Endocrine and Metabolic Disorders

Clinical Lab Testing Manual

Fourth Edition

Endocrine and Metabolic Disorders

Clinical Lab Testing Manual

Fourth Edition

Robert F. Dons and Frank H. Wians, Jr.

CRC Press
Taylor & Francis Group
Boca Raton   London   New York

CRC Press is an imprint of the
Taylor & Francis Group, an **informa** business

Cover art created by Mark Smith of the Pathology Media Services section of the Department of Pathology at UT Southwestern.

CRC Press
Taylor & Francis Group
6000 Broken Sound Parkway NW, Suite 300
Boca Raton, FL 33487-2742

© 2009 by Taylor and Francis Group, LLC
CRC Press is an imprint of Taylor & Francis Group, an Informa business

No claim to original U.S. Government works

Printed in the United States of America on acid-free paper
10 9 8 7 6 5 4 3 2 1

International Standard Book Number: 978-1-4200-7932-6 (Paperback)

Library of Congress Cataloging-in-Publication Data

Dons, Robert F.
    Endocrine and metabolic testing manual / Robert F. Dons and Frank H. Wians Jr. -- 4th ed.
       p. ; cm.
    Includes bibliographical references and index.
    ISBN 978-1-4200-7932-6 (pbk. : alk. paper)
    1. Endocrine glands--Diseases--Diagnosis--Handbooks, manuals, etc. 2.
Metabolism--Disorders--Diagnosis--Handbooks, manuals, etc. I. Wians, Frank H. II. Title.
    [DNLM: 1. Endocrine System Diseases--diagnosis--Handbooks. 2. Metabolic
Diseases--diagnosis--Handbooks. WK 39 D687e 2009]

    RC649.E485 2009
    616.4'075--dc22
                                            2009015402

Visit the Taylor & Francis Web site at
http://www.taylorandfrancis.com

and the CRC Press Web site at
http://www.crcpress.com

Contents

Preface

This book is intended to provide practicing endocrinologists and other allied health professionals with a comprehensive and up-to-date source for all endocrine procedures used in the assessment of patients with potential endocrine disorders. The book is organized by the endocrine organ system into 12 chapters, and all tests within each chapter are organized for efficient use by providing accurate, brief but adequate, and, it is hoped, clear information regarding:

- Indications for the test
- Procedure for performing the test
- Interpretation of test results
- Notes that provide additional information
- Suggested reading
- ICD-9 diagnosis codes that may be useful

Moreover, to assist readers with complex equations and calculations, we have included sample calculations. In addition, our book is laden with tables, and relatively few figures, because in our view, if "a picture is worth a 1000 words" then a good table is worth 2000 words. Tables tend to be easier to use than searching out needed information from a sea of narrative. Nevertheless, users of this book are reminded of several important caveats:

1. No endocrine test or procedure is a substitute for a good history and physical examination of the patient.
2. Clinicians should treat the patient and not the endocrine test result.
3. Accurate and clinically useful endocrine test results are critically dependent on strict adherence to all procedure/test requirements, especially the quality of the specimen submitted for testing, including attention to such issues as patient preparation, timing of specimen collection, specimen collection tubes and additives, and specimen transport and storage conditions prior to laboratory testing. In short, the quality of any endocrine test result is only as good as the quality of the specimen submitted for testing.
4. Test units and reference interval information can differ between different test methodologies and testing laboratories, including the use and reporting of conventional units and/or Systéme International (SI) units. For many, but not all, tests in our book we have included SI units to broaden their applicability.
5. Newer, and perhaps better, endocrine tests may be available other than those listed in our book. These are tests that might have appeared in the interim period of June 2008 to January 2009 between submission of this text and its publication.
6. For the sake of historical interest, we have included purposely older, less often used, endocrine tests in Chapter 12 that may not be entirely obsolete.

We hope that you will find our book informative, useful, and valuable!

Introduction

Important Concepts in Understanding and Correctly Interpreting Biochemical Laboratory Test Result

Because the practice of endocrinology is one of the most clinical laboratory dependent of medical specialties, a word of caution about overzealous reliance on biochemical testing and laboratory test results is appropriate. Because a laboratory test result by itself rarely, if ever, definitively establishes a clinical diagnosis, proper use of laboratory test results requires an understanding of the five principles detailed below.

Laboratory Test Interpretation Principles

- The possibility of *test error* is always present, and errors can occur in any of the three phases of the laboratory test cycle (Figure 1): preanalytical, analytical, or postanalytical (examples of preanalytical sources of error for various lipids are shown in Table 1).
- Most clinical diagnoses can be made by *history and physical examination*; however, biochemical laboratory tests and additional studies (e.g., imaging modalities) can provide useful information in discriminating between the options constituting the differential diagnoses.

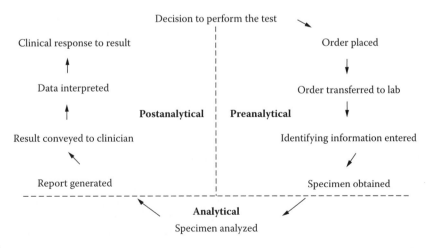

Figure 1
The laboratory test cycle illustrating the preanalytical, analytical, and postanalytical phases of the cycle. (Adapted from Elston, D.M., *MedGenMed*, 8(4), 9, 2006.)

TABLE 1
Preanalytical Variables in Lipid, Lipoprotein, and Apolipoprotein Testing

Variable	Lipids	Lipoproteins	Apolipoproteins
		Effect On	
Biological:			
↑ Age	↑ (TC, TG)	↑ LDL-C	↑ B
Sex/age:			
Just prior to puberty	Transient ↓ (TC, TG)	Transient ↓ (HDL-C, LDL-C)	—
At puberty (males)	slt ↓ TC	↓↓ HDL-C	↓↓ A_I
After puberty	TC: M (20–45 y) > F TC: F (<20 y or PM) > M	M: HDL-C (up to age 55) ME 6 PM: gradual ↑ HDL-C	M: A_I (≤55 y) ME 6 PM: gl ↑ A_I
Race (Blacks ≥ 9 years vs. Caucasians)	—	↑ (HDL-C, VLDL-C); ↓ LDL-C	↑ A_I; ↓ B
Behavioral:			
Diet:			
Cholesterol-rich	↑ TC	—	—
High in mono-/polyunsaturated fatty acids	↓ (TC, TG)	↓ LDL-C	↓ B
High in saturated fatty acids	↑ TC	↑ LDL-C	—
Obesity	↑ (TC, TG)	↑ LDL-C; ↓ HDL-C	—
Smoking	↑ TG	↑ (LDL-C, VLDL-C); ↓ HDL-C	↓ A_I
Alcohol intake	↑ TG	↑ HDL-C; ↓ LDL-C	↑ (A_I, A_{II})
Caffeine intake	↑ TC	↑ LDL-C	—
Exercise:			
↑	↓ TG	↑ HDL-C; ↓ LDL-C	↑ A_I; ↓ B
↓	↑ TG	↑ LDL-C	—
Stress	↑ TC	↓ HDL-C	↓ A_I
Clinical (2° alterations)			
Disease:			
Hypothyroidism	↑ TC	↑ LDL-C	—
Insulin-dependent diabetes mellitus	↑ (TC, TG)	↑ LDL-C	—
Nephrotic syndrome/chronic renal failure	↑ (TC, TG)	↑ LDL-C	↓ A_I; ↑ B

- Clinicians should use the *resources* of appropriately trained and qualified laboratory personnel (e.g., doctoral-level clinical chemists) when interpreting laboratory test results.
- Reference interval and interpretive information, as well as other quantitative analytical and clinical performance characteristics of laboratory tests, usually differ among different methods, instruments, or laboratories for the same analyte or test. Assays for the same analyte are not created equal, and no universal standard or calibrator for any analyte exists; laboratory test results obtained using different methods, instruments, or laboratories are not necessarily interchangeable.
- The selection of which laboratory test is the most appropriate and *medically necessary* to order requires an understanding of the *analytical and diagnostic performance characteristics* of laboratory tests for specific disorders/diseases as detailed below.

TABLE 1 (cont.)
Preanalytical Variables in Lipid, Lipoprotein, and Apolipoprotein Testing

Variable	Effect On		
	Lipids	Lipoproteins	Apolipoproteins
Disease (cont.):			
Biliary tract obstruction	↑↑ TC	↑ Lipoprotein X	—
Acute myocardial infarction	↓↓ TC	↓↓ LDL-C	↓↓ (A$_I$, B)
Drug therapy:			
Diruetics	↑ (TC, TG)	↑ (LDL-C, VLDL-C); ↓ HDL-C	↑ B; ↓ A$_I$
Propanolol	↑↑ TG	↓↓ HDL-C	—
Oral contraceptives with high [progestin]	↑ TC	↑ LDL-C; ↓ HDL-C	—
Oral contraceptives with high [estrogen]	↓ TC	↑ HDL-C; ↓ LDL-C	—
Prednisolone	↑ (TC, TG)	↑ (LDL-C, VLDL-C, HDL-C)	↑ (A$_I$, B)
Cyclosporine	↑↑ TC	↑↑ LDL-C	↑↑ B
Pregnancy	↑↑ (TC, TG)	↑↑ LDL-C	↑↑ (A$_I$, A$_{II}$, B)
Specimen collection and handling			
Nonfasting vs. fasting (12 h)	↑↑ TG	↑ VLDL-C; ↓ LDL-C	↑ B?
Anticoagulants:			
EDTA	—	↓ (3.0-4.7%)	—
Heparin	↓ TG	—	—
Capillary vs. venous blood	TC: ↑ or ↓ (–3%)	—	—
Hemoconcentration (e.g., use of a tourniquet)	↑ TC	↑	↑
Specimen storage (at 0–4°C for up to 4 days)	slt ↓ (TG, PLs)	slt ↓ (LDL-C, HDL-C)	—

Abbreviations: TC, total cholesterol; TG, triglycerides; LDL-C, low-density lipoprotein cholesterol; HDL-C, high-density lipoprotein cholesterol; VLDL-C, very-low-density lipoprotein cholesterol; F, females; M, males; y, years; ME, menarche; PM, postmenopausal; gl, gradual; EDTA, ethylenediaminetetraacetate; slt, slight; PLs, phospholipids. ↑, increased; ↑↑, markedly increased; ↓, decreased; ↓↓, markedly decreased; ↔, no effect (neither decreased nor increased); ?, effect uncertain.

Analytical and Diagnostic Performance Characteristics of Lab Tests

Analytical Performance Characteristics of Lab Tests

These include:

- Analytic sensitivity
- Analytic specificity
- Accuracy (or bias)
- Precision (or reproducibility)
- Functional sensitivity
- Dynamic range
- Reference interval (or "expected values")

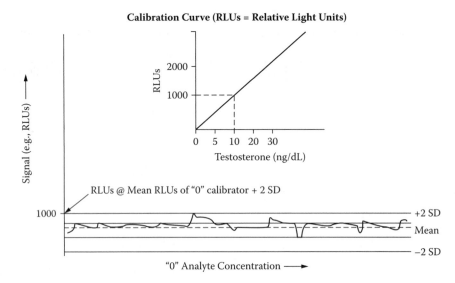

Figure 2
Example illustrating the determination of MDC for testosterone quantified using an immunochemiluminometric assay (ICMA).

Analytical Sensitivity

Analytical sensitivity is the minimum detectable concentration (MDC), which is the concentration of analyte that the test can reliably distinguish from 0 analyte concentration with 95% confidence. MDC is typically determined by performing 20 replicate tests of the zero calibrator, determining the mean and standard deviation (SD) values for the signal provided by all 20 replicates, and calculating the analyte concentration corresponding to the mean signal + 2 standard deviations (SD) (Figure 2). (*Note:* In most cases, the analyte concentration corresponding to the mean signal − 2 SD will be a negative number and therefore cannot be used as the MDC.)

Analytical Specificity

Analytical specificity is a measure of the test's freedom from interferents such as hemolysis, lipemia, icterus, structural analogs, drugs, and any other substance different from the analyte of interest that the test may erroneously quantify as the analyte the test was designed to detect and measure. Analytical specificity is determined by interference studies that assess the magnitude of the interference or cross-reactivity of interferents with the analyte of interest.

Accuracy or Bias

Accuracy or bias is the closeness of the analyte concentration measured in the test with the "true" analyte concentration. The bias is the difference between the observed (O) (or measured) analyte concentration vs. the expected (E) (or true) analyte concentration; that is, bias = $O - E$. Thus, the bias can be either positive or negative. In addition, the true analyte concentration is the analyte concentration obtained by one of the following three methods:

- Definitive method
- Reference method
- Method of known bias

Example: For serum total cholesterol, the definitive method is isotope-dilution mass spectrometry, the reference method is the modified Abell–Kendall method, and a method of known bias is the method currently being used in a particular clinical laboratory for the quantification of serum total cholesterol concentration which has been validated previously against values obtained by one of the three aforementioned methods and, therefore, the magnitude of the bias between these methods is known.

Precision

Precision reflects the reproducibility of the results of the test when the test is performed multiple times on the same sample. Conversely, imprecision refers to a lack of precision. The measure of imprecision is the coefficient of variation (CV), typically expressed as a percent:

$$CV\ (\%) = (Mean/SD) \times 100$$

where the mean and SD are obtained from the replicate values from testing a single sample a minimum of 20 times in a single run (i.e., within-run or "intra-assay" CV) or 1 time in 20 separate runs (i.e., between-run or "interassay" CV). Based on the equation for calculating %CV, values for %CV will necessarily be high when the SD is small if the analyte concentrations being measured are small numbers, such as the imprecision at low thyroid-stimulating hormone (TSH) concentrations (<0.1 mIU/L). Measurement systems with good precision have low imprecision identified by low %CV values. Conversely, measurement systems with poor precision have high imprecision, identified by high %CV values. Typically, the relationship between the magnitude of the %CV and the degree of imprecision is interpreted as: $<2.5\%$, excellent; $<5\%$, very good; $<10\%$, good; and $<15–20\%$, useful for some purposes (e.g., measurement of low levels of TSH using a highly sensitive TSH assay). Moreover, for many laboratory tests (e.g., total cholesterol), serial testing at appropriate time intervals is necessary to determine a patient's true value for a particular test. In addition, for test values near the lower or upper limits of the reference interval or near the cutoff value for analyte concentration associated with optimum diagnostic accuracy (see below), imprecision in the measurement of a single laboratory test result will affect the reproducibility of which side of these limits the result will fall. This is especially true when the imprecision is necessarily high because low levels of analyte concentration are being measured (e.g., TSH concentrations < 0.1 mIU/L); therefore, the functional sensitivity of such tests is critically important to the test's diagnostic performance.

Functional Sensitivity

Also referred to as the lower limit of quantitative (LLQ) measurement, functional sensitivity is the analyte concentration corresponding to an interassay CV $\leq 20\%$. Functional sensitivity is determined from a precision profile (Figure 3), and the LLQ corresponding to an interassay CV $\leq 20\%$ is used over the intra-assay CV because patients' values for any analyte are obtained most often by testing multiple specimens from a patient over several runs on different days.

Dynamic Range

Dynamic range is the range of analyte values over which the test is linear without the need for dilution of the sample. Laboratory tests with a broad dynamic range are desirable because they reduce the need for an offline dilution (and the potential error associated with such a step) of the patient's specimen and the reagent and labor costs associated with repeat testing. The lowest value of the dynamic range is the LLD, and the highest value is the upper limit of the linear range.

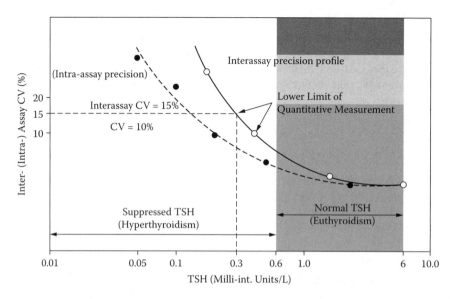

Figure 3

Example of a precision profile for TSH for determining functional sensitivity. (From Wians, Jr., F.H. et al., in *Clinical Chemistry*, Lewandrowski, K., Ed., Lippincott, Williams & Wilkins, Philadelphia, PA, 2000, pp. 81–106. With permission.)

Reference Interval (or Normal Reference Range)

The term "reference interval" is preferred over the older and less accurate term "normal reference range" because the values reported as the lower (LLN) and upper (ULN) limits of normal are not the lowest and highest numbers in the rank-ordered set of all test values on samples from a healthy population of individuals, and the term "normal" does not necessarily mean "healthy." A reference interval for a laboratory test is established by performing the test on a suitable number of apparently healthy individuals and applying parametric or nonparametric statistics to determine the mean and standard deviation from the entire population of values. The reference interval is defined as the 95% confidence interval around the mean such that the lower limit of the interval is the test value corresponding to the mean − 1.96 SD and the upper limit of the interval is the test value corresponding to the mean + 1.96 SD. (*Note:* Because 1.96 rounded to no decimal places is 2, a value of 2 is often used in lieu of the more precise value 1.96 when determining the 95% confidence interval.) Thus, based on this definition of reference interval, 5% (or 1/20) of apparently healthy individuals can have a test value outside these limits; therefore, the lower and upper limits of the reference interval are not necessarily the best cutoff values for identifying individuals with or without disease (see information below on area under the curve (AUC) from a receiver–operator characteristic (ROC) curve analysis). Moreover, the degree of individuality in many analytes, such as thyroid-stimulating hormone (TSH), total triiodothyronine (TT_3), free thyroxine (FT_4), and prolactin, may be such that population-based reference intervals may not correctly identify clinically significant changes in these hormones in individual subjects (Maes et al., 1997).

When clinical assessment of a patient identifies an abnormality and the results of initial, repeat, or serial laboratory tests are abnormal (i.e., outside the reference interval for the test), the clinician must answer the question: *What is the likelihood that the magnitude of the difference in my patient's tests results, compared to the lower or upper limits of the reference interval, is due to the disease process suggested by clinical assessment vs. no disease, a different disease, or biological variation alone?* The answer to this question requires an understanding of the clinical (or diagnostic) performance characteristics of laboratory tests as detailed below.

Clinical (or Diagnostic) Performance Characteristics of Lab Tests

These include:

- Diagnostic sensitivity
- Diagnostic specificity
- Post-test probability of disease given a positive test result (PPD+) or positive predictive value (PPV)
- Post-test probability of disease given a negative test result (PPD–) or negative predictive value (NPV)
- Efficiency
- Area under the curve (AUC) from a receiver–operator characteristic (ROC) curve analysis
- Reference change value (RCV)

Diagnostic Sensitivity

Diagnostic sensitivity is the percentage of *affected* patients who have a *positive* test result:

$$\text{Sensitivity } (\%) = [TP/(TP+FN)] \times 100$$

where TP = number of true positive test results; FN = number of false negative test results.

Highly sensitive tests give a positive result in nearly all diseased subjects and, therefore, are most useful in mass screening for a disease with significant morbidity and/or mortality for which curative therapy is available. The number of FN test results is relatively low, thus avoiding missing individuals with disease.

Diagnostic Specificity

Diagnostic specificity is the percentage of *unaffected* patients who have a *negative* test result:

$$\text{Sensitivity } (\%) = [TN/(TN + FN)] \times 100$$

where TN is the number of true-negative test results, and FP is the number of false-positive test results.

Highly specific laboratory tests give a negative result in most patients without disease and are most useful in confirming the presence of disease in selected populations of individuals already suspected of having the disease on the basis of other diagnostic findings. Such tests can be useful in determining who requires treatment. The number of FP test results is relatively low, thus avoiding the treatment of individuals without disease.

Unfortunately, knowing the sensitivity and specificity of a test is of limited diagnostic value because the questions that clinicians are most concerned with are:

- What is the probability that my patient has disease X when the test is positive?
- What is the probability that my patient does *not* have disease X when the test is negative?

The sensitivity and specificity of a test answer different and less important questions:

- What is the probability of my patient having a positive test result if he or she has disease X?
- What is the probability of my patient having a negative test result if he or she does not have disease X?

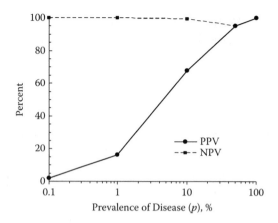

Figure 4
The effect of disease prevalence on positive (PPV) and negative (NPV) predictive value for a laboratory test with 95% sensitivity and 95% specificity.

Pretest Probability of Disease

The pretest probability of disease is the probability that a patient has a particular disease before any laboratory tests are performed; therefore, the pretest probability of disease is the known prevalence (p) of the disease in the population from which the patient comes:

$$p = \text{No. of individuals with disease/no. of individuals in population}$$

When the prevalence of the disease is low, most of the positive laboratory test results associated with a good (i.e., acceptable diagnostic sensitivity and specificity) laboratory test for this disease will be false positives; therefore, to increase the positive predictive power of a laboratory test for a particular disease, clinician's must increase the likelihood of the disease being present by appropriate selection of patients for whom the laboratory test is ordered, based on clinical and other diagnostic test findings.

Post-Test Probability of Disease

The post-test probability of disease is the probability that a patient has, or does not have, a particular disease based on positive (PPD+) or negative (PPD–) test results, respectively. Values for PPD+ and PPD– are calculated using Bayes' equations:

$$\text{PPV} = \left[(p)(\text{Sensitivity})\right] \big/ \left\{\left[(p)(\text{Sensitivity})\right] + \left[(1-p)(1-\text{Specificity})\right]\right\}$$

or, more simply, TP/(TP + FP).

$$\text{NPV} = \left[(1-p)(\text{Sensitivity})\right] \big/ \left\{\left[(p)(1-\text{Sensitivity})\right] + \left[(1-p)(\text{Specificity})\right]\right\}$$

or, more simply, TN/(TN + FN).

Thus, if a laboratory test result for a particular disease is positive, the specificity of that test is the most important determinant of whether or not disease is likely to be present. Conversely, sensitivity of the test is the most important determinant of whether or not disease is likely to be absent. Moreover, the prevalence of the disease in the population being tested markedly affects PPV and has little effect on NPV (Figure 4).

Efficiency

Efficiency is how often the laboratory test provides a positive or negative test result when the patient has, or does not have, the disease the test was designed to identify, respectively. Efficiency is calculated as:

$$\text{Efficiency } (\%) = \left[(TP + TN)/(TP + TN + FP + FN) \right] \times 100$$

To illustrate the calculation and interpretation of the aforementioned diagnostic performance characteristics of laboratory tests, consider Example 1.

Example 1. Sample Calculations for Laboratory Test Diagnostic Performance Characteristics

The prostate-specific antigen (PSA) laboratory test was used to screen 100,000 men for prostate cancer (PCa) and the results of this testing yielded the data in the matrix below:

	No. of Men with PCa[a]	No. of Men without PCa	Total
No. of men with positive PSA test result[b]	175 (TP)	16,287 (FP)	16,462
No. of men with negative PSA test result[c]	25 (FN)	83,513 (TN)	83,538
Total	200[c]	99,800	100,000

[a] Based on prostate biopsy findings.

[b] PSA value ≥ 4.0 ng/mL.

[c] PSA value < 4.0 ng/mL.

Therefore,

$$p = 200/100{,}000 = 0.002 = 0.2\%$$

$$\text{Sensitivity} = \left[TP/(TP + FN) \right] = \left[200/(200 + 25) \right] = \left[200/225 \right] = 0.8888 = 88.9\%$$

$$\text{Specificity} = \left[TN/(TN + FP) \right] = \left[83{,}513/(83{,}513 + 16{,}287) \right]$$

$$= 83{,}513/99{,}800 = 0.08368 = 83.7\%$$

$$PPD+ = \left[(p)(\text{Specificity}) \right] / \left\{ \left[(p)(\text{Sensitivity}) \right] + \left[(1-p)(1-\text{Specificity}) \right] \right\}$$

$$= \left[(0.002)(0.8368) \right] / \left\{ \left[(0.002)(0.8750) \right] + \left[1 - 0.002)(1 - 0.8368) \right] \right\}$$

$$= 0.00153/(0.00153 + 0.16291) = 0.00153/0.16444 = 0.0108 = 1.1\%$$

$$PPV = TP/(TP + FP) = 175/(175 + 16{,}287) = 175/16{,}462 = 0.0106 = 1.1\%$$

$$PPD- = \left[(1-p)(\text{Specificity}) \right] / \left\{ \left[(p)(1 - \text{Sensitivity}) \right] + \left[(1-p)(\text{Specificity}) \right] \right\}$$

$$= \left[(1 - 0.002)(0.8368) \right] / \left\{ \left[(0.002)(1 - 0.8888) \right] + \left[(1 - 0.002)(0.8368) \right] \right\}$$

$$= 0.83513/(0.00022 + 0.83513) = 0.83513/0.83535 = 0.9997 = 100.0\%$$

$$NPV = TN/(TN + FN) = 83{,}513/(83{,}513 + 25)$$

$$= 83{,}513/83{,}538 = 0.9997 = 100.0\%$$

$$\text{Efficiency} = (TP + TN)/(TP + TN + FP + FN) = (175 + 83{,}513)/100{,}000 = 83.7\%$$

Notice the marked effect of low disease prevalence on PPD+, with virtually no effect on PPD–, for a laboratory test with reasonably good sensitivity (89%) and specificity (84%).

Area under the Curve (AUC) from a Receiver–Operator Characteristic (ROC) Curve Analysis

Area under the curve (AUC) from a receiver–operator characteristic (ROC) curve analysis is a measure of the area under the curve of a plot of test values for 1 – specificity (x-axis) vs. sensitivity (y-axis) incorporating the full range of all available data. AUC values range from 0.50 to 1.00, where an AUC value of 0.500 represents a test with no clinical utility (i.e., there is no test cutoff value at which the sensitivity and specificity are adequate for clinical decision making), whereas a test with an AUC value of 1.000 represents a perfect test (i.e., there is a cutoff value for this test at which sensitivity and specificity both equal 100%).

Construction of a receiver–operator characteristic (ROC) curve is the best way to establish the optimal cutoff value for a laboratory test associated with any desired level of diagnostic sensitivity and specificity. Using this approach, two populations of individuals are required: one without and one with the disease the laboratory test was designed to identify. To avoid bias, individuals with disease are identified using tests and diagnostic procedures that do not include the test being evaluated and are selected without obvious bias; for example, using troponin I (TnI) levels to segregate individuals into no-disease or disease groups is not appropriate when evaluating the discriminatory power of TnI in discriminating between individuals with and without an acute coronary syndrome. Moreover, the test and diagnostic procedures used to identify individuals as not having disease should be the best available (e.g., tests and diagnostic procedures accepted as the "gold standard" by the medical scientific community). The quality of the information provided by any ROC curve analysis is only as good as the gold standard used to segregate study subjects into the without-disease and with-disease groups. Once these two populations are created, laboratory testing for the analyte of interest is performed on all individuals in both groups. Subsequently, these data can be entered into a software program for creating a ROC curve, typically as a plot of values for 1 – specificity (x-axis) vs. sensitivity (y-axis) at all test values, and providing relevant statistics (Figure 5).

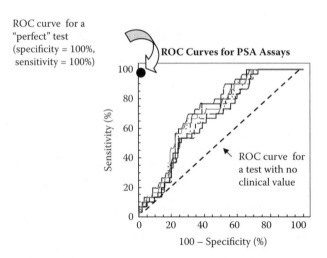

Figure 5

ROC curves of various tPSA assays and cPSA for discriminating between men with benign prostatic hyperplasia (BPH; n = 94) and those with prostate cancer (PCa; n = 30). Curves shown correspond to (from dashed line, the ROC curve for a test with no clinical value, to the point (•) in the upper left-hand corner of the box plot representing a perfect test with 100% sensitivity and 100% specificity) (AUC values; 95% confidence interval): BtPSA2 (0.661; 0.570–0.743), AtPSA (0.676; 0.585–0.757), BtPSA (0.695; 0.606–0.775), BtPSAe (0.698; 0.609–0.777), HtPSA-R (0.701; 0.618–0.785), BcPSA (0.740; 0.661–0.821). AtPSA, Abbott tPSA assay; BcPSA, Bayer complexed PSA assay. BtPSA, older version BtPSA assay; BtPSA2, Bayer total prostate specific antigen assay, newer version; BtPSAe, equimolar Bayer tPSA assay; HtPSA-R, HybriTech tPSA immunoradiometric assay. (From Wians, Jr., F.H. et al., *Clin. Chim. Acta*, 326, 81–95, 2002. With permission.)

The area under the curve (AUC) and the 95% confidence interval for the AUC value evaluate laboratory test performance (i.e., sensitivity and specificity) across all test values and, when multiple laboratory tests are being evaluated, across different tests. Moreover, the optimal cutoff value for the laboratory test being evaluated is at the apex of the ROC curve, corresponding to the test value associated with the highest sum of values for specificity and sensitivity. If clinicians wish to favor sensitivity over specificity, or *vice versa*, the ROC curve analysis makes it easy to identify corresponding test cutoff values; however, there is no objective or absolute level of acceptable sensitivity or specificity. Clinicians must evaluate the sensitivity and specificity of a laboratory test at a specific cutoff value based on their answers to these questions:

- What is the cost (e.g., excess morbidity and/or mortality) of a false-negative test? Is this cost adequate or acceptable to justify using the test based on the magnitude of its *sensitivity* at a particular cutoff value?
- What is the cost (e.g., emotional impact or potential side effects of unnecessary treatment) of a false-positive test? Is this cost adequate or acceptable to justify using the test based on the magnitude of its *specificity* at a particular cutoff value?

Finally, when comparing the discriminatory power of multiple different laboratory tests in the same test populations, AUC values for each test provide an indicator of the rank ordering of these tests with regard to increasing discriminatory power, with higher AUC values representing laboratory tests with higher diagnostic accuracy. To determine if the difference between AUC values for one or more laboratory tests is significant, appropriate statistical testing can and should be performed.

Reference Change Value (RCV)

Reference change value is a reference point representing the magnitude of the inherent error due to both analytical and biological variation alone over the time period between serial laboratory test values. Laboratory tests are requested often at multiple time points (e.g., every day or every other day during a patient's hospital stay). Moreover, values for these tests are inherently affected by test imprecision and biological variability (BV), in addition to any changes in serial test values due to the pathophysiology and course of the patient's disease. The rationale for the applicability of the indices of BV, expressed as a coefficient of variation—intra-individual CV (CV_I), inter-individual (or group) CV (CV_G), and analytical CV (CV_A)—when evaluating the magnitude of the change between serial values for an analyte in individuals with a nonacute pathological process is based on the axiom by Fraser and Williams that "in non-acute pathological processes where new homeostatic steady states are reached, biological variations around the new set points are of the same order as those found in healthy individuals." Therefore, if it was possible to determine a RCV, then any change (positive or negative) greater than this reference point (or RCV) could be interpreted as due to changes in the pathophysiology or course of the patient's disease. If the magnitude of the change between serial values for an analyte was within the limits of change expected based on the combination of analytical and biological variation alone, then clinicians would interpret this finding as providing no evidence of improvement or deterioration in the patient's disease status. With help from Eugene Harris, Fraser distilled these concepts into the equation for RCV:

$$RCV\ (\%) = \pm(2^{0.5})Z\left(CV_A^2 + CV_I^2\right)^{0.5}$$

where $2^{0.5}$ accounts for the fact that two serial analyte values are being compared; Z is 1.96 for a significant change (i.e., a change with 95% probability, or $p = 0.05$) between these two values or 2.58 for a highly significant change (i.e., a change with 99% probability, or $p = 0.01$); CV_A is the analytical variation of the assay used to quantify the analyte; and, CV_I, is the intra-individual biological variation (e.g., within day, day to day, week to week) for the analyte.

Example: Blood samples were taken from a patient upon admission (t_0) to the hospital with signs and symptoms of congestive heart failure (CHF) and 1 week (t_1) later after aggressive treatment. Both samples were tested for the analyte N-terminal propeptide of B-type natriuretic peptide (NT-proBNP), a biochemical marker of the severity of CHF. The pertinent data relevant to this patient were:

$$[\text{NT-proBNP}]_{t0} = 854 \text{ pg/mL}$$

$$[\text{NT-proBNP}]_{t1} = 48 \text{ pg/mL}$$

$$CV_A \text{ for NT-proBNP assay} = 1.6\% \text{ (from in-house quality control data for NT-proBNP)}$$

$$CV_I \text{ week-to-week for NT-proBNP} = 33.3\% \text{ (from published literature data)}$$

Then,

$$RCV\ (\%) = \pm(20.5)(1.96)\left[(1.6)^2 + (33.3)^2\right]^{0.5}$$

$$= \pm(1.414)(1.96)(2.56 + 1108.89)^{0.5}$$

$$= \pm2.77(1111.45)^{0.5} = \pm2.77(33.3) = \pm92.2\%$$

The percent change in the patient's NT-proBNP values was:

$$[(48 - 854)/854]100 = -94.4\%$$

That is, between t_0 and t_1, the patient's NT-proBNP level declined by 94.4%.

Because 94.4% > 92.2%, the magnitude of the patient's decline in NT-proBNP level exceeds what could occur 95% of the time due to analytical and inter-individual variation alone, and the change in the patient's NT-proBNP level may be indicative of improvement (successful treatment) of the patient's CHF.

Note: This example illustrates the application of the concept of RCV, recognizing, of course, that in many clinical situations experienced clinicians recognize when a change is significant and indicative of successful or unsuccessful treatment without the need to calculate RCV. However, when the magnitude of the change is less dramatic than the example provided above, determination of RCV provides a more objective measure of the probability of a statistically significant change in serial laboratory test values that may also be clinically significant.

Summary

The major points to recognize about the use of statistics in laboratory test interpretation are:

- Analytical and functional sensitivity are related but differ in that functional sensitivity invokes an imprecision requirement.
- 5% of healthy individuals can have laboratory test results that exceed the lower or upper limits of the reference interval; therefore, population-based reference intervals are useful in most, but not all, patients.
- The post-test probability of disease is greater when the pretest probability is large. When this concept is applied, the clinician's assessment of the whole patient gets priority. As a result, the danger of relying too heavily on lab test data, in the absence of a thorough history and physical, is avoided.

- Because the differences between PPD+ and PPD– are largest when pretest probability is low, testing is most useful when clinical evaluation is equivocal.
- Laboratory tests with greater sensitivity are less likely to be misleading if negative (i.e., the PPD– will be lower).
- With a positive test result, a test with greater specificity gives the greater post-test probability of disease or diagnosis.
- Be aware of the inability of most diagnostic laboratory tests to rule out disease.
- ROC curves are useful in establishing the cutoff value for laboratory tests that provide the desired level of clinical sensitivity and specificity.
- When interpreting values for RCV, once the trip wire of *statistical* significance has been activated, clinicians should investigate for *clinical* significance, recognizing that there may be none.

Suggested Reading

Bayer MF. Performance criteria for appropriate characterization of "(highly) sensitive" thyrotropin assays. *Clin Chem* 1987; 33:630–1.

Fraser CG. *Biological Variation: From Principles to Practice*. 2001. AACC Press, 151 pp.

Fraser CG, Williams P. Short-term biological variation of plasma analytes in renal disease. *Clin Chem* 1983; 29:508–10.

Lamb CR. Statistical briefing: estimating the probability of disease. *Vet Radiol Ultrasound* 2008; 49:109–10.

Maes M, Mommen K, Hendrickx D, Peeters D, D'Hondt P, Ranjan R, De Meyer F, Scharpe S. Components of biological variation, including seasonality, in blood concentrations of TSH, TT3, FT4, PRL, cortisol and testosterone in healthy volunteers. *Clin Endocrinol* 1997; 46:587–98.

Obuchowski NA, Lieber ML, Wians FH Jr. ROC curves in clinical chemistry: uses, misuses, and possible solutions. *Clin Chem* 2004; 50:1118–25.

Wians FH Jr, Baskin LB. The use of clinical laboratory tests in diagnostic decision-making. In: *Handbook of Clinical Pathology*. 2000. ASCP Press, pp. 9–24.

Wians FH Jr, Cheli CD, Balko JA, Bruzek DJ, Chan DW, Sokoll LJ. Evaluation of the clinical performance of equimolar- and skewed-response total prostate-specific antigen assays versus complexed and free PSA assays and their ratios in discriminating between benign prostatic hyperplasia and prostate cancer. *Clin Chim Acta* 2002a; 326:81–95.

Wians FH Jr, Koch DD, Haara A. Quality control/quality assurance. In: *Clinical Chemistry*, Lewandrowski K (Ed). 2000b. Lippincott, Williams & Wilkins, pp. 81–106.

Wu AHB, Smith A, Wieczorek S, Mather JF, Duncan B, White CM, McGill C, Katten D, Heller G. Biological variation for N-terminal pro- and B-type natriuretic peptides and implications for therapeutic monitoring of patients with congestive heart failure. *Am J Cardiol* 2003; 92:628–31.

The Authors

Robert F. Dons is a clinical endocrinologist practicing in San Antonio, Texas. His focus on both the routine and specialized tests used to make better and more cost-effective diagnoses of endocrine disorders began in 1990 when he served as director and chief of the U.S. Air Force endocrinology training program at Wilford Hall Medical Center, Lackland Air Force Base, in San Antonio. Preparation and publication of both the second and third editions of the *Endocrine and Metabolic Testing Manual* in 1993 and 1997 were under his direction.

He has a long-standing interest in improving the systems of care for patients with diabetes mellitus, particularly those involving life-style and psychosocial aspects of this diverse disease, as well as tests for its diagnosis and complications. He regularly hosts training seminars for diabetes patients on a variety of topics, including glucose monitoring, carbohydrate counting, and exercise. His endocrine specialty areas of expertise include acromegaly, intensive insulin therapy, and thyroid testing in the context of coincident clinical conditions and drug therapies.

Dr. Dons obtained his MD in 1973 and his PhD in biochemistry from the University of Illinois in Chicago. He served 21 years in military medicine (Air Force, Navy, and Public Health Service, National Institutes of Health) before retiring in 1998. He is board certified in Internal Medicine (1977) and in Endocrinology and Metabolism (1987). With the assistance of a fitness trainer, he continues to enhance his balance and core strength in preparation for ever more advanced snow skiing and marathon-level endurance running.

Frank H. Wians, Jr., is currently professor of pathology in the Department of Pathology at The Southwestern Medical Center at Dallas (Texas). His research interests are multifaceted and include endocrine testing, intraoperative hormone testing as an adjunct to imaging studies, evaluation of the analytical and clinical performance characteristics of laboratory test methods, and the role of prostate-specific antigen (PSA) testing in discriminating between prostate cancer and benign prostatic hyperplasia. Past research includes prenatal testing for neural tube defects and fetal aneuploidies, biochemical markers of acute coronary syndromes, and the role of group-specific component (Gc) protein in patients with acute liver failure. Dr. Wians has directed the programs of seven clinical chemistry fellows who have carried out research in clinical chemistry and endocrine topics. He is the author of more than 120 scientific papers and author or coauthor of 9 book chapters on various clinical pathology topics, and he has been the principal or coprincipal investigator for more than 30 projects since 1994.

Dr. Wians obtained a BA degree in advanced mathematics in 1969 from the University of Massachusetts at Boston, a BS degree in biology and chemistry in 1972 from the University of Arizona, an MS degree in biochemistry from the University of Vermont in 1976, an MA degree in health care administration from Rider College in 1979, and a PhD in biochemistry from Harvard University, Division of Medical Sciences, Harvard Medical School, in 1985. Concurrent with his education and training, he served as an officer in the Biomedical Sciences Corps of the U.S. Air

Force, retiring after 22 years of service in 1993 as a Lieutenant Colonel. He served as the Consultant in Clinical Chemistry to the Air Force Surgeon General from 1990 to 1993.

Dr. Wians is board certified in clinical chemistry by the American Board of Clinical Chemistry. He is a fellow of the National Academy of Clinical Biochemistry and an associate member of the American Society of Clinical Pathologists (ASCP), and he has held certification with ASCP as a medical technologist since 1973. He is a member of the American Association for Clinical Chemistry (AACC), a past chair of the Texas Section AACC, and a member of the editorial boards of several scientific journals. He is currently editor-in-chief of *LABMEDICINE*™, a publication of the ASCP. Together with a former pathology resident, Wians is currently in the process of publishing an extensive compilation of clinical and anatomic pathology case studies. Retirement is not an option.

Chapter 1

Thyroid Gland Testing

1.1 Thyroid Hormone Reserve

1.1.1 Thyroid-Stimulating Hormone (Sensitive or Highly Sensitive TSH) as a Test for Hypothyroidism*

Indications for Test

Thyroid-stimulating hormone (TSH) as a test for hypothyroidism is indicated when:

- Patients treated for thyroid disease are being monitored.
- Symptoms of hypothyroidism, including fatigue, cold intolerance, constipation, sleep disturbances, "puffiness," mental slowing, or memory problems, are noted as presented in the "Thyroid Symptom Questionnaire (TSQ)" (Q1.1 in Appendix 1) and "Incidence of Clinical Signs and Symptoms of Hypothyroidism in Patients from Two Different Studies" (Table 1.1).
- Lethargy and fatigue, the most common symptoms, and facial edema, the most prevalent sign of hypothyroidism, are noted.

Procedure

1. Obtain a random blood specimen for TSH testing using an immunometric assay (Figure 1.1). To account for diurnal variation in TSH secretion, obtain a timed blood sample before 0900 hours or after 1500 hours.
2. If a previous result for TSH was >4 mIU/L but <10 mIU/L, consider repeat testing after 1500 hours when the [TSH] is at its nadir.

---

* TSH assays have been variously categorized as 1st, 2nd, 3rd, or 4th generation, based on the magnitude of the assay-specific analytical sensitivity or lower limit of detection (LLD). In general, each generation of TSH assay development has resulted in approximately a 10-fold improvement in the LLD over the previous generation assay. The 1st-generation radioimmunoassays for [TSH] had an LLD of ~0.5 to 1 mIU/L, while 2nd- and 3rd-generation immunometric assays demonstrate LLDs of ~0.05 mIU/L and 0.005 mIU/L, respectively. The principal advantage of 2nd- and 3rd-generation over 1st-generation TSH assays is the ability of TSH values by these assays to discriminate between hyperthyroid and euthyroid individuals, as well as hypothyroid individuals. The terms "sensitive" and "highly sensitive" have been used interchangeably and separately to describe TSH assays with improved analytical sensitivity compared to 1st-generation TSH assays. In our view, it is reasonable to refer to 2nd-generation TSH assays as "sensitive" and to 3rd- and 4th-generation TSH assays as "highly sensitive." In any event, this issue is largely moot because most clinical laboratories in the United States now use a 3rd-generation immunometric assay for the routine measurement of [TSH]; therefore, for simplicity throughout the rest of the text of this publication, the current "highly sensitive TSH assays" will be referred to as "TSH assays."

TABLE 1.1
Incidence of Clinical Signs and Symptoms of Hypothyroidism
in Patients from Two Different Studies

| | % of Patients with Signs and Symptoms of | | | |
| | 1° HypoT Group | | 2° HypoT Group | EuT Control Group |
Signs and Symptoms	Study I[a] (n = 77)	Study II (n = 100)	Study II (n = 15)	Study II (n =100)
Generalized weakness	99	98	100	21
Coarse and dry skin	97	70	7	10
Slow speech	91	56	67	7
Decreased sweating	89	68	80	17
Lethargy/fatigue	91	85	80	17
Facial edema	79	95	53	27
Dry skin	97	79	47	26
Slow movements	n.d.	73	60	14
Edema of eyelids	90	86	40	28
Cold intolerance	89	95	93	39
Peripheral edema	55	57	0	2
Thick tongue	82	60	20	17
Skin cold to touch	83	80	60	33
Constipation	61	54	33	10
Hoarseness	52	74	33	18
Slow mentation/ cerebration	n.d.	49	67	9
Muscle weakness in arm and/or leg	n.d.	61	73	21
Memory impairment	66	65	67	31
Paresthesias	n.d.	56	13	15
Weight gain	59	76	47	36
Coarseness of hair	76	75	40	43
Nervousness/anxiety	35	51	53	42
Anorexia	45	40	0	15
Sparse lateral eyebrow hair	n.d.	84	80	58
Hair loss (from scalp)	57	41	13	21
Depression	n.d.	60	73	41
Deafness	30	40	26	15
Brittle nails	n.d.	41	13	20
Dyspnea/shortness of breath	55	72	73	52
Chest pain, possible angina	25	16	7	9
Exophthalmos	n.d.	11	0	4
Palpitations	31	23	13	20
Heat intolerance	n.d.	2	0	12
Weight loss	13	9	26	23

[a] Study I did not include an euthyroid control group.

Note: 1° HypoT, primary hypothyroidism; 2° HypoT, secondary hypothyroidism; EuT, euthyroidism; n.d., not determined

Source: Means, J.H. et al., in *The Thyroid and Its Diseases*, McGraw-Hill, Chicago, IL, 1963, pp. 321–322. With permission.

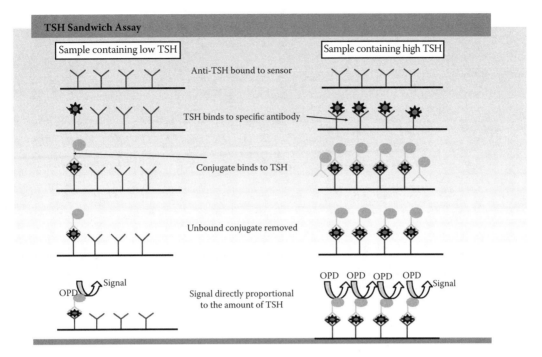

Figure 1.1
Illustration of typical sandwich-type immunometric assay for quantifying serum TSH concentration. OPD, orthophenylene-diamine substrate (converted to a colored product [signal] that absorbs at 492 nm). (Illustration courtesy of DxTech LLC; Melbourne, FL.)

Interpretation

1. Refer to "Non-Drug-Related Causes of Various Patterns of TSH and FT_4 Results" (Table 1.2) and "Drug-Related Causes of Various Patterns of TSH and FT_4 Results" (Table 1.3).
2. Even with the development of laboratory tests with high analytical sensitivity, clinical assessment and judgment remain paramount in making an accurate assessment of thyroid function.
3. Reference intervals for [TSH] vary widely and are typically 0.4 to 4.5 mIU/L.
4. The consequences of subclinical hypothyroidism (serum [TSH] of 4.5–10.0 mIU/L) are often subtle. Thyroid hormone treatment of patients with [TSH] > 6 mIU/L is often recommended, with most experiencing clinical improvement and stability upon achieving a target [TSH] of between 0.3 and 3 mIU/L.
5. Patients with thyroid antibodies, infertility, pregnancy, or imminent pregnancy usually require thyroid hormone treatment at a [TSH] > 5 mIU/L.
6. A [TSH] < 0.3 mIU/L usually indicates hyperthyroidism but may occur in central hypothyroidism secondary to pituitary hypothalamic dysfunction, or such low levels (e.g., 0.1–0.3 mIU/L) may be "normal" for these individuals. Free thyroxine (T_4) will be low (<0.6 ng/dL) in untreated central hypothyroidism. Agents that may directly lower TSH in clinically euthyroid or hypothyroid patients include:
   - Exogenous triiodothyronine (T_3) and/or T_4
   - High-dose glucocorticoids
   - Somatostatin analogs
   - Dopamine agonists
   - Iodides, via the Jod–Basedow effect, if given to relatively iodide-deficient patients
7. [TSH] of 4 to 10 mIU/L may be normal (especially in neonates) but may represent subclinical or early hypothyroidism in adults under 30 years old.
8. Neonates younger than 3 or 4 days of age will have higher [TSH], frequently >20 mIU/L.

TABLE 1.2

Non-Drug-Related Causes of Various Patterns of TSH and FT$_4$ Results

Pattern	TSH	FT$_4$	Non-Drug-Related Thyroid Disease Diagnoses (Dx)
I	L	H	Graves' disease/hyperthyroidism/thyrotoxicosis
			Hyperfunctioning thyroid nodule(s)
			Active thyroiditis
			Choriocarcinoma or struma
			Hyperemesis gravidarum
			Surreptitious (self-treatment) or unsuspected thyroxin intake (ingestion of hamburger contaminated with bovine thyroid gland)
II	L	N	Resolving thyroiditis
			Recent-onset hyperthyroidism
			Multinodular goiter with area of autonomy
			T$_3$ toxicosis or subclinical hyperthyroidism
			Critical illness
			Depression
III	L	L	Severe critical nonthyroidal illness
			Hypothalamic–pituitary dysfunction
			Advanced age
			T$_3$ toxicosis with severe thyroid binding protein deficiency
			Cushing's syndrome
			Use of "fat burner" dietary supplements, such as tiratricol (3,5,3'-triiodothyroacetic acid)
IV	N	H	Increased thyroid binding proteins secondary to pregnancy, acute hepatitis, acute porphyria, or X-linked dominant trait
			Recovery phase of critical illness
			Recent increase in endogenous T$_4$ secretion (spontaneous)
			Acute psychiatric illness
			Blood sampling 2 to 4 hours after T$_4$ dosing
			TSH-secreting tumor
			Peripheral/partial pituitary resistance to T$_4$/T$_3$

V	N	N	N	Normal thyroid–pituitary axis Conditions that may raise or lower TSH and/or free T$_4$ in hypothyroid or hyperthyroid patient (e.g., heparin therapy) Very early hypothyroidism detectable only with TRH testing
VI	N	N	L	Hypothalamic–pituitary dysfunction (i.e., hypopituitarism or second-degree hypothyroidism) Critical illness/hospitalized patient (bioinactive TSH detected by RIA or nonimmunoreactive TSH) Decreased thyroid binding proteins caused by clinical syndromes of acromegaly, nephrosis, hemodialysis, or congenital X-linked recessive trait Malabsorption of oral T$_4$ Rx in hypothyroid patients—non-drug induced (e.g., gastrointestinal hypermotility disorders, new-onset celiac disease) Moderate iodine deficiency
VII	H	H	H	Acute psychoneurosis Recovery phase of critical illness TSH-secreting tumor 5'-Iodinase deficiency Peripheral/pituitary resistance to T$_4$ Analytic error including heterophil antibody or human anti-mouse antibody (HAMA) interference with the TSH assay
VIII	H	H	N	Normal diurnal variation in TSH secretion (sample taken in early morning) Subclinical hypothyroidism in a patient with a TSH < 10 IU/L, but no obvious symptoms; possible increased TBP (e.g., familial dysalbuminemia) and hypothyroidism Overt primary hypothyroidism if TSH > 10 IU/L Recovery phase of critical illness Early phase of thyroid failure in Hashimoto's thyroiditis Secretion of biologically hypofunctional TSH Mild generalized resistance to thyroid hormones Presence of human anti-mouse antibodies (HAMAs) which falsely elevate measured TSH Acute psychiatric illness Hypoadrenalism (70% of cases)
IX	H	H	L	Primary hypothyroidism (idiopathic, congenital, postthyroidectomy) External beam radiation to the neck or ablative dose of radioactive iodine-131 (RAI) Extreme nutritional iodine-deficiency states Increased degradation of T$_4$/T$_3$ by type 3 iodothyronine deiodinase from hemangiomas Hypoadrenalism (30% of cases)

TABLE 1.3
Drug-Related Causes of Various Patterns of TSH and FT_4 Results

Pattern	TSH	FT_4	Drug-Related (Rx) Effects on BTFT[a] Patterns
I	L	H	*Direct T_4 Rx effect in hypothyroid patient:* T_4 overdosage (>6 weeks) *Indirect T_4 Rx effect in hypothyroid patient:* same dose T_4 (>6 weeks) before and after induction of decreased thyroid binding proteins or globulins (TBPs) by agents such as androgens, L-asparaginase, glucocorticoids, nicotinic acid, or weight loss independent of [TBP] Enhancers of T_4 synthesis or release, such as amiodarone, lithium, or iodide treatment of iodine-deficient patient; radioiodine (RAI); or interferon-induced thyroid storm/thyroiditis
II	L	N	*Direct T_4 Rx effect in hypothyroid patient:* T_4 overdosage (>6 weeks) discontinued and adjusted to a lower, correct/therapeutic T_4 dose for <6 weeks TSH suppressants such as glucocorticoids; human growth hormone (hGH), octreotide, dopamine agonists, animal thyroid extracts, dithioerythritol (DTE), or low-dose T_3 Recent RAI as Rx for GD, with positive, initial therapeutic response
III	L	L	*T_4 Rx effect in hypothyroid patient:* T_4 underdosage in cases of central or hypothalamic–pituitary-related hypothyroidism Triiodothyronine Rx usually at a high dose TSH suppressants such as high-dose glucocorticoids, human growth hormone (hGH), octreotide, dopamine agonists, animal thyroid extracts, or DTE
IV	N	H	*T_4 Rx effect in hypothyroid patient:* T_4 overdosage (<6 weeks) *TBP stimulants:* estrogens, tamoxifen, heroin, methadone, fibric acids, mitotane, 5-FU *T_4 to T_3 conversion inhibitors:* beta blockers, high-dose glucocorticoids, dexamethasone *Transient drug effect:* iodinated x-ray contrast media, inhibitors of T_4 binding to proteins or displacers of bound T_4 such as phenytoin, furosemide, NSAIDs, slow-release niacin
V	N	N	*T_4 Rx effect in hypothyroid patient:* correct/therapeutic T_4 replacement dose (>6 weeks) *Drug effect in hyperthyroid patient:* adequate dosages of antithyroid drugs (thionamides) Drugs that may raise or lower TSH and/or free T_4 in hypothyroid or hyperthyroid patient (e.g., heparin therapy) Blood sampling at an interval of days after dosing with inhibitors of T_4 binding to proteins or displacers of bound T_4 (phenytoin, furosemide, NSAIDs, slow-release niacin)

VI	N	L	T_4 Rx effect in hypothyroid patient: same dose T_4 before and after treatment with T_4 absorption interferents or enterohepatic circulation—colestipol, cholestyramine, aluminum hydroxide antacids, calcium or iron supplements, sucralafate, orlistat T_4 Rx effect in hypothyroid patient: same dose T_4 before and after treatment with enhancers of T_4 metabolism including supraphysiologic doses of glucocorticoids or CYP 450 activators such as anticonvulsants or rifampicin Drug effect: triiodothyronine Rx in therapeutic doses Drug effect in hyperthyroid patient: antithyroid agents in excess doses usually during the later phases of treatment (thionamides) Drug effect: high-dose salicylates
VII	H	H	T_4 Rx effect in hypothyroid patient: T_4 overdosage in initial phases of T_4 replacement Drug effect: T_4 to T_3 conversion inhibitors—beta blockers, high-dose glucocorticoids Drug effect: iodinated x-ray contrast media Recombinant TSH (rTSH) stimulation in patients with metastatic thyroid cancer
VIII	H	N	T_4 Rx effect in hypothyroid patient: T_4 underdosage (>6 weeks) T_4 Rx effect in hypothyroid patient: same dose T_4 before and after treatment with TBP stimulants—estrogens, tamoxifen, opiates, fibric acids, mitotane, 5-FU, rifampicin, perphenazine T_4 Rx effect in hypothyroid patient: same dose T_4 before and after treatment with agents that interfere with T_4 absorption or enterohepatic circulation—colestipol, cholestyramine, aluminum hydroxide antacids, calcium or iron supplements, sucralafate, orlistat T_4 Rx effect in hypothyroid patient: same dose T_4 before and after treatment with T_4 metabolism enhancers—CYP 450 activators, high-dose glucocorticoids, rifampicin Drug effect: TSH stimulators such as lithium carbonate, H2 blockers, dopamine antagonists, rTSH, TRH, antithyroid agents
IX	L	H	T_4 Rx effect in hypothyroid patient: T_4 under dosage (e.g., therapeutic miscalculation; onset of celiac disease or other primary T_4 malabsorptive disorder such as impaired gastric acid secretion or nephrotic syndrome in which T_4 bound to serum proteins is lost in the urine with no increase in T_4 dose) T_4 Rx effect in hypothyroid patient: late effects of same dose T_4 before and after treatment with agents that induce increased T_4 metabolism such as CYP 450 activators, high-dose glucocorticoids, rifampicin T_4 Rx effect in hypothyroid patient: same dose T_4 before and after treatment with T_4 absorption or enterohepatic circulation interferents—colestipol, cholestyramine, aluminum hydroxide antacids, calcium or iron supplements, sucralafate, orlistat, H2 blockers, proton pump inhibitors, chromium, phosphate binders (e.g., sevelamer) Drug effect: antithyroid agents (thionamides, ethionamide, ablative-dose RAI); high-dose iodides (amiodarone); inducers of antithyroid antibodies (lithium carbonate, interleukin-2); tyrosine kinase inhibitor (sunitinib maleate)

a BTFT, biochemical thyroid function test.

9. Conditions that may elevate [TSH] include:
   - Thyroid gland failure
   - Psychoneurosis
   - Recovery from severe nonthyroidal illness
   - TSH-secreting pituitary tumors
   - Secretion of biologically hypofunctional TSH
   - Generalized resistance to thyroid hormone
10. Agents that may raise the [TSH] include:
    - Thyrotropin-releasing hormone (TRH)
    - Dopamine antagonists
    - Estrogen therapy, which increases thyroid-hormone-binding proteins in thyroid-hormone-treated patients
    - Agents that impair gastric acid secretion (e.g., cimetidine, H2 blockers, and proton pump inhibitors) in thyroid-hormone-treated patients
    - Lithium
    - Iodides, particularly after ^{131}I therapy
11. Marked elevations of TSH may occur in the presence of human anti-mouse antibodies (HAMAs) or myxedema coma.
12. Elevated [TSH] with elevated free T_4 suggests:
    - A TSH-producing tumor
    - Functionally abnormal TSH
    - Thyroid hormone resistance
    - Psychoneurosis
    - Recovery from severe nonthyroidal illness

Note that none of these conditions necessarily results in clinical hypothyroidism.

Notes

1. TSH originates from the pituitary and is a pituitary test. TSH and thyroid hormones function in a classical endocrine feedback loop.
2. The regular TSH (rTSH) assay became obsolete (*circa* 1992), given its inability to accurately measure levels of TSH below 0.5 mIU/L.
3. A diurnal variation in [TSH] of up to 2 to 3 mIU/L may occur and may complicate interpretation of borderline high [TSH], particularly if obtained in the early morning as is typically done in a hospital setting.
4. [TSH] tends to be highest in the early morning (2200–0400 hours) and lowest in the afternoon (1400–1800 hours).
5. Following a higher, a lower, and no change in treatment dose of T_4, no significant changes in hypothyroid symptoms, well-being, or quality of life occurred despite expected changes in [TSH] (0.3 ± 0.1, 1.0 ± 0.2, and 2.8 ± 0.4 mIU/L, respectively) (Walsh et al., 2006).
6. The signs and symptoms of hypothyroidism are remarkably diverse (Table 1.1). In the thyroid emergency of myxedema coma, blunted ventilatory response with abnormal response to hypoxia and hypercapnia occurs in about a third of cases (Ladenson et al., 1988).
7. The signs that have the greatest association with hypothyroidism include puffiness of the face, coarse skin, bradycardia, and delayed ankle reflexes.
8. When a multiplicity of hypothyroid symptoms occurs in T_4-treated patients whose [TSH] in the morning is ≥ 4 mIU/L, the thyroxine dose may be insufficient for such patients.
9. Prevalence of subclinical hypothyroidism is two- to threefold higher in middle-aged women with high total cholesterol levels than in those with normal (<200 mg/dL) levels. An increase in carotid intima-media thickness (CIMT) (see Test 12.4.5 in Chapter 12) occurs in the hypercholesterolemic hypothyroid patient.
10. Based on studies of twins, genetic, more than environmental, factors appear to control the normal TSH and free T_4 "set point" for an individual.

11. In a 6-year study, the incidence rate of overt hypothyroidism was found to be ~2, 20, and 73 cases per 100 patient-years in subjects with an initial [TSH] of between 5.0 and 9.9, 10.0 and 14.9, and 15.0 and 19.9 mIU/L, respectively (Surks et al., 2004).

12. Offspring of hypothyroid mothers with elevated TSH during pregnancy may suffer from visual contrasting problems resulting in reading and visuospacial difficulties in general.

ICD-9 Codes

Condition(s) that may justify this test *include but are not limited to*:

Hypothyroidism
244.9	acquired, unspecified
244.0	postsurgical
244.1	postablative
243	congenital
648.10	in pregnancy
781.1	with weight gain

Goiter
240.0	simple
240.9	unspecified
241.X	nontoxic nodular
780.79	fatigue

Suggested Reading

Centanni M, Gargano L, Canettieri G, Viceconte N, Franchi A, Delle Fave G, Annibale B. Thyroxine in goiter, *Heliobacter pylori* infection, and chronic gastritis. *N Engl J Med* 2006; 354:1787–95.

Desai J, Yassa L, Marqusee E, George S, Frates MC, Chen MH, Morgan JA, Dychter SS, Larsen PR, Demetri GD, Alexander EK. Hypothyroidism after sunitinib treatment for patients with gastrointestinal stromal tumors. *Ann Intern Med* 2006; 145:660–4.

Diez JJ, Iglesias P. Spontaneous subclinical hypothyroidism in patients older than 55 years: an analysis of natural course and risk factors for the development of overt thyroid failure. *J Clin Endocrinol Metab* 2004; 89:4890–7.

Indra R, Patil SS, Joshi R, Pai M, Kalantri SP. Accuracy of physical examination in the diagnosis of hypothyroidism: a cross-sectional, double-blind study. *J Postgrad Med* 2004; 50:7–11.

Khurram IM, Choudhary KS, Muhammad K, Islam N. Clinical presentation of hypothyroidism, a case control analysis. *J Ayub Med Coll Abbottabad* 2003; 15:45–9.

Ladenson PW, Goldenheim PD, Ridgway EC. Prediction and reversal of blunted ventilatory responsiveness in patients with hypothyroidism. *Am J Med* 1988; 84:877–83.

Mazzaferri EL, Surks MI. Recognizing the faces of hypothyroidism. *Hosp Pract* 1999; 34(3):93–96, 101–105, 109.

Redmond GP. Thyroid dysfunction and women's reproductive health. *Thyroid* 2004; 14(Suppl 1):5–15.

Strickland DM, Whitted WA, Wians FH Jr. Screening infertile women for subclinical hypothyroidism. *Am J Obstet Gynecol* 1990; 163:262–3.

Surks MI, Sievert R. Drugs and thyroid function. *N Engl J Med* 1995; 333:1688–94.

Surks MI, Ortiz E, Daniels GH, Sawin CT, Col NF, Cobin RH, Franklyn JA, Hershman JM, Burman KD, Denke MA, Gorman C, Cooper RS, Weissman NJ. Subclinical thyroid disease: scientific review and guidelines for diagnosis and management. *JAMA* 2004; 291:228–38.

Walsh JP, Ward LC, Burke V, Bhagat CI, Shiels L, Henley D, Gillett MJ, Gilbert R, Tanner M, Stuckey BG. Small changes in thyroxine dosage do not produce measurable changes in hypothyroid symptoms, well-being, or quality of life: results of a double-blind, randomized clinical trial. *J Clin Endocrinol Metab* 2006; 91:2624–30.

1.1.2   TSH as a Screening Test for Hypothyroidism in Adults and Neonates

Indications for Test

Screening a population at risk for hypothyroidism by testing for TSH is indicated:

- When the cost–benefit ratio of testing is sufficiently low and affordable enough to justify testing
- In accordance with the recommendations of various organizations (Table 1.4), depending on the target population to be screened

Procedure

1. Obtain a blood sample from individuals in a high-risk population, such as men over 60 years of age, women over 50 years of age, women in their first trimester of pregnancy with hypothyroid symptoms, or women with thyroid abnormalities, and measure [TSH].
2. For better accuracy in the diagnosis of individuals with nonspecific symptoms and physical findings suggestive of hypothyroidism, proceed directly to a combination of TSH with free T_4 measurement done by "reflex" if the [TSH] is above a specific cutoff value (e.g., 5 mIU/L).
3. In neonates at 6 to 8 days of birth, use a dried blood spotted filter paper specimen to test for [TSH] with confirmation of an abnormal value by serum TSH measurement.
4. Instead of screening 3- to 7-day-old neonates with a TSH test, measure [T_4] and follow up abnormally low values with TSH testing after the first week of life.

Interpretation

1. Adults with [TSH], confirmed by repeat testing, that is >5 mIU/L may be diagnosed with hypothyroidism if they present with signs and symptoms of hypothyroidism and the free [T_4] is low or in the lower range of normal.
2. Adults with [TSH] confirmed to be >5 mIU/L but <10 mIU/L without any signs or symptoms of hypothyroidism (Table 1.1) have subclinical hypothyroidism.
3. An untreated adult with [TSH] confirmed to be >10 mIU/L is unequivocally hypothyroid.
4. Neonates with [TSH] between 25 and 50 mIU/L will require repeat screening. Other authorities consider a TSH result of between 10 and 25 mIU/L borderline and cause for a repeat screen.
5. If the neonatal [TSH] is abnormal on two screenings or is >50 mIU/L on the initial screen, testing for [TSH] and free [T_4] is recommended within 48 hours.
6. Expect a 25 to 33% occurrence of early or transient hypothyroidism on 2-year follow-up of neonates with an initially elevated [TSH], which becomes normal, on repeat, in the absence of thyroid hormone therapy.

Notes

1. The recommendations regarding screening for thyroid disease are diverse. Refer to "Screening for Thyroid Disease in Asymptomatic Adults: Recommendations of Nine Professional Organizations" (Table 1.4) (Helfand, 2004).
2. A portable device (ThyroChek® One-Step Whole Blood Rapid TSH Assay) permits mass screening for hypothyroidism in homes for the elderly, in individuals living in geographically remote areas, or in populations without access to established laboratory testing facilities, such as areas of Asia, Eastern Europe, and South America. The ThyroChek device and 4 drops (100 μL) of serum (or blood) have been used to identify individuals with a TSH > 5 mIU/L. Positive screening tests require follow-up with clinical interview, physical examination, and measurement of TSH.
3. A separate ThyroChek device, incorporating a TSH assay with a cutoff value of 20 mIU/L, is available to identify neonates with a [TSH] > 20 mIU/L.
4. Hospitals routinely screen neonates for hypothyroidism by collecting dried blood specimens on filter paper and testing for either TSH or T_4 as mandated by state law.

TABLE 1.4
Screening for Thyroid Disease in Asymptomatic Adults:
Recommendations of Nine Professional Organizations

Organization	Screen?	Specific Recommendation
American Thyroid Association	Yes	Women and men over age 35; screen every 5 years thereafter with more frequent checks in high-risk or symptomatic individuals
American Association of Clinical Endocrinologists	Yes	Older patients, especially women planning to conceive or in the first trimester of pregnancy
American College of Obstetrics and Gynecology	Yes	Women greater than age 19 in high-risk groups (e.g., individuals with autoimmune disease, a strong family history of thyroid disease, or in the postpartum period)
American Academy of Family Physicians	Yes	May screen individuals at age >60 but do not screen at age <60 if asymptomatic
American College of Physicians	Yes	Screen women over age 50 if they have one or more symptoms suggestive of thyroid disease
College of American Pathologists	Yes	Screen women over age 50 if they have one or more symptoms suggestive of thyroid disease
U.S. Preventive Services Task Force	No	Insufficient evidence for or against screening
Canadian Task Force on the Periodic Health Examination	No	But maintain high index of clinical suspicion on physical examination, particularly in peri- and postmenopausal women
Royal College of Physicians	No	Screening of healthy adults unjustified

5. A 3rd-generation TSH assay (sensitivity, ~0.04 mIU/L; diagnostic accuracy, ~97%), requiring only microliter quantities of blood obtained by finger stick, is offered by Biosafe Laboratories, Inc. (8600 W. Catalpa Ave., Chicago, IL 60656-9958; 888-700-8378 or 773-693-0400). A Biosafe test kit is required for sample collection.

6. Thyroid dysgenesis is the most common (prevalence of 1/4000 live births) cause of neonatal hypothyroidism in the United States, with other causes accounting for less than 10% of the total.

ICD-9 Codes

Condition(s) that may justify this test *include but are not limited to*:

Hypothyroidism
240.9   goiter
243   congenital
244.9   unspecified
780.79   fatigue

Suggested Reading

Foo A, Leslie H, Carson DJ. Confirming congenital hypothyroidism identified from neonatal screening. *Ulster Med J* 2002; 71:38–41.

Helfand M. Screening for subclinical thyroid dysfunction in nonpregnant adults: a summary of the evidence for the U.S. Preventive Services Task Force. *Ann Intern Med* 2004; 140:128–41.

Hershman JM, Berg L. Rapid qualitative TSH test to screen for primary hypothyroidism. *Clin Chem* 1997; 43:1097–8.

Hofman LF, Foley TP, Henry JJ, Naylor EW. Assays for thyroid-stimulating hormone using dried blood spotted filter paper specimens to screen for hypothyroidism in older children and adults. *J Med Screen* 2003; 10:5–10.

King P, Bryant W. Congenital hypothyroidism screening in Oklahoma: a change in follow-up recommendations for 1999. *J Okla State Med Assoc* 1999; 92:42–3.

1.1.3   Free T$_4$ (Free Thyroxine) as a Test for Hypothyroidism

Indications for Test

Free T$_4$ (FT$_4$) as a test for hypothyroidism is indicated:

- In patients with established thyroid or pituitary disorders
- If symptoms including cold intolerance, constipation, fatigue, sleep disturbances, mental slowing, or memory problems are noted in patients with possible hypothyroidism
- When monitoring patients treated for thyroid disease

Procedure

1. Obtain a random blood sample for FT$_4$ testing by a cost-effective, one-step immunometric method (Figure 1.2).
2. Be prepared to order the gold standard FT$_4$ test (i.e., equilibrium dialysis) or a two-step test requiring immunoextraction to clarify a confusing clinical situation, recognizing that this will be needed when thyroid-hormone-binding proteins (TBPs) are markedly increased or decreased.
3. Ordinarily, obtain a [TSH] in conjunction with the FT$_4$ test except in patients with known, well-established, pituitary thyrotrope failure or TSH suppression within the last 6 weeks. In such cases, FT$_4$ testing alone is appropriate.
4. In patients with established hypothyroidism on treatment with thyroxine, identify medications and drugs that may alter or increase TBP (see Table 1.3, Pattern VIII).

Interpretation

1. Refer to "Non-Drug-Related Causes of Various Patterns of TSH and FT$_4$ Results" (Table 1.2) and "Drug-Related Causes of Various Patterns of TSH and FT$_4$ Results" (Table 1.3).
2. The reference interval for [FT$_4$] is 0.6 to 1.8 ng/dL. A narrower interval of 0.8 to 1.5 ng/dL allows identification of an abnormal population with greater diagnostic specificity.
3. Because of intra- and interindividual biological variation, consider that an individual may exhibit a range of [FT$_4$] and [TSH] more typically normal for that individual than what is considered normal for the reference group used to establish the reference interval for these analytes.
4. Increases in circulating T$_4$ result from treatment with agents that inhibit T$_4$ to T$_3$ conversion such as beta blockers and glucocorticoids.
5. An increase in the metabolic breakdown of T$_4$ may occur with the intake of cytochrome P450 activators such as phenytoin and barbiturates.
6. If the [FT$_4$] is in the lower end of the reference range and the [TSH] is normal, the patient may be receiving T$_3$ treatment with triiodothyronine (Cytomel®) or desiccated thyroid extract.

Notes

1. FT$_4$ assays indirectly measure the free, bioactive form of T$_4$.
2. Used in conjunction with TSH, testing for FT$_4$ is preferred over total T$_4$ to monitor and assess thyroid status because [FT$_4$] is less affected by changes in TBP.
3. When the [TSH] is outside an interval of 0.4 to 4 mIU/L, reflex testing for [FT$_4$] is indicated. To save costs, when the TSH is normal, defer measurement of the FT$_4$.
4. The FT$_4$ result, measured using a one-step assay, changes with marked increases and decreases in TBP, particularly albumin. Two-step FT$_4$ assays are practically unaffected by this problem.
5. L-Thyroxine taken orally may be sequestered or malabsorbed in patients taking proton pump inhibitors, H2 blockers, calcium supplements, iron and other metallovitamins, bile acid sequestrants, sucralfate, or aluminum-containing antacids, resulting in lowered [FT$_4$].
6. Patients whose serum contains human anti-mouse antibodies (HAMAs) may have falsely elevated (or decreased) [FT$_4$] if the assay used for FT$_4$ testing uses mouse-derived monoclonal capture or signal antibodies.
7. Equilibrium dialysis is the gold standard method of assay for FT$_4$; however, it is cumbersome, expensive, and not readily available.

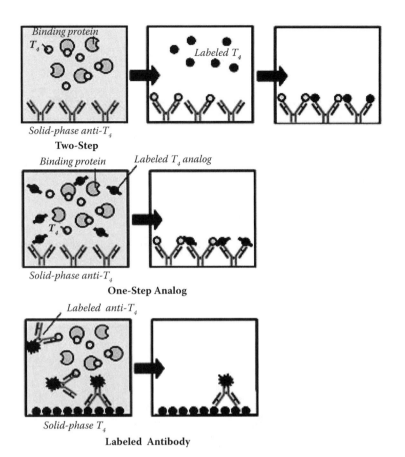

Figure 1.2
Illustration of differences among three different approaches to quantifying serum FT_4 concentration: two-step, one-step analog, and labeled antibody (Ab) immunoassays. The principle of each of these assays is summarized below:

Assay Type	Solid-Phase Ligand	Immunoassay Principle
Two-step	anti-T_4 Ab	FT_4 in sample binds to solid-phase bound capture Ab (first step). Labeled T_4 is then added and saturates any unbound T_4 Ab sites (second step). The amount of signal provided by the labeled T_4 bound to solid-phase Ab is inversely proportional to the concentration of FT_4 in the patient's serum.
One-step	anti-T_4 Ab	A labeled T_4 analog is added to the patient's serum. Competition between unlabeled FT_4 in the patient's serum and labeled T_4 analog for solid-phase anti-T_4 Ab binding sites occurs. After washing to remove all components of the reaction mixture except unlabeled FT_4 from the patient's serum and labeled T_4 analog bound to the solid-phase anti-T_4 Ab, the amount of signal produced is inversely proportional to the concentration of FT_4 in the patient's serum.
Labeled Ab	T_4	Labeled anti-T_4 Ab is added to the patient's serum. FT_4 and solid-phase bound T_4 compete for labeled anti-T_4 Ab. The amount of signal produced is inversely proportional to the concentration of FT_4 in the patient's serum.

Note: The solid phase in these assays could be a plastic bead, paramagnetic particle, test-tube wall, or the surface of a microtiter plate well; the signal could be a radioisotope, a substance capable of chemiluminescence or fluorescence, or an enzyme, requiring the addition of a substrate to measure the conversion of the substrate to a colored or colorless product with a characteristic spectrophotometric absorption maximum.

ICD-9 Codes

Refer to Test 1.1.1 codes.

Suggested Reading

Andersen S, Pedersen KM, Bruun NH, Laurbert P. Narrow individual variations in serum T(4) and T(3) in normal subjects: a clue to the understanding of subclinical thyroid disease. *J Clin Endocrinol Metab* 2002; 87:1068–72.

Arafah BM. Increased need for thyroxine in women with hypothyroidism during estrogen therapy. *N Engl J Med* 2001; 344:1743–9.

Elmlinger MW, Kuhnel W, Lambrecht HG, Ranke MB. Reference intervals from birth to adulthood for serum thyroxine (T_4), triiodothyronine (T_3), free T_3, free T_4, thyroxine binding globulin (TBG), and thyrotropin (TSH). *Clin Chem Lab Med* 2001; 39:973–9.

Ferretti E, Persani L, Jaffrain-Rea ML, Giambona S, Tamburrano G, Beck-Peccoz P. Evaluation of the adequacy of levothyroxine replacement therapy in patients with central hypothyroidism. *J Clin Endocrinol Metab* 1999; 84:924–9.

Singh N, Singh PN, Hershman JM. Effect of calcium carbonate on the absorption of levothyroxine. *JAMA* 2000; 283:2822–5.

Wang R, Nelson JC, Weiss RM, Wilcox RB. Accuracy of free thyroxine measurements across natural ranges of thyroxine binding to serum proteins. *Thyroid* 2000; 10:31–9.

1.1.4   Antimicrosomal (Anti-M) or Antithyroid Peroxidase (Anti-TPO) and Antithyroglobulin (Anti-Tg) Antibodies in Hypothyroidism

Indications for Test

Antithyroid antibody tests are indicated when:

- Screening for thyroiditis and attempting to confirm a diagnosis of Hashimoto's thyroiditis
- Determining type of goiter before thyroidectomy
- Making predictions of thyroid dysfunction when other types of autoimmune disorder are present (type 1 diabetes, polyendocrinopathies, adrenalitis)
- Making a differential diagnosis of the causes for hypothyroidism

Procedure

1. Obtain a random blood sample for testing Tg, which includes anti-Tg and anti-TPO (or anti-M). *Note:* Most laboratories do not perform anti-Tg testing independent of quantitative Tg testing (i.e., all requests for Tg include reflex testing for anti-Tg).

Interpretation

1. Anti-TPO (or anti-M) and anti-Tg concentrations or titers are usually negative or undetectable.
2. Positive anti-M titers occur in up to 25% of normal adults and in patients with no evidence of thyroid disease.
3. In the NHANES III survey of 17,000 adults without reported thyroid disease, 12% had detectable anti-TPO.
4. Very high titers of anti-M (>1:10000) are specific for Hashimoto's thyroiditis.

5. Anti-Tg is usually elevated in conjunction with anti-M, but anti-M may be positive independent of a positive anti-Tg titer.
6. Expect 15 to 20% of patients who are positive for anti-TPO (or anti-M) and have a [TSH] near the upper limit of normal to progress to a hypothyroid state.
7. Thyroid autoimmunity is common in females with type 1 diabetes by the time they are in their second decade of life. When both anti-TPO and anti-Tg antibodies are present in the serum from such patients, their [TSH] is typically >3 mIU/L.

Notes

1. Both anti-Tg and anti-TPO (or anti-M) antibodies are elevated in a variety of thyroid diseases, including 80 to 90% of patients with Hashimoto's thyroiditis.
2. Anti-TPO and anti-M antibodies may be found in the serum from 95% and 99% of patients with Hashimoto's thyroiditis cases, respectively. Diagnostic correlation between anti-TPO and anti-M is high ($r = 0.90$); however, use of an enzyme-linked immunosorbent assay (ELISA) for quantifying anti-TPO antibodies that utilizes calibrators containing recombinant TPO had both a higher diagnostic specificity and sensitivity than anti-M for identification of patients with chronic thyroiditis.
3. More anatomic abnormalities on thyroid ultrasound, specifically heterogeneity and nodules, were present in patients with anti-Tg alone than in those with anti-TPO alone, but hypothyroidism was found more often in the anti-TPO patient group.
4. The presence of anti-TPO antibodies marks a familial predisposition to autoimmune thyroid disease.
5. The presence of anti-Tg antibodies invalidates the results of Tg measurement and obviates its use as a tumor marker.
6. Infertile women have increased thyroid autoimmunity (i.e., anti-TPO antibodies are present in their serum) but no increase in thyroid dysfunction (i.e., TSH and/or free T_4 are within reference range) compared to a fertile peer group.

ICD-9 Codes

Condition(s) that may justify this test *include but are not limited to*:

Thyroiditis
245.1   subacute
245.2   lymphocytic, chronic (Hashimoto's)
648.14   postpartum

Hypothyroidism
244.8   specified, other

Suggested Reading

Dayan CM, Daniels GH. Chronic autoimmune thyroiditis. *N Engl J Med* 1996; 335:99–107.

Kordonouri O, Klinghammer A, Lang EB, Gruters-Kieslich A, Grabert M, Holl RW. Thyroid autoimmunity in children and adolescents with type 1 diabetes: a multicenter survey. *Diabetes Care* 2002; 25:1346–50.

Ruf J, Czarnocka B, Ferrand M, Doullais F, Carayon P. Novel routine assay of thyroperoxidase autoantibodies. *Clin Chem* 1988; 34:2231–4.

Strieder TG, Prummel MF, Tijssen JG, Endert E, Wiersinga WM. Risk factors for and prevalence of thyroid disorders in a cross-sectional study among healthy female relatives of patients with autoimmune thyroid disease. *Clin Endocrinol* 2003; 59:396–401.

Takamatsu J, Yoshida S, Yokozawa T, Hirai K, Kuma K, Ohsawa N, Hosoya T. Correlation of antithyroglobulin and antithyroid-peroxidase antibody profiles with clinical and ultrasound characteristics of chronic thyroiditis. *Thyroid* 1998; 8:1101–6.

Weetman AP, McGregor AM. Autoimmune thyroid disease: developments in our understanding. *Endocr Rev* 1984; 5:309–55.

1.1.5   Thyrotropin-Binding Inhibitory Immunoglobulins (TBIIs) in Hypothyroidism

Indications for Test

Measurement of TBII may be indicated for prediction of:

- The presence and duration of hypothyroidism in neonates of mothers with autoimmune thyroid disease
- Recurrence of Graves' disease after cessation of treatment with antithyroid drugs once a euthyroid or hypothyroid state is achieved in adults
- Which autoimmune disease patients are at high risk for hypothyroidism in cases of postpartum thyroiditis; Hashimoto's thyroiditis with large, potentially failing goiter; or Addison's disease

Procedure

1. Obtain a blood sample for TBII concentration in the high-risk populations noted above.
2. Obtain a blood sample for both TBII and thyroid-stimulating antibody (TS-Ab) in all neonates with a positive screen for neonatal hypothyroidism to help identify those with a transient cause of hypothyroidism.
3. Monitor the concentration of TBII in patients receiving antithyroid drug therapy (i.e., methimazole or propylthiouracil) when deciding on dose adjustment or withdrawal of this therapy.
4. Test for TS-Ab in treated Graves' disease patients who have a normal or low free T_4 level and an unexpectedly low TSH level.

Interpretation

1. The reference interval for TBII is usually reported as negative or less than 130%.
2. The blocking nature of TBII may result in transient hypothyroidism in newborns and an athyreotic appearance on thyroid scan associated with a compensatory elevation in TSH. Prolonged neonatal hypothyroidism, as a result of extremely high TBII levels, is a rare event.
3. A rise and fall in TBII and TS-Ab may be expected to correlate with appropriate changes in thyroid status.
4. The Graves' type of hyperthyroidism caused by increased TS-Ab may be followed immediately by transient hypothyroidism due to coexisting destructive autoimmune thyroiditis during the early postpartum period.
5. In treated patients with Graves' disease who may be euthyroid or even hypothyroid, high titers of TBII may suppress TSH secretion independent of thyroid hormone levels, most likely by binding to the pituitary TSH receptor, resulting in a falsely low TSH.

Notes

1. Elevated TBII titers are seen in most patients with autoimmune thyroid disease. Note that TBII is routinely detected by measuring the ability of test sera to inhibit TSH binding to the receptor and is not a direct measure of TS-Ab.
2. High levels of TBII are found in over 98% of patients with Graves' disease.
3. Even with a highly sensitive TBII assay that uses human recombinant TSH receptor protein as a reagent, the prediction of antithyroid drug treatment outcome in patients with Graves' disease requires assessment of clinical parameters and assays for other types of thyroid antibodies.
4. TBII is only a reflection of the effect of TS-Ab. The titer of TS-Ab is measured in bioassays that quantitate the cAMP response to TS-Ab-containing sera applied to Chinese hamster ovary (CHO) cells. In a chemiluminescent TS-Ab assay using CHO cells, the reference interval for TS-Ab was ≤1.5 relative light units (Evans et al., 1999).
5. The elimination half-life of TS-Ab is about 7.5 days.

ICD-9 Codes

Condition(s) that may justify this test *include but are not limited to*:

Graves' disease
242.00   pregnancy, unspecified
648.10   in pregnancy

Refer also to Test 1.1.4 codes.

Suggested Reading

Brokken LJ, Wiersinga WM, Prummel MF. Thyrotropin receptor autoantibodies are associated with continued thyrotropin suppression in treated euthyroid Graves' disease patients. *J Clin Endocrinol Metab* 2003; 88:4135–8.

Connors MH, Styne DM. Transient neonatal "athyreosis" resulting from thyrotropin-binding inhibitory immunoglobulins. *Pediatrics* 1986; 78:287–90.

Evans C, Morgenthaler NG, Lee S, Llewellyn DH, Clifton-Bligh R, John R, Lazarus JH, Chatterjee VK, Ludgate M. Development of a luminescent bioassay for thyroid stimulating antibodies. *J Clin Endocrinol Metab* 1999; 84:374–7.

Gillis D, Volpe R, Daneman D. A young boy with a thyroid yo-yo. *J Pediatr Endocrinol Metab* 1998; 11:467–70.

Maugendre D, Massart C. Clinical value of a new TSH binding inhibitory activity assay using human TSH receptors in the follow-up of antithyroid drug treated Graves' disease: comparison with thyroid stimulating antibody bioassay. *Clin Endocrinol (Oxf)* 2001; 54:89–96.

Shigemasa C, Mitani Y, Taniguchi S, Ueta Y, Urabe K, Tanaka T, Yoshida A, Mashiba H. Development of postpartum spontaneously resolving transient Graves' hyperthyroidism followed immediately by transient hypothyroidism. *J Intern Med* 1990; 228:23–8.

Usala AL, Wexler I, Posch A, Gupta MK. Elimination kinetics of maternally derived thyrotropin receptor-blocking antibodies in a newborn with significant thyrotropin elevation. *Am J Dis Child* 1992; 146:1074–7.

1.1.6   Thyroid Scanning Scintigraphy Using ^{99m}Tc-Pertechnetate or [^{123}I]Na in Hypothyroidism

Indications for Test

Thyroid scanning scintigraphy using ^{99m}Tc or [^{123}I]Na is indicated:

- In the diagnosis of ectopic or congenitally absent thyroid
- When there is palpable nodularity in a hypothyroid goiter
- In preparation for surgery to debulk a large symptomatic thyroid gland

Procedure

1. Use [^{123}I]Na in preference to ^{99m}Tc as the isotopic scanning agent, particularly in the evaluation and classification of congenital hypothyroidism.
2. Ideally, couple high-resolution thyroid ultrasound imaging with radioisotope scintigraphy on the same day.

Interpretation

1. A warm to hot area on radioisotope scan may be found in 1 to 2% of patients with hypothyroid goiters, usually associated with the presence of thyroid antibodies.
2. Up to 5% of warm to hot areas will contain a primary thyroid cancer.
3. Up to 5% of patients with asymmetry on scan will be found to have hemilobar atrophy, hypogenesis, or hemilobar agenesis.
4. More ectopic thyroid tissue will be found with [^{123}I]Na than ^{99m}Tc scintigraphy in children with congenital hypothyroidism. In such cases, thyroid aplasia may be present in up to 60% of patients scanned with ^{99m}Tc but only about 25% scanned with [^{123}I]Na.

Notes

1. The causes of hot areas on thyroid scintigraphy include hyperfunctioning tumors or nodules, localized functioning thyroid tissue spared from autoimmune destruction, inflammation or tumor invasion, congenital abnormality, clusters of hyperactive follicular cells caused by long-term TSH and/or thyroid-stimulating immunoglobulin (TSI) stimulation, and asymmetry of the thyroid gland.
2. Thyroid scintigraphy is the most accurate diagnostic test to detect thyroid dysgenesis or one of the inborn errors of thyroxine synthesis. Thyroid sonography alone may miss some cases of ectopic glands.

ICD-9 Codes

Condition(s) that may justify this test *include but are not limited to*:

Goiter
240.X   nontoxic goiter

Thyroid nodule(s)
241.0   thyroid nodule, unspecified
242.1   thyroid nodule, toxic
242.20   multinodular, toxic
242.3   unspecified, toxic nodular

Hypothyroidism
243   congenital

Suggested Reading

Iwata M, Kasagi K, Hatabu H, Misaki T, Iida Y, Fujita T, Konishi J. Causes of appearance of scintigraphic hot areas on thyroid scintigraphy analyzed with clinical features and comparative ultrasonographic findings. *Ann Nucl Med* 2002; 16:279–87.

La Franchi S. Congenital hypothyroidism, etiologies, diagnosis, and management. *Thyroid* 1999; 9:735–40.

Panoutsopoulos G, Mengreli C, Ilias I, Batsakis C, Christakopoulou I. Scintigraphic evaluation of primary congenital hypothyroidism: results of the Greek screening program. *Eur J Nucl Med* 2001; 28:529–33.

1.1.7   T$_3$ Resin Uptake (RU) and Thyroid Hormone Binding Ratio (THBR) as Tests to Estimate Thyroid-Hormone-Binding Proteins (TBPs) in Hypothyroidism

Indications for Tests

Measurement of T$_3$ RU and calculation of THBR may be indicated:

- When estimating the variability in protein binding of T_4 by TBPs—albumin, thyroxine-binding globulin (TBG), and thyroxine-binding prealbumin (TBPA)

Procedure

1. Obtain a blood sample for T_3 RU testing.
2. T_3 RU is expressed as a percent of the labeled T_3 added during the test and is inversely related to TBG concentration.
3. The THBR is calculated by dividing the measured T_3 RU value by the mean T_3 RU value of a healthy reference population. It is an assay-dependent variable used to obtain an estimate of the concentration of the THB proteins:

$$THBR = (T_3\text{-}RU)_{measured} / (T_3\text{-}RU)_{mean}$$

Interpretation

1. The reference interval for T_3 RU is typically 25 to 35%. The mean T_3 RU value is laboratory dependent, reflecting the type and precision of the assay used to measure T_3 RU.
2. Low T_3 RU values indicate increased T_4 binding sites in the patient's serum, and high values indicate decreased binding sites.
3. High TBG levels (i.e., low T_3 RU) are seen in pregnancy, use of oral contraceptives, estrogen or perphenazine administration, acute hepatitis, porphyria, and familial syndromes of increased TBG concentration. Hypothyroidism tends to raise the TBG concentration and lower the T_3 RU.
4. Decreased TBG (i.e., high T_3 RU) may be caused by familial TBG deficiency, androgen or glucocorticoid administration, acute and chronic illness, acromegaly, and hyperthyroidism.
5. The reference interval for THBR is 0.83 to 1.15.
6. THBR > 1.15 indicates a decreased availability of TBG binding sites for T_4.
7. THBR < 0.83 indicates an increased availability of TBG binding sites for T_4.

Notes

1. These tests are only needed when FT_4 testing is unavailable, a total T_4 (TT_4) has been obtained, and a free thyroxine index (FTI) is to be calculated.
2. The TT_4 and T_3 RU tests, with calculated FTI, have been made obsolete by the availability of highly analytically sensitive TSH and improved FT_4 assays.

ICD-9 Codes

Refer to Test 1.1.1 codes.

Suggested Reading

Csako G, Zweig MH, Ruddel M, Glickman J, Kestner J. Direct and indirect techniques for free thyroxine compared in patients with nonthyroidal illness. III. Analysis of interference variables by stepwise regression. *Clin Chem* 1990; 36:645–50.

Harman SM, Wehmann RE, Blackman MR. Pituitary–thyroid hormone economy in healthy aging men: basal indices of thyroid function and thyrotropin responses to constant infusions of thyrotropin releasing hormone. *J Clin Endocrinol Metab* 1984; 58:320–6.

Wilke TJ. Free thyroid hormone index, thyroid hormone/thyroxine-binding globulin ratio, triiodothyronine uptake, and thyroxine-binding globulin compared for diagnostic value regarding thyroid function. *Clin Chem* 1983; 29:74–9.

1.1.8   Free Thyroxine Index (FTI or FT_4I) and Total Thyroxine (Total T_4 or TT_4) in Hypothyroidism

Indications for Test

The total T_4 test is indicated:

- When a FT_4 test is unavailable and the FTI is to be calculated from the TT_4 and T_3 resin uptake (T_3 RU) values

Procedure

1. Obtain a random blood sample for TT_4 and T_3 RU only if a FT_4 test cannot be obtained.
2. Calculate the FTI as shown below.

Method 1

$$FTI = TT_4 \times THBR$$

Using this calculation, the reference interval for FTI is similar to that for TT_4.

Method 2

$$FTI \text{ (also called } T_7) = TT_4 \times T_3 \text{ RU}$$

Recognize the limitations of this effort to correct TT_4 for TBP abnormalities, and focus on the patient's clinical symptoms suggestive of hypothyroidism.

Interpretation

1. The reference interval for TT_4 in adults is typically 4.1 to 11.2 mcg/dL. On serial measurement, the TT_4 for an individual patient is characteristic of that patient and falls within the limits of the reference change value (RCV; see Introduction, p. xxxi) for TT_4.
2. A low TT_4 concentration may be explained by:
   - Low levels of TBPs
   - Nonthyroidal illness
   - Low levels of both free and bound T_4 in true hypothyroidism
   - T_3 ingestion.
3. Using Method 1, the reference interval for FTI is 4.1 to 11.2 mcg/dL.
4. Using Method 2, the reference interval for FTI is 1.0 to 4.2 mcg/dL.
5. A decreased FTI is seen in hypothyroidism and in patients who take exogenous T_3.
6. Patients who take exogenous T_3 may have a decrease in calculated FTI but actually be hyperthyroid.

Notes

1. Thyroxine (T_4) acts as a prohormone with minimal intrinsic activity. It is metabolized in peripheral tissues to triiodothyronine (T_3), which is many times more bioactive than thyroxine.
2. TT_4 assays measure both free and bound T_4.
3. Almost all (99.97%) of the circulating T_4 is bound to serum TBPs and is biologically inactive.
4. Oral estrogen treatment stimulates the liver to produce increased amounts of thyroxine-binding globulin (TBG) and results in higher TT_4 levels and falsely high FTI calculations.
5. In the absence of severe illness, a low FTI value may correlate with FT_4 concentration; however, in severe illness, the FTI may be falsely low.
6. The FTI has a lower sensitivity and specificity for diagnosis of hypothyroidism in critical illness than FT_4 by the one-step or two-step method.

7. The FTI helps correct the TT_4 for variation in the concentration of THB proteins; however, an indirect FT_4 measurement is usually preferable to and more cost effective than an FTI, which requires two separate tests and a calculation.
8. In hypothyroid patients with abnormally high TBPs, falsely normal and even elevated TT_4 levels frequently lead to confusion in establishing a correct diagnosis.

ICD-9 Codes

Refer to Test 1.1.1 codes.

Suggested Reading

Bartalena L, Bogazzi F, Brogioni S, Burelli A, Scarcello G, Martino E. Measurement of serum free thyroid hormone concentrations: an essential tool for the diagnosis of thyroid dysfunction. *Horm Res* 1996; 45:142–7.

Midgley JE. Direct and indirect free thyroxine assay methods: theory and practice. *Clin Chem* 2001; 47:1353–63.

Nordyke RA, Reppun TS, Madanay LD, Woods JC, Goldstein AP, Miyamoto LA. Alternative sequences of thyrotropin and free thyroxine assays for routine thyroid function testing: quality and cost. *Arch Intern Med* 1998; 158:266–72.

Spiessens H, Uyttenbroeck F. Evaluation of the changes of *in vitro* thyroid function tests during pregnancy (binding capacity of triiodothyronine, dosage of total thyroxine, free thyroxine and its computed T_7-index). *J Belge Radiol* 1973; 56:217–22.

1.1.9   Thyroid-Binding Globulin (TBG) and Thyroxine-Binding Prealbumin (TBPA) as Tests in the Event of a False-Normal TT_4 in Hypothyroidism

Indications for Tests

Testing for the thyroid-hormone-binding proteins TBG and TBPA in clinically hypothyroid patients is indicated:

- In cases of clinically suspected hypothyroidism when the $[TT_4]$ is normal or elevated
- In the absence of clinical hyperthyroidism when the $[TT_4]$ is normal or elevated and factors (e.g., estrogen therapy) known to elevate THB proteins are not involved and a FT_4 assay is unavailable

Procedure

1. Obtain a random blood sample for both TBG and TBPA in a clinically euthyroid or hypothyroid patient with an elevated or normal $[TT_4]$.

Interpretation

1. The reference intervals for TBG and TBPA are assay dependent.
2. Elevated TBG occurs in pregnancy, with the use of oral contraceptives, the intake of estrogen or perphenazine, acute hepatitis, and porphyria.
3. Elevated TBG and/or TBPA without predisposing factors is found in familial syndromes of thyroxine-binding protein excess.
4. In familial disorders of TBP excess, family counseling may be indicated; however, such disorders rarely result in any clinically significant problem.

Notes

1. Both TBG and TBPA are produced in the liver.
2. TBG and TBPA tests are more expensive than FT_4 determination.
3. If the $[FT_4]$ is within the reference interval, testing for TBG and TBPA is usually unnecessary.

ICD-9 Codes

Condition(s) that may justify this test *include but are not limited to*:

Hypothyroidism
244.9   acquired, unspecified

Suggested Reading

Borst GC, Eil C, Burman KD. Euthyroid hyperthyroxinemia. *Ann Intern Med* 1983; 98:366–78.
Litherland PG, Bromage NR, Hall RA. Thyroxine binding globulin (TBG) and thyroxine binding prealbumin (TBPA) measurement, compared with the conventional T_3 uptake in the diagnosis of thyroid disease. *Clin Chim Acta* 1982; 122:345–52.

1.1.10 Triiodothyronine (T_3), Total T_3 (TT_3), Free T_3 (FT_3), and Reverse T_3 (rT_3) for Determination of Thyroid Status, Especially in Critical Illness or Low T_3 Syndrome

Indications for Tests

The TT_3 and/or FT_3 tests for thyroid status may be indicated:

- As a prognostic factor in chronically and/or critically ill patients
- When symptoms of hypothyroidism appear or persist in thyroxine-treated patients in whom the [TSH] and $[FT_4]$ are within the reference interval
- In hypothyroid patients treated with triiodothyronine (Cytomel®) alone or dessicated thyroid extract

Testing for rT_3 may be indicated:

- When attempting to differentiate true hypothyroidism from nonthyroidal illness (NTI) caused by trauma, infection, starvation, or drug abuse

Procedure

1. Obtain a random blood sample for TT_3, FT_3, and/or rT_3 testing in carefully selected patients as described above.

Interpretation

1. Low $[TT_3]$ and/or $[TT_4]$ or $[FT_4]$ early in the course of critical illness are indicative of a poor prognosis.
2. The $[rT_3]$ reference interval is 30 to 78 ng/dL (0.45–1.16 nmol/L).
3. $[rT_3] > 60$ ng/dL (0.90 nmol/L) suggests NTI or drug effect.
4. Concentrations of rT_3 increase in hyperthyroidism and critical illness.

Notes

1. T_4 is normally deiodinated to produce T_3, the most potent thyroid hormone, or rT_3, which has no biological activity.
2. Reference intervals for $[TT_3]$ and $[FT_3]$ are assay dependent.

3. The rT_3 test does not have a routine clinical use since it does not reliably discriminate among NTI, hyperthyroidism, and hypothyroidism.

4. Withdrawal of prolonged dopamine infusion has been shown to be followed by a 10-fold increase in [TSH]; a 57% and 82% rise in [TT_4] and [TT_3], respectively; and an increase in the T_3/rT_3 ratio, whereas a brief dopamine infusion has a suppressive effect on the thyroid axis within 24 hours.

5. The transient alterations in T_4, T_3, rT_3, and TSH concentrations that occur with acute illness in primary hypothyroid patients and in initially euthyroid patients on appropriate levothyroxine replacement therapy are practically identical to those occurring in normal subjects with acute illness. Thus, frequent changes in T_4 replacement dose during an acute illness in hypothyroid patients are not necessary.

ICD-9 Codes

Condition(s) that may justify this test *include but are not limited to*:

Hypothyroidism
244.9   acquired, unspecified

Suggested Reading

Chopra IJ, Hershman JM, Pardridge WM, Nicoloff JT. Thyroid function in nonthyroidal illnesses. *Ann Intern Med* 1983; 98:946–57.

Custro N, Scafidi V, Costanzo G, Notarbartolo A. Prospective study on thyroid function anomalies in severely ill patients. *Ann Ital Med Int* 1992; 7:13–8.

Girdler SS, Thompson KS, Light KC, Leserman J, Pedersen CA, Prange AJ. Historical sexual abuse and current thyroid axis profiles in women with premenstrual dysphoric disorder. *Psychosom Med* 2004; 66:403–10.

Roti E, Gardini E, Magotti MG, Pilla S, Minelli R, Salvi M, Monica C, Maestri D, Cencetti S, Braverman LE. Are thyroid function tests too frequently and inappropriately requested? *J Endocrinol Invest* 1999; 22:184–90.

Van den Berghe G, de Zegher F, Lauwers P. Dopamine and the sick euthyroid syndrome in critical illness. *Clin Endocrinol (Oxf)* 1994; 41:731–7.

Vasa FR, Molitch ME. Endocrine problems in the chronically critically ill patient. *Clin Chest Med* 2001; 22:193–208.

Wadwekar D, Kabadi UM. Thyroid hormone indices during illness in six hypothyroid subjects rendered euthyroid with levothyroxine therapy. *Exp Clin Endocrinol Diabetes* 2004; 112:373–7.

1.2   Thyroid Hormone Excess

1.2.1   Thyroid-Stimulating Hormone as a Test for Hyperthyroidism

Indications for Test

TSH testing is indicated when:

- Patients treated for thyroid disease are being monitored, particularly if overtreatment with thyroxine or triiodothyroxine is suspected.
- Symptoms suggestive of hyperthyroidism, including heat intolerance, looser or more frequent bowel movements, palpitations, weight loss, tremor, fatigue, sleep disturbances, and/or increased irritability, are noted as presented in the "Thyroid Symptom Questionnaire (TSQ)" (Q1.1 in Appendix 1).

Procedure

1. Obtain a random blood sample for TSH testing.

Interpretation

1. Refer to Tables 1.2 and 1.3.
2. Reference intervals for TSH assays vary and typically range from 0.3 to 6.5 mIU/L. The lower limit of detection varies but is usually about 0.05 to 0.01 mIU/L for the 2nd- and 3rd-generation TSH assays, respectively, and 0.005 mIU/L for the 4th-generation TSH assay.
3. [TSH] < 0.3 mIU/L usually indicates hyperthyroidism or overt pituitary or hypothalamic dysfunction. Low [TSH] in hyperemesis gravidarum may be caused by elevated human chorionic gonadotropin (hCG) in pregnancy or hyperthyroidism. Most patients with frank hyperthyroidism will have an undetectable [TSH] (<0.05 mIU/L). [FT_4] will be high in hyperthyroidism and low (<0.6 mIU/L) in hypopituitarism.
4. A [TSH] interval of 0.1 to 0.3 mIU/L defines subclinical hyperthyroidism with an increased risk for atrial fibrillation associated with [TSH] < 0.1 mIU/L.
5. Agents that reduce the [TSH] include administration of exogenous T_3 and T_4, high-dose glucocorticoids, somatostatin analogs, dopamine agonists, and iodides if given to relatively iodide-deficient patients.
6. Elevated [TSH] with elevated [FT_4] suggests a TSH-producing tumor, functionally abnormal TSH, thyroid hormone resistance, psychoneurosis, or recovery from severe nonthyroidal illness.
7. In unselected general medical, geriatric, or psychiatric inpatients, TSH testing for hyperthyroidism results in a low yield of true positives and many false positives.

Notes

1. The obsolete (*circa* 1992) regular TSH (rTSH) assay was incapable of accurately measuring [TSH] < 0.5 mIU/L and could not specifically identify a hyperthyroid patient.
2. TSH originates from the pituitary and is a pituitary test.
3. A unique role for the 4th-generation TSH assay in the diagnosis of hyperthyroidism has yet to be identified.
4. TSH and thyroid hormones function in a classical endocrine feedback loop.
5. A low or suppressed [TSH] is usually present at any time of the day and does not demonstrate a diurnal cycle.
6. A low serum [TSH] may be a risk factor for atrial fibrillation and perhaps other cardiovascular diseases.
7. The pulse wave arrival time (QKd), which is the time interval between the onset of the Q-wave and the onset of the arrival of Korotkoff sounds by Doppler recording at the antecubital fossa (Kd); the systolic time interval; and the Achilles tendon reflex times are shorter in hyperthyroid patients. These tests are limited to research studies of hyper- and hypothyroidism (Christ-Crain et al., 2004).

ICD-9 Codes

Condition(s) that may justify this test *include but are not limited to*:

Goiter
242.00   diffuse, toxic (Graves')
242.10   uninodular, toxic
242.20   multinodular, toxic
242.90   thyrotoxicosis, unspecified

Thyroiditis
245.0   acute
245.2   Hashimoto's type
648.14   postpartum

Weight loss
783.21

Suggested Reading

Attia J, Margetts P, Guyatt G. Diagnosis of thyroid disease in hospitalized patients: a systematic review. *Arch Intern Med* 1999; 159:658–65.

Christ-Crain M, Meier C, Huber PR, Staub J-J, Muller B. Effect of L-thyroxine replacement therapy on surrogate markers of skeletal and cardiac function in subclinical hypothyroidism. *Endocrinologist* 2004; 14:161–6.

Madeddu G, Spanu A, Falchi A, Nuvoli S. Clinical and laboratory assessment of subclinical thyroid disease. *Rays* 1999; 24:229-42.

Sawin CT. Subclinical hyperthyroidism and atrial fibrillation. *Thyroid* 2002; 12:501–3.

Wians FH Jr, Jacobson JM, Dev J, Heald JI, Ortiz G. Thyrotroph function assessed by sensitive measurement of thyrotropin with three immunoradiometric assay kits: analytical evaluation and comparison with the thyroliberin stimulation test. *Clin Chem* 1988; 34:568–75.

Young RT, Van Herle AJ, Rodbard D. Improved diagnosis and management of hyper- and hypothyroidism by timing the arterial sounds. *J Clin Endocrinol Metab* 1976; 42:330–40.

1.2.2   Free Thyroxine (Free T_4, FT_4) as a Test for Hyperthyroidism

Indications for Test

FT_4 measurement as a test for hyperthyroidism is indicated:

- In patients with established thyroid or pituitary disorders and symptoms of hyperthyroidism including heat intolerance, frequent bowel movements, palpitations, weight loss, tremor, fatigue, sleep disturbances, or irritability

Procedure

1. Obtain a random blood sample for FT_4 testing by a cost-effective, one-step immunometric method.
2. Be prepared to order the gold standard FT_4 test by equilibrium dialysis or by a two-step immunoextraction assay to clarify a confusing clinical situation.
3. FT_4 testing alone is appropriate in patients with known, well-established pituitary thyrotrope failure or TSH suppression within the last 6 weeks.

Interpretation

1. Refer to "Non-Drug-Related Causes of Various Patterns of TSH and FT4 Results" (Table 1.2) and "Drug-Related Causes of Various Patterns of TSH and FT4 Results" (Table 1.3).
2. The $[FT_4]$ reference interval is 0.6 to 1.8 ng/dL. A more narrow range of 0.8 to 1.5 ng/dL may identify an abnormal population of patients with higher specificity.
3. Consider that individuals may exhibit $[FT_4]$ and [TSH] that are outside widely accepted reference intervals for these analytes yet are more typically normal for that individual.
4. An increased $[FT_4]$ may be caused by medications which inhibit T_4 to T_3 conversion such as beta blockers and glucocorticoids.
5. An increase in $[FT_4]$ may occur upon withdrawal of cytochrome P450 activators (e.g., phenytoin, barbiturates) in thyroxine-treated patients.
6. Consider possible intake of T_3 (Cytomel® or desiccated thyroid extract) if a $[FT_4]$ in the lower or normal range is obtained and [TSH] is normal or suppressed.

Notes

1. The FT_4 test measures the free, bioactive form of T_4.
2. FT_4 measured using a one-step assay may be affected by changes in thyroid-hormone-binding proteins, particularly albumin and prealbumin. Two-step free T_4 assays significantly reduce this problem.

3. Patients whose serum contains human anti-mouse antibodies (HAMAs) may have falsely elevated (or decreased) FT_4 levels if the assay uses mouse-derived monoclonal capture or signal antibodies.
4. The gold standard method for quantifying $[FT_4]$ is equilibrium dialysis; however, this test is difficult to perform, expensive, and not readily available.
5. When used in conjunction with TSH, FT_4 is the preferred test to evaluate and monitor thyroid status. When the [TSH] is abnormal, testing for FT_4 may be done reflexively to save the cost of unnecessary testing when the [TSH] is normal.
6. Rarely, hyperthyroidism with low [TSH] and elevated $[FT_4]$ may be associated with elevated free human chorionic gonadotropin beta-subunit (β-hCG) levels, usually from a trophoblastic neoplasm and even more rarely from a nontrophoblastic source. Because of homology between the β-subunits of TSH and β-hCG, very high levels of β-hCG exert TSH-like effects.

ICD-9 Codes

Refer to Test 1.2.1 codes.

Suggested Reading

Andersen S, Pedersen KM, Bruun NH, Laurbert P. Narrow individual variations in serum T(4) and T(3) in normal subjects: a clue to the understanding of subclinical thyroid disease. *J Clin Endocrinol Metab* 2002; 87:1068–72.

Brousse C, Mignot L, Baglin AC, Bernard N, Piette AM, Gepner P, Chapman A. Hyperthyroidism and hypersecretion of chorionic gonadotropin in gastric adenocarcinoma. *Rev Med Interne* 1994; 15:830–3.

Elmlinger MW, Kuhnel W, Lambrecht HG. Ranke MB. Reference intervals from birth to adulthood for serum thyroxine (T_4), triiodothyronine (T_3), free T_3, free T_4, thyroxine binding globulin (TBG), and thyrotropin (TSH). *Clin Chem Lab Med* 2001; 39:973–9.

Uy HL, Reasner CA. Elevated thyroxine levels in a euthyroid patient: a search for the cause of euthyroid hyperthyroxinemia. *Postgrad Med* 1994; 96:195–202.

Wang R, Nelson JC, Weiss RM, Wilcox RB. Accuracy of free thyroxine measurements across natural ranges of thyroxine binding to serum proteins. *Thyroid* 2000; 10:31–9.

1.2.3   Free Thyroxine Index Derived from Total T_4 (TT_4) and T_3 Resin Uptake (T_3 RU) as a Test for Hyperthyroidism

Indications for Test

The FTI calculation as a test for hyperthyroidism is indicated:

- When a FT_4 test is unavailable, but values for TT_4 and T_3 RU can be obtained

Procedure

1. Obtain a random blood sample for TT_4 and T_3 RU testing and calculate the FTI (see Test 1.1.8).

Interpretation

1. The reference interval for [FTI], when calculated using the THBR, is 4.1 to 11.2 mcg/dL.
2. The reference interval for [FTI], when calculated using the T_3 RU, is 1.0 to 4.2 mcg/dL.
3. An elevated [FTI] suggests hyperthyroidism except in the presence of high concentrations of thyroid-hormone-binding proteins, such as in treatment with estrogens, and otherwise normal thyroid status.
4. Patients who take exogenous T_3 may have low FTI values but may actually be hyperthyroid.

Notes

1. Oral estrogen treatment stimulates the liver to produce TBG, resulting in higher $[TT_4]$.
2. In familial dysalbuminemic hyperthyroxinemia (FDH), the FTI may be falsely elevated.
3. The FTI has a lower sensitivity and specificity for diagnosis of hyperthyroidism than does measuring the FT_4.
4. The FTI helps correct the TT_4 for variation in the concentration of serum thyroid-hormone-binding proteins but is not as useful as a directly measured $[FT_4]$.
5. Among euthyroid patients with increased TBG, 12 to 32% and 5 to 20% were found to have an increased or suppressed FTI, respectively, depending on the T-uptake method used.
6. In truly hyperthyroid patients, falsely normal or low $[TT_4]$, not adequately corrected by T_3 RU, may lead to confusion in establishing a correct diagnosis.

ICD-9 Codes

Refer to Test 1.2.1 codes.

Suggested Reading

Abid M, Billington CJ, Nuttall FQ. Thyroid function and energy intake during weight gain following treatment of hyperthyroidism. *J Am Coll Nutr* 1999; 18:189–93.

Faix JD, Rosen HN, Velazquez FR. Indirect estimation of thyroid-hormone-binding proteins to calculate free thyroxine index: comparison of nonisotopic methods that use labeled thyroxine ("T-uptake"). *Clin Chem* 1995; 41:41–7.

1.2.4   Total T_3 to Total T_4 Ratio (TT_3/TT_4 Ratio) in Hyperthyroidism

Indications for Test

The TT_3/TT_4 ratio calculation as a test for hyperthyroidism is indicated:

- When differentiating milder forms of thyroid follicle disruption (as occurs in thyroiditis of pregnancy) from autonomous thyroid hyperfunction (as occurs in patients with a toxic nodule)

Procedure

1. Calculate the ratio in units of ng/mcg by dividing the TT_3 (in ng/dL) by the TT_4 (mcg/dL) value.
2. Calculate the FT_3/FT_4 ratio as a supplemental test.

Interpretation

1. The reference interval for the TT_3/TT_4 ratio is 12 to 20 ng/mcg (mean, 16.0 ng/mcg).
2. A TT_3/TT_4 ratio of <20 ng/mcg is consistent with normal thyroid function or thyroiditis with thyroid hormones released in normal proportions.
3. A TT_3/TT_4 ratio of ≥20 ng/mcg is seen in hyperthyroidism due to autonomous thyroid hormone overproduction, such as in Graves' disease, toxic adenoma, toxic multinodular goiter, or Hashimoto's thyroiditis.
4. An elevated TT_3/TT_4 ratio can also be seen in iodine-deficient states and in patients taking a preparation containing T_3.
5. In subjects taking thyroxine, $[TT_4]$ but not $[TT_3]$ increases with the dose of thyroxine, leading to decreasing TT_3/TT_4 ratios. This decrease occurs regardless of the presence or absence of thyroid tissue and is uninfluenced by antithyroid drug treatment, indicating that peripheral metabolism of T_4 is reduced.
6. A FT_3/FT_4 ratio of 0.3 to 0.5 is typical of Graves' disease, whereas a ratio of <3 is typical of thyroiditis.

Notes

1. Calculation of the TT_3/TT_4 or the FT_3/FT_4 ratio is an alternative to the radioactive iodine uptake (RAIU) imaging test of the thyroid gland in the differential diagnosis of hyperthyroidism when the RAIU test is unavailable or contraindicated, as in pregnancy or breast feeding.
2. Serial change in values for the TT_3/TT_4 ratio has a low specificity for any disorder and probably occurs as the result of a variety of physiological phenomena.

ICD-9 Codes

Refer to Test 1.2.1 codes.

Suggested Reading

Amino N, Yabu Y, Miki T, Morimoto S, Kumahara Y, Mori H, Iwatani Y, Nishi K, Nakatani K, Miyai K. Serum ratio of triiodothyronine to thyroxine, and thyroxine-binding globulin and calcitonin concentrations in Graves' disease and destruction-induced thyrotoxicosis. *J Clin Endocrinol Metab* 1981; 53:113–6.

Erfurth EM, Hedner P. Thyroid hormone metabolism in thyroid disease as reflected by the ratio of serum triiodothyronine to thyroxine. *J Endocrinol Invest* 1986; 9:407–12.

Izumi Y, Hidaka Y, Tada H, Takano T, Kashiwai T, Tatsumi K, Ichihara K, Amino N. Simple and practical parameters for differentiation between destruction-induced thyrotoxicosis and Graves' thyrotoxicosis. *Clin Endocrinol* 2002; 57:51–8.

Rendell M, Salmon D. "Chemical hyperthyroidism": the significance of elevated serum thyroxine levels in L-thyroxine treated individuals. *Clin Endocrinol (Oxf)* 1985; 22:693–700.

Takamatsu J, Kuma K, Mozai T. Serum triiodothyronine to thyroxine ratio: a newly recognized predictor of the outcome of hyperthyroidism due to Graves' disease. *J Clin Endocrinol Metab* 1986; 62:980–3.

1.2.5   Thyroid Autoantibodies: Antimicrosomal (Anti-M), Antithyroid Peroxidase (Anti-TPO), and Antithyroglobulin (Anti-Tg) Antibody Testing in Hyperthyroidism

Indications for Test

Anti-TPO and anti-Tg as tests in hyperthyroid patients are indicated:

*   In cases of acute thyroiditis and for confirmation of the diagnosis of toxic Hashimoto's-type thyroiditis
*   To make a differential diagnosis of the causes for thyrotoxicosis

Procedure

1. Obtain a random blood sample for testing Tg, which includes anti-Tg and anti-TPO (or anti-M). *Note:* Most laboratories do not perform anti-Tg testing independent of quantitative Tg testing (i.e., all requests for Tg include reflex testing for anti-Tg).

Interpretation

1. Anti-Tg and anti-TPO antibodies are elevated in a variety of thyroid diseases, including 80 to 90% of patients with Hashimoto's thyroiditis and 50% of patients with Graves' disease.
2. Positive antibody titers (anti-M, anti-TPO, or anti-Tg) are found in up to 25% of normal adults; however, very high titers (>1:10000) are specific for Hashimoto's thyroiditis.
3. Anti-Tg is almost never elevated except when anti-TPO antibodies are elevated.

Notes

1. Measurement of anti-TPO antibodies suffices as a test for Hashimoto's thyroiditis in the majority of cases.
2. None of the thyroid antibody tests is required for the diagnosis of thyroid disease, but their levels may help confirm or categorize the severity of a condition.
3. There is little interference of TPO antigen in anti-M antibody assays.

ICD-9 Codes

Refer to Test 1.2.1 codes.

Suggested Reading

Kordonouri O, Klinghammer A, Lang EB, Gruters-Kieslich A, Grabert M, Holl RW. Thyroid autoimmunity in children and adolescents with type 1 diabetes: a multicenter survey. *Diabetes Care* 2002; 25:1346–50.

Ruf J, Czarnocka B, Ferrand M, Doullais F, Carayon P. Novel routine assay of thyroperoxidase autoantibodies. *Clin Chem* 1988; 34:2231–4.

Weetman AP, McGregor AM. Autoimmune thyroid disease: developments in our understanding. *Endocr Rev* 1984; 5:309–55.

1.2.6 Thyroid-Stimulating Immunoglobulin (TSI) or Thyroid-Stimulating Antibody (TS-Ab) Testing in Graves' Disease (GD)

Indications for Test

TSI testing in patients with Graves' disease is indicated for:

- The more definitive diagnosis of GD in pregnant patients who are unable to undergo RAIU testing and for the assessment of risk for delivering an infant with neonatal GD
- Neonates of mothers who have GD
- Predicting the response to thionamide treatment of hyperthyroidism

Procedure

1. Obtain a blood sample for TSI testing in carefully selected patients as indicated above.

Interpretation

1. TSI values > 2 mIU TSH equivalents/mL indicate active GD.
2. The level of TSI stimulatory activity may predict the likelihood of neonatal GD and of recurrence of GD after cessation of treatment with antithyroid drugs.

Notes

1. The long-acting thyroid stimulator (LATS) test as a bioassay for TSI is no longer available.
2. The titer of TS-Ab may be measured using bioassays that quantitate the cAMP response to TS-Ab-containing sera applied to Chinese hamster ovary (CHO) cells. In a chemiluminescent assay using CHO cells, the reference interval for TS-Ab was ≤1.5 relative light units (Evans et al., 1999).
3. About 0.2% of pregnant women have GD; however, only 1% of children born to GD women are found to have hyperthyroidism.
4. A mutation in the TSH receptor, resulting in its persistent, constitutive activation, has been found in nonimmune neonatal hyperthyroidism of duration >4 months.

ICD-9 Codes

Refer to Test 1.2.1 codes.

Suggested Reading

Evans C, Morgenthaler NG, Lee S, Llewellyn DH, Clifton-Bligh R, John R, Lazarus JH, Chatterjee VK, Ludgate M. Development of a luminescent bioassay for thyroid stimulating antibodies. *J Clin Endocrinol Metab* 1999; 84:374–7.

Gossage AA, Munro DS. The pathogenesis of Graves' disease. *Clin Endocrinol Metab* 1985; 14:299–330.

Peleg D, Cada S, Peleg A, Ben-Ami M. The relationship between maternal serum thyroid-stimulating immunoglobulins and fetal and neonatal thyrotoxicosis. *Obstet Gynecol* 2002; 99:1040–3.

Polak M. Hyperthyroidism in early infancy: pathogenesis, clinical features and diagnosis with a focus on neonatal hyperthyroidism. *Thyroid* 1998; 8:1171–7.

Schott M, Morgenthaler NG, Fritzen R, Feldkamp J, Willenberg HS, Scherbaum WA, Seissler J. Levels of autoantibodies against human TSH receptor predict relapse of hyperthyroidism in Graves' disease. *Horm Metab Res* 2004; 36:92–6.

1.2.7   Thyroid Scan with Radioactive Iodine and Measurement of Its Uptake (RAIU) in Hyperthyroidism

Indications for Test

Both a thyroid scan and RAIU uptake test in clinical hyperthyroidism are indicated:

- In patients when hyperthyroidism has been biochemically established
- To differentiate between thyroiditis (low uptake), Graves' disease (GD), Hashimoto's thyroiditis, or Pendred's syndrome (high uptake) and a hyperfunctioning or hypofunctioning nodule or area within the thyroid gland

Procedure

1. Administer tracer dose of radiolabeled iodine, preferably iodine-123 (^{123}I).
2. Determine the percent RAIU by the thyroid gland at 4 to 6 hours and at 24 hours after dosing.
3. Obtain only a RAIU at 4 hours and omit the 24-hour RAIU in overtly hyperthyroid patients if the 4-hour RAIU is markedly elevated.
4. Obtain a nuclear scan image of the thyroid and determine the distribution of isotope if there is an anatomic irregularity, palpable on physical exam, or the patient experiences unusual discomfort on palpation of the thyroid.

Interpretation

1. Reference interval for the 24-hour RAIU is 5 to 30% in individuals with adequate dietary iodine intake.
2. Generally elevated, evenly distributed RAIU is found in Graves' disease, the rare cases of TSH-induced hyperthyroidism, iodine deficiency, organification and enzymatic defects, and subsets of patients with chronic lymphocytic (Hashimoto's-type) thyroiditis.
3. Localized areas of increased RAIU are found in autonomous toxic or hot nodules, which tend to suppress uptake in other areas.
4. Generally low RAIU is found in patients with exogenous or cold iodine from any of the following: seafood, prepackaged meals, treatment with saturated solution of potassium iodide (SSKI), amiodarone therapy, recent exposure to radiographic contrast material, thyroid hormone, propylthiouracil or methimazole, or normal thyroid status in patients with adequate dietary iodine intake.
5. Low normal to very low RAIU tends to occur in thyroiditis, hypothyroidism, or delayed (>24 hours) measurements of uptake in rapid-turnover GD.

6. Homogeneous RAIU in patients with hyperthyroidism indicates GD or, rarely, a TSH-producing tumor.
7. Heterogeneous RAIU may be present in patients with toxic or nontoxic nodular disease (Hashimoto's thyroiditis).
8. Nodules without much RAIU on scan (i.e., cold nodules) have up to a 20% chance of harboring a malignancy.
9. Cancerous thyroid nodules occur more frequently in males (8.2%) than in females (4.2%); however, the occurrence of thyroid cancer is more frequent in women since they have eight times as many thyroid nodules as men.

Notes

1. Since the percent uptake of iodine by the thyroid gland only approximates the rate of thyroid hormone production, RAIU measurement plays no role in making a definitive diagnosis of hypothyroidism.
2. The RAIU measures thyroid gland function, whereas the nuclear scan helps to differentiate hot from cold nodules.
3. The RAIU may be within the reference interval for healthy individuals in patients with an autonomously functioning hot nodule if the nodule produces a "normal" amount of thyroid hormone.
4. Multinodular goiters with a dominant cold nodule harbor cancer as often as a thyroid with a solitary nodule. Selection of a cold nodule for biopsy over that of a nodule without decrease in uptake in the same thyroid is not more likely to yield a finding of cancer.
5. RAIU and scan can only be reliably interpreted with concurrent tests for TSH and FT_4.
6. Lithium pretreatment may enhance the effectiveness of radioiodine treatment (i.e., result in more efficient RAIU) of GD by mechanisms not yet determined (Bogazzi et al., 1999).
7. The perchlorate discharge test may be considered the gold standard for diagnosis of Pendred's syndrome, an autosomal recessive condition characterized by a euthyroid or hypothyroid thyroidal iodine organification defect, deafness, and goiter, but it is nonspecific and its sensitivity, independent of clinical findings, is unknown (Reardon et al., 1997).

ICD-9 Codes

Refer to Test 1.2.1 codes.

Suggested Reading

Bogazzi F, Bartalena L, Brogioni S, Scarcello G, Burelli A, Campomori A, Manetti L, Rossi G, Pinchera A, Martino E. Comparison of radioiodine with radioiodine plus lithium in the treatment of Graves' hyperthyroidism. *J Clin Endocrinol Metab* 1999; 84:499–503.

Floyd JL, Rosen PR, Borcert RD, Jackson DE, Weiland FL. Thyroid uptake and imaging with iodine-123 at 4–5 hours: replacement of the 24-hour iodine-131 standard. *J Nucl Med* 1985; 26:884–7.

Reardon W, Coffey R, Phelps PD, Luxon LM, Stephens D, Kendall-Taylor P, Britton KE, Grossman A, Trembath R. Pendred syndrome: 100 years of underascertainment? *QJM* 1997; 90:443–7.

1.2.8   Thyroid Scan with Technetium-99m (^{99m}Tc) Methoxyisobutyl Isonitrile (Sestamibi) Scintigraphy in Hyperthyroidism

Indications for Test

Thyroid scan with ^{99m}Tc-sestamibi scintigraphy in the hyperthyroid patient is indicated:

- For diagnosis of an autonomously functioning thyroid nodule in preference to the use of ^{131}I
- In a euthyroid ophthalmic Graves' disease (GD) patient with thyroid enlargement and proptosis

Procedure

1. Obtain a dual-phase thyroid scan at an early (15–30 minutes) and late (3–4 hours) interval after the administration of 740 to 1000 MBq ^{99m}Tc-MIBI.
2. Measure intensity of uptake and washout of radioactivity from nodules.

Interpretation

1. When a goiter is present and thyroid scan shows a locally hyperfunctioning or a large substernal area, total or subtotal thyroidectomy to remove the hyperfunctioning area or compressive goiter may be indicated.
2. Benign hyperfunctional nodules have intense uptake in the early image and intense uptake to absent retention in the late image with a reported nodular-to-thyroid uptake ratio of 2.94 ± 1.31 and 1.62 ± 0.50 in the early and late images, respectively.
3. Hürthle cell tumor nodules, in contrast to benign hyperfunctioning adenomas, display intense and persistent uptake of ^{99m}Tc-MIBI with a reported nodular-to-thyroid uptake ratio of 2.81 ± 0.52 and 5.53 ± 1.06 in early and late images, respectively.
4. A more uniform distribution of isotope is found in hyperthyroid than in euthyroid GD patients with ophthalmopathy.

Notes

1. The presence of autonomously functioning follicular cells, heterogeneously distributed, has been observed in about half of euthyroid ophthalmic GD patients (16 out of 36 in one study). Chronic stimulation by thyrotropin-receptor-stimulating autoantibody may underlie this phenomenon.
2. Thallium-201 (^{201}Tl) scintigraphy is superior to sestamibi scintigraphy in the visualization of suppressed thyroid tissue in patients with a toxic thyroid nodule.
3. When diffuse, intense thyroid gland radioactivity is observed during bone and lung scintigraphy with ^{99m}Tc-MIBI, the possibility of hyperthyroidism should be considered.
4. If isotope uptake in the stomach is not seen on sestamibi bone or lung scans, hyperthyroidism secondary to diffuse hyperplasia of a thyroid goiter may be present.

ICD-9 Codes

Refer to Test 1.2.1 codes.

Suggested Reading

Campeau RJ, Lichtenstein RJ, Ward TL, Alster DK. Incidental detection of hyperthyroidism during a perfusion lung scan for suspected pulmonary emboli. *Clin Nucl Med* 1991; 16:251–2.

Erdil TY, Onsel C, Kanmaz B, Caner B, Sonmezoglu K, Ciftci I, Turoglu T, Kabasakal L, Sayman HB, Uslu I. Comparison of ^{99m}Tc-methoxyisobutyl isonitrile and ^{201}Tl scintigraphy in visualization of suppressed thyroid tissue. *Nucl Med* 2000; 41:1163–7.

Kasagi K, Hidaka A, Misaki T, Miyamoto S, Takeuchi R, Sakahara H, Sasayama S, Iida Y, Konishi J. Scintigraphic findings of the thyroid in euthyroid ophthalmic Graves' disease. *J Nucl Med* 1994; 35:811–7.

Vattimo A, Bertelli P, Cintorino M, Burroni L, Volterrani D, Vella A, Lazzi S. Hürthle cell tumor dwelling in hot thyroid nodules: preoperative detection with technetium-99m-MIBI dual-phase scintigraphy. *J Nucl Med* 1998; 39:822–5.

1.2.9   Ultrasonograph (USG) Imaging of the Thyroid in Hyperthyroidism

Indications for Test

Ultrasonograph imaging of the thyroid in patients with hyperthyroidism is indicated when:

- A discrete nodule or multiple nodules are palpable.
- Treatment of a toxic thyroid nodule or multinodular goiter with surgery instead of radioiodine is planned.

Procedure

1. Establish the diagnosis of hyperthyroidism with TSH and FT_4 testing.
2. Correlate findings from USG with findings from:
   - Fine-needle aspiration biopsy (FNAB) of the dominant nodule
   - Radioiodine thyroid scan
   - RAIU
3. Use a high-resolution (10 MHz or higher) ultrasound scanning device, preferably with a color Doppler option, with a gray-scale probe covered by gel and held in place with a rubber sheath if a FNAB is to be performed.
4. Apply sonographic gel to the patient's neck and measure the dimensions of the thyroid in the longitudinal and transverse planes.
5. Mark the borders of the thyroid gland and any nodules seen in both planes, looking for heterogeneous features.
6. Distinguish solid nodules from simple and complex cysts, and measure the dimensions of each thyroid lobe.
7. In distinguishing patients with Graves' disease (GD) from those with diffuse toxic goiter, use color duplex USG to measure peak systolic velocity (PSV), volume flow rate (VFR), and color pixel density (CPD).

Interpretation

1. Because toxic multinodular goiters are more radioresistant, it may be possible to bypass the radio-iodine thyroid scan and proceed directly to high-dose radioablation therapy or near-total thyroidectomy in cases of hyperthyroidism with multiple nodules seen on USG.
2. Heterogeneity in the sonographic pattern of the thyroid parenchyma is typical of thyroiditis.
3. Finding a discrete, solitary hypoechoic thyroid nodule on USG prompts thyroid aspiration biopsy, as such a patient is at higher risk for thyroid cancer.
4. Color-flow Doppler (CFD) can be used to show vascular flow and velocity and positively distinguish the presence of a cyst vs. a blood vessel.
5. Nodules outside the thyroid gland may be identified as regional lymphadenopathy with USG of the neck.
6. A diagnosis of goiter can be objectively documented if the USG dimensions of the thyroid lobes exceed the normal of 3.5 to 4.5 cm (length), 2 to 3 cm (transverse width), and 1 to 2 cm (anterior–posterior thickness).
7. Using CFD-duplex sonography, Graves' disease was reliably distinguished from diffuse toxic goiter based on higher PSV (110 ± 49 cm/s), VFR (123 ± 67 mL/min), and CPD (33% ± 12) values in patients with GD vs. the lower PSV (43 ± 9 cm/s), VFR (23 ± 10 mL/min), and CPD (9% ± 6) values observed in patients with diffuse toxic goiter (Vitti et al., 1995).
8. Discovery of markedly increased thyroid blood flow with CFD-duplex USG is characteristic of GD and may help to distinguish GD from Hashimoto's thyroiditis when the echographic pattern seen on conventional USG does not.

Notes

1. Ordinarily, there is no indication for thyroid USG in the case of a smooth goiter of GD.
2. Ultrasound guidance allows the accurate placing of needles into the thyroid for fine-needle aspiration biopsy.
3. Normally, the right thyroid lobe is about 20% larger than the left, regardless of age. Total thyroid volume is normally less than 15 mL and progressively declines with age.

ICD-9 Codes

Refer to Test 1.2.1 codes.

Suggested Reading

Berghout A, Wiersinga WM, Smits NJ, Touber JL. Determinants of thyroid volume as measured by ultra-sonography in healthy adults in a non-iodine deficient area. *Clin Endocrinol* 1987; 26:273–80.

Diez JJ. Hyperthyroidism in patients older than 55 years: an analysis of the etiology and management. *Gerontology* 2003; 49:316–23.

Hegedus L. Thyroid ultrasound. *Endocrinol Metab Clin North Am* 2001; 30:339–60.

Saleh A, Cohnen M, Furst G, Godehardt E, Modder U, Feldkamp J. Differential diagnosis of hyperthyroidism: Doppler sonographic quantification of thyroid blood flow distinguishes between Graves' disease and diffuse toxic goiter. *Exp Clin Endocrinol Diabetes* 2002; 110:32–6.

Vitti P, Rago T, Mazzeo S, Brogioni S, Lampis M, De Liperi A, Bartolozzi C, Pinchera A, Martino E. Thyroid blood flow evaluation by color-flow Doppler sonography distinguishes Graves' disease from Hashimoto's thyroiditis. *J Endocrinol Invest* 1995; 18:857–61.

1.2.10 Thyroglobulin (Tg) Testing in Thyroiditis and Hyperthyroidism

Indications for Test

Thyroglobulin measurement is indicated in hyperthyroid patients when:

- Symptomatic thyroiditis is a predominant feature of their condition.

Procedure

1. Obtain a random blood sample for Tg testing from patients with a painfully sore neck and evidence for thyroid disease based on TSH, free T_4 concentration, thyroid antibodies, or history.
2. If [Tg] is elevated and thyroid tests are consistent with hyperthyroidism, measure RAIU (see Test 1.2.7).

Interpretation

1. The reference interval for [Tg] in healthy individuals with an intact thyroid gland is 30 to 60 ng/mL.
2. In general, the calculated volume (in mL) of a diffuse goiter, as measured from dimensions obtained from ultrasound, correlates with the [Tg].
3. Subacute granulomatous thyroiditis or de Quervain's syndrome is a transient, painful thyroid disorder characterized by elevated erythrocyte sedimentation rate, elevated [Tg], depressed RAIU, and exquisite pain on neck palpation of the thyroid gland.
4. Nontoxic goiters, thyroid cancers, and all forms of endogenous hyperthyroidism may be associated with an elevated [Tg].
5. Low or undetectable [Tg] may be found in patients with exogenous hyperthyroidism caused by T_4 or T_3 ingestion.

Notes

1. Tg is a large glycoprotein produced by the thyroid to store T_4 and T_3.
2. Elevated [Tg] results from disruption of normal thyroid parenchyma or secretion from a thyroid cancer.
3. Anti-Tg antibodies interfere to some extent with all Tg assays; therefore, anti-Tg testing is performed typically in concert with testing for Tg.
4. Low [Tg] does not exclude thyroid diseases.
5. Using special techniques, T_4 or T_3 may be found adherent to Tg. Although increased T_4 and T_3 attached to Tg may occur in patients with Graves' disease or differentiated thyroid cancer, it is found as a common, and probably distinctive, feature in patients with subacute thyroiditis.

ICD-9 Codes

Refer to Test 1.2.1 codes.

Suggested Reading

Druetta L, Bornet H, Sassolas G, Rousset B. Identification of thyroid hormone residues on serum thyroglobulin: a clue to the source of circulating thyroglobulin in thyroid diseases. *Eur J Endocrinol* 1999; 140:457–67.

Rink T, Dembowski W, Schroth HJ, Klinger K. Impact of serum thyroglobulin concentration in the diagnosis of benign and malignant thyroid diseases. *Nuklearmedizin* 2000; 39:133–8.

Slatosky J, Shipton B, Wahba H. Thyroiditis: differential diagnosis and management. *Am Fam Physician* 2000; 61:1047–1052, 1054.

1.2.11 Exophthalmometry in Graves' Disease: Measurement of Ocular Protrusion by Hertel's Method

Indications for Test

Exophthalmometry is indicated in patients with:

- Protrusion of the eyes or unexplained eye irritation, particularly in hyperthyroid patients
- Graves' disease (GD) with or without hyperthyroxinemia

Procedure

1. Examine the patient in the sitting position.
2. Using a hand-held Hertel device, place the notches on the bony lateral orbital ridges, which serve as the fixation base for the instrument and measure the base or distance between the ridges.
3. Record the base (distance along the horizontal bar) measurement for future reference, and set the device to this measurement when performing serial measurements.
4. Ask the patient to fixate on a point on the examiner's forehead.
5. Adjust the device so the apex of the cornea of each eye is superimposed on the millimeter scale on the prisms.
6. Align the two red lines on each prism.
7. Record the distance between the lateral angle of the bony orbit and the most anterior part of the cornea to the nearest millimeter. This distance can be read from a calibrated scale located underneath a mirror in which a side view of the eye is reflected.
8. Subsequent alterations in the orbital rim or ridges will invalidate serial measurements.

TABLE 1.5
Systems for Reporting and Classifying Proptosis in Graves' Class III Ophthalmopathy

Corneal Protrusion[a]				
Absent	Minimal	Moderate	Marked or Severe	Refs.
≤20 mm	21–23 mm	24–27 mm	≥28 mm	1
<3 mm ULN	3–4 mm > ULN	5–7 mm > ULN	≥8 mm over ULN	2, 3

[a] *Important:* Prior to surgical intervention, interpret serial measurements considering the limits of reproducibility for the imaging method (Hertel exophthalmometry, CT, or MR imaging) used to measure corneal protrusion.[4,5]

Note: ULN, upper limit of normal based on ethnicity; CT, computed tomography; MR, magnetic resonance.

1. Werner SC. Classification of the eye changes of Graves' disease. *J Clin Endocrinol Metab* 1969; 29:982–4.
2. Werner SC. Modification of the classification of the eye changes of Graves' disease. *Am J Ophthalmol* 1977; 83:725–7.
3. Van Dyk HJL. Orbital Graves' disease: a modification of the "NO SPECS" classification. *Ophthalmology* 1981; 88:479–83.
4. American Thyroid Association Committee Report. Classification of eye changes of Graves' disease. *Thyroid* 1992; 2:235–6.
5. Feldon S. Classification of Graves' ophthalmology [letter]. *Thyroid* 1993; 3:171.

Interpretation

1. Graves' ophthalmopathy may be defined as a protrusion of the cornea ≥3 mm above the upper limit typical of the patient's ethnic group, or a ≥3-mm difference between the eyes, or a progression of ≥3-mm protrusion under observation. Refer to "Systems for Reporting and Classifying Proptosis in Graves' Class III Ophthalmopathy" (Table 1.5).
2. Ethnicity affects the upper limit of normal (ULN) for ocular protrusion as follows:
   • Caucasian males and females have a mean of 16 mm and a ULN of 20 or 21 mm.
   • Black females have a mean of 17.5 mm and a ULN of 23 mm.
   • Black males have a mean of 18 mm and a ULN of 24 mm.
   • Orientals (Southeast Asians, Chinese, Japanese, and Koreans) have degrees of protrusion that typically do not exceed 18 mm or 19 mm.
   • Mixed-ethnicity patients may have intermediate ocular protrusion values distinctively different from the above levels.
3. Protrusion of 3 to 4 mm above the ethnic-specific ULN in asymptomatic patients may represent clinically insignificant disease, whereas moderate (5 to 7 mm above the ethnic-specific ULN) or marked (≥8 mm above the ethnic-specific ULN) protrusion is usually associated with eye symptoms.
4. A variance of 1 mm in the measurement of protrusion by the same observer may be expected, with normal asymmetry between the eyes rarely exceeding 2 mm. A difference of at least 2 mm, and perhaps 3 to 4 mm, by exophthalmometric measurements is necessary if a pathological difference is to be identified.
5. Blacks have a wider range of ocular protrusion values than Caucasians. Published reference intervals for general, mixed-race populations are not appropriate for adult black patients.

Notes

1. It is best to characterize Graves' ophthalmopathy using objective clinical assessments of soft tissue involvement, optic nerve function, proptosis, ocular motility, and eyelid position rather than arbitrary staging classification or grading systems.
2. Although Hertel exophthalmometry is the most widely used method for measuring ocular protrusion in GD, others include the Luedde device and multislice computed tomography (M-CT) or magnetic resonance (MR) scanning.

3. Orbitometry relies on fixation points on the superior and inferior orbits.
4. Given the wide overlap in ocular protrusion values between ethnic groups and reference populations within the same ethnicity, a single exophthalmometry determination is of limited diagnostic value.
5. Serial measurements of ocular protrusion in an individual patient with orbital disease are more informative and help to assess the activity of disease.
6. The eye disease activity score (range of scores, 0 to 7) may be calculated by assigning one point each to the following signs and symptoms:
   - Spontaneous retrobulbar pain when not moving eyes
   - Pain on eye movement
   - Eyelid erythema
   - Conjunctival injection
   - Chemosis
   - Swelling of the caruncle
   - Eyelid edema or fullness
7. No particular degree or millimeter of corneal protrusion merits a "point value" in staging or grading of the ophthalmopathy of GD.

ICD-9 Codes

Condition(s) that may justify this test *include but are not limited to*:

242.00   Graves' disease
379.90   eye disorder, not otherwise specified

Suggested Reading

American Thyroid Association Committee Report. Classification of eye changes of Graves' disease. *Thyroid* 1992; 2:235–6.
Bogren HG, Franti CE, Wilmarth SS. Normal variations of the position of the eye in the orbit. *Ophthalmology* 1986; 93:1072–7.
Dunsky IL. Normative data for Hertel exophthalmometry in a normal adult black population. *Optom Vis Sci* 1992; 69:562–4.
Feldon S. Classification of Graves' ophthalmology [letter]. *Thyroid* 1993; 3:171.
Fledelius HC. Exophthalmometry and thyroid disease: the value of the Hertel measurement evaluated in a group of patients with thyroid diseases and a control group. *Ugeskr Laeger* 1994; 156:6528–31.
Frueh BR. Why the "NO SPECS" classification of Graves' eye disease should be abandoned, with suggestions for the characterization of this disease. *Thyroid* 1992; 2:85–8.
Van Dyk HJL. Orbital Graves' disease: a modification of the "NO SPECS" classification. *Ophthalmology* 1981; 88:479–483.
Werner SC. Classification of the eye changes of Graves' disease. *J Clin Endocrinol Metab* 1969; 29:982–984.
Werner SC. Modification of the classification of the eye changes of Graves' disease. *Am J Ophthalmol* 1977; 83:725–727.

1.2.12 Criteria for the Objective Diagnosis of Thyroid Storm

Indications for Test

This assessment is indicated when:

- A patient with overt hyperthyroidism presents in critical condition and a decision on the use of aggressive antithyroid therapy must be made on an emergent basis.

TABLE 1.6
Criteria and Point Scores Related to Features of Thyroid Storm

Criterion	Result	Point Value
Body temperature (°F)	99.0–99.9	5
	100.0–100.9	10
	101.0–101.9	15
	102.0–102.9	20
	103.0–103.9	25
	>104.0	30
Central nervous system disturbance	Absent	0
	Mild (*agitation*)	10
	Moderate (*delirium, psychosis, lethargy*)	20
	Severe (*seizure, coma*)	30
Gastrointestinal–hepatic dysfunction	Absent	0
	Moderate (*diarrhea, nausea, vomiting, abdominal pain*)	10
	Severe (*unexplained jaundice*)	20
Tachycardia (heart rate, beats per minute)	90–109	5
	110–119	10
	120–129	15
	130–139	20
	≥140	25
Congestive heart failure	Absent	0
	Mild (*pedal edema*)	5
	Moderate (*bibasilar rales*)	10
	Severe (*pulmonary edema*)	15
Atrial fibrillation	Absent	0
	Present	10
Precipitant factor present (see Table 1.7)	Negative	0
	Positive	10

Procedure

1. Select the thyroid storm risk (TSR) score associated with the highest grade of the patient's signs and symptoms for each criterion indicated in "Criteria and Point Scores Related to Features of Thyroid Storm" (Table 1.6) and "Factors That May Precipitate Thyroid Storm" (Table 1.7).
2. Add point values for each criterion to achieve a composite TSR score within the range of 0 to 140.

Interpretation

TSR score of:

- <25 indicates that thyroid storm is unlikely.
- 25 to 44 is suggestive of impending thyroid storm.
- ≥45 is highly suggestive of thyroid storm.

Case Study

(From Alfardan et al., 2005.)

History

A 60-year-old, cachectic black female presented to the Emergency Department.

TABLE 1.7
Factors That May Precipitate Thyroid Storm

Therapeutic or Diagnostic Misadventure	Medical Conditions
Thyroid surgery	Infection (most common cause)
Nonthyroid surgery including tooth extraction	Diabetic ketoacidosis/hypoglycemia
Radioactive iodine therapy	Congestive heart failure
Iodinated contrast dyes	Toxemia of pregnancy/parturition
Amiodarone (high-iodide) therapy	Severe emotional stress/acute manic crisis
Withdrawal of antithyroid drug therapy	Pulmonary embolism
Thyroid hormone overdose	Cerebral or bowel ischemia or infarction
Vigorous, repeated palpation of thyroid gland	Acute trauma

Source: Sarlis, N.J. and Gourgiotis, L., *Rev. Endocr. Metab. Disord.*, 4, 129–136, 2003. With permission.

Clinical Criteria

- She had a history of overactive thyroid complaining of chills, nausea, vomiting, and diarrhea.
- Vital signs revealed an irregular [*10 points*] pulse of 196 bpm [*25 points*] and a temperature of 36.2°C (97.2°F) [*0 points*].
- Physical examination showed a large goiter, jugular venous distension with shortness of breath consistent with moderate congestive heart failure [*10 points*], and slightly icteric sclerae [*20 points*].
- She was awake, tremulous, and mildly agitated [*10 points*].
- A chest X-ray revealed prominent pulmonary vasculature and small, bilateral effusions; ECG confirmed atrial fibrillation.

Laboratory Values

- Glucose, 38 mg/dL* [*10 points*]
- Total bilirubin, 3.1 mg/dL
- B-type natriuretic peptide (BNP), 408 pg/mL
- [TSH], undetectable
- [FT_4], 9.3 ng/dL (>5× ULN)

Thyroid Storm Risk

Score = 10 + 25 + 0 + 10 + 20 + 10 + 10 = 85, a score highly suggestive of thyroid storm. Despite aggressive management, the patient went into shock and died, as occurs in 25% or more of all patients in thyroid storm.

ICD-9 Codes

Refer to Test 1.2.1 codes.

Suggested Reading

Alfardan J, Wians FH, Dons RF, Wyne K. The "Perfect Storm." *Lab Med* 2005; 36:700–4.

Burch HB, Wartofsky L. Life-threatening thyrotoxicosis: thyroid storm. *Endocrinol Metab Clin North Am* 1993; 22:263–277.

Sarlis NJ, Gourgiotis L. Thyroid emergencies. *Rev Endocr Metab Disord* 2003; 4:129–136.

---

* Hypoglycemia relating to sepsis, suggested by markedly elevated white blood cell count, with sepsis being the most likely precipitating factor contributory to her condition

1.3   Thyroid Malignancies

1.3.1   Fine-Needle Aspiration Biopsy (FNAB) of Thyroid Nodules (TNs)

Indications for Test

Fine-needle aspiration biopsy is indicated for:

- Routine evaluation of solitary or multiple TNs, of size greater than 1 cm by ultrasound, to determine the presence of malignancy
- Cases of possible thyroiditis to document the presence of inflammatory cells
- Patients with TNs with microcalcifications, an irregular or microlobulated margin, or increased density (i.e., hypoechogenicity) or having a shape that is taller than it is wide, regardless of size or palpability

Procedure

1. Examine patient's neck for evidence of lymphadenopathy. Biopsy any enlarged lymph nodes, especially when ultrasonographic evidence of metastatic spread of cancer is suspected (see Test 1.3.4).
2. Explain the procedure, its rationale, and benefits to the patient. Point out its rare but serious risks, which include hematoma formation with airway or vascular compromise.
3. If the patient has ingested any aspirin, nonsteroidal antiinflammatory agents (NSAIDs), or blood thinners 3 days prior to biopsy, do not perform FNAB. In patients who have not ingested these agents, explain the expected side effects of bruising, mild swelling, and mild pain at the site of biopsy and obtain informed consent.
4. Ideally, assemble a three-person operative team consisting of an experienced cytology and ultrasound technician to assist the physician in performing aspiration and slide preparation.
5. Have the patient lie on the examination table with a pillow or rolled-up towel under the shoulders to position the neck for possible hyperextension, if necessary. Prepare the area over the TN with antiseptic in a standard fashion and administer a local anesthetic.
6. Place 1 mL of air as a cushion into a 5- to 10-mL syringe and attach a 27-gauge needle. The length of the needle may have to be varied, with longer (1.5-inch) needles required for deeper lesions or thicker necks.
7. Stabilize the TN between your fingers if it is palpable. Use ultrasound to localize the lesion and help guide the positioning of the needle. Instruct the patient not to talk or swallow while the needle is in the neck.
8. Have patient swallow first and then penetrate the skin with the needle, passing the tip of the needle into the TN. With guidance from ultrasound images, sample various areas on the periphery of the TN as well as its center. The location of calcified areas should be identified and sampled.
9. Apply suction to the syringe, and manipulate the needle (up and down motion with slight changes in direction) to obtain material for cytopathologic analysis. If the operative team is limited to two persons, the physician must hold the ultrasound probe and guide the needle into the TN while the second person applies suction using an extension tube device as described in Dons (1997).
10. If the TN is cystic, drain the cyst using a larger bore needle (18- to 21-gauge) if the fluid is viscous.
11. Release suction on the syringe as soon as the material from the nodule appears in the hub of the needle and prior to removing the needle tip from the nodule. Do not apply excessive suction as the sample may become excessively diluted with blood. After retracting the syringe plunger to about the 1-mL line, remove the needle from the patient's neck.
12. Force the material in the needle onto a slide for fixation. Save one drop per slide if a cyst was drained.
13. Repeat needle aspiration with a new needle. A total of 5 or 6 punctures should be made, depending on the size of the TN, to obtain an adequate sampling of the nodule.
14. After the aspiration procedure has been completed, apply pressure to the biopsy site to prevent bleeding. In most cases, the patient can assist in applying the pressure.
15. A cytopuncture with a 23- to 25-gauge needle alone (i.e., without aspiration) can be performed by:
    - Puncturing the TN and rotating the needle within the tissue

- Observing for material appearing in the needle hub
- Withdrawing the needle from the nodule and attaching the needle hub to a syringe
- Forcing the material in the needle hub onto a slide for fixation

16. Send biopsy specimens to a pathologist specially trained in thyroid cytopathology for interpretation of tissue and fluid obtained at FNAB.

Interpretation

1. If sampling is adequate, the diagnostic sensitivity of positive test approaches 90% (i.e., false negatives approximate 10%).
2. An increased percentage of false-negative biopsies occurs in patients with larger TNs because of inadequate sampling (i.e., the risk of missing abnormal tissue increases when a larger TN is biopsied).
3. In studies of large series of patients, about 15% of solitary TNs > 1 cm contain a cancer. If multiple nodules are present, only 8% of the >1-cm TNs will have a cancer found.
4. In up to 30% of patients with TNs found to contain a "follicular neoplasm" on FNAB, after surgical excision of the nodule, the final diagnosis will be thyroid cancer.
5. In patients in whom anaplastic, sarcomatoid thyroid carcinoma is found on FNAB of TN, expect a median survival of 3 months, with few patients surviving beyond a year following combination therapy with surgery, radiation, and doxorubicin.
6. Patients with family histories of Gardner's or Cowden's syndrome, familial adenomatous polyposis (cribiform variant), or familial papillary carcinoma have an increased risk for papillary or follicular thyroid carcinoma.
7. Patients with hyperparathyroidism, pheochromocytoma, mucosal neuromas, Marfanoid habitus, or Hirschsprung's disease have increased risk for medullary carcinoma.

Notes

1. A frequency of only 3 to 7% of the U.S. population is found to have a TN on palpation vs. 20 to 75% when ultrasound is used, dependent on age and gender. Females have more TNs than males. In general, older patients have more TNs, with an autopsy prevalence of TNs approaching 50%.
2. Expect carotid ultrasound to incidentally detect larger TNs but not necessarily differentiate them from asymmetric goiters. TNs detected by positron emission tomography (PET) are more likely to be malignant (i.e., 1.1% of nonthyroid cancer patients undergoing PET scan will have incidental TNs, of which 42% will be malignant).
3. If a TN contains a cancer and there is evidence of lymphadenopathy with node positive microcarcinomas, expect the tumor to be biologically more aggressive.
4. Risk factors for TN malignancy include obesity in women, breast cancer, late menarche, and gallbladder disease. Negative risk factors are recreational exercise, African-American ethnicity, and (paradoxically) cigarette smoking.
5. FNAB of TNs requires an experienced individual to obtain the tissue sample, an experienced cytopathologist to evaluate the specimen, and a skilled ultrasonographer to help with the imaging studies during the procedure.
6. Ultrasound guidance during performance of FNAB on any size TN is standard procedure.
7. A TN < 1 cm in size is difficult to biopsy by FNA. Always sample nodules of 1 cm or less using ultrasound guidance.
8. A TN > 3 cm may be easy to biopsy with or without ultrasound guidance; however, the size of the nodule limits the ability to sample adequately with only 6 punctures.
9. If the cytologic specimen is inadequate or nondiagnostic and the cytopathologist is not certain that the cells obtained are benign, up to 7% of such patients will harbor a thyroid carcinoma. The occurrence of nondiagnostic biopsies has been greatly reduced with use of ultrasound-guided FNAB but still may be up to 10%.
10. After FNAB, serum thyroglobulin may remain elevated for up to 15 days.
11. In the future, a combination of immunohistochemical studies for tumor markers (e.g., calcitonin, thyroglobulin) and molecular profiling using DNA microarray and proteonomic analysis (e.g., BRAF, RET, RAS mutations) will be used to enhance definitive diagnosis of thyroid malignancies.

ICD-9 Codes

Condition(s) that may justify this test include *but are not limited to*:

Thyroid neoplasm
193   cancer, unspecified
237.4   uncertain behavior
239.7   follicular
241.0   nodule
242.XX   toxic nodule(s)
246.2   cyst

Thyroiditis
245.0   acute
245.2   chronic lymphocytic

Suggested Reading

Blum M. The diagnosis of the thyroid nodule using aspiration biopsy and cytology. *Arch Intern Med* 1984; 149:1140–2.

Chow LS, Gharib H, Goellner JR, van Heerden JA. Nondiagnostic thyroid fine-needle aspiration cytology: management dilemmas. *Thyroid* 2001; 11:1147–51.

Danese D, Sciacchitano S, Farsetti A, Andreoli M, Pontecorvi A. Diagnostic accuracy of conventional versus sonography-guided fine-needle aspiration biopsy of thyroid nodules. *Thyroid* 1998; 8:15–21.

Dons RF (Ed). *Endocrine and Metabolic Testing Manual*, 3rd ed. 1997. CRC Press, Chapter 8, pp 8.13–8.14.

Dwarakanathan AA, Ryan WG, Staren ED, Martirano M, Economou SG. Fine-needle aspiration biopsy of the thyroid: diagnostic accuracy when performing a moderate number of such procedures. *Arch Intern Med* 1989; 149:2007–9.

1.3.2   Thyroglobulin (Tg) Testing in Patients with Known Thyroid Carcinoma after Administration of Recombinant Human Thyrotropin (rhTSH) or Discontinuation of Thyroxine Replacement Therapy

Indications for Test

Thyroglobulin determination before and after administration of rhTSH or after discontinuation of thyroxine replacement therapy is indicated:

* In the follow-up of patients with differentiated thyroid cancer following total or near-total thyroidectomy
* After ^{131}I therapy to ablate residual thyroid tissue in a thyroid cancer patient

General Procedure

1. Do not measure [Tg] if the patient has had a recent thyroid biopsy, thyroid surgery, or high-dose radioiodine therapy, as the [Tg] may be falsely elevated under these circumstances.
2. Known prior elevation of antithyroglobulin (anti-Tg) antibodies is a relative contraindication to this test procedure unless complete thyroid ablation has been achieved and the Tg antigen has been eradicated for a considerable period of time.
3. Obtain a random baseline blood sample for Tg testing. Anti-Tg antibodies (see Test 1.3.3) should be measured simultaneously with Tg.
4. Based on the desired method to stimulate residual or recurrent thyroid cancer tissue, follow Procedure 1 or 2 below.

TABLE 1.8

Normative or Expected Ranges of Thyroglobulin (Tg) for Various Thyroid Conditions, Surgeries, and Thyroid-Stimulating Hormone (TSH) Concentrations in Patients Treated with Thyroid Hormone

Condition/Surgery	[TSH] (mIU/L)	[Tg] (ng/mL)
None (healthy individuals)	0.4–4.0	3–40
On suppressive thyroid hormone treatment:		
No thyroid surgery	<0.1	1.5–20
Thyroid lobectomy	<0.1	<10
Near-total thyroidectomy	<0.1	<2

Procedure 1. Stimulation of Residual or Recurrent Thyroid Cancer Tissue by Administration of rhTSH

1. Administer 2 injections of 0.9 mg rhTSH, 24 hours apart, followed by collection of a blood sample on the third day after the last injection. Use this sample to test for [Tg].
2. If whole-body scanning with radioiodine is to be performed, ^{123}I should be administered on the second day after the last injection of rhTSH.

Procedure 2. Stimulation of Residual or Recurrent Thyroid Cancer Tissue by Discontinuation of Thyroid Hormone Therapy

1. Assuming that the patient has had surgical and radiation therapy appropriate to disease status and is on chronic L-thyroxine suppression therapy, discontinue L-thyroxine (T_4) therapy for 5 to 6 weeks prior to the date of the ^{123}I scan.
2. Upon T_4 withdrawal, administer oral triiodothyronine (T_3) (25 mcg p.o., q.d. or b.i.d.) for 3 to 4 weeks as subreplacement thyroid hormone therapy.
3. Discontinue T_3 for 10 to 14 days prior to collection of a blood sample for Tg testing. In most cases, expect a [TSH] > 40 mIU/L to be achieved at the time the blood sample is collected for Tg testing.
4. If whole-body scanning with radioiodine is to be performed, ^{123}I should be administered on the day prior to collection of the blood specimen for Tg testing.
5. The patient should be on a low-iodine diet (i.e., no seafood or iodized salt) for 1 to 2 weeks prior to the ^{123}I scan and for 24 hours after the ^{123}I dose and should not have iodinated contrast dyes administered for >60 days prior to scanning.
6. If desired, check a 24-hour urine collection for iodine level to validate the patient's compliance with dietary iodine restriction.

Interpretation

1. A low (0.5–2.0 ng/mL) or undetectable (<0.5 ng/mL) [Tg] in a patient with a history of thyroid ablation or surgically absent thyroid gland and a normal or elevated TSH, such as may be induced by administration of rhTSH or by prolonged withdrawal from thyroxine therapy, indicates successful treatment for the thyroid cancer and no residual or recurrent disease. Such findings allow avoidance of more expensive thyroid scanning procedures.
2. Higher Tg concentrations (>2 ng/mL), in the absence of anti-Tg in the patient's serum, particularly if the patient is on adequate thyroid suppression therapy after thyroid ablation, are highly suggestive of the presence of malignant thyroid tissue and dictate further evaluation with body scanning. Refer to "Normative or Expected Ranges of Thyroglobulin (Tg) for Various Thyroid Conditions, Surgeries, and Thyroid-Stimulating Hormone (TSH) Concentrations in Patients Treated with Thyroid Hormone" (Table 1.8).
3. [Tg] may not be increased in patients with undifferentiated or medullary thyroid carcinoma.

4. Anti-Tg interferes with almost all Tg assays, rendering them useless as tools for monitoring recurrence of thyroid cancer. Disparities between serial serum Tg and anti-Tg measurements raise the possibility of anti-Tg interference with the serum Tg measurement and prompt a more cautious use of [Tg] in clinical decision making with regard to the assessment of the presence or absence of residual or recurrent malignant thyroid tissue.

Notes

1. Using rhTSH stimulation, Procedure 1 avoids the morbidity and discomfort associated with the procedure involving withdrawal of thyroid hormone.
2. A baseline Tg measurement is of limited use in the diagnosis of primary thyroid cancer. [Tg] in serum may be elevated for up to 15 days after fine-needle aspiration biopsy.
3. Interfering anti-Tg antibodies are present in the serum of 20 to 50% of thyroid cancer patients and may result in falsely low or high values for [Tg].
4. Because anti-Tg may become undetectable over time in the absence of thyroid antigen, the use of Tg as a cancer marker years after thyroid gland ablation may be possible even if anti-Tg antibodies were present in the patient's serum at a previous time.
5. Newer immunoradiometric assays (IRMAs) for quantifying serum Tg concentration, with a lower limit of Tg detection of 0.2 ng/mL, are not as affected by the presence of most human anti-Tg autoantibodies.
6. Although a large study of 229 thyroid cancer patients on thyroid hormone replacement therapy and with [Tg] $\geq$ 2 ng/mL demonstrated that cancer could be detected by imaging studies in 22% of these patients, stimulation of the thyroid using rhTSH (Procedure 1) or after T_4 withdrawal (Procedure 2) was more effective in detecting cancer in these patients. Using Procedure 1 or 2, 52% and 56%, respectively, of these patients were identified as having residual thyroid cancer (Haugen et al., 1999).

ICD-9 Codes

Condition(s) that may justify this test *include but are not limited to*:

193   malignant neoplasm of thyroid gland
198.89   metastatic thyroid cancer

Suggested Reading

Haugen BR, Pacini F, Reiners C, Schlumberger M, Ladenson PW, Sherman SI, Cooper DS, Graham KE, Braverman LE, Skarulis MC, Davies TF, De Groot LJ, Mazzaferri EL, Daniels GH, Ross DS, Luster M, Samuels MH, Becker DV, Maxon HR, Cavalieri RR, Spencer CA, McEllin K, Weintraub BD, Ridgway EC. A comparison of recombinant human thyrotropin and thyroid hormone withdrawal for the detection of thyroid remnant or cancer. *J Clin Endocrinol Metab* 1999; 84:3877–85.

Lima N, Cavaliere H, Tomimori E, Knobel M, Medeiros-Neto G. Prognostic value of serial serum thyroglobulin determinations after total thyroidectomy for differentiated thyroid cancer. *J Endocrinol Invest* 2002; 25:110–5.

Luboshitzky R, Lavi I, Ishay A. Serum thyroglobulin levels after fine-needle aspiration of thyroid nodules. *Endocr Pract* 2006; 12:264–9.

Mazzaferri EL, Robbins RJ, Spencer CA, Braverman LE, Pacini F, Wartofsky L, Haugen BR, Sherman SI, Cooper DS, Braunstein GD, Lee S, Davies TF, Arafah BM, Ladenson PW, Pinchera A. A consensus report of the role of serum thyroglobulin as a monitoring method for low-risk patients with papillary thyroid carcinoma. *J Clin Endocrinol Metab* 2003; 88:1433–41.

Spencer CA, Takeuchi M, Kazarosyan M, Wang CC, Guttler RB, Singer PA, Fatemi S, Lo Presti JS, Nicoloff JT. Serum thyroglobulin autoantibodies: prevalence, influence on serum thyroglobulin measurement, and prognostic significance in patients with differentiated thyroid carcinoma. *J Clin Endocrinol Metab* 1998; 83:1121–7.

1.3.3 Antithyroglobulin (Anti-Tg) Antibody as a Marker for Thyroid Carcinoma

Indications for Test

Anti-Tg testing is indicated in patients with thyroid cancer as a:

- Concomitant or "reflex" test whenever Tg testing is performed
- Primary test for persistence or reappearance of thyroid cancer in patients with a positive anti-Tg test

Procedure

1. Obtain a blood sample for both anti-Tg and Tg testing.
2. If the anti-Tg test is negative, perform yearly testing for [Tg] and anti-Tg.
3. If the anti-Tg test is positive, consider periodic repeat anti-Tg tests every several years after extirpation of thyroid tissue is accomplished. Perform Tg testing only when the anti-Tg test becomes negative.

Interpretation

1. A positive test for anti-Tg antibodies may invalidate the Tg test result, depending on the assay used to quantify [Tg].
2. A negative anti-Tg test permits use of [Tg] as a tumor marker for evidence of tumor recurrence.
3. In some patients, regardless of serum anti-Tg concentration, [Tg] may be low (<2 ng/mL), while in others very high anti-Tg levels (>1000 IU/mL) may provide no interference in the measurement of [Tg]. This issue is a moot point, however, as referral labs screen serum for anti-Tg level and do not perform quantitative Tg testing when the anti-Tg level exceeds the upper limit of the reference interval (typically, 20 IU/mL).
4. Serial changes in anti-Tg levels (i.e., either increasing or decreasing titers) may correlate with recurrence or resolution of thyroid cancer.

Notes

1. After complete surgical and radioablation of the thyroid and in thyroid malignancy, the level of anti-Tg antibodies may decline or disappear and permit the use of [Tg] as a tumor marker for thyroid cancer recurrence; however, the disappearance of anti-Tg antibodies from patients' sera may take many years after thyroid ablation.
2. Persistently high and/or increasing anti-Tg concentrations with low [Tg] correlate with the recurrence or persistence of differentiated thyroid cancer.
3. Anti-Tg levels decrease or disappear in tumor-free cases, but they may remain unchanged or even increase, in comparison with the preoperative level, in up to 50% of patients with proven metastases.
4. Improved immunometric Tg assays, with a lower limit of detection of 0.2 mcg/L, are largely unaffected by human anti-Tg autoantibodies.
5. Using modern immunometric Tg assays and a [Tg] cutoff value of 1 mcg/L, the diagnostic sensitivity and specificity of Tg tests for disease recurrence in the follow-up of patients with differentiated thyroid carcinoma after total thyroidectomy are 97% and 100%, respectively.

ICD-9 Codes

Condition(s) that may justify this test *include but are not limited to*:

Thyroid neoplasm
193   malignant
198.89   metastatic
226   benign
237.4   uncertain behavior

Thyroiditis
245.0   acute
245.1   subacute
245.2   chronic lymphocytic

Suggested Reading

Marquet PY, Daver A, Sapin R, Bridgi B, Muratet JP, Hartmann DJ, Paolucci F, Pau B. Highly sensitive immunoradiometric assay for serum thyroglobulin with minimal interference from autoantibodies (TG IRMA Pasteur). *Clin Chem* 1996; 42:258–62.

Rubello D, Girelli ME, Casara D, Piccolo M, Perin A, Busnardo B. Usefulness of the combined antithyroglobulin antibodies and thyroglobulin assay in the follow-up of patients with differentiated thyroid cancer. *J Endocrinol Invest* 1990; 13:737–42.

Vincze B, Sinkovics I, Keresztes S, Gergye M, Boer A, Remenar E, Peter I, Szentirmay Z, Kremmer T, Kasler M. Clinical significance of serum thyroglobulin and antithyroglobulin antibody in differentiated thyroid cancer after thyroid ablation. *Magy Onkol* 2004; 48:27–34.

Zophel K, Wunderlich G, Liepach U, Koch R, Bredow J, Franke WG. Recovery test or immunoradiometric measurement of anti-thyroglobulin autoantibodies for interpretation of thyroglobulin determination in the follow-up of different thyroid carcinoma. *Nuklearmedizin* 2001; 40:155–63.

1.3.4   Imaging of Anatomic Abnormalities of the Thyroid Gland and Neck Region Using Ultrasound (USG), Multislice Computed Tomography (M-CT), or Magnetic Resonance Imaging (MRI)

Indications for Test

Thyroid imaging using USG, M-CT, or MR is indicated to:

- Detect and monitor the presence and changes in size of suspicious lesions and nonpalpable abnormalities in high-risk patients such as those who have been exposed to ionizing neck radiation, including radioiodine, before age 20
- Develop a preoperative map of the extent of disease in a thyroid cancer patient
- Guide an FNAB of any thyroid nodule (USG only)
- Differentiate intrinsic thyroid masses from extrinsic neck masses such as thyroglossal duct cysts or lymph nodes

Procedures

Ultrasound (USG)

1. Place the patient in the supine position with the neck hyperextended using a pillow or rolled-up towel under the patient's shoulders and neck.
2. Apply acoustic gel to the patient's neck. Image the thyroid gland, thyroid nodules, and any nearby lymph nodes that appear to be enlarged.
3. For aspiration biopsies of thyroid nodules, first apply gel to the transducer and then cover it with a sterile rubber sheath or condom and sterile gel.
4. Use a high-resolution (7- to 13-MHz probe) scanning device, preferably with a color Doppler option. Use lower megahertz probe energy to image structures deeper below the skin surface.
5. Apply sonographic gel to the patient's neck and measure the dimensions of the thyroid in the longitudinal and transverse planes using the central trachea, carotid arteries, and internal jugular veins as landmarks.
6. Mark the borders, in both planes, of any nodules seen and look for heterogeneous features on the scan.

M-CT and MR

1. Use mild sedation or tranquilizer therapy as appropriate to help the patient remain still during the test.
2. To avoid motion artifact, use multidetector CT (M-CT) and MR devices that can acquire images within short periods of time.
3. Use contrast enhancement cautiously when attempting to define anatomic abnormalities in neck structures, recognizing the effects of iodinated dyes on subsequent iodine uptake.

Interpretation

USG Scan

1. Normally, the thyroid gland is acoustically homogeneous with moderate density echoes.
2. Connective tissue is hyperechoic and muscle is hypoechoic compared to normal thyroid tissue.
3. Fluid-filled thyroid cysts, blood vessels, and the esophagus are hypoechoic, with the esophagus appearing to contain a hyperechoic central area seen best on swallowing. Note that less experienced ultrasonographers can confuse the esophagus with a thyroid nodule unless color Doppler USG is used.
4. Patients with multiple hypoechoic foci or patches scattered throughout an otherwise normoechoic gland often have elevated antithyroid antibodies or biopsy-proven histology consistent with Hashimoto's thyroiditis.
5. Thyroid cancer is less of a concern in the absence of a discrete, solitary dominant hypoechoic nodule of 1 cm or larger size with irregular or blurred borders.
6. Characteristics of malignant thyroid nodules may include:
   - A taller than wider shape of the nodule in anterior–posterior USG views
   - Increased intranodular vascular pattern on Doppler imaging
   - Invasion into surrounding tissues
7. Peripheral calcification (i.e., eggshell-like) around a nodule favors a benign nodule, whereas internal, punctate calcifications within a nodule favor malignancy, particularly in cases of medullary thyroid carcinoma (MTC). "Shadowing" of the echoes is a worse prognostic sign than the appearance of "comet-tail" echoes seen with crystalline colloid densities.
8. The combination of hypoechogenicity, intranodular calcifications, and the absence of the halo sign is characteristic of most cases of MTC, as this combination is only rarely seen in patients with benign thyroid nodules.
9. Papillary thyroid cancer metastatic to lymph nodes presents with one or more of the following findings on ultrasound:
   - Cystic appearance
   - Round rather than oblong shape without a hilar stripe
   - Hyperechogenicity with calcifications
   - Peripheral vascularity
10. Normally, a benign lymph node is oblong with a hilar stripe and a blood vessel.

M-CT and MR Scans

1. Use M-CT and/or MR images as part of preoperative staging of metastatic disease related to a thyroid nodule.
2. Tracheal deviation and compression by the presence of a large goiter and the assessment of risk for tracheomalacia are facilitated with M-CT and MR imaging of the neck.

Notes

1. M-CT imaging of the neck and upper chest gives much better anatomic information about a substernal thyroid gland than imaging of this area with ultrasound.
2. Advanced imaging with M-CT or MR may be limited or unobtainable in claustrophobic individuals. Sedation will often be needed in such patients for an adequate M-CT or MR image to be obtained.
3. In most cases, obtain USG images of thyroid nodules before obtaining M-CT or MR images.
4. Image resolution has improved into the 1- to 2-mm range with the use of modern USG transducers. Image resolution of no better than 2 to 4 mm was typical with use of 1st-generation USG transducers.

5. Spiral CT is required to exclude lung metastases < 1 cm in diameter, as neither PET nor sestamibi scintigraphy is capable of detecting lesions of this size.

6. The halo sign is a hypoechogenic ring seen around a thyroid nodule that represents compressed perinodular blood vessels. The absence of the halo sign may be associated with malignancy (median diagnostic sensitivity, 66%), whereas thick, irregular, or interrupted halos suggest capsular invasion by a thyroid cancer.

7. Many benign thyroid nodules grow slowly over time, independent of volume and function at baseline, prompting discard of the notion that thyroid nodules that grow are malignant and those that do not are benign. In addition, multiple long-term studies have failed to show consistent shrinkage of benign thyroid nodules with thyroxine therapy.

8. In congenital thyroid hemiagenesis, the left lobe of the thyroid gland is nearly always missing and is associated with a slight increase in serum TSH concentration and compensatory enlargement of the remaining lobe.

ICD-9 Codes

Condition(s) that may justify this test include *but are not limited to*:

Thyroid neoplasm
193	cancer, unspecified
198.89	metastatic cancer
237.4	uncertain behavior
239.7	follicular
241.0	nodule
242.XX	toxic nodule(s)
246.2	cyst

Suggested Reading

Alexander EK. Natural history of benign solid and cystic thyroid nodules. *Ann Intern Med* 2003; 138:315–8

Dietlein M, Scheidhauer K, Voth E, Theissen P, Schicha H. Follow-up of differentiated thyroid cancer: what is the value of FDG and sestamibi in the diagnostic algorithm? *Nuklearmedizin* 1998; 37:12–7.

Kim EK, Park CS, Chung WY, Oh KK, Kim DI, Lee JT, Yoo HS. New sonographic criteria for recommending fine-needle aspiration biopsy of nonpalpable solid nodules of the thyroid. *AJR Am J Roentgenol* 2002; 178:687–91.

Lane H, Jones MK. Management of nodular thyroid disease. *Practitioner* 2002; 246:266–9.

Mandel SJ. Diagnostic use of ultrasonography in patients with nodular thyroid disease. *Endocr Pract* 2004; 10:246–252.

Newman E, Shaha AR. Distal tracheal compression. *Head Neck* 1991; 13:251–4.

Papini E, Guglielmi R, Bianchini A, Crescenzi A, Taccogna S, Nardi F, Panunzi C, Rinaldi R, Toscano V, Pacella CM. Risk of malignancy in nonpalpable thyroid nodules: predictive value of ultrasound and color-Doppler features. *J Clin Endocrinol Metab* 2002; 87:1941–6.

Parsons D, Cotton R, Crysdale W. Substernal goiter. *J Surg Oncol* 1995; 60:207–12.

Saller B, Moeller L, Gorges R, Janssen OE. Mann K. Role of conventional ultrasound and color Doppler sonography in the diagnosis of medullary thyroid carcinoma. *Exp Clin Endocrinol Diabetes* 2002; 110:403–7.

Schade G, Krupski G, Leuwer R. Topographic diagnosis in the area of the head-neck: initial experiences comparing ultrasound panorama images with CT and MR. *Laryngorhinootologie* 2001; 80:329–34.

Sostre S, Reyes MM. Sonographic diagnosis and grading of Hashimoto's thyroiditis. *J Endocrinol Invest* 1991; 14:115–21.

Walsh RM, Watkinson JC, Franklyn J. The management of the solitary thyroid nodule: a review of CT and MR. *Clin Otolaryngol* 1999; 24:388–97.

Weber AL, Randolph G, Aksoy FG. The thyroid and parathyroid glands: CT and MR imaging and correlation with pathology and clinical findings. *Radiol Clin North Am* 2000; 38:1105–29.

1.3.5 Thyroid Scanning with ^{131}I, ^{123}I, or ^{99m}Tc-MIBI after Administration of rhTSH or Discontinuation of Thyroxine Replacement Therapy in Patients with or at Risk for Thyroid Cancer Recurrence

Indications for Test

Radioisotopic imaging of thyroid tissue is indicated in thyroid cancer patients:

- When a palpable abnormality in the neck appears
- As a routine follow-up procedure for patients with prior neck irradiation of the thyroid gland
- As a routine and periodic follow-up procedure for patients who have undergone treatment for thyroid cancer

General Procedure

1. Perform a FNAB of a thyroid nodule before thyroid scanning if the [TSH] is normal and the patient is not known to have thyroid cancer.
2. Decide which isotope scan is most appropriate in a treated thyroid cancer patient.
3. If whole-body scanning with radioiodine is to be performed, the patient should be on a low-iodine diet (i.e., no seafood or iodized salt) for 1 to 2 weeks prior to the ^{123}I scan and for 24 hours after the ^{123}I dose and should not have iodinated contrast dyes administered for >60 days prior to scanning.
4. If desired, test a 24-hour urine collection for iodine level to validate the patient's compliance with dietary iodine restriction.
5. Decide whether to stimulate any residual thyroid tissue by increasing the patient's circulating [TSH] by either administering rhTSH (i.e., Thyrogen®) or discontinuing suppressive L-thyroxine therapy for 6 weeks before scanning. Using either of these procedures to stimulate any residual thyroid tissue, [TSH] increase typically to >40 mIU/L, which allows TSH-stimulated RAIU scanning to be performed.

Procedure 1. Stimulation of Thyroid Cells by Increasing Circulating TSH Level by Administration of rhTSH

1. Obtain a random baseline blood sample for Tg testing. Anti-Tg antibodies should be measured simultaneously with Tg.
2. rhTSH administration should be performed using two injections of 0.9 mg rhTSH, 24 hours apart, followed by collection of a blood sample, obtained on the third day after the last injection, for Tg testing.
3. If whole-body scanning with radioiodine is to be performed, ^{123}I should be administered on the second day after the last injection of rhTSH.

Procedure 2. Discontinuation of Thyroid Hormone Replacement Therapy

1. Assuming that the patient has had surgical and radiation therapy appropriate to disease status and is on chronic L-thyroxine (T$_4$) suppression therapy, discontinue L-thyroxine replacement therapy for 5 to 6 weeks prior to the date of the ^{123}I scan.
2. Upon T$_4$ withdrawal, administer oral triiodothyronine (T$_3$) (25 mcg p.o., q.d. or b.i.d.) for 3 to 4 weeks as subreplacement thyroid hormone therapy.
3. Discontinue T$_3$ therapy for 10 to 14 days prior to collection of a blood sample for Tg testing. In most cases, expect a [TSH] > 40 mIU/L to be achieved at the time the blood sample is collected for Tg testing.
4. If whole-body scanning with radioiodine is to be performed, ^{123}I should be administered on the day prior to collection of the blood specimen for Tg testing.

Thyroid Scanning Procedure

1. Do not proceed with radioiodine whole-body scanning in patients who have undergone thyroidectomy and radioiodine ablation if there is no clinical evidence of residual tumor, the [Tg] < 2 ng/mL, and [TSH] is appropriately suppressed by thyroid hormone therapy.
2. For patients who are candidates for radioiodine whole-body scanning, administer an appropriate dose of ^{123}I, ^{131}I, or ^{99m}Tc-MIBI.
3. If radioiodine was administered, scan at 4 to 6 hours and at 24 hours.
4. If ^{99m}Tc-MIBI was administered, scan at 20 to 30 minutes and at 4 hours.

Interpretation

^{123}I Scan and ^{131}I Scan

1. Either an experienced endocrinologist or certified specialist in nuclear medicine should read radio-iodine scans.
2. Increased radioiodine uptake outside the thyroid bed is suspicious for metastatic thyroid cancer.
3. Increased uptake in the liver suggests the presence of thyroid tissue and is abnormal after completion of thyroidectomy and radioiodine ablation.
4. Significant uptake of radioiodine (>1% RAIU) in the neck is suggestive of residual functioning thyroid tissue or metastatic recurrent disease.
5. The higher energy radioiodine isotope ^{131}I gives better images of tissues deeper in the body than can be obtained with ^{123}I.
6. A [Tg] > 2 ng/mL is a more sensitive test for thyroid cancer recurrence than radioisotope scans.

^{99m}Tc-MIBI Scan

1. ^{99}Tc-sestamibi or ^{99m}Tc-MIBI, an isotope similar to ^{201}Tl, has been touted to give better quality images of the thyroid. Parathyroid adenomas may be imaged with ^{99}Tc sestamibi using early and late dual-phase scanning.
2. Heterogeneous isotope uptake may be present in patients with toxic or nontoxic nodular disease (Hashimoto's thyroiditis).
3. Toxic nodules tend to suppress iodine uptake in the remainder of the gland and may not be apparent on ^{99m}Tc-MIBI scan.
4. Significant uptake of ^{99m}Tc-MIBI in the neck is suggestive of residual functioning thyroid tissue or metastatic recurrent disease.
5. Using ^{99m}Tc-MIBI scans, the diagnostic sensitivity for thyroid cancer detection was 94.4% for neck, 78.4% for lung, and 92.8% for skeletal lesions. Positive predictive value (PPV) and negative predictive value (NPV) combinations were 96.3% and 97.7% for head–neck lesions, 94.7% and 50.0% for chest lesions, and 100.0% and 93.1% for lesions in the abdomen–pelvis–extremities regions. For all scan sites taken together, the PPV and NPV were 96.1% (122/127) and 86.5% (109/126), respectively. Only lung metastases were detected < 85% of the time with ^{99m}Tc-MIBI scans (Alam et al., 1998).
6. ^{99m}Tc-MIBI scans may be read as normal even in patients with a nodule if the nodule functions autonomously; therefore, these scans should not be used to confirm a questionable palpable nodule.

Notes

1. Radioisotopic imaging of the thyroid gland may reveal the presence of heterogeneous or uniform, functioning thyroid tissue and should be performed in conjunction with a 24-hour RAIU test if functional assessment of the thyroid is needed.
2. Nodules without uptake on scan (i.e., cold nodules) have up to a 20% chance of harboring a malignancy. These nodules are more frequently cancerous in males (8.2%) than females (4.2%); however, thyroid cancer occurs more frequently in women than men because women have eight times as many thyroid nodules as men.
3. Multinodular goiters with a dominant nodule harbor cancer as often as a thyroid gland with a solitary nodule.
4. ^{99m}Tc-MIBI isotope may be used for imaging of thyroid gland anatomy but has limited use in the assessment of thyroid nodule function or malignancy.

5. rhTSH may be combined with radioiodine treatment to enhance ^{131}I uptake by thyroid cancer cells in subgroups of patients as follows:
   - Those in whom thyroxine withdrawal may be dangerous because of the effects of long-term TSH stimulation on the tumor mass in patients with brain metastases, vertebral metastases, neurological signs, or heart disease
   - Those affected by tumors with marked biological aggressiveness and a low iodine uptake (e.g., variants of follicular carcinoma, insular carcinoma, tall and columnar cell variants of papillary thyroid carcinoma, Hürthle cell carcinoma)
   - Those with hypothalamic–pituitary disturbances resulting in low levels of TSH secreted in response to thyroid hormone withdrawal
6. Lithium treatment combined with administration of rhTSH has not been shown to consistently enhance iodine uptake by residual thyroid tumor.
7. Indium-111 octreotide and positron emission tomography (PET) scans may be found to be positive in treated thyroid cancer patients even when follow-up whole-body radioiodine scans are negative.

ICD-9 Codes

Condition(s) that may justify this test *include but are not limited to*:

Thyroid neoplasm
193   cancer unspecified
198.89   metastatic cancer
237.4   uncertain behavior
239.7   follicular

Suggested Reading

Alam MS, Kasagi K, Misaki T, Miyamoto S, Iwata M, Iida Y, Konishi J. Diagnostic value of technetium-99m methoxyisobutyl isonitrile (^{99m}Tc-MIBI) scintigraphy in detecting thyroid cancer metastases: a critical evaluation. *Thyroid* 1998; 8:1091–100.

Ang ES, Teh HS, Sundram FX, Lee KO. Effect of lithium and oral thyrotropin-releasing hormone (TRH) on serum thyrotropin (TSH) and radioiodine uptake in patients with well differentiated thyroid carcinoma. *Singapore Med J* 1995; 36:606–8.

Belfiore A, La Rosa GL, La Porta GA, Giuffrida D, Milazzo G, Lupo I, Regalbuto C, Vigneri R. Cancer risk in patients with cold thyroid nodules: relevance of iodine intake, sex, age, and multinodularity. *Am J Med* 1992; 93:363–9.

Bombardieri E, Seregni E, Villano C, Aliberti G, Mattavelli F. Recombinant human thyrotropin (rhTSH) in the follow-up and treatment of patients with thyroid cancer. *Tumori* 2003; 89:533–6.

Maxon HR 3rd, Smith HS. Radioiodine-131 in the diagnosis and treatment of metastatic well differentiated thyroid cancer. *Endocrinol Metab Clin North Am* 1990; 19:685–718.

Michael BE, Forouhar FA, Spencer RP. Medullary thyroid carcinoma with radioiodide transport: effects of iodine-131 therapy and lithium administration. *Clin Nucl Med* 1985; 10:274–9.

Pons F, Carrio I, Estorch M, Ginjaume M, Pons J, Milian R. Lithium as an adjuvant of iodine-131 uptake when treating patients with well-differentiated thyroid carcinoma. *Clin Nucl Med* 1987; 12:644–7.

1.3.6   Thyroid Scan, Thallium-201 Whole-Body Scan (^{201}Tl-WBS) in Patients with Thyroid Carcinoma, Particularly Onchyocytic (Hürthle Cell) and Medullary Types

Indications for Test

^{201}Tl-WBS may be indicated for:

- The detection and monitoring of Hürthle cell and medullary thyroid carcinomas that are radioiodine-insensitive tumors when positron emission tomography (PET) scanning is unavailable
- Thyroid cancer patients who are Tg positive (<2 ng/mL) and ^{131}I scan negative

Procedure

1. Obtain a ^{99m}Tc-sestamibi scan before a ^{201}Tl-scan if metastatic lesions of Hürthle cell carcinoma are to be imaged with the greatest efficiency.
2. If a ^{201}Tl scan is to be performed, inject 3 mCi of [201]Tl.
3. Imaging is usually performed between 20 and 60 minutes after the injection of [201]Tl.

Interpretation

1. Scans must be read by an experienced nuclear medicine specialist or endocrinologist to identify areas of abnormal isotope uptake.

Notes

1. [201]Tl chloride is a monovalent radioisotope with biologic properties similar to potassium. It is taken up and concentrated in a variety of tumors including thyroid cancer.
2. ^{99m}Tc-MIBI may be an acceptable alternative to ^{201}Tl scintigraphy when evaluating patients with thyroid carcinoma, especially following thyroidectomy and ^{131}I-Na therapy.
3. Some thyroid malignancies will take up ^{201}Tl and not radioiodine, and *vice versa*.
4. The combination of testing for serum [Tg], ^{131}I-WBS, and ^{201}Tl-WBS increases the sensitivity of screening for metastatic thyroid cancer; thyroid hormone withdrawal prior to scanning is not required, and a same-day result with a low radiation burden is achieved.
5. Because competitive tracer uptake by normal thyroid tissue does not occur with ^{201}Tl, a ^{201}Tl-WBS coupled with a ^{131}I-WBS is useful in excluding metastatic disease.
6. PET scanning (see Test 1.3.7), as a more sensitive and specific test for detection of recurrent thyroid cancer, has practically replaced the ^{201}Tl-WBS since 2004.

ICD-9 Codes

Condition(s) that may justify this test *include but are not limited to*:

Thyroid neoplasm
193   malignant
198.89   metastatic
237.4   uncertain behavior

Suggested Reading

Dadparvar S, Chevres A, Tulchinsky M, Krishna-Badrinath L, Khan AS, Slizofski WJ. Clinical utility of technetium-99m methoxisobutylisonitrile imaging in differentiated thyroid carcinoma: comparison with thallium-201 and iodine-131 Na scintigraphy, and serum thyroglobulin quantitation. *Eur J Nucl Med* 1995; 22:1330–8.

Hoefnagel CA, Delprat CC, Marcuse HR, de Vijlder JJ. Role of thallium-201 total-body scintigraphy in follow-up of thyroid carcinoma *J Nucl Med* 1986; 27:1854–57.

Ramanna L, Waxman A, Braunstein G. Thallium-201 scintigraphy in differentiated thyroid cancer: comparison with radioiodine scintigraphy and serum thyroglobulin determinations. *J Nucl Med* 1991; 32:441–6.

Seabold JE, Gurll N, Schurrer ME, Aktay R, Kirchner PT. Comparison of ^{99m}Tc-methoxyisobutyl isonitrile and 201-Tl scintigraphy for detection of residual thyroid cancer after 131-I ablative therapy. *J Nucl Med* 1999; 40:1434–40.

Yen TC, Lin HD, Lee CH, Chang SL, Yeh SH. The role of technetium-99m sestamibi whole-body scans in diagnosing metastatic Hürthle cell carcinoma of the thyroid gland after total thyroidectomy: a comparison with iodine-131 and thallium-201 whole-body scans. *Eur J Nucl Med* 1994; 21:980–3.

1.3.7 Positron Emission Tomography (PET) Scan and Fusion with MR or M-CT Images in Metastatic Thyroid Carcinoma

Indications for Test

Positron emission tomography scan and fusion with MR or M-CT images are indicated when:

- A persistently elevated thyroglobulin or thyroglobulin antibody concentration is found in an established thyroid cancer patient with no uptake on radioiodine scan
- Planning surgical therapy for patients with possible thyroid tumor metastases, including neuroendocrine or nonendocrine tumors
- Establishing the presence or absence of tumor metastases for prognostic reasons
- Suspicious lesions are seen on MR or M-CT and the metabolic activity of the lesions is to be determined

Procedure

1. In thyroid cancer patients whose thyroglobulin concentration rises to >2 ng/mL, always perform ultrasound imaging of the neck as the first step in a search for metastases.
2. Upon finding neck or thoracic lesions on MR or M-CT imaging of thyroid cancer patients, obtain a follow-up PET scan.
3. If pulmonary lesions are suspected, use spiral-CT of the thorax before PET scan to exclude diffuse lung metastases and lesions < 1 cm in size.
4. Obtain 15 mCi of the PET scan isotope, fluorine-18 (^{18}F)-labeled deoxyglucose (FDG), within less than 1 hour of production. Note that the half-life of ^{18}F is 1.8 hours (108 minutes).
5. Instruct the patient not to exercise heavily for 24 hours prior to a PET scan.
6. Measure the patient's glucose level. Proceed with scanning only if the fasting glucose concentration < 140 mg/dL.
7. To get optimal imaging in a diabetic patient, the patient should fast and abstain from taking any diabetes therapy for at least 6 hours. This is necessary in order to get both the insulin and glucose levels stable enough to avoid insulin-stimulated cardiac muscle uptake of FDG.
8. If the patient's fasting glucose level is elevated, the test must be rescheduled; in the 6-hour period prior to testing, the patient is allowed prescribed diabetes therapy along with a low-carbohydrate meal (e.g., bacon, eggs, water).
9. Consider conscious sedation if the patient is claustrophobic. Routine use of tranquilizers (e.g., Xanax® or Valium®) 5 to 10 minutes before the procedure is recommended.
10. Administer an FDG dose of at least 10 mCi. Instruct the patient not to talk as the larynx will take up the isotope.
11. Image selected body areas 50 to 60 minutes after isotope injection.
12. If scanning the area from the eye to thigh level, acquire images over at least 30 to 40 minutes.
13. Overlay the MR or M-CT image level, or cut with the same cut taken on the PET scan.

Interpretation

1. Highly metabolically active thyroid neoplasms as small as 4 mm in size may be localized with PET, but less active lesions must be larger to be seen.
2. Correlation with lesions seen using other imaging modalities may help to distinguish benign from malignant tumors.
3. After surgical and radioiodine therapy of differentiated thyroid cancer, PET scanning may be expected to detect dedifferentiated lesions as well as residual cervical neck or mediastinal lesions and suspected small metastatic foci in the lung better than a ^{131}I-WBS (Figure 1.3).
4. The ^{131}I-WBS is better than PET in the detection of diffuse pulmonary metastases, lesions < 1 cm in size, and distant bone metastases, especially in the pelvis.

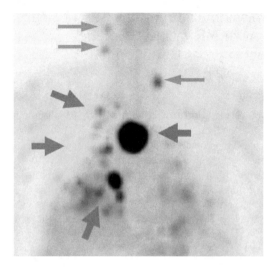

Figure 1.3
FDG–PET (fluorodeoxyglucose–positron emission tomography) scan following a negative radioiodine (^{131}I) uptake scan in a patient with metastatic papillary thyroid cancer and a rising thyroglobulin level. Patient had undergone a total thyroidectomy 1 year prior to both scans. Anterior FDG–PET projection image shows multiple areas of increased tracer uptake consistent with cervical nodal metastases (thin arrows) as well as extensive thoracic metastases (thick arrows).

Notes

1. FDG, an analog of glucose, accumulates in most tumors in greater amount than it does in normal tissue. It also accumulates in inflamed tissues, including gingivitis, adenomatous parathyroids, Graves' disease, and thyrotoxic nodules, as well as Teflon® laryngeal implants.
2. PET scanning is used to localize and identify the presence of primary or metastatic tumors, which metabolize various substrates such as glucose, neurotransmitters, or amino acids differently from normal tissues.
3. Positron emitters other than ^{18}F-deoxyglucose, including ^{11}C-amino acids, H$_2$^{15}O, and ^{13}N, are very short lived. Iodine-124 (^{124}I) has a half-life of 4.2 days and is very difficult to obtain. FDG is now more readily available from a network of cyclotron reactors.
4. Low-dose, continuous insulin infusion or isolated administration may be necessary to correct hyperglycemia, decrease myocardial uptake of isotope, and improve the quality of the FDG–PET image.
5. The availability of PET scanning has been limited by selective reimbursement for this test. In 2004, Medicare considered only studies of thyroglobulin-positive thyroid cancer with no iodine uptake by PET reimbursable.

ICD-9 Codes

Condition(s) that may justify this test *include but are not limited to*:

Thyroid neoplasm
193   malignant
198.89   metastatic
237.4   uncertain behavior
246.0   thyrocalcitonin secretion

Suggested Reading

Dietlein M, Scheidhauer K, Voth E, Theissen P, Schicha H. Follow-up of differentiated thyroid cancer: what is the value of FDG and sestamibi in the diagnostic algorithm? *Nuklearmedizin* 1998; 37:12–7.

Hsu CH, Liu RS, Wu CH, Chen SM, Shih LS. Complementary role of ^{18}F-fluorodeoxyglucose positron emission tomography and ^{131}I scan in the follow-up of post-therapy differentiated thyroid cancer. *J Formos Med Assoc* 2002; 101:459–67.

Rohren EM, Turkington TG, Coleman RE. Clinical applications of PET in oncology. *Radiology* 2004; 231:305–32.

1.3.8 Calcitonin Testing: Serum Screening and Follow-Up in Patients with Medullary Thyroid Carcinoma (MTC)

Indications for Test

Screening for MTC with calcitonin may be indicated when:

- A thyroid biopsy report suggests MTC.
- A patient belongs to a kindred with multiple endocrine neoplasia (MEN) type 2.
- A MTC patient has undergone near-total thyroidectomy, and a test for evidence of residual tumor is required.

Procedure

1. Obtain a blood sample for serum calcitonin testing from high-risk individuals, such as those with multiple mucosal neuromas, Marfanoid habitus, or neurogangliomas of the gastrointestinal tract, or relatives of such individuals even if they have a normal appearance.
2. To confirm the diagnosis of MTC in patients with an elevated serum calcitonin concentration, test for the *RET* protooncogene mutation.

Interpretation

1. Expect falsely elevated calcitonin levels to be found in critical ill, renal insufficient, and renal failure patients.
2. Newly diagnosed patients with MTC may have baseline calcitonin levels ranging from normal (15 pg/mL) to >80,000 pg/mL.
3. A markedly elevated baseline calcitonin concentration does not preclude a favorable short-term outcome in patients with MTC who receive appropriate surgical therapy.
4. Extensive involvement of cervical lymph nodes and high postoperative serum calcitonin levels predict tumor recurrence in patients with MTC.

Notes

1. Calcitonin stimulation with calcium combined with pentagastrin (CPG) infusion is an alternative but less sensitive test than the test for *RET* protooncogene to confirm the diagnosis of MTC.
2. The CPG calcitonin stimulation test may have a place in the assessment of completeness of thyroid extirpation and in follow-up for recurrence of MCT.
3. A negative family history for MTC in a patient presenting with symptoms of MTC does not reliably exclude familial disease.
4. MTC can be detected at an early stage by calcitonin screening in all patients with thyroid nodules; however, this approach is limited by the high cost of testing and the limited specificity of calcitonin for diagnosing MTC, particularly in patients with renal insufficiency.
5. MTCs discovered in a preclinical stage by calcitonin screening are equally distributed between the sexes.
6. Determination of serum calcitonin gene-related peptide (CGRP) adds no information to that of serum calcitonin in MEN 2 family screening for MTC.

7. Pentagastrin stimulation of calcitonin not only is less sensitive than the *RET* protooncogene test but is also less specific, as positive tests are found in patients with C-cell hyperplasia. On the other hand, many sporadic MTCs develop in the presence of C-cell hyperplasia.
8. C-cell hyperplasia as a preneoplastic condition in patients without *RET* protooncogene mutations remains under study.
9. Tests for hyperparathyroidism (e.g., intact PTH, dual-phase sestamibi scanning) and for pheochromocytomas (e.g., plasma metanephrines) are indicated in patients with elevated calcitonin levels.

ICD-9 Codes

Refer to Test 1.3.7 codes.

Suggested Reading

Bieglmayer C, Scheuba C, Niederle B, Flores J, Vierhapper H. Screening for medullary thyroid carcinoma: experience with different immunoassays for human calcitonin. *Wien Klin Wochenschr* 2002; 114: 267–73.

Kaserer K, Scheuba C, Neuhold N, Weinhausel A, Haas OA, Vierhapper H, Niederle B. Sporadic versus familial medullary thyroid microcarcinoma: a histopathologic study of 50 consecutive patients. *Am J Surg Pathol* 2001; 25:1245–51.

Kaserer K, Scheuba C, Neuhold N, Weinhausel A, Vierhapper H, Niederle B. Recommendations for reporting C cell pathology of the thyroid. *Wien Klin Wochenschr* 2002; 114:274–8.

Machens A, Schneyer U, Holzhausen HJ, Dralle, H. Prospects of remission in medullary thyroid carcinoma according to basal calcitonin level. *J Clin Endocrinol Metab* 2005; 2029–34.

Netzloff ML, Garnica AD, Rodgers BM, Frias JL. Medullary carcinoma of the thyroid in the multiple mucosal neuromas syndrome. *Ann Clin Lab Sci* 1979; 9:368–73.

Ponder BA, Ponder MA, Coffey R, Pembrey ME, Gagel RF, Telenius-Berg M, Semple P, Easton DF. Risk estimation and screening in families of patients with medullary thyroid carcinoma. *Lancet* 1988; 1:397–401.

Raue F, Geiger S, Buhr H, Frank-Raue K, Ziegler R. The prognostic importance of calcitonin screening in familial medullary thyroid carcinoma. *Dtsch Med Wochenschr* 1993; 118:49–52.

Saller B, Gorges R, Reinhardt W, Haupt K, Janssen OE, Mann K. Sensitive calcitonin measurement by two-site immunometric assays: implications for calcitonin screening in nodular thyroid disease. *Clin Lab* 2002; 48:191–200.

Schifter S. Calcitonin gene-related peptide and calcitonin as tumour markers in MEN 2 family screening. *Clin Endocrinol (Oxf)* 1989; 30:263–70.

Zangeneh F, Gharib H, Goellner JR, Kao PC. Potential absence of prognostic implications of severe preoperative hypercalcitoninemia in medullary thyroid carcinoma. *Endocr Pract* 2003; 9:284–9.

1.3.9   *RET* Protooncogene Testing in Medullary Thyroid Carcinoma (MTC)

Indications for Test

RET protooncogene determination is indicated to make a genetic diagnosis in patients with:

- Known MTC or who are at high risk for MTC
- MEN 2A or 2B who are at risk for harboring a pheochromocytoma or parathyroid adenoma, in addition to a MTC
- A non-MEN, familial MTC

Procedure

1. Obtain 5 mL of peripheral blood in a lavender-top tube or a swab of the buccal mucosa.
2. Analyze the DNA in the sample collected for the presence and sequencing of a specific *RET* mutation in exon 10, 11, or 16 located within the pericentromeric region of chromosome 10.

Interpretation

1. Determine if the patient is positive for one of the *RET* protooncogene mutations commonly observed in the syndrome present in the kindred under study. In MEN 2A and familial MTC, *RET* protoon-cogene mutations change cysteine residues in the protein normally encoded by codons 609, 611, 618, 620, or 634 on exons 10 or 11.
2. In MEN 2B, a methionine-to-threonine substitution occurs in the protein encoded by codon 918 in exon 16.
3. If no mutation is found, DNA from affected family members should be sequenced to exclude a *RET* protooncogene mutation not previously described.
4. Linkage analysis should be considered if mutations are not found in individuals with clinical evidence of MTC.
5. Generally, a prophylactic thyroidectomy should be offered to patients at a young age if positive for *RET* protooncogene mutations, especially in MEN 2B kindreds.

Notes

1. MEN 2A and 2B and familial MTC are inherited conditions with autosomal dominance and incom-plete penetrance. MTC occurs at a very young age in children with MEN 2.
2. Mutations in exons 10, 11, or 16 of the *RET* protooncogene have been found to be responsible for most (95%) inherited forms of MTC.
3. The DNA from patients with Hirschsprung's disease may also bear a mutation in the *RET* protoon-cogene.
4. Measurement of plasma calcitonin concentration, after stimulation with pentagastrin or calcium (CPG), and urinary excretion of catecholamines and catecholamine metabolites may be used as screening tests for MTC but will lead to false negatives (~20%) and false positives (~10%) when high-risk kindreds (e.g., those with MEN 2A) are tested.
5. Because DNA analysis is less ambiguous than the calcitonin stimulation test with pentagastrin or calcium, it is the diagnostic test of choice in the identification of asymptomatic individuals at risk for MEN 2A.

ICD-9 Codes

Refer to Test 1.3.7 codes.

Suggested Reading

Chi DD, Toshima K, Donis-Keller H, Wells SA. Predictive testing for multiple endocrine neoplasia type 2A (MEN 2A) based on the detection of mutations in the *RET* protooncogene. *Surgery* 1994; 116:124–33.

Goodfellow PJ. Inherited cancers associated with the *RET* proto-oncogene. *Curr Opin Genet Dev* 1994; 4:446–52.

Ledger GA, Khosla S, Lindor NM, Thibodeau SN, Gharib H. Genetic testing in the diagnosis and management of multiple endocrine neoplasia type II. *Ann Intern Med* 1995; 122:118–24.

Lips CJ, Landsvater RM, Hoppener JW, Geerdink RA, Blijham G, van Veen JM, van Gils AP, de Wit MJ, Zewald RA, Berends MJ et al. Clinical screening as compared with DNA analysis in families with multiple endocrine neoplasia type 2A. *N Engl J Med* 1994; 331:828–35.

van Heurn LW, Schaap C, Sie G, Haagen AA, Gerver WJ, Freling G, van Amstel HK, Heineman E. Predictive DNA testing for multiple endocrine neoplasia 2: a therapeutic challenge of prophylactic thyroidectomy in very young children. *J Pediatr Surg* 1999; 34:568–71.

Wells SA, Donis-Keller H. Current perspectives on the diagnosis and management of patients with multiple endocrine neoplasia type 2 syndromes. *Endocrinol Metab Clin North Am* 1994; 23:215–28.

Pituitary Gland Testing

2.1 Corticotroph (Adrenocorticotropic Hormone) Reserve

2.1.1 Cortisol and Adrenocorticotropic Hormone (ACTH) Testing in the Morning: Test Panel for Adrenal Insufficiency Secondary to Central or Isolated ACTH Deficiency

Indications for Test

Measurement of morning cortisol and ACTH is indicated when:

- Features and clinical criteria for the diagnosis of primary hypoadrenalism (Addison's disease) are present (see Table 6.2).
- Adrenal insufficiency is suspected in symptomatically hypoadrenal patients.
- A pituitary or hypothalamic disturbance is known or suspected.

Procedure

1. Use an immunometric assay to measure ACTH because of its speed, cost-effectiveness, improved analytical specificity, and ability to detect low-normal levels compared to high-performance liquid chromatography (HPLC), fluorescence polarization immunoassay (FPIA), and radioimmunoassay (RIA) methods.
2. Screen suspected hypoadrenal patients by obtaining a morning fasting blood sample for ACTH and cortisol. Repeat testing of morning cortisol levels is indicated if results are equivocal.
3. If ACTH and cortisol results are persistently equivocal, proceed to a low-dose cosyntropin stimulation test (see Test 2.1.2).

Interpretation

1. A random, basal plasma ACTH in the reference interval of high to high-normal distinguishes primary adrenal insufficiency from secondary adrenal insufficiency, in which the ACTH is in the low to low-normal portion of the reference interval.
2. A basal cortisol of <3 mcg/dL (83 nmol/L) effectively makes the diagnosis of adrenal insufficiency or hypoadrenalism; levels of >19 mcg/dL (524 nmol/L) exclude it. Most patients will have intermediate cortisol levels and will require dynamic testing (see Test 2.1.2).

3. A single, morning (0800 hours) serum cortisol measurement is of limited value for diagnosis of hypoadrenalism. Although it has a diagnostic sensitivity of up to100%, its specificity is only 33% using a serum cortisol cutoff value of 18 mcg/dL (497 nmol/L). Serial repeat testing for morning cortisol levels is indicated in making the diagnosis of hypoadrenalism.
4. Lack of response to both a low-dose cosyntropin test and ovine corticotropin releasing factor (oCRF) test suggests a pituitary rather than a hypothalamic origin for the rare disorder of isolated ACTH deficiency.
5. The coexistence of positive antithyroglobulin, antimicrosomal, and antiperoxidase autoantibodies in about 25% of cases with secondary hypoadrenalism suggests an autoimmune etiology for some cases of isolated ACTH deficiency.
6. If hypoadrenalism is documented, test the patient for thyroid autoimmunity, as both conditions may be linked (i.e., Schmidt's syndrome).

Notes

1. In secondary hypoadrenalism, patients may have a long history of malaise, fatigue, anorexia, diarrhea, weight loss, and joint and back pain.
2. More severe hypotension and orthostasis occurs in acute or chronic primary hypoadrenalism than in pituitary patients with secondary hypoadrenalism who typically present with vague, nonspecific symptoms (e.g., weakness, fatigue).
3. Positive diagnosis of isolated ACTH deficiency requires demonstration of normal secretory indices for other pituitary hormones.
4. Partial ACTH deficiency may prevent involution of the adrenal cortex and preserve the cortisol response to ACTH stimulation.
5. Using intensive (every 10 minutes) and extended (24-hour) blood sampling, there are normally about 39 to 41 ACTH secretory bursts per 24 hours, with an interburst interval of 37 to 41 minutes. These bursts are discrete, punctuated events, arising without tonic interpulse secretion and evenly distributed throughout the 24-hour period (Veldhuis et al., 1990).
6. The half-life of endogenous ACTH is about 15 minutes, and its daily production rate is 0.8 to 1.1 ng/mL (0.18–0.24 nmol/L) volume of distribution. There is a significant three- to fourfold variance in the mass of ACTH secreted per burst, with maximal amplitude per peak occurring at 0800 to 0900 hours.
7. Significant differences in cortisol levels when determined by different methods (HPLC, FPIA, or RIA) may occur.
8. A normal response to insulin-induced hypoglycemia by insulin tolerance test (ITT) is defined as peak serum cortisol >20 mcg/dL (551 nmol/L) any time during the ITT.
9. Although ITT 5 to 8 days after pituitary surgery is 100% accurate in determining the need for sustained glucocorticoid replacement postoperatively, repeat morning cortisol measurement obviates the need for an ITT in making a diagnosis of central hypoadrenalism in 95% of cases.

ICD-9 Codes

Conditions that may justify this test *include but are not limited to*:

253.2 panhypopituitarism
255.4 corticoadrenal insufficiency

Suggested Reading

Auchus RJ, Shewbridge RK, Shepherd MD. Which patients benefit from provocative adrenal testing after transsphenoidal pituitary surgery? *Clin Endocrinol (Oxf)* 1997; 46:21–7.

de Luis DA, Aller R, Romero E. Isolated ACTH deficiency. *Horm Res* 1998; 49:247–9.

Gordon D, Beastall GH, Thomson C, Thomson JA. ACTH deficiency: hypothalamic or pituitary in origin? *Scott Med J* 1987; 32:49–50.

Kasperlik-Zaluska AA, Czarnocka B, Czech W, Walecki J, Makowska AM, Brzezinski J, Aniszewski J. Secondary adrenal insufficiency associated with autoimmune disorders: a report of twenty-five cases. *Clin Endocrinol (Oxf)* 1998; 49:779–83.

Mukherjee JJ, de Castro JJ, Kaltsas G, Afshar F, Grossman AB, Wass JA, Besser GM. A comparison of the insulin tolerance/glucagon test with the short ACTH stimulation test in the assessment of the hypo-thalamo–pituitary–adrenal axis in the early post-operative period after hypophysectomy. *Clin Endocrinol (Oxf)* 1997; 47:51–60.

Nieman LK. Dynamic evaluation of adrenal hypofunction. *J Endocrinol Invest* 2003; 26(Suppl 7):74–82.

Nye EJ, Grice JE, Hockings GI, Strakosch CR, Crosbie GV, Walters MM, Jackson RV. Comparison of adrenocorticotropin (ACTH) stimulation tests and insulin hypoglycemia in normal humans: low dose, standard high dose, and 8-hour ACTH-(1-24) infusion tests. *J Clin Endocrinol Metab* 1999; 84:3648–55.

Soule SG, Fahie-Wilson M, Tomlinson S. Failure of the short ACTH test to unequivocally diagnose long-standing symptomatic secondary hypoadrenalism. *Clin Endocrinol (Oxf)* 1996; 44:137–40.

Talbot JA, Kane JW, White A. Analytical and clinical aspects of adrenocorticotrophin determination. *Ann Clin Biochem* 2003; 40(Pt 5):453–71.

Veldhuis JD, Iranmanesh A, Johnson ML, Lizarralde G. Amplitude, but not frequency, modulation of adreno-corticotropin secretory bursts gives rise to the nyctohemeral rhythm of the corticotropic axis in man. *J Clin Endocrinol Metab* 1990; 71:452–63.

2.1.2 Cosyntropin Stimulation Test: Short or Rapid (1 Hour) with Low-Dose (1 mcg) Cosyntropin in Central Adrenal Insufficiency or Secondary Hypoadrenalism

Indications for Test

The low-dose cosyntropin stimulation test is indicated in patients with:

- Clinical signs or symptoms of secondary hypoadrenalism (e.g., orthostasis, electrolyte imbalance, fatigue, malaise, anorexia, diarrhea, or weight loss without hyperpigmentation)
- A history of recent exposure to suppressive doses of glucocorticoids associated with subsequent symptoms of hypoadrenalism (e.g., steroid withdrawal)
- A pituitary tumor
- A history of previous pituitary surgery or irradiation

Procedure

1. Discontinue prednisone therapy for at least 24 hours and obtain informed consent before performing this invasive test.
2. To prepare a solution containing a minimum of 1 mcg of cosyntropin, dissolve 250 mcg of cosyntropin in 50 mL of normal saline and withdraw 0.22 mL of the resultant mixture for intravenous infusion. Use the volume of 0.22 mL, instead of 0.20 mL, to compensate for a small loss of fluid volume in the hub of the needle.
3. Obtain blood samples for serum cortisol testing at –15 minutes and at –10 minutes (i.e., the baseline cortisol levels). At zero minutes, inject 1 mcg of cosyntropin.
4. Collect blood samples at 30 and 60 minutes for cortisol testing after the cosyntropin injection.

Interpretation

1. Normally, a rise in serum cortisol to >18 mcg/dL (497 nmol/L) at 30 or 60 minutes or a rise of ≥7 mcg/dL (193 nmol/L) at 30 minutes above the higher of the baseline cortisol values occurs after injection of 1 mcg cosyntropin.

2. In patients whose cortisol response to 1 mcg of exogenously administered cosyntropin fails to meet the aforementioned criteria, it is likely that their adrenal glands are not adequately stimulated by endogenous, pituitary ACTH.

3. Partial ACTH deficiency may prevent involution of the adrenal cortex and preserve the cortisol response to ACTH stimulation.

4. Because the definition of a "normal cortisol response" following the administration of cosyntropin in the low-dose, short cosyntropin stimulation test depends on the cortisol cutoff values chosen to define a normal cortisol response, this test may not detect clinically significant ACTH deficiency in all hypoadrenal patients.

Notes

1. The low-dose cosyntropin stimulation test is physiological and more sensitive than the high-dose test (see Test 6.1.1), especially in cases of mild hypoadrenalism following discontinuation of long-term, suppressive doses of glucocorticoids.

2. A 250-mcg (high-dose) cosyntropin stimulation test (see Test 6.1.1) will fail to detect secondary hypoadrenalism resulting from pituitary or hypothalamic dysfunction.

3. In comparison to a continuous ACTH infusion test, a significant number of falsely abnormal rapid (1 hour) low-dose cosyntropin infusion test results have been observed, indicating that the rapid cosyntropin test is useful as a screening but not diagnostic test.

4. Cortisol immunoassays detect prednisone but not dexamethasone; therefore, do not attempt measurement of cortisol levels during prednisone therapy, unless cortisol measurement by HPLC, a method capable of separating and quantifying both cortisol and prednisone, is available.

5. ACTH levels may be normal even if the adrenal glands are chronically suppressed by glucocorticoids therapy.

6. Use of the low-dose cosyntropin test has been limited by the lack of a ready-to-use, commercially available preparation of dilute cosyntropin and by controversy over the diagnostic criteria for defining a "normal cortisol response" following the administration of 1 mcg of cosyntropin.

7. The metyrapone test and insulin tolerance test (ITT), as alternatives to the low-dose cosyntropin test, are hazardous, poorly tolerated, and cumbersome; nonetheless, in the opinion of some endocrinologists (Soule et al., 2000), they remain gold standard tests for the diagnosis of secondary hypoadrenalism.

8. The diagnostic accuracy of the ITT or metyrapone stimulation test (i.e., the overnight 11-deoxycortisol response following administration of metyrapone) is assumed by some experts to be 100% for the diagnosis of hypoadrenalism. In contrast, the low-dose cosyntropin test is 71 to 100% sensitive and 58 to 93% specific when using cortisol cutoff values of 15 to 21 mcg/dL (414–579 nmol/L) to define a normal cortisol response at 30 or 60 minutes after the administration of cosyntropin.

9. The ITT or metyrapone tests are unnecessary when a normal cortisol peak of >27 mcg/dL (745 nmol/L) occurs after intravenous administration of 1 mcg cosyntropin or when a very low basal cortisol (<3 mcg/dL or 83 nmol/L) is found, indicating overt hypoadrenalism or glucocorticoid treatment.

ICD-9 Codes

Refer to Test 2.1.1 codes.

Suggested Reading

Agwu JC, Spoudeas H, Hindmarsh PC, Pringle PJ, Brook CG. Tests of adrenal insufficiency. *Arch Dis Child* 1999; 80:330–3.

Ambrosi B, Barbetta L, Re T, Passini E, Faglia G. The one microgram adrenocorticotropin test in the assessment of hypothalamic–pituitary–adrenal function. *Eur J Endocrinol* 1998; 139:575–9.

Beishuizen A, van Lijf JH, Lekkerkerker JF, Vermes I. The low dose (1 mcg) ACTH stimulation test for assessment of the hypothalamo–pituitary–adrenal axis. *Neth J Med* 2000; 56:91–9.

May ME, Carey RM. Rapid adrenocorticotropic hormone test in practice: retrospective review. *Am J Med* 1985; 79:679–84.

Nieman LK. Dynamic evaluation of adrenal hypofunction. *J Endocrinol Invest* 2003; 26(Suppl 7):74–82.

Nye EJ, Grice JE, Hockings GI, Strakosch CR, Crosbie GV, Walters MM, Jackson RV. Comparison of adrenocorticotropin (ACTH) stimulation tests and insulin hypoglycemia in normal humans: low dose, standard high dose, and 8-hour ACTH-(1-24) infusion tests. *J Clin Endocrinol Metab* 1999; 84:3648–55.

Soule S, Van Zyl Smit C, Parolis G, Attenborough S, Peter D, Kinvig S, Kinvig T, Coetzer E. The low dose ACTH stimulation test is less sensitive than the overnight metyrapone test for the diagnosis of secondary hypoadrenalism. *Clin Endocrinol (Oxf)* 2000; 53:221–7.

Tordjman K, Jaffe A, Trostanetsky Y, Greenman Y, Limor R, Stern N. Low-dose (1 microgram) adrenocorticotrophin (ACTH) stimulation as a screening test for impaired hypothalamo–pituitary–adrenal axis function: sensitivity, specificity and accuracy in comparison with the high-dose (250 microgram) test. *Clin Endocrinol (Oxf)* 2000; 52:633–40.

2.2 Corticotroph (Adrenocorticotropic Hormone) Excess

2.2.1 Cortisol and Adrenocorticotropic Hormone (ACTH) Testing in the Morning: Test Panel for Cushing's Syndrome

Indications for Test

Measurement of cortisol and ACTH in the morning is indicated when:

- Cushing's syndrome is suspected in patients with signs and symptoms of a pituitary or primary hyperadrenal state.
- Findings suggestive of Cushing's syndrome, which may include weight gain, new-onset diabetes or glucose intolerance, moon facies, abdominal striae, increase in supraclavicular fat pads, bone loss, personality change, poor wound healing, and thinning of the skin with easy bruisability, are noted.

Procedure

1. Screen patients suspected of having Cushing's syndrome by obtaining a fasting morning, or upon arising, blood sample in a serum separator tube (SST) for *serum* cortisol and a separate blood sample collected in a lavender-top (EDTA) tube for *EDTA plasma* ACTH measurement.

Interpretation

1. Typically, both ACTH and cortisol will be elevated in the morning, or in the high-normal reference interval, in patients with Cushing's syndrome resulting from a pituitary or ectopic source of ACTH secretion. Proceed with the collection of a urinary free cortisol concentration (see Test 6.2.2) in such an event.
2. Cushing's syndrome caused by exogenous glucocorticoid administration or adrenal cortisol hypersecretion is characterized by low (<10 pg/mL) to low-normal (10–15 pg/mL) EDTA plasma ACTH concentrations.
3. Extremely high ACTH concentrations are usually associated with aggressive, ectopic tumors such as bronchial carcinoids or small-cell lung cancers, some of which may be secreting corticotropin-releasing hormone (CRH) rather than ACTH.

Notes

1. ACTH measurement is of limited value in distinguishing between the causes of Cushing's syndrome, as there is considerable overlap in ACTH in subjects with either a pituitary or an ectopic ACTH-secreting tumor.
2. Inferior petrosal sinus sampling for ACTH, after administration of ovine CRH, may be required to differentiate between pituitary and ectopic sources of ACTH hypersecretion.

3. A cortisol of <1.8 mcg/dL (50 nmol/L) between 2200 hours and midnight makes active Cushing's syndrome unlikely.
4. The 1-mg overnight dexamethasone cortisol suppression test (1-mg overnight Dex) requires that the tablet of dexamethasone be taken at precisely 2300 hours and that a cortisol is obtained at precisely 0800 hours the next day. Not only does the rate of dexamethasone metabolism vary between individuals, but it is also influenced by other drugs metabolized by the liver.
5. The normal response to the 1-mg overnight Dex test is a morning cortisol of <3 mcg/dL, with a 13% false-positive and a 2% false-negative rate.

ICD-9 Codes

Conditions that may justify this test *include but are not limited to*:

255.X	Cushing's syndrome
278.1	localized adiposity/supraclavicular fat pad/abdomen
701.3	abdominal striae/"stretch marks" (atrophic)
733	osteoporosis
754.0	facial edema/moon facies (puffiness)

Suggested Reading

Newell-Price J, Grossman AB. The differential diagnosis of Cushing's syndrome. *Ann Endocrinol (Paris)* 2001; 62:173–9.

Talbot JA, Kane JW, White A. Analytical and clinical aspects of adrenocorticotrophin determination. *Ann Clin Biochem* 2003; 40(Pt 5):453–71.

2.2.2   Imaging to Find ACTH-Secreting Tumors by High-Resolution Multislice Magnetic Resonance (MR) and Computed Tomographic (M-CT) Scanning

Indications for Test

Magnetic resonance and M-CT imaging of the *sella and parasellar regions* for identification of ACTH-secreting *pituitary* tumors is indicated when:

• Biochemical evidence of hypercortisolism has been found.

Magnetic resonance and M-CT imaging of the *chest and/or abdomen* for identification of ACTH-secreting *ectopic* tumors may be indicated when:

• A non-pituitary tumor source of ACTH is suspected, particularly when the ACTH level is extremely high.

Procedure

1. Perform MR imaging of the pituitary using magnified ≤3-mm coronal cuts of the hypothalamic–hypophyseal (sella and parasellar) region.
2. When a pituitary lesion is identified or highly suspected on noncontrast MR imaging, obtain gadolinium-diethylenetriaminepentaacetic acid contrast (Gd–DTPA)-enhanced images in both the coronal and sagittal planes.
3. Use M-CT with iodinated contrast media to delineate calcified lesions and osseous structures, both of which are better seen with M-CT than MR.

4. If possible, obtain <1.0-mm cuts of the hypothalamic–hypophyseal region with M-CT to provide a higher resolution image.
5. Obtain a M-CT without contrast in the event of a high-density lesion, which may be hypervascular or calcified.
6. If a pituitary tumor mass is not found in an overtly symptomatic, clinically Cushingoid patient with elevated ACTH levels, have a bilateral inferior petrosal sinus sampling (BIPSS) procedure (see Specimen Collection Protocol P5 in Appendix 2) performed when MR and M-CT imaging of the chest and abdomen has been done and failed to show evidence of an ectopic tumor in these areas.

Interpretation

1. Pituitary adenomas are classified as micro (<10 mm) or macro (>10 mm) based on their largest diameter.
2. Autoinfarction of pituitary adenomas, characterized by hyperdense lesions, is often associated with headaches, visual disturbances, and hemorrhage. Catastrophic apoplectic events associated with an infarcted lesion may take 1 to 2 hours to visualize on scan.
3. ACTH-secreting tumors tend to be small, usually <10 mm in size, and may not cause a headache.
4. The pituitary is the most common location for an ACTH-producing tumor.
5. The differential diagnosis of solid lesions in the pituitary sella includes adenomas, chordomas, and gliomas arising from the hypothalamus, optic nerve, metastases, or meningiomas of the tuberculum sella.
6. Angiography may be required to more clearly define supra and parasellar lesions, but it is not of much use in the diagnosis of pituitary adenomas.

Notes

1. The incidental discovery of clinically silent lesions in and around the pituitary is a common occurrence.
2. Magnetic resonance imaging can detect 60 to 70% of all pituitary adenomas in adults, but it is of less use in making the diagnosis of Cushing's disease in children.
3. Computed tomography (CT) imaging is preferable to MR imaging when defining ossified structures prior to trans-sphenoid pituitary surgery.
4. Magnetic resonance imaging better differentiates cystic, solid, and hemorrhagic lesions than does M-CT imaging.

ICD-9 Codes

Refer to Test 2.2.1 codes.

Suggested Reading

Newell-Price J, Grossman AB. The differential diagnosis of Cushing's syndrome. *Ann Endocrinol (Paris)* 2001; 62:173–9.

2.3 Somatotroph (Growth Hormone) Reserve

2.3.1 Insulin-Like Growth Factor 1 (IGF-1) Screening Prior to Use of Growth Hormone Stimulation Testing in Suspected Growth Hormone Deficiency (GHD)

Indications for Test

Insulin-like growth factor 1 (or somatomedin C) screening followed by a GH stimulation test is indicated for:

TABLE 2.1

Adult Age-Specific Total Insulin-Like Growth Factor 1 (IGF-1) Reference Ranges[a] for Quest Diagnostics IGF-1 Assay

	Age (years)													
	18	19	20	25	30	35	40	45	50	55	60	65	70	75–80
Males														
LLN (ng/mL)	187	172	158	112	89	77	70	66	61	56	50	44	38	35
ULN (ng/mL)	554	524	497	402	350	323	307	296	285	271	255	238	223	213
Females														
LLN (ng/mL)	154	140	128	89	71	63	58	54	49	42	35	27	22	21
ULN (ng/mL)	546	515	488	397	352	330	318	307	292	272	248	223	204	199

[a] 95% confidence limits (LLN, lower limit of normal range; ULN, upper limit of normal range)

Source: Adapted from Brabant, G. et al., *Horm. Res.*, 60, 53–60, 2003.

- Children and adults suspected of GHD based on either growth delay in childhood or the presence of hypothalamic pituitary disorders
- Adults with symptoms of fatigue, bone loss, increase in abdominal adiposity, or dysphoric mood disturbances (i.e., social isolation, decrements in energy and sex life) consistent with adult growth hormone deficiency (AGHD)

Procedure

1. Obtain a random blood sample for IGF-1 testing. Repeat a low, low-normal, or borderline normal test result and measure IGF binding protein 3 (IGFBP-3) in children with low or low-normal [IGF-1].
2. If the [IGF-1] is unexpectedly elevated, obtain a fasting blood sample for determination of a resting basal [GH]. Recognize that measurement of a basal [GH] does not help in making a diagnosis of GHD and that greater amounts of GH are produced during normal puberty.
3. Reconfirm a childhood diagnosis of isolated or idiopathic GH deficiency by performing GH provocative testing in these patients when they enter young adulthood.
4. Do not reconfirm childhood GH deficiency by GH provocative testing in young adults who have two or more pituitary hormone deficiencies in addition to GHD, as such testing is not useful or cost effective.
5. In both children and adults, if the [IGF-1] > 80 ng/mL, but the clinical suspicion of GHD is high, perform a combined arginine (ARG) and growth-hormone-releasing hormone (GHRH) or Geref test (ARG–Geref) in preference to other provocative GH stimulation tests.
6. Alternatives to the ARG–Geref GH stimulation test in children include the insulin tolerance test, sleep-induced or exercise-induced GH release test, or the propranolol-augmented glucagon, L-dopamine, arginine-ITT, or clonidine GH stimulation test (see Test 2.3.4). The availability, safety, cost, and performance of these tests vary widely.

Interpretation

1. Refer to the "Adult Age-Specific Total Insulin-Like Growth Factor (IGF-1) Reference Intervals for Quest Diagnostics IGF-1 Concentration (ng/mL) Assay" (Table 2.1) (Brabant et al., 2003).
2. Refer to "Peak Growth Hormone (GH) Response to GH Stimulation Tests" (Table 2.2) for the consensus on the nature of a positive response to various GH stimulation tests.
3. [GH] < 7 ng/mL (<0.32 pmol/L) is consistent with, but not necessarily diagnostic of, GH deficiency in children for the following tests: insulin tolerance test (ITT), sleep-induced or exercise-induced GH release test, or the propranolol-augmented glucagon, L-dopamine, arginine, or clonidine GH stimulation test. With the exception of the ITT, these tests have limited practical value, and their use has almost disappeared.

TABLE 2.2
Peak Growth Hormone Response to
Growth Hormone Stimulation Tests[a]

Type of Subject	Peak Growth Hormone Concentration (ng/mL)	Type of Growth Hormone Response
Children	>10	Normal
	7–10	Equivocal
	<7	Unequivocally low
Adults	>5	Normal
	3–5	Equivocal
	<3	Unequivocally low

[a] Peak GH levels and responses are well established for the insulin tolerance test and less well established for other GH stimulation tests.

4. IGF-1 concentrations may be good predictors of active or cured acromegaly but may be within the normal age-adjusted range in childhood-onset or adult growth hormone deficiency. Approximately 70% of the [IGF-1] and 72% of the [IGFBP-3] in individuals with AGHD secondary to hypopituitarism were within the normal reference interval for the analyte tested (Darendeliler et al., 2004).

5. [IGF-1] < 70 ng/mL on repeat determinations is good evidence for AGHD and obviates the need for stimulation testing, particularly in patients known to have pituitary disorders and hormone deficiencies other than GH.

6. A basal [IGF-1] < 77.2 ng/mL (3.5 pmol/L) was 95% specific for GH deficiency (Biller et al., 2002).

7. In AGHD, a low [IGFBP-3] plus low [IGF-1] suggests a very low degree of residual GH activity. Note that the use of IGFBP-3 testing is reserved, almost exclusively, for children.

8. About 10 to 11% of children diagnosed with complete growth hormone deficiency based on a peak [GH] < 7 ng/mL (0.32 pmol/L) in response to a stimulatory agent will be found to have recovered their GH secretory capacity on retesting.

9. Patients with multiple pituitary hormone deficiencies or a small pituitary or remnant on MR imaging are unlikely to recover normal GH secretory capacity.

10. The ITT and the ARG–Geref tests have the greatest diagnostic accuracy in diagnosing GHD. For reasons of comfort and safety, patients much prefer the latter test.

11. The GH response evoked by the ARG–Geref test in patients with even a mild increase in body mass index (BMI) and a normal pituitary may not be distinctly different from the GH response in AGHD patients, potentially resulting in the erroneous classification of obese subjects as GH deficient.

Notes

1. Growth hormone responses to a variety of stimulatory agents may be reduced in children with decreased growth velocity before puberty but may return to normal afterward, suggesting that transient, functional defects in GH secretion may occur that are reversible with the onset of puberty or sex hormone administration. Up to 70% of idiopathic isolated GHD children are no longer GH deficient when retested at the completion of their linear growth after GH therapy.

2. Studies suggest that even mild degrees of weight gain and those with a BMI of ≥25 can have a markedly decreased GH response in a variety of stimulation tests.

3. Although a significant negative correlation between peak evoked GH response during the ITT test and BMI (Pearson $r = -0.59$; $p < 0.01$) has been reported, additional studies are required to determine the normal reference interval for peak [GH] as a function of BMI.

4. Abnormal, evoked GH responses to ARG–Geref infusion, defined as <9 ng/mL, occurred in 5, 13, 33, and 64% of individuals with BMIs of <25, 25–26.9, 27–29.9, and ≥30, respectively (Bonert et al., 2004).

5. Using peak [GH] cut-points of 5.1 ng/mL for the ITT and 4.1 ng/mL for ARG–Geref stimulation (see Test 2.3.4), both high sensitivity (96% and 95%, respectively) and specificity (92% and 91%, respectively) for AGHD were found (Biller et al., 2002).

6. Ninety-five percent specificity for diagnosis of AGHD using peak serum [GH] cut-points of 3.3 ng/mL and 1.5 ng/mL for the ITT and ARG–Geref tests, respectively, were found (Aimaretti et al., 1998).
7. Tests for the acid-labile subunit and IGF complexes of IGFBP-3 are of limited predictive value in AGHD, as values for these tests are commonly within reference intervals similar to that found for IGF-1.

ICD-9 Codes

Conditions that may justify this test *include but are not limited to*:

Pituitary disorders
253.2   panhypopituitarism
253.3   growth hormone deficiency/pituitary dwarfism

Other
278.1   localized abdominal adiposity
733   osteoporosis, unspecified
758.6   gonadal dysgenesis/Turner syndrome
780.79   fatigue/lethargy

Suggested Reading

Aimaretti G, Corneli G, Razzore P, Bellone S, Baffoni C, Arvat E, Camanni F, Ghigo E. Comparison between insulin-induced hypoglycemia and growth hormone (GH)-releasing hormone + arginine as provocative tests for the diagnosis of GH deficiency in adults. *J Clin Endocrinol Metab* 1998; 83:1615–8.

Biller BM, Samuels MH, Zagar A, Cook DM, Arafah BM, Bonert V, Stavrou S, Kleinberg DL, Chipman JJ, Hartman ML. Sensitivity and specificity of six tests for the diagnosis of adult GH deficiency. *J Clin Endocrinol Metab* 2002: 87:2067–79.

Bonert VS, Elashoff JD, Barnett P, Melmed S. Body mass index determines evoked growth hormone (GH) responsiveness in normal healthy male subjects: diagnostic caveat for adult GH deficiency. *J Clin Endocrinol Metab* 2004; 89:3397–401.

Brabant G, von zur Muhlen A, Wuster C, Ranke MB, Kratzsch J, Kiess W, Ketelslegers JM, Wilhelmsen L, Hulthen L, Saller B, Mattsson A, Wilde J, Schemer R, Kann P. Serum insulin-like growth factor I reference values for an automated chemiluminescence immunoassay system: results from a multicenter study. *Horm Res* 2003; 60:53–60.

Darendeliler F, Spinu I, Bas F, Bundak R, Isguven P, Arslanoglu I, Saka N, Sukur M, Gunoz H. Reevaluation of growth hormone deficiency during and after growth hormone (GH) treatment: diagnostic value of GH tests and IGF-1 and IGFBP-3 measurements *J Pediatr Endocrinol Metab* 2004; 17:1007–12.

Evans AJ. Screening tests for growth hormone deficiency. *J R Soc Med* 1995; 88:161P–165P.

Frasier SD. The diagnosis and treatment of childhood and adolescent growth hormone deficiency: consensus or confusion? [editorial]. *J Clin Endocrinol Metab* 2000; 85:3988–9.

GH Research Society. Consensus guidelines for the diagnosis and treatment of growth hormone (GH) deficiency in childhood and adolescence: summary statement of the GH Research Society. *J Clin Endocrinol Metab* 2000; 85:3990–3.

Gourmelen M, Pham-Huu-Trung MT, Girard F. Transient partial hGH deficiency in prepubertal children with delay of growth. *Pediatr Res* 1979; 13:221–4.

Hoffman DM, O'Sullivan AJ, Baxter RC, Ho KK. Diagnosis of growth-hormone deficiency in adults. *Lancet* 1994; 343:1064–8.

Marzullo P, Di Somma C, Pratt KL, Khosravi J, Diamandis A, Lombardi G, Colao A, Rosenfeld RG. Usefulness of different biochemical markers of the insulin-like growth factor (IGF) family in diagnosing growth hormone excess and deficiency in adults. *J Clin Endocrinol Metab* 2001; 86:3001–8.

Rosenfeld RG, Albertsson-Wikland K, Cassorla F, Frasier SD, Hasegawa Y, Hintz RL, Lafranchi S, Lippe B, Loriaux L, Melmed S et al. Diagnostic controversy: the diagnosis of childhood growth hormone deficiency revisited. *J Clin Endocrinol Metab* 1995; 80:1532–40.

Thorner MO, Bengtsson BA, Ho KY, Albertsson-Wikland K, Christiansen JS, Faglia G, Irie M, Isaksson O, Jorgensen JO, Ranke M et al. The diagnosis of growth hormone deficiency (GHD) in adults. *J Clin Endocrinol Metab* 1995; 80:3097–8.

2.3.2 Total and Free Insulin-Like Growth Factor 1 (IGF-1 or Somatomedin C) Testing for Growth Hormone Deficiency (GHD) in Fasted Adults

Indications for Test

Insulin-like growth factor 1 (or somatomedin C) testing is indicated when:

- Screening for adult GHD
- Monitoring response to therapy in adults with established GHD

Procedure

1. Obtain the appropriate blood samples for liver function tests—aspartate transaminase (AST) and alanine transaminase (ALT)—and tests of nutritional status (e.g., serum albumin, CBC), including calculation of the patient's body mass index (BMI). Note that Chapter 11 presents tests related to nutritional assessment, including calculation of BMI (see Test 11.1.3).
2. In a patient who has fasted overnight and been put at rest (i.e., seated quietly for a minimum of 15 minutes prior to blood collection), obtain a blood sample for total IGF-1 measurement.
3. As a research test, determine the free IGF-1 concentration on a fasting blood sample obtained from an adult with a total [IGF-1] in the lowest quartile of the reference interval for age of patient (Table 2.1).
4. Do not waste resources by attempting to reconfirm childhood GH deficiency by GH provocative testing in young adults who have two or more pituitary hormone deficiencies in addition to GHD.

Interpretation

1. Considerable overlap in [IGF-1] between normal and GHD patients occurs.
2. A normal serum [IGF-1] suggests normal GH secretory dynamics; however, other stimuli for IGF-1 secretion (such as intake of food) may move the [IGF-1] into the normal range in GHD patients. Many GHD adults maintain a serum [IGF-1] within the low-normal portion of the reference interval.
3. Typically, if a patient has been starved (i.e., fasted for 40 hours), there is a 50% reduction in the patient's free IGF-1 concentration, a parallel increase in [IGFBP-1], an increase in the concentration of IGFBP-3–(IGF-1) complex, and a modest reduction in the proteolysis of IGFBP-3.
4. Short-term fasting (i.e., overnight fast) results in only a small percentage drop in total IGF-1 concentration, unless the patient is GH deprived or deficient or has received somatostatin, in which case the free IGF-1 concentration and total IGF-1 concentration may drop by 70% and 35%, respectively.
5. Since age- and sex-related variability is considerable, reference intervals for total IGF-1 are related to the patient's age, sex, Tanner stage, and the assay used to measure [IGF-1] (Table 2.1). IGF-1 concentrations, unlike stimulated GH levels, are uninfluenced by a BMI < 30.
6. In healthy adults, the concentrations of IGF-1 and IGFBP-3 decrease smoothly and steadily with age.
7. An [IGF-1] within the patient-specific reference interval virtually excludes idiopathic GHD in patients with no known pituitary disorder but does not rule out GHD if the patient has a hypothalamic–pituitary disorder.

8. An [IGF-1] below the lower limit of the reference interval in individuals with a normal variant of idiopathic short stature is consistent with GHD, after exclusion of poor nutritional status and/or liver disease as the cause of the decreased [IGF-1].

9. An [IGF-1] below the population-based reference interval, but within the limits of the reference interval for individuals with idiopathic short stature, may or may not be normal even when associated with a low GH peak upon GH stimulation testing.

10. Low [IGF-1] may occur with obesity, dopamine or somatostatin therapy, hypopituitarism, complete or partial GH deficiency, Laron dwarfism (GH resistance), malnutrition, liver failure, psychosocial deprivation, impotency, old age, and conditions that impair GH responsiveness such as hypothyroidism and chronic illnesses.

11. Both free IGF-1 and total IGF-1 concentrations normally increase in adolescence, peak in puberty, and then return to prepubertal levels.

12. In late puberty and early adolescence, 14-year-old girls have a slightly higher mean peak IGF-1 (410 ng/mL) than 16-year-old boys (382 ng/mL). Thereafter, a rapid fall in [IGF-1] occurs in both sexes until approximately 25 years of age, when a slow age-dependent decline begins.

Notes

1. GH binding to the GH-receptor in the liver stimulates IGF-1 production.

2. Circulating IGF-1 has a half-life of ~1 week, and the [IGF-1] in the blood remains relatively constant throughout the day.

3. Although a single [IGF-1] has low diagnostic discriminatory power, values < 77.2 ng/mL have been shown to have 95% specificity for AGHD.

4. Significantly lower IGF-1 concentrations were found in 11 Central African pygmies (69 ng/mL) and in 12 GHD patients (24 ng/mL) compared to those in 31 matched control subjects (193 ng/mL) (Merimee et al., 1981).

5. IGF-2 concentrations in pygmies (mean, 503 ng/mL) and in control subjects (mean, 647 ng/mL) did not differ significantly.

6. Prepubertal pygmy children do not differ from control subjects in linear growth or in [IGF-1] and [IGF-2]; however, in adolescence, both pygmy boys and girls have a mean [IGF-1] level only one third that of age- and Tanner stage-matched control subjects.

7. Growth acceleration in normal males is the result of enhanced sensitivity of the GH receptor–IGF-1 production system to GH secretion triggered by testosterone or other testicular-maturation factors. This phenomenon appears to be absent in male adolescent pygmies with normal testosterone secretion (Merimee et al., 1987, 1991).

8. Growth acceleration in normal females results almost solely from increased production of IGF-1 in response to GH and without sensitization of the system by sex hormones.

9. When measuring free IGF-1, the direct immunoradiometric assay (IRMA), which measures free plus readily dissociable IGF-1, yields higher values than those obtained using the ultrafiltration (UF) method. Thus, reductions in free IGF-1 concentrations are better appreciated using UF than IRMA methods.

10. Unlike total IGF-1 concentrations, free IGF-1 concentrations are higher in serum from obese subjects than nonobese control subjects, probably reflecting the IGF-1 lowering effect of overnutrition and the increased IGF-1 bioactivity resulting from the chronic hyperinsulinemia of obesity (Nam et al., 1997).

ICD-9 Codes

Refer to Test 2.3.1 codes.

Suggested Reading

Bussieres L, Souberbielle JC, Pinto G, Adan L, Noel M, Brauner R. The use of insulin-like growth factor 1 reference values for the diagnosis of growth hormone deficiency in prepubertal children. *Clin Endocrinol (Oxf)* 2000; 52:735–9.

Clemmons DR, Van Wyk JJ. Factors controlling blood concentration of somatomedin C. *Clin Endocrinol Metab* 1984; 13:113–43.

Frystyk J, Ivarsen P, Stoving RK, Dall R, Bek T, Hagen C, Orskov H. Determination of free insulin-like growth factor-I in human serum: comparison of ultrafiltration and direct immunoradiometric assay. *Growth Horm IGF Res* 2001; 11:117–27.

Hoffman DM, Nguyen TV, O'Sullivan AJ, Baxter RC, Ho KK. ITT and IGF-1 in the diagnosis of growth hormone deficiency in adults. *Lancet* 1994; 344(8920):482–3 and 344(8922):613–4.

Juul A, Dalgaard P, Blum WF, Bang P, Hall K, Michaelsen KF, Muller J, Skakkebaek NE. Serum levels of insulin-like growth factor (IGF)-binding protein-3 (IGFBP-3) in healthy infants, children, and adolescents: the relation to IGF-1, IGF-2, IGFBP-1, IGFBP-2, age, sex, body mass index, and pubertal maturation. *J Clin Endocrinol Metab* 1995; 80:2534–42.

Juul A, Holm K, Kastrup KW, Pedersen SA, Michaelsen KF, Scheike T, Rasmussen S, Muller J, Skakkebaek NE. Free insulin-like growth factor I serum levels in 1430 healthy children and adults, and its diagnostic value in patients suspected of growth hormone deficiency. *J Clin Endocrinol Metab* 1997; 82:2497–502.

Marzullo P, Di Somma C, Pratt KL, Khosravi J, Diamandis A, Lombardi G, Colao A, Rosenfeld RG. Usefulness of different biochemical markers of the insulin-like growth factor (IGF) family in diagnosing growth hormone excess and deficiency in adults. *J Clin Endocrinol Metab* 2001; 86:3001–8.

Merimee TJ, Zapf J, Froesch ER. Dwarfism in the pygmy: an isolated deficiency of insulin-like growth factor 1. *N Engl J Med* 1981; 305:965–8.

Merimee TJ, Zapf J, Froesch ER. Insulin-like growth factors (IGFs) in pygmies and subjects with the pygmy trait: characterization of the metabolic actions of IGF-1 and IGF-1I in man. *J Clin Endocrinol Metab* 1982; 55:1081–8.

Merimee TJ, Zapf J, Hewlett B, Cavalli-Sforza LL. Insulin-like growth factors in pygmies: the role of puberty in determining final stature. *N Engl J Med* 1987; 316:906–11.

Merimee TJ, Quinn S, Russell B, Riley W. The growth hormone-insulin-like growth factor I axis: studies in man during growth. *Adv Exp Med Biol* 1991; 293:85–96.

Nam SY, Lee EJ, Kim KR, Cha BS, Song YD, Lim SK, Lee HC, Huh KB. Effect of obesity on total and free insulin-like growth factor (IGF)-1, and their relationship to IGF-binding protein (BP)-1, IGFBP-2, IGFBP-3, insulin, and growth hormone. *Int J Obes Relat Metab Disord* 1997; 21:355–9.

Norrelund H, Frystyk J, Jorgensen JO, Moller N, Christiansen JS, Orskov H, Flyvbjerg A. The effect of growth hormone on the insulin-like growth factor system during fasting. *J Clin Endocrinol Metab* 2003; 88:3292–8.

Rose SR, Ross JL, Uriarte M, Barnes KM, Cassorla FG, Cutler GB. The advantage of measuring stimulated as compared with spontaneous growth hormone levels in the diagnosis of growth hormone deficiency. *N Engl J Med* 1988; 319:201–7.

Tiryakioglu O, Kadiolgu P, Canerolgu NU, Hatemi H. Age dependency of serum insulin-like growth factor (IGF)-1 in healthy Turkish adolescents and adults. *Indian J Med Sci* 2003; 57:543–8.

2.3.3 Free Insulin-Like Growth Factor 1 (Free IGF-1) and Insulin-Like Growth Factor Binding Protein 3 (IGFBP-3) Testing in Growth Hormone Deficiency (GHD) of Children

Indications for Test

Measurement of [free IGF-1] and [IGFBP-3] is indicated:

- In cases of short stature in early adolescence
- Upon detection of GHD by stimulation testing in children
- For pituitary irradiation or surgery in children

Procedure

1. Obtain a random blood sample for free IGF-1 and IGFBP-3 testing from prepubertal children and those in early adolescence with short stature, potentially compromised pituitary–hypothalamic axis, or low GH response to stimulation testing.
2. Obtain objective measures of the patient's age, sex, height, BMI, and stage of pubertal maturation.
3. Do not waste resources by attempting to reconfirm childhood GH deficiency by GH provocative testing in young adults who have two or more pituitary hormone deficiencies in addition to GHD.

Interpretation

1. [IGFBP-3] and [free IGF-1] below the fifth percentile of normal suggest GH deficiency or resistance.
2. In healthy children, free IGF-1 concentrations (mean ± SD) are low (males, 0.71 ± 0.26 ng/mL; females, 1.05 ± 0.49 ng/mL) in infancy (<1 year of age), increase during puberty (males, 5.84 ± 2.18 ng/mL; females, 5.80 ± 1.49 ng/mL), and decline thereafter.
3. Obtain assay-specific reference intervals for [IGFBP-3] and [free IGF-1] in children based on age and Tanner stage.
4. In children and adolescents within Tanner stage 5, [IGF-1] and [IGFBP-3] are significantly greater in females than in males.
5. [IGFBP-3] increases with age in healthy children, with maximal levels occurring in puberty; girls experience peak values approximately 1 year earlier than boys.
6. Children with GHD and Turner syndrome have low IGF-1 and IGFBP-3 concentrations, which increase to within the normal reference interval on therapy with recombinant human growth hormone (rHGH).
7. The increases in sex steroids that occur with puberty correlate with increases in [IGF-1] and IGF-1/IGFBP-3 ratios. Concomitant with these changes are increases in bone growth.
8. Significant lowering in height, growth velocity, target height, [IGF-1] ($p < 0.0001$), and [IGFBP-3] ($p < 0.01$) occurs in familial short-statured patients compared with healthy control subjects (del Valle Nunez et al., 2004).

Notes

1. To block interference from elevated IGF binding proteins on the IGF-1 immunoassay, measure IGF-1 in the presence of an excess of IGF-2.
2. Growth hormone stimulates the synthesis of IGFBP-3, the principal protein that binds IGF-1.
3. A serum [IGFBP-3] within the reference interval does not exclude adult GHD, hence its lack of usefulness in the adult. Measurement of IGFBP-3, combined with free IGF-1, has become the screening test of choice for GHD in children; however, these tests lack specificity in adults.
4. Total IGF-1 concentration is highly dependent on age, sex, and stage of puberty, while [IGFBP-3] is dependent only on age and gender.
5. [IGFBP-3] more reliably separates GH-deficient or -resistant children from healthy or otherwise short children than does total [IGF-1]. Neither [free IGF-1] nor [IGFBP-3] is useful in predicting growth outcomes in reponse to GH therapy.
6. During the growth spurt of puberty, [total IGF-1] increases to relatively higher levels than [IGFBP-3]. This increase leads to an increased molar ratio of [IGF-1] to [IGFBP-3].
7. [Total IGF-1] and [IGFBP-3] are not significantly different between obese and nonobese individuals, even when decreased GH secretion in obesity can be demonstrated.
8. In healthy individuals, most of the somatomedin-binding protein (SmBP) is present in the blood as a large (150-kDa) complex; however, substantial quantities of smaller (32-kDa, 42-kDa, and 60-kDa) fragments can be found by high-performance liquid chromatography (HPLC).
9. Serum [SmBP] is low at birth, rises sharply during the first weeks of life, and has a moderate peak at puberty.
10. High [IGFBP-1], but not [IGFBP-3], may contribute to growth retardation in a subgroup of children with idiopathic short stature through an IGFBP-1-induced decrease in free IGF-1 concentration.
11. In healthy subjects, IGF-1, IGFBP-1, and IGFBP-3 complexes decline with age.

12. Modest correlations have been reported between [IGF-1] and [IGFBP-3] ($r = 0.59$; $p < 0.001$) and between [IGF-1] and [free IGF-1] ($r = 0.40$; $p < 0.05$).

13. In children and young adults reevaluated for GHD, the sensitivities of testing for IGF-1 and IGFBP-3 for confirmation of diagnosis are ~77% and 74%, respectively, when using a stimulated GH cutoff level for GHD of 7 ng/mL (0.32 pmol/L).

14. Free IGF-1 appears to play an essential role in bone formation as evidenced by the high [free IGF-1] in the blood of adolescents when the growth rate accelerates.

ICD-9 Codes

Refer to Test 2.3.1 codes.

Suggested Reading

Blum WF, Ranke MB, Kietzmann K, Gauggel E, Zeisel HJ, Bierich JR. A specific radioimmunoassay for the growth hormone (GH)-dependent somatomedin-binding protein: its use for diagnosis of GH deficiency. *J Clin Endocrinol Metab* 1990; 70:1292–8.

Darendeliler F, Spinu I, Bas F, Bundak R, Isguven P, Arslanoglu I, Saka N, Sukur M, Gunoz H. Reevaluation of growth hormone deficiency during and after growth hormone (GH) treatment: diagnostic value of GH tests and IGF-1 and IGFBP-3 measurements. *J Pediatr Endocrinol Metab* 2004; 17:1007–12.

del Valle Nunez CJ, Lopez-Siguero JP, Lopez-Canti LF, Lechuga Campoy JL, Espigares Martin R, Martinez-Aedo Ollero MJ. GHBP, IGF-1 and IGFBP-3 serum levels in familial short-statured and normal-statured children. *Med Clin (Barc)* 2004; 123:452–5.

Elmlinger MW, Kuhnel W, Weber MM, Ranke MB. Reference intervals for two automated chemiluminescent assays for serum insulin-like growth factor I (IGF-1) and IGF-binding protein 3 (IGFBP-3). *Clin Chem Lab Med* 2004; 42:654–64.

Jaruratanasirikul S, Leethanaporn K, Pradutkanchana S, Sriplung H. Serum insulin-like growth factor-1 (IGF-1) and insulin-like growth factor binding protein-3 (IGFBP-3) in healthy Thai children and adolescents: relation to age, sex, and stage of puberty. *J Med Assoc Thai* 1999; 82:275–83.

Kamoda T, Saitoh H, Hirano T, Matsui A. Serum levels of free insulin-like growth factor (IGF)-I and IGF-binding protein-1 in prepubertal children with idiopathic short stature. *Clin Endocrinol (Oxf)* 2000; 53:683–8.

Kawai N, Kanzaki S, Takano-Watou S, Tada C, Yamanaka Y, Miyata T, Oka M, Seino Y. Serum free insulin-like growth factor I (IGF-1), total IGF-1, and IGF-binding protein-3 concentrations in normal children and children with growth hormone deficiency. *J Clin Endocrinol Metab* 1999; 84:82–9.

Merimee TJ, Russell B, Quinn S. Growth hormone-binding proteins of human serum: developmental patterns in normal man. *J Clin Endocrinol Metab* 1992, 75:852–4.

Rosenfeld RG, Albertsson-Wikland K, Cassorla F, Frasier SD, Hasegawa Y, Hintz RL, Lafranchi S, Lippe B, Loriaux L, Melmed S et al. Diagnostic controversy: the diagnosis of childhood growth hormone deficiency revisited. *J Clin Endocrinol Metab* 1995; 80:1532–40.

2.3.4 Arginine–GHRH (ARG–Geref) Infusion for Stimulation of GH in Growth Hormone Deficiency (GHD)

Indications for Test

Arginine–GHRH stimulation of GH is indicated in:

- Adults in whom there is the suspicion of GHD
- Children with growth failure or short stature
- Children who were treated with GH and have progressed to young adulthood

Procedure

1. Hypoglycemia, especially in children and malnourished adults, is a potential side effect of this test. Obtain informed consent for this invasive test and keep a solution of 10 to 50% dextrose for injection readily available in the event of hypoglycemia (i.e., fingerstick blood glucose ≤ 50 mg/dL). Do not treat if symptoms of hypoglycemia occur but the glucose > 50 mg/dL. Note that a mild decrease in blood glucose is the goal of this test.

2. Treat prepubertal children older than 12 years of age and hypogonadal adults with sex steroids before GH testing. Give ethinyl estradiol (40 mg/m^2/day orally × 2 days) to girls or testosterone enanthate (200 mg i.m.) to boys within the 5 to 10 days prior to testing.

3. After the patient has fasted overnight, obtain a fingerstick blood sample for determination of baseline blood glucose and a venous blood sample for determination of baseline [GH].

4. Give GHRH or Geref (1.0–1.5 mcg/kg, up to 100 mcg total) intravenously before infusion of 0.5 g/kg of L-arginine HCl (up to 30 g total, or one 300-mL bottle of R-Gene® 10) as 10 g arginine/100 mL water within 30 minutes.

5. Because arginine is a hypertonic solution, take special care to avoid extravasation.

6. Infusion of arginine for more than 30 minutes may fail to stimulate a GH response. Too rapid an infusion may result in nausea, vomiting, and flushing.

7. At 30, 60, and 90 minutes after the infusion of arginine, obtain fingerstick blood samples for determination of glucose and venous blood samples for determination of [GH] according to Specimen Collection Protocol P4 in Appendix 2.

Interpretation

1. Following the administration of Geref and L-arginine, an inadequately stimulated [GH] (i.e., <9 ng/mL) identifies patients with probable GHD. Refer to "GH Stimulation Testing Using Geref and Arginine" (Table 2.3).

Notes

1. The stimulated [GH] > 17.8 ng/mL in 97% of healthy individuals.

2. Glucose suppresses GH secretion; therefore, perform GH testing after an overnight fast to minimize the effect of a postprandial rise in plasma glucose.

3. Poorly controlled diabetes mellitus is a relative contraindication to GH testing.

4. GH testing is not diagnostic in patients with liver or renal failure.

5. Based on receiver–operator characteristic (ROC) curve data, ARG–Geref stimulation testing has high sensitivity and specificity for the diagnosis of GH deficiency similar to the ITT and is the preferred test, over the ITT, because of its safety and superior tolerability.

6. Arginine is an insulin secretagog and stimulates GH secretion by inhibiting hypothalamic somatostatin along with a consequent rise in insulin levels (usually >24 mcU/mL) by the end of the arginine infusion.

7. An abnormally low GH response to any GH-secretory agent in prepubertal children and hypogonadal adults without sex steroid "priming" is inconclusive.

8. Dilated cardiomyopathy patients show blunted GH responses to GHRH both alone and when combined with arginine.

9. Substantial overlap in the GH response of AGHD patients to that of control subjects occurs for the ARG alone, L-DOPA alone, and ARG + L-DOPA tests. The recent (2008) limited availability of Geref prompts use of a combined arginine–ITT (0.075 units insulin) but necessitates that testing be done in an infusion center for safety. The cost of effective testing in such settings is markedly increased.

10. Test-specific, peak [GH] cut-points to provide 95% sensitivity for diagnosis of AGHD were reported as 1.4 ng/mL for ARG alone, 0.64 ng/mL for L-dopamine alone, and 1.5 ng/mL for the ARG plus L-dopamine test.

11. A specificity of 95% for diagnosis of AGHD was achieved only with very low peak GH cut-points (0.21 ng/mL for ARG alone; 0.25 ng/mL for ARG + L-dopamine). No GH cutpoint was low enough with the L-dopamine test to achieve a test specificity of 95% (Biller et al., 2002).

12. Stimulation tests for the diagnosis of GHD were more sensitive and clearly superior to the determination of the spontaneous secretion of GH in pooled serum prepared from samples taken every 20 minutes in prepubertal short children over a 24-hour period.

TABLE 2.3
GH Stimulation Testing Using Geref and Arginine[a]

Population	N	Mean Age (years) (±1 SD or range)	Mean Peak GH Concentration (mcg/L) (±1 SD or range)	Refs.
Healthy				
"Short" children	48	12.0 (0.4)	48.8 (22.4–150)	4
Children	10	12.3 (0.9)	61.6 (8.1)	1
	81	11.9 (0.3)	61.8 (19.6–106)	6
Young adult	18	31.1 (1.3)	70.4 (10.1)	1
Adult	178	20–50	65.9[b] (13.8–171)	2, 5
Elderly	12	74.4 (1.8)	57.9 (14.8)	1
GHNSD	21	25.1 (1.6)	65.9 (5.5)	3
i-GHD	23	23.0 (1.5)	18.6 (4.7)	3
o-GHD	15	40.9 (4.1)	2.3 (0.5)	1
	18	26.8 (2.2)	2.8 (0.1–12)	3
	40	36.4 (2.1)	3.0 (0.1–12)	2

[a] Geref (1 mcg/kg) and arginine (30 g/kg for adults > 60 kg or 0.5 g/kg for children < 60 kg).

[b] A cutoff value of ≥9 mcg/L was considered a normal response.

Note: GHD, growth hormone deficiency; GHNSD, GH neurosecretory disorder (only in children); i-GHD, isolated GHD; o-GHD, organic GHD associated with hypopituitarism.

1. Valetto MR et al. *Eur J Endocrinol* 1996; 135:568–72.
2. Aimaretti G et al. *J Clin Endocrinol Metab* 1998; 83:1615–8.
3. Aimaretti G et al. *J Clin Endocrinol Metab* 2000a; 85:3693–9.
4. Bellone J et al. *J Endocrinol Invest* 2000; 23:97–101.
5. Aimaretti G et al. *Eur J Endocrinol* 2000b; 142:347–52.
6. Ghigo E et al. *J Clin Endocrinol Metab* 1996; 81:3323–7.

13. The mean [GH] in pooled serum gathered over 24 hours is below the analytical sensitivity of the GH assay in 80% of hypopituitary subjects and in 16% of a reference population.
14. When GH secretory status is reevaluated during or after GH treatment, especially in patients with idiopathic, partial, or isolated GHD, use of a stimulated [GH] peak cutoff level of <7 ng/mL (0.32 pmol/L) will lower the number of false-positive diagnoses of GHD.

ICD-9 Codes

Refer to Test 2.3.1 codes.

Suggested Reading

Aimaretti G, Corneli G, Razzore P, Bellone S, Baffoni C, Arvat E, Camanni F, Ghigo E. Comparison between insulin-induced hypoglycemia and growth hormone (GH)-releasing hormone + arginine as provocative tests for the diagnosis of GH deficiency in adults. *J Clin Endocrinol Metab* 1998; 83:1615–8.

Aimaretti G, Baffoni C, Bellone S, Di Vito L, Corneli G, Arvat E, Benso L, Camanni F, Ghigo E. Retesting young adults with childhood-onset growth hormone (GH) deficiency with GH-releasing-hormone-plus-arginine test. *J Clin Endocrinol Metab* 2000a; 85:3693–9.

Aimaretti G, Baffoni C, Di Vito L, Bellone S, Grottoli S, Maccario M, Arvat E, Camanni F, Ghigo E. Comparisons among old and new provocative tests of GH secretion in 178 normal adults. *Eur J Endocrinol* 2000b; 142:347–52.

Bellone J, Aimaretti G, Bellone S et al. Sequential administration of arginine and arginine plus GHRH to test somatotroph function in short children. *J Endocrinol Invest* 2000; 23:97–101.

Broglio F, Benso A, Gottero C, Vito LD, Aimaretti G, Fubini A, Arvat E, Bobbio M, Ghigo E. Patients with dilated cardiomyopathy show reduction of the somatotroph responsiveness to GHRH both alone and combined with arginine. *Eur J Endocrinol* 2000; 142:157–63.

Cook M, Fairchild L, Kidder T, Pulaski K, and Solares M. (Eds). *Nursing Handbook for Adult Growth Hormone Deficiency.* 2001. Endocrine Nurses Society.

Ghigo E, Bellone J, Aimaretti G, Bellone S, Loche S, Cappa M, Bartolotta E, Dammacco F, Camanni F. Reliability of provocative tests to assess growth hormone secretory status: study in 472 normally growing children. *J Clin Endocrinol Metab* 1996; 81:3323–7.

Juul A, Kastrup KW, Pedersen SA, Skakkebaek NE. Growth hormone (GH) provocative retesting of 108 young adults with childhood-onset GH deficiency and the diagnostic value of insulin-like growth factor I (IGF-1) and IGF-binding protein-3. *J Clin Endocrinol Metab* 1997; 82:1195–201.

Valetto MR, Bellone J, Baffoni C, Savio P, Aimaretti G, Gianotti L, Arvat E, Camanni F, Ghigo E. Reproducibility of the growth hormone response to stimulation with growth hormone-releasing hormone plus arginine during lifespan. *Eur J Endocrinol* 1996; 135:568–72.

2.3.5 Insulin-Like Growth Factor Binding Protein 3 (IGFBP-3) Test in GHD Patients with Growth Hormone Receptor Defects (GHRD or Laron Syndrome) and/or Hepatic Disorders

Indications for Test

The IGFBP-3 test is indicated when a patient has:

- An [IGF-1] low for age, normal hepatic status, BMI that is <30, and normal response to a GH stimulation test
- History of hypoglycemia associated with micropenis or beta thalassaemia major
- An [IGF-1] low for age and cryptic liver disease, such as fatty infiltration suggested by a low AST to ALT ratio

Procedure

1. Obtain liver enzyme tests (i.e., AST and ALT). Obtain hepatic ultrasound evaluation if the AST/ALT ratio is <0.5.
2. Obtain a fasting blood sample for the measurement of the [IGFBP-3] in patients suspected of having clinical sequelae of GHRD.

Interpretation

1. In adult patients suspected of having GHRD, [IGFBP-3] is typically low (<3 mg/L).

Notes

1. The advent of the IGFBP-3 assay as a marker for GH/GH receptor interaction has largely obviated the need for the low- or high-dose hGH infusion for the diagnosis of Laron dwarfism or GHRD.
2. Laron dwarfs have hypoglycemia in 33% of cases and micropenis in 58% of males.
3. Children with beta thalassaemia major may have a defective GH/IGF-1/IGFBP-3 axis of secretion and response, suggesting the presence of partial, rather than the complete, resistance to GH as is found in the Laron syndrome.
4. Short stature in pygmies does not result from an absolute deficiency of, or defect in, GH receptors as occurs in Laron dwarfism, but from an apparent failure of somatic GH receptor mass to increase appropriately at the time of puberty.
5. In pygmies, there seems to be an alteration in the way the expression of the GH receptor gene is regulated, rather than a structural defect in the coding sequence for it.

ICD-9 Codes

Conditions that may justify this test *include but are not limited to*:

251.2 hypoglycemia
253.3 growth hormone deficiency/pituitary dwarfism (Laron syndrome)
758.6 gonadal dysgenesis/Turner syndrome

Suggested Reading

Blair JC, Camacho-Hubner C, Miraki Moud F, Rosberg S, Burren C, Lim S, Clayton PE, Bjarnason R, Albertsson-Wikland K, Savage MO. Standard and low-dose IGF-1 generation tests and spontaneous growth hormone secretion in children with idiopathic short stature. *Clin Endocrinol (Oxf)* 2004; 60:163–8; discussion 161–2.

Cotterill AM, Camacho-Hubner C, Duquesnoy P, Savage MO. Changes in serum IGF-1 and IGFBP-3 concentrations during the IGF-1 generation test performed prospectively in children with short stature. *Clin Endocrinol (Oxf)* 1998; 48:719–24.

Laron Z, Pertzelan A, Karp M, Kowadlo-Silbergeld A, Daughaday WH. Administration of growth hormone to patients with familial dwarfism with high plasma immunoreactive growth hormone: measurement of sulfation factor, metabolic and linear growth responses. *J Clin Endocrinol Metab* 1971; 33:332–42.

Lieberman SA, Mitchell AM, Marcus R, Hintz RL, Hoffman AR. The insulin-like growth factor I generation test: resistance to growth hormone with aging and estrogen replacement therapy. *Horm Metab Res* 1994; 26:229–33.

Merimee TJ, Baumann G, Daughaday W. Growth hormone-binding protein. II. Studies in pygmies and normal statured subjects. *J Clin Endocrinol Metab* 1990; 71:1183–8.

Moller S, Becker PU. Somatomedins (insulin-like growth factors), growth hormone and chronic liver disease *Ugeskr Laeger* 1990; 152:2022–5.

Savage MO, Blum WF, Ranke MB, Postel-Vinay MC, Cotterill AM, Hall K, Chatelain PG, Preece MA, Rosenfeld RG. Clinical features and endocrine status in patients with growth hormone insensitivity (Laron syndrome). *J Clin Endocrinol Metab* 1993; 77:1465–71.

Seehofer D, Steinmueller T, Graef KJ, Rayes N, Wiegand W, Tullius SG, Settmacher U, Neuhaus P. Pituitary function test and endocrine status in patient with cirrhosis of the liver before and after hepatic transplantation. *Ann Transplant* 2002; 7:32–7.

Soliman AT, El Banna N, Ansari BM. GH response to provocation and circulating IGF-1 and IGF-binding protein-3 concentrations, the IGF-1 generation test and clinical response to GH therapy in children with beta-thalassaemia. *Eur J Endocrinol* 1998; 138:394–400.

2.3.6 Pituitary Imaging by M-CT and MR in GH Deficiency (GHD)

Indications for Test

Pituitary imaging by M-CT, MR, or both is indicated when:

- There is evidence for GHD based on basal and/or GH stimulation tests.
- A search for an anatomic lesion involving the pituitary–hypothalamic axis is required for confirmation of the diagnosis and etiology of primary GHD.

Procedure

1. Image the hypothalamic–pituitary axis with multislice, high-resolution MR in preference to M-CT, as soft tissue abnormalities are better defined on MR.
2. If available, obtain fast-framing dynamic MR with gadolinium (Gd-DTPA) enhancement to quantitate the time course of contrast enhancement within the neurohypophysis, pituitary stalk, posterosuperior adenohypophysis, and anteroinferior adenohypophysis.

Interpretation

1. On MR, a mass in the hypothalamic–pituitary axis, severe hypoplasia of the anterior pituitary lobe (HPAL), hypoplasia or interruption of the pituitary stalk (HPS), or ectopy of the posterior lobe (EPL) in which the infundibulum and posterior bright spot are absent and the neurohypophysis is closely adjacent to the median eminence may be found.

2. The presence of an EPL or HPS and HPAL on MR is highly specific (100% and 89%, respectively) and predictive of GHD (100% and 79% positive predictive value, respectively), indicating that these abnormalities provide strong support for the diagnosis of GHD.

3. An abnormality of the pituitary in patients with multiple pituitary hormone defects or with severe isolated growth hormone deficits may include any of the above (EPL, HPS, HPAL) as well as tumors.

4. A normal pituitary, empty sella, or small, slightly hypoplastic glands occur in individuals with less severe GHD or normal variant short stature (NVSS).

5. Hyperdense lesions seen on M-CT or MR imaging are characteristic of acute autoinfarction of pituitary adenomas, often associated with an unusually severe headache, visual disturbances, and hemorrhage.

6. Autoinfarction of pituitary lesions, even if associated with catastrophic apoplectic events, may take 1 to 2 hours to visualize on scan.

7. Extension of the infundibular stalk to the floor of the sella defines the empty sella syndrome, which may be a benign condition but is often associated with pituitary hormone deficits.

8. Structures of uniform density seen on MR suggest fluid-filled lesions, which include aneurysms and arachnoid cysts and prompt angiographic studies.

9. *Solid hypothalamic lesions* responsible for causing GHD include gliomas, hamartomas, and calcified craniopharyngiomas.

10. The differential diagnosis of *solid pituitary sellar lesions* includes adenomas, chordomas, and gliomas arising from the hypothalamus or optic nerve, metastases, and meningiomas of the tuberculum sella.

11. Complex or *diffuse CNS lesions* in the region of the pituitary include sphenoid sinus mucocoels, dermoids, lymphocytic hypophysitis, Rathke cleft cysts, and granulomas of sarcoid or eosinophilic nature.

12. Sequential time-resolved Gd-enhanced MR may demonstrate reduced contrast enhancement in the pituitary stalk as the only anatomic abnormality related to GHD.

Notes

1. It is common for lesions in and around the pituitary to be clinically silent and discovered only incidentally.

2. Over 40% of patients with isolated GHD have an ectopic posterior lobe of the pituitary, and the absence of the infundibulum. GHD in these patients is most likely a consequence of the absent infundibulum.

3. The presence of any MR abnormality seen on hypothalamic–pituitary imaging has a sensitivity of 79% for diagnosis of GHD but a specificity of only 54%, indicating that the isolated finding of an anatomic abnormality on MR alone does not confirm GHD (Tillmann et al., 2000).

4. Angiography may be needed to more clearly define supra and parasellar lesions such as aneurysms but is not of use in the diagnosis of adenomas confined to the pituitary.

ICD-9 Codes

Conditions that may justify this test *include but are not limited to*:

Pituitary disorders

227.3	pituitary adenomas and cerebral neoplasms pushing on sella, including craniopharyngiomas, meningiomas, gliomas, hamartomas, chordomas, dermoids
253.3	GH deficiency, isolated/pituitary dwarfism
253.4	prolactin deficiency, isolated
253.8	disorders of the pituitary/empty sella/Rathke cleft cysts/lymphocytic hypophysitis

Other

348.0	cerebral cysts

Suggested Reading

Abrahams JJ, Trefelner E, Boulware SD. Idiopathic growth hormone deficiency: MR findings in 35 patients. *AJNR Am J Neuroradiol* 1991; 12:155–60.

Avataneo T, Cirillo S, Cesarani F, Besse F, Vannelli S, Benso L, Bona G. Magnetic resonance in the study of patients of short stature of the hypothalamo–hypophyseal origin: report on 29 cases. *Radiol Med (Torino)* 1994; 88:68–73.

Liotta A, Maggio C, Giuffre M, Carta M, Manfre L. Sequential contrast-enhanced magnetic resonance imaging in the diagnosis of growth hormone deficiencies. *J Endocrinol Invest* 1999; 22:740–6.

Maghnie M, Triulzi F, Larizza D, Scotti G, Beluffi G, Cecchini A, Severi F. Hypothalamic–pituitary dwarfism: comparison between MR imaging and CT findings. *Pediatr Radiol* 1990; 20:229–35.

Tillmann V, Tang VW, Price DA, Hughes DG, Wright NB, Clayton PE. Magnetic resonance imaging of the hypothalamic–pituitary axis in the diagnosis of growth hormone deficiency. *J Pediatr Endocrinol Metab* 2000; 13:1577–83.

2.4 Somatotroph (Growth Hormone) Excess

2.4.1 Basal GH, IGF-1, and IGFBP-3 Testing in Acromegaly

Indications for Test

Measurement of basal GH, IGF-1, and IGFBP-3 levels individually or together is indicated in patients with:

- The clinical syndrome of acromegaly, including acral enlargement and arthropathy, carpal tunnel syndrome, frontal bossing, prognathism with teeth spacing, chronic headaches or visual disturbance with sellar enlargement or tumor, oily skin with skin tags, hyperhidrosis, glucose intolerance progressing to diabetes mellitus, galactorrhea, sleep apnea, and menstrual disturbance
- Persistent elevations of [GH] on repeat testing, suggesting excess GH secretion
- Documented acromegaly after medical or surgical treatment

Procedure

1. To avoid false elevation of GH induced by acute discomfort of venipuncture, insert a small-gauge Hep-Lock® needle intravenously 20 minutes before obtaining a blood sample for GH measurement from a fasting and resting patient. Also, measure the [IGF-1] in this sample.
2. In patients in whom the clinical suspicion for acromegaly is high, determine the [IGF-1] in a random blood sample and proceed directly to an oral glucose tolerance test (see Test 2.4.2) for determination of the nadir of [GH] in response to a glucose load rather than obtaining a basal [GH] alone.
3. If the [IGF-1] in blood is at the upper limit of the reference interval or only minimally elevated, obtain a blood sample for measurement of [IGFBP-3].
4. If testing for GH molecular variants, consult with the research laboratory performing the test to determine specimen requirements.

Interpretation

1. Elevated [IGF-1] occurs in normal adolescence, gigantism, acromegaly, and tumor-associated hypoglycemia.
2. Untreated acromegalic patients mostly have [IGF-1] and [IGFBP-3] above the age-related 97th percentile value for healthy individuals.
3. [IGFBP-3] and [IGF-1] correlate well in patients with acromegaly both before ($r = 0.650$; $p = 0.0162$) and after ($r = 0.644$; $p = 0.0001$) pituitary surgery.

4. An average daytime [GH] < 2.5 ng/mL is evidence for successful treatment of acromegaly, as is a nadir [GH] < 1 ng/mL after an oral glucose tolerance test (see Test 2.4.2).
5. A drop in [IGF-1] to within the age-specific reference interval at one or more months after pituitary surgery is evidence for successful treatment of acromegaly.
6. In acromegaly, a relatively lower (i.e., just above the lower limit of normal) preoperative [GH] is a good predictor of potential responsiveness to selective transsphenoidal surgery.
7. Because normal GH secretion is pulsatile, do not use a random test for [GH] to establish the diagnosis of GH excess or deficiency.
8. Normal basal GH concentrations range from ~0.1 to 5.0 ng/mL but may go higher with stress (e.g., painful venipuncture). During sleep, secretion of GH occurs in a pulsatile fashion in spikes of large amplitude, usually >5 ng/mL.
9. In a small series of untreated acromegaly patients, markedly elevated serum IGFBP-3 concentrations (mean, 6566 mcg/L; range, 4186–10,026 mcg/L) were found (Grinspoon et al., 1995).
10. In healthy subjects, [IGF-1] and [IGFBP-3], as well as IGFBP-3 complexes, decline with age.
11. IGFBP-3 is a sensitive physiological marker of somatotroph function, and [IGFBP-3] is concordant with [GH] after glucose suppression and with [IGF-1] before and after transsphenoidal surgery for removal of a GH-secreting pituitary tumor.

Notes

1. Because circulating GH exists in many molecular forms, reference intervals for [GH] by immunometric assays depend on the specificity of the antibodies used in these assays.
2. IGF-1 is the test of choice for monitoring therapy for acromegaly.
3. Remission achieved with surgery for acromegaly is frequently at the expense of some degree of hypopituitarism.
4. Glucose suppresses [GH]; therefore, do GH testing after an overnight fast to avoid effects of postprandial hyperglycemia.
5. In acromegaly, [IGF-1] correlates well with heel-pad thickness ($r = 0.73$), fasting glucose ($r = 0.74$), and 1-hour postprandial glucose ($r = 0.77$). In contrast, glucose-suppressed [GH] correlated poorly ($r = 0.34, 0.36, 0.34$ respectively) with these clinical indexes of the severity of acromegaly, while fasting [GH] showed no correlation ($r = 0.14$) (Clemmons et al., 1979).
6. When elevated [GH], as found in acromegaly, has been present for an extended period of time and the effect of acute [GH] reduction therapy is to be monitored in the short term (1 or 2 weeks), GH testing serves as a useful alternative to IGF-1 testing.
7. Posttreatment [IGF-1] did not appear to have an impact on mortality; however, after radiotherapy basal [GH] < 2.5 ng/mL in acromegalic patients was associated with a normal lifespan (Kauppinen-Mäkelin et al., 2005).
8. In active cases of acromegaly, IGF-1 bound to IGFBP-3 was 5.4-fold higher than in control subjects and >2 SD above the mean [IGF-1] in control subjects in 95% of cases, whereas IGFBP-3 bound to the acid-labile subunit of IGF-1 was elevated in only 15% of cases.
9. Neither GH nor IGF-1 secretion directly influences leptin concentration in individuals with acromegaly (Bolanowski et al., 2002).

ICD-9 Codes

Conditions that may justify this test *include but are not limited to*:

Visual abnormality
368.4X   visual field defects
368.9   visual disturbance, unspecified

Pituitary disorders
253.0   acromegaly and gigantism
253.1   hyperprolactinemia

Other

227.3	pituitary adenoma
250.X	diabetes mellitus
524.10	unspecified anatomic anomaly/prognathism/teeth spacing
676.6	galactorrhea
713.0	arthropathy associated with other endocrine and metabolic disorders
784.X	headaches

Suggested Reading

Bolanowski M, Milewicz A, Bidzinska B, Jedrzejuk D, Daroszewski J, Mikulski E. Serum leptin levels in acromegaly: a significant role for adipose tissue and fasting insulin/glucose ratio. *Med Sci Monit* 2002; 8:CR685–9.

Clemmons DR, Van Wyk JJ, Ridgway EC, Kliman B, Kjellberg RN, Underwood LE. Evaluation of acromegaly by radioimmunoassay of somatomedin-C. *N Engl J Med* 1979; 301:1138–42.

De P, Rees DA, Davies N, John R, Neal J, Mills RG, Vafidis J, Davies JS, Scanlon MF. Transsphenoidal surgery for acromegaly in Wales: results based on stringent criteria of remission. *J Clin Endocrinol Metab* 2003; 88:3567–72.

Grinspoon S, Clemmons D, Swearingen B, Klibanski A. Serum insulin-like growth factor-binding protein-3 levels in the diagnosis of acromegaly. *J Clin Endocrinol Metab* 1995; 80:927–32.

Kauppinen-Mäkelin R, Sane T, Reunanen A, Välimäki MJ, Niskanen L, Markkanen H, Löyttyniemi E, Ebeling T, Jaatinen P, Laine H, Nuutila P, Salmela P, Salmi J, Stenman UH, Viikari J, Voutilainen E. A nationwide survey of mortality in acromegaly. *J Clin Endocrinol Metab* 2005; 90:4081–6.

Marzullo P, Di Somma C, Pratt KL, Khosravi J, Diamandis A, Lombardi G, Colao A, Rosenfeld RG. Usefulness of different biochemical markers of the insulin-like growth factor (IGF) family in diagnosing growth hormone excess and deficiency in adults. *J Clin Endocrinol Metab* 2001; 86:3001–8.

2.4.2   Oral Glucose Tolerance Test (oGTT) for GH Secretory Dynamics in Acromegaly

Indications for Test

The oGTT for GH secretory dynamics is indicated:

- In the diagnosis of acromegaly
- To document remission after pituitary surgery

Procedure

1. Require the patient to be n.p.o. for food and liquids after midnight before blood sampling.
2. Insert an i.v. catheter into the antecubital fossa. At 15 minutes after catheterization, obtain blood samples for basal GH and glucose determination.
3. Administer a 75-g or 100-g glucose load p.o. within 5 minutes. Use a 100-g load for the average-sized adult.
4. Obtain blood samples for GH and glucose testing at 30, 60, 90, and 120 minutes after administration of the glucose load.
5. Determine the [GH] nadir (i.e., the lowest [GH] after oral glucose administration).

Interpretation

1. A basal [GH] of <1 ng/mL is required to exclude active acromegaly in a patient treated for this disease. Following the administration of an oral glucose load, most healthy males and females suppress [GH] to a nadir of <0.1 ng/mL or <0.8 ng/mL, respectively.

2. The high correlation between baseline and nadir [GH] ($r = 0.82$) and the near equivalent fractional decline in mean [GH] in healthy men vs. women after glucose administration (67% vs. 84%) suggests that lower GH concentrations in men after oGTT are due to lower baseline values and not to a greater suppressive effect of glucose.
3. False-positive tests (i.e., less than expected suppression of GH for healthy individuals after oGTT) may occur in tall adolescents and in patients with diabetes mellitus, liver failure, anorexia nervosa, malnutrition, or GH resistance disorders in the absence of acromegaly.

Notes

1. In healthy individuals, hyperglycemia suppresses GH secretion via substrate-driven IGF-1 secretion.
2. The degree of GH suppression by glucose is usually concordant with the basal [IGFBP-3] and [IGF-1] before and after transsphenoidal surgery to remove a somatotropinoma.
3. Early in the acromegalic disease process and follow-up after treatment, significant predictors for survival were the duration of disease and the postoperative glucose-suppressed [GH]. Later on, a time-dependent effect on survival was observed for serial [IGF-1] (relative risk of 4.78 for an elevated as opposed to a normal [IGF-1]), but not for serial [GH] (Biermasz et al., 2004).

ICD-9 Codes

Refer to Test 2.4.1 codes.

Suggested Reading

Biermasz NR, Dekker FW, Pereira AM, van Thiel SW, Schutte PJ, van Dulken H, Romijn JA, Roelfsema F. Determinants of survival in treated acromegaly in a single center: predictive value of serial insulin-like growth factor I measurements. *J Clin Endocrinol Metab* 2004; 89:2789–96.

Chapman IM, Hartman ML, Straume M, Johnson ML, Veldhuis JD, Thorner MO. Enhanced sensitivity growth hormone (GH) chemiluminescence assay reveals lower postglucose nadir GH concentrations in men than women. *J Clin Endocrinol Metab* 1994; 78:1312–9.

Earll JM, Sparks LL, Forsham PH. Glucose suppression of serum growth hormone in the diagnosis of acromegaly. *JAMA* 1967; 201:628–30.

2.4.3   Growth-Hormone-Releasing Hormone (GHRH): Basal Level in Acromegaly

Indications for Test

Measurement of GHRH is indicated:

- In acromegalic patients particularly those with other known or suspected neoplastic lesions, such as a carcinoid tumor in the lung, pancreas, or other nonpituitary locations

Procedure

1. Identify diffuse, symmetric pituitary enlargement or the presence of a neoplasm in the pancreas, lung, gastrointestinal tract, or elsewhere using appropriate imaging techniques.
2. Obtain a random blood sample for GHRH testing.

Interpretation

1. [GHRH] in healthy individuals is always <1 ng/mL.
2. A major elevation of [GHRH] is typically associated with ectopic secretion of GHRH, a rare cause of acromegaly.

Notes

1. Symmetric enlargement of the sella secondary to somatotrophic hyperplasia may occur in GHRH-secreting patients with acromegaly.
2. In one study of 259 acromegalic patients with typical somatotropinomas of the pituitary, the mean [GHRH] was 23 pg/mL (range, 19–28 pg/mL) (Muller et al., 1999).

ICD-9 Codes

Conditions that may justify this test *include but are not limited to*:

157	malignant neoplasm of pancreas
495.8	malignant neoplasm of trachea, bronchus, or lung

Refer also to Test 2.4.1 codes.

Suggested Reading

Muller B, de Marco D, Burgi U, Mullis PE. What is the value of determining immunoreactive GHRH in acromegaly? *Schweiz Med Wochenschr* 1999; 129:1152–61.

Thorner MO, Frohman LA, Leong DA, Thominet J, Downs T, Hellmann P, Chitwood J, Vaughan JM, Vale W. Extrahypothalamic growth-hormone-releasing factor (GRF) secretion is a rare cause of acromegaly: plasma GRF levels in 177 acromegalic patients. *J Clin Endocrinol Metab* 1984; 59:846–9.

Thorner MO, Vance ML, Evans WS, Rogol AD, Rivier J, Vale W, Blizzard RM. Clinical studies with GHRH in man. *Horm Res* 1986; 24:91–8.

2.4.4 Imaging of Somatotropin-Secreting Pituitary Tumors by M-CT and MR in Acromegaly

Indications for Test

Imaging of the pituitary gland by M-CT and MR is indicated in patients with:

- Obvious acral enlargement
- Various combinations of headaches, prognathism, painful arthropathy, thickening of the palms, oily skin with skin tags, glucose intolerance or diabetes, galactorrhea, or menstrual disturbance
- An elevated fasting [IGF-1] and/or [GH], confirmed by repeat testing
- An elevated prolactin level, even if only mildly (i.e., 20–50 ng/mL) increased

Procedure

1. Because many acromegalic pituitary tumors are macroadenomas associated with sellar enlargement, obtain conventional lateral and anterior-to-posterior (AP) radiographic views of the sella (see Test 2.4.5) as a more readily available, cost-effective baseline study before using advanced imaging techniques.
2. Obtain standard MR views of the brain with enlarged views of the sella to define soft-tissue abnormalities, particularly optic nerve involvement and cavernous sinus invasion. Obtain multidetector or multislice CT (M-CT) views to best define focal asymmetry and any bony abnormalities such as erosion of the sellar wall, particularly in cases of cerebrospinal fluid (CSF) rhinorrhea. Use contrast enhancement, if possible, with both types of imaging.
3. When there is a symmetrically enlarged pituitary without adenoma, found on imaging of the sella, measure a random GHRH (see Test 2.4.3) and proceed to imaging of the pancreas and/or lung to search for a GHRH-secreting tumor.
4. Use digital subtraction angiography in cases of coexistent intrasellar aneurysm and GH-secreting pituitary tumor to define the anatomic nature of both lesions.

Interpretation

1. In general, MR with gadolinium enhancement is more sensitive and specific than M-CT scanning for the detection of pituitary tumors.
2. Typically, primary GH-secreting pituitary tumors erode into the sella asymmetrically.
3. Symmetric sellar enlargement may be the result of pituitary hyperplasia secondary to an ectopic GHRH-secreting tumor, usually of the carcinoid type, originating from a variety of areas in the bronchi, gastrointestinal tract, or pancreas.
4. On MR, carotid artery encasement is a reliable indicator of the cavernous sinus invasion of a pituitary tumor.
5. Heterogeneity seen within an acromegalic pituitary adenoma is consistent with cystic or necrotic change.
6. Asymmetric pituitary adenomas > 10 mm in diameter found on pituitary imaging prompted by an elevated prolactin level may also secrete GH and cause acromegaly (Bohler et al., 1994).

Notes

1. Factors that correlate with high (e.g., >101 ng/mL) preoperative [GH] in acromegaly are the maximum dimension of a pituitary tumor ($r = 0.496$; $p < 0.01$), its volume ($r = 0.439$; $p < 0.05$), sphenoid or cavernous sinus invasion ($p < 0.01$), and intracavernous carotid artery encasement ($p < 0.01$) (Saeki et al., 1999).
2. Findings on imaging that do not correlate with [GH] include suprasellar extension, signal intensity, and contrast enhancement of a pituitary tumor.
3. Lower basal GH concentrations (5–50 ng/mL) are typical of pituitary tumors medial to the intercarotid line.
4. Typically, in population studies, the ratio of pituitary macroadenomas to microadenoma found on diagnosis of acromegaly is 5:1.
5. Histologic staining of resected pituitary adenomas may reveal multiple cell types throughout the tumor. In a unique case of multiple endocrine neoplasia (MEN) type 1, acromegaly that was initially cured by resection of a GHRH-secreting tumor of the pancreas became recurrent secondary to the *de novo* appearance of a GH-secreting tumor of the pituitary.
6. Identify and repair or remove an intrasellar aneurysm before attempting to remove a coexistent pituitary tumor. Application of an endovascular Guglielmi detachable coil facilitates this procedure (Sade et al., 2004).
7. Bilateral inferior petrosal sinus sampling (BIPSS) for GH without GH-releasing hormone (Geref) stimulation in acromegaly patients is plagued by unacceptably high false-negative and false-positive localization study results.
8. A Geref infusion does not improve the sensitivity of BIPSS in patients with small pituitary tumors causing acromegaly and is not helpful or necessary in diagnosis of larger tumors (Frey et al., 1994).

ICD-9 Codes

Refer to Test 2.4.1 codes.

Suggested Reading

Altstadt TJ, Azzarelli B, Bevering C, Edmondson J, Nelson PB. Acromegaly caused by a growth hormone-releasing hormone-secreting carcinoid tumor: case report. *Neurosurgery* 2002; 50:1356–60.

Bohler HC, Jones EE, Brines ML. Marginally elevated prolactin levels require magnetic resonance imaging and evaluation for acromegaly. *Fertil Steril* 1994; 61:1168–70.

Bolanowski M, Schopohl J, Marciniak M, Rzeszutko M, Zatonska K, Daroszewski J, Milewicz A, Malczewska J, Badowski R. Acromegaly due to GHRH-secreting large bronchial carcinoid: complete recovery following tumor surgery. *Exp Clin Endocrinol Diabetes* 2002; 110:188–92.

Frey H, Torjesen PA, Enge IP, Stiris MG, Reinlie S. Analysis of blood from the inferior sinus petrosus in patients with Cushing syndrome and acromegaly. *Tidsskr Nor Laegeforen* 1994; 114:2257–61.

Kannuki S, Matsumoto K, Sano T, Shintani Y, Bando H, Saito S. Double pituitary adenoma: two case reports. *Neurol Med Chir (Tokyo)* 1996; 36:818–21.

TABLE 2.4
**Pituitary Volume and Hardy Grade Estimated from
Dimensions of Sella on Lateral Skull X-Ray**

Surface Area of Sella (mm²)	Estimated Pituitary Volume (mm³)	Hardy Grade
<208 with symmetry or "mild" asymmetry	<1.7	I
>208 with symmetry or "overt" asymmetry	1.3–6.0	II
Erosion through sella floor[a]	2.2–8.0	III
Diffuse destruction of the bony landmarks of the sella[b]	—	IV

[a] Regardless of surface area.

[b] A rare event in a medically served patient population.

Marro B, Zouaoui A, Sahel M, Crozat N, Gerber S, Sourour N, Sag K, Marsault C. MRI of pituitary adenomas in acromegaly. *Neuroradiology* 1997; 39:394–9.

Sade B, Mohr G, Tampieri D, Rizzo A. Intrasellar aneurysm and a growth hormone-secreting pituitary macroadenoma: case report. *J Neurosurg* 2004; 100:557–9.

Saeki N, Iuchi T, Isono S, Eda M, Yamaura A. MRI of growth hormone-secreting pituitary adenomas: factors determining pretreatment hormone levels. *Neuroradiology* 1999; 41:765–71.

2.4.5 Imaging of the Skull, Mandible, and Pituitary Sella by Conventional Radiography and Multislice CT (M-CT) Scanning in Acromegaly

Indications for Test

Imaging of the pituitary sella by conventional radiography and M-CT is indicated:

- In the initial assessment of clinically suspected acromegaly
- When markedly elevated [IGF-1] and/or [GH] are found
- For detection and classification of pituitary macroadenomas to predict response to therapy

Procedure

1. Obtain conventional lateral and anterior-to-posterior (AP) radiographic views of the skull, mandible, and sella turcica
2. Determine sella grade based on its surface projection and area as seen in a lateral radiographic view of the skull (Hardy et al., 1979).
3. In the case of a mildly asymmetric sella, calculate sella volume using the algebraic formula for the volume of an ellipsoid using both antero–posterior and lateral views of the skull. In cases of markedly irregular sellar anatomy, use the method of DiChiro and Nelson (1962) to accommodate for the asymmetry.
4. Obtain M-CT scanning to obtain additional anatomic detail of the sella with dynamic early contrast enhancement to detect the presence of direct arterial supply to tumor vs. those tumors with portal blood supply only.

Interpretation

1. Interpret sella grade from "Pituitary Volume and Hardy Grade Estimated from Dimensions of Sella on Lateral Skull X-Ray" (Table 2.4).
2. Acromegaly is associated with an increase in soft-tissue bulk and overpneumatization of sinuses and air cells, as well as the generalized bony overgrowth (enlarged and prognathic mandible with underbite). Both M-CT and conventional radiographs of the skull show these changes.

Notes

1. Larger, more erosive, acromegalic pituitary tumors are associated with higher preradiotherapy [GH] and higher median [GH] at various intervals after radiotherapy with or without surgical removal of tumor (Figure 2.1) (Dons et al., 1983).
2. Patients with arthritic changes in the neck may be difficult or impossible to position and obtain adequate coronal M-CT scans of the sella. Thus, sellar assessment of patients with severe degenerative joint disease may be limited to conventional radiographs.
3. Preoperative M-CT scans correctly localize microadenomas and macroadenomas with a sensitivity of ~80% (13/16 in one study of acromegalics with microadenomas) and ~95%, respectively (Barmeir et al., 1982).
4. Correct GH-secreting tumor localization and size estimation within 2 mm of that actually found at surgery (range, 2–20 mm) were achieved by preoperative M-CT with an accuracy rate of 91% (Marcovitz et al., 1988).
5. On M-CT scan, about two thirds of all pituitary microadenomas do not show early contrast enhancement before that of the pituitary portal system and appear less enhanced than the rest of the gland. These lesions appear to be under direct control of the hypothalamus.
6. One third of all pituitary microadenomas have early, partial, or complete contrast enhancement before portal system enhancement and are presumably independent of hypothalamic control (Bonneville et al., 1993).

ICD-9 Codes

Refer to Test 2.4.1 codes.

Suggested Reading

Barmeir EP, Lipschitz S, Dubowitz B, Kalk WJ. Diagnosis of acromegaly on cranial computed tomography. *Isr J Med Sci* 1982; 18:830–4.

Bonneville JF, Cattin F, Gorczyca W, Hardy J. Pituitary microadenomas: early enhancement with dynamic CT: implications of arterial blood supply and potential importance. *Radiology* 1993; 187:857–61.

DiChiro G, Nelson KB. The volume of the sella turcica. *Am J Roentgenol* 1962; 87:989–1008.

Dons RF, Rieth KG, Gorden P, Roth J. Size and erosive features of the sella turcica in acromegaly as predictors of therapeutic response to supervoltage irradiation. *Am J Med* 1983; 74:69–72.

Hardy J, Somma M. Acromegaly: surgical treatment by transsphenoidal microsurgical removal of the pituitary adenoma. In: *Clinical Management of Pituitary Disorders*, Tindall GT, Collins WF (Eds). 1979. Raven Press, pp. 209–17.

Marcovitz S, Wee R, Chan J, Hardy J. Diagnostic accuracy of preoperative CT scanning of pituitary somatotroph adenomas. *Am J Neuroradiol* 1988; 9:19–22.

2.5   Gonadotroph Reserve

2.5.1   Follicle-Stimulating Hormone (FSH) and Luteinizing Hormone (LH) Testing: Basal Levels in Central Gonadotropin Deficiency

Indications for Test

FSH and LH as tests for gonadotropin deficiency are indicated in:

- Children or young adults with infertility, hypogonadism, or low testosterone as part of a fertility evaluation
- Patients with anosmia, microphallus, cryptochidism, microtestis, delayed bone maturation, or delayed puberty consistent with hypogonadotropic hypogonadism (HH)

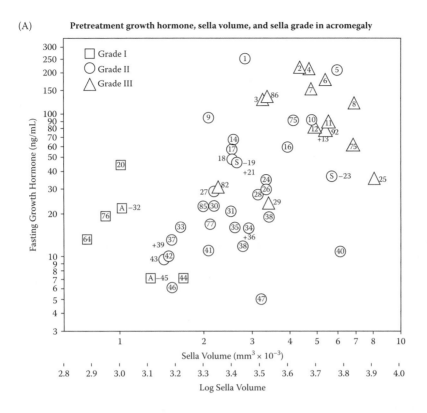

(A) Pretreatment growth hormone, sella volume, and sella grade in acromegaly

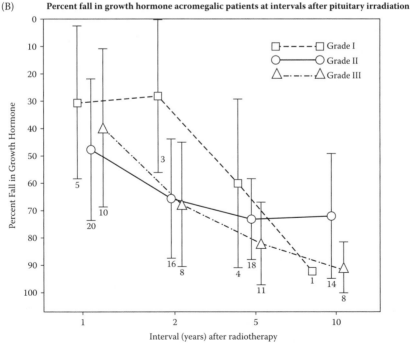

(B) Percent fall in growth hormone acromegalic patients at intervals after pituitary irradiation

Figure 2.1

Growth hormone (GH) levels as a function of (A) sella volume and grade before treatment of acromegaly, and (B) sella grade after radiotherapy to the pituitary in acromegalic patients. In panel A, numbers identify individual patients who participated in a series of NIH-funded endocrine studies; in panel B, numbers refer to the number of patients whose GH data were analyzed. (From Dons, R.F. et al., *Am. J. Med.*, 74, 69–72, 1983. With permission.)

TABLE 2.5
Concentrations of Gonadal Steroids and Gonadotropins

Type of Individual	Gonadal Steroid or Gonadotropin			
	LH (IU/L)	FSH (IU/L)	TT (ng/mL)	Estradiol (pg/mL)
Prepubertal	<2	<2	<1	<1.5
Adult male	<3–10	<1–8	4–9	<35
Adult female				
Follicular	3–11[a]	1–9	0.1–0.6	30–100
Ovulatory	18–70	4–30	Variable	>100
Luteal	2–11	1–7	High normal	70–300
Pregnancy[b]	n.a.	n.a.	Low normal	Up to 35,000
Postmenopausal	>35	>35	↓ with advanced age	<20

[a] Early puberty and anorexia nervosa are characterized by nocturnal LH pulses. Continuous 24-hour sampling may be useful in these cases but is labor intensive and expensive.

[b] The high human chorionic gonadotropin (hCG) concentrations associated with pregnancy suppress LH and FSH.

Note: FSH, follicle-stimulating hormone; LH, luteinizing hormone; n.a., not applicable; TT, total testosterone.

- Patients with pituitary tumors or surgery and findings of hypogonadism
- Early-onset amenorrhea, oligomenorrhea of polycystic ovary syndrome (PCOS), or perimenopause

Procedure

1. In females, identify the patient's menstrual cycle phase and first day of last normal menstrual period.
2. Obtain a blood sample for FSH testing, usually in conjunction with tests for gonadal steroids (i.e., various androgens and estrogens), when attempting to determine if the gonads have been understimulated by FSH, primary gonadal failure has developed, or, in females, only a transient, anovulatory cycle has occurred.
3. Obtain a blood sample for both LH and FSH testing in cases of perimenopause or possible gonadal failure.

Interpretation

1. Refer to "Concentrations of Gonadal Steroids and Gonadotropins" (Table 2.5) for changes in [FSH] and [LH] typical of a normal menstrual cycle.
2. In HH, FSH concentrations are very low (i.e., below assay detectability) with corresponding gonadal steroid deficiency.
3. In primary gonadal failure or early perimenopause, [FSH] may be elevated or in the upper normal portion of the reference interval.
4. In a woman with a transient anovulatory cycle, [FSH] is usually normal or into the lower normal portion of the reference interval.

Notes

1. Isolated deficiencies of LH and FSH are usually the result of beta-subunit gene mutations or defects.
2. Mutations in the gonadotropin-releasing hormone (GnRH) receptor, leptin, and the leptin receptor genes are associated with autosomal recessive syndromes of hypogonadotropic hypogonadism.
3. In the ovary, FSH acts through granulosa cell receptors to stimulate granulosa cell proliferation and differentiation. The follicle appearing at the beginning of the cycle is most responsive to FSH and is the first to produce estrogen and express granulosa cell LH receptors.
4. In the absence of an ovulation-inducing LH surge from the pituitary, terminal differentiation of the granulosa cells stops, leading to failure of the follicle to rupture, release a fertilizable egg, and luteinize.

5. A pulsatile GnRH stimulus from the hypothalamus is required to maintain pituitary gonadotropin synthesis and secretion. Increased pulse frequency (>1 per hour) favors LH secretion, whereas reduced pulse frequencies favor FSH secretion.
6. In hypothalamic amenorrhea, the frequency of GnRH pulses drops below normal. In PCOS, pulses are more frequent and favor LH synthesis, hyperandrogenism, and impaired follicular maturation.
7. Administration of progesterone slows GnRH pulse frequency which favors FSH secretion and follicular maturation.
8. Improved responsiveness of the testes to lower, more physiologic levels of gonadotropins in GnRH-deficient men follows long-term infusion of physiologic pulses of GnRH and the initiation of sexual maturation.

ICD-9 Codes

Conditions that may justify this test *include but are not limited to*:

Ovarian disorders
256.31   premature menopause
256.4   polycystic ovaries

Other
186.0   undescended testis
227.3   pituitary adenoma
257.2   testicular hypofunction
259.0   delayed sexual development and puberty
606.9   male infertility
626.0   absence of menstruation
752.64   microphallus
781.1   disturbance of sensation, smell, and taste

Suggested Reading

Dhindsa S, Prabhakar S, Sethi M, Bandyopadhyay A, Chaudhuri A, Dandona P. Frequent occurrence of hypogonadotropic hypogonadism in type 2 diabetes. *J Clin Endocrinol Metab* 2004; 89:5462–8.

Hillier SG. Gonadotropic control of ovarian follicular growth and development. *Mol Cell Endocrinol* 2001; 179(1–2):39–46.

John H, Schmid C. Kallmann's syndrome: clues to clinical diagnosis. *Int J Impot Res* 2000; 12:121–3 and 269–71.

Layman LC. Genetics of human hypogonadotropic hypogonadism. *Am J Med Genet* 1999; 89:240–8.

Marshall JC, Eagleson CA, McCartney CR. Hypothalamic dysfunction. *Mol Cell Endocrinol* 2001; 183(1–2):29–32.

Spratt DI, Crowley WF. Pituitary and gonadal responsiveness is enhanced during GnRH-induced puberty. *Am J Physiol* 1988; 254(5 Pt 1):E652–7.

2.5.2   Olfactory Function Testing in Gonadotropin Deficiency

Indications for Test

Olfactory function testing is indicated when a patient has:

- A long-term history of anosmia and is suspected of hypogonadotropic hypogonadism (HH), Turner syndrome, or pseudohypoparathyroidism
- Complaints of an acquired loss of sense of smell and one of a variety of common endocrine disorders, including hypothyroidism, adrenal hyperplasia, Addison's disease, Cushing's disease, diabetes mellitus, or primary amenorrhea with ovarian failure

- Complaints of acquired loss of sense of smell and one of a variety of non-endocrine disorders, including nonpituitary CNS neoplasms, cerebral ischemia, Parkinson's disease, Korsakoff psychosis, Alzheimer's disease, cystic fibrosis, or vitamin B_{12} deficiency

Procedure

1. Have the patient close his or her eyes, sniff, and identify several olfactory stimulants. Commonly available substances (e.g., oil of clove, ground coffee, peppermint, and vanilla) with nonirritating odors may be used as test substances with distilled water as a control substance.
2. Lifesaver® candies are also adequate test materials.
3. A do-it-yourself kit can be made by storing olfactory stimulants in corked, unmarked test tubes.
4. Commercially available scratch-and-sniff cards (Smell Identification Test™) can be ordered from Sensonics, Inc., P.O. Box 112, Haddon Heights, NJ 08035; 800-547-8838 (U.S.); 856-547-7702 (outside U.S.); 856-547-5665 (fax); www.sensonics.com.
5. Phenylethylmethylethylcarbinol (PEMEC), which has a floral scent and is normally detectable at very low concentrations (10^{-6} to 10^{-8} M) in water, is the preferred test agent.
6. Do not use substances (e.g., ammonia) with an irritating odor that can stimulate the trigeminal nerve and may produce false-positive results.
7. To determine odor thresholds in cases of incomplete dysfunction, unilateral dysfunction, or distortions of smell, request assistance from specialists in otolaryngology or neurology.

Interpretation

1. Inability to detect or identify any of the above common odors suggests anosmia.
2. The extensive differential diagnosis associated with anosmia, including the endocrine and non-endocrine diseases, is noted above.

Notes

1. Olfactory function test procedures assess the integrity of cranial nerve I.
2. Elaborate quantitative tests of olfactory perception are available from neurologists, but simple office testing as described above is usually adequate.
3. Patients infrequently volunteer a history of anosmia. An interviewer must elicit a history of an absent sense of smell as well as a change in or loss of the sense of smell.
4. Female anosmics with hypogonadotropic hypogonadism fail to develop ovarian follicles beyond the primordial stage, indicating that early stages of follicular maturation require secretion of gonadotropins well above the low levels that occur in these patients.

ICD-9 Codes

Conditions that may justify this test *include but are not limited to*:

Adrenal disorders
255.00   Cushing's syndrome
255.4   corticoadrenal insufficiency

Neurologic disorders
323   Parkinson's disease
331.0   Alzheimer's dementia
435.9   transient cerebral ischemia

Reproductive disorders
256.31   premature menopause
257.2   testicular hypofunction
758.6   gonadal dysgenesis/Turner syndrome

Other
244.9   hypothyroidism
250.X   diabetes mellitus
277   cystic fibrosis
281.1   vitamin B_{12} deficiency anemia
291.1   alcohol-induced persisting amnestic disorder (Korsakoff's type)
781.1   disturbances of sensation of smell and taste

Suggested Reading

Goldenberg RL, Powell RD, Rosen SW, Marshall JR, Ross GT. Ovarian morphology in women with anosmia and hypogonadotropic hypogonadism. *Am J Obstet Gynecol* 1976; 126:91–4.

Nelson LM. Screening military aircrews for anosmia. *Mil Med* 1988; 153:257–8.

Rosen SW, Gann P, Rogol AD. Congenital anosmia: detection thresholds for seven odorant classes in hypogonadal and eugonadal patients. *Ann Otol Rhinol Laryngol* 1979; 88(2 Pt 1):288–92.

2.6   Gonadotroph Excess

2.6.1   Luteinizing Hormone (LH) and Follicle-Stimulating Hormone (FSH) Testing for Gonadotropin Hypersecretion

Indications for Test

As tests for gonadotropin hypersecretion, LH and FSH are indicated in the diagnosis of:

- Primary gonadal failure when the hypothalamic–pituitary axis is intact
- Onset of menopause when the hypothalamic–pituitary axis is intact
- Sexual precocity, on finding of a hypothalamic hamartoma, and in rare cases of pituitary tumors associated with hypergonadism

Procedure

1. Obtain a blood sample for LH and FSH determination within 3 to 6 hours of the patient awakening from sleep. In menstruating women, note the first day of their last menstrual period.
2. Secretion of LH, unlike FSH, occurs in a pulsatile manner. To achieve increased diagnostic accuracy, pool serum from a set of two or three blood samples obtained at 10- to 30-minute intervals and measure [LH] in the pooled sample.

Interpretation

1. In perimenopause, gonadotropin levels are widely variable but tend to be >5 IU/L.
2. In precocious puberty associated with a hypothalamic hamartoma, both gonadotropins and gonadal steroids are elevated.
3. "Concentrations of Gonadal Steroids and Gonadotropins" (Table 2.5) shows the expected normal physiologic changes in gonadotropins and sex steroids.

Notes

1. Mean FSH concentrations vary markedly between individual males, with discrete pulses rarely observed.
2. In healthy men, both mean [LH] and mean LH pulse amplitudes vary up to fourfold between individuals, with LH interpulse intervals varying from 30 to 480 minutes (mean, 119 minutes). Diurnal variation in LH, but not FSH, secretion is typical.

3. After puberty, the testosterone secretory surge lags behind LH pulse secretion by approximately 40 minutes.

4. In one unique case, hypersecretion of functional LH from a pituitary tumor caused a marked increase in testosterone levels (Dizon and Vesely, 2002).

ICD-9 Codes

Conditions that may justify this test *include but are not limited to*:

Pituitary disorders
253.1   Forbes–Albright syndrome
253.2   panhypopituitarism

Other
227.3   pituitary neoplasm/hypothalamic hamartoma
256.31   premature menopause
257.2   testicular hypofunction
259.1   precocious sexual development; not puberty, not elsewhere classified

Suggested Reading

Boyar RM, Finkelstein JW, David R, Roffwarg H, Kapen S, Weitzman ED, Hellman L. Twenty-four hour patterns of plasma luteinizing hormone and follicle-stimulating hormone in sexual precocity. *N Engl J Med* 1973; 289:282–6.

Boyar RM, Kapen S, Finkelstein JW, Perlow M, Sassin JF, Fukushima DK, Weitzman ED, Hellman L. Hypothalamic–pituitary function in diverse hyperprolactinemic states. *J Clin Invest* 1974a; 53:1588–98.

Boyar RM, Katz J, Finkelstein JW, Kapen S, Weiner H, Weitzman ED, Hellman L. Anorexia nervosa: immaturity of the 24-hour luteinizing hormone secretory pattern. *N Engl J Med* 1974b; 291:861–5.

Boyar RM, Rosenfeld RS, Kapen S, Finkelstein JW, Roffwarg HP, Weitzman ED, Hellman L. Human puberty: simultaneous augmented secretion of luteinizing hormone and testosterone during sleep. *J Clin Invest* 1974c; 54:609–18.

Colaco MP, Desai MP, Choksi CS, Shah KN, Mehta RU. Hypothalamic hamartomas and precocious puberty. *Indian J Pediatr* 1993; 60:445–50.

Dizon MN, Vesely DL. Gonadotropin-secreting pituitary tumor associated with hypersecretion of testosterone and hypogonadism after hypophysectomy. *Endocr Pract* 2002; 8:225–31.

Spratt DI, O'Dea LS, Schoenfeld D, Butler J, Rao PN, Crowley WF. Neuroendocrine–gonadal axis in men: frequent sampling of LH, FSH, and testosterone. *Am J Physiol* 1988; 254(5 Pt 1):E658–66.

2.6.2   Thyrotropin-Releasing Hormone (TRH) Stimulation of LH Alpha- and Beta-Subunits in Pituitary Tumors

Indications for Test

Measurement of LH beta-subunit and the common pituitary gonadotropin (PGH) alpha-subunit are indicated in cases of:

- Clinically nonfunctioning or nonsecreting pituitary adenomas (NFPA/NSA)

Procedure

1. In patients with NFPA/NSA, obtain blood samples before and every 15 minutes for up to 2 hours (8 samples) after the injection of standard-dose (usually 500 mcg) TRH for measurement of LH, LH alpha- and beta-subunits, and FSH.

Interpretation

1. Patients may have pituitary tumors in which defects in protein biosynthesis or processing result in complete failure to produce or secrete any detectable gonadotropic hormones or their subunits.
2. Following injection of TRH, a marked rise in free FSH beta- and LH beta-subunits occurs in up to about two thirds of NFPA/NSA patients who have low or normal basal intact [FSH] and [LH], which do not change after TRH.
3. Elaboration of a disproportionate increase in the alpha-subunit, common to pituitary glycoprotein hormones, may occur without concurrent production of the intact hormones in a substantial number of patients with NFPA/NSA.
4. Alpha-subunit responses to stimuli such as TRH are generally parallel with those of the hormones concomitantly produced in adenomas and released in response to TRH.
5. In normal pituitary tissue, but not necessarily in the blood, alpha-subunits are present in excess of beta-subunits at both the mRNA and protein levels, while in NFPA/NSA there may be disproportionate secretion of free beta-subunit relative to alpha-subunit.
6. Patients with pituitary tumors other than NFPA/NSA and healthy subjects exhibit no significant increase in FSH beta- or LH beta-subunit secretion following the administration of TRH.

Notes

1. NFPA/NSA comprise at least 25% of all pituitary tumors.
2. Measurement of baseline serum LH beta-subunit alone is of little value in the diagnosis of NFPA/NSA and in the identification of gonadotroph adenomas among nonfunctioning pituitary adenomas.
3. Most cases of apparent NFPA/NSA in women are macroadenomas that arise from gonadotroph cells and can be recognized, even in postmenopausal women, by hypersecretion of LH beta-subunits or FSH and LH only in response to TRH testing.
4. Excess ACTH production may occur in patients with NFPA/NSA without biochemical hypercortisolism. Coexpression of FSH beta-subunit, LH beta-subunit, human chorionic gonadotropin (hCG) beta-subunit, and TSH beta-subunit molecules is characteristic of patients with NFPA/NSA.
5. Because human luteinizing hormone (hLH) and hCG beta-subunits have high structural homology, immunoassays that use antibodies to epitopes that are different between the hLH and hCG beta-subunits are required to accurately quantify these subunits in human serum.
6. The net increase in serum LH beta-subunit concentration after the administration of TRH in NFPA/NSA patients may range from 0 to 23% over baseline.
7. In patients with NFPA/NSA, TRH elicited a marked rise in [beta FSH] in 29 of 40 patients and [beta LH] in 28 of 36 patients. In a subgroup of eight individuals whose NFPA/NSA were harvested during surgery and cultured for 7 to 21 days, TRH stimulated the release of beta FSH or beta LH and alpha-subunit *in vitro* in all cases, including tumors from subjects not responding to TRH *in vivo* (Somjen et al., 1997).
8. Transsphenoidal decompression remains the treatment of choice for patients with NFPA/NSA, as no effective pharmacotherapy is available as of this writing.

ICD-9 Codes

Conditions that may justify this test *include but are not limited to*:

227.3   pituitary adenomas

Suggested Reading

Birken S, Kovalevskaya G, O'Connor J. Metabolism of hCG and hLH to multiple urinary forms. *Mol Cell Endocrinol* 1996; 125:121–31.

Daneshdoost L, Gennarelli TA, Bashey HM, Savino PJ, Sergott RC, Bosley TM, Snyder PJ. Recognition of gonadotroph adenomas in women. *N Engl J Med* 1991; 324:589–94.

Gil-del-Alamo P; Pettersson KS, Saccomanno K, Spada A, Faglia G, Beck-Peccoz P. Abnormal response of luteinizing hormone beta subunit to thyrotrophin-releasing hormone in patients with non-functioning pituitary adenoma. *Clin Endocrinol (Oxf)* 1994; 41:661–6.

Ishibashi M, Yamaji T, Takaku F, Teramoto A, Fukushima T. Secretion of glycoprotein hormone alpha-subunit by pituitary tumors. *J Clin Endocrinol Metab* 1987; 64:1187–93.

Katznelson L, Alexander JM, Bikkal HA, Jameson JL, Hsu DW, Klibanski A. Imbalanced follicle-stimulating hormone beta-subunit hormone biosynthesis in human pituitary adenomas. *J Clin Endocrinol Metab* 1992; 74:1343–51.

Klibanski A. Nonsecreting pituitary tumors. *Endocrinol Metab Clin North Am* 1987; 16:793–804.

Saccomanno K, Bassetti M, Lania A, Losa M, Faglia G, Spada A. Immunodetection of glycoprotein hormone subunits in nonfunctioning and glycoprotein hormone-secreting pituitary adenomas. *J Endocrinol Invest* 1997; 20:59–64.

Somjen D, Tordjman K, Kohen F, Baz M, Razon N, Ouaknine G, Stern N. Combined beta FSH and beta LH response to TRH in patients with clinically non-functioning pituitary adenomas. *Clin Endocrinol (Oxf)* 1997; 46:555–62.

2.6.3　Imaging of Nonfunctional and Nonsecreting Pituitary–Hypothalamic Tumors (PHTs) by High-Resolution Multislice Magnetic Resonance (MR) and Computed Tomographic (M-CT) Scanning

Indications for Test

Imaging for PHTs by M-CT or MR is indicated when:

- A combination of signs and symptoms such as erectile dysfunction, visual field abnormalities, headaches, and symptoms of hypothyroidism, hypogonadism, or adrenal insufficiency suggests the presence of a PHT.
- Low testosterone with elevated or high-normal serum levels of LH, FSH, and prolactin is found.
- A diagnosis of central precocious puberty (CPP), associated with a seizure disorder, is made.

Procedure

1. After obtaining a blood sample for FSH, LH, and total testosterone testing, obtain a MR, in preference to a M-CT, image of the pituitary.

Interpretation

1. Individuals with PHTs may have low serum testosterone levels similar to those found in non-pituitary tumor patients with idiopathic low testosterone, but the pituitary tumor patients tend to have significantly higher serum levels of LH, FSH, and prolactin.
2. The majority of PHTs occur as macroadenomas about as frequently as prolactinomas and account for 25 to 40% of all pituitary tumors.
3. Distinguish a true CNS neoplasm from benign non-neoplastic hypothalamic hamartomas. Hypothalamic hamartomas frequently appear below the tuber cinereum extending into the supersellar and interpeduncular cistern.
4. More than three quarters of hypothalamic hamartomas are located in parahypothalamic positions not involving the third ventricle. Up to 50% of these are small, pedunculated lesions that are <10 mm in diameter.
5. Characteristically, patients with seizures and a sessile intrahypothalamic hamartoma have a distorted third ventricle.

Notes

1. Most PHTs are gonadotropin producing but do not actively secrete large quantities of hormone(s) into the blood.
2. Small and pedunculated hypothalamic hamartomas are associated with CPP, while large and sessile lesions may be associated with seizures alone and no endocrinopathy.
3. Peduculated hypothalamic hamartomas are amenable to microsurgical ablation. Use a stereotactic Gamma Knife™ to remove sessile lesions.

ICD-9 Codes

Conditions that may justify this test *include but are not limited to*:

Visual abnormality
368.4X   visual field defects
368.9   visual disturbance, unspecified

Other
244.9   hypothyroidism, primary
253.1   hyperprolactinemia/Forbes–Albright syndrome
255.4   Addison's disease/hypoadrenalism
257.2   testicular hypofunction
259.1   precocious sexual development; not puberty not elsewhere classified
302.70   psychosexual dysfunction, unspecified
434.91   stroke, ischemic
784.0   headache

Suggested Reading

Debeneix C, Bourgeois M, Trivin C, Sainte-Rose C, Brauner R. Hypothalamic hamartoma: comparison of clinical presentation and magnetic resonance images. *Horm Res* 2001; 56:12–8.

Dobs AS, El-Deiry S, Wand G, Wiederkehr M. Central hypogonadism: distinguishing idiopathic low testosterone from pituitary tumors. *Endocr Pract* 1998; 4:355–9.

Gsponer J, De Tribolet N, Deruaz JP et al. Diagnosis, treatment, and outcome of pituitary tumors and other abnormal intrasellar masses: retrospective analysis of 353 patients. *Medicine (Baltimore)* 1999; 78:236–69.

Jung H, Neumaier Probst E, Hauffa BP, Partsch CJ, Dammann O. Association of morphological characteristics with precocious puberty and/or gelastic seizures in hypothalamic hamartoma. *J Clin Endocrinol Metab* 2003; 88:4590–5.

Luo S, Li C, Ma Z, Zhang Y, Jia G, Cheng Y. Microsurgical treatment for hypothalamic hamartoma in children with precocious puberty. *Surg Neurol* 2002; 57:356–62.

Unger F, Schrottner O, Feichtinger M, Bone G, Haselsberger K, Sutter B. Stereotactic radiosurgery for hypothalamic hamartomas. *Acta Neurochir Suppl* 2002; 84:57–63.

2.7   Thyrotropin (Thyroid-Stimulating Hormone) Reserve

2.7.1   TSH Testing in Central Hypothyroidism (CH)

Indications for Test

The TSH* test is indicated to evaluate patients suspected of central hypothyroidism (CH) when:

- A pituitary or hypothalamic disturbance or tumor is present or suspected

---

* *Nomenclature footnote:* The current, highly sensitive thyrotropin (hsTSH) assay used by most clinical laboratories in the United States (at a minimum a 2nd-generation TSH assay and commonly a 3rd-generation assay) will be considered as the applicable procedure of choice and throughout this publication is referred to as simply the "TSH" assay.

Procedure

1. Obtain a random blood sample for both TSH and free thyroxine (FT_4) testing in patients with a known or highly suspected hypothalamic–pituitary abnormality.
2. Do not rely on TSH testing alone to screen for, diagnose, or monitor thyroid status in CH. Always obtain the serum FT_4 result.

Interpretation

1. Interpret [TSH] in the context of the [FT_4] and perhaps the [total T_3], as well as the patient's clinical symptoms suggestive of CH and treatment with thyroid hormone.
2. [TSH] within or above the upper limit of the reference interval does not exclude secondary or tertiary hypothyroidism, as this may occur if primary hypothyroidism coexists with partial pituitary or hypothalamic defects in the hypothalamic–pituitary–thyroid (HPT) axis.
3. In patients with CH without primary hypothyroidism, [TSH] is usually within the reference interval for healthy individuals and is typically between 0.3 and 3.0 mIU/L. Coexistence of both primary hypothyroidism and central thyrotrope insufficiency may result in a slightly supranormal [TSH] (>3 mIU/L) but markedly low [FT_4].
4. In patients with CH, [TSH] is usually in the normal range and is responsive to thyroid hormone therapy, as in primary hypothyroidism, but at a lower [TSH] set point.
5. The secretion of TSH molecules with reduced bioactivity is a common alteration in patients with hypothalamic–pituitary lesions, contributing to the pathogenesis of CH as well as impairment of pituitary TSH reserve.
6. In patients with established CH, thyroid hormone therapy resulting in a [TSH] of <0.1 mIU/L predicted euthyroidism in 92% of cases, compared to only 34% when [TSH] remained >1 mIU/L (Shimon et al., 2002).
7. In patients with CH, FT_4 test results may supplant measurement of the [TSH] as an indicator of the adequacy of T_4 replacement therapy.

Notes

1. Typically, the results of a pituitary imaging study as well as tests consistent with multiple pituitary hormone abnormalities lead to the diagnosis a pituitary disorder.
2. Unlike in primary hypothyroidism, several months may pass before the [TSH] in a treated CH patient reaches a new steady state or basal set point.
3. A widely variable delay in the recovery of the hypothalamic–pituitary axis occurs in the majority of patients with Graves' disease treated with radioiodine. Typically, a transient central hypothyroid phase occurs after administration of radioiodine.
4. T_4-mediated negative feedback to the pituitary is preserved in most CH patients, but at a much lower set point than in primary hypothyroidism.
5. In CH, the slopes of plots of [TSH] vs. [free T_4] were unique for individual patients (Shimon et al., 2002).
6. TSH-based newborn screening does not identify hypothyroidism due to TSH beta-subunit mutations.
7. Legally mandated screening programs that measure both [total T_4] and [TSH] have detected newborns with congenital CH usually secondary to rare, compound heterozygous and more common, homozygous (variants C105Vfs114X and Q49X) mutations of the TSH beta-subunit gene (Karges et al., 2004).
8. Algorithms dating back to the 1980s for diagnosing both CH and central thyrotoxicosis using modern free T_4 and TSH assays remain valid today (Hay and Klee, 1988).

ICD-9 Codes

Conditions that may justify this test *include but are not limited to*:

227.3   pituitary adenoma
253.2   hypopituitarism

Suggested Reading

Faglia G. The clinical impact of the thyrotropin-releasing hormone test. *Thyroid* 1998; 8:903–8.

Hay ID, Klee GG. Thyroid dysfunction. *Endocrinol Metab Clin North Am* 1988; 3:473–509.

Karges B, Le Heup B, Schoenle E, Castro-Correia C, Fontoura M, Pfaffle R, Andler W, Debatin KM, Karges W. Compound heterozygous and homozygous mutations of the TSHbeta gene as a cause of congenital central hypothyroidism in Europe. *Horm Res* 2004; 62:149–55.

Persani L, Ferretti E, Borgato S, Faglia G, Beck-Peccoz P. Circulating thyrotropin bioactivity in sporadic central hypothyroidism. *J Clin Endocrinol Metab* 2000; 85:3631–5.

Shimon I, Cohen O, Lubetsky A, Olchovsky D. Thyrotropin suppression by thyroid hormone replacement is correlated with thyroxine level normalization in central hypothyroidism. *Thyroid* 2002; 12:823–7.

Uy HL, Reasner CA, Samuels MH. Pattern of recovery of the hypothalamic–pituitary–thyroid axis following radioactive iodine therapy in patients with Graves' disease. *Am J Med* 1995; 99:173–9.

2.7.2   Thyrotropin (TSH) Surge Test for Diagnosis of Central Hypothyroidism (CH)

Indications for Test

The TSH surge test is indicated:

- To objectify the clinical diagnosis of CH
- In cases of short stature, slow growth velocity (with or without pubertal delay), and a free T_4 in the lowest third of the reference interval for healthy children

Procedure

1. Carefully select patients for testing based on prior free T_4 data and growth studies.
2. Obtain informed consent for testing from the patient (child and parent, if necessary).
3. Admit the patient to the hospital by noon on the day before testing.
4. Between 1300 and 1400 hours, insert a short catheter intravenously into the upper extremity. Maintain patency of catheter with 0.5-mL heparin solution (100 U/mL).
5. Obtain hourly blood samples for [TSH] during the period of the afternoon [TSH] nadir (1500–1800 hours) and the nightly [TSH] peak (2200–0400 hours).
6. Take care not to disturb the patient's sleep, even though the test results are not affected by a failure to sleep well.
7. Calculate the nadir [TSH] as the lowest average of three consecutive [TSH] measured in the afternoon and the peak [TSH] as the highest average of three consecutive nocturnal [TSH].
8. Using the formula below, calculate the nocturnal surge in TSH over the nadir:

$$\text{Nocturnal surge in TSH over nadir } (\%) = \frac{[\text{TSH}]_{\text{peak}} - [\text{TSH}]_{\text{nadir}}}{[\text{TSH}]_{\text{nadir}}} \times 100$$

Interpretation

1. The mean (95% confidence limits) value for the nocturnal percent TSH surge in a healthy reference population of children or adults is 124% (50–300%).
2. Patients with primary hypothyroidism have a normal percent TSH surge.
3. The mean percent TSH surge in patients with various hypothalamic–pituitary disorders is 22 to 53%.
4. The TSH surge test has a sensitivity of 91%, which is higher than the sensitivity of the TRH test for identifying individuals with CH.

Notes

1. TSH undergoes marked diurnal variation of up to 2 or 3 mIU/L between peak (0200 hours) and nadir (1600 hours) [TSH].
2. A loss in the marked rise or surge in TSH, which normally occurs between 2200 and 0400 hours, suggests abnormal hypothalamic–pituitary regulation even though the morning [TSH] may be within the reference interval for healthy individuals.
3. CH may be the cause of short stature children in whom there is a loss in diurnal TSH surge.
4. Use of a direct free T_4 assay to screen short children for possible subtle hypothyroidism has been used to identify a subgroup of these children with a 33% probability of isolated CH.
5. The oral TRH test, as an alternative test for the diagnosis of CH, is invalid in patients with hypopituarism. Typically, 40 mg of TRH administered orally are required to stimulate [TSH] of >5 mIU/L. These large quantities of TRH are not available commercially.
6. Although up to one third of children with GHD may have CH, GH treatment does not appreciably alter thyroid function in children who have no evidence of thyroid axis dysfunction before GH treatment.

ICD-9 Codes

Conditions that may justify this test *include but are not limited to*:

244.9   hypothyroidism, secondary or central
253.3   growth hormone deficiency/pituitary dwarfism

Suggested Reading

Municchi G, Malozowski S, Nisula BC, Cristiano A, Rose SR. Nocturnal thyrotropin surge in growth hormone-deficient children. *J Pediatr* 1992; 121:214–20.

Rose SR. Isolated central hypothyroidism in short stature. *Pediatr Res* 1995; 38:967–73.

Rose SR, Manasco PK, Pearce S, Nisula BC. Hypothyroidism and deficiency of the nocturnal thyrotropin surge in children with hypothalamic–pituitary disorders. *J Clin Endocrinol Metab* 1990; 70: 1750–5.

Staub JJ, Girard J, Mueller-Brand J, Noelpp B, Werner-Zodrow I, Baur U, Heitz P, Gemsenjaeger E. Blunting of TSH response after repeated oral administration of TRH in normal and hypothyroid subjects. *J Clin Endocrinol Metab* 1978; 46:260–6.

2.8 Thyrotropin (Thyroid-Stimulating Hormone) Excess

2.8.1 TSH Testing in Screening for Thyrotrope Hypersecretion

Indications for Test

The TSH test as a screening test for thyrotrope hypersecretion is indicated for:

• Hyperthyroid patients without the autoimmune syndrome of Graves' disease
• Hyperthyroid patients with symptoms of a pituitary mass lesion such as visual field loss, headaches, and loss of other pituitary hormones

Procedure

1. Obtain a random blood sample for TSH and free T_4 testing, as is usually done for any patient suspected of being hyperthyroid.
2. Based on the results of TSH and free T_4 tests, proceed with measurement of free alpha-subunit and pituitary imaging studies.
3. Calculate the alpha-subunit/TSH ratio.

Interpretation

1. If the [TSH] is within the [TSH] reference interval for healthy individuals or is greater than the upper limit of the [TSH] reference interval and the [free T_4] is near the upper limit of the reference interval for healthy individuals or is clearly elevated, hyperthyroidism secondary to thyrotrope hypersecretion is a possibility.
2. Thyrotropinomas may appear in a wide range of sizes from micro- to macroadenomas on M-CT or MR imaging and may contain calcifications.
3. The alpha-subunit concentration and alpha-subunit (ng/mL)/TSH (mIU/L) ratio are usually elevated (i.e., >1.5 ng/mL and >4.0, respectively), but an alpha-subunit concentration and/or alpha-subunit/ TSH ratio within the reference interval for healthy individuals does not categorically exclude the presence of a TSH-secreting pituitary adenoma.

Notes

1. Thyrotropinomas are rare; however, when they do occur, the patient is at high risk for the misdiagnosis of Graves' disease.
2. There may be an association between TSH-secreting pituitary adenomas and thyroid carcinoma or Hashimoto's thyroiditis.
3. High circulating levels of HCG and its beta-subunit component, whether or not of trophoblastic origin, can cause hyperthyroidism by stimulating pituitary thyrotrophs in the absence of a pituitary tumor.
4. There are no descriptions of acquired resistance to thyroid hormone in patients with TSH-secreting tumors. When a pituitary microadenoma is not detected in cases with high [TSH] and [free T_4], selective pituitary resistance to TSH and/or T_3 (i.e., Refetoff's syndrome) may be present.
5. In a study of 14 newly diagnosed patients with thyrotropinomas (Brucker-Davis et al., 1999), each of three criteria (TSH to TRH stimulation, high alpha-subunit concentration, and a high alpha-subunit/TSH ratio) was found in 10, 8, and 12 of the patients to have sensitivity and specificity diagnostic utility of 71% and 96%, 75% and 90%, and 83% and 65%, respectively.

ICD-9 Codes

Conditions that may justify this test *include but are not limited to*:

227.3	pituitary adenoma
242.0	toxic diffuse goiter
242.9	thyrotoxicosis
368.4	visual field defects
784.0	headache

Suggested Reading

Brousse C, Mignot L, Baglin AC, Bernard N, Piette AM, Gepner P, Chapman A. Hyperthyroidism and hypersecretion of chorionic gonadotropin in gastric adenocarcinoma. *Rev Med Interne* 1994; 15:830–3.

Brucker-Davis F, Oldfield EH, Skarulis MC, Doppman JL, Weintraub BD. Thyrotropin-secreting pituitary tumors: diagnostic criteria, thyroid hormone sensitivity, and treatment outcome in 25 patients followed at the National Institutes of Health. *J Clin Endocrinol Metab* 1999; 84:476–86.

Podoba J, Hnilica P, Makaiova I, Kovac A. Thyrotropin-secreting adenomas of the hypophysis. *Vnitr Lek* 1997; 43:611–4.

Yamaguchi S, Sasajima T, Takahashi M, Kinouchi H, Suzuki A, Yoshioka N, Itoh S, Mizoi K. A case of pleomorphic TSH-producing pituitary adenoma with calcification. *No Shinkei Geka* 2004; 32:961–7.

2.8.2 TSH Alpha-Subunit and Intact TSH Testing in TSH-Secreting Pituitary Tumors

Indications for Test

Measurement of TSH alpha-subunit and intact [TSH] is indicated:

- To explain an inappropriately normal or high TSH in a hyperthyroid patient

Procedure

1. Obtain a random blood sample for TSH alpha-subunit and intact TSH testing.
2. Assess for the presence of high circulating levels of alpha-subunits from other causes such as FSH and LH in hypogonadism. Do not use the formula below if [FSH] and [LH] are elevated, as the ratio will be invalid.
3. Using a molecular weight of 28,000 for TSH and 14,000 for the TSH alpha-subunit, a specific activity for TSH of 0.2 µIU/ng, and a conversion factor of 10 derived from:

$$\frac{\left(28{,}000 \text{ g} \cdot \text{mole}^{-1} / 14{,}000 \text{ g} \cdot \text{mole}^{-1}\right)}{0.2 \text{ µIU/ng}}$$

calculate the ratio shown below:

$$\text{TSH/TSH alpha-subunit molar ratio} = \frac{\text{TSH (µIU/mL)}}{\text{TSH alpha-subunit (ng/mL)}} \times 10 \text{ ng/µIU}$$

If the TSH/TSH alpha-subunit ratio is <1, perform an imaging study of the pituitary as well as assays for [LH] and [FSH].

Interpretation

1. Avoid inappropriate (i.e., irreversible radioiodine) antithyroid therapy in hyperthyroid patients with normal or elevated [TSH].
2. In a thyrotoxic patient, a low (<1) TSH/TSH alpha-subunit molar ratio suggests a TSH-secreting pituitary adenoma.
3. A TSH/TSH alpha-subunit molar ratio that is >1 may occur in patients with thyroid hormone resistance.
4. An elevated [TSH] in the presence of a low [free T$_4$] usually indicates primary hypothyroidism, not the presence of a TSH-secreting adenoma.
5. After pituitary surgery, most hyperthyroid patients will become euthyroid, but pituitary imaging, TSH alpha-subunit concentration, and the TSH/TSH alpha-subunit molar ratio may normalize in only 47%, 54%, and 58% of patients, respectively.

Notes

1. [TSH] is rarely elevated (i.e., TSH > 5 mIU/L in patients with a TSH-secreting pituitary adenoma), but it may be inappropriately normal if the patient has hyperthyroidism.
2. Generalized enlargement of the pituitary may occur in cases of severe hypothyroidism and thyrotrope hyperplasia.

3. Hypersecretion of other pituitary hormones may occur in up to one third of patients with a TSH-secreting pituitary adenoma.

4. Free alpha-subunit is a tumor marker for neoplasms of gonadotropic or thyrotropic cell origin but may be negative in isolated instances of TSH-secreting pituitary adenomas.

5. The ratio of biologic (B) TSH activity to immunologic (I) TSH concentration (B/I), determined preoperatively in patients with TSH-secreting tumors, is usually elevated compared with euthyroid subjects.

6. Expect the TRH test to show a blunted TSH response with an elevated baseline [TSH] and the T_3 suppression test (T_3 given at a dose of 60 mcg per day × 10 days) to show no inhibition of TSH secretion in patients with a TSH-secreting adenoma.

7. The majority of TSH-secreting pituitary tumors are macroadenomas by a ratio of >2:1 compared to microadenomas.

8. Isolated cases of primary hyperthyroidism may occur after pituitary surgery for a TSH-secreting tumor.

ICD-9 Codes

Conditions that may justify this test *include but are not limited to*:

227.3 pituitary adenoma
242.9 thyrotoxicosis

Suggested Reading

Fahey P, Sarkos P, Mazzaferri E. An inappropriate thyrotropin-secreting pituitary adenoma in a 30-year-old man. *Arch Fam Med* 1994; 3:190–2.

Gesundheit N, Petrick PA, Nissim M, Dahlberg PA, Doppman JL, Emerson CH, Braverman LE, Oldfield EH, Weintraub BD. Thyrotropin-secreting pituitary adenomas: clinical and biochemical heterogeneity. Case reports and follow-up of nine patients. *Ann Intern Med* 1989; 111:827–35.

Kuo CS, Ho DM, Yang AH, Lin HD. Thyrotropin-secreting pituitary adenoma with growth hormone hypersecretion. *Zhonghua Yi Xue Za Zhi (Taipei)* 2002; 65:489–93.

Losa M, Giovanelli M, Persani L, Mortini P, Faglia G, Beck-Peccoz P. Criteria of cure and follow-up of central hyperthyroidism due to thyrotropin-secreting pituitary adenomas. *J Clin Endocrinol Metab* 1996; 81:3084–90.

Refetoff S, Weiss RE, Usala SJ. The syndromes of resistance to thyroid hormone. *Endocr Rev* 1993; 14:348–99.

Szabolcs I, Czirjak S, Hubina E, Goth M, Tarko M, Bako B, Konrady A, Patkay J, Kovacs L, Gorombey Z, Radacsi A, Kovacs G, Szilagyi G. Thyrotropin-secreting pituitary adenomas: diagnosis and treatment in five cases. *Orv Hetil* 2002; 143(Suppl 19):1074–7.

Wu YY, Chang HY, Lin JD, Chen KW, Huang YY, Jung SM. Clinical characteristics of patients with thyrotropin-secreting pituitary adenoma. *J Formos Med Assoc* 2003; 102:164–71.

2.8.3 Imaging of TSH-Secreting Central Nervous System (CNS) Tumors by High-Resolution Multislice Magnetic Resonance (MR) and Computed Tomographic (M-CT) Scanning

Indications for Test

Imaging for detection of a TSH-secreting CNS tumor by M-CT or MR is indicated in cases with:

- Hyperthyroidism and symptoms related to a pituitary mass effect such as hemianopsia
- An inappropriately elevated [TSH] for elevated [free T_4]

Procedure

1. Image the nasopharyngo–hypothalamic–pituitary space with contrast-enhanced MR in preference to M-CT, as both larger and smaller lesions are better defined with MR.

Interpretation

1. TSH-secreting adenomas are usually large tumors at initial presentation with hypodense features compared with normal pituitary tissue.
2. TSH-secreting adenomas tend to be invasive and, rarely, may be ectopic with locations in the nasopharynx or hypothalamus.

Notes

1. In a series of 21 patients, TSH-secreting tumor volumes ranged from 0.42 to 94.2 cm^3, with a mean $\pm$ SD of 16 $\pm$ 18 cm^3 and a tumor invasion score of 4.8 $\pm$ 2.1 out of 9.0. Higher staging scores predicted less favorable responses to surgery (Sarlis et al., 2003).
2. In a series of 43 cases, TSH-secreting pituitary microadenomas, which make up almost one third of cases observed since 1990, tended to be medially located without associated TSH hypersecretion (Socin et al., 2003).
3. ^{111}In-pentetreotide scanning was found to visualize somatostatin receptors in five patients with TSH-secreting pituitary adenomas, confirming their frequent presence in these rare tumors (Losa et al., 1997).
4. Consider the possibility of an ectopic TSH-secreting tumor in cases of inappropriate secretion of TSH with hyperthyroidism but no evidence of pituitary tumor by M-CT and/or MR. Ectopic TSH-secreting pituitary tumors are usually located along the migration path of the Rathke's pouch, usually through the nasal cavity and the nasopharynx.
5. Pituitary radiotherapy and somatostatin analog therapy may allow shrinkage of tumor mass not completely removed at surgery.

ICD-9 Codes

Conditions that may justify this test *include but are not limited to*:

Visual abnormality
368.4X   visual field defects
368.9   visual disturbance, unspecified

Other
227.3   pituitary adenoma
242.9   thyrotoxicosis

Suggested Reading

Losa M, Magnani P, Mortini P, Persani L, Acerno S, Giugni E, Songini C, Fazio F, Beck-Peccoz P, Giovanelli M. Indium-111 pentetreotide single-photon emission tomography in patients with TSH-secreting pituitary adenomas: correlation with the effect of a single administration of octreotide on serum TSH levels. *Eur J Nucl Med* 1997; 24:728–31.

Pasquini E, Faustini-Fustini M, Sciarretta V, Saggese D, Roncaroli F, Serra D, Frank G. Ectopic TSH-secreting pituitary adenoma of the vomerosphenoidal junction. *Eur J Endocrinol* 2003; 148:253–7.

Sarlis NJ, Gourgiotis L, Koch CA, Skarulis MC, Brucker-Davis F, Doppman JL, Oldfield EH, Patronas NJ. MR imaging features of thyrotropin-secreting pituitary adenomas at initial presentation. *Am J Roentgenol* 2003; 181:577–82.

Socin HV, Chanson P, Delemer B, Tabarin A, Rohmer V, Mockel J, Stevenaert A, Beckers A. The changing spectrum of TSH-secreting pituitary adenomas: diagnosis and management in 43 patients. *Eur J Endocrinol* 2003; 148:433–42.

2.9 Prolactotroph (Prolactin) Reserve

2.9.1 Prolactin (PRL) and Growth Hormone (GH) Testing in Lactation Failure in Prolactotroph and Somatotroph Insufficiency

Indications for Test

Measurement of [PRL] and/or [GH] is indicated for:

- Women unable to lactate normally postpartum
- Failure to resume menses after delivery, usually after a severe postpartum hemorrhage
- Enlargement of the pituitary with pregnancy combined with small sella size and postpartum vaginal hemorrhage
- The event of disseminated intravascular coagulation (DIC) or thrombotic thrombocytopenic purpura (TTP) or the presence of an autoimmune syndrome and failure to lactate

Procedure

1. Inquire as to timing of last intercourse in a female subject.
2. Test for [PRL] only if 4 or more hours have elapsed since last coitus and orgasm.
3. Obtain a blood sample for PRL testing from an individual in the fasting, nonstressed state.
4. In the event of complete lactation failure, also measure [GH] and [IGF-1] and image the pituitary, looking for an empty sella or a neoplastic, cystic, or inflammatory lesion.

Interpretation

1. In lactation failure, [PRL] and [GH] are usually undetectable but may fall into the low end of the reference range.
2. Lymphocytic hypophysitis, Sheenan's syndrome, and adult GH deficiency are in the differential diagnosis of lactation failure when the [PRL] is normal or elevated.
3. A postpartum pituitary infarction may impair secretion of one or more pituitary hormones in addition or as an alternative to GH and PRL.

Notes

1. Patients with Sheehan's syndrome develop an abrupt-onset, severe hypopituitarism immediately after delivery of an infant, but most will have partial hypopituitarism and may go undiagnosed and untreated for an extended period.
2. Growth hormone is one of the hormones lost earliest after a severe postpartum hemorrhage and, together with PRL deficiency, contributes to lactation failure.

ICD-9 Codes

Conditions that may justify this test *include but are not limited to*:

253.2	hypopituitarism
446.6	thrombotic microangiopathy
626.0	absence of menstruation
666.1	postpartum hemorrhage
676.4	failure of lactation

Suggested Reading

Kelestimur F. Sheehan's syndrome. *Pituitary* 2003; 6:181–8.

Milewicz A, Bohdanowicz-Pawlak A, Bednarek-Tupikowska G. Importance of prolactin level indication in detection of Sheehan syndrome. *Ginekol Pol* 1992; 63:232–5.

2.9.2　Prolactin (PRL) Testing: Monitoring after Treatment of a Prolactinoma

Indications for Test

Monitoring of [PRL] after treatment of a prolactinoma is indicated when:

- Persistence of galactorrhea and infertility or impotence with decreased libido are noted after surgery or dopamine receptor agonist therapy.
- A pituitary–hypothalamic disorder or anatomic lesion has appeared after interventional therapy for the pituitary tumor.
- A periodic check on the status of putatively effective pituitary tumor therapy is due, usually at a yearly interval.

Procedure

1. Inquire as to timing of last intercourse in a female subject.
2. Test for PRL only if 4 or more hours have elapsed since last coitus and orgasm.
3. Obtain a blood sample for PRL testing from an individual in the fasting, nonstressed state.

Interpretation

1. Reference intervals reported for [PRL] vary widely. Usually, [PRL] in healthy individuals is <15 ng/mL.
2. A normal or elevated [PRL] may occur with adequate or excessive, but benign, PRL production.
3. A [PRL] > 250 ng/mL usually indicates a pituitary prolactinoma; lesser elevations may result from hypothalamic–pituitary stalk compression or damage such as in lymphocytic hypophysitis or granulomatous disease or from an adenoma that primarily secretes GH and causes acromegaly.
4. Diseases such as renal failure, hypothyroidism, cirrhosis, epilepsy, or pseudocyesis may elevate [PRL].
5. Estrogen, antiandrogen, or neuroleptic agent treatment can cause a secondary rise in [PRL].
6. Minor elevations in [PRL] occur commonly as result of common physiologic conditions such as pregnancy, lactation, sleep, postcoital state, or major psychological stress.
7. Mildly elevated [PRL] (15–20 ng/mL) may disrupt ovulation and impair fertility.

Notes

1. There is no established indication for use of a provocative test to stimulate PRL in patients with PRL excess or recurrent prolactinoma. Baseline testing for PRL is sufficient.
2. Prolactinomas may secrete a multimeric macroprolactin form of PRL.

ICD-9 Codes

Conditions that may justify this test *include but are not limited to*:

227.3	pituitary adenoma
302.7	psychosexual dysfunction
606.9	male infertility
676.6	galactorrhea
799.81	decreased libido

Suggested Reading

Gsponer J, De Tribolet N, Deruaz JP, Janzer R, Uske A, Mirimanoff RO, Reymond MJ, Rey F, Temler E, Gaillard RC, Gomez F. Diagnosis, treatment, and outcome of pituitary tumors and other abnormal intrasellar masses: retrospective analysis of 353 patients. *Medicine (Baltimore)* 1999; 78:236–69.

Mounier C, Trouillas J, Claustrat B, Duthel R, Estour B. Macroprolactinaemia associated with prolactin adenoma. *Hum Reprod* 2003; 18:853–7.

2.10 Prolactotroph (Prolactin) Excess

2.10.1 Monomeric Prolactin (PRL) and Multimeric Macroprolactin (MPRL) Testing in Apparent States of Hyperprolactinemia

Indications for Test

Measurement of [PRL] is indicated for:

- Women with amenorrhea, decreased libido, and/or infertility
- Men with decreased libido, infertility, gynecomastia, low testosterone, and/or impotence
- Patients with a pituitary tumor, empty sella, acromegaloid or Cushingoid appearance, headaches, possible cirrhosis, or galactorrhea

Procedure

1. Inquire as to timing of last intercourse in a female subject, nipple stimulation, chest wall trauma, thyroid and renal status.
2. Inquire as to intake of phenothiazines, metoclopramide, estrogens, opiates, methyldopa, and other drugs that are dopamine depleting or dopamine receptor antagonists.
3. Test for PRL only if 4 or more hours have elapsed since last coitus and orgasm.
4. Obtain a blood sample for PRL and TSH testing from an individual in the fasting, nonstressed state. Save any unused serum for pooling or dilution (see Test 2.10.2), as described below.
5. Given that PRL is a pulsatile hormone, ideally obtain blood samples on three separate days (or 3 samples at 30-minute intervals from a resting patient) and pool the serum for PRL testing.
6. If the observed [PRL] is unexpectedly low, given the patient's signs and symptoms (i.e., galactorrhea), evaluate the patient for a falsely low [PRL] due to the high-dose hook effect. Request that the patient's serum sample be retested for [PRL] using a minimum of two carefully selected dilutions of the patient's serum.
7. If unexpectedly low PRL concentrations obtained on the diluted samples are in good agreement, after laboratory personnel have corrected the measured values for the dilution factor, and these values are appreciably higher than the initial value, then the initial value may be discounted as falsely low as a result of the high-dose hook effect.
8. If a measured [PRL] is unexpectedly high, reassay the patient's sample for [PRL] using an immunoassay less affected by [MPRL], or screen for the presence of MPRL in the sample using the polyethylene glycol (PEG) precipitation–RIA method (Cavaco et al., 1999; Smith et al., 2002).

Interpretation

1. The upper limit of the [PRL] reference interval for healthy individuals is typically <15 ng/mL.
2. A prolactinoma cannot be excluded when [PRL] is <100 ng/mL, and idiopathic hyperprolactinemia or macroprolactinemia may be present at a [PRL] >100 ng/mL.
3. A [PRL] of >250 ng/mL usually indicates a pituitary prolactinoma, whereas less elevated values (usually <85 ng/mL but >15 ng/mL) may result from stress, hypothalamic–pituitary stalk compression, or damage such as in lymphocytic hypophysitis or granulomatous disease.
4. Diseases such as renal failure, hypothyroidism, cirrhosis, epilepsy, and pseudocyesis may elevate PRL concentration.
5. Medications that may cause a secondary rise in [PRL] include treatment with estrogens, antiandrogens, or neuroleptic agents such as metoclopramide.
6. The PRL measured by RIA may be up to 30% MPRL such that up to 25% of patients thought to have hyperprolactinemia actually have macroprolactinemia with a much lower [PRL] following treatment of the serum with PEG and repeat immunoassay measurement.
7. In patients not found to have typical symptoms of hyperprolactinemia, MPRL may comprise up to 90% of the PRL detected.

Notes

1. MPRL (or big-big prolactin), as measured by gel chromatography, has reduced bioactivity compared to PRL and may be a significant cause of misdiagnosis, unnecessary investigation, and inappropriate treatment.
2. When macroprolactinemic samples are assayed for [PRL], different assay systems yield results that may vary by two- to eightfold.
3. In studies of different immunoassay methods used to measure [PRL] in macroprolactinemic sera, the effect of MPRL on [PRL] by these methods decreased in the order:
    *   *Study 1.* ADVIA Centaur® < ACS:180™ < ACCESS < IMMULITE® 2000 < ARCHITECT < AxSym® < Immuno 1™ < DELFIA® < Elecsys™
    *   *Study 2.* PROL-CTK < ACS:180™ < DELFIA® = RIAgnost
    Thus, the ADVIA Centaur® and PROL-CTK PRL assays were least affected by the presence of MPRL.
4. Bioassay of macroprolactinemic serum shows more activity than the MPRL extracted by gel chromatography, suggesting that either a change in the structure of MPRL or a removal of substances that potentiate the bioactivity of PRL occurs with extraction.
5. Causes of hyperprolactinemia include physiologic, pharmacologic, and pathologic factors such as pituitary adenoma. Chronic kidney disease and PRL secretion-enhancing drugs such as metoclopramide may elevate [PRL].
6. [PRL] may be mildly elevated (15–50 ng/mL) in 25 to 30% of patients with PCOS.
7. Macroprolactinemia is not known to be a genetically transmitted trait in first-degree relatives.
8. Infertile females with a pituitary tumor and a [PRL] of >20 ng/mL demonstrate a better response to transsphenoidal surgical removal of the tumor than those with lower [PRL].

ICD-9 Codes

Conditions that may justify this test *include but are not limited to*:

227.3	pituitary adenoma
253.1	hyperprolactinemia
302.7	psychosexual dysfunction
606.9	male infertility
611.1	hypertrophy of breast
626.0	absence of menstruation
676.6	galactorrhea
799.81	decreased libido

Suggested Reading

Alfonso A, Rieniets KI, Vigersky RA. Incidence and clinical significance of elevated macroprolactin levels in patients with hyperprolactinemia. *Endocr Pract* 2006; 12:275–80.

Cavaco B, Prazeres S, Santos MA, Sobrinho LG, Leite V. Hyperprolactinemia due to big big prolactin is differently detected by commercially available immunoassays. *J Endocrinol Invest* 1999; 22:203–8.

Ciccarelli E, Camanni F. Diagnosis and drug therapy of prolactinoma. *Drugs* 1996; 51:954–65.

Leite V, Cosby H, Sobrinho LG, Fresnoza MA, Santos MA, Friesen HG. Characterization of big, big prolactin in patients with hyperprolactinaemia. *Clin Endocrinol (Oxf)* 1992; 37:365–72.

Smith TP, Suliman AM, Fahie-Wilson MN, McKenna TJ. Gross variability in the detection of prolactin in sera containing big big prolactin (macroprolactin) by commercial immunoassays. *J Clin Endocrinol Metab* 2002; 87:5410–5.

Suliman AM, Smith TP, Gibney J, McKenna TJ. Frequent misdiagnosis and mismanagement of hyperprolactinemic patients before the introduction of macroprolactin screening: application of a new strict laboratory definition of macroprolactinemia. *Clin Chem* 2003; 49:1504–9.

2.10.2 PRL Testing and the Hook Effect in Extreme Hyperprolactinemia

Indications for Test

Dilution of serum with an elevated [PRL] for detection of the hook effect in extreme hyperprolactine-mia is indicated in:

- Known hypothalamic disease–pituitary tumor patients
- Patients whose serum may contain only a moderately elevated [PRL] but have disproportionately severe clinical symptoms of hyperprolactinemia (e.g., galactorrhea)

Procedure

1. Clearly identify the severity of a patient's signs and symptoms of hyperprolactinemia if any (i.e., perform breast exam and obtain detailed history of menses and recent sexual activity).
2. Using a properly obtained serum sample, perform a 1:1 and a 1:10 dilution of sample with saline before assay of [PRL] in all hypothalamic–pituitary tumor patients.

Interpretation

1. When an analyte such as PRL is present in extremely high levels, a falsely low determination may result unless the sample assayed is diluted. Usually the falsely "low" test result remains above the reference interval for the PRL assay.
2. The actual [PRL] of a sample may be over 100-fold that of the undiluted sample.

Notes

1. The high-dose hook effect occurs when attempting to measure an extremely high level of an analyte such as PRL, without appropriate dilution of the sample, resulting in a falsely low value.
2. Decisions as to response to treatment of moderately elevated [PRL] should be based on diluted samples.

ICD-9 Codes

Refer to Test 2.10.1 codes.

Suggested Reading

Frieze TW, Mong DP, Koops MK. "Hook effect" in prolactinomas: case report and review of literature. *Endocr Pract* 2002; 8:296–303.

2.10.3 Imaging of Prolactin-Secreting Pituitary Tumor by High-Resolution Multislice Magnetic Resonance (MR) and Computed Tomographic (M-CT) Scanning

Indications for Test

M-CT and MR imaging studies to search for prolactin-secreting pituitary tumors may be indicated for:

- Patients when the [PRL] is elevated with or without chronic headache or signs such as visual disturbance suggestive of a mass effect in the pituitary or when acromegaly is suspect
- Males when galactorrhea, decreased libido, gynecomastia, or impotence is noted
- Females when galactorrhea, decreased libido, amenorrhea, or problems with fertility are noted

Procedure

1. Proceed directly to MR imaging of the pituitary, if available, rather than M-CT imaging.
2. Use dynamic, rapid-acquisition pituitary MR imaging during bolus contrast infusion to visualize small microadenomas suspected, but not seen, on M-CT imaging.
3. Upon finding a vascular, rather than solid, lesion, proceed directly to cerebral angiography.

Interpretation

1. Prolactinomas differ greatly in size, degree of PRL production, and aggressiveness.
2. A carotid aneurysm can result in a prolactin level that is >300 ng/mL. Although there is no reliable cut-off [PRL] for distinguishing PRL-secreting adenomas from other causes of elevated [PRL], the great majority of tumors are associated with a [PRL] that is >200 ng/mL.
3. Nonsecreting pituitary adenomas may be associated with elevated [PRL] (up to ~85 ng/mL) if the PRL inhibitory factor pathway from the hypothalamus is interrupted by the tumor mass.

Notes

1. Prolactinoma is the most frequently found pituitary tumor, accounting for approximately 40% of cases in men and women. Microprolactinomas (<10 mm) are found more often in women than in men.
2. Immunostaining for PRL was positive in 24 of 25 adenomas from acromegalic patients with hyperprolactinemia, but no PRL was found in the tumor tissue of 10, matched normoprolactinemic patients with macroadenomas.
3. Malignant prolactinomas are extremely rare events with metastasis to extrasellar sites reported in less than a dozen cases (Popadic et al., 1999).
4. After full-term pregnancy, prolactinomas may be expected to decrease in size in >75% of cases.

ICD-9 Codes

Refer to Test 2.10.1 codes.

Suggested Reading

Gsponer J et al. Diagnosis, treatment, and outcome of pituitary tumors and other abnormal intrasellar masses: retrospective analysis of 353 patients. *Medicine (Baltimore)* 1999; 78:236–69.

Kahn SR, Leblanc R, Sadikot AF, Fantus IG. Marked hyperprolactinemia caused by carotid aneurysm. *Can J Neurol Sci* 1997; 24:64–6.

Maroldo TV, Dillon WP, Wilson CB. Advances in diagnostic techniques of pituitary tumors and prolactinomas. *Curr Opin Oncol* 1992; 4:105–15.

Popadic A, Witzmann A, Buchfelder M, Eiter H, Komminoth P. Malignant prolactinoma: case report and review of the literature. *Surg Neurol* 1999; 51:47–55.

Serri O, Robert F, Comtois R, Jilwan N, Beauregard H, Hardy J, Somma M. Distinctive features of prolactin secretion in acromegalic patients with hyperprolactinaemia. *Clin Endocrinol (Oxf)* 1987; 27:429–36.

2.11   Posterior Pituitary Arginine Vasopressin Reserve

2.11.1 Measurement of Serum and Urine Osmolality Followed by Determination of Plasma or Urine AVP in Diabetes Insipidus (DI)

Indications for Test

Measurement of arginine vasopressin (AVP) and osmolalities in serum and urine are indicated when:

- Polyuria and polydipsia appear in a patient without diabetes mellitus (DM)
- Screening for possible cases of primary polydipsia (PPd), central or neurogenic diabetes insipidus (CDI), and nephrogenic diabetes insipidus (NDI)

TABLE 2.6
Causes of Diabetes Insipidus (DI)

Central/Neurogenic (CDI)	Nephrogenic (NDI)
Vascular—cerebral aneurysms, hemorrhage, or infarction	*Vascular*—sickle cell disease
Gene mutations—X-linked recessive (Xq28); arginine vasopressin (AVP)–neurophysin II abnormality (multiple forms), AVP precursor protein (autosomal recessive and dominant)	*Gene mutations*—X-linked V_2 vasopressin receptor; aquaporin 2 vasopressin-responsive water channel (autosomal dominant and recessive)
Trauma—head injury or any type of neurosurgery	*Kidney*—chronic renal disease; renal washout secondary to polydipsia from psychogenic or dipsogenic (i.e., reset thirst threshold)
Infections—tuberculosis, encephalitis or meningitis, toxoplasmosis; rarely sepsis	*Electrolyte/osmotic imbalances (severe)*—hypokalemia; hypercalcemia; hyperglycemia
Tumors—metastases to hypothalamus, dysgerminoma, craniopharyngioma, suprasellar pituitary macroadenoma	*Drugs*—lithium, demeclocycline, aminoglycosides, cisplatin, rifampin, methoxyflurane
Autoimmune inflammatory—antibodies to vasopressin, lymphocytic infiltration into hypothalamus/pituitary, systemic lupus erythematosus (SLE), Wegener's granulomatosis (WG)	*Infiltrative diseases*—sarcoid, amyloid
Pregnancy-related—postpartum hemorrhage with hypothalamic/pituitary necrosis; placental vasopressinase with increased vasopressin catabolism	

Procedure

1. From a patient who has fasted overnight (i.e., >8 hours), obtain both a first-morning void urine and simultaneous blood sample for urine (U_{osm}) and serum (S_{osm}) osmolality as a screening test.
2. If the U_{osm} is >300 mOsm/kg and the S_{osm} is <310 mOsm/kg, proceed with a water deprivation test (see Test 2.11.2) with measurement of [AVP] to help differentiate among PPd, CDI, and NDI.
3. If the U_{osm} is <300 mOsm/kg and S_{osm} is >310 mOsm/kg, proceed with a therapeutic trial of vasopressin or DDAVP as treatment for presumptive, usually overt, CDI.
4. Plot the plasma [AVP] (linearly on the vertical axis) vs. S_{osm} (linearly on the horizontal axis) and U_{osm} (linearly on the vertical axis) vs. log plasma [AVP] (horizontal axis).

Interpretation

1. Refer to "Causes of Diabetes Insipidus (DI)" (Table 2.6) and Figure 2.2.
2. The inability to concentrate urine (i.e., U_{osm} < 300 mOsm/kg in a first-morning urine void with simultaneous S_{osm} > 310 mOsm/kg) is diagnostic of a severe or complete form of DI, either CDI or NDI, and usually obviates the need for a water deprivation test.
3. CDI is not present in a dehydrated, water-deprived patient who has a proportionally parallel increase in S_{osm} and plasma [AVP].
4. Lesser degrees of DI or partial CDI may occur even in the absence of overt pituitary or neurohypophyseal disease and require water deprivation testing and vasopressin infusion for definition of the severity of the disorder.
5. Refer to Nomogram 7.1 and 7.2 in Dons (1997) for the reference ranges (shaded areas) of plasma [AVP] in pmol/L, U_{osm}, and S_{osm} (Robertson et al., 1988).

Notes

1. Plasma AVP determination without measurement of both serum and urine osmolalities in context of specific interval from last fluid intake is not informative or diagnostic of any one condition. Similarly, determination of osmolalities alone is not sufficient for definitive diagnosis of DI.
2. DI results from disturbances of a complex neurosecretory physiologic pathway involving (in order) central osmoreceptors; vasopressinergic hypothalamic magnocellular nuclei; the median eminence;

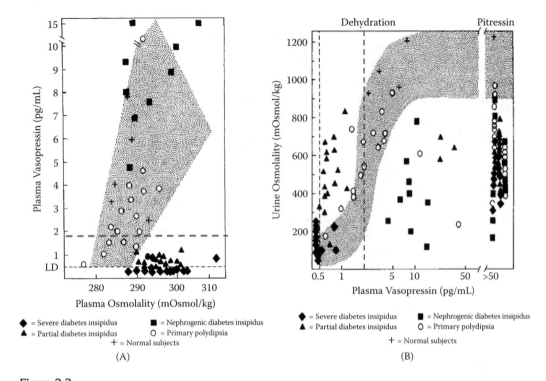

Figure 2.2

Relationship between plasma arginine vasopressin levels, [AVP], and (A) plasma osmolality (P_{osm}) and (B) urine osmolality (U_{osm}) during water deprivation testing. Panel A: [AVP] in patients with severe or partial neurogenic diabetes insipidus (DI) are near to or below the limit of detection (LD; 0.5 pg/mL) of the AVP assay over a broad range of P_{osm}. [AVP] from patients with nephrogenic ($n = 9$) or dipsogenic ($n = 17$) DI are usually within or above the expected range of normal values for P_{osm} (shaded area). Patients with the syndrome of inappropriate antidiuresis (SIAD) have $P_{osm} < 280$ mOsmol/kg and [AVP] ranging from about 2 to 50 pg/mL (data not shown). Panel B: U_{osm} in patients with nephrogenic DI fall uniformly below that expected for [AVP]. In most cases, [AVP] in patients with dipsogenic DI are normal. A few patients with dipsogenic DI have subnormal U_{osm}, probably because of washout of the medullary concentration gradient. Treatment with pitressin results in a markedly diverse range of U_{osm} values in all groups. Dotted lines indicate the ULN for plasma vasopressin concentration ([AVP]; 1.7 pg/mL) for usual P_{osm} of 280 to 292 mOsmol/kg. (From Robertson, G.L., in *Endocrinology and Metabolism*, Felig, P., Baxter, J.D., Broadus, A.E., and Frohman, L.A., Eds., McGraw-Hill, New York, 1987. With permission.)

    pituitary stalk and neurohypophysis, which releases and inactivates AVP; and, finally, the concentrating renal tubules, which respond to AVP.

3. Simultaneous measurement of plasma [AVP] and S_{osm} in a dehydration test is the most powerful diagnostic tool in the differential diagnosis of polyuria/polydipsia.

4. Measurement of urinary [AVP] with commercially available assays has almost the same sensitivity as measurement of plasma [AVP] (Diederich et al., 2001).

5. A plot of [AVP] in urine (pg/mL) vs. U_{osm} with interpretive reference ranges and areas is shown on page 315 of the *Quest Diagnostics Manual* (Fisher, 2007). Note that pg/mL of AVP × 0.926 = pmol/L of AVP.

ICD-9 Codes

Conditions that may justify this test *include but are not limited to*:

253.5	central diabetes insipidus
588.1	nephrogenic diabetes insipidus
783.5	polydipsia
788.42	polyuria

Suggested Reading

Chadha V, Garg U, Alon US. Measurement of urinary concentration: a critical appraisal of methodologies. *Pediatr Nephrol* 2001; 16:374–82.

Diederich S, Eckmanns T, Exner P, Al-Saadi N, Bahr V, Oelkers W. Differential diagnosis of polyuric/polydipsic syndromes with the aid of urinary vasopressin measurement in adults. *Clin Endocrinol (Oxf)* 2001; 54:665–71.

Dons RF, *The Endocrine and Metabolic Testing Manual*, 3rd ed. 1997. CRC Press, pp. 7–32

Fisher DA (Ed). *Quest Diagnostics Manual: Endocrinology Test Selection and Interpretation*, 4th ed. 2007. Quest Diagnostics Nichols Institute.

Robertson GL. Differential diagnosis of polyuria *Ann. Rev. Med* 1988; 39:425–42.

Zerbe RL, Robertson GL. A comparison of plasma vasopressin measurements with a standard indirect test in the differential diagnosis of polyuria. *N Engl J Med* 1981; 305:1539–46.

2.11.2 Water Deprivation Test (WDT) for Adult Diabetes Insipidus (DI)

Indications for Test

The WDT is indicated:

- In cases of polyuria without DM or other overt cause for an osmotic diuresis and polydipsia
- To differentiate among nephrogenic diabetes insipidus (NDI) secondary to renal medullary washout in a nondehydrated patient, central neurogenic DI (CDI), or primary polydipsia (PPd)

Procedure

1. Perform the WDT only in the hospital with close observation of the patient. In extreme cases, shutting off water sources near the patient may be necessary. Do not do this test if the patient is treated with vasopressin (i.e., DDAVP).
2. Instruct the nursing staff to alert the endocrinologist if the patient feels ill, develops orthostatic hypotension, or develops a reduced level of consciousness. As in the case of all pituitary disease patients, test for thyroid and adrenal function. The patient must have normal thyroid and adrenal function using effective replacement therapy if necessary before water deprivation.
3. Coordinate sample handling and stat assays with the laboratory in advance of the test.
4. Initiate the timing of the WDT such that the patient is most likely to become dehydrated between 0800 and 1600 hours when maximum laboratory and nursing support are available.
5. Obtain baseline parameters, including body weight, supine and standing blood pressure and heart rates, and both serum and urine osmolality (S_{osm} and U_{osm}) and a blood sample for sodium [Na^+].
6. Arrange to obtain serial [Na^+] determinations "stat" to assess immediately the presence of and risk for clinically significant dehydration.
7. Obtain blood samples for simultaneous determination of [Na^+] and plasma [AVP] and S_{osm} at baseline, toward the middle, and at the end of the dehydration period.
8. Be prepared to obtain parameters noted above at least every 4 hours. These include body weight, blood pressure, pulse, serum [Na^+], S_{osm}, U_{osm}, and urine volume.
9. At a S_{osm} of 280 mOsm/kg, obtain the above parameters every 2 hours, then hourly when S_{osm} is ~290 mOsm/kg, [Na^+] is >145 mmol/L, and the patient's weight loss is nearing 3%.
10. Be sure to stop the test if hypotension or mental status changes occur, or if the patient's weight falls by >3%. In children, it is customary to stop the test after 7 hours.
11. When U_{osm} increases by <30 mOsm/kg on two (or better, three) consecutive hourly samples or the S_{osm} rises to >300 mOsm/kg with a serum [Na^+] >145 mmol/L, repeat the measurement of all baseline parameters and obtain a blood sample for measurement of the final plasma [AVP].
12. Elect to administer low-dose (5 units) aqueous vasopressin subcutaneously and repeat parameters at 1 and 2 hours after vasopressin injection (VI).

13. Two hours after VI, stop the test and let the patient drink water *ad lib*.
14. Plot the plasma [AVP] (linearly on the vertical axis) vs. S_{osm} (linearly on the horizontal axis) and U_{osm} (linearly on the vertical axis) vs. log plasma [AVP] (horizontal axis).

Interpretation

1. A decline in S_{osm} and/or U_{osm} during the WDT suggests patient noncompliance with fluid restriction.
2. The criterion for severe or complete CDI is a baseline $S_{osm} > 310$ mOsm/kg before vasopressin administration. Do not do a WDT in such a case, as it may be hazardous.
3. Differentiation between types of severe DI based on plasma [AVP] and U_{osm} after vasopressin injection (VI) is made as follows:
   - CDI—[AVP] is undetectable with a change in U_{osm} of >50% after VI.
   - NDI—[AVP] is high (usually >6 pg/mL) with a change in U_{osm} of <9% after VI.
4. Primary polydipsia and partial CDI cannot be distinguished from each other if both the U_{osm} and S_{osm} are >300 mOsm/kg.
5. Diagnosis of partial CDI is based on simultaneous plasma [AVP] and S_{osm} with the [AVP] being low but detectable (usually <1.5 pg/mL), and S_{osm} being >290 mOsm/kg (Robertson et al., 1995).
6. Base the diagnosis of PPd on simultaneous plasma [AVP] and S_{osm} for which the [AVP] is high or normal and all data points lie in or above shaded area of Nomogram 7.1 (Dons, 1997).
7. In patients with NDI, all data points lie to right of the shaded area in Nomogram 7.2 (Dons, 1997) (i.e., [AVP] is disproportionately high in relation to the U_{osm}), whereas in PPd or partial CDI the data points generally lie within the shaded area of the plot but may lie to the right if urine volumes are high.
8. If a diagnosis for polyuria based on the tests above remains equivocal, a hypertonic saline load (HSLT) and/or extended vasopressin infusion test may be helpful; however, these tests are hazardous and rarely necessary.
9. Patients treated with vasopressin or its analogs can develop antivasopressin antibodies that may spuriously alter, and usually lower, the measured plasma [AVP].
10. It is not always possible to differentiate PPd from partial forms of DI based on the results of the WDT alone (Figure 2.2).

Case History 1

(Adapted from Fukagawa et al., 2001.)

History

A 32-year-old female had meningitis at 4 years of age and insertion of a ventriculoperitoneal shunt at age 13 because of normopressure hydrocephalus. Persistent hypernatremia (150–166 mmol/L) was noted since age 14. Pituitary–adrenocortical function was normal.

Test Results

- U_{osm}, S_{osm}, AVP, and urine volume (U_{vol}) with *ad libitum* water drinking (baseline):

Total U_{vol}	750–1700 mL/day
U_{osm}	446–984 mmol/kg
AVP	0.4–1.2 pmol/L (normal)
S_{osm}	298–343 mmol/kg (elevated)

  S_{osm} was positively correlated with U_{osm} ($r = 0.545$; $p < 0.05$) but not with AVP.
- Hypertonic saline load test (HSLT) results:

U_{osm} increase	377 mOsm/kg to 679 mOsm/kg
AVP increase	0.2 pmol/L to a subnormal concentration of 1.3 pmol/L

  S_{osm} and plasma AVP were positively correlated ($r = 0.612$; $p < 0.05$) after saline.
- Water load test (acute, oral water load of 20 mL/kg body weight):

Minimum U_{osm}	710 mOsm/kg (elevated)

  Excretion of the water load reduced to only 8.5%.

Clinical Assessment

Marked hypernatremia as a result of partial CDI characterized by a subnormal increase in [AVP] with an elevated thirst threshold and reduced renal free water clearance permitting maintenance of body water.

Case History 2

(Adapted from Okayasu et al., 1990.)

History

Two brothers presented with fever and vomiting and failure to gain weight.

Test Results in Brothers

- WDT (4-hour):
U_{osm}	<115 mOsm/kg
Body weight change	decline from baseline of 4.6%

- U_{osm} response to vasopressin:
Desamino-8-D-arginine vasopressin (intranasal)	no significant elevation in U_{osm}
Vasopressin injection (low-dose VI)	no significant elevation in U_{osm}

Test Results in Parents

- U_{osm} response to vasopressin:
Mother	subnormal
Father	normal

Clinical Assessment

The brothers had congenital NDI, and their mother carries the genetic abnormality responsible for the insensitivity of their distal nephrons to AVP.

Notes

1. In patients with either CDI or NDI, the WDT is potentially dangerous, particularly if the patient starts out dehydrated.
2. Severe complications of dehydration can occur after greater than 3 to 5% total body water depletion.
3. The HSLT is potentially even more hazardous than the WDT, and its use should be strictly limited primarily to research settings.
4. Subclinical CDI may appear in late pregnancy because of a combination of physiological vasopressinase secretion with reduced AVP secretory capacity and reduction in the thirst threshold that accompanies normal pregnancy.
5. The sensitivity for vasopressin secretion in response to dehydration as well as the osmotic thresholds for thirst and vasopressin demonstrates significant polygenetic variance among healthy adults.
6. Maximal AVP concentrations during the WDT do not show any correlation with age.
7. In a healthy reference population, the minimal [AVP] (~6 pg/mL) at the point of maximal U_{osm} during the HSLT is higher than the [AVP] (2.41 ± 1.37 pg/mL) at maximal U_{osm} during the WDT.
8. Normally, the maximal U_{osm} in the WDT (1040 ± 154 mOsm/kg) is significantly higher than that for the HSLT (713 ± 109 mOsm/kg).
9. A water load suppresses S_{osm}, plasma vasopressin, and U_{osm} regardless of cause for polydipsia.
10. In polydipsic psychiatric patients with hyponatremia, U_{osm} and free water clearance are impaired after water loading, even though plasma [AVP] and solute clearance are similar to those found in nonpolydipsic, eunatremic psychiatric patients.

ICD-9 Codes

Refer to Test 2.11.1 codes.

Suggested Reading

Dons RF (Ed). *The Endocrine and Metabolic Testing Manual*, 3rd ed. 1997. CRC Press, pp. 7–32.

Fukagawa A, Ishikawa SE, Saito T, Kusaka I, Nakamura T, Higashiyama M, Nagasaka S, Kusaka G, Masuzawa T, Saito T. Chronic hypernatremia derived from hypothalamic dysfunction: impaired secretion of arginine vasopressin and enhanced renal water handling. *Endocrinol Jpn* 2001; 48:233–9.

Goldman MB, Luchins DJ, Robertson GL. Mechanisms of altered water metabolism in psychotic patients with polydipsia and hyponatremia. *N Engl J Med* 1988; 318:397–403.

Hasegawa Y. The relationship between antidiuretic hormone and plasma or urine osmolalities during water restriction test and hypertonic saline loading test in normal children: a change in the apparent tubular response to AVP during these two tests. *Endocrinol Jpn* 1991; 38:451–6.

Miller M, Dalakos T, Moses AM, Fellerman H, Streeten DH. Recognition of partial defects in antidiuretic hormone secretion. *Ann Intern Med* 1970; 73:721–9.

Okayasu T, Shigihara K, Kobayashi N, Ishikawa A, Fukushima N, Takase A, Hattori S, Nakajima T, Shishido T, Agatsuma Y. A family case of nephrogenic diabetes insipidus. *Tohoku J Exp Med* 1990; 162:137–45.

Robertson GL. Differential diagnosis of polyuria. *Ann. Rev. Med* 1988; 39:425–42.

Robertson GL. Posterior pituitary. In: *Endocrinology and Metabolism*, Felig P, Baxter J, Frohman L, Eds. 1995. McGraw-Hill, pp. 385–432.

Vokes TJ, Robertson GL. Disorders of antidiuretic hormone. *Endocrinol Metab Clin North Am* 1988; 17:281–99.

Williams DJ, Metcalfe KA, Skingle L, Stock AI, Beedham T, Monson JP. Pathophysiology of transient cranial diabetes insipidus during pregnancy. *Clin Endocrinol (Oxf)* 1993; 38:595–600.

Zerbe RL, Miller JZ, Robertson GL. The reproducibility and heritability of individual differences in osmoregulatory function in normal human subjects. *J Lab Clin Med* 1991; 117:51–9.

2.11.3 Water Deprivation Test (WDT) Using Urinary [AVP] as a Test for Diabetes Insipidus in Pediatric Patients

Indications for Test

The WDT using urinary [AVP] is useful:

- In small children with polyuria where repeated phlebotomy is impractical
- To differentiate between nephrogenic diabetes insipidus (NDI) secondary to renal medullary washout, central neurogenic DI (CDI), or primary polydipsia (PPd) in a nondehydrated pediatric patient

Procedure

1. Collect a series of urine samples after water deprivation during sleep and measure U_{osm} and [AVP] in these specimens.
2. If possible, obtain blood samples for plasma [AVP] at baseline, toward the middle, and at the end of the dehydration period.

Interpretation

1. Normally, U_{osm} and urinary [AVP] are closely correlated ($r = 0.89$; $p < 0.001$).
2. The normal maximal U_{osm} during WDT (1040 ± 154 mOsm/kg) may be age-dependent ($r = 0.52$; $p < 0.05$), although maximal [AVP] during WDT (2.41 ± 1.37 pg/mL) did not correlate with age (Hasegawa et al., 1991).
3. The urinary [AVP] relative to the U_{osm} varies depending on AVP secretory capacity of the pituitary and the ability of the kidney to concentrate urine.
4. Compulsive water drinkers with PPd may not be discriminated from normal using this test.

Notes

1. The standard WDT with measurement of plasma [AVP] is preferred over this test, the use of which was reported in only one study (Shimura et al., 1993).
2. Measurement of [AVP] and U_{osm} alone is too insensitive for differentiating between primary polydipsia and partial or complete CDI in adult patients. (Diederich et al., 2001).

ICD-9 Codes

Refer to Test 2.11.1 codes.

Suggested Reading

Diederich S, Eckmanns T, Exner P, Al-Saadi N, Bahr V, Oelkers W. Differential diagnosis of polyuric/polydipsic syndromes with the aid of urinary vasopressin measurement in adults. *Clin Endocrinol (Oxf)* 2001; 54:665–71.

Hasegawa Y. The relationship between antidiuretic hormone and plasma or urine osmolalities during water restriction test and hypertonic saline loading test in normal children: a change in the apparent tubular response to AVP during these two tests. *Endocrinol Jpn* 1991; 38:451–6.

Shimura N. Urinary arginine vasopressin (AVP) measurement in children: water deprivation test incorporating urinary AVP *Acta Paediatr Jpn* 1993; 35:320–4.

2.11.4 Imaging of the Pituitary and Hypothalamus in Diabetes Insipidus (DI)

Indications for Test

Imaging of the pituitary and hypothalamus is indicated in:

- Polyuria and/or polydipsia without DM or a cause for an osmotic diuresis
- Patients with a family history of polydipsia
- Patients with DI associated with sarcoidosis, metastatic carcinoma, or Langerhans cell histiocytosis

Procedure

1. Obtain T1-weighted magnetic resonance (MR) images of the posterior pituitary in preference to M-CT images.

Interpretation

1. In pituitary diabetes insipidus (CDI), there is the loss of the so-called "bright spot" characteristic of a normal posterior pituitary containing an adequate store of AVP.
2. Displacement of the stalk by an asymmetric pituitary macroadenoma may be associated with erosion of the sella floor.
3. A thickened pituitary stalk is frequently found in patients with DI, prompting evaluation for a variety of neoplastic and inflammatory diseases.
4. Idiopathic autoimmune hypophysitis, lymphocytic-type, most commonly affects the adenohypophysis without DI and less commonly the pituitary stalk with resultant DI.

Notes

1. Deletion of nucleic acid C59 and substitution of A60W in the AVP-NP II precursor is predicted to disrupt one of the seven disulfide bridges required for correct folding of the neurophysin moiety.
2. Disturbances in the function of neurophysin, the AVP transport protein, will lead to a decrease in AVP storage and a reduced density in the posterior pituitary.

3. Primary pituitary tumors such as craniopharyngioma and germ-cell tumors can affect the pituitary stalk and result in DI.
4. Lymphocytic hypophysitis typically presents in young pregnant and postpartum women with symmetric pituitary enlargement and diffuse contrast enhancement extending into the stalk. It may extend into the basal hypothalamus in a tongue-like fashion without sellar erosion.

ICD-9 Codes

Refer to Test 2.11.1 codes.

Suggested Reading

Fluck CE, Deladoey J, Nayak S, Zeller O, Kopp P, Mullis PE. Autosomal dominant neurohypophyseal diabetes insipidus in a Swiss family, caused by a novel mutation (C59Delta/A60W) in the neurophysin moiety of prepro-vasopressin-neurophysin II (AVP-NP II). *Eur J Endocrinol* 2001; 145:439–44.

2.12　Posterior Pituitary (Arginine Vasopressin) Excess

2.12.1　Sodium [Na⁺] Testing in Syndrome of Inappropriate Antidiuresis (SIAD)

Indications for Test

Measurement of [Na^+] is indicated:

- In patients with central nervous system (CNS) symptoms, including lethargy, apathy, confusion, dementia, disorientation, agitation, depression, or acute psychosis
- As routine screening in a wide variety of pulmonary and CNS conditions, including infections, inflammatory and demyelinating diseases, stroke, trauma, artificial ventilation, hypoadrenal, neoplastic (both intra- and extrathoracic), or paraneoplastic syndromes
- In patients with one of the porphyria syndromes (e.g., acute intermittent, variegate, coproporphyria)

Procedure

1. Obtain a blood sample for [Na^+] testing in a patient who has been drinking fluids *ad libitum* and has fasted overnight, but not necessarily kept n.p.o.
2. Differentiate patients with non-euvolemic hyponatremia from euvolemic hyponatremia based on history and measurement of a random urine [Na^+], thyroid, adrenal, and kidney function.
3. Monitor blood [Na^+] closely in patients during the first 2 weeks after transsphenoidal surgery, along with provision of dietary sodium supplements, mild fluid restriction, and observation for symptoms of hyponatremia.
4. Differentiate mineralocorticoid-responsive hyponatremia from SIAD secondary to AVP excess in elderly, hyponatremic subjects by a therapeutic trial of oral fludrocortisone acetate (Fluorinef®) (Ishikawa et al., 2001).

Interpretation

1. The reference interval for [Na^+] in a healthy population is 135 to 145 mmol/L.
2. Refer to "Causes of SIAD-Associated Hyponatremia" (Table 2.7) for identification of the variety of factors and conditions that may underlie the euvolemic hyponatremia of SIAD usually associated with a urine [Na^+] > 20 mmol/L.

TABLE 2.7
Causes of Syndrome of Inappropriate
Antidiuresis (SIAD)-Associated Hyponatremia

Drug therapy

Nonsteroidal antiinflammatory drugs (NSAIDs), prostaglandin synthesis inhibitors, opiates

Selective serotonin reuptake inhibitors (SSRIs), tricyclic antidepressants, phenothiazines, carbamazepine

Chlorpropamide, desmopressin, oxytocin

Cyclophosphamide, vincristine

Neoplasia

Lung cancer

Mediastinal chest tumors

Tumors outside the chest

Central nervous system disorders

Acute psychoses, senile cerebral atrophy

Brain/pituitary tumors with mass effect

Cerebral ischemia or hemorrhage

Closed or open head trauma

Encephalitis, meningitis, or inflammatory/demyelinating diseases

Pulmonary problems

Pneumonias

Acute respiratory failure

Positive-pressure ventilation

Miscellaneous

Severe nausea upon prolonged exercise

Postoperative state

Human immunodeficiency virus (HIV) infection

3. Marked hyponatremia may appear as a result of either primary AVP excess, causing classical SIAD, or excess AVP secondary to stimulation from baroreceptor activation as occurs in liver cirrhosis, cardiac failure, and volume contraction with circulatory compromise.

4. Patterns of plasma [AVP] response to an osmotic stimulus vary, but plasma [AVP] neither increases significantly nor become undetectable with normal changes in S_{osm}.

5. An appropriate increase in [AVP] occurs to compensate for any significant (>7%) loss of body fluid in elderly patients with primary adrenal insufficiency and mineralocorticoid-responsive hyponatremia.

Notes

1. Hyponatremia may be associated with nonspecific muscle cramps, anorexia, nausea, and weakness.

2. Hypovolemic hyponatremia is the result of extrarenal losses of Na^+ from diarrhea, vomiting, gastrointestinal suction, surgical drainage tubes, draining fistulas, burns, pancreatitis, or bowel obstruction in which the urine $[Na^+]$ is <10 mmol/L. It may also be caused by renal salt wasting (i.e., diuretics, mineralocorticoid deficiency) in which the urine $[Na^+]$ is >20 mmol/L.

3. Hypervolemic hyponatremia is marked by edema. When this is associated with renal failure, the urine $[Na^+]$ is >20 mmol/L. When associated with congestive heart failure, nephrotic syndrome, or liver cirrhosis, the urine $[Na^+]$ is <10 mmol/L.

4. SIAD secondary to paraneoplasia is most frequently associated with small-cell lung cancer (SCLC) even with a normal chest x-ray. Use of a chest M-CT scan and bronchoscopy may detect SCLC at an early stage in the hyponatremic patient (Gunther et al., 2003).

5. Unusual cases of inappropriate antidiuresis in the absence of excess AVP secretion have been found secondary to a mass effect from a nonsecreting pituitary tumor.

6. Hyponatremic, salt-depleted patients with urinary-to-plasma creatinine ratio values that are >140, fractional sodium excretion (FENa) < 0.15%, and fractional urea excretion (FEurea) < 45% do not have SIAD. In SIAD patients, the FENa falls to <0.5%, when the urinary to plasma creatinine ratio exceeds 180.

7. In SIAD, low levels of serum creatinine and urea with hypouricemia, hypokalemia, hyponatremia, hypomagnesemia, hypophosphatemia, and hypercalciuria occur because of the associated volume expanded state. All of these conditions tend to resolve with water restriction.

8. Within a week of pituitary surgery, expect about 25% of patients to develop hyponatremia rather than dehydration from CDI. Lack of AVP suppression occurs after water-load testing in hyponatremic pituitary surgery patients, but only one third of normonatremic patients will excrete the water load and suppress AVP normally.

9. In severe hyponatremia (serum [Na$^+$] < 115 mmol/L), its rate of correction should not exceed 1 mmol/L/hr with a short-term, corrected serum [Na$^+$] target of ≤130 mmol/L.

10. Inappropriate secretion of AVP in patients with SIAD and hypopituitarism persists while urinary excretion of aquaporin-2 (AQP-2) and renal water excretion normalize on replacement with hydrocortisone for secondary hypoadrenalism. This indicates that urinary AQP-2 may be a more sensitive measure of AVP effects on renal collecting tubules than plasma [AVP] themselves (Ishikawa et al., 2001).

11. Hypersecretion of AVP may occur with hyponatremia of adrenal insufficiency, but the increase in urine concentration in such cases may reflect a variety of AVP-dependent or -independent phenomena.

12. Rarely, a congenital gain-of-function mutation causing constitutive activation of the V2 AVP receptor may occur resulting in a recognized (Feldman et al., 2005) hyponatremic syndrome termed *nephrogenic syndrome of inappropriate antidiuresis*.

13. Excessive sun exposure in patients with porphyria may trigger a hyponatremic episode.

ICD-9 Codes

Conditions that may justify this test *include but are not limited to*:

006.5	amebic brain abscess
255.4	corticoadrenal insufficiency
277.1	porphyria
290	dementias
311	depressive disorder, not elsewhere classified
340	multiple sclerosis
436	stroke (cerebrovascular)
518.81	acute respiratory failure
780.7	malaise and fatigue
959.01	head injury, unspecified

Suggested Reading

Feldman BJ, Rosenthal SM, Vargas GA, Fenwick RG, Huang EA, Matsuda-Abedini M, Lustig RH, Mathias RS, Portale AA, Miller WL, Gitelman SE. Nephrogenic syndrome of inappropriate antidiuresis. *N Engl J Med* 2005; 352:1884–90.

Gross P, Wehrle R, Bussemaker E. Hyponatremia: pathophysiology, differential diagnosis and new aspects of treatment. *Clin Nephrol* 1996; 46:273–6.

Gunther A, Rauch M, Krumpelmann U, Driessen M. Hyponatraemic delirium as an early symptom of small-cell bronchial carcinoma. *Nervenarzt* 2003; 74:1016–9.

Hung SC, Wen YK, Ng YY, Yang WC. Inappropriate antidiuresis associated with pituitary adenoma: mechanisms not involving inappropriate secretion of vasopressin. *Clin Nephrol* 2000; 54:157–60.

Ishikawa S, Saito T, Fukagawa A, Higashiyama M, Nakamura T, Kusaka I, Nagasaka S, Honda K, Saito T. Close association of urinary excretion of aquaporin-2 with appropriate and inappropriate arginine vasopressin-dependent antidiuresis in hyponatremia in elderly subjects. *J Clin Endocrinol Metab* 2001; 86:1665–71.

Iwasaki Y, Kondo K, Hasegawa H, Oiso Y. Osmoregulation of plasma vasopressin in three cases with adrenal insufficiency of diverse etiologies. *Horm Res* 1997; 47:38–44.

Kamoi K, Toyama M, Takagi M, Koizumi T, Niishiyama K, Takahashi K, Sasaki H, Muto T. Osmoregulation of vasopressin secretion in patients with the syndrome of inappropriate antidiuresis associated with central nervous system disorders. *Endocr J* 1999; 46:269–77.

Musch W, Hedeshi A, Decaux G. Low sodium excretion in SIADH patients with low diuresis. *Nephron Physiol* 2004; 96:P11–8.

Olson BR, Gumowski J, Rubino D, Oldfield EH. Pathophysiology of hyponatremia after transsphenoidal pituitary surgery. *J Neurosurg* 1997; 87:499–507.

Rizos E, Liamis G, Elisaf M. Multiple metabolic abnormalities in a patient with the syndrome of inappropriate antidiuresis. *Nephron* 2002; 91:339–40.

2.12.2 Arginine Vasopressin (AVP) with Serum and Urine Osmolality Tests for Syndrome of Inappropriate Anti-Diuresis (SIAD)

Indications for Test

Tests for AVP, along with urine and serum osmolality, may be indicated:

- In patients fulfilling all of the following five criteria: (1) hypotonic hyponatremia, (2) natriuresis, (3) urine osmolality (U_{osm}) greater than serum osmolality (S_{osm}), (4) absence of edema and volume depletion, and (5) normal renal and adrenal function
- When other physiologic stimuli to free water clearance blockade have been excluded, including volume depletion/dehydration, postoperative pain, hypothyroidism, congestive heart failure, hepatic failure, nausea, and emesis
- In severely hyponatremic patients not treated with pharmacological agents that interfere with free water clearance, including NSAIDs, phenothiazines, chlorpropamide, carbamazepine, nicotine, opiates, vincristine, cyclophosphamide, clofibrate, selective serotonin reuptake inhibitors, tricyclic antidepressants, or high-dose insulin for uncontrolled diabetes
- In hyponatremic patients with neoplasia, neurological diseases, and/or lung diseases

Procedure

1. In an established euglycemic, euadrenal hyponatremic patient, obtain urine and blood samples for simultaneous determination of U_{osm} and S_{osm}.
2. Measure the plasma [AVP] when true hyponatremia or hyposmolality, without hyperglycemia, is established and urine is found to be more concentrated than serum.
3. Do not measure [AVP] in a patient with pituitary macroadenoma, craniopharyngioma, or a large aneurysm affecting the sella or empty sella, as central hypoadrenalism (CH) may be present.
4. Perform a low-dose ACTH test followed by appropriate imaging studies in CH patients rather than measuring [AVP].

Interpretation

1. Refer to Nomograms 7.1 and 7.2 in Dons (1997) for the reference ranges (shaded areas) of plasma [AVP] in pmol/L, U_{osm} and S_{osm} (Robertson et al., 1988).
2. The plot of plasma [AVP] (pg/mL) vs. S_{osm} on page 314 of the *Quest Diagnostics Manual* (Fisher, 2007) shows the interpretive reference ranges and areas consistent with SIAD. Refer to Figure 2.2A.

3. Abnormal patterns of AVP secretion result from:
   - Erratic AVP release
   - Resetting of the CNS osmostat
   - Persistent AVP release at low plasma osmolality
4. If true hyponatremia and inappropriately concentrated urine for plasma are found, even if urine is not hypertonic, a diagnosis of SIAD can be made by exclusion and measurement of plasma [AVP] is unnecessary.
5. Extremely high plasma [AVP] occurs in cases of small-cell carcinoma of the lung, reflecting the ectopic production of this neurohormone. AVP may be a useful tumor marker.

Notes

1. SIAD is usually a clinical diagnosis not requiring [AVP] measurement.
2. In the majority of cases of hyponatremia, non-neoplastic thoracic or intracranial disturbances or a variety of medications can trigger the secretion of excess AVP originating from the hypothalamic–posterior pituitary axis.
3. SIAD is associated with strikingly low plasma concentrations of uric acid, creatinine, and urea nitrogen. In most patients, the blood pressure is normal and there is no edema.
4. In the hyponatremia of liver cirrhosis or heart failure, the blood pressure tends to be low and the patient is edematous.
5. A local pituitary tumor may cause exaggerated secretion of AVP.
6. Interleukin-6 (IL-6) secreted during an aseptic inflammatory state, such as head trauma, is quantitatively correlated with [AVP], suggesting that this cytokine is directly or indirectly involved in the pathogenesis of SIAD.
7. Hyponatremia in central hypoadrenalism is promptly corrected by glucocorticosteroid replacement. Thus, it is important to ascertain adrenal status before exploring other causes of hyponatremia.

ICD-9 Codes

Conditions that may justify this test *include but are not limited to*:

199.1	neoplasm of lung
276.1	hyposmolality and/or hyponatremia
791.9	nonspecific findings on examination of urine

Suggested Reading

Baylis PH. The syndrome of inappropriate antidiuretic hormone secretion. *Int J Biochem Cell Biol* 2003; 35:1495–9.

Dons RF (Ed). *The Endocrine and Metabolic Testing Manual*, 3rd ed. 1997. CRC Press, pp. 7–32.

Fisher DA (Ed). *Quest Diagnostics Manual: Endocrinology Test Selection and Interpretation*, 4th ed. 2007. Quest Diagnostics Nichols Institute.

Gionis D, Ilias I, Moustaki M, Mantzos E, Papadatos I, Koutras DA, Mastorakos G. Hypothalamic–pituitary–adrenal axis and interleukin-6 activity in children with head trauma and syndrome of inappropriate secretion of antidiuretic hormone. *J Pediatr Endocrinol Metab* 2003; 16:49–54.

Kanda M, Omori Y, Shinoda S, Yamauchi T, Tamemoto H, Kawakami M, Ishikawa SE. SIADH closely associated with non-functioning pituitary adenoma. *Endocr J* 2004; 51:435–8.

Olchovsky D, Ezra D, Vered I, Hadani M, Shimon I. Symptomatic hyponatremia as a presenting sign of hypothalamic–pituitary disease: a syndrome of inappropriate secretion of antidiuretic hormone (SIADH)-like glucocorticosteroid responsive condition. *J Endocrinol Invest* 2005; 28:151–6.

Robertson GL. Differential diagnosis of polyuria. *Ann. Rev. Med* 1988; 39:425–42.

2.13 Detection of the Leakage of Cerebrospinal Fluid (CSF) in Pituitary Disease or after Transsphenoidal Surgery

2.13.1 Beta-2 Transferrin and Beta-2 Microglobulin Testing in Nasal Secretions of Patients with a Possible CSF Leak

Indications for Test

Measurement of beta-2 transferrin in nasal fluid is indicated in:

- Any patient with unexplained, persistent rhinorrhea associated with a pituitary macroadenoma or status after pituitary surgery

Procedure

1. Collect a fluid sample from the nose of patients with suspected CSF leak and assay immediately for beta-2 transferrin, preferably with an immunofixation electrophoresis technique. If necessary, store the sample frozen prior to testing.
2. In addition, collect a blood sample for simultaneous testing of both the patient's serum and nasal fluid for presence of beta-2 transferrin.
3. As an alternative to beta-2 transferrin testing, measure nasal fluid beta-2 microglobulin concentration by enzyme-linked immunosorbent assay (ELISA) in patients without neuroinflammatory disorders, if this test is more readily available than beta-2 transferrin.

Interpretation

1. The test for beta-2 transferrin in nasal fluid will be positive (i.e., a band consistent with beta-2 transferrin is present in the nasal fluid but not in the patient's serum) if the fluid contains CSF, but not in normal or hemorrhagic nasal secretions (Figure 2.3).

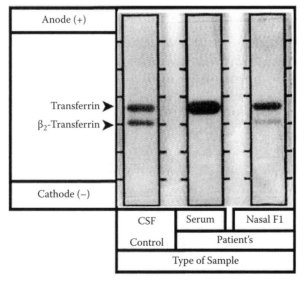

Figure 2.3
Beta-2 transferrin by immunofixation electrophoresis (IFE) in the diagnosis of a cerebrospinal fluid (CSF) leak after transsphenoidal surgery (TSS). Nasal Fl, nasal fluid obtained from rhinorrheal secretions emanating from the nose of a patient with a persistent defect in the cribiform plate after TSS. (Figure courtesy of Dr. Frank H. Wians, Jr.)

2. Beta-2 microglobulin may be present in hemorrhagic nasal secretions or in patients with inflammatory disorders without CSF leaks; thus, this test is of limited use in patients with any leakage of blood into the nasal cavity.

Notes

1. CSF and serum beta-2 microglobulin are elevated in patients with neuro-Behçet's syndrome compared to control patients and may decrease with successful treatment of this and other neuroinflammatory disorders such as multiple sclerosis and CNS multiple myeloma.
2. Patients with idiopathic normal-pressure hydrocephalus showed significantly lower CSF beta-2 trans-ferrin concentration (17.5 mg/L) than patients with Alzheimer's disease (23.8 mg/L) or depression (24.2 mg/L) or normal control subjects (25.3 mg/L).

ICD-9 Codes

Conditions that may justify this test *include but are not limited to*:

227.3   pituitary adenoma
349.81   cerebrospinal fluid rhinorrhea

Suggested Reading

Bateman N, Jones NS. Rhinorrhoea feigning cerebrospinal fluid leak: nine illustrative cases. *J Laryngol Otol* 2000; 114:462–4.

Chan DT, Poon WS, IP CP, Chiu PW, Goh KY. How useful is glucose detection in diagnosing cerebrospinal fluid leak? The rational use of CT and beta-2 transferrin assay in detection of cerebrospinal fluid fistula. *Asian J Surg* 2004; 27:39–42.

Kawai M, Hirohata S. Cerebrospinal fluid beta (2)-microglobulin in neuro-Behçet's syndrome. *J Neurol Sci* 2000; 179(Suppl 1–2):132–9.

Rice DH. Cerebrospinal fluid rhinorrhea: diagnosis and treatment. *Curr Opin Otolaryngol Head Neck Surg* 2003; 11:19–22.

2.13.2 Imaging Studies to Localize Site of CSF Leakage Following Pituitary Surgery

Indications for Test

Imaging of the sphenoid sinuses and base of skull is indicated in:

• Any patient with proven CSF rhinorrhea associated with a pituitary macroadenoma or after pituitary surgery

Procedure

1. Obtain a positive test for beta-2 transferrin in nasal secretions of the patient with pituitary disease or head trauma before imaging the nasopharyngeal sphenoid sinus.
2. Use M-CT rather than MR to identify disruptions of the bony floor of the sphenoid.
3. When obtaining M-CT images, get high-resolution isoviews of the skull base and sphenoids to assess for evidence of basilar skull fracture or other bony defects separate from the sphenoid floor.

Interpretation

1. Discontinuity in course of sphenoid bone marks defect and source of leakage related to pituitary tumor or pituitary surgery.

Notes

1. Lumbar puncture with infusion of fluorescein may be useful in establishing the exact site of an intracranial CSF leak. This procedure can confirm the absence of a leak when clinical suspicion of a leak is high, but beta-2 transferrin is negative.
2. Transnasal endoscopy can be used to repair anterior cranial and sphenoid leaks, but not if the defects are found in the posterior wall of the frontal sinus or when the defect is larger than 5 cm.

ICD-9 Codes

Conditions that may justify this test *include but are not limited to*:

227.5   pituitary adenoma
349.81   cerebrospinal fluid rhinorrhea

Suggested Reading

Marshall AH, Jones NS, Robertson IJ. CSF rhinorrhoea: the place of endoscopic sinus surgery. *Br J Neurosurg* 2001; 15:8–12.

Pancreas Gland Testing (Hypoglycemia and Diabetes)

3.1 Tests for Nonfasting or Random Hypoglycemia and Fasting Hypoglycemia

3.1.1 Random or Casual Plasma Glucose (RPG or CPG) Concentration in Reactive Hypoglycemia and Neuroglycopenia

Indications for Test

A RPG is indicated when a patient experiences symptoms of:

- Symptoms of hypoglycemia, including dizziness, tachycardia, diaphoresis, anxiety, hunger, or feeling cold or weak
- Neuroglycopenia, including confusion, loss of consciousness, or seizures

Procedure

1. Obtain a blood sample for plasma (preferably, using a collection tube containing sodium fluoride anticoagulant to inhibit glycolysis) or peripheral capillary (i.e., fingerstick) blood glucose [BG] testing when a patient, particularly a diabetic, experiences symptoms or signs of hypoglycemia such as those noted above.
2. Depending on circumstances, but particularly after an insulin overdose, repeat testing for blood glucose every 15 to 30 minutes until the patient exhibits a sustained response to treatment with food intake or glucose administration.
3. Obtain a blood sample for serum glucose testing and other metabolites to screen for hypoglycemia or other causes of nonspecific symptoms suggestive of hypoglycemia, recognizing the risk of a falsely low glucose in serum obtained from clotted blood samples left too long at room or refrigerator temperature prior to separation of the serum.

Interpretation

1. The lower limits of normal for RPG vary with gender. For males, the lower limit rarely falls below 50 mg/dL (2.8 mmol/L). For females, plasma glucose of 45 mg/dL (2.5 mmol/L) may occur normally after an overnight fast, particularly in pregnancy. In general, a RPG of ≥62 mg/dL is not associated with any symptoms of hypoglycemia.

2. Female patients may drop their glucose levels into the 20- to 30-mg/dL (1.1–1.7 mmol/L) range without hypoglycemic symptoms after prolonged fasting.

3. The rate of decline, rather than the absolute glucose nadir, is the trigger for symptoms of hypoglycemia.

4. The presence of Whipple's triad verifies an episode of severe hypoglycemia:
   • Neuroglycopenic symptoms
   • Blood glucose ≤ 40 mg/dL (plasma glucose < 50 mg/dL) at time of symptoms
   • Resolution of symptoms by carbohydrate administration

5. Causes of hypoglycemia include medications, usually insulin or sulfonylureas; toxins capable of decreasing blood glucose; metabolic or neoplastic disorders associated with fasting hypoglycemia; and postprandial hypoglycemic disorders.

6. Insulinoma may be the most common of all the hormone-secreting islet cell tumors but is a very rare cause of hypoglycemia.

7. Postprandial glucagon-like peptide 1 (GLP-1)-induced insulin release and inhibition of pancreatic glucagon both contribute to the reactive hypoglycemia encountered in some patients following gastrectomy. Rapid emptying of the reconstructed gastric pouch seems to be one causative factor for the exaggerated GLP-1 release in these subjects.

Notes

1. The blood glucose concentration as measured by some hand-held devices used by diabetes patients is lower than the plasma or serum glucose concentration. Multiply by a constant factor of 1.11 to convert blood glucose to an estimate of plasma glucose.

2. Because consumption of glucose by metabolically active white or younger red cells may occur in blood samples from which the serum or plasma has not been removed promptly, spuriously low plasma glucose levels (pseudohypoglycemia) may be seen in patients with reticulocytosis, leukemia, or other conditions that result in increased levels of young red blood cells (RBCs) or a markedly elevated white cell count.

3. In emergency settings, hastily obtained fingerstick blood samples, particularly when assayed with hand-held glucose monitoring devices using glucose oxidase test methodology, may lead to inaccurate (usually low) glucose test results.

4. Ideally, blood samples for glucose testing should be collected in tubes containing sodium fluoride anticoagulant to inhibit glycolysis and immediately centrifuged to separate the plasma from both red and white cells.

5. The plasma glucose at which there is activation of an adrenergic response, as marked by the sudden and rapid increase in heart rate and diaphoresis, is <18 mg/dL ($<1.0 \pm 0.1$ mmol/L) in type 1 diabetics vs. < 25 mg/dL (1.4 ± 0.1 mmol/L) ($p < 0.05$) in a nondiabetic control group (Hepburn et al., 1991).

6. In normal subjects, an episode of prolonged, moderate, hypoglycemia substantially blunts the neuroendocrine and symptomatic responses to hypoglycemia induced a few hours later.

7. In recent and repeated episodes of hypoglycemia, posthypoglycemic insulin resistance develops and can compensate for the resultant blunted adrenergic neuroendocrine responses and protect the patient from severe neuroglycopenia by limiting glucose flux and glucose oxidation.

ICD-9 Codes

Conditions that may justify this test *include but are not limited to*:

Mentation disorders
780.4   dizziness/giddiness
780.93   memory loss

Other
250.X   diabetes mellitus
251.2   hypoglycemia, unspecified
785.0   tachycardia, unspecified

Suggested Reading

Bartlett D. Confusion, somnolence, seizures, tachycardia? Question drug-induced hypoglycemia. *J Emerg Nurs* 2005; 31:206–8.

Davis SN, Tate D. Effects of morning hypoglycemia on neuroendocrine and metabolic responses to subsequent afternoon hypoglycemia in normal man. *J Clin Endocrinol Metab* 2001; 86:2043–50.

Hepburn DA, Patrick AW, Brash HM, Thomson I, Frier BM. Hypoglycaemia unawareness in type 1 diabetes: a lower plasma glucose is required to stimulate sympatho–adrenal activation. *Diabet Med* 1991; 8:934–45.

Miholic J, Orskov C, Holst JJ, Kotzerke J, Meyer HJ. Emptying of the gastric substitute, glucagon-like peptide-1 (GLP-1), and reactive hypoglycemia after total gastrectomy. *Dig Dis Sci* 1991; 36:1361–70.

Pourmotabbed G, Kitabchi AE. Hypoglycemia. *Obstet Gynecol Clin North Am* 2001; 28:383–400.

3.1.2　Mixed Meal Tolerance Test in Idiopathic or Postprandial Reactive Hypoglycemia and Postprandial Syndrome

Indications for Test

A mixed meal tolerance test is indicated:

- When adrenergic symptoms (e.g., diaphoresis, anxiety, hunger with food cravings) occur after ingestion of larger carbohydrate loads, usually associated with a postprandial glucose of <60 mg/dL (<3.34 mmol/L)
- In patients with hypopituitarism who experience hypoglycemia with few or no warning symptoms or gastric bypass with possible "dumping" syndrome

Procedure

1. Obtain the patient's history of any previous gastrointestinal surgery, panic attacks, or anxiety disorders.
2. Prepare the patient in the same way as for the 100-g oral glucose tolerance test (oGTT) (Test 3.5.3). Use a food log and patient recall to document that the patient has ingested >150 g carbohydrates per day for the 3 days prior to testing and has not engaged in weight-loss dieting in the past 7 days.
3. Following an 8- to 14-hour overnight fast, insert the sampling catheter, wait 5 to 10 minutes, and obtain a baseline blood sample for plasma glucose testing and, if necessary, a second tube of blood for later measurement of insulin, glucagon, cortisol, and growth hormone (GH) concentrations.
4. Instruct the patient to ingest, in less than 5 minutes, 70 to 80 g carbohydrates as a commercial, high-fiber drink such as Boost® or Ensure® Fiber with FOS having 500 calories (usually 2 cans) distributed as approximately 34 g simple sugars, 6 g dietary fiber, 2 g soluble fiber, 15 to 16 g fat, and 18 g protein.
5. Obtain a minimum of six fingerstick blood samples for glucose testing at 0 (baseline), 30, 60, 90, 120, and 180 minutes after ingestion of the "meal." Note that this testing can be done by the patient with minimal supervision or availability of special assistance.
6. Obtain additional fingerstick blood samples for glucose at 240 and 300 minutes if symptoms of hypoglycemia are noted by the observer or reported by the patient during the first 180 minutes of the test.
7. Instruct the patient to keep a log of symptoms occurring during the test, noting time of onset for each symptom.
8. If hypoglycemia is documented, usually based on a fingerstick blood glucose test, obtain insulin, epinephrine, and cortisol levels from two successive blood samples within the hour following the onset of symptoms to identify the cause of the hypoglycemia.

Interpretation

1. A confirmed diagnosis of postprandial hypoglycemia is made when a plasma glucose of <40 mg/dL (2.2 mmol/L) is documented; when adrenergic or, on rare occasions, neuroglycopenic symptoms consistent with hypoglycemia occur; and when symptoms are relieved upon achieving a glucose > 62 mg/dL (> 3.47 mmol/L).
2. Rapid gastric emptying, secondary to dumping syndrome, may explain postprandial hypoglycemia, and performance of a gastric emptying rate study (Test 5.3.3) may be indicated.
3. Glucagon receptor downregulation and impaired glucagon sensitivity and secretion may contribute to postprandial hypoglycemia but are rarely the sole cause of this condition.
4. Over half of the cases of postprandial hypoglycemia may result from increased sensitivity to insulin accompanied by persistence of high insulin levels, usually occurring in very lean subjects, in lean patients with polycystic ovary disease, or after massive weight loss as occurs after gastric exclusion surgery.
5. Causes of postprandial hypoglycemia, other than rapid gastric emptying, include:
   - An exaggerated and delayed insulin secretory response secondary to insulin resistance or hyper-secretion of glucagon-like peptide 1 (GLP-1)
   - Carbohydrate "drain" or deficit from renal glycosuria
6. A doubling of the serum cortisol concentration at 30 to 90 minutes after a fall in plasma glucose suggests activation of the hypothalamic–pituitary–adrenal axis and helps to confirm the diagnosis of a more severe hypoglycemic episode.
7. Up to 30% of patients with documented insulinoma present with postprandial symptoms only.

Notes

1. Because of the high rate of false positives with the 75-g or 100-g oGTT, a mixed meal tolerance test is a more appropriate way to establish the diagnosis of reactive hypoglycemia.
2. If possible, have the patient keep a diary of food and drink with correlation of any symptoms suggestive of low blood sugar.
3. Patients with meal-related hypoglycemic disorders may have symptoms that result from ingestion of excessive quantities of refined carbohydrate. To determine the clinical relevance of such a patient's symptoms, documentation of hypoglycemia in the home setting with a food log and measurement of blood glucose during a symptomatic episode after a carbohydrate-loaded meal are required. Standardization of such a meal has been described (Brun et al., 1995).
4. Reactive hypoglycemia may occur in patients with prediabetes (reactive hypoglycemia of glucose intolerance), gastrointestinal dysfunction (alimentary reactive hypoglycemia or gastric dumping after weight loss surgery or possibly secondary to GLP-1 deficiency), or deficiency of counter-regulatory hormones, including glucagon and GH.
5. Patients with true postprandial hypoglycemia are frequently confused with patients having an under-lying psychiatric illness or idiopathic postprandial syndrome. Patients with either true postprandial hypoglycemia or idiopathic postprandial symptoms of hypoglycemia have similar adrenergic-medi-ated responses and personality traits, as noted on Minnesota Multiphasic Personality Inventory (MMPI) testing.
6. During an oGTT, patients with idiopathic postprandial symptoms of hypoglycemia have a significantly lower mean glucose nadir, and plasma epinephrine, cortisol, GH, glucagon, and norepinephrine responses are significantly higher than controls. Only the epinephrine hyper-response was found to separate patients with idiopathic postprandial symptoms of hypoglycemia from controls without overlap (Chalew et al., 1984).

ICD-9 Codes

Conditions that may justify this test *include but are not limited to*:

251.2   hypoglycemia, unspecified
253.2   hypopituitarism
790.29   other abnormality of glucose tolerance

Suggested Reading

Ahmadpour S, Kabadi UM. Pancreatic alpha-cell function in idiopathic reactive hypoglycemia. *Metabolism* 1997; 46:639–43.

Altuntas Y, Bilir M, Ucak S, Gundogdu S. Reactive hypoglycemia in lean young women with PCOS and correlations with insulin sensitivity and with beta cell function. *Eur J Obstet Gynecol Reprod Biol* 2005; 119:198–205.

Brun JF, Fedou C, Bouix O, Raynaud E, Orsetti A. Evaluation of a standardized hyperglucidic breakfast test in postprandial reactive hypoglycaemia. *Diabetologia* 1995; 38:494–501.

Brun JF, Fedou C, Mercier J. Postprandial reactive hypoglycemia. *Diabetes Metab* 2000; 26:337–51.

Chalew SA, McLaughlin JV, Mersey JH, Adams AJ, Cornblath M, Kowarski AA. The use of the plasma epinephrine response in the diagnosis of idiopathic postprandial syndrome. *JAMA* 1984; 251:612–5.

Charles MA, Hofeldt F, Shackelford A, Waldeck N, Dodson LE, Bunker D, Coggins JT, Eichner H. Comparison of oral glucose tolerance tests and mixed meals in patients with apparent idiopathic postabsorptive hypoglycemia: absence of hypoglycemia after meals. *Diabetes* 1981; 30:465–70.

Hofeldt FD. Reactive hypoglycemia. *Endocrinol Metab Clin North Am* 1989; 18:185–201.

3.1.3 Imaging of the Pancreas and Abdomen in a Patient with Suspected Tumor-Associated Hypoglycemia

Indications for Test

Imaging of the pancreas and abdomen in the hypoglycemic patient is indicated:

- Upon establishment of a biochemical diagnosis of an insulinoma

Procedure

1. Use multidetector (16 to 64 slices) computed x-ray tomography (M-CT) with high-resolution, three-dimensional images or endoscopic ultrasonography (EUS) to localize pancreatic tumors preoperatively.
2. Use intraoperative ultrasonography (IUS) combined with palpation as a highly sensitive method for localizing pancreatic masses for excision from insulinoma patients.

Interpretation

1. Review radiologist's interpretation of any imaging studies.
2. In IUS studies, insulinomas appear sonolucent compared to the more echo-dense surrounding pancreas.

Notes

1. Preoperative M-CT and EUS may detect pancreatic masses in 85 to 100% of patients with pancreatic cancer.
2. EUS has been reported to be more accurate than M-CT (<16 slice) in detecting tumor spread to surrounding tissues and blood vessels.
3. EUS more often detected pancreatic masses in patients shown subsequently to have pancreatic cancer than did M-CT. Both tests performed similarly in identifying cancers that were resectable.
4. IUS alone can identify 75 to 90% of insulinomas without palpation and can help define the relationship between the tumor and splenic artery and vein, pancreatic duct, common bile duct, and superior mesenteric vessels.
5. Intraoperative palpation of the pancreas alone, without use of IUS, has the same sensitivity for tumor localization as IUS (75–90%).

6. When the biochemical diagnosis of an insulinoma has been established, the combined use of IUS and palpation has almost 100% sensitivity for identifying insulinomas. M-CT and EUS had similar accuracy for detecting the spread of a variety of gastrointestinal cancers to lymph nodes and correctly identified ~90% of the resectable and ~65% of the unresectable tumors.

7. The accuracy of intraoperative palpation of a pancreatic or abdominal tumor improves markedly following appropriate preoperative localization.

8. The high density of glucagon-like peptide 1 (GLP-1) receptors in insulinomas as well as the high specific uptake of [Lys^{40}(Ahx-DTPA-^{111}In)NH$_2$]-exendin-4 in the tumor of Rip1Tag2 mice indicate that targeting of GLP-1 receptors in insulinomas may become a useful imaging method to localize insulinomas in patients, either preoperatively or intraoperatively (Wild et al., 2006).

ICD-9 Codes

Refer to Test 3.2.6 codes.

Suggested Reading

Bottger TC, Junginger T. Is preoperative radiographic localization of islet cell tumors in patients with insulinoma necessary? *World J Surg* 1993; 17:427–32.

DeWitt J, Devereaux B, Chriswell M, McGreevy K, Howard T, Imperiale TF, Ciaccia D, Lane KA, Maglinte D, Kopecky K, LeBlanc J, McHenry L, Madura J, Aisen A, Cramer H, Cummings O, and Sherman S. Comparison of endoscopic ultrasonography and multidetector computed tomography for detecting and staging pancreatic cancer. *Ann Intern Med* 2004; 141:753–63.

Doppman JL, Chang R, Fraker DL, Norton JA, Alexander HR, Miller DL, Collier E, Skarulis MC, Gorden P. Localization of insulinomas to regions of the pancreas by intra-arterial stimulation with calcium. *Ann Intern Med* 1995; 123:269–73 (*erratum* appears in *Ann Intern Med* 1995; 123:734).

Goode PN, Farndon JR, Anderson J, Johnston ID, Morte JA. Diazoxide in the management of patients with insulinoma. *World J. Surg* 1986; 10:586–92.

Hashimoto LA, Walsh RM. Preoperative localization of insulinomas is not necessary. *J Am Coll Surg* 1999; 189:368–73.

Mansour JC, Chen H. Pancreatic endocrine tumors. *J Surg Res* 2004; 120:139–61.

Sheppard BC, Norton JA, Doppman JL, Maton PN, Gardner JD, Jensen RT. Management of islet cell tumors in patients with multiple endocrine neoplasia: a prospective study. *Surgery* 1989; 106:1108–17.

Wild D, Behe M, Wicki A, Storch D, Waser B, Gotthardt M, Keil B, Christofori G, Reubi JC, Mäcke HR. [Lys40(Ahx-DTPA-^{111}In)NH$_2$]exendin-4, a very promising ligand for glucagon-like peptide-1 (GLP-1) receptor targeting. *J Nucl Med* 2006; 47:2025–33.

Zeiger MA, Shawker TH, Norton JA. Use of intraoperative ultrasonography to localize islet cell tumors. *World J Surg* 1993; 17:448–54.

3.2 Tests for Fasting Hypoglycemia in Adult Patients

3.2.1 Insulin/Glucose (I/G) Ratio and the I/G Amended Ratio in the Hypoglycemic Patient

Indications for Test

The insulin/glucose (I/G) ratio and the I/G amended ratio are indicated:

- To differentiate hypoglycemia caused by an insulinoma from that secondary to other causes in patients with fasting hypoglycemia

Procedure

1. Following a 12- to 14-hour overnight fast, obtain a blood sample for concurrent glucose and insulin levels.
2. Document the presence or absence of hypoglycemic symptoms.
3. Obtain a blood sample to screen for the presence of sulfonylureas and other hypoglycemic agents if symptomatic hypoglycemia occurs and the clinical suspicion of surreptitious ingestion of these agents is high.
4. Document the type of hypoglycemic symptoms occurring in patients with a plasma glucose < 50 mg/dL (< 2.8 mmol/L).
5. Calculate the I/G ratio as follows:

$$I/G = \frac{\text{Plasma insulin (mcU/mL)}}{\text{Plasma glucose (mg/dL)}}$$

Interpretation

1. In nonobese subjects, the I/G ratio before meals is usually <0.04. In insulinoma patients, the I/G ratio may rise to 1.0.
2. In patients with hypoglycemia, I/G ratios greater than 0.25 are very specific for insulinoma, but false negatives may occur using this criterion alone (low sensitivity).
3. A more sensitive test parameter, known as the amended I/G ratio, may be calculated:

$$\text{Amended I/G ratio} = \frac{\text{Plasma insulin (mcU/mL)}}{\text{Plasma glucose (mg/dL)} - 30} \times 100$$

4. An amended I/G ratio that is >50 suggests the presence of an insulinoma; however, because this ratio is associated with more false-positive results (low specificity), it must be confirmed by another test of insulin suppressibility, such as a prolonged fast, before diagnosis of an insulinoma can be made.
5. The presence of anti-insulin antibodies invalidates the use of the I/G ratio by interfering with the insulin immunoassay, often causing spuriously high insulin values. Moreover, the presence of anti-insulin antibodies in the absence of overt autoimmune disorder(s) suggests that the patient has taken exogenous insulin previously and supports the presumptive diagnosis of surreptitious insulin administration.

Notes

1. Approximately 80% of patients with known insulinomas will have hypoglycemia following an overnight (~12-hour) fast.
2. Anomalously normal I/G ratios may be found during supervised fasts of patients with an insulinoma, prompting performance of a broader range of diagnostic tests to establish a correct diagnosis.
3. The optimal timing for confirming the presence and accurately diagnosing the type of hypoglycemic disorder a patient may have is at the time of a spontaneous hypoglycemic event.

ICD-9 Codes

Conditions that may justify this test *include but are not limited to*:

Mentation disorders
780.4   dizziness / giddiness
780.93   memory loss

Pancreatic islet cells
157.4   malignant neoplasm
211.7   benign neoplasm

Other
251.2   hypoglycemia, unspecified

Suggested Reading

Dons RF, Hodge J, Ginsberg BH, Brennan MF, Cryer PE, Kourides IA, Gorden P. Anomalous glucose and insulin responses in patients with insulinoma: caveats for diagnosis. *Arch Intern Med* 1985; 145:1861–3.

Fajans SS, Floyd JC. Fasting hypoglycemia in adults. *N Engl J Med* 1976; 294:766–72.

Frerichs H, Creutzfeldt W. Hypoglycaemia 1. Insulin secreting tumours. *Clin Endocrinol Metab* 1976; 5:747–67.

Service FJ (Ed). *Hypoglycemia Disorders: Pathogenesis, Diagnosis, and Treatment.* 1983. G.K. Hall, p. 111.

Service FJ. Diagnostic approach to adults with hypoglycemic disorders. *Endocrinol Metab Clin North Am* 1999; 28:519–32, vi.

3.2.2   Extended Fasting and Monitoring of Glucose in the Hypoglycemic Patient

Indications for Test

Extended monitoring of fasting glucose in adults is indicated when:

- A patient develops symptoms of hypoglycemia confirmed by measurement of a random plasma glucose < 62 mg/dL (3.47 mmol/L) with no obvious cause for the hypoglycemia.
- Tumor-related hypoglycemia or surreptitious hypoglycemic agent administration is suspected.

Procedure

1. Admit the patient to the hospital and supervise the patient closely, usually in a private room with one-on-one nursing support.
2. Secure a reliable intravenous line with infusion of normal saline, and have a syringe of 25% dextrose solution, diluted from a 50% solution, on hand for emergency administration as necessary.
3. Allow only water, calorie-free, low-caffeine drinks (e.g., unsweetened tea) during the test. Inspect the patient's room for hidden "contraband" sources of calories and hypoglycemic agents (e.g., sulfonylureas or insulin).
4. Document that 4 or more hours have elapsed since the patient last ingested calories, then obtain baseline blood samples for glucose, insulin, cortisol, and C-peptide determinations.
5. Begin fingerstick capillary glucose monitoring every 2 hours. Increase the frequency of monitoring to every 1 hour when the glucose level is <70 mg/dL (3.89 mmol/L) and every 30 minutes when it is <60 mg/dL (3.34 mmol/L).
6. Routinely, obtain blood samples for glucose and insulin as well as for measurement of cortisol and C-peptide at intervals of 4 to 8 hours, at the physician's discretion, if the patient remains asymptomatic. Carefully document the time of blood sampling.
7. Use frequent fingerstick blood glucose monitoring for interpreting symptoms suggestive of hypoglycemia. If the fingerstick blood glucose is <50 mg/dL (2.78 mmol/L), confirm the result using a venous blood sample collected in a sodium fluoride tube and sent to a central laboratory for quantitative blood glucose measurement.
8. Provide general supervision when the patient is in the room. Always have the patient accompanied when outside of the room. Beyond this, it is not necessary to restrict the patient's activity.
9. Avoid administering any medications unless they are absolutely necessary.
10. If the patient has or reports adrenergic symptoms compatible with hypoglycemia, obtain blood samples for glucose, insulin, cortisol, and C-peptide determinations as soon as possible.
11. If the patient has symptoms of neuroglycopenia, such as impaired cognitive function, disorientation, unusual behavior, or loss of consciousness associated with a blood glucose < 40 mg/dL (2.22 mmol/L), obtain blood samples for glucose, insulin, cortisol, and C-peptide determinations, then administer 25% dextrose intravenously and terminate the fast.
12. Systematically evaluate the patient's cognitive function by a Mini-Mental Status Exam (MMSE) at the beginning and at the termination of the fast to objectively determine and document a deterioration of the cognitive state in patients with presumed neuroglycopenia even if the fingerstick blood glucose is <40 mg/dL (<2.22 mmol/L) (Wiesli et al., 2005).

13. Do not stop the test if the patient has a low blood glucose (i.e., <62 mg/dL or 3.47 mmol/L) but is asymptomatic. Administer the MMSE before terminating the test if uncertain as to the patient's mental status. Avoid terminating the test because of adrenergic symptoms alone.

14. If the patient is asymptomatic following 72 hours of fasting, have the patient exercise vigorously for 15 to 30 minutes and then obtain blood samples for insulin, glucose, and C-peptide.

Interpretation

1. An insulin level of >6 mcU/mL at the time of hypoglycemia (plasma glucose < 50 mg/dL or 2.78 mmol/L) during a fast is virtually diagnostic for insulinoma.

2. A baseline I/G ratio of >0.25 is usually diagnostic for insulinoma but may occur in a patient without this tumor.

3. Serial estimates of the I/G ratio tend to decrease during a fast in normal subjects but will increase in patients with an insulinoma.

4. The presence of insulin antibodies invalidates tests for insulin levels and suggests either surreptitious administration of insulin or the presence of autoantibody to insulin, which may be responsible for reactive hypoglycemia in patients with autoimmune diseases.

5. A C-peptide that is ≥1.2 ng/mL at the time of hypoglycemia indicates abnormal endogenous insulin secretion whereas a value of <1.2 ng/mL with hypoglycemia and a high circulating insulin level are highly suggestive of surreptitious administration of insulin.

6. Serum cortisol should increase by >7 mcg/dL (193 mmol/L) over baseline to at least 20 mcg/dL (552 mmol/L) when the plasma glucose is <40 mg/dL (2.22 mmol/L). Failure to meet these criteria is consistent with adrenal insufficiency or inaccurate determination of glucose and/or cortisol.

7. Rarely, females without hyperinsulinemia may have a fasting plasma glucose as low as 30 mg/dL (1.67 mmol/L) during a 72-hour fast without development of hypoglycemic symptoms (Merimee and Tyson, 1979).

8. During a fast, 70% of patients with an insulinoma will have a drop in their blood glucose to <38 mg/dL (2.13 mmol/L) within the first 24 hours. The sensitivity of this degree of hypoglycemia for prediction of insulinoma is 75% at 24 hours, 90% at 48 hours, and 98% at 72 hours.

9. A plasma beta-hydroxybutyrate of ≤2.7 mmol/L at the end of a prolonged fast is indicative of hyperinsulinemia of insulinoma when the plasma glucose is ≤60 mg/dL (3.34 mmol/L).

Notes

1. An extended fast is the gold standard test for the diagnosis of suspected insulinoma.

2. Fasting hypoglycemia secondary to an insulin-secreting neoplasm may occur more than 6 hours after a meal, may be precipitated by exercise, and is often associated with profound symptoms of neuroglycopenia.

3. The etiologies of neuroglycopenia include:
   - Hypoglycemic medications
   - Liver and/or kidney disease
   - Hormonal deficiencies of cortisol and growth hormone
   - Insulinoma or non-islet cell tumors
   - Autoimmune diseases associated with anti-insulin or anti-insulin receptor antibodies

4. The diagnosis of insulinoma depends on the demonstration of a failure of the blood insulin level to suppress appropriately in the setting of symptomatic fasting hypoglycemia (Figure 3.1).

5. A subnormal plasma beta-hydroxybutyrate and glucose response to glucagon helped to distinguish 20 out of 40 hypoglycemic insulinoma patients from 13 out of 25 hypoglycemic non-insulinoma subjects when the glucose was ≤60 mg/dL (3.3 mmol) at the same time as the insulin and C-peptide levels were within the expected range for a fasting patient.

6. Changes in fasting plasma free fatty acids, insulin, or C-peptide levels do not clearly differentiate normal subjects from insulinoma patients when the glucose is ≤60 mg/dL (3.3 mmol).

7. Monitoring glucose levels during surgery for removal of an insulinoma is of little use in determining the effectiveness of surgery, as anesthesia and the stresses of surgery alone tend to raise the blood glucose (Schwartz et al., 1979).

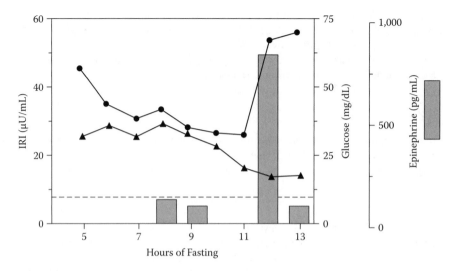

Figure 3.1
Counter-regulatory hormone surge with reversal of hypoglycemia during an extended fast in an insulinoma patient. Note the correction in glucose as the epinephrine levels (hatched bars) increased sevenfold after the glucose (closed circles) nadir at 12 hours of fasting, while immunoreactive insulin (IRI) levels (closed triangles) remained unchanged but persistently elevated (dotted line indicates upper limit of normal for fasting IRI levels). (From Dons, R.F. et al., *Arch. Intern. Med.*, 145, 1861–1863, 1985. With permission. Copyright © 2009. American Medical Association. All rights reserved.)

ICD-9 Codes

Refer to Test 3.2.1 codes.

Suggested Reading

Dons RF, Hodge J, Ginsberg BH, Brennan MF, Cryer PE, Kourides IA, Gorden P. Anomalous glucose and insulin responses in patients with insulinoma: caveats for diagnosis. *Arch Intern Med* 1985; 145:1861–3.

Joffe B, Kew M, Beaton G, Kusman B, Seftel H. Serum somatomedin and insulin levels in tumor hypoglycemia. *J Endocrinol Invest* 1978; 1:269–71.

Merimee TJ, Tyson JE. Stabilization of plasma glucose during fasting: normal variations in two separate studies. *N Engl J Med* 1974; 291:1275–8.

O'Brien T, O'Brien PC, Service FJ. Insulin surrogates in insulinoma. *J Clin Endocrinol Metab* 1993; 77:448–51.

Schwartz SS, Horwitz DL, Zehfus B, Langer BG, Kaplan E. Continuous monitoring and control of plasma glucose during operation for removal of insulinomas. *Surgery* 1979; 85:702–7.

Wiesli P, Schwegler B, Schmid B, Spinas GA, Schmid C. Mini-Mental State Examination is superior to plasma glucose concentrations in monitoring patients with suspected hypoglycaemic disorders during the 72-hour fast. *Eur J Endocrinol* 2005; 152:605–10.

3.2.3 C-Peptide Suppression with Low-Dose Insulin as a Test for Abnormal Endogenous Insulin Secretion in the Hypoglycemic Patient

Indications for Test

C-Peptide suppression with insulin infusion is indicated:

TABLE 3.1
Interpretation of C-Peptide Suppression with Insulin Test for Insulinoma

[Glucose] nadir, mg/dL (mmol/L)	[C-Peptide], ng/mL (nmol/L)	Interpretation
≤40 (2.2)	Not applicable	Valid test
<40 (2.2)	<1.2 (0.40)	Normal response[a]
<40 (2.2)	>1.2 (0.40)	Abnormal endogenous insulin secretion consistent with an insulinoma[b]

[a] Similar findings can be seen in individuals with factitious hypoglycemia.

[b] A post-insulin infusion [C-peptide] decline of <50% over baseline C-peptide concentration is also highly suggestive of an insulinoma.

- As an alternative to the extended fasting and glucose monitoring test for evaluation of a patient with normal renal function who is suspected of abnormal endogenous insulin secretion consistent with an insulinoma based on confirmed hypoglycemic episode(s)
- To avoid the prolonged hospitalization in the evaluation of the hypoglycemic patient during the performance of the extended (i.e., 3-day) fast fasting and glucose monitoring test

Procedure

1. Because severe hypoglycemia may occur during this test and be poorly tolerated by patients with cardiac or neurological conditions, choose candidates for study carefully and always obtain informed consent for this potentially hazardous test.
2. Instruct patient to ingest at least 300 g/day of carbohydrates for 3 days prior to the test.
3. Following an 8- to 14-hour overnight fast, establish intravenous heparin locks for blood sampling and insulin administration.
4. Obtain baseline blood samples for glucose and C-peptide level at –20 and 0 minutes prior to initiation of the insulin infusion.
5. For emergent treatment of hypoglycemia, maintain availability of a syringe containing a 20 to 50% dextrose solution for injection of a minimum of 12.5 g of glucose.
6. Administer an insulin dose, 0.125 unit/kg body weight, as a continuous infusion over 1 hour beginning at time zero.
7. Obtain blood samples for glucose and C-peptide at 10, 30, 60, 90, and 120 minutes from the start of the infusion of insulin.
8. Observe the patient closely for symptoms of neuroglycopenia. If such symptoms occur or the glucose level is <40 mg/dL (2.2 mmol/L) at any time during the test, obtain another glucose and C-peptide level and give food or intravenous dextrose, followed by termination of the test.

Interpretation

1. Refer to "Interpretation of C-Peptide Suppression with Insulin Test for Insulinoma" (Table 3.1) for typical changes in glucose and C-peptide during and after insulin infusion.
2. In insulinoma patients, mean serum [C-peptide] was suppressed to a nadir of 1.81 ± 0.87 ng/mL (median, 1.83 ng/mL) vs. 0.40 ± 0.15 ng/mL (median, 0.30 ng/mL) in non-insulinoma patients.
3. Maximal suppression of C-peptide with insulin was $37 \pm 24\%$ of the basal [C-peptide] in insulinoma patients vs. $75 \pm 9\%$ in non-insulinoma patients ($p < 0.001$) (Figure 3.2).
4. The following are suggestive of insulinoma:
   - Development of severe symptomatic hypoglycemia with plasma glucose levels in the range of <30 mg/dL (1.8 mmol/L)
   - Any rise in the C-peptide concentration coupled with a decline in plasma glucose

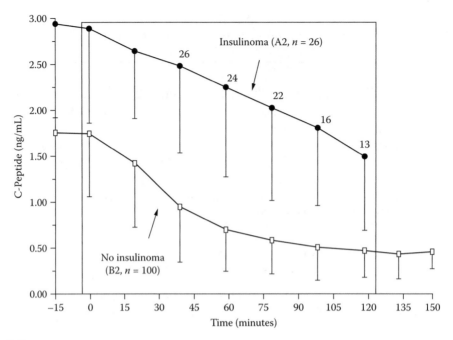

Figure 3.2
Serum C-peptide levels in response to 120 minutes of insulin infusion (indicated by box) in patients with an insulinoma (A2, n = 26) and without an insulinoma (B2, n = 100). Mean glycemia (30.8 ± 3.3 mg/dL vs. 47.5 ± 8.3 mg/dL, p < 0.001) and glucose/C-peptide ratio (21.9 ± 14.6 vs. 139.2 ± 43.8; p < 0.001) during the test were lower in patients with an insulinoma compared to those without an insulinoma. C-peptide values shown are the mean – 1 SD. (From Saddig, C. et al., *J. Pancreas*, 3, 16–25, 2002. With permission.)

5. This test has a specificity of 96% to rule out the diagnosis of insulinoma at a 1% probability threshold with sensitivity of 100%.
6. A false-positive test result occurs in patients with chronic kidney disease Stage 4 or 5 and those taking sulfonylurea agents surreptitiously.

Notes

1. A significantly lower percent decrease in C-peptide level from baseline occurs during insulin infusion in cases of insulin resistance (i.e., older patients; individuals with obesity) independent of gender (Service et al., 1992).
2. The C-peptide suppression test does not supplant the extended fast test; however, it may be useful in diagnosing factitious hypoglycemia as well as the rare case of coexisting diabetes mellitus and insulinoma.
3. In 11 of 12 insulinoma patients, the mean C-peptide level could not be suppressed to <1.9 ng/mL by this test (Service et al., 1977).
4. This test differs from the glucose reduction challenge test (GRCT). In the GRCT, a rapid reduction in plasma glucose is induced by a continuous low-dose insulin infusion. In insulinoma patients, C-peptide levels remain high at 0.96 to 4.8 ng/mL (0.32–1.6 nmol/L) during the GRCT (Ipp et al., 1990).

ICD-9 Codes

Refer to Test 3.2.1 codes.

Suggested Reading

Ipp E, Sinai Y, Forster B, Cortez C, Baroz B, Nesher R, Cerasi E. A glucose reduction challenge in the differential diagnosis of fasting hypoglycemia: a two-center study. *J Clin Endocrinol Metab* 1990; 70:711–7.

Ravnik-Oblak M, Janez A, Kocijanicic A. Insulinoma induced hypoglycemia in a type 2 diabetic patient. *Wien Klin Wochenschr* 2001; 113:339–41.

Saddig C, Bender R, Starke AA. A new classification plot for the C-peptide suppression test. *J Pancreas* 2002; 3:16–25.

Service FJ, Horwitz DL, Rubenstein AH, Kuzuya H, Mako ME, Reynolds C, Molnar GD. C-peptide suppression test for insulinoma. *J Lab Clin Med* 1977; 90:180–6.

Service FJ, O'Brien PC, Kao PC, Young WF. C-peptide suppression test: effects of gender, age, and body mass index; implications for the diagnosis of insulinoma. *J Clin Endocrinol Metab* 1992; 74:204–10.

3.2.4   Proinsulin (PI) Level Test for Insulinoma

Indications for Test

Measurement of [PI] is indicated:

- For the evaluation of patients with fasting hypoglycemia
- When an insulinoma is suspected but insulin levels are not clearly elevated

Procedure

1. Note the patient's height, weight, and waist circumference.
2. Following an 8- to 14-hour overnight fast, obtain blood samples for basal glucose, insulin, and proinsulin determination.

Interpretation

1. PI concentrations in normal-weight individuals without an insulinoma are <0.2 ng/mL.
2. PI concentrations > 0.2 ng/mL are seen in patients with conditions where insulin release is abnormal or when the metabolism of proinsulin is decreased.
3. In addition to patients with insulinoma, an elevated [PI] may be seen in familial hyperproinsulinemia, insulin resistance, hypokalemia, hyperthyroidism, chronic kidney disease, and hepatic cirrhosis.
4. In both type 2 diabetes mellitus and insulinoma patients, the [PI] is significantly higher than in healthy subjects.
5. The proinsulin (ng/mL)/insulin (mcU/mL) ratio is interpreted as follows:
   - <0.20, normal
   - 0.20–0.30, suspicious for insulinoma or marked increase in waist circumference
   - >0.30, consistent with presence of an insulinoma

Notes

1. Proinsulin is the single-chain precursor to insulin.
2. Proinsulin usually comprises 10 to 15% (and, normally, always <20%) of the total immunoreactive insulin.
3. Although proinsulin has a direct biologic effect one tenth that of insulin, C-peptide has no biologic activity of significance.

4. The processing of proinsulin to insulin is less efficient in neoplastic cells such that patients with an insulinoma usually have higher proinsulin/insulin ratios than lean, non-insulinoma subjects.
5. In a prospective analysis of a native Canadian population, baseline waist circumference was positively associated with [PI] related to the effect of abdominal obesity on beta-cell function (Hanley et al., 2002).

ICD-9 Codes

Refer to Test 3.2.1 codes.

Suggested Reading

Clark PM. Assays for insulin, proinsulin(s) and C-peptide. *Ann Clin Biochem* 1999; 36(Pt 5):541–64.

Gorden P, Skarulis MC, Roach P, Comi RJ, Fraker DL, Norton JA, Alexander HR, Doppman JL. Plasma proinsulin-like component in insulinoma: a 25-year experience. *J Clin Endocrinol Metab* 1995; 80:2884–7.

Hanley AJ, McKeown-Eyssen G, Harris SB, Hegele RA, Wolever TM, Kwan J, Zinman B. Cross-sectional and prospective associations between abdominal adiposity and proinsulin concentration. *J Clin Endocrinol Metab* 2002; 87:77–83.

Houssa P, Dinesen B, Deberg M, Frank BH, Van Schravendijk C, Sodoyez-Goffaux F, Sodoyez JC. First direct assay for intact human proinsulin. *Clin Chem* 1998; 44:1514–9.

Kitabchi AE. Proinsulin and C-peptide: a review. *Metabolism* 1977; 26:547–87.

Sherman BM, Pek S, Fajans SS, Floyd JC, Conn JW. Plasma proinsulin in patients with functioning pancreatic islet cell tumors. *J Clin Endocrinol Metab* 1972; 35:271–80.

3.2.5 Provocative Insulin Secretagog Tests for Insulinomas Using Infusion of Calcium, Tolbutamide, Glucagon, or Secretin to Stimulate Insulin

Indications for Test

Provocative insulin secretagog tests using glucagons, secretin, or tolbutamide may be indicated in:

- Evaluation of hypoglycemic patients in whom other tests for insulinoma are equivocal
- Patients with symptomatic fasting neurohypoglycemia, associated with apathy, amnesia, dizziness, confusion, personality change, diplopia, seizures, stroke, coma, substupor, mutism, unusual mannerisms, restlessness, or incoherence

Intraarterial calcium infusion may be indicated when:

- Attempting to better localize an insulinoma not adequately localized by imaging techniques (see Test 3.2.8)

General Procedure

1. Obtain informed consent for these invasive tests.
2. Following an 8- to 14-hour overnight fast, start a normal saline infusion.
3. Keep a 20 to 50% dextrose solution available for emergency use as necessary. A physician must be in attendance during testing.
4. Determine baseline glucose and insulin levels in peripheral blood samples taken before intravenous or intraarterial infusion of a secretagog, usually at −30, −15, and 0 minutes.

Glucagon Stimulation Procedure

1. Administer 2 mg of glucagon intravenously as a bolus into the brachial vein over 30 to 40 sec.
2. Obtain blood samples for glucose and insulin levels at 3, 6, 15, 20, 25, 30, and 60 minutes after the administration of glucagon.

Secretin Stimulation Procedure

1. Administer 0.2 mcg secretin/kg body weight intravenously as a bolus into the brachial vein over 2 minutes.
2. Obtain blood samples for glucose and insulin levels at 3, 5, 10, 15, 30, 45, and 60 minutes after the intravenous administration of secretin.
3. Calculate the percent increase above the mean basal insulin level (mean of insulin values on samples collected at −30, −15, and 0 minutes prior to the administration of secretin) and identify the peak insulin response.

Tolbutamide Stimulation Procedure

1. Administer 1 gram of tolbutamide (Orinase Diagnostic®, Pfizer) intravenously over 2 minutes into the brachial vein.
2. Obtain blood samples for glucose and insulin levels at 2, 5, 10, 15, and 30 minutes after the intravenous administration of tolbutamide.
3. If an initial hypersecretory insulin response is to be confirmed, obtain blood samples at 120, 150, and 180 minutes after the intravenous administration of tolbutamide.

Calcium Stimulation Procedure

1. Selectively inject boluses of calcium gluconate (0.025 mEq calcium per kg body weight) along the length of the superior mesenteric artery using Imamura's technique (Brandle et al., 2001).
2. Follow each selective calcium injection by rapidly taking samples for insulin testing from the right hepatic vein before and at 30, 60, and 120 seconds after injection of calcium gluconate (i.e., four blood samples per each calcium injection).
3. Higher dose intraarterial calcium injection (up to 0.05 mEq calcium per kg) may be needed to evoke insulin secretion and localize an insulinoma.

Interpretation of Glucagon Stimulation Test

1. A normal response is a transient rise in plasma glucose followed by a reactive rise in serum insulin levels and a prompt return of the glucose toward pretest levels as glucagon stimulates liver glycogenolysis. Patients with reactive hypoglycemia respond normally to glucagon.
2. In hepatocellular carcinoma (HCC) patients without hypoglycemia, plasma insulin and C-peptide responses to glucagon injection are normal.
3. Expect the glucose response to glucagon administration to be blunted (glucose increment of <25 mg/dL or <1.4 mmol) in patients with insulin-like substances from HCC as well as in patients with insulinoma.
4. A plasma glucose of <40 mg/dL, when found 30 to 180 minutes following a glucagon injection and followed by persistent hyperinsulinism, has a sensitivity of 55% for the diagnosis of insulinoma. Characteristically, insulinoma patients have increased insulin and C-peptide responses to glucagon inappropriate for their low glucose levels.
5. A peak insulin response of >130 mcU/mL is seen in 70 to 80% of insulinoma cases and has a sensitivity of 72% for detection of tumor.
6. False-positive results using the insulin criteria above (i.e., >130 mcU/mL) can be seen in type 2 diabetes and obesity and with sulfonylurea or aminophylline therapy.
7. False-negative results can be seen with hydrochlorothiazide, diazoxide, or diphenylhydantoin therapy.
8. If a glucagon stimulation test were to be done at the end of a prolonged fast, the responses of plasma glucose to glucagon would be significantly greater in insulinoma patients (mean, 3.0 mmol/L; range, 1.4–5.4) than in a reference population (mean, 0.7 mmol/L; range, 0.0–1.3).

Interpretation for Secretin Stimulation Test

1. A <50% increase in insulin above the basal concentration is consistent with either a normal response or presence of a solitary insulinoma in patients with hypoglycemia and documented hyperinsulinism.
2. A >1000% (>10-fold) increase in insulin above the basal concentration suggests multiple beta-cell adenomas and/or hyperplasia.
3. Multiple insulinomas may be unusually responsive to secretin, whereas a lack of a several-fold insulin secretory response to secretin in patients with single insulinomas is typical.

Interpretation for Tolbutamide Stimulation Test

1. A peak insulin response of >190 mcU/mL associated with hypoglycemia occurs in 90% of insulinoma patients.
2. The percent drop in glucose from baseline induced by tolbutamide is 30% greater in insulinoma patients than in healthy non-insulinoma subjects. Moreover, the tolbutamide-induced glucose nadir in insulinoma patients is usually in the 20- to 40-mg/dL range vs. 50 to 60 mg/dL in normal subjects.
3. The fifth percentiles of the mean of plasma glucose levels at 120, 150, and 180 minutes after injection of tolbutamide were 55 mg/dL for lean and 62 mg/dL for obese control subjects. At 95% specificity, the sensitivity of these criteria for detecting patients with an insulinoma was 95% for lean and 100% for obese patients (McMahon et al., 1989).
4. In insulinoma patients, persistent (i.e., 2.5 to 3.0 hours) hypoglycemia is induced by tolbutamide vs. only 1.5 to 2.0 hours in healthy, non-insulinoma subjects. This test is 80% sensitive for insulinoma.

Interpretation of Calcium Stimulation Test

1. In patients without an insulinoma, all insulin values on all three samples obtained from the hepatic vein after injection of calcium at each site along the mesenteric artery will not be significantly different from the insulin level on the corresponding sample obtained prior to calcium injection. No significant difference in insulin values is defined as a ±2 times within-run CV for the insulin assay used for testing.

   Example—Just prior to the injection of calcium at the first site along the mesenteric artery, the blood sample taken from the hepatic vein of a patient had a baseline insulin level of 50 mcU/mL:

   Baseline [insulin] = 50 mcU/mL

   Insulin values on each of the three samples taken from the hepatic vein at 30, 60, and 120 seconds after the injection of calcium were:

   30-sec sample [insulin] = 52 mcU/mL
   60-sec sample [insulin] = 54 mcU/mL
   120-sec sample [insulin] = 46 mcU/mL

   If the within-run coefficient of variation (CV) for the insulin assay used for testing is 5%, then, the 2× within-run CV = 2 × 5 = 10%.

   10% × baseline insulin of 50 mcU/mL = 5.0 mcU insulin/mL

   Thus, in the absence of an insulinoma, all insulin values for the three samples taken along with this baseline sample should be between 45 and 55 mcU/mL; therefore, the patient's insulin values (46, 52, and 54 mcU/mL) indicated above are indicative of the absence of an insulinoma.

2. In patients with an insulinoma, injection of calcium into the mesenteric artery feeding the tumor will cause a rise in hepatic vein insulin concentration within 2 minutes of injection. For example, in a blood sample taken from the hepatic vein *prior* to the injection of calcium at a site near the location of an insulinoma in the pancreas the insulin concentration may be elevated, and in subsequent blood samples taken at 30, 60, and 120 seconds *after* the injection of calcium the insulin levels may rise sharply in one or more of these samples over an already elevated baseline insulin level.
3. Arterial calcium infusion and hepatic venous sampling is a very sensitive technique (up to 100%) for preoperative localization of insulin-producing tumors.

Notes

1. Each of the secretagog tests described above has no known differences in diagnostic accuracy in identifying patients with an insulinoma. Moreover, it should be noted that no provocative test is capable of making a definitive diagnosis of insulinoma.

2. In up to 20% of insulinoma patients, sympathoadrenal symptoms, not neuroglycopenic phenomena, may predominate.

3. Induction of profound and/or prolonged hypoglycemia with tolbutamide is hazardous and should be protected against.

4. Both insulinomas and gastrinomas may be found in patients with multiple endocrine neoplasia type 1 (MEN 1), and gene mutation analysis may be required for diagnosis of MEN 1. Multiple insulinomas are present in only about 10% of insulinoma patients, many of whom have MEN 1.

5. Selective intraarterial injection of secretin, combined with hepatic venous sampling for gastrin, can be used for the detection of small gastrinomas, as well as insulinomas.

6. In some insulinoma patients, the long-term use of a calcium channel blocker may be associated with an exaggerated insulin response and systemic hypoglycemia during selective intraarterial calcium stimulation (Moreira et al., 2003).

7. Noninvasive localization modalities such as ultrasound, multislice (<16) computed tomography (CT), or magnetic resonance (MR) imaging may fail to localize insulinomas smaller than 2 cm.

8. In a National Institute of Health (NIH) study of 25 patients, the sensitivities for detection of insulinoma were as follows: arterial stimulation with calcium followed by hepatic venous sampling, 88%; percutaneous transhepatic portal venous sampling, 67%; magnetic resonance, 43%; arteriography, 36%; CT scan, 17%; and transabdominal ultrasound, 9% (Doppman et al., 1995).

9. Most (97%) insulinomas are located in the pancreas with the rest typically located close to it. The size of the tumor typically ranges from 4 to 25 mm. Forty percent are <1 cm in diameter, 66% are <1.5 cm, and 90% are <2 cm.

10. Acute insulin responses to peripherally infused calcium and tolbutamide were not sufficient to differentiate focal adenomatous hyperplasia from diffuse beta-cell hypersecretion (nesidioblastosis) in pediatric hypoglycemia patients.

11. Anti-insulin antibodies that bind and release insulin in an unregulated manner may cause hypoglycemia in multiple myeloma or systemic lupus erythematosus patients. Testing for anti-insulin antibodies should be done in these patients if they become hypoglycemic.

ICD-9 Codes

Refer to Test 3.2.1 codes.

Suggested Reading

Brandle M, Pfammatter T, Spinas GA, Lehmann R, Schmid C. Assessment of selective arterial calcium stimulation and hepatic venous sampling to localize insulin-secreting tumours. *Clin Endocrinol (Oxf)* 2001; 55:357–62.

Doppman JL, Chang R, Fraker DL, Norton JA, Alexander HR, Miller DL, Collier E, Skarulis MC, Gorden P. Localization of insulinomas to regions of the pancreas by intra-arterial stimulation with calcium. *Ann Intern Med* 1995; 123:269–73.

Fajans SS, Vinik AI. Insulin-producing islet cell tumors. *Endocrinol Metab Clin North Am* 1989; 18:45–74.

Giurgea I, Laborde K, Touati G, Bellanne-Chantelot C, Nassogne MC, Sempoux C, Jaubert F, Khoa N, Chigot V, Rahier J, Brunelle F, Nihoul-Fekete C, Dunne MJ, Stanley C, Saudubray JM, Robert JJ, de Lonlay P. Acute insulin responses to calcium and tolbutamide do not differentiate focal from diffuse congenital hyperinsulinism. *J Clin Endocrinol Metab* 2004; 89:925–9.

Glaser B, Shapiro B, Glowniak J, Fajans SS, Vinik AI. Effects of secretin on the normal and pathological beta-cell. *J Clin Endocrinol Metab* 1988; 66:1138–43.

Gutman RA, Lazarus NR, Penhos JC, Fajans S, Recant L. Circulating proinsulin-like material in patients with functioning insulinomas. *N Engl J Med* 1971; 284:1003–8.

Honda M, Ishibashi M. The diagnosis and treatment of insulinoma and gastrinoma. *Gan To Kagaku Ryoho* 2004; 31:337–41.

Marks V, Alberti KG. Selected tests of carbohydrate metabolism. *Clin Endocrinol Metab* 1976; 5:805–20.

McMahon MM, O'Brien PC, Service FJ. Diagnostic interpretation of the intravenous tolbutamide test for insulinoma. *Mayo Clin Proc* 1989; 64:1481–8.

Moreira RO, Vaisman M, Peixoto PC, Azevedo F. Is long-term use of a calcium channel blocker associated with an exacerbated response of an insulinoma during selective intraarterial calcium stimulation? *Acta Radiol* 2003; 44:354.

Nakamura Y, Doi R, Kohno Y, Shimono D, Kuwamura N, Inoue K, Koshiyama H, Imamura M. High-dose calcium stimulation test in a case of insulinoma masquerading as hysteria. *Endocrine* 2002; 19:127–30.

O'Brien T, O'Brien PC, Service FJ. Insulin surrogates in insulinoma. *J Clin Endocrinol Metab* 1993; 77:448–51.

Pun KK, Young RT, Wang C, Tam CF, Ho PW. The use of glucagon challenge tests in the diagnostic evaluation of hypoglycemia due to hepatoma and insulinoma. *J Clin Endocrinol Metab* 1988; 67:546–50.

3.2.6 Human Chorionic Gonadotropin (hCG) and Its Subunits as Tumor Markers in Insulinoma Patients

Indications for Test

Testing for hCG and its subunits is indicated:

- In patients with an established diagnosis of an insulinoma
- For the purpose of finding a tumor marker in cases of metastatic or functioning islet-cell carcinomas

Procedure

1. Obtain blood sample for hCG and alpha- and beta-subunit hCG testing.

Interpretation

1. Assay-specific reference intervals for hCG and its subunits in males, premenopausal females, and postmenopausal females should be consulted.
2. Any elevation in alpha- or beta-subunit hCG in a patient with a known insulinoma supports the possibility that the tumor is a functioning islet-cell carcinoma. Secretion of hCG and its subunits is often discordant in recurrent insulinomas.
3. Low or normal hCG subunit levels do not exclude the diagnosis of malignancy particularly in postop patients.
4. Closely monitoring the clinical course of tumor patients with detectable alpha-subunit levels, the most frequently occurring abnormality in recurrent insulinomas, is recommended.

Notes

1. Increased levels of hCG and its subunits may be found in patients with malignant insulinoma as well as those with testicular cancer and pregnancy.
2. In 14 patients with malignant insulinoma, the hCG alpha-subunit (common to all glycoprotein hormones) was elevated in 8 (57%) of these patients.
3. Beta-subunit hCG and intact hCG were elevated in only 21% and 25% of malignant insulinoma patients, respectively.
4. None of 41 patients with benign insulinoma had elevated levels of alpha-subunit, beta-subunit, or intact hCG.
5. HCG and its subunits are specific markers for islet-cell carcinoma. Their ectopic secretion appears to result from malignant depression of the genome rather than overproduction by an aberrant "cell rest."

ICD-9 Codes

Conditions that may justify this test *include but are not limited to*:

Pancreatic islet cells
157.4   malignant neoplasm
211.7   benign neoplasm

Suggested Reading

Graeme-Cook F, Nardi G, Compton CC. Immunocytochemical staining for human chorionic gonadotropin subunits does not predict malignancy in insulinomas. *Am J Clin Pathol* 1990; 93:273–6.

Kahn CR, Rosen SW, Weintraub BD, Fajans SS, Gorden P. Ectopic production of chorionic gonadotropin and its subunits by islet-cell tumors: a specific marker for malignancy. *N Engl J Med* 1977; 297:565–9.

3.2.7   Insulin-Like Growth Factor 2 (IGF-2) and IGF-1 (Somatomedin C) as Tests for Non-Insulin Causes of Hypoglycemia

Indications for Test

Testing for IGF-2 and IGF-1 is indicated:

- In cases of tumor-associated hypoglycemia when insulin or C-peptide levels are not elevated
- When a diagnosis of hemangiopericytoma is established

Procedure

1. Obtain a blood sample for insulin, IGF-2, and IGF-1 testing during a hypoglycemic episode in patients with a large mesenchymal tumor, particularly of the liver.
2. Obtain a tissue sample for fluorescence *in situ* hybridization (FISH) analysis to detect expression of IGF-2 mRNA in patients who are not yet known to be hypoglycemic but are known to have a hemangiopericytoma or a mesenchymal tumor.

Interpretation

1. Interpret [IGF-2] against the assay-specific reference interval for healthy individuals. IGF-2 concentrations are typically elevated in patients with IGF-2–mediated hypoglycemia.
2. Interpret [IGF-1] against the age- and gender-specific reference interval for healthy individuals. [IGF-1] may or may not be elevated in patients with hemangiopericytoma and mesenchymal tumors.
3. If FISH is positive for IGF-2 mRNA, the patient's hypoglycemia may be explained by increased production of IGF-2 by hemangiopericytoma or mesenchymal tumors.

Notes

1. Although not all hemangiopericytomas produce hypoglycemia in patients with these tumors, those arising from the liver and pancreas are most likely to do so.
2. Hemangiopericytoma tumors are known to secrete high levels of IGF-2, usually the high-molecular-weight (HMW) variant, which can cause hypoglycemia. Although HMW IGF-2 has been well identified as a cause of tumor-associated hypoglycemia, other IGFs may be involved as well.
3. Three out of 19 hemangiopericytoma tumors (15.8%) were found to be positive for IGF-1 mRNA, 11 were positive for IGF-2 mRNA (57.9%), and 17 (almost 90%) expressed IGF-1 receptor mRNA (Pavelic et al., 1999).

4. It has been proposed that concomitant overproduction of IGF-1 and IGF-2 ligands, together with the presence of tumor IGF-1 receptor, could be responsible for hypoglycemia in hemangiopericytoma patients. Alternatively, the hypoglycemia may be mediated via nonspecific binding of IGF-2 or IGF-1 to insulin receptors throughout the body.

5. In some mesenchymal tumors present for an extended period of time, the actual clinical manifestation of hypoglycemia may be blocked by counter-regulatory factors.

ICD-9 Codes

Conditions that may justify this test *include but are not limited to*:

| 155 | malignant neoplasm of liver and intrahepatic bile ducts/tumor-associated hypoglycemia/hemangiopericytoma |
| 251.2 | hypoglycemia, unspecified |

Suggested Reading

Hoog A, Sandberg Nordqvist AC, Hulting AL, Falkmer UG. High-molecular-weight IGF-2 expression in a haemangiopericytoma associated with hypoglycaemia. *APMIS* 1997; 105:469–82.

Kruskal JB, Kane RA. Paraneoplastic hypoglycemia associated with a hepatic hemangiopericytoma. *J Ultrasound Med* 2002; 21:927–32.

Pavelic K, Cabrijan T, Hrascan R, Vrkljan M, Lipovac M, Kapitanovic S, Gall-Troselj K, Bosnar MH, Tomac A, Grskovic B, Karapandza N, Pavelic LJ, Kurslin B, Spaventi S, Pavelic J. Molecular pathology of hemangiopericytomas accompanied by severe hypoglycemia: oncogenes, tumor-suppressor genes and the insulin-like growth factor family. *J Cancer Res Clin Oncol* 1998; 124:307–14.

Pavelic K, Pavelic ZP, Cabrijan T, Karner I, Samarzija M, Stambrook PJ. Insulin-like growth factor family in malignant haemangiopericytomas: the expression and role of insulin-like growth factor I receptor. *J Pathol* 1999; 188:69–75.

3.3 Tests for Hypoglycemia in Pediatric Patients

3.3.1 Pediatric Hypoglycemia: Routine Serum and Urine Testing (Glucose, Electrolytes, Ketones, and Lactate)

Indications for Test

Blood and urine testing for glucose, electrolytes, ketones, and lactate is indicated:

- Upon encountering a child suspected of hypoglycemia, typically in the critical-care setting of the newborn nursery or emergency department
- In a nonemergent consultative setting in children with non-life-threatening hypoglycemia

Procedure

1. In an emergency setting, obtain a random blood sample for glucose to confirm the clinical impression of hypoglycemia (glucose < 62 mg/dL or < 3.47 mmol/L) on an urgent basis before obtaining a complete history and physical.

2. In an emergency, if capillary blood glucose by dry reagent strip testing is <62 mg/dL (<3.47 mmol/L), then urine and blood should be collected and saved frozen as contingency samples for additional testing. Next, screen a urine sample for ketones followed by quantitative measurement of electrolytes (Na^+, K^+, Cl^-, CO_2), beta-hydroxybutyrate, and lactate in blood if clinically indicated.

3. In nonemergent situations involving children who are referred by school personnel or parents for behaviors that may be attributed to hypoglycemia (e.g., inattention, tiredness, transient seizures, or lack of responsiveness), complete a history and physical exam before obtaining blood and urine specimens from these patients.

4. When suspected hypoglycemia is not proven, provide a letter to parents or guardians outlining the diagnostic tests recommended when the patient presents with acute symptoms suggestive of hypoglycemia.

5. To avoid artifactual hypoglycemia resulting from consumption of glucose by cellular elements, collect blood for glucose in sodium-fluoride-containing (i.e., gray-top) tubes and separate the cells from plasma as soon as possible.

6. When blood samples are collected in a red-top tube, separate cells immediately from serum or refrigerate until separation within no more than an hour.

7. When hypoglycemia is confirmed, evaluate the patient for various endocrine abnormalities, such as hyperinsulinemia, poorly treated diabetes, hypopituitarism, growth hormone deficiency (see tests in Section 2.3), and adrenal insufficiency (see tests in Section 6.1). After reviewing the history and physical exam findings, obtain the following blood tests as appropriate:
   - Baseline cortisol, IGF-1, or growth hormone levels
   - Adrenocorticotropic hormone (ACTH) level if cortisol is low (a normal cortisol makes ACTH testing more elective)
   - Glucose and insulin levels, particularly when the patient is symptomatic of hypoglycemia
   - C-peptide level
   - Glucagon level (when considering the extremely rare case of isolated glucagon deficiency)

8. Contingency samples of urine (10 mL) and blood (10 mL) should be collected and frozen as serum (5 mL) for later analysis (i.e., metabolic screening).

Interpretation

1. If whole blood, serum, or plasma glucose concentration is low, ensure that the blood sample was collected, processed, and stored optimally prior to testing.

2. If both cortisol and GH levels are elevated in a blood sample, spontaneous hypoglycemia may be the cause for these hormone abnormalities.

3. Low cortisol or IGF-1 levels suggest hypopituitarism. Low GH levels alone are usually not diagnostic.

4. A high C-peptide level may be found in children with a solitary islet-cell tumor or islet-cell hyperplasia (i.e., nesidioblastosis).

5. A low C-peptide level may be found if factitious hyperinsulinemia is uncovered, usually in the context of:
   - Surreptitious use of insulin in child abuse or assault cases (Munchausen by proxy)
   - Overtreatment in adolescents with type 1 diabetes

6. Based on history and physical exam findings, explore the possibility of various toxic exposures such as ethanol or oral hypoglycemic agent ingestion in hypoglycemic children.

7. Based on history, physical exam findings, and the results of testing for urine ketones, serum acetone, and plasma lactate, evaluate patients for various metabolic defects of fatty acid, amino acid, or carbohydrate metabolism by obtaining additional metabolic screening tests, such as plasma amino acid analysis by liquid chromatography–tandem mass spectrometry (LC-MS/MS).

Notes

1. In anticipation of an encounter with a hypoglycemic pediatric patient, have a predetermined list of laboratory tests to be obtained as appropriate.

2. Be careful to use only the proper types of specimen collection tubes (i.e., red-top for serum and green-top or gray-top for plasma).

3. Keep an updated list of the proper methods of blood and urine sample preparation and storage for the tests required.

4. Recognize that symptomatic hypoglycemia is often intermittent, and abnormalities on testing are much more likely when the patients are symptomatic than when they are not.

5. Getting a quantitative venous plasma or serum glucose test from a central laboratory is preferred over finger- or heelstick samples whenever testing for glucose in children is necessary. In emergency situations, however, dry reagent strip testing of capillary blood for glucose is an acceptable and rapid alternative.
6. Confirm an abnormal finger- or heelstick capillary glucose value with a subsequent or simultaneous quantitative glucose determination using a venous blood sample and central laboratory testing.

ICD-9 Codes

Conditions that may justify this test *include but are not limited to*:

251.2   hypoglycemia, unspecified

Suggested Reading

Bartlett D. Confusion, somnolence, seizures, tachycardia? Question drug-induced hypoglycemia. *J Emerg Nurs* 2005; 31:206–8.
Dekelbab BH, Sperling MA. Hyperinsulinemic hypoglycemia of infancy: the challenge continues. *Diabetes Metab Res Rev* 2004; 20:189–95.
Hussain K, Thornton PS, Otonkoski T, Aynsley-Green A. Severe transient neonatal hyperinsulinism associated with hyperlactataemia in non-asphyxiated infants. *J Pediatr Endocrinol Metab* 2004; 17:203–9.
Valencia I, Sklar E, Blanco F, Lipsky C, Pradell L, Joffe M, Legido A. The role of routine serum laboratory tests in children presenting to the emergency department with unprovoked seizures. *Clin Pediatr (Phila)* 2003; 42:511–7.

3.3.2   Pediatric Hypoglycemia: Nonroutine Testing (Arterial Blood Gases, Blood Alcohol, Sulfonylureas, Ammonia, Uric Acid, and Lipids)

Indications for Test

Testing for arterial blood gases, blood alcohol, sulfonylureas, uric acid, and lipids is indicated in the evaluation of children:

*   Confirmed to have transient, extended, or severe hypoglycemia (typical glucose much lower than 62 mg/dL) and hypoglycemic symptoms
*   With poorly controlled diabetes and those suspected of hypopituitarism with growth hormone (GH) deficiency and/or adrenal insufficiency or surreptitious insulin abuse
*   For suspected poisoning (e.g., salicylates, ethanol, oral hypoglycemic agents) or rare, hereditary syndromes causing hypoglycemia

Procedure

1. Obtain or retrieve the appropriate specimen for testing of:
    *   Arterial blood gases
    *   Blood alcohol
    *   Sulfonylureas
    *   Salicylates
2. If sepsis appears possible, obtain:
    *   Blood, urine, and spinal fluid cultures
    *   Collect and freeze additional serum (5 mL) and urine (10 mL) samples for culture and later analysis of special metabolites, substrates, and hormones if hypoglycemia is persistent or recurrent.

TABLE 3.2
Laboratory Abnormalities Associated
with Specific Pediatric Hypoglycemic Disorders

Abnormality	Associated Diagnoses
Hyperlipidemia	Type I glycogen storage disease (GSD) secondary to glucose-6-phosphatase deficiency Fructose-1,6-diphosphatase (F-1,6-DiPase) deficiency
Hyperuricemia	Type I GSD Medium-chain acyl dehydrogenase (MCAD) deficiency
Hyperammonemia	Carnitine deficiency, including MCAD deficiency Reye's syndrome
Elevated creatine kinase and/or lactate dehydrogenase	Carnitine deficiency, including MCAD deficiency

3. For advanced diagnosis of inborn errors of metabolism causing hypoglycemia, elect to obtain blood and urine for the following special tests:
   • Plasma free fatty acids (PFFAs)
   • Plasma amino acids (alanine and branched-chain amino acids first, others second)
   • Serum beta-hydroxybutyrate
   • Serum carnitine/acylcarnitine ratio
   • Urine organic acids
4. Obtain additional routine and specialized blood tests if rare syndromes causing hypoglycemia are suspected based on the finding of hypoglycemia associated with one or more of the laboratory abnormalities listed in "Laboratory Abnormalities Associated with Specific Pediatric Hypoglycemic Disorders" (Table 3.2) or a family history of sudden infant death syndrome (SIDS) or Reye's syndrome:
   • Ammonia, creatine kinase (CK), and lactate dehydrogenase (LD)
   • Uric acid
   • Cholesterol and triglycerides

Interpretation

1. Based on both clinical and laboratory findings, evaluate for causes of hypoglycemia characteristic for:
   • Age at onset
   • Infants with neonatal intrauterine growth retardation
   • Infants of diabetic mothers
   • Childhood ketosis
2. Evaluate for causes of *ketotic hypoglycemia* if insulin secretion is normal. *Nonketotic or hypoketotic* hyperinsulinemia may be a cause of hypoglycemia in patients with fatty acid oxidation defects.
3. Thyrotoxicosis or hyperinsulinemia may be the cause of glucose *overutilization* in children with fasting hypoglycemia. Disorders of glycogenolysis or gluconeogenesis may be the cause of glucose *underutilization* in children with hypoglycemia.
4. If hypoglycemia occurs at an *early* age, evaluate for defective glycogenolysis. If the onset of hypoglycemia occurs *later* in childhood, evaluate for fatty acid oxidation defects.
5. If GH, glucagon, or cortisol levels are low, evaluate for counter-regulatory hormone deficiency or deficiencies.
6. If hepatomegaly is present, evaluate for glycogen storage diseases (GSDs).

Notes

1. The differential diagnosis of hypoglycemia in pediatric individuals represents a very large and diverse group of conditions that defy easy categorization.
2. Hypoglycemia secondary to hyperinsulinemia is the most common cause of persistent hypoglycemia in infants and children.

3. Rare syndromes causing hypoglycemia include type I GSD (i.e., glucose-6-phosphatase or fructose-1,6-diphosphatase [F-1,6-DiPase] deficiency) or medium-/long-chain acyl–CoA dehydrogenase deficiency (M/LCADD) (Table 3.2).

ICD-9 Codes

Conditions that may justify these tests *include but are not limited to*:

251.2   hypoglycemia, unspecified
253.2   hypopituitarism
E932.3 insulins and antidiabetic agents toxicity/drug abuse

Suggested Reading

Gregg AR, Weiner CP. "Normal" umbilical arterial and venous acid–base and blood gas values. *Clin Obstet Gynecol* 1993; 36:24–32.

Henriquez H, el Din A, Ozand PT, Subramanyam SB, al Gain SI. Emergency presentations of patients with methylmalonic acidemia, propionic acidemia and branched chain amino acidemia (MSUD). *Brain Dev* 1994; 16(Suppl):86–93.

Little GL, Boniface KS. Are one or two dangerous? Sulfonylurea exposure in toddlers. *J Emerg Med* 2005; 28:305–10.

Salhab WA, Wyckoff MH, Laptook AR, Perlman JM. Initial hypoglycemia and neonatal brain injury in term infants with severe fetal acidemia. *Pediatrics* 2004; 114:361–6.

3.3.3   Extended Fasting and Monitoring of Glucose in Pediatric Patients

Indications for Test

Extended monitoring of fasting glucose in pediatric patients is indicated:

*   In suspected cases of hyperinsulinemic hypoglycemia when an episode of hypoglycemia has been confirmed with a random blood glucose determination
*   When the facilities with adequate staff to respond to both the diagnostic and resuscitative demands of patients with potentially severe hypoglycemia undergoing this monitoring are available

Procedure

1. Because of the potential risk of provoking an episode of severe encephalopathy in a child with a defect in fatty acid metabolism, measure plasma carnitine and acylcarnitine before initiating this elective fast. Initiate fast only if these analytes are normal.
2. Admit the patient to the hospital.
3. Insert a normal saline or heparin lock i.v. for blood sampling and administration of 20 to 25% dextrose and/or glucagon as necessary.
4. Begin the fast after the evening meal, timed so the anticipated hypoglycemia will occur between 1000 and 1800 hours when the patient is normally awake and physician and laboratory staff availability are optimal.
5. Obtain blood samples at 8, 16, 20, 21, 22, 23, and 24 hours into the fast and whenever symptoms of hypoglycemia occur.
6. Assay blood samples for plasma glucose, ketones, or beta-hydroxybutyrate, lactate, alanine, and insulin.
7. Obtain blood samples at the end of the fast (either at 24 hours or when symptomatic hypoglycemia occurs) for cortisol, growth hormone, free fatty acid, free and total plasma carnitine, and quantitative plasma amino acids.

8. Collect urine at the end of the fast for quantitative amino acid determination and organic acid analysis.
9. At the time of hypoglycemia, administer glucagon, 30 mcg/kg (maximum 1 mg) i.v., and obtain a blood sample for plasma glucose determination 20 minutes later.
10. If hypoglycemia has not resolved following parenteral glucagon, administer oral carbohydrate (10–20 g) if the patient is without neurologic impairment or administer i.v. dextrose (10 or 25%) as a bolus (0.25–0.50 g/kg) followed by a continuous infusion (4–6 mg/kg/min) until the patient is stable and euglycemic.

Interpretation

1. A plasma glucose < 40 mg/dL (2.2 mmol/L) with hypoglycemic symptoms during fasting is confirmation of the diagnosis of hypoglycemia in pediatric patients.
2. The presence of fasting ketonemia with hypoglycemia is seen in individuals with galactosemia and defects of gluconeogenesis or amino acid metabolism.
3. Endocrine deficiencies such as hypopituitarism with isolated growth hormone or ACTH deficiency and abnormalities in substrate availability such as ketotic hypoglycemia result in hypoglycemia during fasting.
4. The absence of ketosis, associated with hypoglycemia, suggests hyperinsulinism, disorders of fatty acid oxidation (FAO), hereditary fructose intolerance (HFI), or certain glycogen storage diseases (GSDs) types I, III, VI, and IX.
5. Lactic acidosis is associated with HFI, disorders of gluconeogenesis, disorders of FAO, and the glycogen storage diseases (especially glucose-6-phosphatase deficiency in type I GSD).
6. Quantitative amino acids will demonstrate the characteristic elevations of leucine, isoleucine, valine, alanine, and glutamine in maple syrup urine disease (MSUD) or branched-chain alpha-ketoacid dehydrogenase deficiency.
7. Hyperalaninemia is seen in deficiencies of glucose-6-phosphatase, pyruvate carboxylase, and phosphoenolpyruvate (PEP) carboxykinase.
8. Plasma insulin is normally suppressed to <10 mcU/mL during hypoglycemia in all infants and children. A plasma insulin > 10 mcU/mL or an insulin (mcU/mL)/glucose (mg/dL) ratio > 0.3 is presumptive evidence for hyperinsulinism.
9. Normally, in the nonfasted state, administration of glucagon should result in a >50% increase in serum glucose within 30 minutes. If this occurs when glucagon is administered for fasting hypoglycemia, hyperinsulinism or glucagon deficiency should be suspected. A totally absent glycemic response to glucagon is seen in GSD (especially type I varient) or with depletion of hepatic glycogen stores.
10. Elevations in free fatty acid (FFA), accompanied by low concentrations of free and total carnitine, suggest a defect in fatty acid metabolism. Reye's syndrome gives a similar picture but is also associated with defects in ureagenesis. FFAs are also increased in fructose-1,6-diphosphatase deficiency.
11. Urine organic acid analysis is necessary when specific enzyme defects are suspected, such as in the rare syndromes of M/LCADD, isovaleric acidemia, propionic acidemia, or methylmalonic aciduria.
12. Hypopituitarism should be suspected if, in a hypoglycemic serum sample, the cortisol is <20 mcg/dL, GH is <7 ng/mL, or ACTH is low.
13. The most common endocrine or metabolic disorders identified during a fast in children are hyperinsulinism and beta-oxidation (i.e., FAO) defects, which are usually discovered in 20 to 25% of hypoglycemic children undergoing extended fasting.

Notes

1. Because the symptoms and signs of hypoglycemia are nonspecific, particularly in children, extended fasting should not be initiated until hypoglycemia is confirmed.
2. Because hypoglycemia is often intermittent, the detection of laboratory test abnormalities that suggest a specific pathophysiologic cause for hypoglycemia are more likely to be found when the patient is symptomatic.

3. Symptoms can occur either spontaneously, when precipitated by an intercurrent illness or stress, or in response to a monitored fast of 8 to 24 hours.
4. Although significant morbidity may occur during a fast, when compared to the unpredictable timing and circumstances of spontaneous hypoglycemia, monitored fasting has the advantage of readily available medical personnel prepared to respond as needed.

ICD-9 Codes

Conditions that may justify this test *include but are not limited to*:

251.1   hypoglycemia, specified/hyperinsulinism
251.2   hypoglycemia, unspecified

Suggested Reading

Morris AA, Thekekara A, Wilks Z, Clayton PT, Leonard JV, Aynsley-Green A. Evaluation of fasts for investigating hypoglycaemia or suspected metabolic disease. *Arch Dis Child* 1996; 75:115–9.

3.3.4   Beta-Hydroxybutyrate (BHB) Testing in Hypoglycemic Pediatric Patients

Indications for Test

Beta-hydroxybutyrate testing is indicated in cases of:

- Fasting hypoglycemia (e.g., due to alcohol ingestion)
- Disorders of intermediary carbohydrate metabolism

Procedure

1. Use a rapid, commercially available assay (e.g., Precision Xtra® by Abbot Labs) for measurement of beta-hydroxybutyrate in a drop of capillary blood.

Interpretation

1. The absence of ketosis in a patient with hypoglycemia is abnormal and suggests the diagnosis of either hyperinsulinism or a rare inborn error of fatty acid metabolism.
2. Elevated BHB occurs in severe ethanol toxicity with coma, usually at a lower blood alcohol level in children than in adults.
3. The differential diagnosis of ketotic hypoglycemia of childhood includes corticosteroid or growth hormone deficiency, salicylate poisoning, glycogen synthase deficiency, and other inborn errors of carbohydrate metabolism.

Notes

1. In fatalities where there is a history of alcohol intoxication, a raised level of BHB in postmortem samples of vitreous humor and urine can be used as an indicator for possible alcoholic ketosis-related hypoglycemia.
2. In the rare case of complete absence of insulin receptors (i.e., Donohue's syndrome), fasting plasma BHB is low, falls during the early postprandial period, and fails to rise in response to hypoglycemia consistent with persistent insulin-like effects on the liver secondary to the insulinomimetic activity of grossly elevated circulating insulin levels on the IGF-1 receptor (Ogilvy-Stuart et al., 2001).

3. At the end of a prolonged fast, plasma BHB (in mmol/L) was lower in insulinoma patients (median, 0.3; range 0.1–2.7) compared to normals (median, 4.5; range 1.2–7.0) ($p < 0.0001$) (O'Brien et al., 1993).

4. Ketosis may be expected during fasting, after prolonged exercise, and when a high-fat diet is consumed. BHB, the principal "ketone" body in starving humans, normally displaces glucose as the predominant fuel for brain and decreases the need for glucose synthesis from liver, kidney, and muscle-derived amino acids.

5. In acute diabetic ketoacidosis, the ratio of BHB to acetoacetate (AcAc) rises from normal (i.e., 1:1) to as high as 10:1. In response to insulin therapy, BHB levels commonly decrease long before AcAc levels.

ICD-9 Codes

Conditions that may justify this test *include but are not limited to*:

251.2 hypoglycemia, unspecified
270.3 disturbances of branched chain amino-acid metabolism
305.0 acute alcohol intoxication

Suggested Reading

Denmark LN. The investigation of beta-hydroxybutyrate as a marker for sudden death due to hypoglycemia in alcoholics. *Forensic Sci Int* 1993; 62:225–32.

Laffel L. Ketone bodies: a review of physiology, pathophysiology and application of monitoring to diabetes. *Diabetes Metab Res Rev* 1999; 15:412–26.

Lamminpaa A, Vilska J, Korri UM, Riihimaki V. Alcohol intoxication in hospitalized young teenagers. *Acta Paediatr* 1993; 82:783–8.

Mitchell GA, Kassovska-Bratinova S, Boukaftane Y, Robert MF, Wang SP, Ashmarina L, Lambert M, Lapierre P, Potier E. Medical aspects of ketone body metabolism. *Clin Invest Med* 1995; 18:193–216.

O'Brien T, O'Brien PC, Service FJ. Insulin surrogates in insulinoma. *J Clin Endocrinol Metab* 1993; 77: 448–51.

Ogilvy-Stuart AL, Soos MA, Hands SJ, Anthony MY, Dunger DB, O'Rahilly S. Hypoglycemia and resistance to ketoacidosis in a subject without functional insulin receptors. *J Clin Endocrinol Metab* 2001; 86:3319–26.

Rutledge SL, Atchison J, Bosshard NU, Steinmann B. Case report: liver glycogen synthase deficiency—a cause of ketotic hypoglycemia. *Pediatrics* 2001; 108:495–7.

3.3.5   Organic Acids: Urine Testing in Hypoglycemic Pediatric Patients

Indications for Test

Testing for organic acids in urine is indicated:

- In families where sudden infant death syndrome (SIDS), near-miss SIDS, or Reye's-like disease has occurred
- When screening for cases of medium-chain acyl–CoA dehydrogenase deficiency (MCADD)
- When screening for peroxisomal disorders including Zellweger syndrome, neonatal adrenoleuko-dystrophy, 3-methylcrotonyl–CoA carboxylase (3-MCC) deficiency, and isolated, single deficiencies of peroxisomal beta-oxidation enzymes

Procedure

1. Collect a newborn urine sample on filter paper for assay within 28 days.
2. Analyze the urine sample for organic acids by gas chromatography–mass spectrometry (GC-MS) or LC-MS/MS.
3. Measure urinary organic acids before and after loading with 3-phenylpropionate when making a diagnosis of MCADD.
4. Screen the urine sample for 3-hydroxyisovalerylcarnitine levels with LC-MS/MS in newborns to detect 3-MCC deficiency.

Interpretation

1. Markedly elevated excretion of urinary organic acids, particularly phenylpropionylglycine, after the 3-phenylpropionate load test is consistent with, but not completely specific for, primary MCADD.

Notes

1. MCADD is an autosomal, recessively transmitted mitochondrial beta-oxidation defect involving C6 to C12 carboxylic acid metabolism, resulting in episodic attacks of life-threatening hypoketotic hypoglycemia commonly provoked by infections or prolonged fasting.
2. Most (80%) MCADD patients are homozygous for a common mutation, 985A→G, with another 18% of patients known to be heterozygous. In addition, many, rare disease-causing mutations have been identified and characterized, but no clear genotype–phenotype correlation has been found for any enzyme abnormality.
3. Recurrent hypoglycemic episodes in MCADD can be avoided by application of dietary measures.
4. Patients with short-chain acyl–CoA dehydrogenase deficiency do not have problems with hypoglycemia.
5. Using GC-MS, nine types of alpha-ketoacids, including fumaric, methylmalonic, *N*-acetylaspartic, pyroglutamic, and homogentisic acids, as well as glutaric acid but not phenylpyruvate, 2-ketoadipate, or *p*-OH-phenylpyruvate, are recoverable from filter paper urine samples.
6. Pyruvate, branched-chain alpha-ketoacids, alpha-ketoadipate, and alpha-ketoglutarate are stable on filter paper for at least 28 days.
7. When collected on filter paper, alpha-ketoacids, such as succinylacetone, are unstable, resulting in difficulties in the diagnosis of tyrosinemia type 1.

ICD-9 Codes

Conditions that may justify this test *include but are not limited to*:

Metabolic disorders
277.85   fatty acid oxidation
277.86   peroxisome-related/adrenoleukodystrophy

Other
253.3   pituitary dwarfism
331.81   Reye's syndrome
798.0   sudden infant death syndrome (Reye's-like)

Suggested Reading

Andresen BS, Dobrowolski SF, O'Reilly L, Muenzer J, McCandless SE, Frazier DM, Udvari S, Bross P, Knudsen I, Banas R, Chace DH, Engel P, Naylor EW, Gregersen N. Medium-chain acyl–CoA dehydrogenase (MCAD) mutations identified by MS/MS-based prospective screening of newborns differ from those observed in patients with clinical symptoms: identification and characterization of a new, prevalent mutation that results in mild MCAD deficiency. *Am J Hum Genet* 2001; 68:1408–18.

Barbas C, Garcia A, de Miguel L, Simo C. Evaluation of filter paper collection of urine samples for detection and measurement of organic acidurias by capillary electrophoresis. *J Chromatogr B Analyt Technol Biomed Life Sci* 2002; 780:73–82.

Bhala A, Willi SM, Rinaldo P, Bennett MJ, Schmidt-Sommerfeld E, Hale DE. Clinical and biochemical characterization of short-chain acyl-coenzyme A dehydrogenase deficiency. *J Pediatr* 1995; 126:910–5.

Koeberl DD, Millington DS, Smith WE, Weavil SD, Muenzer J, McCandless SE, Kishnani PS, McDonald MT, Chaing S, Boney A, Moore E, Frazier DM. Evaluation of 3-methylcrotonyl–CoA carboxylase (3-MCC) deficiency detected by tandem mass spectrometry newborn screening. *J Inherit Metab Dis* 2003; 26:25–35.

Lehnert W. Experience with the 3-phenylpropionic acid loading test for diagnosis of medium-chain acyl–CoA dehydrogenase deficiency (MCADD). *Padiatr Padol* 1993; 28:9–12.

Ohie T, Fu X, Iga M, Kimura M, Yamaguchi S. Gas chromatography–mass spectrometry with *tert*-butyldimethylsilyl derivation: use of the simplified sample preparations and the automated data system to screen for organic acidemias. *J Chromatogr B Biomed Sci Appl* 2000; 746:63–73.

Yamaguchi S, Iga M, Kimura M, Suzuki Y, Shimozawa N, Fukao T, Kondo N, Tazawa Y, Orii T. Urinary organic acids in peroxisomal disorders: a simple screening method. *J Chromatogr B Biomed Sci Appl* 2001; 758:81–6.

Zschocke J, Penzien JM, Bielen R, Casals N, Aledo R, Pié J, Hoffmann GF, Hegardt FG, Mayatepek E. The diagnosis of mitochondrial HMG-CoA synthase deficiency. *J Pediatr* 2002; 140:778–80.

3.3.6   Free Fatty Acid (FFA) Testing in Hypoglycemic Pediatric Patients

Indications for Test

FFA testing is indicated in:

- Children with hypoglycemia, vomiting, lethargy, encephalopathy, respiratory arrest, hepatomegaly, seizures, apnea, cardiac arrest, and/or coma
- Children with developmental and behavioral disability, chronic muscle weakness, failure to thrive, cerebral palsy, and attention-deficit disorder (ADD)
- A variety of syndromes of impaired mitochondrial FFA beta-oxidation
- Systemic carnitine deficiency, carnitine transferase deficiency (CTD), and medium- and long-chain acyl–CoA dehydrogenase deficiency (MCAD and LCAD deficiency)
- Multiple acyl–CoA dehydrogenase deficiency (MACADD), including glutaric aciduria, ethylmalonic–adipic aciduria, and riboflavin-responsive MACADD

Procedure

1. Using five blood spots on filter paper, each 3 mm in diameter, measure the following FFAs:
   - Octanoate
   - Decanoate
   - *cis*-4-Decenoic acid (C10:1)
   - *cis*-5-Tetradecenoic acid (C14:1)
2. Use one-step transmethylation followed by GC-MS or LC-MS/MS as the current methods of choice for assay of FFAs.
3. When ordering quantitative nonesterified FFA (NEFA) testing, be sure that the method used has a lower limit of detection (LLD) of 100 μmol/L and an upper limit of detection (ULD) of 2000 μmol/L.
4. As appropriate, measure carnitine and ammonia in blood samples found to have elevated FFAs.

Interpretation

1. In CTD, precursors of the blocked enzyme reaction (FFAs, carnitine, ammonia) are elevated in blood with lack, or deficiency, of the metabolites unique to this reaction (i.e., triglycerides and urine acetyl carnitine relative to longer acyclic derivatives).
2. In MCADD, only C10:1 fatty acid is increased.
3. In very-long-chain acyl–CoA dehydrogenase (VLCAD) deficiency, only C14:1 fatty acid is increased.
4. In MACADD, both C10:1 and C14:1 fatty acids are increased

Notes

1. FAO plays a major role in energy production during prolonged fasting and in working cardiac and skeletal muscle. Life-threatening episodes of coma and hypoglycemia induced by fasting are common presenting features in most FAO disorders.
2. At least 12 fatty acid oxidation disorders are known to be responsible for cases of sudden and unexpected death in early childhood (Oakes and Furler, 2002).
3. Advances in tandem MS technology are likely to result in the increased identification of patients with milder variant FAO disorders.
4. Deficient states of the gene for MCAD deficiency, located on chromosome one (1p31), are inherited in an autosomal recessive manner with a high frequency among people of Northern European descent (Wang et al., 1999).
5. Simultaneous infusion of ^{14}C-palmitate and the non-beta-oxidizable FFA analog [9,10-^{3}H]-(R)-2-bromopalmitate (^{3}H-R-BrP) and measurement of their disappearance from plasma into tissues have been used to demonstrate tissue-level control of FFA utilization.
6. FFA utilization is not simply supply driven but is highly dependent on the rate or capability of tissue metabolism via mitochondrial FFA beta-oxidation.
7. In short-chain acyl–CoA dehydrogenase (SCAD) deficiency, chronic CNS toxicity is a dominant feature reflecting the fact that short-chain fatty acids more readily cross the blood–brain barrier than longer chain fatty acids.

ICD-9 Codes

Conditions that may justify this test *include but are not limited to*:

Metabolic disorders
277.85   fatty acid oxidation
277.87   mitochondrial metabolism

Neurodysfunction
314.0   attention deficit disorder
343   infantile cerebral palsy
348.30   encephalopathy, unspecified
780.01   coma
780.39   convulsions

Other
251.2   hypoglycemia unspecified
427.5   cardiac arrest
728.87   muscle weakness (generalized)
783.41   failure to thrive in childhood
786.03   apnea
789.1   hepatomegaly
799.1   respiratory arrest

Suggested Reading

Kimura M, Yoon HR, Wasant P, Takahashi Y, Yamaguchi S. A sensitive and simplified method to analyze free fatty acids in children with mitochondrial beta oxidation disorders using gas chromatography/mass spectrometry and dried blood spots. *Clin Chim Acta* 2002; 316:117–21.

Oakes NE, Furler SM. Evaluation of free fatty acid metabolism *in vivo. Ann NY Acad Sci* 2002; 967:158–75.

Rufini S, Bragetti P, Brunelli B, Campolo G, Lato M. Non-ketotic hypoglycemia caused by carnitine palmitoyl transferase 1 deficiency. *Pediatr Med Chir* 1993; 15:63–6.

Wang SS, Fernhoff PM, Hannon WH, Khoury MJ. Medium chain acyl–CoA dehydrogenase deficiency human genome epidemiology review. *Genet Med* 1999; 1:332–9.

3.3.7   Carnitine/Acylcarnitine Ratio in Hypoglycemic Pediatric Patients

Indications for Test

Measurement of the carnitine/acylcarnitine ratio is indicated in:

- Hypoglycemic infants with Reye's-like syndrome suspected of having VLCAD deficiency (VLCADD) or carnitine palmitoyltransferase (CPT)-I or -II deficiency
- Patients with exercise intolerance and hypoglycemia

Procedure

1. Screen for acylcarnitines in blood spots on filter paper (i.e., Guthrie card) using LC-MS/MS.
2. Measure plasma free carnitine and acylcarnitine levels as well as tetradecenoylcarnitine, octanoyl-carnitine, and dodecanoylcarnitine in patients who are positive on screening with LC-MS/MS.
3. Calculate the carnitine/acylcarnitine and tetradecenoylcarnitine/dodecanoylcarnitine ratios.
4. Check for myoglobinuria.

Interpretation

1. A low serum or plasma free carnitine and low carnitine/acylcarnitine ratio are characteristic of these disorders of FAO, but some overlap with normals may occur.
2. There is a lesser degree of overlap in the ratio of tetradecenoylcarnitine/dodecanoylcarnitine (T/D), characteristic of VLCADD, compared to the T/D ratio found in hypoglycemic infants without disorders of peroxisomal FAO.
3. In atypical cases, specific enzyme assay may be necessary to make a diagnosis of a disorder of FAO.
4. Diagnosis of MCAD deficiency by acylcarnitine analysis is established on finding an elevated C8 acylcarnitine level (>0.3 μmol), a C8/C10 acylcarnitine ratio > 5, and low or normal levels of acylcarnitines with fatty acid chain lengths > C10.
5. The elevated C8 acylcarnitine (octanoylcarnitine)/acylcarnitine ratios clearly differentiate MCAD deficiency from normal controls.
6. CPT-II deficiency is invariably associated with repetitive episodes of myoglobinuria triggered by exercise, cold exposure, fever, or fasting, with the diagnosis dependent on demonstration of the CPT-II enzyme deficiency in muscle.
7. If urine myoglobin testing is positive, it provides evidence suggestive of defects in mitochondrial respiratory chain DNA, particularly complexes I, III, and IV.
8. Rarely, exercise-intolerant patients have an alteration in the assembly of the complex IV of the mitochondrial respiratory chain. When these patients are treated with statins in combination with fibrates, myoglobinuria and hypoglycemia can result.

Notes

1. Carnitine is derived from protein in the diet and can be synthesized from lysine and methionine.
2. Carnitine is a pivotal substrate in:
   - Branched-chain amino acid metabolism
   - Removal of excess acyl groups
   - Peroxisomal FAO, in addition to buffering of the acyl–coenzyme A (CoA)/nonacylated CoA ratio
3. Primary and secondary forms of carnitine deficiency result in increased lipolysis, increased lipid peroxidation, accumulation of acylcarnitines, and altered membrane permeability.
4. Metabolic consequences of defective carnitine–acylcarnitine translocase (CACT) are fasting hypoketotic hypoglycemia, hyperammonemia, elevated creatine kinase and transaminases, dicarboxylic aciduria, very low free carnitine, and marked elevation of the long-chain acylcarnitines.
5. Because acylcarnitine profiles observed in CACT-deficient patients are identical to those in CPT-II-deficient patients, definitive identification of CACT deficiency requires determination of the enzymatic activity of CACT.
6. A characteristic pattern of change in serum free carnitine levels during and after attacks of Reye's-like disease in 1- to 3-year-olds with MCADD has been reported.
7. Spontaneous hypoglycemic episodes may occur in the rare patient with riboflavin-responsive, mild MCADD of the ethylmalonic–adipic aciduria type when treated with supplemental L-carnitine (Knudsen et al., 2001).
8. Hypoglycemia and/or exercise intolerance may be the only clinical presentation of defects in mitochondrial respiratory chain DNA, particularly those involving complexes I, III, or IV.

ICD-9 Codes

Conditions that may justify this test *include but are not limited to*:

251.2	hypoglycemia, unspecified
277.85	disorders of fatty acid oxidation
331.81	Reye's syndrome

Suggested Reading

Andresen BS, Dobrowolski SF, O'Reilly L, Muenzer J, McCandless SE, Frazier DM, Udvari S, Bross P, Knudsen I, Banas R, Chace DH, Engel P, Naylor EW, Gregersen N. Medium-chain acyl–CoA dehydrogenase (MCAD) mutations identified by MS/MS-based prospective screening of newborns differ from those observed in patients with clinical symptoms: identification and characterization of a new, prevalent mutation that results in mild MCAD deficiency. *Am J Hum Genet* 2001; 68:1408–18.

Arenas J, Martin MA. Metabolic intolerance to exercise. *Neurologia* 2003; 18:291–302.

Bzduch V, Behulova D, Salingova A, Ponec J, Fabriciova K, Kozak L. Serum free carnitine in medium-chain acyl–CoA dehydrogenase deficiency. *Bratisl Lek Listy* 2003; 104:405–7.

Green A, Preece MA, de Sousa C, Pollitt RJ. Possible deleterious effect of L-carnitine supplementation in a patient with mild multiple acyl–CoA dehydrogenation deficiency (ethylmalonic–adipic aciduria). *J Inherit Metab Dis* 1991; 14:691–7.

Haut S, Brivet M, Touati G, Rustin P, Lebon S, Garcia-Cazorla A, Saudubray JM, Boutron A, Legrand A, Slama A. A deletion in the human QP-C gene causes a complex III deficiency resulting in hypoglycaemia and lactic acidosis. *Hum Genet* 2003; 113:118–22.

Hoppel C. The role of carnitine in normal and altered fatty acid metabolism. *Am J Kidney Dis* 2003; 41(Suppl 4):S4–12.

Rubio-Gozalbo ME, Bakker JA, Waterham HR, Wanders RJ. Carnitine–acylcarnitine translocase deficiency: clinical, biochemical, and genetic aspects. *Mol Aspects Med* 2004; 25:521–32.

Schmidt-Sommerfeld E, Penn D, Duran M, Rinaldo P, Bennett MJ, Santer R, Stanley CA. Detection and quantitation of acylcarnitines in plasma and blood spots from patients with inborn errors of fatty acid oxidation. *Prog Clin Biol Res* 1992; 375:355–62.

Shigematsu Y, Hirano S, Hata I, Tanaka Y, Sudo M, Tajima T, Sakura N, Yamaguchi S, Takayanagi M. Selective screening for fatty acid oxidation disorders by tandem mass spectrometry: difficulties in practical discrimination. *J Chromatogr B Analyt Technol Biomed Life Sci* 2003; 792:63–72.

Van Hove JL, Zhang W, Kahler SG, Roe CR, Chen YT, Terada N, Chace DH, Iafolla AK, Ding JH, Millington DS. Medium-chain acyl–CoA dehydrogenase (MCAD) deficiency: diagnosis by acylcarnitine analysis in blood. *Am J Hum Genet* 1993; 52:958–66.

3.3.8 Quantitative Amino Acid (AA), Alanine, and Branched-Chain AA Levels in Plasma and Urine in Hypoglycemic Pediatric Patients

Indications for Test

Quantitative measurements of alanine and branched-chain amino acids are indicated in:

- Mild or severe hypoglycemia associated with any of the various organic acidemias
- Branched-chain amino acidemia (e.g., maple syrup urine disease [MSUD]), particularly in people from the Middle East

Procedure

1. Measure branched-chain L-amino acids in plasma and urine by quantitative amino acid analysis using LC-MS/MS.
2. Specifically quantitate L-alloisoleucine, isoleucine, valine, leucine, and alanine in blood and urine.
3. In families of children diagnosed with organic acidemias, use home monitoring methods to determine branched-chain ketoacid excretion and blood glucose levels on a frequent basis and report results to a specialist for assistance with management of diet and other therapy.
4. Monitor branched-chain amino acid levels in plasma at the time of specialty clinic visits to promptly diagnose metabolic derangements in established hypoglycemia patients with known organic acidemias and treat with modified or special diets.

Interpretation

1. Mild hypoglycemic events tend to occur in the rare neonatal syndromes of classic MSUD, ethylmalonic aciduria, and isovaleric acidemia (IVA). Fewer than 50% of MSUD patients older than 8 months with pyruvate carboxylase deficiency, methylmalonic acidemia, or propionic acidemia will experience a hypoglycemic event.
2. Patients with 3-hydroxy-3-methyl glutaryl–CoA lyase deficiency, holocarboxylase synthetase deficiency, MCAD or LCAD deficiency, neonatal-onset 3-methylglutaconic aciduria, glutaric aciduria type 2, or disorders of fructose metabolism have moderate to severe hypoglycemia associated with metabolic crises.
3. Plasma L-alloisoleucine above the cutoff value of 5 µmol/L is the most specific and sensitive diagnostic marker for all forms of MSUD. The frequency of diagnostically significant increases in other branched-chain amino acids was reported to be <45%.
4. In MSUD, the molar ratio of leucine to alanine in plasma may range from 1.3 to 12.4 compared with a control range of 0.12 to 0.53.
5. In patients with diabetes mellitus, ketotic hypoglycemia, phenylketonuria, and, in obligate heterozygous parents of MSUD patients, alloisoleucine are not significantly elevated compared to healthy subjects.

Notes

1. Patients with organic acidemias that primarily manifest with neurologic signs, including 4-hydroxy-butyric aciduria, infantile-onset 3-methylglutaconic aciduria, and glutaric aciduria type 1, as well as beta-ketothiolase deficiency, biotinidase deficiency, or intermittent or intermediate MSUD, do not experience hypoglycemia.
2. The most common branched-chain organic acidurias include MSUD, IVA, propionic aciduria, and methylmalonic aciduria, which result from abnormalities of specific enzymes involving the catabolism of the branched-chain amino acids leucine, isoleucine, and valine.
3. Molecular testing for the Y393N mutation of the E1 alpha subunit of the branched-chain alpha-ketoacid dehydrogenase enzyme allows early identification of infants who are at high risk for MSUD.

ICD-9 Codes

Conditions that may justify this test *include but are not limited to*:

251.2   hypoglycemia, unspecified
270.3   disturbances of branched-chain amino-acid metabolism
271.1   galactosemia

Suggested Reading

Morton DH, Strauss KA, Robinson DL, Puffenberger EG, Kelley RI. Diagnosis and treatment of maple syrup disease: a study of 36 patients. *Pediatrics* 2002; 109:999–1008.

Ogier de Baulny H, Saudubray JM. Branched-chain organic acidurias. *Semin Neonatol* 2002; 7:65–74.

Schadewaldt P, Bodner-Leidecker A, Hammen HW, Wendel U. Significance of L-alloisoleucine in plasma for diagnosis of maple syrup urine disease. *Clin Chem* 1999; 45:1734–40.

Worthen HG, al Ashwal A, Ozand PT, Garawi S, Rahbeeni Z, al Odaib A, Subramanyam SB, Rashed M. Comparative frequency and severity of hypoglycemia in selected organic acidemias, branched chain amino acidemia, and disorders of fructose metabolism. *Brain Dev* 1994; 16(Suppl):81–5.

3.3.9   Testing for Reducing Substances in Urine in Screening for Galactosemia and Hereditary Fructose Intolerance (HFI) in Hypoglycemic Pediatric Patients

Indications for Test

Screening for reducing substances in urine as a test for galactosemia or HFI in pediatric patients is indicated:

- In cases of hypoglycemia in infants when sucrose feedings are begun after weaning
- Based on a detailed nutritional history of gastrointestinal symptoms after high fructose intake

Procedure

1. Obtain random urine for glucose measurement by dry chemical strip analysis or quantitative lab method.
2. Test urine for the presence of reducing substances (usually detected with Clinitest®).

Interpretation

1. In a patient without glucosuria, but a positive Clinitest® reaction, even if trace, consider galactosemia, benign fructosuria, or HFI.
2. This test has poor specificity and results in a high number of false positives.

TABLE 3.3
Fructose Tolerance Test Protocol

Time	Tests to Be Performed
0[a]	Glu, K+, LA, Pi, UA

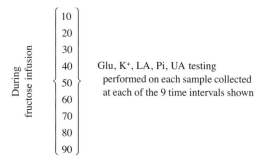

Time after beginning fructose infusion (minutes)

During fructose infusion

$\left\{\begin{array}{c}10\\20\\30\\40\\50\\60\\70\\80\\90\end{array}\right.$ Glu, K+, LA, Pi, UA testing performed on each sample collected at each of the 9 time intervals shown

Time after stopping fructose infusion (minutes)

After stopping fructose infusion

$\left\{\begin{array}{c}10\\20\\30\end{array}\right.$ Glu, K+, LA, Pi, UA testing performed on each sample collected at each of the 3 time intervals shown

[a] Just prior to fructose infusion.

Note: Glu, glucose; K+, potassium; LA, lactic acid; Pi, inorganic phosphorus; UA, uric acid.

Notes

1. Congenital aldolase B deficiency causes the HFI syndrome, characterized by hypoglycemia and severe abdominal symptoms.
2. The two most common aldolase B mutations, A149P and A174D, account for more than 70% of HFI cases. These mutations can be detected by polymerase chain reaction (PCR) analysis of a 224-base-pair segment of exon 5.
3. Infants with galactosemia usually present with an undiagnosed acute illness and may have only a trace of non-glucose-reducing substances in the urine on screening.
4. The prevalence of HFI has been estimated to be as high as 1 per 20,000 individuals and is an autosomal recessive trait that is equally distributed between the sexes.
5. A breath test with measurement of hydrogen and methane after receiving 150 mL of a 33% fructose solution is a widely available screening test with low specificity for HFI (Choi et al., 2003).
6. The "Fructose Tolerance Test Protocol" (Table 3.3) in which a 25% solution of 250 mg fructose per kg ideal body weight is infused for the purpose of inducing hypoglycemia, hypokalemia, hypophosphatemia, hyperuricemia, and lactic acidosis is a less palatable and more hazardous method of definitive diagnosis of HFI (Burmeister et al., 1991) than PCR analysis for aldolase B mutations.
7. Typically, in HFI there is a prompt fall in [phosphorus] and an increase in [uric acid] over baseline at 30 minutes after infusion of fructose. In addition, serum [K+] declines at 30 to 120 minutes with a rise in [lactate] between 60 and 120 minutes into the FTT.
8. Patients with benign, essential fructosuria will not become hypoglycemic in response to the FTT.
9. Definitive diagnosis of HFI may require laparoscopic liver biopsy with assay for reduced activity or deficiencies of fructaldose, fructose-1,6-diphosphatase, and glucose-6-phosphatase and for characteristic histologic changes (Steinmann and Gitzelmann, 1981).

ICD-9 Codes

Conditions that may justify this test *include but are not limited to*:

251.2 hypoglycemia, unspecified
271.1 galactosemia
789.0 abdominal pain

Suggested Reading

Ali M, Cox TM. Diverse mutations in the aldolase B gene that underlie the prevalence of hereditary fructose intolerance. *Am J Hum Genet* 1995; 56:1002–5.

Burmeister LA, Valdivia T, Nuttall FQ. Adult hereditary fructose intolerance. *Arch Int Med* 1991; 151:773–6 (see comment in *Arch Intern Med* 1992; 152:881).

Choi YK, Johlin FC Jr, Summers RW, Jackson M, Rao SSC. Fructose intolerance: an under-recognized problem. *Am J Gastroenterol* 2003; 98:1348–53.

Item C, Hagerty BP, Muhl A, Greber-Platzer S, Stockler-Ipsiroglu S, Strobl W. Mutations at the galactose-1–*p*-uridyltransferase gene in infants with a positive galactosemia newborn screening test. *Pediatr Res* 2002; 51:511–6.

Kirby LT, Norman MG, Applegarth DA, Hardwick DF. Screening of newborn infants for galactosemia in British Columbia. *Can Med Assoc J* 1985; 132:1033–5.

Kullberg-Lindh C, Hannoun C, Lindh M. Simple method for detection of mutations causing hereditary fructose intolerance. *J Inherit Metab Dis* 2002; 25:571–5.

Steinmann B, Gitzelmann R. The diagnosis of hereditary fructose intolerance. *Helv Paediatr Acta* 1981; 36:297–316.

3.4   Insulin Reserve

3.4.1   C-Peptide: Basal Level as a Test for Insulin Deficiency and Type 1 Diabetes Mellitus (T1DM)

Indications for Test

C-peptide testing is indicated:

- To assess pancreatic capacity for insulin secretion in suspected T1DM
- As preliminary to initiation of continuous subcutaneous insulin infusion or insulin pump therapy

Procedure

1. Following an 8- to 14-hour overnight fast, obtain a blood sample for C-peptide concentration and plasma glucose determinations.

Interpretation

1. A [C-peptide] can only be interpreted when the fasting glucose is ≤224 mg/dL (16.7 mmol/L).
2. The normal basal [C-peptide] is typically 0.9 to 4.0 ng/mL (0.297–1.32 nmol/L).
3. In T1DM without renal insufficiency, a [C-peptide] < 1.0 ng/mL, or 110% of the lower limit of normal (LLN) for the C-peptide assay, after fasting indicates inadequate insulin reserve or diminished beta-cell responsiveness to glucose as a secretagogue (i.e., glucose toxicity).
4. Beta-cell recovery may occur in T2DM, and [C-peptide] may return to normal if chronic hyperglycemia is corrected with a course of insulin therapy.

5. In cases of chronic kidney disease stage ≥ 3 (creatinine clearance < 50 mL/min), insulin dependency is considered to be present at a [C-peptide] ≤ 200% of the LLN for its assay.

6. A [C-peptide] > 3.0 ng/mL suggests recent carbohydrate ingestion or insulin resistance as is found in T2DM.

Notes

1. Qualification for insulin pump therapy by Medicare criteria is dependent on a sufficiently low [C-peptide] as defined above or the presence of beta-cell antibodies or antigenic markers for pancreatic failure such as antibodies to glutamic acid decarboxylase (GAD65) or glutamic acid decarboxylase antibody (GADA).

2. C-peptide is produced when proinsulin is cleaved to form one fully active insulin molecule and one C-peptide molecule.

3. C-peptide does not undergo first-pass hepatic metabolism and thus is metabolized more slowly than insulin. It serves as an integrated measure of endogenous insulin secretion.

4. In new-onset T1DM, it may be useful to determine the area under the curve (AUC) for [C-peptide] vs. time following an oGTT or mixed meal tolerance test (Iwasaki et al., 1994). Based on the magnitude of the AUC value, patients can be categorized as having *fulminant* (markedly low AUC) or *acute-onset* (less reduced AUC compared to healthy controls) T1DM. Thus, characteristics associated with fulminant T1DM are:
   - Hemoglobin A_{1c} (HbA_{1c}) ≤ 8.0%
   - Undetectable fasting [C-peptide]
   - [C-peptide] AUC ≤ 0.54 after oGTT

5. The fulminant T1DM patient has extremely reduced beta-cell function at onset of disease that rarely recovers later (Tanaka et al., 2004).

ICD-9 Codes

Conditions that may justify this test *include but are not limited to*:

250.X   diabetes mellitus

Suggested Reading

Iwasaki Y, Kondo K, Hasegawa H, Oiso Y. C-peptide response to glucagon in type 2 diabetes mellitus: a comparison with oral glucose tolerance test. *Diabetes Res* 1994; 25:129–37.

Tanaka S, Endo T, Aida K, Shimura H, Yokomori N, Kaneshige M, Furuya F, Amemiya S, Mochizuki M, Nakanishi K, Kobayashi T. Distinct diagnostic criteria of fulminant type 1 diabetes based on serum C-peptide response and HbA$_{1c}$ levels at onset. *Diabetes Care* 2004; 27:1936–41.

Umpaichitra V, Bastian W, Taha D, Banerji MA, Avruskin TW, Castells S. C-peptide and glucagon profiles in minority children with type 2 diabetes mellitus *J Clin Endocrinol Metab* 2001; 86:1605–9.

Vendrame F, Zappaterreno A, Dotta F. Markers of beta-cell function in type 1 diabetes mellitus. *Minerva Med* 2004; 95:79–84.

3.4.2   Glucagon-Stimulated C-Peptide Levels in Testing for Insulin Deficiency and Type 1 Diabetes (T1DM)

Indications for Test

Glucagon-stimulated C-peptide levels as a test for insulin deficiency is indicated to:

- Estimate endogenous insulin reserves in insulin-treated diabetics.
- Determine whether insulin can be safely discontinued in an insulin-treated diabetic.
- Assist in determining if insulin should be started in a newly diagnosed diabetic.
- Distinguish between T1DM and T2DM in investigational settings.

Procedure

1. Obtain baseline blood samples for fasting plasma glucose and serum C-peptide testing from insulin-dependent diabetic patients not treated with insulin for at least 2 days.
2. Abort test if the glucose is >224 mg/dL (16.7 mmol/L) in the baseline blood sample.
3. Administer glucagon, 1 mg i.v. over 10 seconds.
4. To reduce the risk of aspiration, turn the patient on his or her side, as vomiting may occur after intravenous glucagon administration.
5. Obtain blood samples for plasma glucose and serum [C-peptide] at 5 and 10 minutes after the injection of glucagon.

Interpretation

1. An increase in glucose from baseline should be observed for this test to be considered valid.
2. At similar glucose concentration, the fasting [C-peptide] of T1DM significantly overlaps that of T2DM.
3. Normally, glucagon-stimulated peak [C-peptide] can be as high as 9.0 ng/mL (2.97 nmol/L).
4. Most patients with a glucagon-stimulated peak serum [C-peptide] > 1.8 ng/mL (0.6 nmol/L) and an elevated fasting glucose > 110 mg/dL (6.1 mmol/L) or post-glucagon glucose > 200 mg/dL (11.1 mmol/L) appear to have T2DM.
5. An absent or very low [C-peptide] is indicative of T1DM. If the peak glucagon-stimulated [C-peptide] is <1.5 ng/mL (0.495 nmol/L), the patient will definitely require insulin therapy.
6. Glucagon-stimulated peak [C-peptide] of 1.5 to 4.5 ng/mL (0.495–1.485 nmol/L) indicates impaired insulin reserve.
7. This test is not very helpful in determining whether or not a patient with T2DM will require insulin treatment but will identify those whose pancreatic reserves of insulin are sufficient to accommodate ideal levels of activity and calorie intake. A clinical assessment of a patient's adherence to proper nutritional intake, degree of insulin resistance, and beta-cell function are of more use in making an insulin therapy decision.

Notes

1. Anti-insulin or islet-cell antibodies or antigenic markers for pancreatic failure such as glutamic acid decarboxylase antibodies (GADA) are good indicators of the T1DM insulin deficiency syndrome.
2. In addition to vomiting, side effects of intravenous glucagon administration include flushing and nausea.
3. Only 2 out of 8 T2DM using insulin therapies had pancreatic reserves poor enough to actually require insulin. Of 19 patients who were on diet and/or oral hypoglycemic agents, only one newly diagnosed DM patient had a definitive indication for insulin therapy based on glucagon-stimulated C-peptide (Rajasoorya et al., 1990).

ICD-9 Codes

Refer to Test 3.4.1 codes.

Suggested Reading

Madsbad S, Krarup T, McNair P, Christiansen C, Faber OK, Transbol I, Binder C. Practical clinical value of the C-peptide response to glucagon stimulation in the choice of treatment in diabetes mellitus. *Acta Med Scand* 1981; 210:153–6.

Rajasoorya C, Tan YT, Chew LS. The value of C-peptide level measurements in diabetes mellitus. *Ann Acad Med Singapore* 1990; 19:463–6.

3.4.3 First-Phase Insulin Response (FPIR) to Intravenous Glucose as a Test for Pancreatic Beta-Cell Function

Indications for Test

Measurement of FPIR to intravenous glucose may be indicated to:

- Help predict the onset of clinical diabetes mellitus (DM) in an individual with a positive test for DM-related autoantibodies, such as insulin autoantibody (IAA), GADA, and islet-cell autoantibody (ICA) (see Test 3.5.9).
- Identify children with transient hyperglycemia and apparent type 2 diabetes mellitus who will go on to, or are less likely to, develop type 1 diabetes mellitus.

Procedure

1. Following a 10- to 14-hour overnight fast, insert an antecubital catheter to be kept open with normal saline and used for glucose infusion and blood sampling.
2. Obtain a baseline blood sample for insulin testing.
3. Infuse glucose intravenously (i.v.) at a dose of 0.5 g/kg body weight as a 25% solution over precisely 3 minutes ± 15 seconds starting at time zero.
4. Obtain blood samples for insulin testing at 1 and 3 minutes following completion of the glucose infusion (i.e., 4 and 6 minutes into the test).

Interpretation

1. The FPIR is defined as the sum of the 1- and 3-minute post-glucose infusion insulin levels.
2. In 99% of normal individuals, the FPIR > 48 mcU/mL.
3. In individuals whose serum contains 1 or more DM-related autoantibodies and an FPIR < 48 mcU/mL, the risk of developing T1DM within 4 years is 100%.
4. In children with transient hyperglycemia, which is associated with a 10 to 27% risk of developing overt T1DM, a FPIR < 48 mcU/mL on two separate occasions is both 100% sensitive and specific for predicting the clinical onset of T1DM, usually within 9 months (Herskowitz et al., 1988).
5. The combination of various DM-related autoantibodies with the FPIR may more accurately predict the risk of developing T1DM, as shown in "Sensitivity and Positive Predictive Value (PPV) of the First-Phase Insulin Response (FPIR) in Conjunction with the Number of Positive Autoantibody Tests" (Table 3.4).

TABLE 3.4
Sensitivity and Positive Predictive Value (PPV) of the First-Phase Insulin Response (FPIR) in Conjunction with the Number of Positive Autoantibody Tests

FPIR (μU/mL)	Autoantibody Tests[a] (No. of Positive Tests)	Sensitivity (%)	PPV (%) 3-Year Risk	PPV (%) 5-Year Risk
<48	≥1	51	79	100
48–81	≥1	28	22	66
>81	≥1	18	16	21
<48	≥2	41	87	100
48–81	≥2	21	13	82
>81	≥2	15	19	29

[a] Insulin (IAA), glutamic acid decarboxylase (GADA), and islet-cell (ICA) autoantibodies.

6. The combination of (1) high-titer cytoplasmic islet-cell antibodies, (2) insulin autoantibodies, and (3) FPIR < 48 mcU/mL after i.v. glucose tolerance test helps to identify first-degree relatives of diabetics who will develop T1DM within <10 years (Bingley et al., 1996).

Notes

1. The oGTT is the least accurate predictor of T1DM, and FPIR has the highest negative predictive value and the greatest overall accuracy of predicting impending T1DM.
2. Immunological abnormalities resulting in high levels of DM-related autoantibodies have a much higher positive predictive value for development of T1DM than FPIR.
3. The presence of two or more DM-related autoantibodies is highly predictive of the development of T1DM among younger age relatives of diabetics.
4. Overt T1DM can usually be diagnosed with the use of a combination of tests, including measurement of fasting and postprandial glucose, calculation of glucose/insulin ratio, and [HbA$_{1c}$] determination.
5. An intravenous glucose tolerance test can be used to assess the adequacy of the FPIR and the degree of insulin resistance of a subject who has gastrointestinal disturbances that interfere with normal absorption of glucose or who are prone to the dumping syndrome (e.g., individuals after gastric bypass surgery) (Bingley et al., 1992; Galvin et al., 1992).

ICD-9 Codes

Refer to Test 3.4.1 codes.

Suggested Reading

Bingley PJ. Interactions of age, islet cell antibodies, insulin autoantibodies, and first-phase insulin response in predicting risk of progression to IDDM in ICA+ relatives: the ICARUS data set. Islet Cell Antibody Register Users Study. *Diabetes* 1996; 45:1720–8.

Bingley PJ, Colman P, Eisenbarth GS, Jackson RA, McCulloch DK, Riley WJ, Gale EA. Standardization of IVGTT to predict IDDM. *Diabetes Care* 1992; 15:1313–6.

Bruce DG, Chisholm DJ, Storlein LH, Kraegen EW. Physiological importance of deficiency in early prandial insulin secretion in non-insulin-dependent diabetes. *Diabetes* 1988; 37:736–44.

Bruttomesso D, Pianta A, Mari A, Valerio A, Marescotti MC, Avogaro A, Tiengo A, Del Prato S. Restoration of early rise in plasma insulin levels improves the glucose tolerance of type 2 diabetic patients. *Diabetes* 1999; 48:99–105.

Galvin P, Ward G, Walters J, Pestell R, Koschmann M, Vaag A, Martin I, Best JD, Alford F. A simple method for quantitation of insulin sensitivity and insulin release from an intravenous glucose tolerance test. *Diabet Med* 1992; 9:921–8.

Herskowitz RD, Wolfsdorf JI, Ricker AT, Vardi P, Dib S, Soeldner JS, Eisenbarth GS. Transient hyperglycemia in childhood: identification of a subgroup with imminent diabetes mellitus. *Diabetes Res* 1988; 9:161–7.

Lundgren H, Bengtsson C, Blohmè G, Lapidus L, Waldeström J. Fasting serum insulin concentration and early insulin response as risk determinants for developing diabetes. *Diabet Med* 1990; 7:407–13.

Luzi L, DeFronzo RA. Effect of loss of first-phase insulin secretion on hepatic glucose production and tissue glucose disposal in humans. *Am J Physiol* 1989; 257:E241–6.

Mitrakou A, Kelley D, Mokan M et al. Role of reduced suppression of glucose production and diminished early insulin release in impaired glucose tolerance. *N Engl J Med* 1992; 326:22–9.

Polonsky KS, Given BD, Hirsch L, Shapiro ET, Tillil H, Beebe C, Galloway JA, Frank BH, Karrison T, Van Cauter E. Quantitative study of insulin secretion and clearance in normal and obese subjects. *J Clin Invest* 1988; 81:435–41.

Steiner KE, Mouton SM, Williams PE, Lacy WW, Cherrington AD. The relative importance of first- and second-phase insulin secretion in countering the action of glucagon on glucose turnover in the conscious dog. *Diabetes* 1982; 31:964–72.

Verge CF, Gianani R, Kawasaki E, Yu L, Pietropaolo M, Jackson RA, Chase HP, Eisenbarth GS. Prediction of type I diabetes in first-degree relatives using a combination of insulin, GAD, and ICA512bdc/IA-2 autoantibodies. *Diabetes* 1996; 45:926–33.

Weyer C, Bogardus C, Mott DM, Pratley RE. The natural history of insulin secretory dysfunction and insulin resistance in the pathogenesis of type 2 diabetes mellitus. *J Clin Invest* 1999; 104:787–94.

Ziegler AG, Herskowitz RD, Jackson RA, Soeldner JS, Eisenbarth GS. Predicting type I diabetes. *Diabetes Care* 1990; 13:762–5.

3.5 Diabetes Mellitus: Diagnosis and Control of Hyperglycemia

3.5.1 Glucose Testing: Fasting Plasma Glucose (FPG) in the Detection of Diabetes Mellitus (DM)

Indications for Test

The use of a FPG is indicated when:

- Screening for DM in a symptomatic individual or in high-risk groups such as African- and Hispanic-Americans
- Diagnosing DM in an asymptomatic individual at age ≥45 years or at high risk given family history of DM and signs of insulin resistance including hypertension
- Assessing metabolic control in DM as an adjunct to HbA_{1c} testing
- Monitoring patients not known to have DM who are receiving medications, other than insulin, that can alter glucose metabolism such as niacin, thiazide diuretics, beta blockers, and glucocorticoids

Procedure

1. Following an 8- to 14-hour overnight fast, obtain a venous blood sample in a sodium fluoride (gray-top) tube.
2. Separate plasma immediately to prevent metabolism of the glucose by cellular elements and a falsely lowered glucose. Measure the plasma glucose concentration.
3. Never rely on a routine fasting or random serum glucose as adequate validation of the nondiabetic state. Use only a plasma glucose determination, obtained as above, to make this distinction.

Interpretation

1. A FPG in nonpregnant adults of ≤99 mg/dL (5.5 mmol/L) is in the non-DM range.
2. Compared to non-DM with a FPG of <85 mg/dL (4.76 mmol/L), those with a FPG of 95 to 99 mg/dL (5.32–5.5 mmol/L) are more than twice as likely to develop DM within the next 108 months (Nichols et al., 2008).
3. A single FPG or random plasma glucose of ≥200 mg/dL (11.2 mmol/L) plus symptoms of polyuria, polydipsia, and polyphagia establishes the diagnosis of overt DM.
4. A FPG of ≥126 mg/dL (7 mmol/L) on two separate occasions is diagnostic of overt DM.
5. If the FPG is >99 mg/dL (5.5 mmol/L) but is <126 mg/dL (7 mmol/L), the patient has early DM (i.e., prediabetes) or impaired glucose tolerance (IGT).
6. In early DM, a 75-g or 100-g oral glucose tolerance test (oGTT) may be considered as a research study, but it is not acceptable as a diagnostic test.
7. If the FPG is <40 mg/dL (2.4 mmol/L) and the patient has hypoglycemic symptoms, evaluation for fasting hypoglycemia is indicated.
8. The finding of a FPG of >140 mg/dL (7.8 mmol/L) is a sensitive, but not specific, test for identifying poor glycemic control in DM especially if found on three or more separate occasions.

Notes

1. The blood glucose concentration, [BG], is lower than the plasma glucose concentration. Multiply by a constant factor of 1.11 to convert [BG] to an estimate of plasma glucose. Although plasma and serum glucose might be equivalent, the serum glucose tends to be lower depending on the interval between phlebotomy and separation from red cells.
2. If the FPG is >140 mg/dL (7.8 mmol/L), up to 90% of patients with HbA_{1c} > 7% will be detected.
3. In the Australian Diabetes, Obesity, and Lifestyle Study, elevated FPG ≥ 110 mg/dL (6.1 mmol/L) but < 126 mg/dL (7 mmol/L) with a 2-hour postprandial glucose of <140 mg/dL (7.8 mmol/L) was associated with increased risk for all-cause mortality and twofold higher CVD mortality at an average of 5 years' follow-up (Barr et al., 2006).
4. The "dawn phenomenon" is a rise in [BG] as a result of the counter-regulatory hormone surge that occurs toward the last few hours of sleep or just before the dawn of the day. This hormone surge as well as a decrease in responsiveness of the liver to suppression of hepatic gluconeogenesis by insulin promotes an increase in the [FBG] in the DM patient.

ICD-9 Codes

Conditions that may justify this test *include but are not limited to*:

250.XX	diabetes mellitus
277.7	metabolic syndrome
401	essential hypertension

Suggested Reading

American Diabetes Association. Position statement: diagnosis and classification of diabetes mellitus. *Diabetes Care* 2007; 30(Suppl 1):S42–7.

Barr ELM, Magliano DJ, Zimmet PZ, Polkinghorne KR, Atkins RC, Dunstan DW, Murray SG, Shaw JE. The Australian Diabetes, Obesity and Lifestyle Study: tracking the accelerating epidemic—its causes and outcomes. In: *The AusDiab 2005 Report*. 2006. International Diabetes Institute, p. 71.

Genuth S, Alberti KG, Bennett P, Buse J, Defronzo R, Kahn R, Kitzmiller J, Knowler WC, Lebovitz H, Lernmark A, Nathan D, Palmer J, Rizza R, Saudek C, Shaw J, Steffes M, Stern M, Tuomilehto J, Zimmet P. Follow-up report on the diagnosis of diabetes mellitus. *Diabetes Care* 2003; 26:3160–7.

Nichols GA, Hillier TA, Brown JB. Normal fasting plasma glucose and risk of type 2 diabetes diagnosis, *Am J Med* 2008; 121:519–24.

3.5.2   Glucose Testing: Random Plasma Glucose (RPG) in the Detection of DM and Beta-Hydroxybutyrate (BHB) for Detection of Diabetic Ketoacidosis (DKA)

Indications for Test

A RPG determination, combined with BHB testing, is indicated to:

- Look for new-onset DM with ketosis in an individual with overtly hyperglycemic symptoms (i.e., polyuria, polydipsia, polyphagia, weight loss, blurred vision, or fatigue).
- Detect uncontrolled diabetes with the possibility of diabetic ketoacidosis in a patient with known diabetes (T2DM or T1DM).
- Provide early detection of ketosis in patients on insulin pump therapy.

Procedure

1. Obtain a random venous or capillary blood sample for glucose (RPG) without regard to time of day or last meal.
2. If the [RPG] > 250 mg/dL (13.9 mmol/L), obtain a blood sample for BHB in a clinically ill patient with evidence of ketosis (i.e., serum or urine ketones present or ketones on the breath).
3. Ordinarily, use a hand-held test meter with a LLD of 0.1 mmol/L (e.g., Precision Xtra® by Abbot Labs) to measure capillary blood [BHB].
4. Recognize the limitations of qualitatively estimating acetoacetate (AcAc) in urine with dipstick (Keto-stick) or acetone in serum using tablet-based methods (see Test 5.4.1).

Interpretation

1. Elevation of [RPG] to >140 mg/dL (7.8 mmol/L), but <200mg/dL (11.2 mmol/L), and a [BHB] < 0.4 mmol/L are highly suggestive of impaired glucose tolerance, inadequate glycemic control, or the new onset of DM, but not ketoacidosis.
2. [RPG] > 200 mg/dL (11.2 mmol/L) in the context of hyperglycemic symptoms unequivocally confirms the diagnosis of DM.
3. [RPG] > 250 mg/dL (13.9 mmol/L) is consistent with very poorly controlled DM possibly complicated by DKA.
4. [RPG] > 250 mg/dL (13.9 mmol/L) and a [BHB] > 1.0 mmol/L are consistent with advanced DKA, usually requiring prompt medical intervention.
5. [BHB] ≥ 0.4 mmol/L is significantly abnormal and should prompt more intensive glucose monitoring, therapy for hyperglycemia, and prevention of more overt DKA.
6. An elevated [RPG] and a positive urine AcAc or serum ketones test may reflect starvation in a diabetic, but neither is adequate for the diagnosis of DKA.

Notes

1. The RPG is sometimes referred to as casual plasma glucose (CPG).
2. The sensitivity/specificity of urine dipstick testing for AcAc and of capillary blood testing for BHB in diagnosis of DKA were 66%/78% and 72%/82%, respectively (Bektas et al., 2004).
3. In a study of 139 DM patients, there was a positive correlation between capillary and venous blood BHB ($r = 0.488$; $p < 0.001$) (Bektas et al., 2004).
4. Detection of ketosis in DM is readily accomplished regardless of the method used to detect ketones but has low specificity (<60%) for the diagnosis of DKA.
5. RPG > 270 mg/dL (15 mmol/L) is often accompanied by [BHB] > 0.2 mmol/L (Samuelsson and Ludvigsson, 2002).
6. In 14 DKA patients seen as emergencies, the mean plasma [BHB] on presentation was 7.4 mmol/L (range, 3.9–12.3 mmol/L). Once the hyperglycemia was corrected, the median half-life of the circulating [BHB] was estimated to be 1.6 hours (Wallace et al., 2001).

ICD-9 Codes

Conditions that may justify this test *include but are not limited to*:

Nutrition disorders
783.21   weight loss
783.5   polydipsia
783.6   polyphagia

Other
250.2   diabetic ketoacidosis (DKA)
368.8   visual disturbances
780.7   malaise and fatigue

Suggested Reading

Bektas F, Eray O, Sari R, Akbas H. Point of care blood ketone testing of diabetic patients in the emergency department. *Endocr Res* 2004; 30:395–402.

Samuelsson U, Ludvigsson J. When should determination of ketonemia be recommended? *Diabetes Technol Ther* 2002; 4:645–50.

Wallace TM, Meston NM, Gardner SG, Matthews DR. The hospital and home use of a 30-second hand-held blood ketone meter: guidelines for clinical practice. *Diabet Med* 2001; 18:640–5

3.5.3 Oral Glucose Tolerance Tests (oGTTs) in the Diagnosis of Gestational Diabetes Mellitus (GDM): 50-g and 100-g Glucose Load

50-g Oral Glucose Load oGTT in Screening for GDM

Indications

A 50-g oGTT is indicated:

- Between 14 and 18 weeks' gestation to screen and diagnose GDM in pregnant patients without known pregestation diabetes mellitus
- At 28 weeks' gestation to screen for GDM if the pregnant woman was not screened at an earlier time point
- As a repeat test if the result of the initial 14- to 16-week gestation 50-g glucose oGTT was in the nondiagnostic range of 111 to 134 mg/dL (6.12–7.38 mmol/L)

Procedure

1. Elect to do this screening test only in asymptomatic, low-risk (e.g., Europid) ethnic groups. Measure [HbA$_{1c}$] (see Test 3.5.4) first and do not perform the oGTT if the HbA$_{1c}$ is ≥6% or <5.3%.
2. Have pregnant patients with a positive family history of DM, obesity, glycosuria, or previous history of GDM or macrosomia or ethnicity other than northern European proceed directly to the 100-g oGTT for the most cost-effective screening for GDM.
3. Administer a 50-g oral glucose load without regard to the time of the last meal or time of day.
4. Obtain a blood sample for plasma glucose at 1 hour after administration of the oral glucose load.

Interpretation

1. The negative predictive value of a 1-hour glucose concentration <110 mg/dL (6.06 mmol/L) at 16 weeks' gestation approaches 100%, as does a value of <130 mg/dL obtained at 24 to 28 weeks' gestation.
2. The positive predictive value of a 1-hour glucose concentration of >135 mg/dL (7.44 mmol/L) for GDM is 55% at 16 weeks' gestation vs. <23% at 28 weeks' gestation.
3. "Typical Ethnicity-Related Reference Range Screening Glucose Values for Gestational Diabetes Mellitus (GDM)" (Table 3.5) shows the typical glucose response at 1 hour after a 50-g glucose load at 28 weeks' gestation in different ethnic populations.
4. Using a cutoff value of >140 mg/dL (7.71 mmol/L) for the 1-hour specimen at 24 to 28 weeks' gestation, 80 to 90% of all women with GDM will be detected; this is thought to be the most cost-effective cutoff value, as it results in the requirement for 100-g oGTT in only 15% of pregnancies.
5. Using a cutoff value of >135 mg/dL (7.38 mmol/L) for the 1-hour specimen to be followed up with a formal 100-g oGTT may be a reasonable compromise, resulting in the requirement for 100-g oGTT in only 20% of pregnancies while improving the detection rate for GDM.

TABLE 3.5
Typical Ethnicity-Related Reference Range Screening Glucose Values for Gestational Diabetes Mellitus (GDM)[a]

Race	1-Hour Glucose Value, mg/dL (mmol/L)
Caucasian	≤140 (7.8)
African-American or Filipina	≤145 (8.1)
Asian	≤150 (8.3)
Mixed or unknown	≤140 (7.8)[b]
	≤130 (7.3)[c]

[a] After 50-g oral glucose load during third trimester of pregnancy.

[b] For 80% sensitivity in the diagnosis of GDM.

[c] For 90% sensitivity in the diagnosis of GDM.

Notes

1. The plasma glucose cutoff value for the decision to perform a 100-g oGTT after the 1-hour 50-g GDM screen was lowered from >143 mg/dL to >135 mg/dL (Carpenter and Coustan, 1982).
2. A practical test consisting of glucose measurement in the fasting state and 2 hours after a "usual" breakfast has been proposed as being more sensitive (54.3% vs. 20.1%; $p = 0.0001$) but less specific (70.0% vs. 80.6%; $p = 0.0001$) in the prediction of macrosomia than the reference 50-g screen between 24 and 28 weeks of gestation (Chastang et al., 2003).

100-Gram Oral Glucose Load in the Diagnosis of GDM

Indication for Test

A 100-g oGTT is indicated for diagnosis of possible GDM in a pregnant patient when:

- The 50-g oGTT screening test is abnormal.
- There is a positive family history of DM, obesity, glycosuria, previous history of GDM, macrosomia, or ethnicity other than northern European.

Procedure

1. Do not proceed with this test if the fasting plasma glucose is >125 mg/dL (6.95 mmol/L). The diagnosis of DM is established.
2. Have patient prepare for this test by ingestion of at least 150 g of carbohydrates for 3 days, ideally using a food intake log to make this determination.
3. Following a 10- to 14-hour overnight fast, obtain a baseline blood sample to determine plasma glucose.
4. Administer a 100-g oral glucose load over 5 minutes.
5. Obtain blood samples for plasma glucose testing at 1, 2, and 3 hours after administration of the oral glucose load.
6. When the suspicion for GDM is high, repeat the 100-g oGTT 4 weeks after the initial test if the initial test, done at 14 to 18 gestational weeks, was abnormal at only one time point.

Interpretation

1. Refer to "Reference Glucose Values During 3-Hour Glucose Tolerance Test in Pregnancy" (Table 3.6) for the cutoff values used to interpret the results of the 100-g oGTT.
2. Two or more patient's glucose values greater than or equal to these cutoffs are diagnostic for GDM.
3. One abnormal value indicates a significant risk (34%) for developing GDM.

TABLE 3.6
Reference Glucose Values During 3-Hour Glucose Tolerance Test in Pregnancy

Time[a] (hr)	[Glucose], mg/dL (mmol/L)			
		Historical Data		
	Current ADA[b]	Sacks[c]	Carpenter and Coustan[d]	NDDG[e]
0	<95 (5.3)	<96 (5.3)	<95 (5.3)	<105 (5.8)
1	<180 (10.0)	<172 (9.6)	<180 (10.0)	<190 (10.6)
2	<155 (8.6)	<152 (8.4)	<155 (8.6)	<165 (9.2)
3	<140 (7.8)	<131 (7.3)	<140 (7.8)	<145 (8.1)

[a] After 100-g oral glucose load.

[b] American Diabetes Association, *Diabetes Care*, 27(Suppl. 1), S88–S90, 2004.

[c] Sacks, D.A. et al., *Am. J. Obstet. Gynecol.*, 161, 638–645, 1989.

[d] Carpenter, M.W. and Coustan, D.R., *Am. J. Obstet. Gynecol.*, 144, 768–773, 1982.

[e] National Diabetes Data Group, *Diabetes*, 28, 1039–1057, 1979.

Notes

1. The diagnosis of GDM is controversial. The criteria for diagnosis originally proposed by O'Sullivan and Mahan in 1964 were based on methods for determining blood glucose concentrations that are no longer in use.
2. In 1979, the National Diabetes Data Group (NDDG) proposed numerical conversions for 1964 post-oGTT glucose data that are still endorsed by the American College of Obstetrics and Gynecology.
3. A set of glucose cutoff values for interpreting the results of oGTTs based on modern methods (glucose oxidase or hexokinase) for determining plasma glucose concentration was proposed (Coustan et al., 1982) and later verified prospectively (Sacks et al., 1989).
4. If the HbA_{1c} (see Test 3.5.4) is ≤5.2%, GDM does not develop during the entire course of pregnancy and oGTT screening is unnecessary.
5. During pregnancy, a significant association between 1-hour glucose level after a 75-g glucose load and abnormal neonatal anthropometric features (macrosomia) was documented at a glucose cutoff value of 150 mg/dL at 16 to 20 weeks' gestation and 160 mg/dL at 26 to 30 weeks' gestation (Mello et al., 2003).
6. For women with diet-treated GDM, the odds ratio of having a child with cryptorchidism was 3.98. The risk of cryptorchidism was increased by more than twofold when all mothers with only an abnormal oGTT result were added to the group with GDM (Virtanen et al., 2006).

ICD-9 Codes

Conditions that may justify this test *include but are not limited to*:

648.8	abnormal glucose tolerance
V22.2	pregnancy, incidental

Suggested Reading

American Diabetes Association. Gestational diabetes mellitus. *Diabetes Care* 2004; 27(Suppl 1):S88–90.

Carpenter MW, Coustan DR. Criteria for screening tests for gestational diabetes. *Am J Obstet Gynecol* 1982; 144:768–73.

Chastang N, Hartemann-Heurtier A, Sachon C, Vauthier D, Darbois Y, Bissery A, Golmard JL, Grimaldi A. Comparison of two diagnostic tests for gestational diabetes in predicting macrosomia. *Diabetes Metab* 2003; 29(2 Pt 1):139–44.

Expert Committee on the Diagnosis and Classification of Diabetes Mellitus. Report. *Diabetes Care* 2003; 26:S5–S20.

Mello G, Parretti E, Cioni R, Lucchetti R, Carignani L, Martini E, Mecacci F, Lagazio C, Pratesi M. The 75-gram glucose load in pregnancy: relation between glucose levels and anthropometric characteristics of infants born to women with normal glucose metabolism. *Diabetes Care* 2003; 26:1206–10.

Nahum GG, Wilson SB, Stanislaw H. Early-pregnancy glucose screening for gestational diabetes mellitus. *J Reprod Med* 2002; 47:656–62.

National Diabetes Data Group. Classification and diagnosis of diabetes mellitus and other categories of glucose intolerance. *Diabetes* 1979; 28:1039–57.

Neiger R, Coustan DR. The role of repeat glucose tolerance tests in the diagnosis of gestational diabetes. *Am J Obstet Gynecol* 1991; 165:787–90.

O'Brien K, Carpenter M. Testing for gestational diabetes. *Clin Lab Med* 2003; 23:443–56.

Sacks DA, Abu-Fadil S, Greenspoon JS, Fotheringham N. Do the current standards for glucose tolerance testing in pregnancy represent a valid conversion of O'Sullivan's original criteria and how reliable is the 50-gram, one-hour glucose screening test? *Am J Obstet Gynecol* 1989; 161:638–5.

Virtanen HE, Tapanainen AE, Kaleva MM, Suomi AM, Main KM, Skakkebaek NE, Tooppari J. Mild gestational diabetes raises risk of cryptorchidism. *J Clin Endocrinol Metab* 2006; 91:4862–5.

3.5.4 Hemoglobin A_{1c} (HbA_{1c}) Testing and A_{1c}-Derived Average Glucose (ADAG) in Monitoring Patients with Diabetes Mellitus

Indications for Test

Measurement of percent HbA_{1c} (%HbA_{1c}) is indicated:

- Every 3 to 6 months in the assessment of long-term glycemic control in patients with diagnosed DM

Procedure

1. Identify confounding factors that may alter the HbA_{1c} test result, including hemoglobinopathies, chronic alcohol ingestion, treatment with medications such as atorvastatin, carbamylation products in uremia of chronic kidney disease, iron-deficiency anemias, and hematologic conditions associated with shortened red blood cell life span.
2. Obtain a random, anticoagulated blood sample for HbA_{1c} testing. A fasting sample is not required.
3. Alternatively, use fingerstick capillary blood samples (<10 μL) for HbA_{1c} testing using a point-of-care testing device (e.g., DRx HbA_{1c} method developed by Metrika, Inc.). This test incorporates microelectronics, optics, and dry reagent chemistry inside a self-contained, integrated, single-use device with results obtainable within 8 minutes after sample application.
4. At appropriate intervals, usually 3 to 4 months, measure the %HbA_{1c} at the time of the patient's routine DM clinic visit when the focus is on longer term glycemic control.
5. Convert the %HbA_{1c} to an A_{1c}-derived average glucose value in either mg/dL or mmol/L (Table 3.7).

Interpretation

1. The criteria for good vs. poor glycemic control can vary markedly between different assays for quantifying %HbA_{1c}.
2. Nondiabetic cutoff values for HbA_{1c} depend on the laboratory and method used. Ion-exchange, high-performance liquid chromatography (HPLC) is the reference method. Other methods for measuring %HbA_{1c} can be calibrated to this reference method.
3. Most HbA_{1c} assays today use a reference interval of 3 to 6% after calibration to results obtained by ion-exchange HPLC. To convert National Glycohemoglobin Standardization Group (NGSP) %HbA_{1c}

TABLE 3.7
Assessment of Degree of Glycemic Control
Based on Hemoglobin A_{1c} (HbA_{1c}) Values
by Ion-Exchange Chromatography

HbA$_{1c}$ (%)	Plasma Glucose (mg/dL)[a]	Qualitative Interpretation of Glycemic Control[b] 2004	1996
<5.9	<133	Excellent	Excellent
<6.5	<154	On target	Excellent
6.5–7.0	154–172	Good	Excellent
7.1–7.5	175–190	Fair	Good
>7.5	190	Poor	Fair

[a] Estimated as the mean plasma glucose over previous 2 to 3 months.

[b] Authors' interpretation of Association of Clinical Endocrinologists (2004) and American Diabetes Association (1996) information guidance.

to International Federation of Clinical Chemistry (IFCC) %HbA$_{1c}$ and then to mmol HbA$_{1c}$/mol Hb, use the following formulas (http://www.ngsp.org/prog/index.html):

$$\text{NGSP \%HbA}_{1c} = \left(0.915 \times \text{IFCC \%HbA}_{1c}\right) + 2.15\%$$

or

$$\text{IFCC \%HbA}_{1c} = \frac{\text{NGSP \%HbA1c} - 2.15\%}{0.915}$$

and,

$$\left(\text{IFCC \%HbA}_{1c}\right) \times 10 = \frac{\text{mmol HbA}_{1c}}{\text{mol Hb}}$$

To convert NGSP HbA$_{1c}$ (%) to estimated average glucose (eAG), use the formulas shown in Figure 3.4.

Example—If the NGSP %HbA$_{1c}$ = 7.0%, then the IFCC %HbA$_{1c}$ = (7.0 − 2.15)/0.915 = 5.3%, or 53 mmol HbA$_{1c}$/mol Hb; eAG (mg/dL) = 28.7(7.0) − 46.7 = 154.2 mg/dL; and eAG (mmol/L) = 1.59(7.0) − 2.59 = 8.54 mmol/L.

4. In patients with normal hematologic status, the %HbA$_{1c}$ appears to correlate with the mean blood glucose concentration over the past 2 to 3 months. Note that this interval is shorter than the RBC life span (120 days).

5. The %HbA$_{1c}$ can be used to determine the degree of glycemic control in a treated diabetic and correlates to a high degree of statistical significance with the risk of developing microvascular complications of DM, including retinopathy, neuropathy, and nephropathy. A HbA$_{1c}$ < 6.5% is considered the goal of therapy in most patients with established diabetes, and a HbA$_{1c}$ > 7% in a DM patient usually requires pharmacologic intervention.

6. Patients with an elevated %HbA$_{1c}$ but no diagnosis of DM may need more careful follow-up and possibly aggressive treatment to reduce the risk of DM. Non-DM patients with high-normal %HbA$_{1c}$, in the 5.7 to 6.0% interval, may require follow-up for assessment of glycemic status earlier than 3 years, especially if they are significantly overweight or obese.

7. Hemoglobin mutations Graz, Sherwood Forest, O Padova, C, D, and S have differing effects on a HbA$_{1c}$ result, either increasing or lowering the result based on the type of assay method used. The boronate affinity HbA$_{1c}$ assay appears to be the least affected by these hemoglobin variants.

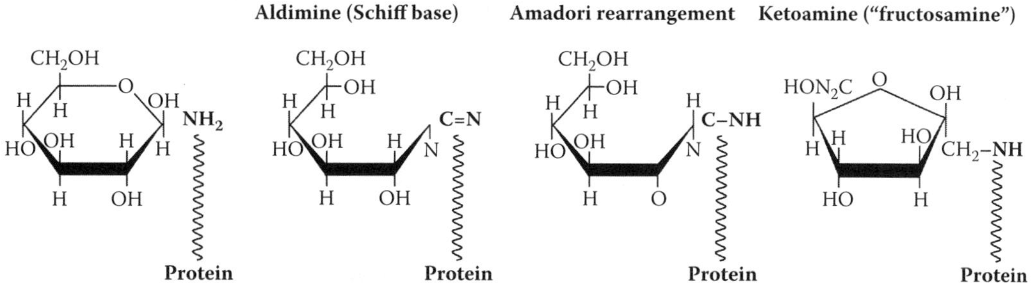

| Aldimine (Schiff base) | Amadori rearrangement | Ketoamine ("fructosamine") |

Figure 3.3

Glycation of amino groups of circulating serum proteins by glucose. Glucose reacts with free amino groups on proteins to form a labile Schiff base that rapidly undergoes rearrangement ("Amadori rearrangement") to form a stable ketoamine. When the protein is albumin, fructosamine (or glycated albumin) is formed; when it is hemoglobin, the modified product, with an altered electrophoretic pattern, is called hemoglobin A_{1c} (HbA_{1c}).

Notes

1. The glycosylated form of hemoglobin A is called HbA_{1c}. It is formed continuously over the life span of the red cell because of modification of free amino groups by adducts of glucose (Figure 3.3).
2. Laboratory reporting of the HbA_{1c} may change from a percentage to mmol HbA_{1c}/mol Hb (see Figure 3.4) based on the 2007 recommendations of the International Diabetes Federation and the International Federation of Clinical Chemistry (Consensus Committee, 2007). This recommendation is prompted by proposals to:
   * Create an international standard for measurement of gkycosylated hemoglobin.
   * Report the estimate of glycemic control as HbA_{1c}-derived average glucose (ADAG) as a uniform, internationally recognized measure.
   Note that these proposals have yet to be adopted as of 2009 and will require development of a broad consensus before they are implemented
3. A proposal for the interconversion of HbA_{1c} and mean plasma blood glucose using the equations:

$$HbA_{1c} = \frac{\text{Plasma blood glucose} + 77.3}{35.6}$$

   and

$$\text{Plasma blood glucose} = (HbA_{1c} \times 35.6) - 77.3$$

   has been presented (Rohlfing et al., 2002).
4. The sensitivity, specificity, and receiver operating characteristic (ROC) curves for the diagnosis of DM using $\%HbA_{1c}$ are similar, but not identical, to the fasting plasma glucose (FPG) and 2-hour blood glucose after a 75-g oral glucose load.
5. The variability in $\%HbA_{1c}$ (intraclass correlation coefficient [ICC] = 0.35) measurements was much greater than the FPG (ICC = 0.65) in a non-DM reference population as a result of hematologic factors that influenced $\%HbA_{1c}$ to a greater extent than [FPG] and 2-hour glucose after 75-g oGTT.
6. Neither HbA_{1c} nor fructosamine appears to be a cost-effective screening test for diagnosis of DM. In patients with stable T2DM not on insulin, the [FPG] correlates fairly well with $\%HbA_{1c}$.
7. An appreciation of the significance of the $\%HbA_{1c}$ may help motivate patients to improve their overall glycemic control and should be carefully taught and thoroughly explained to all DM patients. Since 1996, the qualitative criteria for interpreting glycemic control based on $\%HbA_{1c}$ have changed, as shown in Table 3.7. HbA_{1c} testing at point of care can improve a DM patient's glycemic control when the results are provided rapidly and accurately (Petersen et al., 2007).
8. For every 1% reduction in $\%HbA_{1c}$, the relative risk for microvascular complications in DM drops by 25 to 37%, for myocardial infarction by 14%, and for DM-related death by 21%.

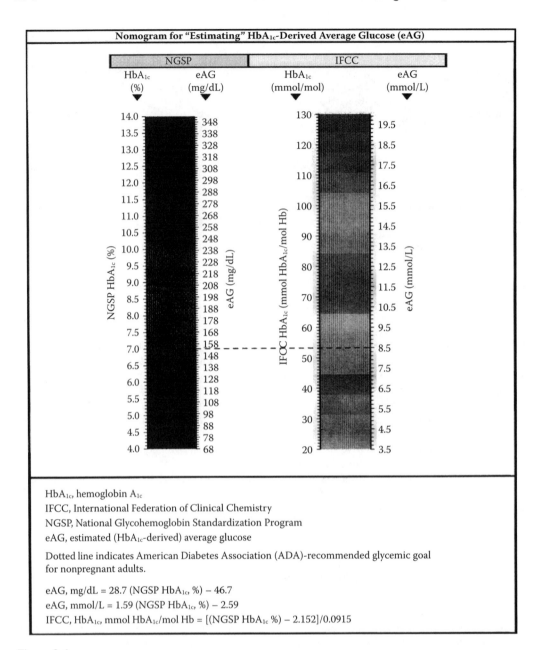

Figure 3.4

Nomogram for HbA$_{1c}$-derived estimated average glucose (eAG) values. (Figure courtesy of Dr. Frank H. Wians, Jr.)

9. When interpreting an HbA$_{1c}$ result, identify which of the four principal glycohemoglobin assay techniques (ion-exchange HPLC, electrophoresis, affinity chromatography, or immunoassay methods) was used and determine if that assay was calibrated to the ion-exchange HPLC method.

10. Clinicians ordering an HbA$_{1c}$ test should be aware of the type of assay method used, its analytical and clinical performance characteristics, including the reference interval and potential assay interferents. Note that more than 20 methods are currently available that measure one or more Hb adducts, in addition to HbA$_{1c}$.

11. In a retrospective study of 154 Japanese T2DM patients, a predisposition to deterioration in glycemic control was found upon treatment with atorvastatin (%HbA$_{1c}$ increased from 6.8 ± 0.9 to 7.2 ± 1.1) in comparison to no effect from pravastatin treatment (Takano et al., 2006).

ICD-9 Codes

Conditions that may justify this test *include but are not limited to*:

250.XX diabetes mellitus/hypoglycemia in diabetes mellitus
251.2   hypoglycemia
277.7   metabolic syndrome

Suggested Reading

Consensus Committee: The American Diabetes Association, European Association for the Study of Diabetes, International Federation of Clinical Chemistry and Laboratory Medicine, and the International Diabetes Federation. Consensus statement on the worldwide standardization of the hemoglobin A_{1c} measurement. *Diabetes Care* 2007; 30:2399–400.

Edelman D, Olsen MK, Dudley TK, Harris AC, Oddone EZ. Utility of hemoglobin A_{1c} in predicting diabetes risk. *J Gen Intern Med* 2004; 19:1175–80.

Hanson RL, Nelson RG, McCance DR, Beart JA, Charles MA, Pettitt DJ, Knowler WC. Comparison of screening tests for non-insulin-dependent diabetes mellitus. *Arch Intern Med* 1993; 153:2133–40.

Knowler WC. Screening for NIDDM: opportunities for detection, treatment, and prevention. *Diabetes Care* 1994; 17:445–50.

Little RR, Wiedmeyer HM, England JD, Wilke AL, Rohlfing CL, Wians FH, Jacobson JM, Zellmer V, Goldstein DE. Interlaboratory standardization of measurements of glycohemoglobins. *Clin Chem* 1992; 38:2472–78.

Nathan, DM et al. Translating the A_{1c} assay into estimated average glucose values. *Diabetes Care* 2008; 31:1473–1478.

Petersen JR, Finley JB, Okorodudu AO, Mohammad AA, Grady JJ, Bajaj M. Effect of point-of-care on maintenance of glycemic control as measured by A1C. *Diabetes Care* 2007; 30:713–715.

Peterson CM, Jovanovic L. Glycosylated proteins in normal and diabetic pregnancy. *Acta Endocrinol Suppl* 1986; 277:107–111.

Rohlfing CL, Wiedmeyer H-M, Little RR, England JD, Tennill A, Goldstein DE. Defining the relationship between plasma glucose and HbA_{1c}: analysis of glucose profiles and HbA_{1c} in the Diabetes Control and Complications Trial. *Diabetes Care* 2002; 25:275–8.

Schnedl WJ, Krause R, Halwachs-Baumann G, Trinker M, Lipp RW, Krejs GJ. Evaluation of HbA_{1c} determination methods in patients with hemoglobinopathies. *Diabetes Care* 2000; 23:339–44.

Stettler C, Mueller B, Diem P. What you always wanted to know about HbA_{1c}. *Schweiz Med Wochenschr* 2000; 130:993–1005.

Stratton IM, Adler AI, Neil HA, Matthews DR, Manley SE, Cull CA, Hadden D, Turner RC, Holman RR: Association of glycaemia with macrovascular and microvascular complications of type 2 diabetes (UKPDS 35): prospective observational study. *BMJ* 2000; 321:405–412.

Takano T, Yamakawa T, Takahashi M, Kimura M, Okamura A. Influences of statins on glucose tolerance in patients with type 2 diabetes mellitus. *Atheroscler Thromb* 2006; 13:95–100.

The Diabetes Control and Complications Trial Research Group. The effect of intensive treatment of diabetes on the development and progression of long-term complications in insulin-dependent diabetes mellitus. *N Engl J Med* 1993; 329:977–86.

Tsuji I, Nakamoto K, Hasegawa T, Hisashige A, Inawashiro H, Fukao A, Hisamichi S. Receiver operating characteristic analysis on fasting plasma glucose, HbA_{1c}, and fructosamine on diabetes screening. *Diabetes Care* 1991; 14:1075–7.

United Kingdom Prospective Diabetes Study (UKPDS). Intensive blood-glucose control with sulphonylureas or insulin compared with conventional treatment and risk of complications in patients with type 2 diabetes (UKPDS 33). *Lancet* 1998; 352:837–53.

3.5.5   Fructosamine (Glycosylated Albumin) Testing in Patients with Diabetes Mellitus

Indications for Test

Testing for fructosamine is indicated in patients with DM to assess:

- Short-term glycemic control and effectiveness of recent glucose-lowering interventions in patients with DM
- Glycemic control in patients with fluctuating anemias and/or hemoglobinopathies that may interfere with the accuracy of HbA_{1c} testing (e.g., patients on hemodialysis)

Procedure

1. Obtain a random blood sample for fructosamine testing.
2. Reconsider or avoid fructosamine testing in patients with hyperbiliribinemia or hyperuricemia or who are on heparin therapy. Bilirubin, uric acid, and heparin are known to interfere with fructosamine assays.
3. Assay serum albumin in patients with conditions known to be associated with a low circulating albumin concentration (e.g., liver or chronic kidney disease [CKD]). Abort this test if the serum albumin is <3 g/L, as the results of fructosamine testing will be uninterpretable.
4. Calculate the glycosylation gap (GG) from a measured HbA_{1c} and fructosamine value as follows:
   - Measure the HbA_{1c}.
   - Obtain the predicted HbA_{1c} from the regression equation that describes the correlation between measured fructosamine and HbA_{1c} concentration in a large reference population (Cohen et al., 2003).
   - Calculate the GG from the equation shown below:

$$GG = \text{Measured } HbA_{1c} - \text{Predicted } HbA_{1c}$$

Interpretation

1. A reference interval of 1.8 to 2.9 mmol/L is in common use for fructosamine.
2. In 153 DM patients, the GG was found to range from −3.2 to 5.5%, with 40% of samples having values indicating major differences in the prediction of risk for complications of DM (Cohen et al., 2003).
3. A positive GG correlates better with DM nephropathy than either HbA_{1c} or fructosamine alone in patients with T1DM of >15 years' duration.
4. Fructosamine levels decrease rapidly during the initial 2 to 3 weeks after acute glycemic normalization, with a half-time of 12 ± 5 days in contrast to a half time of 35 ± 10 days for the HbA_{1c}.

Notes

1. The fructosamine assay detects a nonenzymatically glycosylated serum protein, mostly albumin, which has a much shorter serum half-life (1 to 2 weeks) than HbA_{1c} (Figure 3.3).
2. In T2DM, short-term increases in fructosamine during winter holiday months were observed that resolved in the post-holiday months, while a trend toward higher HbA_{1c} levels occurred over the next year.
3. Fructosamine appears to have clinical utility similar to that of HbA_{1c} for monitoring long-term glycemic control and prediction of DM complications in most patients with the possible exception of those with T1DM of long duration.
4. The elevated glycated albumin/HbA_{1c} ratio found in DM CKD Stage 5 patients relative to DM subjects without nephropathy is consistent with a falsely reduced HbA_{1c} in patients with shortened erythrocyte survival.

5. Using weighted regression analysis, the lengths of time over which HbA_{1c} and fructosamine reflect ambient glucose levels extend back to ~100 and 30 days, respectively. Thus, HbA_{1c} and fructosamine do not reflect the simple mean but reflect the weighted mean of the preceding plasma glucose levels over a considerably longer period than the red cell life span or albumin clearance divided by 2.

ICD-9 Codes

Conditions that may justify this test *include but are not limited to*:

Chronic kidney disease
585.4   Stage 4
585.5   Stage 5

Other
250. X   diabetes mellitus
282.7   hemoglobinopathies
285   unspecified anemias

Suggested Reading

Chen HS, Jap TS, Chen RL, Lin HD. A prospective study of glycemic control during holiday time in type 2 diabetic patients. *Diabetes Care* 2004; 27:326–30.

Cohen RM, Holmes YR, Chenier TC, Joiner CH. Discordance between HbA_{1c} and fructosamine: evidence for a glycosylation gap and its relation to diabetic nephropathy. *Diabetes Care* 2003; 26:163–7.

Tahara Y, Shima K. Kinetics of HbA_{1c}, glycated albumin, and fructosamine and analysis of their weight functions against preceding plasma glucose level. *Diabetes Care* 1995; 18:440–7.

3.5.6   Glucose Monitoring by Devices for Use at Home and Work in Patients with Diabetes Mellitus

Indications for Test

Glucose monitoring by devices used in the home or workplace is indicated:

- For all patients with diabetes (T1DM and T2DM) who are capable of self-care outside of the hospital
- By competent family members when DM patients are incapable of self-testing

Procedure

1. Instruct DM patient or the patient's family member on the proper technique for self blood glucose monitoring (SBGM) and on the limitations of urine glucose and ketone testing even when done on second-void specimens.
2. Optimally, obtain blood sample from the lateral aspect of a fingertip. Apply sample to glucose monitoring test device (i.e., strip). Note that there may be a lag time in measurement of the core blood glucose levels [BG] when an alternative site, such as the forearm, is accessed. The palm test site appears to be equivalent to any fingertip.
3. If more than one family member has DM, be sure that each member has his or her own meter and counsel each member on the importance of not sharing meters with other family members as individual glucose readings are not erasable or transferable.
4. Be aware of meters that can be unintentionally switched from one unit of measurement to another (English vs. metric), as use of relatively higher [BG] in mg/dL vs. relatively lower [BG] in mmol/L can lead to severe hypo- or hyperglycemia when SBGM is used to adjust therapy for diabetes.

5. Decide on the frequency of SBGM based on the current degree of glycemic control exhibited by the patient.

6. Increase the frequency of monitoring (i.e., testing up to 4 to 10 times per day) if the patient has an elevated HbA$_{1c}$, is pregnant, or is undergoing major changes in diet and other therapy. More frequent testing will better detect the true rate of hyper- and hypoglycemic episodes.

7. Advise patients with T1DM to periodically perform ketone monitoring (via urine dipstick or Precision Xtra®), particularly during sick days or in the event of injury or unusual stress.

8. Routinely advise patients to obtain a fasting [BG] measurement on a daily basis regardless of the type of DM that the patient may have.

9. For most patients with DM, particularly T2DM, advise the performance of 1.5- to 2.5-hour postprandial [BG] checks on a daily basis, with postprandial testing being rotated between different meals.

10. Utilize continuous glucose monitoring for 48 to 72 hours using a sensor device inserted under the patient's skin for recording subcutaneous glucose concentration every few minutes in patients with unusually labile glucose levels or unpredictable hypoglycemic episodes.

11. For patients with vision impairment, recommend meters with audible read-outs of [BG] results.

Interpretation

1. Fasting [BG] < 100 mg/dL and postprandial [BG] < 140 mg/dL are indicative of good glycemic control in patients with DM who conscientiously perform SBGM.

2. In patients with DM and normal to near-normal indices of glycemic control, reduce the frequency of SBGM tests to once per day or less, on occasion, but not less than once per week.

3. Expect negative second-void urine glucose and ketone results if the patient's [BG] has been less than the renal threshold for glucose, usually less than 140 mg/dL, and if the patient has not been fasting for over 12 hours.

Notes

1. Glucose monitoring by DM patients in both the home and work environment is routine, with determination of [BG], rather than glucose in the urine, being the standard of care.

2. Order urine glucose testing as an acceptable, but limited, alternative in nonpregnant patients with DM, normal renal function (creatinine clearance of >100 mL/min), and a generally well-established level of metabolic control.

3. Glucose monitoring devices should be accurate (within ±5–10% of laboratory values) and easy to use; they should include programming of date and time, should be shock resistant, and should not be prone to breaking down easily or frequently.

4. Optional, but highly desirable, features of meters include:
   - Memory for more than 1 week of data (20+ test results)
   - Test strips that do not have to be specially coded and are more stable in warm humid environments
   - Rapid testing (<20 seconds to result)
   - Simple to use with easy access to customer support
   - Highly visible number display
   - Computer download and data processing capability
   - Use of coulometric rather than amperometric assay technology so as to avoid interference from the presence of drugs or metabolites in small blood samples, particularly at low glucose concentrations, and to get better accuracy over a broader range of hematocrits as well as smaller blood sample sizes
   - Use of glucose dehydrogenase rather than glucose oxidase enzyme-based test strips when the patient is undergoing hyperbaric oxygen treatment or is at high altitude

5. The use of BG monitoring test strips and devices did not affect HbA$_{1c}$, [BG] results, number of laboratory tests ordered, or number and type of treatment interventions in patients with T2DM taking a sulfonylurea agent in the primary-care setting.

6. When electrochemical glucose test strips employ glucose oxidase, there is generally interference by oxygen. High oxygen concentration leads to lower signal and underestimation of [BG], while low oxygen leads to a higher signal and overestimation of [BG].

7. Automated [BG] testing using electric current (e.g., GlucoWatch®) to obtain samples of subcutaneous fluid for glucose analysis is limited by problems with accuracy, time mismatch in readout with actual core [BG], skin irritation, the requirement for frequent calibration, and cost.
8. Up to one in four patients do not test their [BG] correctly. Pain on testing affects more than 40% of patients, and, along with having to restick, doubts about test accuracy and trouble getting a large enough blood sample are barriers to patient testing.

ICD-9 Codes

Conditions that may justify this test *include but are not limited to*:

250.X	diabetes mellitus
277.7	metabolic syndrome

Suggested Reading

American Diabetes Association. Postprandial blood glucose, *Diabetes Care* 2001; 24:775–8

American Diabetes Association. Standards of medical care in diabetes. Part V.C.1a. Self-monitoring of blood glucose. *Diabetes Care* 2005; 28:S10.

Bohme P, Floriot M, Sirveaux A, Durain D, Ziegler O, Drouin P, Guerci B. Evolution of analytical performance in portable glucose meters in the last decade. *Diabetes Care* 2003; 26:1170–5.

El-Kebbi IM, Ziemer DC, Cook CB, Gallina DL, Barnes CS, Phillips LS. Utility of casual postprandial glucose levels in type 2 diabetes management. *Diabetes Care* 2004; 27:335–9.

Goldstein DE, Little RR. Monitoring glycemia in diabetes. Short-term assessment. *Endocrinol Metab Clin North Am* 1997; 26:475–86.

Goldstein DE, Little RR, Lorenz RA, Malone JI, Nathan D, Peterson CM, Sacks DB. Tests of glycemia in diabetes. *Diabetes Care* 2004; 27:1761–73.

Gross TM, Bode BW, Einhorn D, Kayne DM, Reed JH, White NH, Mastrototaro JJ. Performance evaluation of the MiniMed continuous glucose monitoring system during patient home use. *Diabetes Technol Ther* 2000; 2:49–56.

Khan AI, Vasquez Y, Gray J, Wians FH Jr, Kroll MH. The variability of results between point-of-care testing glucose meters and the central laboratory analyzer. *Arch Pathol Lab Med* 2006; 130:1527–32.

Rindone JP, Austin M, Luchesi J. Effect of home blood glucose monitoring on the management of patients with non-insulin dependent diabetes mellitus in the primary care setting. *Am J Manag Care* 1997; 3:1335–8.

3.5.7 Glucose Monitoring During Intravenous Insulin Therapy for Postoperative and Post-Myocardial Infarction (MI) Glycemic Control: The Van den Berghe and Diabetes Mellitus Insulin–Glucose Infusion in Acute Myocardial Infarction (DIGAMI) Study Protocols

Indications for Test

Glucose monitoring during intravenous insulin therapy is indicated to:

*   Guide adjustment of insulin doses.
*   Optimize blood glucose (BG) levels postoperatively and post-MI to reduce mortality, prevent infections, enhance wound healing, and shorten hospital stays.

Procedure

1. Do not use this protocol in patients with diabetic ketoacidosis or hyperosmolar coma.
2. Specially trained personnel without close physician supervision are to be authorized to use this protocol, which is medically supervised but nurse directed.
3. Ensure that the patient receives a steady source of calories, such as an infusion of 5% dextrose with appropriate potassium supplement into a large vein at a rate sufficient to keep the vessel open.
4. Obtain a stat venous blood sample for plasma glucose determination and send to lab prior to beginning bedside BG monitoring. Begin infusion of insulin only if plasma glucose is >110 mg/dL and usually >140 mg/dL.
5. Use bedside, fingertip capillary BG monitoring results to adjust the rate of insulin infusion so the desired BG target is achieved, optimally 80 to 110 mg/dL.
6. Estimate the degree of the patient's insulin resistance or sensitivity based on factors that may alter insulin requirements such as obesity, prior history of DM, anesthesia used, criticality of illness or stress of surgery, steroid therapy, vomiting, infection, chronic kidney disease, or change in nutritional supplements.
7. Based on the clinical estimate of the degree of the patient's insulin resistance and ambient BG, begin an initial rate of insulin infusion usually ranging from 2 to 8 units/hour.
8. Upon starting insulin infusion, check BG every 1 or 2 hours and increase the rate of insulin infusion as per "Insulin Infusion Rate Algorithm for Postoperative Glycemic Control Based on Patient's Blood Glucose Concentration" (Table 3.8) to achieve and maintain the BG within an optimal range.
9. The BG may be checked at shorter time intervals (i.e., every 30 minutes) at the discretion of the infusion operator and upon a rapid drop in BG with insulin infusion (e.g., decline in BG of >40% in 1 hour).
10. Be prepared to activate a hypoglycemia treatment protocol in the event of a BG < 62 mg/dL.
11. Under no circumstances should any patient receiving an insulin infusion go longer than 2 hours without a blood glucose check.
12. Expect higher insulin dose requirements on the day of surgery and for at least 2 postoperative days for optimal BG control. Longer duration of insulin infusions may be necessary if complications (e.g., infection) develop.
13. Expect achievement of target BG within 2 to 4 hours using an insulin infusion program algorithm (Table 3.8) vs. 4 hours or more if an *ad hoc* insulin infusion sliding scale is used.

Interpretation

1. The infusion should be stopped at a glucose of <70 mg/dL and increased for a glucose 40 to 50 mg/dL above target unless a rapid fall in glucose has occurred in the previous hour.
2. In patients with severe insulin resistance, it may be necessary to use insulin infusion rates as high as 40 units/hour or more.
3. If the BG is 46 to 70 mg/dL, increasing the rate of i.v. glucose infusion is usually sufficient for correction of the patient's BG after suspending insulin infusion.
4. If the BG < 46 mg/dL, a bolus infusion of 10 g of glucose is indicated.
5. Maintenance of target range BG will still require 1 to 2 hours of monitoring even if the target BG level has been reached for a period of ≥8 hours.

Notes

1. Achieving and maintaining euglycemia in a postoperative patient has been shown to reduce mortality, the incidence of infection, length of hospital stays, and time to recovery as per "Outcomes in Surgical Intensive Care Patients Treated with Insulin Infusions to Maintain Glucose Levels Between 80 and 110 mg/dL" (Table 3.9).
2. Improved long-term outcomes in the event of critical illness have been achieved with insulin infusions, with risk of increased mortality seen only in the first 3 days in critically ill nonsurgery patients. Results of a recent NICE–SUGAR (2009) study, however, indicate that mortality was reduced with "intensive control" of the glucose only in trauma and steroid-treated nonsurgery patients but not better than to a $p = 0.06$ level of significance.
3. An insulin infusion is standard therapy for postoperative BG control when a diabetic faces surgery and may be indicated when a diabetic must be kept n.p.o. for a prolonged diagnostic or therapeutic procedure. Adjustments in the rate of an insulin infusion are dependent on the BG determined during the infusion and clinical estimates of the severity of the patient's insulin resistance.

TABLE 3.8
Insulin Infusion Rate Algorithm for Postoperative Glycemic Control Based on Patient's Blood Glucose Concentration

Program 1[a]		Program 2[a]		Program 3[a]		Program 4[a]		Program 5[a]		Program 6[a]	
Glucose (mg/dL)	Insulin (Units/hr)	Glucose (mg/dL)	Insulin (Units/hr)	Glucose (mg/dL)	Insulin (Units/hr)	Glucose (mg/dL)	Insulin (Units/hr)	Glucose (mg/dL)	Insulin (Units/hr)	Glucose (mg/dL)	Insulin (Units/hr)
<80	Off[b]	<70	Off[b]	<70	Off[b]	<70	Off[b]	<70	Off[b]	<70	Off[b]
80–109	0.2	70–109	0.2–0.4	70–109	1.0	70–109	1.5	70–109	2.0	70–109	3.0
110–119	0.5	110–119	0.5	110–119	2.0	110–119	3.0	110–119	4.0	110–119	5.0
120–149	1.0	120–149	1.0	120–149	3.0	120–149	5.0	120–149	6.0	120–149	7.0
150–179	1.5	150–179	1.5	150–179	4.0	150–179	7.0	150–179	8.0	150–179	10.0
180–209	2.0	180–209	2.0	180–209	5.0	180–209	9.0	180–209	12.0	180–209	14.0
210–239	2.0	210–239	4.0	210–239	6.0	210–239	12.0	210–239	16.0	210–239	18.0
240–269	3.0	240–269	5.0	240–269	8.0	240–269	16.0	240–269	20.0	240–269	24.0
270–299	4.0	270–299	6.0	270–299	10.0	270–299	20.0	270–299	24.0	270–299	28.0
300–329	4.0	300–329	7.0	300–329	12.0	300–329	24.0	300–329	28.0	300–329	32.0
330–359	6.0	330–359	8.0	330–359	14.0	330–359	28.0	330–359	32.0	330–359	36.0
≥360		≥360	12.0	≥360	16.0	≥360	32.0	≥360	36.0	≥360	40.0

[a] Move *up* to the next program: (1) if the [glucose] does not correct by at least 50 mg/dL in 1 hour for the first 4 hours of infusion and remains >50 mg above the chosen target range and (2) if the [glucose] does not correct by at least 20 mg/dL in the fifth and each subsequent 2 hours of infusion and remains >40 mg above chosen target range.

Move *down* to the next program if [glucose] is <80 mg/dL two times within 2 hours.

[b] No insulin infusion (insulin pump is off).

Source: Adapted from Markovitz, L.J. et al., *Endocr. Pract.*, 8, 10–18, 2002; Moghissi, E., *Cleve. Clin. J. Med.*, 71, 801–808, 2004.

TABLE 3.9
Outcomes in Surgical Intensive Care Patients Treated with Insulin Infusions to Maintain Glucose Levels Between 80 and 110 mg/dL

Critical Illness Occurrence	Percent Reduction in Occurrence (%)	p Value
Intensive care unit (ICU) mortality	43	$p = 0.036$
Antibiotics required for >10 days	35	Not available
Septicemia	46	$p = 0.003$
Hospital mortality	34	Not available

Source: Adapted from Van den Berghe, G et al., *N. Engl. J. Med.*, 345, 1359–1367, 2001.

4. Once the patient begins to aliment, initiation of a supplemental bolus of rapidly acting insulin before a feeding will be required to maintain the target glucose, as adjustments in rate of insulin infusion alone will not suffice.

5. Transition from insulin infusion to basal/bolus types of insulin and maintenance of euglycemia is possible with q4h or a.c. and h.s. BG monitoring. A reduction of approximately 10 to 20% in total insulin dose on transition from insulin infusion may be expected.

6. The use of an inpatient insulin sliding scale to get adequate control of the BG has been discredited (Robbins et al., 1963). Use of the basal bolus insulin approach to inpatient management of hyperglycemia gives much better results.

7. Intensive perioperative glycemic control reduced the long-term relative mortality by 25% in insulin-treated coronary care patients. Improved long-term survival was especially evident in a prestratified group of patients without prior insulin treatment, in whom the 3.4-year mortality reduction was 45% in the DIGAMI study.

8. In the DIGAMI-2 study, it was shown that glycemic control, regardless of the method used to achieve it, was more important than use of initial insulin–glucose infusion or long-term subcutaneous treatment with insulin in achieving reduced mortality and morbidity.

9. Use and effectiveness of intraoperative control of BG to levels of between 4.4 (80 mg/dL) and 5.6 mmol/L (100 mg/dL) remain experimental (Gandhi et al., 2007).

ICD-9 Codes

Conditions that may justify this test *include but are not limited to*:

250.XX   diabetes mellitus
410   acute myocardial infarction
958.3   posttraumatic wound infection, not elsewhere classified

Suggested Reading

Brown G, Dodek P. Intravenous insulin nomogram improves blood glucose control in the critically ill. *Crit Care Med* 2001; 29:1714–9.

Gandhi GY, Nuttall GA, Abel MD, Mullany CJ, Schaff HV, O'Brien PC, Johnson MG, Williams AR, Cutshall SM, Mundy LM, Rizza RA, McMahon MM. Intensive intraoperative insulin therapy versus conventional glucose management during cardiac surgery. *Ann Intern Med* 2007; 146:233–43.

Kaufman FR, Devgan S, Roe TF, Costin G. Perioperative management with prolonged intravenous insulin infusion versus subcutaneous insulin in children with type I diabetes mellitus. *J Diabetes Complications* 1996; 10:6–11.

Malmberg K. Role of insulin–glucose infusion in outcomes after acute myocardial infarction: the diabetes and insulin-glucose infusion in acute myocardial infarction (DIGAMI) study. *Endocr Pract* 2004; 10(Suppl 2):13–6.

TABLE 3.10
Gender-Related 1,5-Anhydro-D-Glucitol
(1,5-ADG) Reference Ranges
for Healthy Individuals

Gender	Mean SD (mcg/mL)	Reference Interval (mcg/mL)
Male	22.5 (5.8)	10.7–32.0
Female	17.7 (6.2)	6.8–29.3

Malmberg K, Ryden L, Wedel H, Birkeland K, Bootsma A, Dickstein K, Efendic S, Fisher M, Hamsten A, Herlitz J, Hildebrandt P, Macleod K, Laakso M, Torp-Pedersen C, Waldenstrom A. Intense metabolic control by means of insulin in patients with diabetes mellitus and acute myocardial infarction (DIGAMI 2): effects on mortality and morbidity *Eur Heart J* 2005; 26:650–61.

Markovitz LJ, Wiechmann RJ, Harris N, Hayden V, Cooper J, Johnson G, Harelstad R, Calkins L, Braithwaite SS. Description and evaluation of a glycemic management protocol for patients with diabetes undergoing heart surgery. *Endocr Pract* 2002; 8:10–8.

Moghissi E. Hospital management of diabetes: beyond the sliding scale. *Cleve Clin J Med* 2004; 71:801–8.

NICE–SUGAR Study Investigators. Intensive versus conventional glucose control in critically ill patients. *N Engl J Med* 2009; 360, 1283–1297.

Robbins L. Let's get the sliding scale out of medicine. *Med Rec Ann* 1963; 56:201.

van den Berghe G, Wouters P, Weekers F, Verwaest C, Bruyninckx F, Schetz M, Vlasselaers D, Ferdinande P, Lauwers P, Bouillon R. Intensive insulin therapy in the critically ill patients. *N Engl J Med* 2001; 345:1359–67.

3.5.8 Serum 1,5-Anhydro-D-Glucitol (1,5-ADG) as a Test for Degree of Postprandial Hyperglycemia

Indications for Test

Measurement of 1,5-ADG in serum is indicated for:

- Intermediate-term monitoring of postprandial glycemia in patients with diabetes mellitus
- Early detection of potential abnormalities of postprandial glucose tolerance and diagnosis of early diabetes

Procedure

1. Obtain a random blood sample for determination of plasma or serum [1,5-ADG] .
2. Store plasma or serum sample at room temperature or 2 to 8°C if testing will occur within 1 week.
3. Store plasma or serum sample by freezing if assay is delayed for more than 1 week. Sample may be freeze–thawed up to three times before accurate assay of [1,5-ADG].

Interpretation

1. Refer to "Gender-Related 1,5-Anhydro-D-Glucitol (1,5-ADG) Reference Ranges for Healthy Individuals" (Table 3.10); also, the expected glucose level after meals as a function of [1,5–ADG] is shown in "1,5-Anhydro-D-Glucitol (1,5-ADG) Level and Glycemic Status" (Table 3.11).
2. Refer to "Longitudinal Changes in Biochemical Measures of Glycemic Control with Multiple Modality Therapy Designed to Achieve Euglycemia in Type 1 and Type 2 Diabetes Patients" (Table 3.12). Note the rapid, short-term response of [1,5-ADG], compared to the delayed response of HbA_{1c} and fructosamine, to effective diabetes therapy.

TABLE 3.11
1,5-Anhydro-D-Glucitol (1,5-ADG)
Level and Glycemic Status

1,5-ADG (mcg/mL)	Assessment	Expected Glucose Levels after Meals (mg/dL)
≥14	Nondiabetic	<160
10–13.9	Acceptable glycemic control	<200 (prediabetes)
6–9.9	Moderate glycemic control	200–300
2–5.9	Poor glycemic control	Mostly >300
≤1.9	Very poor glycemic control	Almost always >300

*Source:*Adapted from Yamanouchi, T. et al., *Diabetes*, 40, 52–57, 1991.

3. A falsely low serum [1,5-ADG], suggestive of poor glycemic control, may occur in:
    - Persistent glucosuria
    - Reactive hypoglycemia after gastrectomy
    - Pregnancy
    - Chronic kidney disease (CKD), particularly stage 5
    - Advanced cirrhosis
    - Steroid therapy
    - Prolonged inanition
4. Short-term intravenous hyperalimentation associated with a mild to moderate degree of glucosuria may falsely increase the [1,5-ADG] as a result of an overall increase in blood glucose and faster production of circulating 1,5-ADG than its loss in the urine.
5. Severe, prolonged hyperglycemia may deplete the internal pool of 1,5-ADG as a result of persistent glucosuria (nephrogenic polyuria) leading to delay in its response to blood glucose lowering.
6. In the context of abnormal glomerular filtration rates (i.e., CKD Stages ≥ 3) and variations in the renal threshold for glucose, the [1,5-ADG] will be less reflective of changes in control of blood glucose.

Notes

1. Glucosuria prevents the reabsorbtion of 1,5-ADG, a monosaccharide that has structural similarity to glucose, resulting in a fall in its concentration in the blood and loss in the urine.
2. When glucose is present in the renal tubule, there is a resulting net urinary excretion of 1,5-ADG and hence a progressively lower level in the blood.
3. Due to the direct relationship between glucosuria and blood [1,5-ADG], there is a rapid decrease in [1,5-ADG] in response to blood glucose levels that exceed the patient's renal threshold for glucosuria. Upon return to urine glucose levels below this threshold, [1,5-ADG] in blood increases at a steady rate.
4. The GlycoMark™ test is available from Esoterix, Inc. (1-800-444-9111; www.esoterix.com). This test is reflective of postmeal glucose spikes, as described at www.glycomark.com (212-397-5443).

ICD-9 Codes

Conditions that may justify this test *include but are not limited to*:

    250.XX   diabetes mellitus

Suggested Reading

Buse J, Freeman J, Edelman S, Jovanovic L, McGill JB. Serum 1,5-anhydroglucitol (GlycoMark™): a short-term glycemic marker. *Diabetes Technol Ther* 2003; 5:355–63.

Dworacka M, Winiarska H, Szymanska M, Kuczynski S, Szczwinska K, Wierusz-Wysocka B. 1,5-Anhydro-D-glucitol: a novel marker of glucose excursions. *Int J Clin Pract* 2002; 129(Suppl):40–4.

TABLE 3.12
Longitudinal Changes in Biochemical Measures of Glycemic Control with Multiple Modality Therapy Designed to Achieve Euglycemia in Type 1 and Type 2 Diabetes Patients (n = 77)

Time after Therapy	Mean ± 1 SD			
	Blood Glucose (mg/dL)	HbA$_{1c}$ (%)	Fructosamine (mcmol/L)	1,5-ADG (mcg/mL)
Baseline	225.0 ± 105.6	9.5 ± 1.7	410.6 ± 108.6	1.9 ± 1.9
2 weeks	187.4 ± 91.0	9.1 ± 1.5	362.4 ± 76.5	3.0 ± 2.2
% Change from baseline	−16.7*	−4.2	−12.2	57.9*
4 weeks	181.4 ± 102.4	8.8 ± 1.4	340.0 ± 79.1	3.7 ± 2.5
% Change from baseline	−19.4*	−7.4	−17.7*	94.7*
8 weeks	172.6 ± 100.5	8.2 ± 1.2	317.5 ± 75.4	5.0 ± 3.6
% Change from baseline	−23.3*	−13.7*	−22.7*	163.2*

Note: HbA$_{1c}$, hemoglobin A$_{1c}$; 1,5-ADG, 1,5-anhydro-D-glucitol; *p < 0.05 vs. baseline.

Source: McGill, J.B. et al., *Diabetes Care*, 27, 1859–1865, 2004. With permission.

Fukumura Y, Tajima S, Oshitani S, Ushijima Y, Kobayashi I, Hara F, Yamamoto S, Yabuuci M. Fully enzymatic method for determining 1,5-anhydro-D-glucitol in serum. *Clin Chem* 1994; 40:2013–6.

Kilpatrick ES, Keevilt BG, Richmond KL, Newland P, Addison GM. Plasma 1,5-anhydroglucitol concentrations are influenced by variations in the renal threshold for glucose. *Diabet Med* 1999; 16:496–499.

Kishimoto M, Yamasaki Y, Kubota, M, Arai K, Morishima T, Kawamori R, Kamada T. 1,5-Anhydro-D-glucitol evaluates daily glycemic excursions in well-controlled NIDDM. *Diabetes Care* 1995; 18:1156–9.

McGill JB, Cole TG, Nowatzke W, Houghton S, Ammirati EB, Gautille T, Sarno M. Circulating 1,5-anhydroglucitol levels in adult patients with diabetes reflect longitudinal changes of glycemia: a U.S. trial of the GlycoMark assay. *Diabetes Care* 2004; 27:1859–65

Tam WH, Rogers MS, Lau TK, Arumanayagam M. The predictive value of serum 1,5-anhydro-D-glucitol in pregnancies at increased risk of gestational diabetes mellitus and gestational impaired glucose tolerance. *BJOG* 2001; 108:754–6.

Yamanouchi T, Akanuma Y, Toyota T, Kuzuya T, Kawai T, Kawazu S, Yoshioka S, Kanazawa Y, Ohta M, Baba S, Kosaka K. Comparison of 1,5-anhydroglucitol, A$_{1c}$, and fructosamine for detection of diabetes mellitus. *Diabetes* 1991; 40:52–7.

3.5.9 Autoantibody Screening Tests for Type 1 Diabetes Mellitus: Insulin Autoantibodies (IAAs), Protein Tyrosine Phosphatase-Like Protein (IA-2) Autoantibodies, and Glutamic Acid Decarboxylase Autoantibodies (GADAs)

Indications for Test

Autoantibody screening is indicated:

- In all first-degree relatives of a proband with T1DM
- To predict risk for diabetes in family members of T1DM and T2DM patients
- When deciding if interventions should be made to prevent the development of T1DM
- In ostensibly T2DM patients when onset of hyperglycemia is after age 50 years, BMI is over 25, acute symptoms of diabetes are present, and there is a personal history of autoimmune disease

TABLE 3.13
Typical Values for Negative Type 1
Diabetes Autoantibody Screening Tests

Test	Negative Result
IAA (nU/mL)	<42
GADA (index)	<0.032
ICA512bdcAA (index)	<0.071
ICA (JDF units)	<20

Note: IAA, insulin autoantibodies; GADA, glutamic acid decarboxylase antibody; ICA, islet cell autoantibody; JDF, Juvenile Diabetes Foundation.

Procedure

1. Obtain the appropriate blood samples for IAA, IA-2, and the GAD 65-kDa isoform (GAD65) autoantibodies (GADAs) indicated by the referral laboratory where testing for these antibodies will be performed.
2. Be sure the laboratory provides the GAD65 index based on the ratio of GAD65 autoantibody concentration in the test sample vs. the concentration found in positive and negative control samples.

Interpretation

1. IA-2 is responsible for most of the islet-cell autoantibody (ICA) immunofluorescent tissue staining and is measured in serum using a simple radioligand assay.
2. "Typical Values for Negative Type 1 Diabetes Autoantibody Screening Tests" (Table 3.13) shows the negative cutoff values for autoantibodies associated with T1DM.
3. "Sensitivity and Positive Predictive Value (PPV) of Positive Autoantibody (AA) Findings in First-Degree Relatives of Probands with Type 1 Diabetes Mellitus (DM)" (Table 3.14) shows the potential use of autoantibodies alone and in combination for prediction of risk for diabetes.
4. The presence of three autoantibodies (IAA + GAD + IA-2) by age 14 has a 50% positive predictive value (PPV) for the onset of T1DM within 8 years (LaGasse et al., 2002) and 100% PPV within 5 years (Verge et al., 1996).
5. If none of the aforementioned autoantibodies is present, the 5-year risk of T1DM is 0%.
6. The diagnostic accuracy of the aforementioned autoantibodies can be improved further by combining the autoantibody results with the results of the first-phase insulin response (see Test 3.4.3 and Table 3.4).
7. The high sensitivity of ICAs, measured by the original methodology, prevents their use in the early detection of T1DM but permits their use in follow-up monitoring of patients with known T1DM.

Notes

1. The clinical onset of T1DM occurs as the end-stage of an immunological process that runs its course over months to years during which time the presence of autoantibodies against various islet-cell antigens can be detected.
2. Insulin and GAD autoantibodies are found in patients at high risk for developing T1DM.
3. When the presence of defined autoantibodies to human GAD and insulin was assessed in 4505 school children, 6 (0.13%) of these children had IAA and GAD autoantibodies by age 14 years and developed T1DM within 8 years. The sensitivity and PPV of the presence of these autoantibodies and development of T1DM were 50% (95% CI, 25–75%) and 100% (95% CI, 58–100%), respectively (La Gasse et al., 2002).
4. The finding of isolated ICA using the more complicated, labor-intensive, and operator-dependent methodologies did not predict development of T1DM with sufficiently high diagnostic specificity to

TABLE 3.14
Sensitivity and Positive Predictive Value of
Positive Autoantibody Findings in First-Degree Relatives
of Probands with Type 1 Diabetes Mellitus

AA Test Positive	Sensitivity (%)	PPV (%) 3-Year Risk of DM	5-Year Risk of DM
IAA	76	33	59
GADA	90	28	52
ICA512bdcAA	64	40	81
ICA	74	31	51
GADA + IAA	68	41	68
GADA + ICA512bdcAA	62	45	86
IAA + ICA512bdcAA	54	47	100
GADA or IAA	98	25	48
GADA or ICA521bdcAA	92	27	50
IAA or ICA512bdcAA	86	31	58
No. of positive AA tests:			
2	98	24	46
>2	80	39	68

Note: AA, autoantibody; DM, diabetes mellitus; GADA, glutamic acid decarboxylase antibodies;
         IAA, insulin autoantibodies; ICA, islet cell autoantibodies; PPV, positive predictive value.

Source: Verge, C.F. et al., *Diabetes*, 45, 926–933, 1996. With permission.

make this a useful test in ruling in T1DM. In general, testing for ICA has been replaced by testing for IA-2 autoantibodies.

5. Out of 538 patients undergoing chronic dialysis therapy, 52 (9.7%) had T1DM, 434 (80.6%) had T2DM, and 52 (9.7%) had latent autoimmune diabetes of adulthood (LADA). The prevalence of positive GADA tests was 17.3% in the T1DM patients and 26.9% in the LADA patients. None of the T2DM patients was positive for GADA (Biesenbach et al., 2005).

6. Retrospectively, five clinical features were more frequent in LADA compared with T2DM at diagnosis (Fourlanos et al., 2006):
   - Age of onset < 50 years ($p < 0.0001$)
   - Acute symptoms ($p < 0.0001$)
   - BMI < 25 kg/m^2 ($p = 0.0004$)
   - Personal history of autoimmune disease ($p = 0.011$)
   - Family history of autoimmune disease ($p = 0.024$)

7. Prospectively, the presence of at least two distinguishing clinical features (see note 6 above) had a 90% sensitivity and 71% specificity for identifying LADA, with a negative predictive value of 99% if one or no feature was present (Fourlanos et al., 2006).

8. Four major autoantigens (IA-2, IA-2β, GAD65, and insulin) of T1DM are all associated with dense core or synaptic vesicles. Secretory vesicle-associated membrane protein (VAMP2) and neuropeptide Y (NPY), found in dense core or synaptic vesicles, have been identified as new minor autoantigens (Hirai et al., 2008), with 21% and 9%, respectively, of 200 T1DM sera reacting positively.

ICD-9 Codes

Conditions that may justify this test *include but are not limited to*:

250.XX   diabetes mellitus
277.7   metabolic syndrome

Suggested Reading

Biesenbach G, Auinger M, Clodi M, Prischl F, Kramar R. Prevalence of LADA and frequency of GAD antibodies in diabetic patients with end-stage renal disease and dialysis treatment in Austria. *Nephrol Dial Transplant* 2005; 20:559–65.

Bingley PJ, Williams AJ, Gale EA. Optimized autoantibody-based risk assessment in family members: implications for future intervention trials. *Diabetes Care* 1999; 22:1796–801.

Fourlanos S, Perry C, Stein MS, Stankovich J, Harrison LC, Colman PG. A clinical screening tool identifies autoimmune diabetes in adults. *Diabetes Care* 2006; 29:970–5.

Hirai H, Miura J, Hu Y, Larsson H, Larsson K, Lernmark A, Ivarsson S-A, Wu T, Kingman A, Tzioufas AG, Notkins AL. Selective screening of secretory vesicle-associated proteins for autoantigens in type 1 diabetes: VAMP2 and NPY are new minor autoantigens. *Clin Immunol* 2008; 127:366–74.

La Gasse JM, Brantley MS, Leech NJ, Rowe RE, Monks S, Palmer JP, Nepom GT, McCulloch DK, Hagopian WA. Successful prospective prediction of type 1 diabetes in schoolchildren through multiple defined autoantibodies: an 8-year follow-up of the Washington State Diabetes Prediction Study. *Diabetes Care* 2002; 25:505–11.

Verge CF, Gianani R, Kawasaki E, Yu L, Pietropaolo M, Jackson RA, Chase HP, Eisenbarth GS. Prediction of type I diabetes in first-degree relatives using a combination of insulin, GAD, and ICA512bdc/IA-2 autoantibodies. *Diabetes* 1996; 45:926–33.

Verge CF, Stenger D, Bonifacio E, Colman PG, Pilcher C, Bingley PJ, Eisenbarth GS. Combined use of autoantibodies (IA-2 autoantibody, GAD autoantibody, insulin autoantibody, cytoplasmic islet cell antibodies) in type 1 diabetes: Combinatorial Islet Autoantibody Workshop. *Diabetes* 1998; 47:1857–66.

3.6   Criteria for Identification of Insulin Resistance or Early Diabetes Mellitus

3.6.1   Five Criteria for Identification of Insulin Resistance or Early Diabetes Mellitus (EDM)

Indications for Test

Determination of five specific criteria may be used to identify individuals at high risk for insulin resistance and EDM:

- Features of EDM, such as sedentary lifestyle, obesity, hypertension, hyperlipidemia, a family history of diabetes mellitus
- High risk for cardiovascular disease, based on clinical and biochemical findings, including chest pain or increased blood levels of biochemical markers of inflammatory disease

Procedure

1. Determine which of the criteria presented in "Criteria for Identifying Patients as Having the Metabolic Syndrome" (Table 3.15) are met by the patient being evaluated for insulin resistance syndrome and EDM.
2. Reassess patients on at least a yearly basis particularly if their body mass index changes by more than 2 points.

Interpretation

1. If three or more of the above criteria are met, the patient is highly likely to have an insulin resistance syndrome.
2. If three criteria are met, the risk for DM and/or heart disease is increased by at least twice that of those who meet none of the above criteria.

TABLE 3.15
Criteria for Identifying Patients as Having the Metabolic Syndrome

Criterion	Cutpoint Values
Elevated fasting glucose (mg/dL)	≥100 or diagnosed with DM[a]
Elevated [TG] (mg/dL)	≥150 or treated for hyperTG[b]
Reduced [HDL-C] (mg/dL)	<40 (M), <50 (F), or currently being treated for low [HDL-C][c]
Elevated blood pressure (SBP/DBP) (mmHg)	≥130/≥85[d]

Ethnicity	Elevated Waist Circumference, cm (in.)[e]	
	M	F
European		
Sub-Saharan African		
Eastern-Mediterranean	≥94 (37)	≥80 (31)
Middle-Eastern Arab		
Japanese		
Chinese or Korean		
South Asian	≥90 (35)	≥80 (31)
South Central American		

[a] Same as AHA and NHLBI.

[b] Same as AHA, NHLBI, NCEP ATP III, and WHO.

[c] Same as AHA, NHLBI, and NCEP ATP III.

[d] Same as NCEP ATP III.

[e] Measured 2.5 cm (1 in.) above the umbilicus and parallel to the ground.

Note: AHA, American Heart Association; ATP, Adult Treatment Panel; DBP, diastolic blood pressure; DM, diabetes mellitus; F, females; HDL-C, high-density lipoprotein cholesterol; hyperTG, hypertriglyceridemia; M, males; NCEP, National Cholesterol Education Program; NHLBI, National Heart Lung Blood Institute; SBP, systolic blood pressure; TG, triglyceride; WHO, World Health Organization.

Source: IDF, *The IDF Consensus Worldwide Definition of the Metabolic Syndrome*, International Diabetes Federation, 2005 (http://www.idf.org/webdata/docs/Metac_syndrome_def.pdf).

Notes

1. In the initial description of syndrome X, fewer and less stringent criteria were presented for definition of an entity eventually called "metabolic syndrome"; for example, BP ≥ 140 mmHg systolic and/or ≥ 90 mmHg diastolic and a fasting plasma glucose (FPG) > 109 mg/dL were used in lieu of the values for these criteria indicated above (Reaven, 1988).
2. Formal testing for EDM may be indicated when the FPG exceeds 99 mg/dL.
3. Clinicians should evaluate and treat all cardiovascular risk factors without regard as to whether a patient meets the more formal or artificial criteria of three findings for diagnosis of the "metabolic syndrome" (Kahn et al., 2005).

ICD-9 Codes

Conditions that may justify this test *include but are not limited to*:

272.4	hyperlipidemia
277.7	metabolic syndrome
278.00	obesity, unspecified
401	essential hypertension
429.2	cardiovascular disease

Suggested Reading

Alberti K, Zimmet PZ. Definition, diagnosis and classification of diabetes mellitus and its complications. Part 1. Diagnosis and classification of diabetes mellitus provisional report of a WHO consultation. *Diabet Med* 1998; 15:539–53.

Expert Panel on Detection, Evaluation, and Treatment of High Blood Cholesterol in Adults. Executive summary of the Third Report of the National Cholesterol Education Program (NCEP) Expert Panel on Detection, Evaluation, and Treatment of High Blood Cholesterol in Adults (Adult Treatment Panel III). *JAMA* 2001; 285:2486–97.

Grundy SM, Cleeman JI, Daniels SP et al.; American Heart Association; National Heart, Lung, and Blood Institute. Diagnosis and management of the metabolic syndrome: an American Heart Association/ National Heart, Lung, and Blood Institute Scientific Statement. *Circulation* 2005; 112:2735–52.

IDF. *The IDF Consensus Worldwide Definition of the Metabolic Syndrome*. 2005. International Diabetes Federaion (http://www.idf.org/webdata/docs/Metac_syndrome_def.pdf).

Kahn R, Buse J, Ferrannini E, Stern M. The metabolic syndrome: time for a critical appraisal. Joint statement from the American Diabetes Association and the European Association for the Study of Diabetes. *Diabetes Care* 2005; 28:2289–304.

Reaven GM. Role of insulin resistance in human disease. *Diabetes* 1988; 37:1595–1607.

Sone H, Tanaka S, Ishibashi S, Yamasaki Y, Oikawa S, Ito H, Saito Y, Ohashi Y, Akanuma Y, Yamada N. The new worldwide definition of metabolic syndrome is not a better diagnostic predictor of cardiovascular disease in Japanese diabetic patients than the existing definitions: additional analysis from the Japan Diabetes Complications Study. *Diabetes Care* 2006; 29:145–7.

3.6.2   Insulin Sensitivity Index (ISI) in Gestational Diabetes Mellitus (GDM)

Indications for Test

Calculation of the ISI is indicated:

- To more objectively determine which patients with GDM may need insulin therapy vs. those who can be managed with diet alone with or without oral hypoglycemic agents

Procedure

1. Select candidates for study based on an abnormal 50- or 100-g oGTT between 24 and 28 weeks' gestation (see Test 3.5.3)
2. Following an 8- to 10-hour overnight fast by the patient, start a normal saline infusion and obtain a blood sample for basal plasma glucose determination at 5 minutes before insulin injection.
3. Inject insulin, 0.1 units/kg body weight, as a rapid i.v. bolus. Use pre- or early pregnancy weight for this calculation. Note that edema may significantly contribute to the patient's weight gain during pregnancy.
4. Obtain blood samples for glucose testing at precisely 0, 3, 5, 7, 10, and 15 minutes after the i.v. administration of insulin.
5. Monitor the patient closely for the development of hypoglycemic symptoms. A syringe containing 25 mL of 25% dextrose should be available for use if necessary, and the patient fed at the end of test.
6. Calculate from the data the:
   - Baseline glucose (G_0) concentration by averaging the –5-minute and 0-minute glucose values
   - Change in plasma glucose from G_0 at each time point (ΔG_{t3}–ΔG_{t15})
   - Ratio of the change from G_0 divided by G_0 at each time point; for example, $ISI_{t=3} = \Delta G_{t3}/G_0$
   - Maximal ISI_t during the test (usually found at 15 minutes)

Interpretation

1. Gestational diabetics requiring insulin will have a lower maximal ISI, usually <0.15.
2. Only 10% of nondiabetics will have a maximal ISI < 0.4.
3. The higher the ISI, the more insulin sensitive the patient.

Notes

1. Calculation of an ISI helps direct therapy toward either insulin therapy or diet alone in the treatment of patients with GDM.
2. Use of sulfonylureas as therapy for GDM remains investigational in most patients but is appropriate if the ISI is in the range of 0.14 to 0.20.
3. The use of other indices, such as HbA_{1c}, fasting glucose, and changes in triglycerides, are less helpful in deciding therapy for GDM than an ISI.
4. The *minimal model technique for estimation of insulin sensitivity* is a research technique involving the infusion of insulin for the estimation of two key indices of glucose/insulin dynamics: fractional glucose clearance and insulin sensitivity (Pacini et al., 1982).
5. The *euglycemic and hyperglycemic insulin clamp procedure* (De Fronzo et al., 1979) quantitates whole-body insulin sensitivity, and the *intravenous insulin tolerance test to determine insulin sensitivity in non-pregnant adults* (Del Prato, 1986) is a method of estimating insulin dose requirements in type 1 diabetes patients. Both of these tests are used in advanced research studies.

ICD-9 Codes

Conditions that may justify this test *include but are not limited to*:

648.8   abnormal glucose tolerance in pregnancy (gestational diabetes mellitus)

Suggested Reading

Bartha JL, Comino-Delgado R, Martinez-Del-Fresno P, Fernandez-Barrios M, Bethencourt I, Moreno-Corral L. Insulin-sensitivity index and carbohydrate and lipid metabolism in gestational diabetes. *J Repro Med* 2000; 45:185–9.

De Fronzo RA, Tobin JD, Andres R. Glucose clamp technique: a method for quantifying insulin secretion and resistance. *Am J Physiol* 1979; 237:E214–23.

Del Prato S. *Methods in Diabetes Research*, Vol II. 1986. John Wiley & Sons, pp. 35–76.

Pacini G, Finegood DT, Bergman RN. A minimal-model-based glucose clamp yielding insulin sensitivity independent of glycemia. *Diabetes* 1982; 31(5 Pt 1):432–41.

3.6.3   Fasting Glucose/Fasting Insulin (Glu/Ins) Ratio as an Estimate of Insulin Sensitivity or Deficiency

Indications for Test

Calculation of the ratio of fasting glucose (Glu) to fasting insulin (Ins) as an estimate of insulin sensitivity or deficiency may be indicated if three or more of the following are present:

- Fasting plasma glucose (FPG) concentration >99 mg/dL but <126 mg/dL
- Hypertension (BP that is >120/>80 mmHg)
- Gender-dependent, low high-density lipoprotein (HDL) levels and/or hypertriglyceridemia (>149 mg/dL)
- Hemoglobin A_{1c} >5.7% but <6.1%

- Acanthosis nigricans (hyperpigmentation) around the neck and other intertrigonous areas
- Large number (>5) of visible skin tags
- Coronary artery disease

Procedure

1. Following a 10- to 14-hr overnight fast, obtain a blood sample for both FPG and insulin determination.
2. Calculate the Glu/Ins ratio:

$$\text{Glu (mg/dL)/Ins (}\mu\text{U/mL)}$$

3. Do not perform calculation of the Glu/Ins ratio if the glucose is >126 mg/dL, as the patient is either overtly diabetic or nonfasting.

Interpretation

1. Recognize that the imprecision of the insulin assay, pulsatility of insulin secretion, high proinsulin levels, stress, and the significant overlap of insulin values between normal subjects and those with insulin resistance limit the utility of the Glu/Ins ratio in a clinical setting.
2. Determination of plasma insulin concentration provides, at best, only a qualitative estimate of insulin resistance as based on a population study of 490 nondiabetic volunteers with fasting insulins that varied between about 5 and 24 μU/mL (Yeni-Komshian et al., 2000).
3. A fasting Glu/Ins ratio < 6.0 is consistent with significant insulin resistance. A Glu/Ins ratio between 6 and 10 suggests possible insulin resistance.
4. Neither the homeostasis model assessment (HOMA) insulin sensitivity index (see Test 12.3.3) nor the ratio of Glu/Ins correlated any better with a specific measure of insulin resistance (i.e., steady-state plasma glucose) than did fasting insulin alone (Yeni-Komshian et al., 2000).
5. A very high Glu/Ins ratio (e.g., >50) is consistent with insulin deficiency and type 1 diabetes if the fasting glucose is >110 mg/dL.
6. Extreme forms of insulin resistance affect patients with insulin receptor mutations (e.g., leprechaunism), autoantibodies to the insulin receptor, generalized and partial lipodystrophy, protease inhibitor-associated lipodystrophy, myotonic dystrophy, or ataxia-telangectasia.

Notes

1. Refer to Test 12.3.3
2. A fairly good relationship between the Glu/Ins ratio and insulin resistance estimated using the frequently sampled intravenous glucose tolerance test (FSIGT) in 55 women, 40 of whom had polycystic ovary syndrome, was found ($r = 0.73$) (Legro et al., 1998).
3. The research-oriented *intravenous glucose tolerance test for determination of insulin sensitivity* procedure tests the adequacy of the first-phase insulin response (FPIR) or secretion as well as the degree of insulin resistance. This test is particularly useful in subjects who have gastrointestinal disturbances that interfere with normal absorption of glucose or who are prone to the dumping syndrome (e.g., after gastric bypass surgery patients).

ICD-9 Codes

Conditions that may justify this test *include but are not limited to*:

277.7   metabolic syndrome

Suggested Reading

Galvin P, Ward G, Walters J, Pestell R, Koschmann M, Vaag A, Martin I, Best JD, Alford F. A simple method for quantitation of insulin sensitivity and insulin release from an intravenous glucose tolerance test. *Diabet Med* 1992; 9:921–8.

Haffner SM, D'Agostino R, Festa A, Bergman RN, Mykkanen L, Karter A, Saad MF, Wagenknecht LE. Low insulin sensitivity (S(i) = 0) in diabetic and nondiabetic subjects in the insulin resistance athero-sclerosis study: is it associated with components of the metabolic syndrome and nontraditional risk factors? *Diabetes Care* 2003; 26:2796–803.

Legro RS, Finegood D, Dunaif A. A fasting glucose to insulin ratio is a useful measure of insulin sensitivity in women with polycystic ovary syndrome. *J Clin Endocrinol Metab* 1998; 83:2694–8.

Matthews DR, Hosker JP, Rudenski AS, Naylor BA, Treacher DF, Turner RC. Homeostasis model assessment: insulin resistance and beta-cell function from fasting plasma glucose and insulin concentrations in man. *Diabetologia* 1985; 28:412–9.

Yeni-Komshian H, Carantoni M, Abbasi F, Reaven GM. Relationship between several surrogate estimates of insulin resistance and quantification of insulin-mediated glucose disposal in 490 healthy nondiabetic volunteers. *Diabetes Care* 2000; 23:171–5.

Risk Assessment for Complications of Diabetes Mellitus

4

4.1 Nonlipid Markers of Cardiovascular Disease

4.1.1 Uric Acid (UA) as a Risk Factor for Cardiovascular Disease

Indications for Test

Uric acid determination may be indicated in patients who:

- Are at high risk for cardiovascular disease (CVD) such as those with diabetes mellitus (DM)
- Have coronary artery disease (CAD) or are to be treated for dyslipidemia
- Are obese; have had a recent rapid weight loss; have a history of daily alcohol intake or recent alcohol binge, particularly beer (not wine); or have a purine-rich (i.e., meat, seafood) diet, high fructose intake, hypertriglyceridemia, hypertension, congestive heart failure (CHF), or gout, either as an acute problem or by history
- Are insulin-resistant, black, preeclamptic, postmenopausal immigrants to Western cultures; individuals who have moved from rural to urban communities; or individuals who have been exposed to low doses of lead

Procedure

1. Obtain a random blood sample for UA testing after gathering the patient's history of medications; fructose, alcohol, and animal protein intake; and risk factors for CVD.
2. Note season of year when uric acid test is performed and patient's gender.
3. If serum [UA] is elevated, obtain a blood sample for TSH, calcium, creatinine, blood urea nitrogen (BUN), and complete blood count (CBC) testing to identify hypothyroidism, hyperparathyroidism, renal insufficiency, or hematologic proliferative disorders as possible underlying causes for the hyperuricemia.

Interpretation

1. Elevated (>6 mg/dL in females and >7 mg/dL in males) serum [UA] may predispose the patient to gouty attacks, nephrolithiasis, and hypertension and may serve as an additional nonlipid marker for cardiovascular disease.
2. Because of seasonal changes in UA production, higher [UA] may be found in colder, winter months.
3. In an acute gouty attack, the serum [UA] may be normal and actually increase in response to aspirin therapy of <3 g, allopurinol therapy, or dietary restrictions.

4. In patients with CHF, high [UA] may predict the need for heart transplant.
5. Hypertensive patients with hyperuricemia have a three- to fivefold higher risk of developing acute coronary syndrome or CVD compared to those with a normal [UA].
6. Hyperuricemia is linked to endothelial dysfunction with adverse platelet adhesion, aggregation, and blood flow.
7. Hyperuricemia may occur with recent, rapid (>1 kg/wk) weight loss.

Notes

1. The prevalence of features of the metabolic syndrome, which include increased abdominal girth, hypertriglyceridemia, low high-density lipoprotein (HDL), high blood pressure (BP) or BP medication use, and hyperglycemia (fasting glucose ≥ 100 mg/dL), increases steadily with increasing [UA] (Choi et al., 2007).
2. Intake of purine-rich vegetables is not associated with gout; however, a higher intake of meat and seafood is associated with gouty attacks.
3. Niacin may increase [UA] and precipitate a gouty attack, while a higher intake of dairy products may decrease the risk of gout.
4. Medications that can increase [UA] include nicotinic acid, thiazide diuretics, levodopa, cyclosporine, tacrolimus, ethambutol, pyrazinamide, and low-dose salicylates (<2 g).
5. High-dose salicylate (>3 g) and losartan therapy are uricosuric. Fenofibrates also increase renal urate excretion and reduce [UA], but do so much more than atorvastatin. Simvastatin has no effect on UA excretion.
6. Treatment with allopurinol is the most direct method of reducing serum uric acid levels.
7. Lowering [UA] in hyperuricemic patients has not been shown to lower rates of death from CVD.
8. Hyperuricemia is frequently found as a primary idiopathic condition without a well-defined predisposing factor or overt complication such as gout.

ICD-9 Codes

Conditions that may justify this test *include but are not limited to*:

Body weight changes
783.1   weight gain
783.2   weight loss

Vascular disorders
274.9   gout, unspecified
278.0   obesity
303.9   alcohol dependence
305.0X   alcohol abuse
401   hypertension
414.0   coronary atherosclerosis
428.0   CHF
437.9   CVD

Suggested Reading

Alderman M, Aiyer KJ. Uric acid: role in cardiovascular disease and effects of losartan. *Curr Med Res Opin* 2004; 20:369–79.

Anker SD, Doehner W, Rauchhaus M et al. Uric acid and survival in chronic heart failure: validation and application in metabolic, functional, and hemodynamic staging. *Circulation* 2003; 107:1991–7.

Choi HK, Ford ES. Prevalence of the metabolic syndrome in individuals with hyperuricemia. *Am J Med* 2007; 120:442–7.

Daskalopoulou SS, Athyros VG, Elisaf M, Mikhailidis DP. Uric acid levels and vascular disease. *Curr Med Res Opin* 2004; 20:951–4.

Elisaf M. Effects of fibrates on serum metabolic parameters. *Curr Med Res Opin* 2002; 18:269–76.

TABLE 4.1
High-Sensitivity C-Reactive
Protein (hsCRP) as a Marker
for Risk of Cardiovascular Events

[hsCRP] (mg/L)	Risk
<0.49	Low
>4.19	Twofold higher[a]
>10	Nonspecific[b]

[a] Compared to healthy individuals with [hsCRP] < 0.49 mg/L.

[b] Additional testing is required for evidence of cryptic infection or inflammation (e.g., unrecognized periodontal disease)

Feig DI, Kang D-H, Johnson RJ. Uric acid and cardiovascular risk. *New Engl J Med* 2008; 359:1811–1821.

Hare JM, Johnson RJ. Uric acid predicts clinical outcomes in heart failure: insights regarding the role of xanthine oxidase and uric acid in disease pathophysiology. *Circulation* 2003; 107:1951–3.

Johnson RJ, Kang DH, Feig D, Kivlighn S, Kanellis J, Watanabe S, Tuttle KR, Rodriguez-Iturbe B, Herrera-Acosta J, Mazzali M. Is there a pathogenetic role for uric acid in hypertension and cardiovascular and renal disease? *Hypertension* 2003; 41:1183–90.

Milionis HJ, Kakafika AI, Tsouli SG, Athyros VG, Bairaktari ET, Seferiadis KI, Elisaf MS. Effects of statin treatment on uric acid homeostasis in patients with primary hyperlipidemia. *Am Heart J* 2004; 148:635–40.

4.1.2 High-Sensitivity C-Reactive Protein (hsCRP) as a Risk Factor for CVD

Indications for Test

Determination of hsCRP may be indicated in patients with:

- Risk factors for atherosclerotic cardiovascular disease (ASCVD), including obesity, sedentary lifestyle, smoking, hypertension, or insulin resistance/early diabetes mellitus syndrome
- Normal lipid profiles and intermediate risk for ASCVD based on family history or other factors

Procedure

1. Because CRP is an acute-phase reactant, obtain a blood sample for hsCRP testing from a non-DM patient who is not obviously infected or acutely ill.
2. Do not measure [hsCRP] in established DM patients, as they already have a known twofold or greater increased risk for ASCVD.

Interpretation

1. Relative risk of ASCVD as a function of [hsCRP] is shown in "High Sensitivity C-Reactive Protein (hsCRP) as a Marker for Risk of CV Events" (Table 4.1).
2. Patients with any combination of high and low low-density lipoprotein (LDL) concentration and [hsCRP] have a 50% higher risk (relative risk [RR] = 1.5) of CV events compared to those with both low [LDL] and [hsCRP] (RR = 1.0).
3. Patients with both high [LDL] and [hsCRP] have a high relative risk (= 2.1) of CV events.

4. [hsCRP] can be lowered with a low-fat diet, weight loss, and exercise, as well as with lipid-lowering drugs (e.g., statins) and antihypertensive drugs (e.g., ACE inhibitors).
5. [hsCRP] tends to be higher in obese and sedentary individuals, smokers, and those with hypertension.
6. Significant race and gender differences may exist in the population distribution of CRP values (Khera et al., 2005)

Notes

1. The relative risk of an initial CV event in women increases with increasing quintile of [hsCRP]. Compared to women with a [hsCRP] in the lowest quintile (RR = 1.0), RR values increased to 1.4, 1.6, 2.0, and 2.3 ($p < 0.001$), respectively, for each successive increase in quintile, whereas the corresponding RR values with increasing quintile of [LDL-C] were 0.9, 1.1, 1.3, and 1.5 ($p < 0.001$), respectively.
2. hsCRP combined with LDL analysis is superior to either test alone in the prediction of CV events.
3. hsCRP alone may be superior to LDL as a predictor of CV events, even after adjusting for other CV risk factors such as smoking, age, DM, and hypertension.
4. As a marker for inflammation, [hsCRP] may reflect damage to vascular endothelium caused by white blood cell toxins. Chronic inflammation increases plasma [TNFα] (see Test 12.4.2), which in turn appears to mediate an increase in [hsCRP].
5. [hsCRP] is significantly correlated with the calculated 10-year Framingham Coronary Heart Disease Risk in men and in women not taking hormone replacement therapy (HRT) but is correlated minimally, if at all, with most individual risk factors of the Framingham Coronary Heart Disease Risk Score (Albert et al., 2003).
6. For patients with coronary artery disease, the reduced rate of progression of ASCVD associated with intensive statin treatment, as compared with moderate statin treatment, is significantly related to greater reductions in the levels of both atherogenic lipoproteins and hsCRP.
7. Fenofibrates activate peroxisome proliferator-activated receptor alpha (PPAR-α) agonist pathways and inhibit the inflammatory response within aortic smooth-muscle cells and decrease the concentration of plasma acute-phase proteins, including interleukin-6, fibrinogen, and hsCRP.
8. The PPAR-γ pathway agonists have been shown to reduce [hsCRP] by >25% over placebo in patients with type 2 DM (Haffner et al., 2002).
9. Currently, it has not been proven that lowering [hsCRP] reduces the risk of developing DM or the complications of CV disease in individuals who already have DM and other CVD risk factors.

ICD-9 Codes

Conditions that may justify this test *include but are not limited to*:

278.0	obesity
305.1	smoking
401	hypertension
414.0	coronary artery disease (CAD)
440	atherosclerosis
783.1	weight gain

Suggested Reading

Albert MA, Glynn RJ, Ridker PM. Plasma concentration of C-reactive protein and the calculated Framingham Coronary Heart Disease Risk Score. *Circulation* 2003; 108:161–5.

Haffner SM, Greenberg AS, Weston WM, Chen H, Williams K, Freed MI. Effect of rosiglitazone treatment on nontraditional markers of cardiovascular disease in patients with type 2 diabetes mellitus. *Circulation* 2002; 106:679–84.

Khera A, McGuire DK, Murphy SA, Stanek G, Das SR, Vongpatanasin W, Wians FH Jr, Grundy SM, de Lemos JA. Race and sex differences in C-reactive protein levels. *J Am Coll Cardiol* 2005; 46:464–9.

Khera A, de Lemos JA, Peshock RM, Lo HS, Stanek HG, Murphy SA, Wians FH Jr, Grundy SM, McGuire DK. Relationship between C-reactive protein and subclinical atherosclerosis: the Dallas Heart Study. *Circulation* 2006; 113:38–43.

Nissen SE, Tuzcu EM, Schoenhagen P, Crowe T, Sasiela WJ, Tsai J, Orazem J, Magorien RD, O'Shaughnessy C, Ganz P. Statin therapy, LDL cholesterol, C-reactive protein, and coronary artery disease. *N Engl J Med* 2005; 352:29–38.

Ridker PM, Rifai N, Rose L, Buring JE, Cook NR. Comparison of C-reactive protein and low-density lipoprotein cholesterol levels in the prediction of first cardiovascular events. *N Engl J Med* 2002; 347:1557–65.

Ridker PM. C-reactive protein: a simple test to help predict risk of heart attack and stroke. *Circulation* 2003a; 108:e81–5.

Ridker PM. Clinical application of C-reactive protein for cardiovascular disease detection and prevention. *Circulation* 2003b; 107:363–9, 391–7.

Ridker PM, Cannon CP, Morrow D, Rifai N, Rose LM, McCabe CH, Pfeffer MA, Braunwald E. C-reactive protein levels and outcomes after statin therapy. *N Engl J Med* 2005; 352:20–8.

Staels B, Koenig W, Habib A, Merval R, Lebret M, Torra IP, Delerive P, Fadel A, Chinetti G, Fruchart JC, Najib J, Maclouf J, Tedgui A. Activation of human aortic smooth-muscle cells is inhibited by PPAR alpha but not by PPAR gamma activators. *Nature* 1998; 393(6687):790–3.

4.1.3   Homocysteine (Hcy) as a Risk Factor for CVD

Indications for Test

Hcy determination may be indicated in patients with:

- Increased risk for premature atherosclerosis or known premature CAD and normal calculated (Friede-wald formula) or direct LDL cholesterol (LDL-C) level
- Age >70 years, who are at higher risk for osteoporosis or those with an osteoporotic fracture, diabetes mellitus (DM), or cognitive decline

Procedure

1. Obtain a fasting blood sample for plasma Hcy determination.
2. If the plasma [Hcy] is elevated (>10 mcmol/L), elect to test for folate and vitamin B_{12} levels.

Interpretation

1. [Hcy] is typically <10 mcmol/L in healthy individuals. [Hcy] > 15 mcmol/L is considered to independently increase the risk for CVD.
2. Elevated plasma [Hcy] can occur in patients with inborn errors of Hcy metabolism (as either the more common heterozygous or the rare homocysteinuric homozygous variant), chronic kidney disease, or vitamin (i.e., folic acid or B_{12}) deficiency.
3. Elevated plasma [Hcy] may develop on treatment with niacin even at lower doses of 500 mg. These elevations are reversible with folic acid therapy.

Notes

1. Hyperhomocysteinemia is attributed to nutritional or genetic deficiencies of folate and vitamin B_{12}.
2. A metaanalysis of 27 studies relating [Hcy] to arteriosclerotic vascular disease and 11 studies of folic acid effects on [Hcy] suggested that, at most, only 10% of the total risk for CAD may be attributable to Hcy (Boushey et al., 1995).
3. The odds ratio (OR) for cerebrovascular disease for each 5-mcmol/L incremental increase in [Hcy] was 1.5 (95% CI, 1.3 to 1.9), with peripheral arterial disease also strongly associated with [Hcy].

4. Increased (≥200 mcg/day) folic acid intake reduces [Hcy] by about 4 mcmol/L such that serum [Hcy] is negatively correlated with serum folate levels.

5. Independent of bone density, older individuals with an increased [Hcy] are almost twice as likely to have osteoporotic fractures compared to their peers with [Hcy] in the lowest quartile for this population (van Meurs et al., 2004).

6. In clinical practice, the direct Hcy test has supplanted the use of the *methionine load for stimulation of homocysteine* test (Stampfer et al., 1992).

7. Plasma [Hcy] was inversely ($p < 0.0001$) associated with serum folate and plasma vitamin B_{12} concentrations regardless of the presence of genetic variants of methylene tetrahydrofolate reductase (MTHFR) or methionine synthase reductase.

8. The MTHFR TT genotype of the C677T polymorphism was significantly associated with log-transformed [Hcy] even after adjustment for age, gender, smoking status, ethnicity, serum folate, and vitamin B_{12} concentrations ($p < 0.01$) (Vaughn et al., 2004).

9. Regardless of adjustments for sex, gender, vitamin cofactors, and CVD risk factors, statistically significant inverse associations between [Hcy] and multiple cognitive domains were observed for older individuals (≥60 years of age). High [Hcy] is associated with a decline in recall memory while increased dietary folate seemed to protect against decline in verbal fluency.

10. A combination of folic acid, vitamin B_6, and B_{12} supplements administered for an average of 5 years failed to significantly reduce the risk of major CVD events in patients with vascular disease (HOPE 2 Investigators, 2006).

11. High-dose folic acid, vitamin B_6, and B_{12} supplements lower [Hcy] and enhance wound healing. Higher [Hcy] appears to be a risk factor for slower wound healing in DM via an associated lower level of nitric oxide bioactivity in wound fluid (Boykin et al., 2006).

12. The relative risk for vascular disease in the top fifth compared with the bottom four fifths of the control fasting total homocysteine distribution was 2.2 (95% CI, 1.6–2.9). Methionine loading identified an additional 27% of at-risk cases (Graham et al., 1997).

13. Moderately high levels of plasma homocyst(e)ine are associated with a subsequent risk of myocardial infarction (MI) independent of other coronary risk factors based on a study of 14,916 male physicians (Stampfer et al., 1992); however, no prospective evidence suggests that treatment of elevated homocyst(e)ine levels will reduce the risk of ASCVD.

14. Frequency of heterozygosity for homocystinuria, an autosomal recessive disorder, affects 1 in 70 members of the normal population and may be a risk factor for premature occlusive arterial disease (Boers et al., 1985), including intermittent claudication, renovascular hypertension, and ischemic cerebrovascular disease.

ICD-9 Codes

Conditions that may justify this test *include but are not limited to*:

250.XX   diabetes mellitus
414.0   coronary atherosclerosis
733   osteoporosis

Suggested Reading

Anon. Assessment of laboratory tests for plasma homocysteine: selected laboratories, July–September 1998. *MMWR* 1999; 48:1013–5.

Boers GH, Smals AG, Trijbels FJ, Fowler B, Bakkeren JA, Schoonderwaldt HC, Kleijer WJ, Kloppenborg PW. Heterozygosity for homocystinuria in premature peripheral and cerebral occlusive arterial disease. *N Engl J Med* 1985; 313:709–15.

Boushey CJ, Beresford SA, Omenn GS, Motulsky AG. A quantitative assessment of plasma homocysteine as a risk factor for vascular disease: probable benefits of increasing folic acid intakes. *JAMA* 1995; 274:1049–57.

Boykin JV, Baylis C. Homocysteine—a stealth mediator of impaired wound healing: a preliminary study. *Wounds* 2006; 18:101–16.

Elias MF, Sullivan LM, D'Agostino RB, Elias PK, Jacques PF, Selhub J, Seshadri S, Au R, Beiser A, Wolf PA. Homocysteine and cognitive performance in the Framingham offspring study: age is important. *Am J Epidemiol* 2005; 162:644–53.

Graham IM, Daly LE, Refsum HM, Robinson K, Brattstrom LE, Ueland PM, Palma-Reis RJ, Boers GH, Sheahan RG, Israelsson B, Uiterwaal CS, Meleady R, McMaster D, Verhoef P, Witteman J, Rubba P, Bellet H, Wautrecht JC, de Valk HW, Sales Luis AC, Parrot-Rouland FM, Tan KS, Higgins I, Garcon D, Andria G, et al. Plasma homocysteine as a risk factor for vascular disease: the European Concerted Action Project. *JAMA* 1997; 277:1775–81.

HOPE 2 Investigators. Homocysteine lowering and cardiovascular events after acute myocardial infarction. *N Engl J Med* 2006; 354:1567–77.

Refsum H, Ueland PM, Nygard O, Vollset SE. Homocysteine and cardiovascular disease. *Annu Rev Med* 1998; 49:31–62.

Stampfer MJ, Malinow MR, Willett WC, Newcomer LM, Upson B, Ullmann D, Tishler PV, Hennekens CH. A prospective study of plasma homocyst(e)ine and risk of myocardial infarction in US physicians. *JAMA* 1992; 268:877–81.

van Meurs JB, Dhonukshe-Rutten RA, Pluijm SM, van der Klift M, de Jonge R, Lindemans J, de Groot LC, Hofman A, Witteman JC, van Leeuwen JP, Breteler MM, Lips P, Pols HA, Uitterlinden AG. Homocysteine levels and the risk of osteoporotic fracture. *N Engl J Med* 2004; 350:2033–41.

Vaughn JD, Bailey LB, Shelnutt KP, Dunwoody KM, Maneval DR, Davis SR, Quinlivan EP, Gregory JF, Theriaque DW, Kauwell GP. Methionine synthase reductase 66A→G polymorphism is associated with increased plasma homocysteine concentration when combined with the homozygous methylenetetrahydrofolate reductase 677C→T variant. *J Nutr* 2004; 134:2985–90.

Vrentzos GE, Papadakis JA, Malliaraki N, Zacharis EA, Mazokopakis E, Margioris A, Ganotakis ES, Kafatos A. Diet, serum homocysteine levels and ischaemic heart disease in a Mediterranean population. *Br J Nutr* 2004; 91:1013–9.

4.1.4 B-Type Natriuretic Peptide (BNP) and N-Terminal proBNP (NT-proBNP): Tests for Congestive Heart Failure (CHF) or Risk for CHF in DM Patients Treated with Oral Hypoglycemic Agents

Indications for Test

Measurement of a B-type natriuretic peptide (BNP or NT-proBNP) is indicated in the assessment of:

- Older patients with type 2 diabetes mellitus (T2DM) who may be candidates for treatment with hypoglycemic drugs, such as the thiazolidinediones (TZDs), that might result in overt cardiac failure if given unknowingly to a patient predisposed to CHF (e.g., those with prior cardiac ischemia, hypertension, or alcoholism)
- Patients presenting to urgent-care clinics with unexplained dyspnea or edema in whom a cardiologic evaluation is indicated
- Outpatients in cardiac or primary-care clinics when therapies for CHF are titrated

Procedure

1. Identify older T2DM patients, usually over age 60, with a prior history of cardiovascular disease (CVD), lower extremity edema, or any possible symptoms of CHF such as edema and shortness of breath who are:
   - Being considered for therapy with the TZD class of hypoglycemic agents
   - Potential candidates for CHF therapy (e.g., diuretics or cardioselective beta blockers)
2. Obtain a blood sample for BNP or NT-proBNP testing. Similar BNP results have been obtained with EDTA-plasma or whole blood (Wieczorek et al., 2002).

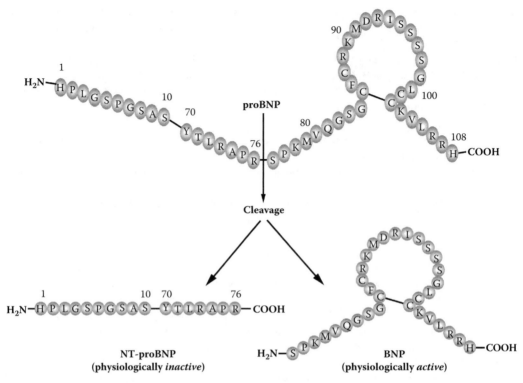

Figure 4.1
Formation of NT-ProBNP and BNP from ProBNP. Both BNP, the physiologically active peptide, and NT-proBNP, which is physiologically inactive but with a longer half-life than BNP, are derived in equimolar amounts from the enzymatic cleavage of a precursor propeptide. Letters in circles represent the standard single-letter abbreviations for the amino acids constituting the primary structures of these proteins and peptides.

Interpretation

Important: Both BNP, the physiologically active peptide, and NT-proBNP, which is physiologically inactive but has a longer half-life than BNP, are derived in equimolar amounts from the enzymatic cleavage of a precursor propeptide (Figure 4.1). The magnitude of values associated with the analytical and clinical performance characteristics of BNP and NT-proBNP in discriminating between individuals with and without CHF differs. For brevity, the information below pertains predominantly to BNP.

1. [BNP] in healthy individuals is typically <30 pg/mL, and this upper limit of normal will vary by BNP assay. The upper limit of linearity of most BNP assays, without sample dilution, is typically 4000 to 5000 pg/mL.
2. [BNP] cutoff values of 49 to 75 pg/mL have a high sensitivity (91–75%) and specificity (82–97%) for detection of left ventricular (LV) dysfunction, including ejection fraction (LVEF) <50%, regional wall motion abnormality, impaired relaxation, and a restrictive or pseudonormal pattern of LV filling. The risk of precipitating CHF in DM patients who have [BNP] in this range and are treated with TZD hypoglycemic agents is unknown.
3. Patients with both systolic and diastolic dysfunction have the highest [BNP] (675 ± 423 pg/mL).
4. The median [BNP] (25th to 75th quartile values) in patients with clinical signs and symptoms of CHF fitting the various New York Heart Association (NYHA) classes were (Figure 4.2):
   - NYHA 1 = 83.1 pg/mL (49.4–137)
   - NYHA 2 = 235 pg/mL (137–391)
   - NYHA 3 = 459 pg/mL (200–871)
   - NYHA 4 = 1119 pg/mL (728–>1300)

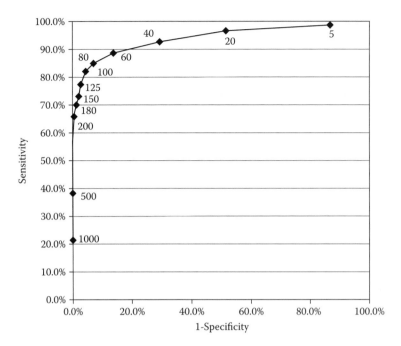

Figure 4.2

Receiver operator characteristic (ROC) curve for B-type natriuretic peptide (BNP) values for discriminating between patients without congestive heart failure (CHF) vs. those with various classes of CHF (New York Heart Association classes 1 to 4). Area under the curve = 0.971 (0.96–0.99) (95% CI; $p < 0.001$). (From Wieczorek, S.J. et al., *Am. Heart J.*, 144, 834–839, 2002. With permission.)

5. Overlap in [BNP] between the NYHA classes reflects the variability of [BNP] with disease states but also reflects the subjectivity of the NYHA classification system.

6. TZD (Avandia®, Actos®) therapy is contraindicated in patients with signs and symptoms suggestive of NYHA CHF classes 3 and 4 and should probably be used with great caution, or not at all, with any clinical sign of CHF regardless of NYHA CHF class.

7. [BNP] is significantly correlated with LV wall thickness and mass.

8. [BNP] increases in Stages 3 and 4 of chronic kidney disease and in dialysis patients. [BNP] may decrease after dialysis of end stage renal disease patients but usually not to normal, indicating that only part of the increase is due to fluid overload.

9. Beta-blocker, but not angiotensin-converting enzyme (ACE) inhibitor, therapy affects [BNP]. Non-cardioselective beta-blocker treatment may raise the [BNP].

10. [BNP] is unaffected by uncomplicated hypertension, DM, emphysema, asthma, and pneumonia. The effect of cor pulmonale on [BNP] is not clear.

Notes

1. Both [BNP] and [NT-proBNP]:
   - Reflect the severity of heart failure in the later stages of CHF.
   - Are useful in the detection of early stages of impaired cardiac function, as LV end diastolic pressure (LVEDP) correlates with [BNP] and [NT-proBNP] across a broad range of LVEDP values.
   - Reliably predict the presence or absence of LV dysfunction on an echocardiogram. A normal BNP may preclude the need for echocardiography; however, [NT-proBNP] can be <100 pg/mL in patients with LVEF < 50%.
2. BNP is a mild diuretic produced by cardiac myocytes. Distension of the heart ventricles results in the release of BNP to help restore normal fluid and electrolyte balance by decreasing aldosterone, blocking sodium reabsorption, and increasing renal filtration and blood flow.

3. [BNP] reliably predicts the presence or absence of LV dysfunction on an echocardiogram. A normal BNP may preclude the need for echocardiography; however, [BNP] can be <100 pg/mL in patients with LVEF < 50%.
4. When 30 T2DM patients with no sign of LV dysfunction were treated with Actos® (15 mg/day), a total of 12 developed a BNP of >100 pg/mL. The basal BNP in these 12 exceeded 18 pg/mL but did not in the 18 patients with BNPs of <100 pg/mL after TZD treatment (Ogawa et al., 2003).
5. In Cox proportional-hazards models adjusting for conventional CVD risk factors, [BNP] most strongly predicted the risk of death (adjusted hazard ratio [AHR] per 1 SD increment in the log values) with a ratio of 1.4 and a major CV event with AHR of 1.25, suggesting that [BNP] may be an emerging biomarker for CVD (Wang et al., 2006).
6. A [BNP] ≥ 100 pg/mL misidentified CHF as a cause for dyspnea in 30% of emergency room patients complaining of shortness of breath after exclusion of cases with acute coronary syndrome. If the [BNP] cutoff was raised to 400 pg/mL, its positive predictive value increased to 77%, but the negative predictive value was lowered to 79% (Chung et al., 2006).
7. The coefficients of variation for intraassay and total imprecision at BNP concentrations of 67 pg/mL and 508 pg/mL, respectively, are typically 4.5% and 5.8% (intraassay) and 3.3% and 3.3% (total) (Wians et al., 2005).

ICD-9 Codes

Conditions that may justify this test *include but are not limited to*:

428.0   congestive heart failure
782.3   edema
786   dyspnea

Suggested Reading

Chung T, Sindone A, Foo F, Dwyer A, Paoloni R, Janu MR, Wong H, Hall J, Freedman SB. Influence of history of heart failure on diagnostic performance and utility of B-type natriuretic peptide testing for acute dyspnea in the emergency department. *Am Heart J* 2006; 152:949–55.

Das SR, Drazner MH, Dries DL, Vega GL, Stanek HG, Abdullah SM, Canham RM, Chung AK, Leonard D, Wians FH Jr, de Lemos JA. Impact of body mass and body composition on circulating levels of natriuretic peptides: results from the Dallas Heart Study. *Circulation* 2005; 112:2163–8.

Hirata Y, Matsumoto A, Aoyagi T, Yamaoki K, Komuro I, Suzuki T, Ashida T, Sugiyama T, Hada Y, Kuwajima I, Nishinaga M, Akioka H, Nakajima O, Nagai R, Yazaki Y. Measurement of plasma brain natriuretic peptide level as a guide for cardiac overload. *Cardiovasc Res* 2001; 51:585–91.

Hussey SM, Wians FH Jr. Shortness of breath in a 74-year-old woman. *Lab Med* 2004; 35:408–12.

Krishnaswamy P, Lubien E, Clopton P, Koon J, Kazanegra R, Wanner E, Gardetto N, Garcia A, De Maria A, Maisel AS. Utility of B-natriuretic peptide levels in identifying patients with left ventricular systolic or diastolic dysfunction. *Am J Med* 2001; 111:274–9.

Maisel A. B-type natriuretic peptide in the diagnosis and management of congestive heart failure. *Cardiol Clin* 2001; 19:557–71.

Ogawa S, Takeuchi K, Ito S. Plasma BNP levels in the treatment of type 2 diabetes with pioglitazone. *J Clin Endocrinol Metab* 2003; 88:3993–6.

Wang TJ, Gona P, Larson MG, Tofler GH, Levy D, Newton-Cheh, Jacques PF, Rifai N, Selhub J, Robins SJ, Benjamin EJ, D'Agostino RB, Vasan RS. Multiple biomarkers for the prediction of first major cardiovascular events and death. *N Engl J Med* 2006; 355:261–9.

Wians FH Jr, Apple FS. Cardiac markers: the current, the past, and potential [guest editorial]. *J Clin Ligand Assay* 2002; 25:129–31.

Wians FH Jr, Wilson BA, Grant A, Bailey J, Gheorghiu I, Conarpe C, Mindicino H, Thakur K, Belenky A, Despres N, Bluestein B. Multi-site evaluation of the analytical performance characteristics of the Bayer ACS:180 B-type natriuretic peptide (BNP) assay. *Clin Chim Acta* 2005; 353:147–55.

Wieczorek SJ, Wu AHB, Christenson R, Krishnaswamy P, Gottlieb S, Rosano T, Hager D, Gardetto N, Chiu A, Bailly KR, Maisel A. A rapid B-type natriuretic peptide assay accurately diagnoses left ventricular dysfunction and heart failure: a multicenter evaluation. *Am Heart J* 2002; 144:834–9.

4.1.5 Measurement of Blood Pressure (BP) and Calculation of Mean Arterial Pressure (MAP) and Pulse Pressure (PP) in Diabetes Mellitus (DM) and Non-DM Patients

Indications for Test

Determination of BP and calculation of MAP and PP are indicated:

- At the time of all routine visits by patients for endocrine or medical evaluation
- In the event of any emergent condition, including stroke, heart attack, or loss of consciousness

Procedure

1. The patient should be comfortably seated, with the legs uncrossed, and the back and arm supported, such that the middle of the BP cuff when placed on the upper arm is at the level of the right atrium (the midpoint of the sternum).
2. The patient should be instructed to relax as much as possible and not to talk during the measurement procedure; ideally, 10 minutes should elapse after sitting down before the first BP reading is taken, because 75% of a standing to sitting BP drop occurs in the first 10 minutes of being seated.
3. Because extremes of room temperature, recent exercise, recent alcohol or nicotine consumption, positioning of the arm, muscle tension, bladder distension, talking, and background noise affect the BP, neutralize or minimize these as much as possible before taking the BP.
4. Securely wrap an appropriately sized blood pressure cuff around the patient's uncovered upper arm. Based on clinical judgment, it may be appropriate to obtain the patient's BP using both a regular and large cuff such that the pressure bladder encircles 80 to 100% of the arm circumference. Because of their lack of reproducibility and correlation when compared to BP measured at the arm, avoid wrist cuffs for determination of BP.
5. Inflate the cuff to a level above the normal systolic blood pressure (SBP), usually 150 mmHg but much higher if necessary.
6. Allow the mercury column to fall at 2 to 3 mm/sec and read the BP to the nearest 2 mmHg.
7. Use either an electronic BP monitoring device or a stethoscope to detect the appearance of a steady tapping (phase I), which marks the SBP, and the disappearance of the tapping (phase V) to mark the diastolic blood pressure (DBP).
8. Use phase V to determine the DBP, except in situations in which the disappearance of sounds cannot reliably be determined because sounds are audible even after complete deflation of the cuff (for example, in pregnant women), in which case phase IV (muffling) may be used to signal the DBP.
9. At an initial encounter, the BP should be taken in both arms.
10. Calculate the MAP and PP as follows:

$$\text{Mean arterial pressure (MAP)} = (\text{SBP} + \text{DBP})/2$$

$$\text{Pulse pressure (PP)} = \text{SBP} - \text{DBP}$$

Interpretation

1. "Classification of Blood Pressure for Adults Ages >18 Years: JNC 7 vs. JNC 6 Guidelines" (Table 4.2) compares Joint National Committee (JNC) on Prevention, Detection, and Treatment of High Blood Pressure guidelines before and after 2003.

TABLE 4.2
Classification of Blood Pressure for Adults
Ages >18 Years: JNC 7 vs. JNC 6 Guidelines

Blood Pressure Category		Systolic Blood Pressure (SBP)	and/or	Diastolic Blood Pressure (DBP)
JNC 7	JNC 6	(mmHg)		(mmHg)
Normal	Optimal	<120	and	<80
Prehypertension	—	120–139	or	80–89
—	Normal	<130	and	<85
—	High-normal	130–139	or	85–89
Stages of hypertension				
1	1	140–159	or	90–99
2	—	≥160	or	≥100
—	2	161–179	or	101–109
—	3	≥180	or	≥110

Note: JNC, Joint National Committee on Prevention, Detection, and Treatment of High Blood Pressure.

2. An SBP between 120 and 140 mmHg or a DBP between 80 and 90 mmHg is considered prehypertensive in untreated non-DM patients. Serially repeated measures of BP are indicated in such cases to help identify candidates for antihypertensive pharmacotherapy.
3. An SBP of >129 mmHg and a DBP >84 mmHg are abnormal and justify the addition of or an adjustment in antihypertensive drug therapy in DM and non-DM patients.
4. Normally, the MAP should be <100 mmHg.
5. Normally, the pulse pressure should be ≤40 mmHg.
6. In DM patients, the target BP is <130/80 mmHg. Patients with BPs slightly above normal or even normal by these criteria who have one or more CV risk factors should be treated with antihypertensive agents (Kostis et al., 2005).
7. Use of too large a cuff (pressure bladder exceeding 100% of arm circumference) will give a falsely lower BP. Use of too small a cuff (pressure bladder encircling <80% of arm) will give a falsely higher BP.
8. Starting at a BP of 115/75 mmHg, each increase of 20 mmHg in SBP or of 10 mmHg in DBP is associated with a doubling of the death rate due to stroke myocardial ischemia or other vascular disease, up to an eightfold higher risk of death in patients with a BP of 175/105 mmHg. A BP ≥ 150/90 may be an independent risk factor for developing DM.
9. On 24-hour BP monitoring, a normal diurnal variation occurs in both SBP and DBP such that lower readings occur during deep sleep with increases occurring toward the end of the sleep cycle (i.e., the BP "dawn" phenomenon).
10. Secondary causes of hypertension (i.e., endocrine hypertension) should be evaluated when the BP is:
    • Unassociated with obesity or high salt intake
    • Intermittently elevated, of unexpectedly sudden onset or lability
    • Associated with hypokalemia or physical findings consistent with Cushing's syndrome
    • Uncontrolled with use of three medications including a diuretic

Notes

1. In the Modification of Diet in Renal Disease (MDRD) Study, a target MAP of <92 mmHg resulted in a slower progression of non-DM kidney disease in patients with moderately to severely decreased glomerular filtration rates.
2. Throughout middle and old age, a BP that is >115/75 mmHg is directly related to vascular and overall mortality.

3. The onset of phase I of the Korotkoff sounds tends to underestimate the true SBP, while the disappearance of sounds (phase V) tends to occur before DBP as determined by direct intraarterial measurement.

4. BP determinations by a physician or medical assistant are typically higher than the patient's average daytime level, particularly in hypertensive patients (the "white coat" effect). Thus, physician-determined BPs should not be used exclusively in the routine management of patients with hypertension.

5. Masked hypertension, defined as a clinic BP < 140/90 mmHg and daytime ambulatory BP ≥ 135/85 mmHg, with microvascular and macrovascular end organ damage may affect more than 40% of diabetics not thought to be hypertensive without 24-hour BP monitoring.

6. In most hypertensive patients, home BP measurements are consistently lower than those measured in the clinic. Consider a home or ambulatory BP of 135/85 mmHg as roughly equivalent to a clinic pressure of 140/90 mmHg. Ambulatory monitoring of BP for the first 48 hours is associated with a pressor response that mostly likely reflects a novelty effect in the use of the monitoring device.

7. Hypertension in African-Americans tends to be exacerbated with higher sodium intake and is less responsive to antihypertensive monotherapy, particularly angiotensin-converting enzyme (ACE) inhibitors, than in other ethnic groups.

8. Endocrine hypertension is found to be associated with a variety of adrenal abnormalities, hyperthyroidism (i.e., widened pulse pressure), hypothyroidism, acromegaly, insulin resistance of pregnancy, renal disease of hyperparathyroidism, and extreme hypercalcemia, which has a vasoconstrictor effect.

ICD-9 Codes

Conditions that may justify this test *include but are not limited to*:

401	hypertension
436	stroke
440.9	myocardial infarction, acute (MI)

Suggested Reading

Calvo C, Hermida RC, Ayala DE, Lopez JE, Fernandez JR, Dominguez MJ, Mojon A, Covelo M. The "ABPM effect" gradually decreases, but does not disappear in successive sessions of ambulatory monitoring. *J Hypertens* 2003; 21:2265–73.

ESH/ESC. 2003 European Society of Hypertension–European Society of Cardiology guidelines for the management of arterial hypertension. *J Hypertens* 2003; 21(6):1011–53.

Hermida RC, Calvo C, Ayala DE, Dominguez MJ, Covelo M, Fernandez JR, Mojon A, Lopez JE. Administration time-dependent effects of valsartan on ambulatory blood pressure in hypertensive subjects. *Hypertension* 42:283–90.

JNC. The Seventh Report of the Joint National Committee on Prevention, Detection, and Treatment of High Blood Pressure (JNC 7). *JAMA* 2003; 289:2560–72.

Kostis JB, Messerli F, Giles TD. Hypertension: definitions and guidelines. *J Clin Hypertens (Greenwich)*. 2005; 7:538–9.

Lewington S, Clarke R, Qizilbash N, Peto R, Collins R. Age-specific relevance of usual blood pressure to vascular mortality: a meta-analysis of individual data for one million adults in 61 prospective studies. *Lancet* 2002; 360:1903–13.

Pickering TG, Hall JE, Appel LJ, Falkner BE, Graves JW, Hill MN, Jones DW, Kurtz T, Sheps SG, Roccella EJ. Recommendations for blood pressure measurement in humans: an AHA scientific statement from the Council on High Blood Pressure Research Professional and Public Education Subcommittee. *J Clin Hypertens* 2005; 7:102–9.

Sarnak MJ, Greene T, Wang X, Beck G, Kusek JW, Collins AJ, Levey AS. The effect of a lower target blood pressure on the progression of kidney disease: long-term follow-up of the MDRD study. *Ann Intern Med.* 2005; 142:342–51.

4.1.6   Ankle–Brachial Index (ABI) Using Segmental Blood Pressures as a Measure of Blood Flow to the Extremities and the Degree of Peripheral Artery Disease (PAD)

Indications for Test

Measurement of the ABI is indicated in:

- The presence of claudication or exertional leg discomfort, atherosclerotic risk factors such as hyperlipidemia and smoking, nonhealing foot ulcer, hypertension, and diabetes mellitus
- Screening those patients at risk for lower extremity wounds (i.e., those with peripheral neuropathy) and other problems associated with poor peripheral arterial circulation
- In conjunction with pulse volume recordings, lower extremity segmental systolic limb pressures, and transcutaneous oximetry mapping (see Test 4.1.8) in high-risk PAD patients

Procedure

1. Place patient in a resting supine position.
2. Use appropriate sized arm and ankle BP cuffs and a 5- or 7-mHz hand-held Doppler device.
3. Allow the patient to rest for a few minutes before pumping up the cuff.
4. Determine the systolic blood pressure (SBP) in both arms followed by determination of the SBP in both ankles, measuring pressures in both the posterior tibial and dorsalis pedis arteries. Repeat measurement until readings stabilize, usually two or three times.
5. Calculate the ABI for each leg by dividing the higher of the two SBP readings in the leg arteries by the higher of the two readings in the arms.
6. Use the lower of the two leg SBP readings for diagnostic purposes.

Interpretation

1. ABI ratios > 0.9 are normal, and ratios < 0.4 indicate severe obstruction as per "Guidelines for Assessment of the ABI Ratio" (Table 4.3).
2. ABI ratios ≤ 0.9, when associated with more advanced lower extremity wounds, are highly predictive of limb ischemia.
3. ABI ratios < 0.8 are associated with 70% survival at 6 years from testing and a twofold risk of total and CVD mortality, even after the exclusion of those with clinical CVD at initial presentation (Figure 4.3).

TABLE 4.3
Guidelines for Assessment
of the ABI Ratio[a]

Ankle–Brachial Index (ABI) Ratio	Degree of Obstruction
>0.90	Normal to no significant obstruction
0.71–0.90	Mild
0.41–0.71	Moderate
0.00–0.40	Severe

[a] Recommendations of the Education Committee of the Society of Vascular Medicine and Biology.

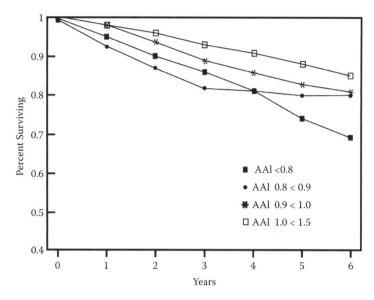

Figure 4.3

Ankle–brachial index (ABI), or ankle–arm index (AAI), and percent survival over 6 years among subjects ($n = 4268$) enrolled in the Cardiovascular Health Study at risk for cardiovascular disease (CVD). Survival is stratified based on ABI at initial presentation: <0.8, severely impaired; 0.8 to <0.9, moderately impaired; 0.9 to <1.0, slightly impaired; 1.0 to <1.5, normal. (From Newman, A.B. et al., *Arterioscler. Thromb. Vasc. Biol.*, 19(3), 538–545, 1999. With permission.)

Notes

1. If the ankle SBP is artifactually high (>30 mmHg higher than the arm SBP), the arteries in the leg may be so stiff or calcified that they cannot be compressed by the BP cuff. In such patients, use the services of a specialized vascular laboratory for diagnosis of PAD using Doppler ultasonographic studies.
2. An absent pedal pulse may not be abnormal; thus, ABI measurements using the lowest ankle artery pressure reading are not recommended.
3. The lowest variance in ABI measurements was observed when the highest pressure between posterior tibial and dorsalis pedis arteries of each leg was divided by the mean of systolic pressures of both arms.
4. ABI is more closely associated with leg function (i.e., ability to walk distances) in persons with PAD than is intermittent claudication or other leg symptoms.
5. Baseline ABI values and the nature of leg symptoms predict the degree of functional decline in walk performance at 2-year follow-up. Specifically, lower baseline ABI values were associated with a greater mean [95% CI] annual decline in 6-minute walk performance:
   - ABI <0.50: –73.0 feet [–142 to –4.2]
   - ABI 0.50 to <0.90: –58.8 feet [–83.5 to –34.0]
   - ABI 0.90 to 1.50: –12.6 feet [–40.3 to 15.1]
6. Compared with participants without PAD, asymptomatic PAD was associated with a greater mean annual decline in 6-minute walk performance (–76.8 feet [–135 to –18.6] vs. –8.67 feet [–36.9 to 19.5]; $p = 0.04$) and an increased odds ratio for becoming unable to walk for 6 minutes continuously (3.63 [1.58 to 8.36], $p = 0.002$).
7. An ABI < 0.50 was associated with a shorter distance walked in 6 minutes (beta regression coefficient = –523 feet; 95% CI, –592 to –454; $p < 0.001$), less physical activity (beta regression coefficient = –514.8 activity units; CI, –657 to –373; $p < 0.001$), slower 4-m walking velocity (beta regression coefficient = –0.21 m/sec; CI, –0.27 to –0.15; $p < 0.001$), and less likelihood of maintaining a tandem stand for 10 seconds compared to an ABI of 1.10 to 1.50 (McDermott et al., 2002).
8. The ABI is not a useful screening test for early disease; rather, it is a very specific test for advanced disease. ABIs sensitivity and specificity for CVD mortality are 30% and 91%, respectively, similar to those of left ventricular hypertrophy by echocardiogram at 43% and 70%, respectively, as a predictor of CVD events (Newman et al., 1999).

ICD-9 Codes

Conditions that may justify this test *include but are not limited to*:

250.7X diabetes with peripheral circulatory disorders
272.4   hyperlipidemia, other and unspecified
305.1   smoking
440   atherosclerosis
682.6   ulcer of leg, except foot
707.15 ulcer of other part or foot, toes

Suggested Reading

Aboyans V, Lacroix P, Preux PM, Vergnenegre A, Ferrieres J, Laskar M. Variability of ankle–arm index in general population according to its mode of calculation. *Int Angiol* 2002; 21:237–43.

Holland T. Utilizing the ankle-brachial index in clinical practice. *Ostomy Wound Manage* 2002; 48:38–49.

McDermott MM, Greenland P, Liu K, Guralnik JM, Celic L, Criqui MH, Chan C, Martin GJ, Schneider J, Pearce WH, Taylor LM, Clark E. The ankle–brachial index is associated with leg function and physical activity: the Walking and Leg Circulation Study. *Ann Intern Med* 2002; 136:873–83.

McDermott MM, Liu K, Greenland P, Guralnik JM, Criqui MH, Chan C, Pearce WH, Schneider JR, Ferrucci L, Celic L, Taylor LM, Vonesh E, Martin GJ, Clark E. Functional decline in peripheral arterial disease: associations with the ankle–brachial index and leg symptoms. *JAMA* 2004; 292:453–61.

Newman AB, Shemanski L, Manolio TA, Cushman M, Mittelmark M, Polak JF, Powe NR, Siscovick D. Ankle–arm index as a predictor of cardiovascular disease and mortality in the Cardiovascular Health Study. The Cardiovascular Health Study Group. *Arterioscler Thromb Vasc Biol.* 1999; 19(3):538–45.

4.1.7 Transcutaneous Oxygen Tension (TC pO$_2$) Measurement with 100% O$_2$ Challenge for Prediction of Wound Healing in Lower Extremity Ulcers of Diabetes Mellitus (DM) Patients*

Indications for Test

TC pO$_2$ measurement is indicated when a patient:

- Has a nonhealing wound for more than 6 weeks
- Requires mapping of the area of viability in an extremity prior to amputation
- Is being considered for hyperbaric oxygen therapy
- May have compromised perfusion to a wounded extremity
- Has Rutherford category 4 ischemic disease marked by minor tissue loss; a nonhealing ulcer; focal gangrene with diffuse pedal ischemia; resting ankle pressure less than 50 to 70 mmHg; plethysmographic, Doppler, or photoplethysmography waveforms demonstrating flat or barely pulsatile flow; or toe pressure less than 30 to 50 mmHg

Procedure

1. Assess the status of parameters associated with lower extremity wound or infection as per "Wound Classification: Depth of Wound (0, I, II, III) vs. Stage (A, B, C, D) Based on Ischemia and Infection" (Table 4.4).
2. Photograph or make a detailed sketch of the lesion for baseline purposes. Measure the dimensions of the wound in both longitudinal and horizontal planes, to include the depth in centimeters (Figure 4.4).

* Contributed by Dr. Jayesh Shah (www.wounddoctors.com) and Robert Babiak, RN (wound care nurse).

TABLE 4.4
Wound Classification: Depth of Wound (0, I, II, III) vs.
Stage (A, B, C, D) Based on Ischemia and Infection[a]

| Stage | Depth of Wound | | | |
	0	I	II	III
A	Pre- or post-ulcerative lesion completely epithelialized	Superficial wound not involving tendon, capsule, or bone	Wound penetrating to tendon or capsule	Wound penetrating to bone or joint
B	Pre- or post-ulcerative lesion completely epithelialized with infection	Superficial wound not involving tendon, capsule, or bone with infection	Wound penetrating to tendon or capsule with infection	Wound penetrating to bone or joint with infection
C	Pre- or post-ulcerative lesion completely epithelialized with ischemia	Superficial wound not involving tendon, capsule, or bone with ischemia	Wound penetrating to tendon or capsule with ischemia	Wound penetrating to bone or joint with ischemia
D	Pre- or post-ulcerative lesion completely epithelialized with infection and ischemia	Superficial wound not involving tendon, capsule, or bone with infection and ischemia	Wound penetrating to tendon or capsule with infection and ischemia	Wound penetrating to bone or joint with infection and ischemia

[a] University of Texas, San Antonio, wound classification system.

Source: Adapted from Lavery, L.A. et al., *J. Foot Ankle Surg.*, 35, 528–531, 1996.

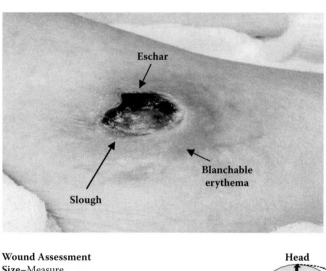

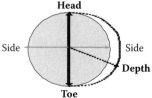

Wound Assessment
Size–Measure
in centimeters.
From **head to
toe**, side to side,
and **depth** (if any).

Figure 4.4
Diagram for assessing the size of a left medial malleous ulcer in a diabetic patient. The characteristics of the ulcer shown were full-thickness skin loss; size, 3 cm × 4 cm × 0.3 cm; 2 cm of peri-wound blanchable erythema; wound bed <15% eschar crescent at the apex; <0.5% slough crescent at the distal base; >80% red base in the middle of the wound; no undermining or tunneling; minimal serous drainage; no odor. (Image courtesy of R. Babiak, RN.)

3. List identifying characteristics, in percentages, of the wound bed (e.g., 10% eschar, 20% slough, and 70% granulating).
4. Note any undermining (lip) or tunnels in the wound bed (in centimeters), using a clock face to demonstrate direction with the head of the patient being at the 12 o'clock position.
5. Note any drainage and its characteristics (e.g., serous, purulent, malodorous, or pungent).
6. Describe the skin surrounding the wound (e.g., erythemic, ischemic, warm to touch), and record the size of affected areas in centimeters.
7. Access oxygen tension sensing equipment supplied by Radiometer™ TCM3, TCM30, TCM400 (www.radiometer.com), Perimed AB (www.perimed.se), or Novametrix® (www.novametrix.com).
8. For measurement of TC pO_2, select six sites on the affected extremity, avoiding major vessels, plantar surfaces, and bony, tendinous, and calloused areas. Choose at least two sites proximate to the wound.
9. Clean the skin with soap and water, rinsing thoroughly to prepare the area for testing.
10. Position the patient such that there is no restriction in circulation or pressure on the head of the electrode. No supplemental oxygen should be administered during the baseline test.
11. Apply a self-adhesive fixation ring securely to the skin area.
12. Moisten the fixation ring with contact fluid and the electrode head with TC pO_2 fresh electrolyte solution.
13. Secure the electrode to the fixation ring such that there is no leakage of fluid beyond the adhesive barrier.
14. Anchor the electrode cable from the oxygen sensor to the test subject.
15. Let 15 minutes elapse after the electrode is put in place to allow baseline TC pO_2 values to stabilize.
16. Elevate the extremity for 10 minutes, and mark the location of each TC pO_2 measurement on a diagram of the skin area being evaluated.
17. Move the extremity back to a neutral or supine position and obtain another set of TC pO_2 measurements.
18. Measure TC pO_2 as above after giving a normobaric 100% oxygen challenge for 10 minutes and after a hyperbaric oxygen challenge for ≥10 minutes.

Interpretation

1. Interpret baseline TC pO_2 measurements as per "Baseline Transcutaneous (TC) Oxygen Tension (pO_2) and Severity of Hypoxia in a Wounded Extremity and Prediction of Response to 10-Minute Normobaric 100% O_2 Challenge and to Hyperbaric Oxygen (HBO) Challenge" (Table 4.5).
2. There is a high likelihood of healing without adjunctive hyperbaric oxygen (HBO) therapy in diabetics with a TC pO_2 > 40 mmHg and in nondiabetics with a TC pO_2 > 30 mmHg.
3. Poor wound healing may be predicted in those with a TC pO_2 of <20 mmHg without HBO intervention.
4. The normobaric 100% oxygen challenge test should be considered for any patient with a TC pO_2 < 40 mmHg.
5. Interpret TC pO_2 values after normobaric 100% oxygen challenge from Table 4.5.
6. Assess response to HBO challenge if there is a suboptimal response (TC pO_2 of 51 to 100 mmHg) to a normobaric 100% oxygen challenge (see Table 4.5).
7. Failure to achieve a TC pO_2 of >100 mmHg after HBO challenge usually means that the blood supply to the wound area is insufficient to allow healing without revascularization of the extremity.
8. Adjunctive HBO therapy may promote wound healing if the TC pO_2 close to a diabetic foot ulcer in 2.5 atmospheres of hyperbaric oxygen is over 200 mmHg.
9. Falsely high TC pO_2 values may be caused by bubbles in the electrode electrolyte; leaks around the fixation ring caused by pulling away from the skin related to hair, skin wrinkles, loose skin, etc.; or by faulty attachment of the electrode head at the wound site being evaluated.
10. Falsely low TC pO_2 values may be caused by bubbles in the electrode electrolyte; excess dead skin, barrier wipes, dirt, or wound care products under the electrode head; outdated, cracked, or an otherwise dysfunctional electrode membrane; absent, insufficient, or outdated electrolyte solution; or incorrect positioning of the patient that restricts circulation or allows pressure on the electrode head.

TABLE 4.5
Baseline Transcutaneous (TC) Oxygen Tension (pO_2) and Severity of Hypoxia in a Wounded Extremity and Prediction of Response to 10-Minute Normobaric 100% O_2 Challenge and to Hyperbaric Oxygen (HBO) Challenge

TC pO_2 (mmHg)	Severity of Hypoxia
>40	No significant hypoxia in diabetic
>30	No significant hypoxia in nondiabetic
39–20	Moderate in diabetic
29–20	Moderate in nondiabetic
19–10	Intermediate
<10	Very severe

TC pO_2 (mmHg)	Predicted Response to Therapy
TC pO_2 after 10-minute normobaric 100% oxygen challenge	
>300	Excellent
>200–299	Good
>100–200	Adequate
51–100	Suboptimal (consider HBO challenge)
0–50	Vascular consultation indicated
TC pO_2 after HBO challenge as a predictor of response to HBO therapy	
>1000	Excellent
>100–1000	Adequate–intermediate
<100	Poor

Notes

1. Causes of DM foot ulcers include neuropathy, infection, and small-vessel angiopathy with ischemia, leading to functional and structural microcirculatory disturbances in perfusion to the skin and deeper structures.

2. Of patients with baseline TC pO_2 < 40 mmHg that could be increased to ≥100 mmHg during inhalation of pure oxygen, 76% responded to HBO therapy, with healing of foot ulcers to the point of intact skin at 3 years compared to 48% healing for conventional, non-HBO-treated subjects (Fife et al., 2002).

3. Pooled data from five trials on DM skin ulcers (118 patients) suggested a reduction in the risk of major amputation with HBO therapy (RR = 0.31; CI, 0.13–0.71). Data supporting the use of HBO therapy in the treatment of venous, arterial, or pressure ulcers are insufficient (Roeckl-Wiedmann et al., 2005).

4. TC pO_2 is a better predictor for healing of chronic DM foot ulcers than toe blood pressure measurement, with failure to heal in ~85% of cases with a normobaric TC pO_2 < 25 mmHg.

5. TC pO_2 and laser Doppler fluxometry (LDF), a skin blood flow measurement technique, yield similar information in diabetics with macrovascular/neuropathic foot syndromes. In contrast, capillary density by video microscopy is widely variable and does not yield any insights beyond that obtainable from clinical exam.

6. A significant trend toward increased amputations as wounds increased in both depth (χ^2 trend = 143.1; $p < 0.001$) and stage (χ^2 trend = 91.0; $p < 0.001$) was found using the University of Texas Wound Classification System (see Table 4.4). Outcomes deteriorated with increasing grade and stage of wounds regardless of modern interventions.

TABLE 4.6
Clinical Description and Objective Criteria for Determining Rutherford Category

Category	Clinical Description	Objective Criteria
0	Asymptomatic; no hemodynamically significant occlusive disease	Normal treadmill/stress test
1	Mild claudication	Completes treadmill exercise (>250 m); post-exercise ankle pressure > 50 mmHg but > 25 mmHg less than normal
2	Moderate claudication	Treadmill not completed (100–250 m)
3	Severe claudication	Failed treadmill test (<100 m); post-exercise ankle pressure < 50 mmHg
4	Ischemic rest pain	Resting ankle pressure < 50–70 mmHg; plethysmographic or Doppler waveforms demonstrating flat or barely pulsatile flow; toe pressure < 30–50 mmHg
5	Minor tissue loss; nonhealing ulcer; focal gangrene with diffuse pedal ischemia	Resting ankle pressure < 50–70 mmHg; plethysmographic or Doppler waveforms demonstrating flat or barely pulsatile flow; toe pressure < 30–50 mmHg
6	Major tissue loss extending above transmetatarsal level; functional foot no longer salvageable.	Resting ankle pressure < 50–70 mmHg; plethysmographic or Doppler waveforms demonstrating flat or barely pulsatile flow; toe pressure < 30–50 mmHg

7. No significant associations have been shown between development of foot ulceration and large-vessel vascular disease, level of formal education (grade school to advanced degrees), nephropathy, retinopathy, impaired vision, or obesity.

8. By definition, Rutherford categories 1 through 4 are nonulcerative ischemic limb disease; category 5 ischemia involves so much tissue loss as to render the limb unsalvageable as per "Clinical Description and Objective Criteria for Determining Rutherford Category" (Table 4.6).

ICD-9 Codes

Conditions that may justify this test *include but are not limited to*:

Diabetes*
| 250.7X | with peripheral circulatory disorders |
| 250.8X | with any associated ulceration |

Vascular*
440.X	atherosclerosis, renal artery and arteries of the extremities
441–442.X	aneurysms
443	peripheral vascular disease
444.XX	arterial embolism and thrombosis
447.X	stricture/rupture of artery
454.X	venous stasis/dermatitis
459.81	venous insufficiency
747.6X	anomalies of peripheral vascular system

Infection
| 682.X | cellulitis/abscesses |

Skin—various locations/organisms
| 707.XX | skin ulcers/various locations |
| 708 | pressure ulcer (i.e., decubitus ulcer), any site |

---

* When diabetes or vascular disease is the cause of the ulcer or gangrene, the ICD code for the underlying condition is to be used in preference to the code(s) for ulcers or gangrene.

785.4	gangrene, lower extremity
862.7	cellulitis/abscessed skin/foot, except toes
891.X	wound leg, traumatic
892.X	wound foot
996.XX	complications of vascular, prosthetic, orthopedic device, implant, or graft
997.62	infection of amputation stump, chronic

Suggested Reading

Fife CE, Buyukcakir C, Otto GH, Sheffield PJ, Warriner RA, Love TL, Mader J. The predictive value of transcutaneous oxygen tension measurement in diabetic lower extremity ulcers treated with hyperbaric oxygen therapy: a retrospective analysis of 1144 patients. *Wound Repair Regen* 2002; 10:198–207.

Kalani M, Brismar K, Fagrell B, Ostergren J, Jorneskog G. Transcutaneous oxygen tension and toe blood pressure as predictors for outcome of diabetic foot ulcers. *Diabetes Care* 1999; 22:147–51.

Kalani M, Jorneskog G, Naderi N, Lind F, Brismar K. Hyperbaric oxygen (HBO) therapy in treatment of diabetic foot ulcers: long-term follow-up. *J Diabetes Complications* 2002; 16:153–8.

Lavery LA, Armstrong DG, Vela SA, Quebedeaux TL, Fleischli JG. Practical criteria for screening patients at high risk for diabetic foot ulceration. *Arch Intern Med* 1998; 158:157–62.

Lawall H, Amann B, Rottmann M, Angelkort B. The role of microcirculatory techniques in patients with diabetic foot syndrome. *Vasa* 2000; 29:191–7.

Niinikoski J. Hyperbaric oxygen therapy of diabetic foot ulcers, transcutaneous oxymetry in clinical decision making. *Wound Repair Regen* 2003; 11:458–6.

Roeckl-Wiedmann I, Bennett M, Kranke P. Systematic review of hyperbaric oxygen in the management of chronic wounds. *Br J Surg.* 2005; 92:24–32.

Sheffield PJ. Measuring tissue oxygen tension: a review. *Undersea Hyperb Med* 1998; 25:179–88.

4.1.8 Coronary Artery Imaging Using Electron Beam Tomography (EBT) or Multidetector (16- to 64-Slice) Computerized Tomography (M-CT) as a Test for Coronary Artery Disease (CAD)

Indications for Test

Coronary artery imaging using EBT or M-CT is indicated in patients:

- At high risk for CAD such as diabetics who may be asymptomatic yet have advanced CAD that might be reversed with earlier diagnosis and intervention
- Without cardiac symptoms who in the judgment of the clinician may require treatment and CAD prevention, especially individuals with LDL cholesterol (LDL-C) values > 190 mg/dL at an age of >50 years and for whom the Framingham model may incorrectly predict low risk of CAD
- With angina or other, possibly cardiac, types of chest pain
- To visualize the patency of native coronary arteries and coronary artery stents

Procedure

1. Obtain informed consent before subjecting the patient to radiation exposure. Pregnancy is a contraindication to testing.
2. Determine the coronary artery calcium score as a noninvasive, noncontrast agent requiring a screening test for atherosclerotic disease in the coronary arteries.
3. For optimal imaging, slow the heart rate with oral or intravenous administration of a beta-blocker agent to get the pulse to <80 bpm with approval of the physician ordering the test.

TABLE 4.7
Image Acquisition Speed as a Function of CT Device Capability

Imaging Device	Number of Detectors	Milliseconds per Image[a]
4-Slice CT	4	250
EBT	20–10	50–100
16-Slice M-CT	16	62
Cineangiographic camera[b]	Not applicable	33
40-Slice M-CT	40	25
64-Slice M-CT	64	15

[a] Cut, slice, or frame; shorter sweep speeds correlate with higher resolution and less radiation per image acquired.

[b] Nonradiographic device for comparison purposes only.

Note: CT, computed tomography; EBT, electron beam tomography; M-CT, multislice computed tomography.

4. Assess renal status before subjecting the patient to an iodinated contrast agent, which is usually administered at a dose of <100 mL using a bolus tracking computer program. Avoid using this test in patients with Stage 3 or 4 chronic kidney disease (CKD).

5. In general, use of iodinated contrast agents puts patients at risk for kidney damage; thus, use of renoprotective doses of *N*-acetylcysteine before and after administration of intravenous contrast media is recommended in most patients.

6. Instruct the patient to hold his or her breath for 20 seconds.

7. Noninvasively image the heart and coronary arteries with subsecond ECG-gated scanning and image reconstruction techniques using an EBT or M-CT device with 16 to 64 detectors, such as SOMATOM Sensation 16 (Siemens, Forchheim, Germany) or Brilliance CT 40-Slice or 64-Slice (Philips, Netherlands).

8. With sweep speeds of <100 msec/cut (superfast scan), heart rates up to about 110 bpm can be tracked and two- to three-dimensional images generated. See "Image Acquisition Speed as a Function of CT Device Capability" (Table 4.7).

9. Calculate the percent blockage of arteries in the same way as with coronary cineangiography.

10. Determine the coronary artery calcium score using linear mass–volume averaging, based on 2.5- to 3.0-mm pixels computed to have an average density distribution of >130 Hounsfield units.

Interpretation

1. Values for the percent blockage of arteries can be used to help direct the most appropriate intervention, ranging from therapeutic lifestyle change or lipid-lowering drug therapy to coronary artery surgery, angioplasty, or placement of stents.

2. With M-CT three-dimensional imaging of the arteries, evidence for buildup of atherosclerotic plaque within and outside the vessel wall may be observed, thus providing the earliest detection of a significant atherosclerotic process in individuals at risk for occlusive disease at a later date.

3. Values for calcium scores range from 1 to 10,000 and can be converted into the percentile corresponding to scores obtained in an age- and gender-matched population of apparently healthy individuals.

4. When arterial calcification is clearly identified, it provides incontrovertible evidence of the presence of moderately advanced CAD, often over a decade before cardiac symptoms and events might occur. Coronary calcium is present in most patients who suffer acute coronary events.

5. A calcium score ≥ 200 among those ≥50 years old and a calcium score ≥ 100 among those <50 years old is evidence for variably obstructive CAD with the potential for myocardial ischemia even without chest pain. A calcium score of zero is good evidence for the absence of obstructive CAD.

6. Based on calcium scoring, earlier diagnosis of CAD, more aggressive intervention, and interval checking of response to intervention may be more effective in slowing or reversing the atherosclerotic process before the onset of symptoms and clinical events.

7. Calcium scoring can be a very sensitive test for the detection of CAD, but its specificity for obstructing lesions is much less clear, prompting high-resolution (<2-mm) imaging of vessels with M-CT.
8. Calcium scoring does not reveal information about plaque instability in the short term or, without the additional use of contrast media, coronary lumen stenosis; thus, a false-negative result may be obtained with inappropriate reassurance and delay in necessary treatment.
9. When used alone, the specificity and sensitivity of the calcium score for diagnosis of low-risk vs. high-risk CAD is unknown.
10. As of 2008, no published evidence suggested that earlier detection of CAD with calcium scoring alone leads to better patient outcomes.

Notes

1. The radiation dose from this test is dependent on the number of slices taken but is usually <10 mSv, and even lower (4–5 mSv) with ECG-controlled dose modulation (i.e., ECG pulsing). Nonetheless, radiation exposure is considerable and in the range of that received cumulatively during two or three standard cardiac catheterizations.
2. In populations at low risk for coronary heart disease (CHD) events, there is no evidence that detection of CAD with M-CT, electrocardiography, or exercise treadmill tests results in improved health outcomes. Indeed, false-positive tests are likely to cause harm because of the unnecessary invasive procedures, overtreatment, and disease labeling that can occur in these patients.
3. The noninvasive nature of M-CT is a great advantage over conventional cardiac catherization, but contrast dye is still necessary, even though at much lower volumes (i.e., 100 mL). Because the use of iodinated contrast agents puts patients with compromised renal status at risk for further kidney damage, the use of renoprotective doses of N-acetylcysteine before and after administration of intravenous contrast media is strongly recommended.
4. The Agatston-based calcium scoring system is poorly reproducible and nonlinear. It has been largely replaced by the linear mass–volume scoring approach, in which scoring is based on 2.5- to 3.0-mm pixels computed to have an average density distribution in the coronary vessels of >130 Hounsfield units
5. Calcium scoring varies considerably, especially in the lower calcium score ranges of <200, between standard, spiral single-detector CT scanners and EBT/M-CT devices, related to the number of detectors in the device.
6. Multivariate clinical risk assessment using the Framingham model and measurements of lipid fractions, homocysteine, hsCRP, and blood pressure are likely to provide better prediction of clinically significant CAD than the Agatston coronary artery calcium score.
7. High-resolution M-CT imaging of arteries to the kidneys and legs, the carotids, and aorta may be indicated in patients with renovascular hypertension, Stage 4 or 5 CKD, claudication, and other vasculopathic phenomena including cerebrovascular ischemia.
8. Based on likelihood ratios, specificity, and negative predictive value, coronary artery imaging using M-CT appeared to be superior to MR for ruling out CAD, but neither test reliably excluded CAD if the pretest odds were high, as in typical angina pectoris (Dewey et al., 2006).

ICD-9 Codes

Conditions that may justify this test *include but are not limited to*:

413	angina pectoris or chest pain
414.0	coronary atherosclerosis

Suggested Reading

Bielak LF, Rumberger JA, Sheedy PF, Schwartz RS, Peyser PA. Probabilistic model for prediction of angiographically defined obstructive coronary artery disease using electron beam computed tomography calcium score strata. *Circulation* 2000; 102:380–5.

Blacher J, Guerin AP, Pannier B, Marchais SJ, London GM. Arterial calcifications, arterial stiffness, and cardiovascular risk in end-stage renal disease. *Hypertension* 2001; 38:938–42.

TABLE 4.8
Major (Positive) Risk Factors
for Cardiovascular Disease[a]

Male ≥ age 45 years; female ≥ age 55 years or with premature
  menopause without estrogen replacement
Family history of premature (before age 60 years) atherosclerosis
Current cigarette smoking
Hypertension (± treatment) or BP > 140/90 mmHg (untreated)
HDL cholesterol < 35 mg/dL (0.9 mmol/L)
Diabetes mellitus (counts as two risk factors)

[a] The presence of documented peripheral vascular disease is
  prima facie evidence for coronary artery disease.

Dewey M, Teige F, Schnapauff D, Laule M, Borges AC, Wernecke K-D, Schink T, Baumann G, Rutsch W, Rogalla P, Taupitz M, Hamm B. Noninvasive detection of coronary artery stenoses with multislice computed tomography or magnetic resonance imaging. *Ann Intern Med* 2006; 145:407–15.

Flohr T, Kuttner A, Bruder H, Stierstorfer K, Halliburton SS, Schaller S, Ohnesorge BM. Performance evaluation of a multi-slice CT system with 16-slice detector and increased gantry rotation speed for isotropic submillimeter imaging of the heart. *Herz* 2003; 28:7–19.

He ZX, Hedrick TD, Pratt CM, Verani MS, Aquino V, Roberts R, Mahmarian JJ. Severity of coronary artery calcification by electron beam computed tomography predicts silent myocardial ischemia. *Circulation* 2000; 101:244–51.

Raggi P, Callister TQ, Cooil B, He ZX, Lippolis NJ, Russo DJ, Zelinger A, Mahmarian JJ. Identification of patients at increased risk of first unheralded acute myocardial infarction by electron-beam computed tomography. *Circulation* 2000; 101:850–5.

US Preventative Services Task Force. Screening for coronary heart disease: recommendation statement. *Ann Intern Med* 2000; 140:569–72.

4.2   Lipid Markers of Cardiovascular Disease

4.2.1   Lipid Screening: Conventional Lipid Profile* as a Predictor of Cardiovascular Disease (CVD) in DM Patients

Indications for Test

A conventional lipid panel is indicated when:

- Screening and monitoring individuals ages 20 and up per National Cholesterol Education Program (NCEP) guidelines for HDL-C and LDL-C at a minimum of 5-year intervals
- Deciding if intervention to reduce the risk of atherosclerotic cardiovascular disease (ASCVD) is warranted
- Monitoring the status of patients with known lipid abnormalities regardless of age
- Evaluating patients with two or more major CVD risk factors as per "Major (Positive) Risk Factors for Cardiovascular Disease" (Table 4.8)
- Assessing individuals, regardless of age, per the American Heart Association (AHA) guidelines for insulin resistance syndrome or early DM, characterized by the occurrence of any three of the following five criteria:

---

* Components of the conventional lipid profile include total cholesterol (TC), triglycerides (TG), high-density lipoprotein cholesterol (HDL-C), TC/HDL ratio, calculated low-density lipoprotein cholesterol (LDL-C), and calculated non-HDL-C.

1. Waist circumference >40 in. (102 cm) in men or >35 in. (88 cm) in women measured by determining abdominal circumference 1 inch above the umbilicus
2. BP ≥ 130/≥80 mmHg
3. Fasting plasma glucose > 99 mg/dL (5.5 mmol/L)
4. [TG] >149 mg/dL (1.68 mmol/L)
5. [HDL-C] < 40 mg/dL (1.04 mmol/L) in men or < 50 mg/dL (1.30 mmol/L) in women

Procedure

1. Instruct the patient to fast for a minimum of 9 to 12 hours before presenting for phlebotomy.
2. Obtain a blood sample for lipid panel testing from an appropriately fasted patient.

Interpretation

1. If the patient's [TG] < 400 mg/dL (4.52 mmol/L), the estimate of [LDL-C] using the Friedewald formula is valid:

$$LDL\text{-}C \ (mg/dL) = TC - HDL - (TG/5)$$

   where TG/5 provides an estimate of very-low-density lipoprotein cholesterol (VLDL-C) concentration
2. If the patient's [TG] ≥ 400 mg/dL (4.52 mmol/L), obtain a direct LDL-C measurement, as a calculated [LDL-C] is not accurate.
3. If an obese patient's [TC] < 200 mg/dL (5.18 mmol/L), but the [TG] > 149 mg/dL (1.68 mmol/L) and the [HDL-C] meets or exceeds the criterion indicated above, the patient may have the insulin resistance syndrome. A fasting glucose measurement in blood should be obtained to establish the diagnosis.
4. Because an [HDL-C] > 60 mg/dL (1.55 mmol/L) is a negative risk factor for CVD, one risk factor may be subtracted from factors calculated using "Framingham Coronary Disease Risk Prediction Score Sheet for Men and Women" (Table 4.9A,B) when making a CVD risk assessment.
5. In the evaluation of patients with premature (onset at <60 years of age) atherosclerosis, elect to analyze HDL-C and LDL-C particle size; lipoprotein little a, or Lp(a); apolipoprotein B (ApoB); homocysteine; and other nonlipid markers of atherosclerotic disease to best identify the potential causes for the CVD and design appropriate preventive interventions.

Notes

1. The American College of Physicians (ACP) guidelines presented in 1996 did not recommend lipid screening for men younger than 35, women younger than 45, and anyone 75 or older. These guidelines are no longer supported by current data.
2. Based on American Association of Clinical Endocrinologists (AACE) advocacy, diagnosis of dysmetabolism or early DM may be made and coded (ICD-9 Code 277.7) for purposes of billing based on professional opinion rather than the AHA guideline that the patient meets three out of the five criteria indicated above.
3. Dysmetabolism is intrinsically an insulin-resistant state; thus, elevated fasting insulin levels as well as nonalcoholic fatty liver disease (NAFLD) and development of acanthosis nigricans and multiple acrochordons are to be considered part of this syndrome.
4. Agents and treatments that reduce insulin resistance (e.g., weight-loss surgery, rimonabant, and thiadiazolidines) will lower [TG] and help to raise [HDL].
5. The definition and implications of the so-called metabolic syndrome remain controversial. In 2006, it was proposed that this syndrome be renamed *insulin resistance syndrome* or *early diabetes mellitus*.
6. In 2005, the IDF proposed criteria for dysmetabolism different from that of the NCEP:
   • Waist circumference ≥ 94 cm (37 in.) for Europid males
   • Waist circumference ≥ 80 cm (32.6 in.) for Europid females
   • Lower, ethnic-specific waist circumference in Chinese, Japanese, and South Asians as per Table 3.15, "Criteria for Identifying Patients as Having the Metabolic Syndrome"

TABLE 4.9A
Framingham Coronary Disease Risk Prediction Score Sheet for Men

Step 1

Age	
Years	Points
30-34	-1
35-39	0
40-44	1
45-49	2
50-54	3
55-59	4
60-64	5
65-69	6
70-74	7

Step 2

LDL - Cholesterol		
(mg/dl)	(mmol/L)	Points
<100	≤2.59	-3
100-129	2.60-3.36	0
130-159	3.37-4.14	0
160-189	4.15-4.91	1
≥190	≥4.92	2

Key	
Color	Risk
green	Very low
white	Low
yellow	Moderate
rose	High
red	Very high

Step 3

HDL - Cholesterol		
(mg/dl)	(mmol/L)	Points
<35	≤0.90	2
35-44	0.91-1.16	1
45-49	1.17-1.29	0
50-59	1.30-1.55	0
≥60	≥1.56	-1

Step 4

Blood Pressure					
Systolic	Diastolic (mmHg)				
(mmHg)	<80	80-84	85-89	90-99	≥100
<120	0				
120-129		0 pts			
130-139			1		
140-159				2	
≥160					3 pts

Note: When systolic and diastolic pressures provide different estimates for point scores, use the higher number

Step 5

Diabetes	
	Points
No	0
Yes	2

Step 6

Smoker	
	Points
No	0
Yes	2

Risk estimates were derived from the experience of the NHLBI's Framingham Heart Study, a predominantly Caucasian population in Massachusetts, USA

Step 7 (sum from steps 1-6)

Adding up the points	
Age	_____
LDL Cholesterol	_____
HDL Cholesterol	_____
Blood Pressure	_____
Diabetes	_____
Smoker	_____
Point Total	_____

Step 8 (determine CHD risk from point total)

CHD Risk	
Point Total	10 Yr CHD Risk
≤-3	1%
-2	2%
-1	2%
0	3%
1	4%
2	4%
3	6%
4	7%
5	9%
6	11%
7	14%
8	18%
9	22%
10	27%
11	33%
12	40%
13	47%
≥14	≥56%

Step 9 (compare to man of the same age)

Comparative Risk		
Age (years)	Average 10 Yr CHD Risk	Low* 10 Yr CHD Risk
30-34	3%	2%
35-39	5%	3%
40-44	7%	4%
45-49	11%	4%
50-54	14%	6%
55-59	16%	7%
60-64	21%	9%
65-69	25%	11%
70-74	30%	14%

*Low risk was calculated for a man the same age, normal blood pressure, LDL cholesterol 100-129 mg/dL, HDL cholesterol 45 mg/dL, non-smoker, no diabetes

TABLE 4.9B
Framingham Coronary Disease Risk Prediction Score Sheet for Women

Step 1

Age	
Years	Points
30-34	-9
35-39	-4
40-44	0
45-49	3
50-54	6
55-59	7
60-64	8
65-69	8
70-74	8

Step 2

LDL - Cholesterol		
(mg/dl)	(mmol/L)	Points
<100	≤2.59	-2
100-129	2.60-3.36	0
130-159	3.37-4.14	0
160-189	4.15-4.91	2
≥190	≥4.92	2

Key	
Color	Risk
green	Very low
white	Low
yellow	Moderate
rose	High
red	Very high

Step 3

HDL - Cholesterol		
(mg/dl)	(mmol/L)	Points
<35	≤0.90	5
35-44	0.91-1.16	2
45-49	1.17-1.29	1
50-59	1.30-1.55	0
≥60	≥1.56	-2

Step 4

Blood Pressure					
Systolic		Diastolic (mmHg)			
(mmHg)	<80	80-84	85-89	90-99	≥100
<120	-3 pts				
120-129		0 pts			
130-139			0 pts		
140-159				2 pts	
≥160					3 pts

Note: When systolic and diastolic pressures provide different estimates for point scores, use the higher number

Step 5

Diabetes	
	Points
No	0
Yes	4

Step 6

Smoker	
	Points
No	0
Yes	2

Risk estimates were derived from the experience of the NHLBI's Framingham Heart Study, a predominantly Caucasian population in Massachusetts, USA

Step 7 (sum from steps 1-6)

Adding up the points	
Age	_____
LDL Cholesterol	_____
HDL Cholesterol	_____
Blood Pressure	_____
Diabetes	_____
Smoker	_____
Point Total	_____

Step 8 (determine CHD risk from point total)

CHD Risk	
Point Total	10 Yr CHD Risk
≤-2	1%
-1	2%
0	2%
1	2%
2	3%
3	3%
4	4%
5	5%
6	6%
7	7%
8	8%
9	9%
10	11%
11	13%
12	15%
13	17%
14	20%
15	24%
16	27%
≥17	≥32%

Step 9 (compare to women of the same age)

Comparative Risk		
Age (years)	Average 10 Yr CHD Risk	Low* 10 Yr CHD Risk
30-34	<1%	<1%
35-39	1%	<1%
40-44	2%	2%
45-49	5%	3%
50-54	8%	5%
55-59	12%	7%
60-64	12%	8%
65-69	13%	8%
70-74	14%	8%

*Low risk was calculated for a woman the same age, normal blood pressure, LDL cholesterol 100-129 mg/dL, HDL cholesterol 55 mg/dL, non-smoker, no diabetes

TABLE 4.10
Lipid Value Conversion Table

Total Cholesterol		LDL		Triglycerides		HDL	
mmol/L	mg/dL	mmol/L	mg/dL	mmol/L	mg/dL	mmol/L	mg/dL
3.2	124	1.5	58	0.5	44	0.25	9.5
3.4	131.5	2.0	77	0.7	62	0.30	11.5
3.6	139	2.3	89	1.0	88.5	0.35	13.5
3.8	147	2.6	100.5	1.1	97.5	0.40	15.5
3.9	151	2.9	112	1.2	106	0.45	17.5
4.0	154.5	3.1	120	1.3	115	0.50	19
4.2	162.5	3.2	124	1.4	124	0.55	21.0
4.4	170	3.3	127.5	1.5	133	0.60	23
4.6	178	3.4	131.5	1.6	142	0.65	25
4.8	185.5	3.5	135	1.7	151	0.70	27
4.9	189.5	3.6	139	1.8	159.5	0.75	29
5.0	193	3.7	143	1.9	168	0.80	31
5.1	197	3.8	147	2.0	177	0.85	33
5.2	201	3.9	151	2.1	186	0.90	35
5.3	205	4.0	154.5	2.2	195	0.95	38
5.4	209	4.1	158.5	2.3	204	1.00	39
5.5	213	4.2	162.5	2.4	213	1.05	40.5
5.6	216.5	4.3	166	2.5	221.5	1.10	42.5
5.7	220.5	4.4	170	2.6	230	1.15	44.5
5.8	224	4.5	174	2.7	239	1.20	46.5
5.9	228	4.6	178	2.8	248	1.29	50
6.0	232	4.7	182	2.9	257	1.40	54
6.1	236	4.8	186	3.0	266	1.50	58
6.2	240	4.9	189.5	3.2	283.5	1.55	60
6.3	244	5.0	193	3.4	301	1.60	62
6.5	251	5.2	201	3.6	318	1.70	66
6.7	259	5.4	209	3.8	337	1.80	70
6.9	267	5.6	216.5	4.0	354	1.90	73.5
7.0	271	5.8	224	4.2	372	2.00	77
7.2	278.5	6.0	232	4.6	407.5	2.50	97
7.3	282	6.5	251	5.0	443		
7.5	290	7.0	271	5.4	478		
7.7	298	7.5	290	6.0	531		
7.8	302	8.0	309	10.0	886		
8.0	309.5			15.0	1329		
8.2	317			20.0	1771		
8.5	329						

Note: For total cholesterol, LDL, and HDL, mg/dL × 0.02586 = mmol/L. For triglycerides, mg/dL × 0.01129 = mmol/L.

Plus at least *two* of the following:
- [TG] ≥ 150 mg/dL (1.70 mmol/L)
- [HDL-C] < 40 mg/dL (1.04 mmol/L) in males or < 50 mg/dL (1.30 mmol/L) in females
- BP ≥ 130/85 mmHg
- Fasting plasma glucose ≥ 100 mg/dL (5.6 mmol/L) or previous diagnosis of DM or impaired glucose tolerance

7. Refer to Table 4.10 for the conversion of conventional and SI units used to measure the components of a lipid profile.

ICD-9 Codes

Conditions that may justify these tests *include but are not limited to*:

Hyperlipidemia
272.0   pure hypercholesterolemia
272.1   pure hypertriglyceridemia
272.2   mixed
272.3   hyperchylomicronemia
272.4   other and unspecified

Other
250.XX   diabetes mellitus
272.5   lipoprotein deficiencies
277.7   dysmetabolism or early DM
278.0   obesity
783.1   weight gain, abnormal

Suggested Reading

ADA. Standards of medical care for patients with diabetes mellitus. *Diabetes Care*. 2003; 26(Suppl 1):S33–50.

NCEP. Executive Summary of the Third Report of the National Cholesterol Education Program (NCEP) Expert Panel on Detection, Evaluation and Treatment of High Blood Cholesterol in Adults. *JAMA* 2001; 285:2486–97.

NCEP. *High Blood Cholesterol: Detection, Evaluation, Treatment*. 2002. National Cholesterol Education Program. (http://www.nhlbi.nih.gov/guidelines/cholesterol/atp3full.pdf).

Sharrett AR, Ballantyne CM, Coady SA, Heiss G, Sorlie PD, Catellier D, Patsch W, Atherosclerosis Risk in Communities Study Group. Coronary heart disease prediction from lipoprotein cholesterol levels, triglycerides, lipoprotein (a), apoliporoteins A-I and B, and HDL density subfractions: the Atherosclerosis Risk in Communities (ARIC) Study. *Circulation* 2001; 104:1108–13.

4.2.2   Total Cholesterol (TC) Fasting Test

Indications for Test

Fasting TC determination is indicated:

- When screening patients for hyperlipidemia

Procedure

1. Obtain a blood sample for TC analysis after a fast of 9 to 12 hours.
2. In women, obtain blood sample during the follicular phase of the menstrual cycle, as TC levels drop up to 20% in the luteal phase.
3. Avoid TC testing during acute illness, including myocardial infarction, which may artifactually lower levels by up to 40%.

Interpretation

1. [TC] < 200 mg/dL (5.18 mmol/L) is desirable.
   [TC] = 200 to 239 mg/dL (5.18–6.19 mmol/L) is borderline high.
   [TC] ≥ 240 mg/dL (6.2 mmol/L) is high.
2. Given the influence of [HDL] on [TC], TC measurement alone has a low positive predictive value for CAD.

Notes

1. [TC] reflects the combination of LDL-C, VLDL-C, and HDL-C concentrations.
2. A higher [TC] may result from a high [HDL], which is usually a favorable condition.
3. [TC] is lower in the presence of an elevated bilirubin concentration and treatment with vitamin C or methyldopa.

Suggested Reading

NCEP. *High Blood Cholesterol: Detection, Evaluation, Treatment*. 2002. National Cholesterol Education
   Program (http://www.nhlbi.nih.gov/guidelines/cholesterol/atp3full.pdf).
Sharrett AR, Ballantyne CM, Coady SA, Heiss G, Sorlie PD, Catellier D, Patsch W, Atherosclerosis Risk in
   Communities Study Group. Coronary heart disease prediction from lipoprotein cholesterol levels,
   triglycerides, lipoprotein (a), apolipoproteins A-I and B, and HDL density subfractions: the Atheroscle-
   rosis Risk in Communities (ARIC) Study. *Circulation* 2001; 104:1108–13.

4.2.3 Triglyceride (TG) Fasting and Nonfasting Tests

Indications for Test

TG testing is indicated when:

- Screening for patients with hypertriglyceridemia-associated insulin resistance or early DM
- There is evidence for diabetes (T1DM or T2DM), presence of atherosclerotic cardiovascular disease
  (ASCVD) or high risk for it (two or more risk factors), or pancreatitis

Procedure

1. Obtain a fasting (minimum 9- to 12-hour) blood sample for TG testing in a patient who has not
   smoked tobacco recently.
2. Avoid blood sampling at mid-cycle in an ovulating woman at the time when the [TG] is usually the
   highest.
3. Obtain a nonfasting blood sample for TG testing in a patient who might be at a higher risk for
   ASCVD based on the judgment of the clinician ordering the test.

Interpretation

1. Fasting:
   [TG] < 75 mg/dL (0.83 mmol/L) is the physiologic ideal.
   [TG] < 150 mg/dL (1.69 mmol/L) is acceptable.
   [TG] = 150 to 199 mg/dL (1.69–2.25 mmol/L) is elevated.
   [TG] = 200 to 499 mg/dL (2.26–5.63 mmol/L) is very high.
   [TG] ≥ 500 mg/dL (5.64 mmol/L) is extremely high, and the patient is at risk for chylomicronemia.
2. [TG] ≥ 150 mg/dL (169 mmol/L) is an element of the insulin resistance syndrome.
3. Nonfasting [TG] measured 2 to 4 hours postprandially normally does not exceed 200 mg/dL. A cutoff
   level has not been established for the nonfasting [TG]. When nonfasting TG concentrations are divided
   into quintiles (i.e., ≤85, 86–113, 114–154, 155–214, ≥215), the CV event rate per 1000 person-years
   steadily increased from 1.28 to 6.27 with a rate of about 3 events in the 114- to 154-mg/dL group
   (Bansal et al., 2007).
4. [TG] levels increase up to 60% with stress, smoking, and starvation and are highest in females at
   time of ovulation. Expect up to a 30% daily variation in [TG].
5. Elevated [TG] occurs with acquired medical conditions, including insulin-resistant T2DM, early DM,
   hypothyroidism, uremia, nephrosis, obstructive liver disease, excessive ingestion of alcohol or carbohydrates, Cushing's syndrome, sarcoidosis, acute spinal cord injury, myeloma, lupus, or pregnancy.
6. Elevated [TG] may result from treatment with medications, including estrogens, tamoxifen, corticosteroids, bile acid sequestrants, diuretics (e.g., thiazides, spironolactone), nonselective beta blockers, cimetidine, protease inhibitors, cyclosporine, clozapine, olanzapine, and isotretinoin.
7. A lowering of [TG] results from treatment with various classes of drugs as follows: fibrates >
   pharmacologic doses of niacin > dietary supplements of omega-3 fatty acids (eicosapentaenoic acid
   [EPA] and docosahexaenoic acid [DHA]) > statins (15–30%).

8. Pharmacologic oral doses of EPA (2–4 g) and DHA are required to lower [TG] comparable to levels achieved with fibrates.

9. Approximately equal elevations of [TG] and [TC], in the range of 6.0 ± 1.0 mmol/L, occur in familial dysbetalipoproteinemia, mostly from decreased clearance; in familial combined hyperlipidemia, there is mostly an overproduction of TG.

10. A relatively benign, autosomal dominant disorder known as familial hypertriglyceridemia (FHT) accounts for 10 to 20% of all patients with TG concentrations between 250 and 1000 mg/dL (2.83–11.3 mmol/L). These patients do not develop coronary disease unless some other risk factor such as diabetes, low [HDL], high [LDL], or family history of CVD at age <60 is present.

11. Risk of pancreatitis increases markedly with [TG] > 700 mg/dL (7.91 mmol/L), even in those with an otherwise benign syndrome of FHT.

12. Severe hypertriglyceridemia with a [TG] >1000 mg/dL (11.3 mmol/L) is associated with chylomicronemia and occurs most commonly in DM patients as well as alcohol abuse.

13. A subset of children with hyperchylomicronemia, usually associated with a [TG] >1000 mg/dL (11.3 mmol/L), has a recessive familial disorder of apolipoprotein CII (ApoCII) deficiency and lipoprotein lipase (LPL) deficiency. Measurement of LPL activity is an investigative tool used to identify defects in LPL as a cause of severe hypertriglyceridemia (Hokanson et al., 1999).

Notes

1. Triglycerides are comprised of a complex aggregate of molecules, including chylomicrons, their remnants, VLDL-C, and intermediate-density lipoprotein cholesterol (IDL-C).

2. Mild hypertriglyceridemia, as well as high-normal [TG], may convey additional risk for CVD in hypertensive patients, even though the rest of the results in the screening lipid profile are within or near the normal range (Rubies-Prat et al., 2001).

3. Fasting [TG] did not predict CVD in women; however, a very significant trend ($p < 0.001$) was observed for the CVD hazard ratio comparing the highest to the lowest tertiles of TG levels postprandially in the Women's Health Study (Bansal et al., 2007).

4. On combination statin plus fenofibrate therapy vs. statin therapy alone, [TG] decreased 43.0% vs. 20.1%, [LDL-C] decreased 31.2% vs. 25.8%, and [HDL–C] increased 18.6% vs. 9.7% ($p < 0.001$ for all) (Grundy et al., 2005).

5. Clinical trials of fibrate therapy, such as the Helsinki Heart Study, Veterans Affairs High-Density Lipoprotein Intervention Trial, Diabetes Atherosclerosis Intervention Study, and Bezafibrate Infarction Prevention trial, suggest but have not proven the superiority of fibrates over statins in the management of obese, insulin-resistant, and DM patients presenting with near-goal [LDL-C] and lower [HDL] and higher [TG] (Fazio and Linton, 2004).

6. The NIH study Action to Control Cardiovascular Risk in Diabetes (ACCORD), due for completion in 2010, will compare the benefits of simvastatin plus fenofibrate to simvastatin alone.

7. The risk of recurrent cardiovascular events (CVEs) in post-acute coronary syndrome patients on treatment with statins was lower for those with [TG] < 150 mg/dL (1.69 mmol/L). Using a hazard ratio (HR) of 1.0 for CVE in those patients with [LDL-C] ≥ 70 mg/dL and [TG] ≥ 150 mg/dL, the HR for CVE was significantly lowered only in those with [LDL-C] ≤ 70 mg/dL and [TG] ≤ 150 mg/dL (Miller et al., 2008).

Suggested Reading

Bansal S, Buring JE, Rifai N, Mora S, Sacks FM, Ridker PM. Fasting compared with nonfasting triglycerides and risk of cardiovascular events in women. *JAMA* 2007; 298:309–16.

Fazio S, Linton MF. The role of fibrates in managing hyperlipidemia: mechanisms of action and clinical efficacy. *Curr Atheroscler Rep* 2004; 6:148–57.

Grundy SM, Vega GL, Yuan Z, Battisti WP, Brady WE, Palmisano J. Effectiveness and tolerability of simvastatin plus fenofibrate for combined hyperlipidemia (the SAFARI trial). *Am J Cardiol* 2005; 95:462–8.

Hokanson JE, Brunzell JD, Jarvik GP, Wijsman EM, Austin MA. Linkage of low-density lipoprotein size to the lipoprotein lipase gene in heterozygous lipoprotein lipase deficiency. *Am J Hum Genet* 1999; 64:608–18.

Miller M, Cannon CP, Murphy SA, Qin J, Kausik K, Ray KK, Braunwald E. Impact of triglyceride levels beyond low-density lipoprotein cholesterol after acute coronary syndrome in the PROVE IT-TIMI 22-trial. *J Am Coll Cardiol* 2008; 51:724–30.

Rubies-Prat J, Ordonez-Llanos J, Martin S, Blanco-Vaca F, Molina L, Goday A, Pedro-Botet J. Low-density lipoprotein particle size, triglyceride-rich lipoproteins, and glucose tolerance in non-diabetic men with essential hypertension. *Clin Exp Hypertens* 2001; 23:489–500.

Sharrett AR, Ballantyne CM, Coady SA, Heiss G, Sorlie PD, Catellier D, Patsch W, Atherosclerosis Risk in Communities Study Group. Coronary heart disease prediction from lipoprotein cholesterol levels, triglycerides, lipoprotein (a), apolipoproteins A-I and B, and HDL density subfractions: the Atherosclerosis Risk in Communities (ARIC) Study. *Circulation* 2001; 104:1108–13.

Talmud PJ, Hawe E, Miller GJ, Humphries SE. Nonfasting apolipoprotein B and triglyceride levels as a useful predictor of coronary heart disease risk in middle-aged UK men. *Arterioscler Thromb Vasc Biol* 2002; 22:1918–23.

4.2.4   Very-Low-Density Lipoprotein (VLDL): VLDL Cholesterol (VLDL-C) and VLDL Particles (VLDL-p)

Indications for Test

VLDL-C and/or VLDL-p testing is indicated when:

- Triglyceride (TG) levels are elevated.
- A patient has diabetes (T1DM or T2DM) or cardiovascular disease.

Procedure

1. Obtain a blood sample after an 8- to 12-hour fast for [TG] determination.
2. If the [TG] < 400 mg/dL (4.52 mmol/L), estimate the [VLDL-C] from the [TG] with the equation:

$$[VLDL\text{-}C] = [TG]/5$$

3. If the [TG] is between 400 mg/dL and 800 mg/dL (4.52–9.04 mmol/L), elect to obtain an estimate of the VLD concentration using the NMR LipoProfile® or VAP® (see Test 12.4.7) to better define the CVD risk based on lipid particle size.
4. If the [TG] > 800 mg/dL (9.04 mmol/L), obtain a measure of [VLDL-C] from a laboratory that performs this test by ultracentrifugation (beta-quantification), as neither the NMR LipoProfile® nor VAP® is capable of accurately estimating [VLDL] in extreme hypertriglyceridemia.

Interpretation

1. [VLDL-C] ≤ 30 mg/dL is normal (i.e., not elevated).
2. The calculated [VLDL-C] can be related to [TG] as follows:
   - [TG] < 150 mg/dL, [VLDL-C] is normal (i.e., <30 mg/dL)
   - [TG] ≥ 150 mg/dL, [VLDL-C] is elevated (i.e., ≥30 mg/dL)
3. The [VLDL-p] measured by the NMR LipoProfile® relates to risk for CVD as follows:
   - [VLDL-p] < 7 mg/dL (0.18 mmol/L), low risk for CVD
   - [VLDL-p] ≥ 7 but ≤ 27 mg/dL, intermediate risk for CVD
   - [VLDL-p] > 27 mg/dL (0.70 mmol/L), high risk for CVD

Notes

1. The postprandial increase in large VLDL and remnant-like particles contributes to the formation of small dense atherogenic LDL-C particles in patients with CVD (Koba et al., 2003).

2. Larger VLDL particles measured by the NMR LipoProfile® or VAP® are considered more atherogenic than smaller ones, presumably because they are more efficient at transferring triglycerides into the arterial wall.

3. Increased postprandial concentrations of TG-rich lipoprotein in the large VLDL-1 fraction on ultracentrifugation are present in early DM and probably contribute to the excess risk of future coronary events in insulin-resistant men (Johanson et al., 2004).

4. Patients with T1DM on a diet higher (17 to 20%) in monounsaturated fats compared to those on a diet lower (10% to 13%) in monounsaturated fats and higher in carbohydrates had lower plasma TG (by 18%; $p = .027$), lower VLDL triglycerides (by 26%; $p = .043$), lower VLDL cholesterol (by 48%; $p = .043$), but higher ApoA1 (by 7%; $p = .018$) and smaller LDL particle size (by 1%; $p = .043$) (Strychar et al., 2003).

5. Oxidative modification of LDL-C particles has been implicated in its abnormal deposition in the arterial wall, and accelerated development of atherosclerosis in DM patients based on significant correlations between the increased susceptibility to *in vitro* oxidation of LDL-C + VLDL-C and LDL-C ($r = 0.54$; $p < 0.001$) and TG ($r = 0.34$; $p < 0.05$) independent of the HbA_{1c} level (Jain et al., 2000).

6. VLDL triglycerides were increased ($p < 0.05$) in non-DM, untreated hypertensive men (0.89 ± 0.79 mmol/L) compared with controls (0.54 ± 0.35 mmol/L). Adjustment for body mass index and abdomen-to-hip circumference ratio did not alter the positive relationship between hypertension and VLDL triglycerides.

7. Mild hypertriglyceridemia, as well as high-normal [TG], may convey additional (i.e., residual) risk for CVD in hypertensive patients even though the rest of the results in the screening lipid profile are within or near the normal range (Rubies-Prat et al., 2001).

Suggested Reading

Blom DJ, O'Neill FH, Marais AD. Screening for dysbetalipoproteinemia by plasma cholesterol and apolipoprotein B concentrations. *Clin Chem* 2005; 51:904–7.

Jain SK, McVie R, Meachum ZD, Smith T. Effect of LDL+VLDL oxidizability and hyperglycemia on blood cholesterol, phospholipid and triglyceride levels in type-I diabetic patients. *Atherosclerosis* 2000; 149:69–73.

Johanson EH, Jansson PA, Gustafson B, Lonn L, Smith U, Taskinen MR, Axelsen M. Early alterations in the postprandial VLDL1 ApoB-100 and ApoB-48 metabolism in men with strong heredity for type 2 diabetes. *J Intern Med* 2004; 255:273–9.

Koba S, Hirano T, Murayama S, Kotani T, Tsunoda F, Iso Y, Ban Y, Kondo T, Suzuki H, Katagiri T. Small dense LDL phenotype is associated with postprandial increases of large VLDL and remnant-like particles in patients with acute myocardial infarction. *Atherosclerosis* 2003; 170:131–40.

Rubies-Prat J, Ordonez-Llanos J, Martin S et al. Low-density lipoprotein particle size, triglyceride-rich lipoproteins, and glucose tolerance in non-diabetic men with essential hypertension. *Clin Exp Hypertens* 2001; 23:489–500.

Strychar I, Ishac A, Rivard M et al. Impact of a high-monounsaturated-fat diet on lipid profile in subjects with type 1 diabetes. *J Am Diet Assoc* 2003; 103:467–74.

Suter PM, Marmier G, Veya-Linder C, Hanseler E, Lentz J, Vetter W, Otvos J. Effect of orlistat on postprandial lipemia, NMR lipoprotein subclass profiles and particle size. *Atherosclerosis* 2005; 180:127–35.

4.2.5 Non-High-Density Lipoprotein Cholesterol (Non-HDL-C) as an Index of Atherogenic Hyperlipidemia

Indications for Test

Non-HDL-C (or atherogenic cholesterol) testing is indicated:

- When attempting to develop an index of CVD risk in hypertriglyceridemic T1DM and T2DM patients
- As a secondary target of lipid-lowering therapy in any patient with markedly elevated triglycerides (TG) ($\geq$200 mg/dL, or 2.26 mmol/L)

Procedure

1. Obtain a blood sample after an 8- to 12-hour fast for TG, TC, and HDL-C testing.
2. Calculate the [non-HDL-C] and the TC/HDL-C and non-HDL-C/HDL-C ratios:

$$[\text{Non-HDL-C}] \ (mg/dL) = [TC] \ (mg/dL) - [HDL\text{-}C] \ (mg/dL)$$

$$TC/HDL\text{-}C \ \text{ratio} = [TC] \ (mg/dL)/[HDL\text{-}C] \ (mg/dL)$$

$$\text{Non-HDL-C}/HDL\text{-}C \ \text{ratio} = [\text{Non-HDL-C}] \ (mg/dL)/[HDL\text{-}C] \ (mg/dL)$$

Interpretation

1. Non-HDL-C:
   - [Non-HDL-C] > 85 mg/dL (2.20 mmol/L) may help to identify individuals not otherwise at high risk for coronary heart disease (CHD).
   - [Non-HDL-C] > 115 mg/dL (2.98 mmol/L) may help to identify DM patients at greater risk of CHD when no risk factors other than DM are present.
   - [Non-HDL-C] > 160 mg/dL (4.14 mmol/L) may help to identify non-DM patients at higher risk of CHD.
2. TC/HDL-C ratios < 4.5 are favorable with regard to decreased risk of CAD compared to higher ratios; however, current Adult Treatment Panel (ATP) III guidelines do not define the TC/HDL-C ratio as either a primary or secondary target of lipid-lowering therapy. LDL-C is retained as the primary target of lipid-lowering therapy.
3. Typically, a non-HDL-C/HDL-C ratio < 4.0 is associated with a low risk of CHD.

Notes

1. Non-HDL-C is made up of LDL, IDL, and VLDL particles.
2. In epidemiologic studies, a hazard ratio (HR) for CVD of 2.23 was found for the highest tertile of non-HDL-C in DM.
3. Hazard ratios for the highest tertile of non-HDL-C in American Indian men with DM (HR = 2.23) and women (HR = 1.80) without known CAD were higher than those for either LDL-C or TG alone in both men and women.
4. Differences of 30 mg/dL (0.78 mmol/L) in [non-HDL-C] and [LDL-C] corresponded to increases in CVD risks of 19% and 15%, respectively, in men and increases in CVD risks of 11% and 8%, respectively, in women.
5. Almost 50% of T2DM with [TG] <200 mg/dL (2.26 mmol/L) have elevated [non-HDL-C].
6. About 50% of T2DM with normal [non-HDL-C] have an increased apolipoprotein B (ApoB) concentration, whereas all hypertriglyceridemic individuals with T2DM have an elevated [non-HDL-C] > 130 mg/dL, and 95% of them have an elevated [ApoB].
7. Use of a target [LDL-C] of <70 mg/dL (1.81 mmol/L) during lipid-lowering treatment of DM with very high CVD risk has been justified by the observation of a greater risk for atherosclerosis in these individuals when the [non-HDL-C] is >100 mg/dL (2.59 mmol/L).
8. Homozygous sitosterolemia is characterized by hyperabsorption and decreased biliary excretion of dietary plant sterols (i.e., campesterol and sitosterol) as well as LDL. This is a rare autosomal recessive disorder in which non-HDL-C may be markedly elevated.

Suggested Reading

Cui Y, Blumenthal RS, Flaws JA, Whiteman MK, Langenberg P, Bachorik PS, Bush TL. Non-high-density lipoprotein cholesterol level as a predictor of cardiovascular disease mortality. *Arch Intern Med* 2001; 161:1413–9.

Jiang R, Schulze MB, Li T, Rifai N, Stampfer MJ, Rimm EB, Hu FB. Non-HDL cholesterol and apolipoprotein B predict cardiovascular disease events among men with type 2 diabetes. *Diabetes Care* 2004; 27:1991–7.

Lu W, Resnick HE, Jablonski KA, Jones KL, Jain AK, Howard WJ, Robbins DC, Howard BV. Non-HDL cholesterol as a predictor of cardiovascular disease in type 2 diabetes: the strong heart study. *Diabetes Care* 2003; 26:16–23.

Miettinen TA. Phytosterolaemia, xanthomatosis and premature atherosclerotic arterial disease: a case with high plant sterol absorption, impaired sterol elimination and low cholesterol synthesis. *Eur J Clin Invest* 1980; 10:27–35.

Wagner AM, Perez A, Zapico E, Ordonez-Llanos J. Non-HDL cholesterol and apolipoprotein B in the dyslipidemic classification of type 2 diabetic patients. *Diabetes Care* 2003; 26:2048–51.

4.2.6   High-Density Lipoprotein Cholesterol (HDL-C), Non-HDL-C, Total Cholesterol (TC)/HDL-C Ratio, and the Risk for Coronary Artery Disease (CAD) in Diabetes Mellitus (DM) Patients

Indications for Test

HDL-C testing with calculation of the [non-HDL-C] and the TC/HDL-C ratio are indicated in:

- Assessment of risk for CAD
- Sedentary patients and those with a family history of premature heart disease who are at higher risk for CAD
- Male hypogonad patients

Procedure

1. Obtain a fasting (minimum of 9- to 12-hour) blood sample for TG, TC, and HDL-C testing.
2. Calculate the [non-HDL-C] and TC/HDL-C ratio:

$$[\text{Non-HDL-C}]\ (\text{mg/dL}) = [\text{TC}]\ (\text{mg/dL}) - [\text{HDL-C}]\ (\text{mg/dL})$$

$$\text{TC/HDL-C ratio} = [\text{TC}]\ (\text{mg/dL})/[\text{HDL-C}]\ (\text{mg/dL})$$

Interpretation

1. HDL-C < 40 mg/dL (1.04 mmol/L) indicates low [HDL-C] in men and women and increased risk of CAD.
2. HDL-C < 50 mg/dL (1.30 mmol/L) indicates low [HDL-C] in women and increased risk of CAD.
3. HDL-C ≥ 60 mg/dL (1.56 mmol/L) indicates high [HDL-C] and reduced risk of CAD.
4. Non-HDL-C is a secondary target (after LDL-C) of therapy in individuals with TG levels ≥ 200 mg/dL (2.26 mmol/L).
5. The target for the non-HDL-C level is the LDL-C goal for the patient + 30 mg/dL.
6. A TC/HDL-C ratio > 4.5 is a significant predictor of CAD and cardiovascular events in T2DM (Gimeno-Orna et al., 2005).

Notes

1. HDL-C:
   - HDL-C reference intervals in African-Americans and females are generally higher than for Caucasians and males.
   - HDL-C is protective against atherosclerotic disease, as it acts via reverse cholesterol transport as a scavenger of particles injurious to the vascular endothelium.
   - HDL-C may be reduced along with other circulating lipid fractions following effective lipid-lowering therapy, including diet therapy. Such reductions are not known to be adverse.
   - HDL-C may be lower in the nonfasted state or when triglycerides are elevated.
   - HDL-C may be lowered by high-dose testosterone therapy (e.g., via injection), which may promote atherogenesis (Isidori et al., 2005). This is in contrast to the physiologic replacement of testosterone with gels, patches, or buccal pellets which may raise [HDL-C] in male hypogonad patients.
   - HDL-C levels that are <5 mg/dL (0.13 mmol/L) occur in patients with Tangier disease as per "Hypolipidemic Disorders" (Table 4.11).

TABLE 4.11
Hypolipidemic Disorders

Disorder	Defect	Inheritance Pattern	Physical Exam Findings	Laboratory Findings
Tangier's disease	Markedly ↑ HDL catabolism	Autosomal recessive	Orange tonsils Cloudy corneas Organomegaly	Low [HDL-C] (2–5% of normal)
Abetalipoproteinemia	Defect in packaging of lipoproteins containing B-100 or B-48, but normal B-100 and B-48 genes	Autosomal recessive	Fat malabsorption Retinitis pigmentosa Hemolytic anemia Neurologic dysfunction Friedrich's ataxia	TC < 50 mg/dL TG < 50 mg/dL Normal HDL-C
Hypobetalipoprotein-emia (heterozygotes)	Defective B-100 gene with a (±) B-48 defect	Autosomal codominant	No disease or characteristic physical finding	Low TC Low or normal TG Normal HDL-C ($HDL_2 > HDL_3$)
Hypobetalipoprotein-emia (homozygotes)	Defective B-100 gene with a (±) B-48 defect	Autosomal codominant	Same as in abetalipoproteinemia, but less severe	Very low TC Low or normal TG Normal HDL-C ($HDL_2 > HDL_3$)

Note: HDL-C, high-density lipoprotein cholesterol; TC, total cholesterol; TG, triglycerides.

- HDL-C levels that are <25 mg/dL (0.65 mmol/L) are typical of patients with markedly impaired liver function with cholestasis.
- HDL-C levels that are <25 mg/dL (0.65 mmol/L) in males and <35 mg/dL (0.91 mmol/L) in females have been associated with twice the average risk of CAD in the Framingham study, primarily related to an associated increase in [LDL-C].
- HDL-C levels that are <30 mg/dL (0.78 mmol/L) are characteristic of patients with familial hypoalphalipoproteinemia and a family history of premature CAD; however, many patients with [HDL-C] in this range have no family history or evidence of CAD, possibly because of their favorable levels of larger sized HDL-p and LDL-p.
- HDL-C is raised significantly by estrogen hormone replacement therapy (HRT) in women, often into a range > 60 mg/dL (1.55 mmol/L) in a fashion similar to pharmacologic doses of niacin.
- A high [HDL-C] phenotype occurs in the human genetic deficiency of cholesteryl ester transfer protein (CETP) associated with substantially enhanced cholesterol efflux from macrophages. Subjects with the CETP VV genotype have lower levels of CETP (1.73 ± 0.11 vs. 2.12 ± 0.10 g/mL; $p = 0.01$), higher [HDL-C] ($p = 0.02$), and larger LDL particles ($p = 0.03$) (Barzilai et al., 2006).
- HDL-C was raised significantly by the CETP inhibitor torcetrapib but resulted in an excess of deaths, myocardial infarction, angina, revascularization procedures, and heart failure in patients receiving torcetrapib plus atorvastatin, as compared with those receiving atorvastatin alone (Tall et al., 2007).
- Higher levels may be associated with longevity. [HDL-C] > 50 mg/dL (1.30 mmol/L) for females and >45 mg/dL (1.17 mmol/L) in males have been associated with a reduced risk of CAD.
- The larger, less dense fractions (i.e., HDL-2), measured directly by VAP® or nuclear magnetic resonance (NMR) spectroscopy, are favorable and reduce CAD risk if > 30 mg/dL (0.78 mmol/L).
- Deficient levels of HDL-2 (<11 mg/dL or 0.29 mmol/L) may be a positive risk factor for CAD.
- The magnitude of increase following statin therapy has been linked to a polymorphism in the promoter region for ApoA1 synthesis.
- In patients with T2DM, a low HDL/triglyceride ratio ($p = 0.011$), but not a higher [LDL] ($p = 0.597$), was associated with ≥50% coronary stenoses and significant CAD.
- Low HDL/triglyceride ratios ($p = 0.044$) were more predictive of an increased incidence of CAD and prevalence of ASCVD events than elevated [LDL] ($p = 0.061$).
- HDL-C can be quantitatively separated by proton NMR spectroscopy into a small, dense fraction and a larger and lighter fraction in blood samples with a total TG < 800 mg/dL (9.04 mmol/L).

- HDL-C can be estimated by VAP® as a large, light (HDL$_2$) and a small, dense (HDL$_3$) fraction.
- HDL-C can be separated and quantified as HDL$_3$ and HDL$_2$ by the following methods: (a) dextran sulfate/Mg^{2+} double precipitation (Northwest Lipid Research Laboratories; Seattle, WA), (b) 4 to 30% polyacrylamide–agarose nondenaturing gradient gel electrophoresis, and (c) analytical ultracentrifugation, which yields results similar to those obtained by VAP®.

2. Non-HDL-C:
   - Reduction of this fraction is a secondary target of therapy in persons with high TG (≥200 mg/dL or 2.26 mmol/L).
3. TC/HDL-C ratio:
   - Ratios < 4.5 are favorable for a decreased risk of CAD.
   - The ratio is neither a primary nor secondary target of lipid-lowering therapy per the Adult Treatment Panel (ATP) III guidelines (i.e., LDL-C lowering is the primary and non-HDL-C lowering is the secondary target of therapy).
   - In T2DM patients, those with a high TC/HDL ratio (cutoff 4.5) had a 2.5 relative risk (RR) for ASCVD events vs. those with an LDL > 135 mg/dL (3.5 mmol/L) with a much higher RR of 4.

Suggested Reading

Barzilai N, Atzmon G, Derby CA, Bauman JM, Lipton RB. A genotype of exceptional longevity is associated with preservation of cognitive function *Neurology* 2006; 67:2170–5.

Caslake MJ, Packard CJ. Phenotypes, genotypes and response to statin therapy. *Curr Opin Lipidol* 2004; 15:387–92.

Drexel H, Aczel S, Marte T, Benzer W, Langer P, Moll W, Saely CH. Is atherosclerosis in diabetes and impaired fasting glucose driven by elevated LDL cholesterol or by decreased HDL cholesterol? *Diabetes Care* 2005; 28:101–7.

Gimeno-Orna JA, Faure-Nogueras E, Sancho-Serrano MA. Usefulness of total cholesterol/HDL-cholesterol ratio in the management of diabetic dyslipidaemia. *Diabet Med* 2005; 22:26–31.

Isidori AM, Giannetta E, Greco EA, Gianfrilli D, Bonifacio V, Isidori A, Lenzi A, Fabbri A. Effects of testosterone on body composition, bone metabolism and serum lipid profile in middle-aged men: a meta-analysis *Clin Endocrinol* 2005; 63:280–93.

Kulkarni KR, Marcovina SM, Krauss RM, Garber DW, Glasscock AM, Segrest JP. Quantification of HDL$_2$ and HDL$_3$ cholesterol by the Vertical Auto Profile II (VAP-II) methodology. *J Lipid Res* 1997; 38:2353–64.

NCEP. *Adult Treatment Panel (ATP) III Guidelines*, NIH Publ. No. 02-5215. 2002. National Cholesterol Education Program, National Heart, Lung, and Blood Institute, National Institutes of Health.

Sacks FM. The role of high-density lipoprotein (HDL) cholesterol in the prevention and treatment of coronary heart disease: expert group recommendations. *Am J Cardiol* 2002; 90:139–43.

Tall AR, Yvan-Charvet L, Wang N. The failure of torcetrapib: was it the molecule or the mechanism? *Arterioscler Thromb Vasc Biol* 2007; 27:257–60.

4.2.7 Low-Density Lipoprotein Cholesterol (LDL-C) and LDL Particle (LDL-p) Testing

Indications for Test

LDL-C testing is indicated:

- When screening for CAD in patients with T1DM or T2DM or suspected atherosclerotic cardiovascular disease (ASCVD)
- In patients, regardless of age, with a family history of premature heart disease, known or suspected familial hypercholesterolemia (FH), or familial combined hyperlipidemia (FCH)

LDL-p testing is indicated in patients with:

- Advanced or advancing ASCVD on hypolipidemic therapy and an [LDL-C] at, or slightly above, the target range
- Markedly elevated [LDL-C] and few or no ASCVD risk factors

Procedure

1. Obtain a fasting (minimum of 9- to 12-hour) blood sample from the patient for lipoprotein profile testing.
2. If the patient's triglyceride concentration is <400 mg/dL (4.52 mmol/L), the laboratory performing the lipoprotein profile will report the patient's LDL-C concentration as a calculated value based on the Friedewald equation:

$$[LDL\text{-}C] \ (mg/dL) = [TC] \ (mg/dL) - [HDL\text{-}C] \ (mg/dL) - ([TG] \ (mg/dL)/5)$$

3. If the [TG] ≥ 400 mg/dL (4.52 mmol/L), elect to treat with a fibrate or niacin, repeat testing after an extended period of fasting, and/or obtain better glycemic control in a DM patient.
4. If the [TG] > 800 mg/dL (9.04 mmol/L), obtain a direct measure of [LDL-C].
5. If LDL particle size testing (see Test 12.4.7) is desired, obtain the appropriate fasting blood specimen for testing, per the specimen requirements of the method (e.g., NMR or VAP®) being used for LDL-p testing.
6. In patients with an elevated [TG], estimate the [non-HDL-C] by subtracting the [HDL-C] from the [TC] (see Test 4.2.5).
7. In patients highly suspected of FH, as per "Dutch Lipid Clinic Network Diagnostic Criteria for Familial Hypercholesterolemia (FH)" (Table 4.12), refer for analysis of possible mutations in the LDL receptor gene, especially in young adults with [LDL-C] >190 mg/dL (4.92 mmol) in whom early initiation and life-long therapy with LDL-C-lowering drugs may be indicated.

Interpretation

1. LDL-C:
   - Undetectable LDL-C is consistent with abetalipoproteinemia and may be associated with fat malabsorption and polyneuropathy characteristic of patients with acanthosis and Bassen–Kornzweig syndrome.
   - LDL-C < 70 mg/dL (1.81 mmol/L) may be associated with longevity and is the goal of therapy in very high-risk patients with established CAD or DM (Grundy et al., 2004).
   - LDL-C < 100 mg/dL (2.6 mmol/L) is optimal.
   - LDL-C = 100 to 129 mg/dL (2.6–3.34 mmol/L) is near to above optimal.
   - LDL-C = 130 to 159 mg/dL (3.36–4.11 mmol/L) is borderline high.
   - LDL-C = 160 to 189 mg/dL (4.14–4.89 mmol/L) is high.
   - LDL-C ≥ 190 mg/dL (4.91 mmol/L) is very high.
   - LDL-C > 200 mg/dL (5.18 mmol/L) is consistent with heterozygous FH or Fredrickson's Type II hyperlipidemia, in which about 50% of patients develop obstructive CAD before age 60. Refer to "Hyperlipidemic Disorders" (Table 4.13).
2. LDL-p (by NMR particle analysis):
   - LDL-p < 1100 nmol/L and size = 20.6 to 22.0 nm indicates a low risk for CHD.
   - LDL-p > 1399 nmol/L and size = 19.0 to 20.5 nm indicates a high risk for CHD.

Notes

1. The LDL-C from the Friedewald formula has been the method used to identify the reference intervals used for clinical interpretation of the LDL-C as a CVD risk factor.
2. LDL-C has limited diagnostic utility because up to a third of patients achieving an [LDL-C] ≤ 100 mg/dL (2.59 mmol/L) may still be at a higher risk for ASCVD.
3. A definitive diagnosis of FH can be made by genetic analysis, although mutations are currently only detected in 30 to 50% of patients with a clinical diagnosis. Underdiagnosis of FH has been reported to range from <1% to 44% (Marks et al., 2003).

TABLE 4.12
Dutch Lipid Clinic Network Diagnostic Criteria
for Familial Hypercholesterolemia (FH)

Criterion	Points[a]
Family history	
First-degree relative with:	
Premature (men, <55 years; women, <60 years) coronary artery and vascular disease	1
Known LDL-C above the 95th percentile in an adult relative	1
Tendinous xanthomata and/or arcus cornealis	2
Children <18 years with LDL-C above the 95th percentile	2
Clinical history	
Patient with:	
Premature (men, <55 years; women, <60 years) coronary artery disease	2
Cerebral or peripheral vascular disease	1
Physical exam findings	
Tendinous xanthomata	6
Arcus cornealis prior to age 45 years	4
LDL-C (mmol/L):	
4.0–4.9	1
5.0–6.4	3
6.5–8.4	5
≥8.5	8
Functional mutation in the LDLR gene	
Present	8
Absent	0

[a] Diagnosis of FH is based on total points: 3–5, possible FH; 6–8, probable FH; >8, definite FH.

Note: LDL-C, low-density lipoprotein cholesterol; *LDLR*, LDL receptor gene.

Source: WHO, *Familial Hypercholesterolemia—Report of a Second WHO Consultation*, Publ. No. WHO/HGN/FH/CONS/99.2, World Health Organization, Geneva, Switzerland, 1999; Austin, M.A. et al., *Am. J. Epidemiol.*, 160, 407–420, 2004.

4. The mutations in the LDL-C receptor gene of patients with heterozygous FH are markedly heterogeneous with over 700 variants identified (Austin et al., 2004).
5. Familial combined hyperlipidemia is characterized by a variable and delayed appearance of elevated [LDL-C] and [TG]. These patients do not have xanthomas.
6. A marked increase in [LDL-C] may occur in moderate to severe hypothyroidism or primary biliary cirrhosis and may be the underlying cause of accelerated atherosclerosis in these patients.
7. Patients with polygenic hypercholesterolemia have a less severe form of Type IIa hyperlipidemia but are still at increased risk for premature atherosclerosis.
8. Because the extent to which the numbers of small LDL particles predict CHD independently of other risk factors is unresolved and standard and inexpensive methods for quantifying [LDL-p] are not currently available, the ATP III guidelines do not recommend measurement of [LDL-p] in routine practice.
9. Individuals with the rare D19H genotype exhibit the greatest statin-induced reduction in LDL-C (Caslake and Packard, 2004).
10. Patients who hyperabsorb LDL from the intestine will respond to LDL absorption inhibitors such as ezetimibe, orlistat, and cholestyramine.

TABLE 4.13
Hyperlipidemic Disorders

Disorder	Inheritance Pattern	Defect	Physical Exam Findings	Lipoprotein Classification (Fredricksen System)
Familial hyperlipidemia (FH)	Autosomal codominant	LDL receptor	Tendinous xanthoma Xanthelasma Arcus juvenilis Planar xanthoma	Type IIA
Familial combined hyperlipidemia (FCH)	Autosomal dominant	Increased LDL/ApoB ratio	Usually normal Rarely xanthelasma/ arcus	Types IIA, IIB, or IV
Familial dysfunctional betalipoproteinemia (FDB)	Autosomal codominant (?)	ApoB-100 mutation	Tendinous xanthoma Xanthelasma Arcus juvenilis Planar xanthoma	Type IIA
Dysbetalipoproteinemia	Autosomal codominant	ApoE (E_2/E_2)	Palmar xanthoma Tuberous xanthoma	Type III
Apolipoprotein CII (ApoCII) deficiency	Autosomal recessive	ApoCII mutation	Eruptive xanthomas Pancreatitis Lipemia retinalis Organomegaly	Types I or IV
Lipoprotein lipase (LPL) deficiency	Autosomal recessive	LPL defect	Eruptive xanthomas Pancreatitis Lipemia retinalis Organomegaly	Types I or IV
Familial hypertriglyceridemia	Autosomal dominant	Unknown	Eruptive xanthomas Pancreatitis	Type IV

Suggested Reading

Austin MA, Hutter CM, Zimmern RL, Humphries SE. Genetic causes of monogenic heterozygous familial hypercholesterolemia: a HuGE prevalence review. *Am J Epidemiol* 2004; 160:407–20.

Bodamer OA, Bercovich D, Schlabach M, Ballantyne C, Zoch D, Beaudet AL. Use of denaturing HPLC to provide efficient detection of mutations causing familial hypercholesterolemia. *Clin Chem* 2002; 48:1913–8.

Caslake MJ, Packard CJ. Phenotypes, genotypes and response to statin therapy. *Curr Opin Lipidol* 2004; 15:387–92.

Grundy SM, Cleeman JI, Merz CN, Brewer HB, Clark LT, Hunninghake DB, Pasternak RC, Smith SC, Stone NJ. Implications of recent clinical trials for the National Cholesterol Education Program Adult Treatment Panel III Guidelines. *Circulation* 2004; 110:2227–39.

Hayward RA, Hofer TP, Vijan S. Narrative review: lack of evidence for recommended low-density lipoprotein treatment targets—a solvable problem. *Ann Intern Med* 2006; 145:520–30.

Marks D, Thorogood MH, Neil AW, Humphries SE. A review on the diagnosis, natural history, and treatment of familial hypercholesterolaemia. *Atherosclerosis* 2003; 168:1–14.

Muller PY, Miserez AR. Large heterogeneity of mutations in the gene encoding the low-density lipoprotein receptor in subjects with familial hypercholesterolaemia. *Atheroscler Suppl* 2004; 5:1–5.

O'Keefe JF, Cordain L, Harris WH, Moe RM, Vogel R. Optimal low-density lipoprotein is 50 to 70 mg/dL: lower is better and physiologically normal. *J Am Coll Cardiol* 2004; 43:2142–46.

WHO. *Familial Hypercholesterolemia—Report of a Second WHO Consultation*, Publ. No. WHO/HGN/FH/ CONS/99.2. 1999. World Health Organization.

4.3 Apolipoprotein (ApoB, ApoC, ApoE, and Lp(a)) Testing in Disorders of Lipid Metabolism

4.3.1 Apolipoprotein B (ApoB)

Indications for Test

Tests for apolipoprotein B (ApoB) may be indicated if a patient with a normal or only slightly abnormal lipid profile has:

- A moderate, high, or very high risk for atherosclerotic cardiovascular disease (ASCVD)
- Xanthomas or a history of premature, familial, or recurrent ASCVD
- Advancing ASCVD on treatment

Procedure

1. Instruct patient not to take in any alcohol for 24 hours before testing.
2. Obtain a >12-hour fasting blood sample for ApoB and lipid profile testing so a calculated LDL-C value can be obtained and the LDL-C/ApoB ratio can be determined.
3. Alternatively, obtain an NMR lipoprotein particle analysis and measure the particle size and concentration of the VLDL-p and LDL-p lipid fractions, which are the main components of ApoB as measured by the VAP®.

Interpretation

1. Reference intervals for ApoB are:
   - 55 to 151 mg/dL in males
   - 44 to 148 mg/dL in females
2. Severe hyperapobetalipoproteinemia is suggested by any ApoB concentration > 130 mg/dL. An ApoB > 60 mg/dL in a very high risk CVD individual is abnormal.
3. If the LDL-C/ApoB ratio is:
   - Elevated, then familial combined hyperlipidemia (FCH) is likely.
   - Reduced in children, then there is a positive relationship to the parental incidence of myocardial infarction (Srinivasan and Berenson, 1995).
4. If NMR particle analysis is performed, patients with [LDL-p] > 1399 nmol/L (the 40th percentile population cutoff value) and those with smaller (<20.6 nm), denser LDL particles may be at a higher risk for ASCVD, dependent on age and other established CVD risk factors.

Notes

1. ApoB is a marker of LDL particle concentration, which, when increased, confers a higher risk for ASCVD. The relative LDL particle size is not related to ApoB concentration.
2. The B-48 subgroup of ApoB lipoproteins is found in chylomicrons, whereas the B-100 subgroup is found in all hepatic-produced beta-lipoproteins, including IDL, LDL, Lp(a), and VLDL-C. Unlike B-100 subgroup ApoB lipoproteins, the B-48 subgroup is not recognized or cleared by LDL receptors.
3. Refer to Test 12.4.8 for interpretive information regarding ApoA and the ApoA-I/ApoB ratio. This ratio has been used only in population studies and is not typically used to assess the risk of ASCVD in an individual patient.

ICD-9 Codes

Conditions that may justify these tests *include but are not limited to*:

Hyperlipidemia
272.2   mixed
272.4   other and unspecified
272.7   lipidoses (xanthomas, xanthomatosis, tuberoeruptive, or palmar xanthomas)

Other

272.5	lipoprotein deficiencies/abnormalities
374.51	xanthelasma of eyelid
413	angina pectoris or chest pain
414.0	coronary atherosclerosis
440	atherosclerosis

Suggested Reading

Contois J, McNamara JR, Lammi-Keefe C. Reference intervals for plasma ApoB determined with standardized commercial immunoturbidimetric assay: results from the Framingham Offspring Study. *Clin Chem* 1996; 42:515–23.

Estonius M, Kallner A. How do conventional markers of lipid disorders compare with apolipoproteins? *Scand J Clin Lab Invest* 2005; 65:33–44.

Jiang R, Schulze MB, Li T, Rifai N, Stampfer MJ, Rimm EB, Hu FB. Non-HDL cholesterol and apolipoprotein B predict cardiovascular disease events among men with type 2 diabetes. *Diabetes Care* 2004; 27:1991–7.

Sattar N, Williams K, Sniderman AD, D'Agostine R Jr, Haffner SM. Comparison of the associations of apolipoprotein B and non-high-density lipoprotein cholesterol with other cardiovascular risk factors in patients with the metabolic syndrome in the insulin resistance study. *Circulation* 2004; 110:2687–93.

Srinivasan SR, Berenson GS. Serum apolipoproteins A-I and B as markers of coronary artery disease risk in early life: the Bogalusa Heart Study. *Clin Chem* 1995; 41:159–64.

4.3.2 ApoCII and ApoCIII

Indications for Test

Tests for ApoCII and ApoCIII may be indicated in patients with:

- Diabetes mellitus (T1DM and T2DM)
- Extreme hypertriglyceridemia (e.g., >1000 mg/dL or 11.3 mmol/L) in a patient with a poor response to lipid-lowering therapy)
- Lipemia retinalis, eruptive xanthomas, or pancreatitis

Procedure

1. Obtain a fasting blood sample (minimum 9 to 12 hours) for measurement of [ApoCII] and [ApoCIII].
2. In patients with marked hypertriglyceridemia, elect to do ApoCII and ApoCIII gene sequencing by polymerase chain reaction (PCR) to identify individuals with mutations of the ApoCII and/or ApoCIII gene. Proceed with a family study to find others in kindred at high risk for complications of hypertriglyceridemia.

Interpretation

1. In normolipidemic individuals, ApoCIII concentration is directly and inversely linked with the [HDL].
2. In lipoprotein subclass analysis (NMR LipoProfile®), an elevated [ApoCIII] is positively and highly correlated with increases in [VLDL-C] and hypertriglyceridemia.
3. Elevation of [ApoCIII] in patients with uremia can alter the metabolism of TG-rich lipoproteins, leading to an elevation in TG and [ApoB].
4. High concentrations of ApoCIII-containing VLDL molecules are associated with hypertriglyceridemia and lower [ApoCII].
5. An elevated [ApoCIII] may increase the risk for ASCVD in normolipidemic T1DM patients through adverse changes in lipoprotein subfraction distributions, independent of ApoCIII genotype.
6. The chylomicronemia associated with patients having the Fredrickson Type V mixed hyperlipidemic phenotype may be the result of a rare, recessive trait involving lipoprotein lipase (LPL) or ApoCII deficiencies in which TG cannot be cleared and may rise to levels > 26,500 mg/dL (300 mmol/L).

Notes

1. ApoCIII decreases TG-rich lipoprotein (TRL) catabolism by inhibiting LPL activity and reducing ApoE-dependent hepatic uptake of TRLs and their remnants.
2. In prospective studies of coronary artery events and angiographically demonstrated progression of occlusive vessel disease, patients who progressed with or without statin therapy had significantly higher TG ($p = .003$), VLDL-C ($p = .005$), ApoCIII in VLDL + LDL molecules ($p = .008$), total ApoCIII ($p = .01$), ApoB ($p = .03$), and TC ($p = .04$) than nonprogressors (Alaupovic et al., 1997).
3. The [VLDL-C] and [ApoCIII] in VLDL and LDL molecules may be more specific measures of CHD risk than plasma [TG], but prospective studies are needed before their use as a screening tool for CHD can be recommended.
4. A patient's status as having DM *per se* does not appear to be associated with high TG-rich [ApoCIII]; thus, although [ApoCIII] appears to be closely linked with metabolic glucose control in T2DM, it is not necessarily correlated with the [TG].
5. Within 4 weeks of adequate thyroid hormone replacement therapy (HRT) in markedly hypothyroid patients, a significant decrease occurs in total cholesterol, TG, ApoB, and ApoCIII concentrations.
6. Plasma [ApoCIII] of 11 ± 3 mg/dL has been found in healthy individuals vs. 20 ± 7 mg/dL in predialysis, 18 ± 5 mg/dL in hemodialysis, and 22 ± 8 mg/dL in peritoneal dialysis patients.
7. The ApoCIII gene on chromosome 11 encodes an 8.8-kDa polypeptide synthesized mainly by the liver in three isoforms. Several ApoCIII gene polymorphisms have been associated with hypertriglyceridemia and/or CAD.
8. Seven known mutations of the ApoCII gene have been identified (Klein et al., 2004). The lipid-binding domain of ApoCII appears to be a requirement for the full activation of LPL against chylomicrons (Olivecrona and Beisiegel, 1997).
9. Familial ApoCII deficiency is a rare autosomal recessive inborn error of metabolism clinically resembling LPL deficiency.
10. Homozygosity for a donor splice-site mutation in the second intron (ApoCII-Hamburg) caused ApoCII deficiency in a 9-year-old boy presenting with chylomicronemia, eruptive xanthoma, and pancreatitis (Nauck et al., 1998).

ICD-9 Codes

Conditions that may justify these tests *include but are not limited to:*

Hyperlipidemia
272.1   pure hypertriglyceridemia
272.3   hyperchylomicronemia (lipemia retinalis)
272.7   lipidoses (xanthomas, xanthomatosis, tuberoeruptive, or palmar xanthomas)

Pancreatitis
577.0   acute
577.1   chronic

Other
250.XX   diabetes mellitus (T1DM and T2DM)
9374.51   xanthelasma of eyelid

Suggested Reading

Alaupovic P, Mack WJ, Knight-Gibson C, Hodis HN. The role of triglyceride-rich lipoprotein families in the progression of atherosclerotic lesions as determined by sequential coronary angiography from a controlled clinical trial. *Arterioscler Thromb Vasc Biol* 1997; 17:715–22.

Fredenrich A. Role of apolipoprotein CIII in triglyceride-rich lipoprotein metabolism. *Diabetes Metab* 1998; 24:490–5.

Gerber Y, Goldbourt U, Segev S, Harats D. Indices related to ApoCII and CIII serum concentrations and coronary heart disease: a case-control study. *Prev Med* 2003; 37:18–22.

Klein RL, McHenry MB, Lok KH, Hunter SJ, Le NA, Jenkins AJ, Zheng D, Semler AJ, Brown WV, Lyons TJ, Garvey WT. Apolipoprotein CIII protein concentrations and gene polymorphisms in type 1 diabetes: associations with lipoprotein subclasses. *Metabolism* 2004; 53:1296–304.

Lee SJ, Moye LA, Campos H, Williams GH, Sacks FM. Hypertriglyceridemia but not diabetes status is associated with VLDL containing apolipoprotein CIII in patients with coronary heart disease. *Atherosclerosis* 2003; 167:293–302.

Moberly JB, Attman PO, Samuelsson O, Johansson AC, Knight-Gibson C, Alaupovic P. Apolipoprotein CIII, hypertriglyceridemia, and triglyceride-rich lipoproteins in uremia. *Miner Electrolyte Metab* 1999; 25:258–62.

Nauck MS, Nissen H, Hoffmann MM, Herwig J, Pullinger CR, Averna M, Geisel J, Wieland H, Marz W. Detection of mutations in the apolipoprotein CII gene by denaturing gradient gel electrophoresis: identification of the splice site variant apolipoprotein CII-Hamburg in a patient with severe hypertriglyceridemia. *Clin Chem* 1998; 44:1388–96.

Olivecrona G, Beisiegel U. Lipid binding of apolipoprotein CII is required for stimulation of lipoprotein lipase activity against apolipoprotein CII-deficient chylomicrons. *Arterioscler Thromb Vasc Biol* 1997; 17:1545–9.

Roselli della Rovere G, Lapolla A, Sartore G, Rossetti C, Zambon S, Minicuci N, Crepaldi G, Fedele D, Manzato E. Plasma lipoproteins, apoproteins and cardiovascular disease in type 2 diabetes patients: a nine-year follow-up study. *Nutr Metab Cardiovasc Dis* 2003; 13:46–51.

Sacks FM, Alaupovic P, Moye LA, Cole TG, Sussex B, Stampfer MJ, Pfeffer MA, Braunwald E. VLDL, apolipoproteins B, CIII, and E, and risk of recurrent coronary events in the Cholesterol and Recurrent Events (CARE) trial. *Circulation* 2000; 102:1886–92.

4.3.3 ApoE Phenotype

Indications for Test

Determination of ApoE phenotype may be indicated:

- In patients with tuberoeruptive or palmar xanthomas
- When approximately equal elevations of fasting triglycerides and total cholesterol are found, with both in the range of 5.0 to 7.0 mmol/L (triglycerides, 440 to 620 mg/dL; total cholesterol, 190 to 270 mg/dL)
- In the diagnosis of patients with familial dysbetalipoproteinemia

Procedure

1. Obtain the appropriate blood specimen for ApoE phenotype analysis by polymerase chain reaction (PCR) and isoelectric focusing (IEF) assay.

Interpretation

1. Of the four different isoforms (E_1, E_2, E_3, and E_4) of ApoE, E_3/E_3 is the predominant normal phenotype.
2. The homozygous E_2/E_2 phenotype in a patient with an abnormal lipid profile is associated with familial (Type III) dysbetalipoproteinemia, a disorder of VLDL remnant clearance (see Table 4.13).
3. When compared to allele E_3, the E_2 allele is associated with a lower [LDL-C] and no significant association with CHD risk (OR, 0.98; CI, 0.66 to 1.46).
4. Allele E_4 is associated with higher [LDL-C] and a 42% higher risk for CVD compared to individuals with the normal E_3/E_3 phenotype.
5. The Framingham data demonstrated an association between the E_4 allele and an increased risk for late-onset familial and sporadic forms of Alzheimer's disease (AD); however, the positive predictive value of the E_4 allele as a marker for this disorder is low (i.e., 0.10).

Notes

1. Evidence suggests an association of $ApoE_4$ with dementias other than AD, primarily multi-infarct dementia and stroke. The use of ApoE genotyping as a screening test for AD is not supported by Framingham data.
2. $ApoE_2$ and possibly $ApoE_3$ genotype individuals respond to HMG–CoA reductase inhibition therapy (i.e., statins) with lower [LDL-C] than do $ApoE_4$ individuals, suggesting that patients might be directed to optimal therapy for improving lipid profiles and ASCVD risk based in part on ApoE phenotype.
3. A negative association ($p < 0.05$) between alcohol intake and [LDL-C] was found in men with the E_2 allele. In contrast, a positive association between alcohol intake and [LDL-C] was found in men with the E_4 allele (Corella et al., 2001).
4. Synergistic effects of the E_4 allele and smoking on carotid atherosclerosis were found with odds ratios of 1.7 (95% CI, 0.8 to 3.6) for smoking alone, 1.0 (95% CI, 0.6 to 1.8) for the E_4 allele alone, and 3.7 (95% CI, 1.1 to 3.6) for the joint presence of the E_4 allele and smoking (Djoussé et al., 2002).

ICD-9 Codes

Conditions that may justify these tests *include but are not limited to:*

Hyperlipidemia
272.1   pure hypertriglyceridemia
272.2   mixed
272.3   hyperchylomicronemia
272.4   other and unspecified
272.7   lipidoses (xanthomas, xanthomatosis, tuberoeruptive or palmar xanthomas, familial dysbetalipoproteinemia)

Other
374.51   xanthelasma of eyelid

Suggested Reading

Corella D, Tucker K, Lahoz C, Coltell O, Cupples LA, Wilson PWF, Schaefer EJ, Ordovas JM. Alcohol drinking determines the effect of the APOE locus on LDL-cholesterol concentrations in men: the Framingham Offspring Study. *Am J Clin Nutr* 2001; 73:736–45.

Djoussé L, Myers RH, Province MA, Hunt SC, Eckfeldt JH, Evans G, Peacock JM, Ellison RC. Influence of apolipoprotein E, smoking, and alcohol intake on carotid atherosclerosis. *Stroke*. 2002; 33:1357–61.

Hagberg JM, Wilund KR, Ferrell RE. APO E gene and gene-environment effects on plasma lipoprotein-lipid levels. *Physiol Genomics* 2000; 4:101–108.

Horejsi B, Ceska R. Apolipoproteins and atherosclerosis: apolipoprotein E and apolipoprotein(a) as candidate genes of premature development of atherosclerosis. *Physiol Res* 2000; 49(Suppl 1):S63–9.

Myers RH, Schaefer EJ, Wilson PW, D'Agostino R, Ordovas JM, Espino A, Au R, White RF, Knoefel JE, Cobb JL, McNulty KA, Beiser A, Wolf PA. Apolipoprotein E epsilon4 association with dementia in a population-based study: the Framingham study. *Neurology* 1996; 46:673–7.

Song Y, Stampfer MJ, Liu S. Meta-analysis: apolipoprotein E genotypes and risk for coronary heart disease. *Ann Intern Med* 2004; 141:137–47.

4.3.4   Lipoprotein Little a

Indications for Test

Measurement of Lp(a) may be indicated if a patient with a normal or only slightly abnormal lipid profile has:

TABLE 4.14
Guidelines for Interpreting Lipoprotein Little a Concentration, [Lp(a)]

[Lp(a)] (mg/dL)[a]	Risk of Atherosclerotic Cardiovascular Disease (ASCVD)[b]
≤ 30	Low
30–40	2-fold increase
>40	3-fold increase
>40 + TC/HDL-C > 5.8	25-fold increase
>40 + ≥2 CVD risk factors	122-fold increase

[a] To convert mg/dL to mmol/L, multiply mg/dL by 2.333.

[b] Compared to low-risk individuals with [Lp(a)] < 30 mg/dL (70 nmol/L).

Note: CVD, cardiovascular disease; HDL-C, high-density lipoprotein cholesterol; TC, total cholesterol.

- Atherosclerotic cardiovascular disease (ASCVD)
- Xanthomas, a history of premature, familial, or recurrent ASCVD
- Advancing ASCVD on treatment

Procedure

1. Obtain the appropriate blood specimen for lipoprotein electrophoresis (LPE) testing as a qualitative screening test for an elevated Lp(a) concentration or proceed directly to quantitative Lp(a) testing.
2. If the LPE is positive (i.e., the pattern demonstrates an increased prebeta band) or the patient has a history and physical exam findings strongly suggestive of a high risk for ASCVD (e.g., xanthomas), proceed to quantitative Lp(a) testing.

Interpretation

1. Note "Guidelines for Interpreting Lipoprotein Little a Concentration, [Lp(a)]" (Table 4.14). [Lp(a)] increases with menopause and falls with estrogen and estrogen/progesterone hormone replacement therapy (HRT).
2. Guidelines (i.e., reference interval data) for interpreting [Lp(a)] are method dependent, and values obtained by different assay methods cannot be used interchangeably unless a method validation study was performed that demonstrated adequate agreement between the methods.

Notes

1. Lp(a) is an LDL molecule covalently linked to apolipoprotein(a).
2. Increased [Lp(a)] is associated with an increased incidence of ASCVD, specifically ischemic heart disease.
3. Lp(a) interferes with plasminogen activation, which may result in susceptibility to thrombogenesis.
4. In the Framingham study, elevated plasma [Lp(a)] was an independent risk factor for the development of premature CHD in men, comparable in magnitude and prevalence (i.e., attributable risk) to a total cholesterol concentration ≥ 240 mg/dL (6.2 mmol/L) or [HDL-C] < 35 mg/dL (0.91 mmol/L).

ICD-9 Codes

Conditions that may justify tests *include but are not limited to:*

Hyperlipidemia
272.0   pure hypercholesterolemia
272.1   pure hypertriglyceridemia
272.2   mixed
272.3   hyperchylomicronemia
272.4   other and unspecified

Other
272.5   lipoprotein deficiencies
413   angina pectoris or chest pain
414.0   coronary atherosclerosis
440   atherosclerosis

Suggested Reading

Bostom AG, Cupples LA, Jenner JL, Ordovas JM, Seman LJ, Wilson PW, Schaefer EJ, Castelli WP. Elevated plasma lipoprotein(a) and coronary heart disease in men aged 55 years and younger: a prospective study. *JAMA* 1996; 276:544–8.

Marcovina SM, Albers JA, Gabel B, Koschinsky ML, Gaur VP. Effect of number of apolipoprotein(a) kringle 4 domains on immunochemical measurements of lipoprotein(a). *Clin Chem* 1995; 41:246–55.

Chapter 5

Tests for Complications of Hyperglycemia

5.1 Retinopathy of Diabetes Mellitus

5.1.1 Indirect Ophthalmoscopy and Stereoscopic Slit-Lamp Biomicroscopy of the Retina with Multi-Field Stereophotography of the Fundus for Detection and Monitoring of Diabetic Retinopathy (DR) and Macular Edema (ME)*

Indications for Test

Indirect retinoscopy and stereoscopic slit-lamp biomicroscopy of the retina are indicated for:

- Early detection of DR and ME in diabetes mellitus (DM) patients (T1DM and T2DM)

Fundus photography with fluorescein angiography is indicated for:

- Clinically significant macular edema (CSME), preliminary to laser treatment if needed
- Monitoring clinical status and the effect of laser therapy for DR or ME

Procedure

1. Time DM patient referrals to an ophthalmologist or, preferably, a retina specialist at appropriate intervals as per "Recommended Screening with Retinoscopy in Diabetes Mellitus Patients" (Table 5.1). Carefully document referrals for eye exams for medicolegal purposes.
2. Ideally, have the retinal specialist perform an ophthalmologic examination of the retina and lens with a stereoscopic slit-lamp biomicroscope with the capability of seven-field stereophotography of the fundus through a fully dilated pupil.
3. In locations without advanced optical imaging capabilities, screen for early DR using a nonmydriatic camera and obtain two 45° images centered on the optic disc and on the macula. Alternative to this approach in remote locations with telemedicine capability, obtain two-field 50° nonstereo digital images and forward them for review by an ophthalmologist or retina specialist.

---

* Reviewed by ophthalmologist Dr. Thomas Oei, San Antonio, Texas.

TABLE 5.1
Recommended Screening with Retinoscopy in Diabetes Mellitus Patients

Patient	Initial Examination	Frequency of Follow-Up[a]
<30 years	Within 5 years of diagnosis	Yearly
≤30 years	At diagnosis	Yearly
Planning pregnancy	Preconception	No retinopathy to mild or moderate NPDR, every
Pregnant	1st trimester	3 to 4 months
		Severe NPDR or worse, every 1 to 3 months

[a] Abnormal findings may prompt follow-up examinations at more frequent intervals.
Note: NPDR, nonproliferative diabetic retinopathy.

Interpretation

1. Diabetic ME is defined as a collection of intraretinal fluid in the macular area with or without lipid exudates and with or without cystoid changes.
2. Criteria for staging nonproliferative (NPDR) and proliferative diabetic retinopathy (PDR) are provided in "Criteria for Interpreting the Results of Retinoscopy, Slit-Lamp Biomicroscopy, and Multi-Field Stereophotography for CSME, NPDR, and PDR" (Table 5.2).

Notes

1. Patients with T2DM are at significant risk for developing DR, but at a lower incidence and severity than in the T1DM patient.
2. PDR affects 25% of patients with T1DM after 15 years and often remains asymptomatic beyond the optimal stage for treatment.
3. Both moderate to severe vision loss from DR are essentially preventable with timely detection and treatment, careful long-term follow-up, and comprehensive, evidence-based care.
4. Because timely laser photocoagulation can reduce severe visual loss from high-risk PDR by 60% and moderate visual loss from CSME by 50%, early detection and management are essential.
5. CSME and cataract disease are common and important causes of visual loss in patients with DM. CSME may be seen even with achievement of tight glycemic control in patients who have previously undergone a period of chronically elevated blood glucose levels.
6. Standard direct ophthalmoscope examination has only moderate sensitivity (~80% in research settings) and specificity (>90% for PDR, but lower for CSME).
7. Early subcapsular or more advanced cataract disease may be apparent on finding lenticular clouding or opacity with the use of the slit lamp.
8. Slit-lamp biomicroscopy is the reference method for detection of CSME, but this technique is less sensitive than optical coherence tomography.
9. Detection of CSME using biomicroscopy is superior to digital or standard nonstereo photographs, and some patients with sight-threatening DR will be missed by photography alone.
10. DR, assessed with seven-field stereo fundus photography, as done in the Wisconsin Epidemiologic Study of Diabetic Retinopathy (WESDR), was detected in 7.6% of patients with impaired glucose tolerance (IGT) and in 12.5% of those with early T2DM 6 to 12 months after making the transition from IGT to DM (Hamman et al., 2005).
11. Somatostatin (SS), an endogenous antiangiogenic factor, is reduced in the vitreous fluid of patients with PDR. The SS-28 isoform appears to be the predominant molecular variant in the vitreous fluid of patients with PDR (Hernandez et al., 2005).
12. Blood flow velocity and arterial diameters were identical in both T1DM patients without more advanced DR and in non-DM controls, clearly indicating that blood flow is not disturbed in well-controlled T1DM early in the course of the disease (Caglieero et al., 2005).
13. Although numerous biochemical factors are thought to play a role in the development of retinopathy, activation of protein kinase C (PKC), specifically its beta isoform (PKC-β), is implicated in the pathogenesis of both the early and late-stage manifestations of DR (Aiello, 2002).

TABLE 5.2
Criteria for Interpreting the Results of Retinoscopy, Slit-Lamp Biomicroscopy, and Multi-Field Stereophotography for CSME, NPDR, and PDR

	One or More of the Following Criteria Are Present
Clinically significant macular edema (CSME)	Retinal thickening at or within 500 μm of the center of the macula Hard exudates at or within 500 μm of the center of the macula if associated with thickening of the adjacent retina A zone or zones of retinal thickening 1 disc area or larger, any part of which is within 1 disc diameter of the center of the macula
Nonproliferative diabetic retinopathy (NPDR)	
Mild stage	Few scattered retinal microaneurysms and hemorrhages Hard exudates
Moderate stage	More extensive hemorrhages and/or microaneurysms Mild IRMAs Early venous beading
Severe stage	>20 intraretinal hemorrhages in each of 4 quadrants Definite venous beading in at least 2 quadrants Prominent IRMA in 1 or more quadrants
Proliferative diabetic retinopathy (PDR)	
Early stage	Minimal NVD involving <1/4 disc area without preretinal or vitreous hemorrhage, or NVE without preretinal or vitreous hemorrhage
High-risk stage	NVD involving >1/4 disc area without preretinal or vitreous hemorrhage Less extensive NVE or NVD ≥ 1/2 disc area in size with preretinal vitreous hemorrhage New vessels within 1 disc diameter of the optic nerve head that are larger than 1/3 disc area

Note: IRMA, intraretinal microvascular abnormality; NVD, new vessels on the disc; NVE, new vessels elsewhere (on the retina).

14. In Diabetes Control and Complications Trial (DCCT) patients, glycation and advanced glycation end-product parameters from skin biopsies obtained in 1992 were correlated with retinopathy data from 2004. In a multivariate analysis, glycation and glycoxidation parameters predicted the progression of DR, independent of hemoglobin A_{1c}, which was itself a strong predictor. Thus, glycation of long-lived matrix molecules may explain the effect of intensified insulin treatment on DR progression (Genuth et al., 2005).

ICD-9 Codes

Conditions that may justify this test *include but are not limited to*:

Retinopathy
362   other retinal disorders (i.e., macular edema)
362.01   background diabetic
362.02   proliferative diabetic

Other
250.5   diabetes with ophthalmic manifestations

Suggested Reading

ADA. *Therapy for Diabetes Mellitus and Related Disorders*, 2nd ed. 1994. American Diabetes Association, pp. 235–48.

Aiello LP. The potential role of PKC beta in diabetic retinopathy and macular edema. *Surv Ophthalmol* 2002; 47(Suppl 2):S263–9.

Boucher MC, Gresset JA, Angioi K, Olivier S. Effectiveness and safety of screening for diabetic retinopathy with two nonmydriatic digital images compared with the seven standard stereoscopic photographic fields. *Can J Ophthalmol* 2003; 38:557–68.

Browning DJ, McOwen MD, Bowen RM, O'Marah TL. Comparison of the clinical diagnosis of diabetic macular edema with diagnosis by optical coherence tomography. *Ophthalmology* 2004; 111:712–5.

Cagliero E, Feke GT, Pitler L et al. Retinal blood flow abnormalities do not precede detectable retinopathy in subjects with well-controlled T1DM: program and abstracts of the 65th Scientific Sessions of the American Diabetes Association, June 10–14, 2005, San Diego, CA.

Early Treatment Diabetic Retinopathy Study (ETDRS) Research Group. Photocoagulation for diabetic macular edema. ETDRS report number 1. *Arch Ophthalmol* 1985; 103:1796–806.

Genuth S, Monnier V, Sun W, DCCT/EDIC Research Group. Glycation products in skin collagen predict future progression of retinopathy over ten years in T1DM: program and abstracts of the 65th Scientific Sessions of the ADA, June 10–14, 2005, San Diego, CA.

Hamman R. Late breaking clinical trials: program and abstracts of the 65th Scientific Sessions of the ADA, June 10–14, 2005, San Diego, CA.

Hernandez C, Carrasco E, Casamitjana R, Deulofeu R, Mesa J, Simo R. Somatostatin molecular variants in the vitreous fluid: a comparative study between diabetic patients with proliferative diabetic retinopathy and non-diabetic control subjects: program and abstracts of the 65th Scientific Sessions of the ADA, June 10–14, 2005, San Diego, CA.

Liesenfeld B, Kohner E, Piehlmeier W, Kluthe S, Aldington S, Porta M, Bek T, Obermaier M, Mayer H, Mann G, Holle R, Hepp KD. A telemedical approach to the screening of diabetic retinopathy: digital fundus photography. *Diabetes Care* 2000; 23:345–8.

Singer DE, Nathan DM, Fogel HA, Schachat AP. Screening for diabetic retinopathy. *Ann Intern Med* 1992; 116:660–71.

5.2   Nephropathy of Diabetes Mellitus (DM)

5.2.1   Glomerular Filtration Rate (GFR) Estimation by Creatinine Clearance (Cl_{cr}), the Modification of Diet in Renal Disease (MDRD) Equations, and the Cockcroft–Gault Formula

Indication for Tests

Determination of estimated GFR (GFR_e) from a timed or 24-hour urine collection is indicated:

- As a baseline test to estimate GFR in all DM patients capable of and willing to collect a 24-hour urine specimen
- For routine monitoring of renal status every 2 years in adequately hydrated DM patients with a Cl_{cr} 100 mL/min/1.73 m^2
- Annually in all DM patients when Cl_{cr} drops to <100 mL/min/1.73 m^2

Determination of GFR_e by the MDRD equations is indicated:

- As a low cost, simple estimate of renal status in DM patients unable or unwilling to accurately collect a 24-hour urine
- As a screening test of renal function in adequately hydrated adult (≥18 years old) individuals

Determination of GFR_e by the Cockcroft–Gault formula is indicated:

- As a crude estimate of renal status with abnormality to be followed up by GFR_e using the MDRD equation or collection of a timed urine

Procedure

Timed or 24-Hour Urine Specimen for Cl$_{cr}$

1. Calculate the patient's body surface area (BSA) in m^2 using the equation:

$$BSA\ (m^2) = Wt^{0.5378} \times Ht^{0.3964} \times 0.024265$$

where Wt = weight (lb), and Ht = height (in.).
Or, the equation:

$$BSA\ (m^2) = Wt^{0.425} \times Ht^{0.725} \times 0.20247$$

where Wt = weight (kg), and Ht = height (m).
2. Alternatively, use a nomogram to accurately estimate BSA.
3. Because of the reduced extracellular fluid volume of children compared to adults, determine BSA in children using the formula:

$$BSA\ (m^2) = Wt^{0.6469} \times Ht^{0.7236} \times 0.02154$$

where Wt = weight (kg), and Ht = height (m).
4. Obtain a 24-hour or shorter term, precisely timed urine specimen for creatinine and volume determination, with collection time to start at the point when the patient empties his or her bladder.
5. Use the following formula to calculate the BSA-corrected Cl$_{cr}$ from a timed urine collection:

$$Cl_{cr}\left(mL/min/1.73\ m^2\right) = \frac{U_{vol}}{(T \times 60)} \times \frac{U_{creat}}{S_{creat}} \times \frac{1.73\ m^2}{BSA}$$

where:

U$_{vol}$ = volume (mL) of urine collected.
T = time (hr) of urine collection period (for a 24-hour urine collection, T × 60 = 1440 minutes).
U$_{creat}$ = urine creatinine concentration (mg/dL).
S$_{creat}$ = serum creatinine concentration (mg/dL), preferably on serum from a blood specimen obtained at the point when the timed urine sample is delivered to the clinical laboratory for testing.
BSA = body surface area in m^2.

MDRD Equations

1. Assess the hydration status of the patient and obtain blood sample for serum creatinine and blood urea nitrogen (BUN). More than one sample may be required if hydration status is uncertain (i.e., patient is dehydrated or treated with diuretics).
2. Apply the *standard* MDRD equation using S$_{creat}$, BUN, and albumin values and demographic variables (Levey et al., 1999):

$$GFR\left(mL/min/1.73\ m^2\right) = \left(170S_{creat}^{-0.999}\right) \times \left(BUN^{-0.170}\right) \times \left(Alb^{0.318}\right) \times \left(Age^{-0.176}\right) \times GEF$$

where:

S$_{creat}$ = serum creatinine concentration (mg/dL).
BUN = serum blood urea nitrogen concentration (mg/dL).
Alb = serum albumin concentration (mg/dL).
Age = chronological age in years
GEF = gender/ethnicity factor: 0.762, non-African-American female; 1.000, non-African-American male; 1.180, African-American male; 0.899, African-American female.

3. Apply the *modified* or abbreviated MDRD equation from www.nkdep.nih.gov/professionals/gfr_calculators/index.htm. (*Note:* The GFR_e using the MDRD equation below is most accurate if it is ≤60 mL/min/1.73 m²):

$$GFR\left(mL/min/1.73\ m^2\right) = 186\left(S_{creat}^{-1.154}\right) \times \left(Age^{-0.203}\right) \times GEF$$

where:

S_{creat} = serum creatinine concentration (mg/dL) using a creatinine method that is not traceable to isotope dilution mass spectrometry (IDMS).
Age = chronological age in years, for patients ≥ 18 years.
GEF = gender/ethnicity factor: 0.742, female; 1.212, male.

Cockcroft–Gault Formula

1. Calculate Cl_{cr} from a random S_{creat} value using the Cockcroft–Gault formula:

Men

$$Cl_{cr}\ (mL/min) = \frac{[140 - Age\ (years)] \times IBW_{men}}{(72 \times S_{creat})}$$

Women

$$Cl_{cr}\ (mL/min) = [140 - Age\ (years)] \times \left(\frac{IBW_{women}}{72S_{creat}}\right) \times 0.85$$

where:

Ideal body weight for men (IBW_{men}) = 50 kg + [2.3 kg × (no. of inches > 60 inches of height)].
Ideal body weight for women (IBW_{women}) = 45.5 kg + [2.3 kg × (no. of inches > 60 inches of height)].

2. The Cockcroft–Gault formula:
   - Requires use of actual body weight (kg) in lieu of IBW when the patient weighs less than the estimated IBW.
   - Does not apply to individuals < 60 inches (5 feet) in height.

Interpretation

Guidelines for interpreting GFR_e are shown in "Stages of Chronic Kidney Disease (CKD)" (Table 7.5).

Example of the Use of the MDRD Equations

A patient has the following demographic and serum values:

Age = 40 years
Ethnicity = African-American
Gender = Female
S_{creat} = 1.5 mg/dL
BUN = 25 mg/dL
Alb = 3.5 g/dL

Standard MDRD equation:

$$GFR\ (mL/min/1.73\ m^2) = \left[170(1.5)^{-0.999}\right] \times (25)^{-0.170} \times (3.5)^{0.318} \times (40)^{-0.176} \times 0.899 = 45.9$$

Modified MDRD equation:

$$\text{GFR (mL/min/1.73 m}^2) = \left[186(1.5)^{-1.154}\right] \times (40)^{-0.203} \times 0.899 = 49.5$$

Note that the modified MDRD equation tends to overestimate GFR in hypoalbuminemic patients.

Notes

1. The average Cl_{cr} is 15% less in females compared to males.
2. With or without DM, individuals with moderate to severe gingival disease have twice the risk of chronic kidney disease (CKD). Toothless individuals have more than 2.5 times the risk of developing CKD. Periodontitis predicts the development of Stage 5 CKD.
3. Creatinine production is related to muscle mass, degree of physical activity, and amount of cooked meat ingested. These factors affect values for Cl_{cr} by any method that uses S_{creat} values.
4. The MDRD equation may systematically underestimate GFR_e by as much as 29% in healthy persons and 6.2% in CKD patients. Use of a complex quadratic equation (Rule et al., 2004) to overcome these biases is controversial (Stevens, 2004).
5. The Cockcroft–Gault equations are preferred over the equations of Jelliffe, Mawer, or Hull, which were presented in the 1970s and 1980s.
6. The advantages of a single serum determination of cystatin C over the Cockcroft–Gault equation estimate of Cl_{cr} are controversial (Hoek, 2003).
7. The reference interval for GFR by gold standard methods, such as infusion of inulin or measurement of isotope clearance, is 80 to 130 mL/min/1.73 m² in young individuals and typically declines at about 10 mL/min/decade after age 50.
8. The correlation coefficient between predicted Cl_{cr} by the Cockcroft–Gault formula and the mean of duplicate values for measured Cl_{cr} was shown to be 0.83, with the difference between them being no greater than that between paired values for measured clearances.

ICD-9 Codes

Conditions that may justify this test *include but are not limited to*:

Chronic kidney disease (CKD)

585.1	Stage 1
585.2	Stage 2
585.3	Stage 3
585.4	Stage 4
585.5	Stage 5

Nephrosis

581	nephrotic syndrome
581.8	specified pathological lesion
581.81	nephrotic syndrome in diseases classified elsewhere

Nephropathy

250.4	diabetes with renal manifestations
583	nephritis not specified as acute or chronic
583.8	specified pathological lesion

Suggested Reading

ADA. *Therapy for Diabetes Mellitus and Related Disorders,* 2nd ed. 1994. American Diabetes Association, pp. 256–269.

Bertino JS Jr. Measured versus estimated creatinine clearance in patients with low serum creatinine values. *Ann Pharmacother* 1993; 27:1439–42 (erratum in *Ann Pharmacother* 1994; 28:811).

Cockcroft DW, Gault MH. Prediction of creatinine clearance from serum creatinine. *Nephron* 1976; 16:31–41.

Go AS, Chertow GM, Fan D, McCulloch CE, Hsu CY. Chronic kidney disease and the risks of death, cardiovascular events, and hospitalization. *N Engl J Med* 2004; 351:1296–305.

Granerus G, Aurell M. Reference values for 51Cr-EDTA clearance as a measure of glomerular filtration rate. *Scand J Clin Lab Invest* 1981; 41:611–6.

Haycock GB, Schwartz GJ, Wisotsky DH. Geometric method for measuring body surface area: a height–weight formula validated in infants, children, and adults. *J Pediatr* 1978; 93:62–6.

Hoek FJ, Kemperman FA, Krediet RT. A comparison between cystatin C, plasma creatinine and the Cockcroft and Gault formula for the estimation of glomerular filtration rate. *Nephrol Dial Transplant* 2003; 18:2024–31.

Hull JH, Hak LJ, Koch GG, Wargin WA, Chi SL, Mattocks AM. Influence of range of renal function and liver disease on predictability of creatinine clearance. *Clin Pharmacol Ther* 1981; 29:516–21.

Jelliffe RW. Estimation of creatinine clearance when urine cannot be collected. *Lancet* 1971; 1:975–6.

Jelliffe RW. Creatinine clearance: bedside estimate [letter]. *Ann Intern Med* 1973; 79:604–5.

Levey AS, Bosch JP, Breyer Lewis J, Greene T, Rogers N, Roth D. A more accurate method to estimate glomerular filtration rate from serum creatinine: a new prediction equation. *Ann Intern Med* 1999; 130:461–70.

Mawer GE, Lucas SB, Knowles BR, Stirland RM. Computer-assisted prescribing of kanamycin for patients with renal insufficiency. *Lancet* 1972; 1(740):12–15.

National Kidney Foundation. K/DOQI clinical practice guidelines for chronic kidney disease: evaluation, classification and stratification. *Am J Kidney Dis* 2002; 39(Suppl 1):S1–76.

Peters AM. The kinetic basis of glomerular filtration rate measurement and new concepts of indexation to body size. *Eur J Nucl Med Mol Imaging* 2004; 31:137–49.

Rule AD, Larson TS, Bergstralh EJ, Slezak JM, Jacobsen SJ, Cosio FG. Using serum creatinine to estimate GFR: accuracy in good health and in chronic kidney disease. *Ann Intern Med* 2004; 141:929–37.

Stevens LA. Clinical implications of estimating equations for glomerular filtration rate [editorial]. *Ann Intern Med* 2004; 141:959–61.

5.2.2   Microalbumin (MicroAlb) in the Diagnosis of Nephropathy of DM Patients

Indications for Test

Semiquantitative and quantitative microalbumin determination in a timed (i.e., minimum of 12- up to 24-hour) urine collection is indicated in:

- All patients with DM initially upon achieving glycemic control (i.e., within 3 months of diagnosis) and capable of accurately and completely collecting a timed urine specimen
- The identification of patients with incipient DM-related chronic kidney disease (CKD) and differentially diagnosing those with microalbuminuria vs. those with macroalbuminuria
- Monitoring DM patients on a yearly basis, or more often (i.e., up to every 2 months), when adjusting therapy to reduce albuminuria of any degree
- DM patients with autonomic neuropathy suggested by symptoms such as tachycardia at rest, gustatory sweating, erectile dysfunction, or chronic diarrhea, as indicated in the "Autonomic Neuropathy Symptom" questionnaire" (Q5.1 in Appendix 1)

Procedure

1. Be sure that none of the following factors, known to give a false-positive test, is present:
   - Strenuous exercise within the previous 72 hours
   - Urinary tract infection
   - Acute febrile illness

TABLE 5.3
Albumin Concentration or Albumin Excretion Rate
in the Assessment of Diabetic Nephropathy

Type of Urine Specimen	Normal	Nephropathy	
		Incipient	Overt
Timed (24-hour)[a]	<25 mg/day <20 mcg/minute	25–250 mg/day 20–200 mcg/minute	>250 mg/day >200 mcg/minute
Random, semiquantitative (dipstick, such as Micral-Test®)	0 mg/L[b]	50–100 mg/L	≥300 mg/L
Random, quantitative	<20 mcg/mg creatinine	20–250 mcg/mg creatinine	>250 mcg/mg creatinine

[a] Reference gold standard.

[b] 10–20 mg/L is indeterminate.

- Uncontrolled (i.e., malignant) hypertension
- Heart failure
- Pronounced hyperglycemia (>450 mg/dL), even if short-term

2. Instruct the patient to collect either random ("spot") or timed (minimum of 12- up to 24-hour) urine specimens. All urine samples must be refrigerated after collection or frozen, if testing will not be performed promptly. Frozen urine specimens should be thawed only once prior to microalbumin testing (i.e., avoid repeated freeze–thaw cycles).

3. Proceed with one or more of the following three methods for detecting microalbuminuria:

Method 1. Quantitative Microalbumin on a Random ("Spot") Urine Sample

- Obtain a random (spot) urine specimen for quantitative microalbumin and creatinine determination.
- Request urine microalbumin and creatinine testing so the microalbumin (mcg)/creatinine (mg) ratio can be calculated by laboratory personnel.
- If possible, confirm abnormal results for the microalbumin/creatinine ratio by Method 2.

Method 2. Quantitative Microalbumin on a "Timed" Urine Specimen

- Obtain a 12- to 24-hour urine collection for microalbumin and creatinine determination.
- The laboratory performing the microalbumin testing should report both the microalbumin concentration (mg/L) in the urine sample and microalbumin excretion rate (MAER) in mg/day.
- For a 12-hour urine collection, be sure that the laboratory provides the MAER in mcg/min.
- High-performance liquid chromatography (HPLC) measures total albumin, including immunoreactive and non-immunoreactive forms, and may allow better early detection of incipient diabetic nephropathy. Check to see what detection method the lab performing the assay employs.

Method 3. Semiquantitative Microalbumin on a Random ("Spot") Urine

- Obtain a random urine specimen.
- Test the urine for microalbumin using a commercially available, semiquantitative dipstick method (e.g., Micral®).
- To ensure reliable results when using dipstick microalbumin methods, follow the manufacturer's instructions precisely.
- Confirm abnormal results by Method 1 or, preferably, by Method 2.

Interpretation

1. Refer to "Albumin Concentration or Albumin Excretion Rate in the Assessment of Diabetic Nephropathy" (Table 5.3). Note that the ranges presented differ from and are lower than those identified by the American Diabetes Association (ADA, 2004).

2. Other causes of renal disease can occur in DM and should be suspected if:
   - Hematuria is present.
   - Azotemia occurs in the absence of proteinuria.
   - Nephrotic-range proteinuria is found in patients with relatively new-onset DM.

- There is evidence of urinary tract obstruction and/or infection.
- Retinopathy is absent (as nephropathy of DM is usually seen in the context of concurrent retinopathy).
- The patient is taking drugs known to be nephrotoxic.
- The patient has malignant hypertension.

3. Persistently overt proteinuria or macroproteinuria (>500 mg/day) is irreversible and heralds the inevitable progression to end-stage renal disease (CKD Stage 5).
4. A cutoff value of 17 to 20 mg/L in a random urine specimen has a sensitivity of 100% and a specificity of 80% or better for the diagnosis of microalbuminuria when compared to a 24-hour timed urine collection as a reference standard.
5. Because of the known day-to-day variability in urinary albumin excretion, abnormal microalbumin tests by Methods 1 or 2 require confirmation by repeat testing.
6. An abnormal microalbumin test by Method 3 is not diagnostic and requires confirmation.

Notes

1. Collecting a 12-hour overnight urine specimen is more convenient for most patients and has been shown to be about as accurate as a 24-hour collection.
2. Conventional immunochemical-based assays for microalbuminuria may not detect an unreactive fraction of albumin in urine resulting in an underestimate of protein excreted. High-performance liquid chromatography may allow early detection of incipient nephropathy by measuring all urinary albumins present.
3. Normal urinary microalbumin excretion is approximately 10 mg/day (range, 2.5 to 25 mg/d).
4. Microalbumin excretion must be ~500 mg/day or more before the qualitative urine dipstick test for microalbumin becomes positive.
5. CKD Stage 3 or incipient nephropathy (MAER = 25 to 250 mg/day) may antedate the onset of overt proteinuria (>250 mg/day) by 5 years and the onset of azotemia by 10 years.
6. The risk of microalbuminuria in patients with T1DM increases abruptly above a hemoglobin A_{1c} value of 8.1%.
7. Detection of diabetic nephropathy at an earlier stage (i.e., incipient nephropathy) allows for earlier intervention utilizing strict glycemic control or blood pressure normalization (i.e., BP <120/<80 mmHg), often with modest dietary protein restriction.
8. Gustatory sweating was found to occur in 69% of patients with nephropathy of DM, 36% with neuropathy of DM, 5% in DM controls, and 5% in non-DM renal failure patients.
9. Proteinuria occurs in 15 to 40% of T1DM patients, with a peak incidence around 15 to 20 years of DM. In T2DM patients, the prevalence is highly variable, depending on population studied, ranging from 5% to 20%.
10. Diabetic nephropathy is more prevalent among African-Americans, Asians, and Native Americans than Caucasians.
11. In 140 T1DM patients with persistent microalbuminuria (20–200 mcg/min), 11% of patients on 1.25 mg ramipril, an angiotensin-converting enzyme (ACE) inhibitor, regressed to normoalbuminuria (<20 mcg/min), while 20% regressed on 5 mg ramipril, and 4% regressed on placebo ($p = 0.053$) (O'Hare et al., 2000).
12. Combination therapy with medications from both the ACE inhibitor and angiotensin receptor blocker classes of agents work synergistically to reduce proteinuria.

ICD-9 Codes

Refer to Test 5.2.1 codes.

Suggested Reading

Agarwal R, Acharya M, Tian J, Hippensteel RL, Melnick JZ, Qiu P, Williams L, Batlle D. Antiproteinuric effect of oral paricalcitol in chronic kidney disease. *Kidney Int* 2005; 68:2823–8.

Alberti, KGMM et al. (Eds). *International Textbook of Diabetes Mellitus*, 1992. John Wiley & Sons, pp. 1267–1329.

American Diabetes Association. Nephropathy in diabetes (position statement). *Diabetes Care* 2004; 27(Suppl 1): S79–83.

American Diabetes Association. Standards of medical care in diabetes. *Diabetes Care* 2005; 28(Suppl 1):S4–S36.

Eknoyan G, Hostetter T, Bakris GL, Hebert L, Levey AS, Parving HH, Steffes MW, Toto R. Proteinuria and other markers of chronic kidney disease: a position statement of the National Kidney Foundation (NKF) and the National Institute of Diabetes and Digestive and Kidney Diseases (NIDDK). *Am J Kidney Dis* 2003; 42:617–22.

Hovind P, Tarnow L, Rossing P, Jensen BR, Graae M, Torp I, Binder C, Parving HH. Predictors of the development of microalbuminuria and macroalbuminuria in patients with type 1 diabetes: inception cohort study. *BMJ* 2004; 328:1105–8.

Krolewski AS, Warram JH, Christlieb AR, Busick EJ, Kahn CR. The changing natural history of nephropathy in type I diabetes. *Am J Med* 1985; 78:785–94.

Krolewski AS, Laffel LM, Krolewski M, Quinn M, Warram JH. Glycosylated hemoglobin and the risk of microalbuminuria in patients with insulin-dependent diabetes mellitus. *N Engl J Med* 1995; 332:1251–5.

O'Hare P, Bilbous R, Mitchell T, O' Callaghan CJ, Viberti GC. Low-dose ramipril reduces microalbuminuria in type 1 diabetes patients without hypertension: results of a randomized controlled trial. *Diabetes Care* 2000; 23:1823–9.

Shaw JE, Parker R, Hollis S, Gokal R, Boulton AJ. Gustatory sweating in diabetes mellitus. *Diabet Med* 1996; 13:1033–7.

Young BA, Maynard C, Boyko EJ. Racial differences in diabetic nephropathy, cardiovascular disease, and mortality in a national population of veterans. *Diabetes Care* 2003; 26:2392–9.

Zelmanovitz T, Gross JL, Oliveira JR, Paggi A, Tatsch M, Azevedo MJ. The receiver operating characteristics curve in the evaluation of a random urine specimen as a screening test for diabetic nephropathy. *Diabetes Care* 1997; 20:516–19.

5.2.3 Fractional Excretion of Sodium (FENa) for Differentiation of Prerenal Azotemia from Acute Tubular Necrosis (ATN) in the Diagnosis of Hyperosmolar Nonketotic Hyperglycemia in DM Patients

Indications for Test

Determination of the FENa is indicated:

- In DM patients with indeterminate or no known renal dysfunction whose intravascular volume status (high or low) is clinically difficult to ascertain, particularly in the context of hyperglycemia
- To discriminate between the prerenal azotemia and oliguric phase of ATN in the context of hyperglycemic crisis

Procedure

1. If possible, discontinue patient's diuretic therapy for more than 24 hours.
2. Simultaneously obtain a random urine and blood sample.
3. Request glucose, sodium (Na), chloride (Cl), and creatinine testing on the blood sample and Na, Cl, and creatinine on the urine specimen.
4. Calculate FENa (%):

$$\frac{\left(U_{Na}/S_{Na}\right)}{\left(U_{Cr}/S_{Cr}\right)} \times 100$$

where U_{Na}, S_{Na}, U_{Cr}, and S_{Cr} are the urine sodium, serum sodium, urine creatinine, and serum creatinine concentrations, respectively.

TABLE 5.4
Laboratory Tests Used to Differentiate Prerenal Azotemia from Acute Tubular Necrosis

Test	Prerenal Azotemia	Acute Tubular Necrosis
U_{Na} (mEq/L)	<20	>40
U_{Cr}/S_{Cr} ratio	>40:1	<20:1
U_{Osm}	100 mOsm > S_{Osm}	<S_{Osm}
Urinary sediment	Normal	Casts, cellular debris present
FENa	<1%	>2%

Note: U_{Na}, urinary sodium (Na) concentration; U_{Cr}/S_{Cr}, urine creatinine/serum creatinine ratio; U_{Osm}, urine osmolality; S_{Osm}, serum osmolality; FENa, fractional excretion of sodium = $[(U_{Na}/S_{Na})/(U_{Cr}/S_{Cr})] \times 100$.

5. In extremely hyperglycemic, hyperosmolar states (e.g., serum glucose concentration > 600 mg/dL), most likely associated with prerenal azotemia, obtain serum (S_{Osm}) and urine (U_{Osm}) osmolality and a urine microscopic analysis for detection of casts in urinary sediment in lieu of FENa testing.

Interpretation

1. See "Laboratory Tests Used to Differentiate Prerenal Azotemia from Acute Tubular Necrosis" (Table 5.4).
2. FENa values:
   - 1–2% is typical in healthy individuals free of renal disease.
   - <1% indicates prerenal azotemia. Note that iron saturation (FeS) values < 1% can occur in acute renal failure from myoglobinuria or hemoglobinuria, iodinated radiocontrast nephropathy, hepatorenal syndrome, renal allograft rejection, burns, sepsis, urinary tract obstruction, acute glomerulonephritis, and drug-related alterations in renal hemodynamics. Occasionally, patients with oliguric and nonoliguric ATN will have an FeS < 1%.
   - >2% indicates sodium leak associated with CKD Stages 3 to 5 and chronic or acute tubular damage (i.e., ATN).
3. Abnormal renal and adrenal function and the use of diuretics will interfere with assessments of intravascular volume made using urine Na and Cl concentrations.
4. Urine osmolality can be used to assess action of antidiuretic hormone (see Test 2.11.1) and to determine the etiology of polyuria or hypernatremia when diabetes insipidus is suspected.

Notes

1. The results of the FENa test in the differential diagnosis of acute renal failure must be interpreted in conjunction with the patient's clinical course and the use of other urine and serum tests such as urine output and serum albumin and creatinine concentrations.
2. There are no "normal values" for urine electrolyte concentrations and osmolality, only "expected values" relative to specific clinical situations.

ICD-9 Codes

Conditions that may justify this test *include but are not limited to*:

Diabetes
250.2   with hyperosmolarity
250.3   with coma

Other
584.5   with lesion of tubular necrosis/renal failure with (acute) tubular necrosis
593.9   unspecified disorder of kidney and ureter/prerenal azotemia

Refer also to Test 5.2.1 codes.

Suggested Reading

Kamel KS, Ethier JH, Richardson RM, Bear RA, Halperin ML. Urine electrolytes and osmolality: when and how to use them. *Am J Nephrol* 1990; 10:89–102.

Zarich S, Fang LS, Diamond JR. Fractional excretion of sodium: exceptions to its diagnostic value. *Arch Intern Med* 1985; 145:108–12.

5.3   Neuropathy of Diabetes Mellitus

5.3.1   Semmes–Weinstein Monofilament (S-WMF) Semiquantitative Test for Neurosensory Deficits in Diabetes Mellitus (DM) Patients

Indications for Test

S-WMF testing is indicated as a semiquantitative test for neurosensory deficits in:

- Protective light touch or pressure sensation in patients with DM, scleroderma, and other conditions (i.e., calluses) that increase the risk for ulceration of the extremities
- Patients with foot lesions (e.g., ulcers) or deformity; patients with autonomic neuropathy (see Q5.1 in Appendix 1)
- High-risk DM patients who require self-monitoring as part of daily foot care
- Patients who may not volunteer symptoms but on inquiry admit that their feet feel numb or "dead"
- Patients with symptoms of burning pain, electrical or stabbing sensations, paresthesias, hyperesthesias

Procedure

1. Obtain an array of varying sizes of monofilaments (e.g., 2.83, 3.61, 4.17, 4.31, 4.56, 5.07, 6.1, and 6.65 units).
2. Inspect both feet and note areas of deformity, redness, swelling, warmth, dryness, and maceration.
3. Before S-WMF testing, obstruct the patient's view of the foot (i.e., have the patient close his eyes or lie flat and look at ceiling).
4. Apply the S-WMF perpendicular to the skin surface with sufficient force to cause the filament to bend. The total duration of contact should be approximately 1.5 seconds.
5. Do not allow the filament to slide across the skin surface or make overly repetitive contact at a test site.
6. Reduce the potential for patient guessing by randomizing the test sites and the time between tests. Ask the patient to respond with the words 'touch", "now," or "yes" each time the filament is felt. Do *not* prompt the patient by asking: "Do you feel this?"
7. Determine whether the patient has a normal cutaneous sensation threshold by testing with the smaller and lighter S-WMFs (2.83 to 4.17 units) on dorsal and proximal areas of the lower extremities (i.e., above the ankle). Apply the stimulus at random intervals up to three times in the same general location to elicit a response.
8. Begin with use of the 4.17-unit S-WMF to test sensation on the plantar foot surface. If the patient does not respond to this lower stimulus, advance to the next higher strength S-WMF, which would be the 5.07 unit (10 g).
9. Test the plantar surfaces of the first and fifth toes and the metatarsal heads of these toes (four sites), test the medial and lateral mid-foot and heel (three sites), and dorsal mid-foot and area proximal to the first and second toes (two sites). Do not apply S-WMFs to ulcer sites, calluses, scars, or necrotic tissue.
10. Use S-WMFs of sizes greater than 6.0 units to test for deep pressure sensation only.
11. Before making a diagnosis of diabetic peripheral neuropathy, other neuropathic conditions, including chronic inflammatory demyelinating polyneuropathy (CIDP), B_{12} deficiency, hypothyroidism, and uremia should be excluded. Thus, obtain tests for serum B_{12}, blood urea nitrogen, and creatinine concentration as well as thyroid function. Consult with a neurologist for diagnosis of CIDP.

TABLE 5.5
Categorization of Neuropathy in Relation to Protective Sensation, Foot Deformity, and ABI

Category of Neuropathy	Principal Characteristic	Protective Sensation	Wound Stage[a]	Criteria				
				Ankle–Brachial Index (ABI)	Toe Systolic Pressure (SP) (mmHg)	Ulceration	Neuro-arthropathy	Foot Deformity
0	No neuropathy in patient with DM	Intact[f]	n.a.	>0.80	>45	–	–	±
1	Neuropathy with no deformity	–	n.a.	>0.80	>45	–	–	–
2	Neuropathy with deformity	–	n.a.	>0.80	>45	–	–	+
3	History of prior pathologic process	–	n.a.	>0.80	>45	+[b]	+	+
4A	Neuropathic wound	–	A	>0.80	>45	±	–	+
4B	Acute Charcot's joint	–	n.a.	>0.80	>45	±[c]	+	±
5	Infected diabetic foot	±	B[d]	n.a.	n.a.	±	±	±
6	Ischemic limb	–	C or D	<0.80	<45[e]	±	±	±

[a] See Table 4.4, "Wound Classification: Depth of Wound (0, I, II, III) vs. Stage (A, B, C, D) Based on Ischemia and Infection."

[b] Or amputation.

[c] For noninfected neuropathic ulceration.

[d] Infected.

[e] Or pedal transcutaneous O_2 tension < 40 mmHg.

[f] By Semmes–Weinstein monofilament (S-WMF) testing at 4.56 monofilament strength.

Note: +, present; –, absent or none; ±, may be present or absent; n.a., not applicable.

Source: Adapted from Armstrong, D.G. et al., *Diabetes Care,* 21, 855–859, 1998.

Interpretation

1. Interpret response to S-WMF testing from "Categorization of Neuropathy in Relation to Protective Sensation, Foot Deformity, and ABI" (Table 5.5).

Notes

1. Loss of sensation is the primary reason for the initial development of a foot wound and its failure to heal. Secondary factors are chronically elevated blood glucose concentrations, usually >140 mg/dL.
2. Up to 50% of diabetic neuropathy patients may experience symptoms, most frequently burning pain, electrical or stabbing sensations, parasthesias, hyperesthesias, and deep aching neuropathic pain that is typically worse at night.
3. The S-WMF test for sensory neuropathy requires a portable, hand-held device widely available commercially as well as through government agencies, voluntary organizations, and pharmaceutical companies, although usually only a single-strength device (i.e., 4.56 or 5.07 [10g]) is offered gratis, and its strength may not be clearly specified.
4. Abnormal sensation on testing with the 4.56 unit is a sensitive and highly predictive indicator of risk for ulceration of the extremities but is positive only with more advanced neuropathy.
5. Quantitative vibratory sensory testing may be more sensitive than the monofilament test for detection of early sensory neuropathy; however, at similar sensitivities, specificities for foot ulcer patient identification were lower with vibration and thermal threshold determination compared to S-WMF testing.
6. A systematic, detailed lower-extremity examination for every DM patient who is admitted to a hospital, particularly those who are admitted with a primary diagnosis that involves a foot complication, is required, but less than half actually get tested for the presence or absence of protective sensation (Edelson et al., 1996).
7. Symptomatic neuropathy with increase in vibration perception threshold, foot deformity, high plantar pressures, a history of amputation, male sex, lengthy duration of DM (>10 years), foot deformities (hallux rigidus or hammer toes), poor DM control (glycosylated hemoglobin [HbA$_{1c}$] > 9%) are all significantly associated with the presence of foot ulceration.
8. Primary-care physicians may be expected to miss at least a third of cases of peripheral neuropathy without actual performance of a test for its presence.
9. Compared to neuropathy, vascular or renal disease is not as significant a risk factor for the development of foot ulceration (Abbott et al., 2002).
10. Monofilament sets may be obtained from:
    - Sensory Testing Systems, 1815 Dallas Dr., Suite 11A, Baton Rouge, LA 70806 (504-923-1297; fax 504-923-3670)
    - North Coast Medical, Inc., 18305 Sutter Blvd., Morgan Hill, CA 95037 (800-821-9319; fax 877-213-9300)
    - Lower Extremity Amputation Prevention Program, Bureau of Primary Health Care (BPHC), Division of Programs for Special Populations, 4350 East West Highway, 9th Floor, Bethesda, MD 20814 (888-ASK-HRSA; 888-275-4772)

ICD-9 Codes

Conditions that may justify this test *include but are not limited to*:

Neuritis

354	upper limb and multiplex
355	lower limb

Other

250.6	diabetes mellitus with neurological manifestations
337	disorders of the autonomic nervous system
357	inflammatory and toxic neuropathy

Suggested Reading

Abbott CA, Carrington AL, Ashe H, Bath S, Every LC, Griffiths J, Hann AW, Hussein A, Jackson N, Johnson KE, Ryder CH, Torkington R, Van Ross ER, Whalley AM, Widdows P, Williamson S, Boulton AJM. The North-West Diabetes Foot Care Study: incidence of, and risk factors for, new diabetic foot ulceration in a community-based patient cohort. *Diabet Med* 2002; 19:377–84.

Armstrong DG, Lavery LA, Harkless LB. Treatment-based classification system for assessment and care of diabetic feet. *J Am Pod Med Assn* 1996; 86:311–6.

Armstrong DG, Lavery LA, Harkless LB. Validation of a diabetic wound classification system: the contribution of depth, infection, and ischemia to risk of amputation. *Diabetes Care* 1998a; 21:855–9.

Armstrong DG, Lavery LA, Vela SA, Quebedeaux TL, Fleischli JG. Choosing a practical screening instrument to identify patients at risk for diabetic foot ulceration. *Arch Intern Med* 1998b; 158:289–92.

Ayyar DR, Sharma KR. Chronic demyelinating polyradiculoneuropathy in diabetes. *Curr Diab Rep* 2004; 4:409–12.

Bell-Krotoski J, Weinstein S, Weinstein C. Testing sensibility, including touch-pressure, two-point discrimination, point localization, and vibration. *J Hand Therapy* 1993; 6:114–23.

Birke JA, Sims DS. Plantar sensory threshold in the ulcerative foot. *Lepr Rev* 1986; 57:261–7.

Duffy JC, Patout CA. Management of the insensitive foot in diabetes: lessons learned from Hansen's disease. *Mil Med* 1990; 155:575–9.

Edelson GW, Armstrong DG, Lavery LA, Caicco G. The acutely infected diabetic foot is not adequately evaluated in an inpatient setting. *Arch Intern Med* 1996; 156:2373–78.

Lavery LA, Armstrong DG, Harkless LB. Classification of diabetic foot wounds. *J Foot Ankle Surg* 1996; 35:528–31.

Lavery LA, Armstrong DG, Vela SA, Quebedeaux TL, Fleischli JG. Practical criteria for screening patients at high risk for diabetic foot ulceration. *Arch Intern Med* 1998; 158:157–162.

Sosenko JM, Kato M, Soto R, Bild DE. Comparison of quantitative sensory-threshold measures for their association with foot ulceration in diabetic patients. *Diabetes Care* 1990; 13:1057–61.

Thomas PK. Classification, differential diagnosis, and staging of diabetic peripheral neuropathy. *Diabetes* 1997; 46(Suppl 2):S54–S57.

Vinik A, Mehrabyan A, Colen L, Boulton AJM. Focal entrapment neuropathies in diabetes [review]. *Diabetes Care* 2004; 27:1783–8.

5.3.2 Vibratory Sensation Testing in the Screening and Diagnosis of Peripheral Neuropathy in DM Patients

Indications for Test

Vibratory sensation testing is indicated in:

- The initial screening evaluation and annual physical examination of all DM patients
- Any patient with neurologic disturbance of gait, cutaneous sensation, or proprioception
- Risk assessment of DM patients to allow effective prevention of damage to lower extremities and consequent ulcers and amputations
- Patients with distended dorsal foot veins, dry skin, absent ankle reflexes, and the presence of calluses under pressure-bearing areas

Screening Procedure

1. Screen all DM patients for DM neuropathy with qualitative neuropathy questionnaires (see Q5.1 and Q5.2 in Appendix 1). Obtain a Semmes–Weinstein monofilament (S-WMF) semiquantitative test for sensory deficits (see Test 5.3.1).

2. Place a 128-Hz tuning fork over the bony prominences of the upper and lower extremities (i.e., medial malleolus) and lift and reapply the tuning fork along the extremity, beginning proximally and moving distally to the great toe.
3. With each application of the tuning fork, ask the patient to respond to the vibration, not to the pressure or the sound.
4. Compare your (normal) response to the vibration with that of the patient by quickly transferring the fork to one of your bony prominences.
5. Compare the strength of the vibratory sense in the distal vs. proximal extremities.
6. To improve the assessment of vibration sensation, take into account the patient's age, height, and weight (or body surface area) when judging vibration abnormalities.

Quantitative Procedure Options

1. Use biothesiometry to determine the patient's vibration perception threshold (VPT) in the feet (e.g., Diabetica Solutions VPT device). Choose vibration testing by the on–off or timed method. Recognize that this more objective methodology is still dependent on a subjective response from the patient.
2. To measure patient's vibration and temperature sensitivity in the hands and feet, use Physitemp's Vibratron II device (Physitemp Instruments, Inc.; 154 Huron Avenue, Clifton NJ 07013; www.physitemp.com/sensory.htm#vibration).
3. Measure the patient's response to vibratory stimuli and heat- or cold-induced pain using the Medoc TSA-II or Case IV devices (www.medoc-web.com/tsa.html; www.wrmed.com/neurophys/neuro/caseiv.htm).

Interpretation

1. Diminution or loss of vibratory sensation in the more *distal* areas is an early indicator of DM sensory polyneuropathy.
2. Diminution or loss of vibratory sensation in the more *proximal* areas is an indicator of more advanced DM neuropathy.
3. Both the qualitative or semiquantitative tuning fork vibration and S-WMF pressure tests fail to detect the early phases of neuropathy (i.e., they lack adequate diagnostic sensitivity).
4. In general, vibratory thresholds correlate best with the function of large myelinated nerve fibers.
5. Vibratory sensation is typically diminished at the toes and ankles with advancing age.
6. With the Vibratron II, in an 18- to 65-year-old reference population, the vibration threshold in vibration units (mean ± SD) is 0.7 ± 0.4 for the index finger vs. 1.20 ± 0.5 for the great toe.
7. With the Diabetica Solutions VPT, insensitivity of the hallux to vibration intensity of <26 volts (6.5 µm of tip vibration displacement) is clearly abnormal.
8. DM patients with a VPT < 15 have a cumulative incidence of foot ulceration sevenfold that of those with a VPT > 25 (Young et al., 1994).

Notes

1. DM neuropathies are complex, heterogeneous disorders encompassing a wide range of abnormalities affecting both peripheral, somatic, and autonomic nervous systems. Neuropathies may be focal or diffuse, proximal or distal.
2. Diminution of vibratory sensation or defective proprioception occurs early in the course of diabetic sensory polyneuropathy and usually precedes loss of light touch, thermal, and pain sensation, as well as loss of deep tendon reflexes. Alcoholic neuropathies may be superimposed on diabetic ones, resulting in additive hypoesthesia.
3. Instruct DM patients to inspect their feet every day, to get an annual test for cutaneous sensation, and to be professionally evaluated promptly in case of foot injury.
4. Up to 75% of DM patients may not experience symptoms of numbness, prickling, aching, or burning pain, either mild or lancinating, yet still have clearly reduced vibratory or pressure sensation and are at risk for foot injury.

5. Expect sustained improvements in glycemic control to lessen the severity of DM sensorimotor polyneuropathy based on the reduction in myelinated fiber density loss as shown in sural nerve biopsies from subjects with better controlled DM.
6. Acute sensory neuropathy may follow an episode of ketoacidosis or sudden improvement in glycemic control following insulin therapy. This rare condition is characterized by the acute onset of severe, painful sensory symptoms.
7. Patients with an abnormal vibration test tend to be older and to have a duration of DM and HbA$_{1c}$ greater than subjects with a normal vibration sense (Kastenbauer et al., 2004).
8. Although the S-WMF test identifies patients with the highest risk of foot complications, up to two times as many patients can be identified to be at risk by using vibration threshold perception testing, which is clearly the more sensitive test for neuropathy (Gin et al., 2002).
9. Combination testing using the 10-g S-WMF and vibration tests adds little value, at most a slight increase in specificity, to the screening test for vibratory sensation.
10. A quantitative method for assessment of vibratory sensation using tuning fork extinction times in the early detection of neuropathy has been described (London et al., 2000). Delays in application of the tuning fork to the base of the medial malleolus of 3 seconds or more were associated with average increases in the extinction time of 1.7 seconds.
11. With the graduated tuning fork screening test, within-test variation at big toes reached 8.4% in DM patients vs. 2.2% in control subjects. Mean contralateral variation was 7.5% in DM patients vs. 2.5% in control subjects.
12. In a 12-year study of the progression of DM peripheral neuropathy, assessed with VPT, in patients with average glycemic control (i.e., 7–8% HbA$_{1c}$ when tested), only about 20% of cases experienced a significant deterioration in VPT (14.2 ± 3.7 volts at baseline to 35.9 ± 9.5 volts). No significant change in VPT (10.1 ± 3.7 volts at baseline to 14.2 ± 4.7 volts) occurred in the remainder. These data suggest that only a subset of DM patients are predisposed to neuropathy or have higher average glucose levels than was apparent with tests used for assessment of glycemic control in the 1990s (Coppini et al., 2001).
13. Refinements in the quantitative measurement of the defects that occur in neuropathies are needed so that therapy can be directed to specific nerve fiber types. Tests for neuropathy must be validated and standardized so studies can be more directly compared.
14. Because malingering can influence quantitative sensory tests results, their use is very limited when trying to settle medicolegal disability issues.
15. When specificity is maintained at >90%, the sensitivities of tests for cutaneous perception in DM neuropathy patients are:
    • Vibration test, 88%
    • Thermal test—warm, 78%; cold, 77%
    • Tactile-pressure test, 77%
    • Cutaneous perception test (CPT)—5 Hz, 52%; 250 Hz, 48%; 2000 Hz, 56%
16. The Center for Medicare and Medicaid Services (CMS) has maintained its noncoverage policy on the CPT test.

ICD-9 Codes

Refer to Test 5.3.1 codes.

Suggested Reading

Apelqvist J, Bakker K, van Houtum WH, Nabuurs-Franssen MH, Schaper NC. International consensus and practical guidelines on the management and the prevention of the diabetic foot. International Working Group on the Diabetic Foot. *Diabetes Metab Res Rev* 2000; 16(Suppl 1):S84–92.

Burns TM, Taly A, O'Brien PC, Dyck PJ. Clinical versus quantitative vibration assessment: improving clinical performance. *J Peripher Nerv Syst* 2002; 7:112–7.

Coppini DV, Wellmer A, Weng C, Young PJ, Anand P, Sonksen PH. The natural history of diabetic peripheral neuropathy determined by a 12-year prospective study using vibration perception thresholds. *J Clin Neurosci* 2001; 8:520–4.

Gin H, Rigalleau V, Baillet L, Rabemanantsoa C. Comparison between monofilament, tuning fork and vibration perception tests for screening patients at risk of foot complication. *Diabetes Metab* 2002; 28(6 Pt 1):457–61.

Kastenbauer T, Sauseng S, Brath H, Abrahamian H, Irsigler K. The value of the Rydel–Seiffer tuning fork as a predictor of diabetic polyneuropathy compared with a neurothesiometer. *Diabet Med* 2004; 21:563–7.

London L, Thompson ML, Capper W, Myers JE. Utility of vibration sense testing for use in developing countries: comparison of extinction time on the tuning fork to vibration thresholds on the Vibratron II. *Neurotoxicology* 2000; 21:743–52.

Perkins BA, Greene DA, Bril V. Glycemic control is related to the morphological severity of diabetic sensori-motor polyneuropathy. *Diabetes Care* 2001a; 24:748–52.

Perkins BA, Olaleye D, Zinman B, Bril V. Simple screening tests for peripheral neuropathy in the diabetes clinic. *Diabetes Care* 2001b; 24:748–52.

Thivolet C, el Farkh J, Petiot A, Simonet C, Tourniaire J. Measuring vibration sensations with graduated tuning fork: simple and reliable means to detect diabetic patients at risk of neuropathic foot ulceration. *Diabetes Care* 1990; 13:1077–80.

van Deursen RW, Sanchez MM, Derr JA, Becker MB, Ulbrecht JS, Cavanagh PR. Vibration perception threshold testing in patients with diabetic neuropathy: ceiling effects and reliability. *Diabet Med* 2001; 18:469–75.

Vinik AI, Mehrabyan A. Diagnosis and management of diabetic autonomic neuropathy. *Compr Ther* 2003; 29:130–45.

Vinik AI, Park TS, Stansberry KB, Pittenger GL. Diabetic neuropathies. *Diabetologia* 2000; 43:957–73.

Vinik AI, Suwanwalaikorn S, Stansberry KB, Holland MT, McNitt PM, Colen LE. Quantitative measurement of cutaneous perception in diabetic neuropathy. *Muscle Nerve* 1995; 18:574–84.

Young MJ, Veves A, Breddy JL, Boulton AJM. The prediction of diabetic neuropathic foot ulceration using vibration perception thresholds: a prospective study. *Diabetes Care* 1994; 17:557–60.

5.3.3 Calculation of Gastric Emptying Rate by Conventional Radioscintigraphic Measurement of the Clearance of ^{99m}Tc-Albumin from the Stomach in DM Patients

Indications for Test

Gastric emptying rate measurement is indicated in DM (T1DM or T2DM) patients when:

- Gastroparesis-associated symptoms of early satiety, epigastric discomfort, bloating, abdominal distention, constipation, reflux, or hypermotility with diarrhea are noted, particularly in context of retinopathy, nephropathy, neuropathy (autonomic and/or peripheral), and chronically poor metabolic control of DM with loss of appetite and weight loss, as per the "Autonomic Neuropathy Symptom Questionnaire" (see Q5.1 in Appendix 1).
- Nausea or vomiting lasts for days to months or occurs in episodic cycles.
- Glucose control is erratic, with symptoms of reactive hypoglycemia or dumping syndrome.
- Symptoms of fecal incontinence are associated with anal sphincter incompetence or reduced rectal sensation.
- Treatment is planned with incretin modifiers, such as incretin degradation blockers or exenatide (Byetta®) for T2DM or amylin analog (Symlin®) for T1DM.
- Other medical problems are present, including amyloidosis, scleroderma, Parkinson's disease, hypothyroidism, hypoadrenalism, or chronic narcotic analgesia.

Procedure

1. Obtain a history of the use of agents that can slow gastric emptying and discontinue intake of these substances if possible, including narcotics, tricyclic antidepressants, calcium channel blockers, clonidine, dopamine agonists, lithium, nicotine, and progesterone.

2. Consultation with a nuclear medicine facility is required to ensure that the protocol used is adequate to detect rapid as well as delayed gastric emptying. Do not accept a report of "percent retention at 2 hours" as an adequate assessment of gastric emptying.

3. Measure the patient's blood glucose level following an overnight fast. Instruct the patient to take his or her usual diabetes medications.

4. Note that hyperglycemia slows gastric emptying in DM patients. Gastric contractions were nearly absent at a serum glucose level of 250 mg/dL and markedly reduced at 175 and 140 mg/dL (Barnett and Owyang, 1988).

5. If the fasting glucose level is <200 mg/dL, proceed with the test by having the patient ingest a low-carbohydrate, ^{99m}Tc-coated solid meal. If available, insulin should be administered prior to the meal in a dose calculated to keep the blood glucose concentration as close to normal as possible (i.e., 1 unit regular insulin for each 10 to 15 g carbohydrate in the test meal plus 1 unit for each 25 g of blood sugar over 125 mg/dL). Instruct the patient in the use of insulin to accomplish this part of the procedure as most radiology centers do not give insulin.

6. Use radioscintigraphy to determine the rate of gastric emptying after ingestion of a standard, radio-labeled test meal (e.g., a 310-kcal omelet with 12 to 15 MBq ^{99m}Tc-labeled macroaggregated albumin).

7. Place the patient under a gamma counter and measure the radioactivity retained in the stomach at 10- to 15-minute intervals following ingestion of the meal. Measurements at hourly intervals for up to 4 hours are an acceptable alternative to the gold standard of measurements at 15-minute intervals.

8. Calculate the half-life ($t_{1/2}$) of ^{99m}Tc activity in the stomach from a plot of the amount of radioactivity retained at each 15-minute interval up to 2.5 hours.

9. If available, use your local radiologic facility's protocol for measurement of:
   - The rate of appearance of [^{13}C] in CO_2 expired after ingestion of [^{13}C]-enriched octanoic acid (100 mg) or acetate (150 mg sodium [^{13}C]-acetate)
   - Serum levels of acetaminophen sampled 6 to 20 times over the 1.5 to 8 hours after ingestion of paracetamol (Willems et al., 2001)

10. Obtain appropriate imaging studies or referral for endoscopy to determine causes of diarrhea and constipation other than gastroparesis.

Interpretation

1. Median values (95th percentile) for percent gastric retention at 60, 120, and 240 minutes were 69% (90%), 24% (60%), and 1.2% (10%), respectively. A power exponential model yielded similar emptying curves and estimated T_{50} when using images taken only at 1, 2, and 4 hours or with imaging taken every 10 minutes.

2. Gastric emptying is initially more rapid in men but is similar in men and women at 4 hours; it is faster in older subjects ($p < 0.05$) but is independent of body mass index (Tougas et al., 2000). Most gastroparesis patients are female.

3. A $t_{1/2}$ clearance of ^{99m}Tc activity in a liquid test meal was reported to be 34 minutes for non-neuropathic DM patients and 65 minutes for a non-DM control group ($p = 0.0005$) (Phillips et al., 1992). Accelerated gastric emptying may occur as a result of DM neuropathy (Couturier et al., 2004).

4. DM patients without gastrointestinal (GI) symptoms or orthostatic hypotension may have a $t_{1/2}$ clearance of ^{99m}Tc activity after a solid meal test twofold (70 minutes) faster than a control group of DM patients with gastropathy symptoms or orthostatic hypotension (148 minutes) (Nowak et al., 1995).

5. Gastric retention of >10% at 4 hours is indicative of delayed emptying, a value comparable to those provided by more intensive (every 10 minutes) scanning approaches.

6. Percent retention of a solid meal at 100 minutes ($p = 0.032$) and the $t_{1/2}$ for its clearance ($p = 0.032$) increased during hyperglycemia (Fraser et al., 1990). Severe hyperglycemia (>300 mg/dL or 16.7 mmol/L) delays gastric emptying and increases retention of liquids and solids in the absence of autonomic neuropathy, resulting in a false-positive test result in both T1DM and T2DM patients (Barnett and Owyang, 1988).

7. The $t_{1/2}$ for clearance of [^{13}C]-enriched octanoic acid may be much slower in neuropathic DM vs. control subjects—for example, 100 ± 35 vs. 77 ± 21 minutes ($p < 0.003$), with 2-hour retention values of 31% ± 17% vs. 20% ± 10% ($p < 0.006$) (Braden et al., 1995).

8. The differential diagnosis of DM-related diarrhea includes reduced GI tract motility, reduced receptor-mediated fluid absorption, bacterial overgrowth, pancreatic insufficiency, coexistent celiac disease, and abnormalities in bile salt metabolism.

9. The finding of retained food in the stomach after an 8- to 12-hour fast, in the absence of obstruction, is diagnostic of severe gastroparesis.

Notes

1. Constipation is the most common symptom of DM gastropathy, but it can alternate with episodes of diarrhea. Constipation affects nearly 60% of DM patients and is more often associated with calcium channel blocker use in T1DM than with symptoms of autonomic neuropathy.

2. Loss of incretin function via deficiency of glucagon-like peptide 1 (GLP-1) and amylin can lead to more rapid gastric emptying early in the course of both T1DM and T2DM, resulting in higher postprandial glucose levels. However, delayed gastric emptying may be observed in up to 28% of unselected, but poorly controlled, DM patients.

3. Gastric emptying depends largely on vagus nerve function, which can be severely disrupted in DM or after surgical vagotomy. Hyperglycemia may contribute to dysfunction of intrinsic enteric neurons and inhibition of antral motility independent of plasma motilin concentrations, probably via alterations in the interdigestive migrating motor complex.

4. Although dyspeptic lower GI symptoms scored during gastric emptying tests may predict the rate of gastric emptying, upper GI sensations experienced during daily life do not (Samsom et al., 2003).

5. Partial gastrectomy usually leads to rapid gastric emptying but may lead to stasis if performed along with vagotomy.

6. The half-emptying times for the [^{13}C]-acetate breath test correlates closely to those measured by radioscintigraphy both for semisolid ($r = 0.87$) and liquid ($r = 0.95$) test meals.

7. Gastric emptying results using the paracetamol absorption test correlate well with those obtained using scintigraphy; however, further standardization of this test is needed to optimize its use in clinical and research studies (Willems et al., 2001).

8. In neuropathic DM, erythromycin therapy can shorten abnormally prolonged gastric-emptying times for both liquids and solids. In these patients, mean gastric retention of solids at 2 hours was 63% with placebo and 4% with erythromycin. Two hours after ingestion of liquids, retentions were 32% and 9%, respectively. In 10 healthy control subjects, 2-hour gastric retention was 9% for solids and 4% for liquids (Janssens et al., 1990).

9. Many patients with symptoms suggestive of gastroparesis have no definable delays in gastric emptying as measured by the relatively insensitive scintigraphic techniques used to determine gastric emptying or because other mechanisms, besides gastroparesis, are responsible for the symptoms.

ICD-9 Codes

Conditions that may justify this test *include but are not limited to*:

Gastrointestinal symptoms
536.2   persistent vomiting
564.0   constipation
787.3   flatulence, eructation, gas pain, bloating
787.6   incontinence of feces
787.91 diarrhea

Other
251.2   hypoglycemia, reactive
536.3   gastroparesis
780.94 early satiety
789.3   abdominal or pelvic swelling, mass

Refer also to Test 5.3.1 codes.

Suggested Reading

Barnett JL, Owyang C. Serum glucose concentration as a modulator of interdigestive gastric motility. *Gastroenterology* 1988; 94:739–44.

Braden B, Adams S, Duan LP, Orth KH, Maul FD, Lembcke B, Hor G, Caspary WF. The [^{13}C] acetate breath test accurately reflects gastric emptying of liquids in both liquid and semisolid test meals. *Gastroenterology* 1995; 108:1048–55.

Bytzer P, Talley NJ, Leemon M, Young LJ, Jones MP, Horowitz M. Prevalence of gastrointestinal symptoms associated with diabetes mellitus: a population-based survey of 15,000 adults. *Arch Intern Med* 2001; 161:1989–96.

Couturier O, Bodet-Milin C, Querellou S, Carlier T, Turzo A, Bizais Y. Gastric scintigraphy with a liquid:solid radiolabelled meal: performances of solid and liquid parameters. *Nuc Med Commun* 2004; 25:1143–50.

Fraser RJ, Horowitz M, Maddox AF, Harding PE, Chatterton BE, Dent J. Hyperglycaemia slows gastric emptying in type 1 (insulin-dependent) diabetes mellitus. *Diabetologia* 1990; 33:675–80.

Janssens J, Peeters TL, Vantrappen G, Tack J, Urbain JL, De Roo M, Muls E, Bouillon R. Improvement of gastric emptying in diabetic gastroparesis by erythromycin: preliminary studies. *N Engl J Med* 1990; 322:1028–31.

Kong MF, Macdonald IA, Tattersall RB. Gastric emptying in diabetes. *Diabet Med* 1996; 13:112–9.

Kong MF, Horowitz M, Jones KL, Wishart JM, Harding PE. Natural history of diabetic gastroparesis. *Diabetes Care* 1999; 22:503–7.

Lysy J, Israeli E, Goldin E. The prevalence of chronic diarrhea among diabetic patients. *Am J Gastroenterol* 1999; 94: 2165–70.

Maleki D, Locke GR, Camilleri M, Zinsmeister AR, Yawn BP, Leibson C, Melton LJ. Gastrointestinal tract symptoms among persons with diabetes mellitus in the community. *Arch Intern Med* 2000; 160:2808–16.

Mistiaen W, Van Hee R, Blockx P, Bortier H, Harrisson F. Gastric emptying rate for solid and for liquid test meals in patients with dyspeptic symptoms after partial gastrectomy and after vagotomy followed by partial gastrectomy. *Hepatogastroenterology* 2001; 48:299–302.

Nowak TV, Johnson CP, Kalbfleisch JH, Roza AM, Wood CM, Weisbruch JP, Soergel KH. Highly variable gastric emptying in patients with insulin dependent diabetes mellitus. *Gut* 1995; 37:23–9.

Phillips WT, Schwartz JG, McMahan CA. Rapid gastric emptying of an oral glucose solution in type 2 diabetic patients. *J Nucl Med* 1992; 33:1496–500.

Samsom M, Vermeijden JR, Smout AJ, Van Doorn E, Roelofs J, Van Dam PS, Martens EP, Eelkman-Rooda SJ, Van Berge-Henegouwen GP. Prevalence of delayed gastric emptying in diabetic patients and relationship to dyspeptic symptoms: a prospective study in unselected diabetic patients. *Diabetes Care* 2003; 26:3116–22.

Tougas G, Eaker EY, Abell T et al. Assessment of gastric emptying using a low fat meal: establishment of international control values. *Am J Gastroenterol* 2000; 95:1456–62.

Vinik A, Erbas T, Stansberry K. Gastrointestinal, genitourinary, and neurovascular disturbances in diabetes. *Diabetes Rev* 1999; 7:358–78.

Willems M, Quartero AO, Numans ME. How useful is paracetamol absorption as a marker of gastric emptying? A systematic literature study. *Dig Dis Sci* 2001; 46:2256–62.

5.3.4 Assessment of Erectile Function (EF) in Male DM Patients with Possible Neuropathy, Vasculopathy, and/or Hypogonadism

Indications for Test

Assessment of erectile function is indicated in male patients with DM who:

- Are at high risk for erectile dysfunction (ED) secondary to neuropathy, vasculopathy, renal insufficiency, or low testosterone levels as per "Risk Factors for Erectile Dysfunction" (Table 5.6)
- Report ED, suggesting an abnormality of penile erectile function (PEF)

TABLE 5.6
Risk Factors for Erectile Dysfunction

Risk Factor	Odds Ratio for Risk Factor ($p < 0.05$)	95% Confidence Interval
Age	Steady increase with age	Not applicable
Diabetes	3.00	1.53–5.87
Hyperlipidemia	2.29	1.42–3.70
Lower urinary tract symptoms	2.20	1.76–2.76
Hypertension	2.05	1.61–2.60
Psychological stress	1.68	1.43–1.98
Low physical activity	1.35	1.15–1.60

Source: Ponholzer, A. et al., *Eur. Urol.*, 47, 80–86, 2005. With permission.

Procedure

1. Obtain a complete history of drug and medication intake, especially of alcohol, cocaine, antidepressants, antihypertensives (i.e., thiazides, calcium channel and beta blockers), tranquilizers, immunomodulators, estrogens, glucocorticoids, H2 antagonists, anticholinergics.
2. Screen patients for irritative and obstructive urinary tract voiding symptoms (McVary et al., 2005). Obtain a detailed history of smoking, depression, social or marital stress, and central neurological and spinal cord conditions (e.g., Parkinson's disease, stroke, Alzheimer's disease, brain tumor).
3. Use the simplified five-item International Index of Erectile Function (IIEF-5) questionnaire to screen for ED (see Q5.3 in Appendix 1). Administer the IIEF-5 without pressuring or making demands on the patient.
4. Follow a positive screening test with the IIEF-5 questionnaire with administration of the full 15-question IIEF questionnaire (IIEF-15) (see Q5.4 in Appendix 1).
5. For the IIEF-15 questionnaire, calculate the scores for each of the five domains (i.e., erectile function, orgasmic function, sexual desire, intercourse satisfaction, and overall satisfaction; total scores range from 5 to 75).
6. Obtain a measurement of positive and negative mood scores using the "Mood Score Questionnaire" (see Q8.2 in Appendix 1) in cases of suspected hypogonadism with or without suspected neuropathy, DM, or specific complaint of ED.
7. In those with negative mood scores, obtain a blood sample for total and free testosterone concentrations.

Interpretation

1. For Q5.3, the maximum score is 25; a score ≤ 21 suggests ED and prompts testing using the IIEF-15 questionnaire (see Q5.4 in Appendix 1).
2. For Q5.4, the maximum score is 75; men with a score ≤ 25 for the PEF factor alone are classified as having ED.
3. Those scoring above 25 on the EF factor of the Q5.4 do not have ED (sensitivity, 97%; specificity, 88%).
4. Apart from age, the most important risk factors for ED are DM, hyperlipidemia, lower urinary tract symptoms, hypertension, and psychological stress.
5. The results of the IIEF questionnaire cannot be used for diagnosis or comparison of specific vascular causes of ED, including arterial insufficiency or venous leakage.
6. Administer the "Mood Score Questionnaire" (see Q8.2 in Appendix 1) in cases of hypogonadism or as a screening tool for hypogonadism, which is not always associated with ED.

TABLE 5.7
Sample Results of Scoring the 15-Item International Index of Erectile Function (IIEF) before and after Interventional Pharmacotherapy

IIEF Question Topic[a]	Mean Score ± (Standard Deviation)		p Value[b]
	Baseline	Follow-Up	
1. Frequency of erections during sexual activity	1.3 (1.5)	2.5 (1.8)	<0.001
2. *Frequency of erections hard enough to penetrate*	1.3 (1.4)	2.3 (1.7)	<0.001
3. Frequency of penetration during intercourse	1.3 (1.4)	2.3 (1.8)	<0.001
4. *Frequency of penetration and maintenance of erection during intercourse*	1.4 (1.6)	2.4 (1.8)	0.002
5. *Maintaining erection to completion of intercourse*	1.3 (1.5)	2.4 (1.9)	<0.001
6. Frequency of attempts at intercourse	1.6 (1.7)	2.2 (1.2)	0.013
7. *Frequency of satisfaction with intercourse*	1.5 (1.7)	2.3 (1.8)	0.013
8. Enjoyment of intercourse	1.8 (1.8)	2.5 (1.7)	0.032
9. Frequency of ejaculation	1.5 (1.7)	2.3 (1.8)	0.011
10. Frequency of orgasm or climax	2.7 (1.9)	3.5 (1.6)	0.011
11. Frequency of sexual desire	3.4 (1.1)	3.5 (1.1)	0.631
12. Rating of sexual desire	3.1 (1.0)	3.3 (1.0)	0.287
13. Satisfaction with sex life	2.2 (1.4)	2.6 (1.6)	0.128
14. Satisfaction with sexual relationship	2.9 (1.6)	3.4 (1.6)	0.078
15. *Rating of confidence in achieving/maintaining erections*	2.0 (1.1)	2.3 (1.4)	0.273

[a] Italicized question topics constitute the IIEF-5 questionnaire.

[b] From t-test.

Source: Lowentritt, B.H. et al., *J. Urol.*, 162, 1614, 1999. With permission.

Notes

1. As many as 50% of men over age 40 will suffer some degree of ED. More than 80% of these are not treated because they do not seek medical attention or their physicians do not initiate an assessment or dialogue about sexual problems during their visits.

2. Once identified, ED should prompt careful physical examination particularly of the penis, testes, and rectum. Laboratory studies for investigation of possible hypogonadism (i.e., testosterone levels), which occurs in at least 6% of those with ED, vascular obstruction to penile vessels, prostate disease (e.g., PSA testing), and neuropathology (Tests 5.3.1 to 5.3.3) should be done.

3. The IIEF-15 questionnaire addresses all five factors of male sexual function (erectile function, orgasmic function, sexual desire, intercourse satisfaction, and overall satisfaction), is psychometrically sound, and has been linguistically validated in 10 languages. Refer to "Sample Results of Scoring the 15-Item IIEF before and after Interventional Pharmacotherapy" (Table 5.7).

4. A moderate-to-high correlation exists between a patient's self-assessment of EF (as severe, moderate, minimal/mild, or no problem) and the PEF factor score of the IIEF-15 questionnaire (Cappelleri et al., 2000).

5. A self-report measure, the Erection Quality Scale (EQS), provides an overall index of erection quality. The EQS correlated well with the results of the IIEF-15 questionnaire and differentiated between men with untreated or treated ED and those without ED in a healthy control group (Wincze et al., 2004).

6. ED is clearly associated with aging, but not total testosterone levels, which have no consistent correlation with erectile function.

7. The age-related prevalence of ED in Japan was 40 to 49 years, 36.4%; 50 to 59 years, 42.5%; 60 to 69 years, 58.1%; 70 to 79 years, 79.4%; ≥80 years, 100% (Marumo et al., 2001). In Western cultures,

the age-related prevalence of moderate and severe cases of ED was 23 to 29 years, 1.8% and 0%; 30 to 39 years, 2.6% and 0%; 40 to 49 years, 7.6% and 1.0%; 50 to 59 years, 14.0% and 6.0%; 60 to 69 years, 25.9% and 15.9%; 70 to 79 years, 27.9% and 36.4% (Rhoden et al., 2002).

8. In a study of the effect of the antihypertensive agent valsartan, at baseline, 75.4% of the total group of 3502 hypertensive patients investigated and 65.0% of the subgroup of patients without previous antihypertensive treatment ($n = 952$) were diagnosed by the IIEF-15 questionnaire as having ED. Valsartan therapy markedly reduced ED in these groups to 53% and 45% ($p < 0.0001$), respectively (Dusing, 2003).

9. Diabetic autonomic neuropathy (DAN) may be associated with sexual dysfunction. DAN can cause loss of penile erection or retrograde ejaculation.

10. A comprehensive work-up for impotence in men includes history (medical and sexual); psychological evaluation; measurement of sex hormone levels and nocturnal penile tumescence; tests to assess penile, pelvic, and spinal nerve function; cardiovascular autonomic function tests; and measurement of penile and brachial blood pressure.

11. "The Sexual Encounter Profile (SEP) Questionnaire" (see Q5.5 in Appendix 1) may be used as an alternative or supplement to Q5.3 in screening for ED and in the assessment of response to therapy for ED.

ICD-9 Codes

Conditions that may justify this test *include but are not limited to*:

250.6	diabetes mellitus with neurological manifestations
337	disorders of the autonomic nervous system
357	inflammatory and toxic neuropathy

Suggested Reading

Bacon CG, Hu FB, Giovannucci E, Glasser DB, Mittleman MA, Rimm EB. Association of type and duration of diabetes with erectile dysfunction in a large cohort of men. *Diabetes Care* 2002; 25:1458–63.

Cappelleri JC, Rosen RC, Smith MD, Mishra A, Osterloh IH. Diagnostic evaluation of the erectile function domain of the International Index of Erectile Function. *Urology* 1999; 54:346–51.

Cappelleri JC, Siegel RL, Osterloh IH, Rosen RC. Relationship between patient self-assessment of erectile function and the erectile function domain of the International Index of Erectile Function. *Urology* 2000; 56:477–81.

Carson CC. Erectile dysfunction in the 21st century: whom we can treat, whom we cannot treat and patient education. *Int J Impot Res* 2002; 14(Suppl 1):S29–34.

Carson CC. Erectile dysfunction: evaluation and new treatment options. *Psychosom Med* 2004; 66:664–71.

Carson CC, Giuliano F, Goldstein I, Hatzichristou D, Hellstrom W, Lue T, Montorsi F, Munarriz R, Nehra A, Porst H, Rosen R. The "effectiveness" scale: therapeutic outcome of pharmacologic therapies for ED—an international consensus panel report. *Int J Impot Res*. 2004; 16:207–13.

Dusing R. Effect of the angiotensin II antagonist valsartan on sexual function in hypertensive men. *Blood Press Suppl* 2003; 2:29–34.

Fisher WA, Rosen RC, Eardley I, Niederberger C, Nadel A, Kaufman J, Sand M. The multinational Men's Attitudes to Life Events and Sexuality (MALES) Study Phase II: understanding PDE5 inhibitor treatment seeking patterns, among men with erectile dysfunction. *J Sex Med*. 2004; 1:150–60.

Fugl-Meyer AR, Lodnert G, Branholm IB, Fugl-Meyer KS. On life satisfaction in male erectile dysfunction. *Int J Impot Res* 1997; 9:141–8.

Kassouf W, Carrier S. A comparison of the International Index of Erectile Function (IIEF-5) and erectile dysfunction studies. *BJU Int* 2003; 91:667–9.

Lewis R, Bennett CJ, Borkon WD, Boykin WH, Althof SE, Stecher VJ, Siegel RL. Patient and partner satisfaction with Viagra (sildenafil citrate) treatment as determined by the Erectile Dysfunction Inventory of Treatment Satisfaction Questionnaire. *Urology*. 2001; 57:960–5.

Lowentritt BH, Scardino PT, Miles BJ, Orejuela FJ, Schatte EC, Slawin KM, Elliott SP, Kim ED. Sildenafil citrate after radical retropubic prostatectomy. *J Urol* 1999; 162:1614–17.

Marumo K, Nakashima J, Murai M. Age-related prevalence of erectile dysfunction in Japan: assessment by the International Index of Erectile Function. *Int J Urol* 2001; 8:53–9.

McVary KT. Erectile dysfunction and lower urinary tract symptoms secondary to BPH. *Eur Urol* 2005; 47:838–45.

Ponholzer A, Temml C, Mock K, Marszalek M, Obermayr R, Madersbacher S. Prevalence and risk factors for erectile dysfunction in 2869 men using a validated questionnaire *Eur Urol* 2005; 47:80–6.

Rhoden EL, Teloken C, Sogari PR, Souto CA. The relationship of serum testosterone to erectile function in normal aging men. *J Urol* 2002a; 167:1745–48.

Rhoden EL, Teloken C, Sogari PR, Vargas Souto CA. The use of the simplified International Index of Erectile Function (IIEF-5) as a diagnostic tool to study the prevalence of erectile dysfunction. *Int J Impot Res* 2002b; 14:245–50.

Rosen RC, Riley A, Wagner G, Osterloh IH, Kirkpatrick J, Mishra A. The International Index of Erectile Function (IIEF): a multidimensional scale for assessment of erectile dysfunction. *Urology* 1997; 49:822–30.

Wincze J, Rosen R, Carson C, Korenman S, Niederberger C, Sadovsky R, McLeod L, Thibonnier M, Merchant S. Erection Quality Scale: initial scale development and validation. *Urology* 2004; 64:351–6.

5.4 Ketoacidopathy of Diabetes Mellitus

5.4.1 Ketone Testing for Diabetic Ketoacidosis (DKA) in Patients with Type 1 Diabetes Mellitus (T1DM)

Indications for Test

Testing for ketones—3-beta-hydroxybutyrate (BHB) in blood or acetoacetate (AcAc) in urine—is indicated:

- In T1DM with any infection or severe illness
- If the blood glucose concentration is markedly elevated (e.g., > 300 mg/dL or 16.7 mmol/L) in a clinically ill patient with T1DM, usually with evidence of dehydration (i.e., rapid weight loss)
- When symptoms of vomiting, deep breathing, stomach ache, dry mouth or tongue, frequent urination, or fruity odor to the breath are noted
- When insulin doses have been omitted, the insulin used is old or expired, the insulin has been heated or freeze–thawed, or an insulin pump device fails or its catheter becomes blocked

BHB testing alone is indicated to:

- Monitor the progress of treatment for DKA

Procedure

1. Use a handheld device to test for blood [BHB] (e.g., Precision Xtra®) such that results are available quickly. *Blood* BHB testing is preferred over *urine* testing for AcAc by dipstick (Keto-Diastix®).
2. Do not measure for the presence of serum acetone as a way of monitoring the progress of treatment or resolution of acidosis in patients with DKA. Confine the use of serum acetone determination to the initial diagnosis of DKA only.
3. If BHB testing is not readily available, obtain a second-void urine specimen for AcAc testing to get a more current estimate of the degree of acidosis.

TABLE 5.8
Implications of Blood Beta-Hydroxybutyrate (BHB) and
Urine Acetoacetate (AcAc) Concentrations

[BHB] (mmol/L)	Qualitative Dipstick Reagent Strip Color and AcAc Level in a Second Void Urine Specimen		Implications
	Reagent Strip Color	[AcAc]	
<0.6	Slight change	Trace	Normal/negative for DKA
0.6–1.0	Purple	Small to moderate	At risk for DKA[a]
1.1–3.0	Dark purple	Large	Probable DKA
>3.0	Very dark purple	Very large	Critical DKA illness

[a] If blood glucose concentration < 150 mg/dL, the possibility of starvation ketosis can be determined by having the patient ingest carbohydrate and observing a reversal of the acidosis based on [BHB] or [AcAc].

Note: DKA, diabetic ketoacidosis.

Interpretation

1. Urine [AcAc] lags behind blood [BHB] and may give the false impression of worsening rather than resolving DKA as per "Implications of Blood Beta-Hydroxybutyrate (BHB) and Urine Acetoacetate (AcAc) Concentrations" (Table 5.8). On the other hand, children recovering from DKA may have continuing elevations of blood [BHB] after the urine becomes free of ketones evidencing a recurrence of DKA without ketonuria (Nadgir et al., 2001).
2. With effective insulin therapy, time to resolution of DKA may range from a few hours to more than 2 days.

Notes

1. In DKA, the body mobilizes free fatty acids for energy, resulting in a major increase in the blood BHB/AcAc ratio. Acetone production is derived from AcAc, and its levels increase later in the course of DKA. Thus, serum acetone reflects neither the early phase of DKA nor its resolution as it lags way behind the [BHB].
2. Measurement of [BHB] in blood is more accurate than urine AcAc testing and is preferred. Use of acetone test tablets (Acetest®) for measurement of serum acetone is even more inaccurate than urine testing and poorly reflects the clinically acidotic state.
3. Urine [AcAc] may be inaccurate because the test results are affected by urine concentration, the freshness of the test strip if the container was opened more than 6 months previously, medications such as captopril, and urine coloring agents such as phenazopyridine.
4. Urine [AcAc] may not reflect the current levels of ketones in the blood and may be inaccurate or be difficult to obtain if the patient is a small child or too ill and exhausted to accurately test a second-void specimen.
5. In 14 DM patients in DKA, the median time from the initiation of treatment to resolution of acidosis based on a [BHB] < 1 mmol/L was 8.46 hours (range, 5 to 58 hours) (Wallace et al., 2001).
6. In the treatment of DKA, once near-normoglycemia is achieved with insulin, clearance of BHB and hence resolution of the DKA can be achieved promptly. Measurement of [BHB] permits cost-effective assessment of DKA resolution and expeditious release of the patient from critical care settings.
7. In young people with T1DM, routine monitoring of [BHB] in the management of sick days can potentially reduce emergency visits and hospitalization for DKA compared with urine ketone testing and is more cost effective (Laffel et al., 2006).

ICD-9 Codes

Conditions that may justify this test *include but are not limited to*:

Diabetes
250.1   with ketoacidosis
250.2   with hyperosmolarity
250.3   with coma

Suggested Reading

Guerci B, Benichou M, Floriot M, Bohme P, Fougnot S, Franck P, Drouin P. Accuracy of an electrochemical sensor for measuring capillary blood ketones by fingerstick samples during metabolic deterioration after continuous subcutaneous insulin infusion interruption in type 1 diabetic patients. *Diabetes Care* 2003; 26:1137–41.

Laffel LM, Wentzell K, Loughlin C, Tovar A, Moltz K, Brink S. Sick day management using blood 3-hydroxybutyrate (3-OHB) compared with urine ketone monitoring reduces hospital visits in young people with T1DM: a randomized clinical trial. *Diabet Med* 2006; 23:278–84.

Nadgir UM, Silver FL, MacGillivray MH. Unrecognized persistence of beta-hydroxybutyrate in diabetic ketoacidosis. *Endocr Res* 2001; 27:41–6.

Vanelli M, Chiari G, Capuano C, Iovane B, Bernardini A, Giacalone T. The direct measurement of 3-beta-hydroxy butyrate enhances the management of diabetic ketoacidosis in children and reduces time and costs of treatment. *Diabetes Nutr Metab* 2003; 16:312–6.

Wallace TM, Meston NM, Gardner SG, Matthews DR. The hospital and home use of a 30-second hand-held blood ketone meter: guidelines for clinical practice. *Diabet Med* 2001; 18:640–5.

Wiggam MI, O'Kane MJ, Harper R, Atkinson AB, Hadden DR, Trimble ER, Bell PM. Treatment of diabetic ketoacidosis using normalization of blood 3-hydroxybutyrate concentration as the endpoint of emergency management: a randomized controlled study. *Diabetes Care* 1997; 20:1347–52.

5.4.2   Blood pH in Diabetic Ketoacidosis (DKA) and Hyperglycemic Hyperosmolar Nonketoacidosis (HHNK) Syndrome

Indications for Test

A blood pH determination is indicated in the evaluation of the critically ill T1DM patient who:

- Becomes confused, delirious, and dehydrated
- Experiences deep prolonged expiratory phase (Kussmall) hyperventilatory breathing with a fruity odor noted on the breath
- Has markedly elevated blood glucose concentrations and develops a combination of the signs and symptoms of DKA including vomiting, diarrhea, and/or complaints of nausea, abdominal pain, and a cold or clammy sensation

Procedure

1. Observe the critically ill T1DM patient for signs of tachycardia, nature of breathing, level of mentation, and signs of dehydration.
2. In the critically ill T1DM patient, obtain a random blood sample for glucose testing to identify a glucose level > 250 mg/dL.
3. In the hyperglycemic, critically ill T1DM patient, obtain an arterial (or venous) blood sample for blood pH testing.
4. If the blood pH < 7.30, obtain a blood sample for beta-hydroxybutyrate testing (see Test 5.4.1).

Interpretation

1. A random plasma glucose level of >250 mg/dL (13.9 mmol/L), arterial blood pH < 7.30, and serum found to be positive for ketones on screening (see Test 5.4.1) are diagnostic of DKA in the critically ill T1DM patient.
2. In HHNK, the clinical signs and symptoms are similar to those of DKA but acidosis is not present. Moreover, the blood glucose level is usually higher (700–1000 mg/dL) than in DKA (600–800 mg/dL).
3. Because the pH of arterial and venous blood is highly correlated, either may be used for the diagnosis of acidosis.
4. Six variables have been found to be significant independent predictors ($p < 0.05$) of mortality in DKA:
   - Severe coexisting diseases at presentation
   - pH < 7.0 at presentation
   - ≥50 units of insulin required in the first 12 hours of treatment
   - Serum glucose >300 mg/dL (16.7 mmo/L) after 12 hours of therapy
   - Depressed mental state persistent at 24 hours from diagnosis
   - Fever persisting or developing after 24 hours of therapy

Notes

1. Patients with HHNK syndrome may have less severe abdominal pain than patients with DKA.
2. Treatment of DKA with intravenous administration of bicarbonate is most appropriately reserved for occasions when the blood pH < 7.1.
3. Hydration alone can frequently reverse mild to moderate acidotic states.
4. Expect a mean and maximum difference between arterial and venous pH values of about 0.03 and 0.11 units, respectively.
5. Venous blood pH correlates well with arterial blood pH ($r = 0.951$ to 0.969). Arterial and venous blood HCO_3^- levels are also highly correlated ($r = 0.9543$) (Brandenburg and Dire, 1998).

ICD-9 Codes

Refer to Test 5.4.1 codes.

Suggested Reading

Brandenburg MA, Dire DJ. Comparison of arterial and venous blood gas values in the initial emergency department evaluation of patients with diabetic ketoacidosis. *Ann Emerg Med* 1998; 31:459–65.

Efstathiou SP, Tsiakou AG, Tsioulos DI et al. A mortality prediction model in diabetic ketoacidosis. *Clin Endocrinol (Oxf)* 2002; 57:595–601.

Ma OJ, Rush MD, Godfrey MM, Gaddis G. Arterial blood gas results rarely influence emergency physician management of patients with suspected diabetic ketoacidosis. *Acad Emerg Med* 2003; 10:836–41.

5.5   Hepatopathy as a Complication of Diabetes Mellitus

5.5.1   Aspartate Aminotransferase (AST) and Alanine Aminotransferase (ALT) in Screening for Hepatic Steatosis, Nonalcoholic Fatty Liver Disease (NAFLD), and Nonalcoholic Steatohepatitis or Steatohepatopathy (NASH) in the Obese Diabetes Mellitus (DM) Patient*

Indications for Test

Measurement of AST and ALT is indicated in overweight and obese T2DM patients with:

---

* Reviewed by Dr. Eric J Lawitz, Medical Director, Gastroenterology/Hepatology, Alamo Medical Research, San Antonio, Texas.

TABLE 5.9
Guidelines for Interpreting AST, ALT, AST/ALT Ratio, and Fat Fraction (FF) Values

Test	Mean	Range	Possible Interpretation
Aspartate aminotransferase	66	31–74	Nonalcoholic fatty liver disease (NAFLD)
(AST) (U/L)	152	58–222	Alcoholic liver disease (ALD)
Alanine aminotransferase	M, <30 (500 nkat/L)[a–c]	—	
(ALT) (U/L)	F, <19 (317 kat/L)[a–c]		
	>40	—	Steatosis (sensitivity, 45%; specificity, 100%)[d]
	91	52–118	NAFLD
	70	10–91	ALD
AST/ALT ratio	0.9	1.0–2.0	Typical of individuals without liver disease
		0.3–2.8	NAFLD
		1.1–11.2	ALD
	2.0[e]	—	Alcoholic hepatitis (AH)
Fat fraction (%)	≤18[f]	—	ALT within normal limits (WNL)
	>18	—	ALT > upper limit of normal (ULN)

[a] These values are significantly lower than previously thought (Prati et al., 2002).

[b] Baseline elevations of ALT and AST above the upper limit of the reference interval suggest the presence of NAFLD or other hepatic pathologies such as alcohol-related or infectious hepatitis in blood donors or historical intravenous drug users.

[c] Most subjects with NAFLD have an ALT value of <250 U/L, and values over 300 U/L warrant a search for additional or alternative causes of ALT elevation.

[d] In morbidly obese individuals undergoing bariatric surgery.

[e] Median ratio value in individuals with NASH = 0.7.

[f] Fat fraction measured using PMR; hepatic triglyceride content was found to vary over a wide range (0.0–41.7%; median, 3.6%) in a multiethnic, population-based sample of 2287 subjects (Browning et al., 2004).

- Dyslipidemia, hypertension, cardiovascular disease, or liver disease with or without a history of drug or alcohol abuse
- Possible liver disease associated with the cardiometabolic syndrome or early DM, including NAFLD, as well as differentiating alcoholic hepatitis from NASH and NAFLD

Procedure

1. Obtain the patient's history of both current and former moderate to heavy alcohol drinking and illegal drug use. Note intake of oral agents for treatment of hyperlipidemia and DM.
2. Measure the patient's waist circumference and examine the patient's abdomen for the size and palpability of the liver and spleen.
3. Obtain a random blood sample for AST and ALT testing, including calculation of the AST/ALT ratio. If both the AST and ALT values are elevated, request testing for γ-glutamyltransferase (GGT).
4. Obtain a fasting blood sample for triglyceride, high-density lipoprotein cholesterol (HDL-C), and glucose testing.

Interpretation

1. Refer to "Guidelines for Interpreting AST, ALT, AST/ALT Ratio, and Fat Fraction (FF) Values" (Table 5.9). (Note that these guidelines represent a summary of reports taken from peer-reviewed scientific literature and are presented as the authors' own synthesis of the data.)

Notes

1. NAFLD is defined as the spectrum of histology associated with significant lipid deposition in the hepatocytes of patients without a history of excessive alcohol ingestion. The definitive diagnosis of NAFLD is based on histologic examination of liver biopsy samples.

2. The AST/ALT ratio, albumin/gamma-globulin ratio, and GGT level do not identify patients with severe chronic liver damage with enough sensitivity to avoid liver biopsy for the specific diagnosis and staging of these forms of liver disease.

3. Severe steatosis and bridging fibrosis on liver biopsy have been associated with the cardiometabolic syndrome.

4. When entered into the same model with adjustment for demographic variables, both C-reactive protein (CRP) and ALT independently predicted onset of T2DM.

5. Both AST and ALT were positively associated with incident T2DM after excluding former and moderate to heavy drinkers (Hanley et al., 2004). A multivariate analysis suggested that associations between liver enzyme markers and risk of developing T2DM were independent of directly measured insulin sensitivity. Odds ratios (ORs) for this risk were adjusted for age, gender, clinical center, ethnicity, and alcohol intake. Quartile ranges (means) were as follows (see Figure 5.1):
   - AST—Q1 = 6 to 16 (14.0); Q2 = 17 to 20 (18.5); Q3 = 21 to 25 (22.9); Q4 ≥ 25 (41.2) units/L
   - ALT—Q1 = 0 to 11 (8.1); Q2 = 12 to 16 (14.0); Q3 = 17 to 23 (19.5); Q4 ≥ 23 (44.7) units/L
   - ALK—Q1 = 23 to 50 (42.6); Q2 = 51 to 61 (56.0); Q3 = 62 to 74 (67.5); Q4 ≥ 74 (93.2) units/L

6. The mean AST/ALT ratio of <1.0 in the fourth quartile group is consistent with nonalcoholic steatohepatitis (NASH).

7. A longitudinal study of Pima Indians reported that ALT was a significant predictor of DM in these individuals over an average follow-up of 7 years, with the association remaining significant after adjustment for percentage of body fat and direct measures of insulin sensitivity and secretion.

8. Serum ALT activity is independently related to body mass index and to laboratory indicators of abnormal lipid or carbohydrate metabolism.

9. Juvenile obesity involves a high risk of liver steatosis, the degree of which correlates with increases in AST, ALT, GGT, and lipids but not abnormalities of glucose metabolism (Guzzaloni et al., 2000).

10. Hepatic fat has been associated with defects in the suppression, by insulin, of endogenous glucose production.

11. Increased high-sensitivity C-reactive protein (hsCRP) and liver marker enzyme concentrations predicted T2DM independent of each other with associations of similar magnitude.

12. In patients with a history of hepatitis or intravenous drug use, screening for hepatitis C should be undertaken regardless of ALT or AST levels.

ICD-9 Codes

Conditions that may justify this test *include but are not limited to*:

Liver disease
571.1   acute alcoholic
571.8   chronic nonalcoholic

Other
250.XX diabetes mellitus
272.4   dyslipidemia
277.7   dysmetabolic syndrome
305.0   alcohol abuse
429.2   cardiovascular disease
997.91 hypertension
V77.8   obesity

Suggested Reading

Angelico F, Del Ben M, Conti R, Francioso S, Feole K, Maccioni D, Antonini TM, Alessandri C. Non-alcoholic fatty liver syndrome: a hepatic consequence of common metabolic diseases. *J Gastroenterol Hepatol* 2003; 18:588–94.

Angulo P, Lindor KD. Non-alcoholic fatty liver disease. *J Gastroenterol Hepatol* 2002; 17(Suppl.):S186–90.

Bajaj M, Suraamornkul S, Pratipanawatr T, Hardies LJ, Pratipanawatr W, Glass L, Cersosimo E, Miyazaki Y, DeFronzo RA. Pioglitazone reduces hepatic fat content and augments splanchnic glucose uptake in patients with type 2 diabetes. *Diabetes* 2003; 52:1364–70.

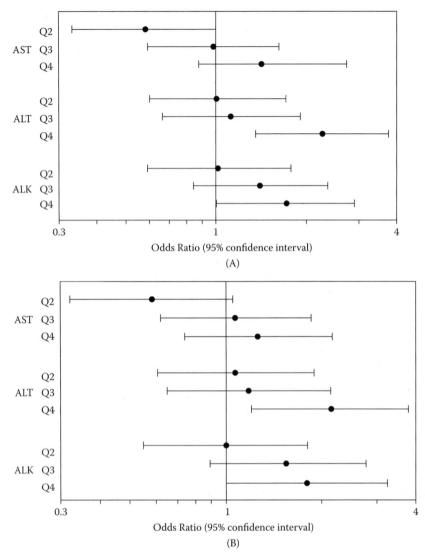

Figure 5.1
Liver enzymes—aspartate aminotransferase (AST), alanine aminotransferase (ALT), alkaline phosphatase (ALK)—and risk of developing diabetes mellitus (DM) in: (A) a large cohort of individuals with no exclusions for former or current level of alcohol intake, and (B) this same cohort of individuals excluding participants who were either former drinkers or moderate/heavy drinkers. Odds ratios (ORs) and 95% confidence intervals (CIs) for individuals in Q2, Q3, and Q4 of each marker are in relation to individuals in the lowest (i.e., reference) quartile (Q1). Quartile ranges (means) were as follows:

Enzyme (U/L)	Q1 (Mean)	Q2 (Mean)	Q3 (Mean)	Q4 (Mean)
AST	6–16 (14)	17–20 (18.5)	21–25 (22.9)	>25 (41.2)
ALT	0–11 (8.1)	12–16 (14)	17–23 (19.5)	>23 (44.7)
ALK	23–50 (42.6)	51–61 (56)	62–74 (67.5)	>74 (93.2)

Source: Hanley, A.J. et al., *Diabetes*, 53, 2623–32, 2004. With permission. Copyright © 2007 American Diabetes Association.

Clark JM, Diehl AM. Nonalcoholic fatty liver disease: an underrecognized cause of cryptogenic cirrhosis. *JAMA* 2003; 289:3000–4.

Fishbein MH, Miner M, Mogren C, Chalekson J. The spectrum of fatty liver in obese children and the relationship of serum aminotransferases to severity of steatosis. *J Pediatr Gastroenterol Nutr* 2003; 36:54–61.

Guzzaloni G, Grugni G, Minocci A, Moro D, Morabito F. Liver steatosis in juvenile obesity: correlations with lipid profile, hepatic biochemical parameters and glycemic and insulinemic responses to an oral glucose tolerance test. *Int J Obes Relat Metab Disord* 2000; 24:772–6.

Hanley AJ, Williams K, Festa A, Wagenknecht LE, D'Agostino RB, Kempf J, Zinman B, Haffner SM. Elevations in markers of liver injury and risk of type 2 diabetes: the insulin resistance atherosclerosis study. *Diabetes* 2004; 53:2623–32.

Mulhall BP, Ong JP, Younossi ZM. Non-alcoholic fatty liver disease: an overview. *J Gastroenterol Hepatol* 2002; 17:1136–43.

Neuschwander-Tetri BA, Brunt EM, Wehmeier KR, Oliver D, Bacon BR. Improved nonalcoholic steatohepatitis after 48 weeks of treatment with the PPAR-gamma ligand rosiglitazone. *Hepatology* 2003; 38:1008–17.

Nyblom H, Berggren U, Balldin J, Olsson R. High AST/ALT ratio may indicate advanced alcoholic liver disease rather than heavy drinking. *Alcohol Alcohol* 2004; 39:336–9.

Papadia FS, Marinari GM, Camerini G, Murelli F, Carlini F, Stabilini C, Scopinaro N. Liver damage in severely obese patients: a clinical-biochemical-morphologic study on 1000 liver biopsies *Obes Surg* 2004; 14:952–8.

Prati D, Taioli E, Zanella A, Della Torre E, Butelli S, Del Vecchio E, Vianello L, Zanuso F, Mozzi F, Milani S, Conte D, Colombo M, Sirchia G. Updated definitions of healthy ranges for serum alanine aminotransferase levels. *Ann Intern Med* 2002; 137:1–10.

Seppala-Lindroos A, Vehkavaara S, Hakkinen AM, Goto T, Westerbacka J, Sovijarvi A, Halavaara J, Yki-Jarvinen H. Fat accumulation in the liver is associated with defects in insulin suppression of glucose production and serum free fatty acids independent of obesity in normal men. *J Clin Endocrinol Metab* 2002; 87: 3023–8.

Sorbi D, Boynton J, Lindor KD. The ratio of aspartate aminotransferase to alanine aminotransferase: potential value in differentiating nonalcoholic steatohepatitis from alcoholic liver disease. *Am J Gastroenterol* 1999; 94:1018–22.

Vozarova B, Stefan N, Lindsay RS, Saremi A, Pratley RE, Bogardus C, Tataranni PA. High alanine aminotransferase is associated with decreased hepatic insulin sensitivity and predicts the development of type 2 diabetes. *Diabetes* 2002; 51:1889–95.

Zamin JI, de Mattos AA, Perin C, Ramos GZ. The importance of AST/ALT rate in nonalcoholic steatohepatitis diagnosis. *Arq Gastroenterol* 2002; 39:22–6.

5.5.2 Liver Ultrasonography (USG), Computed Tomography (CT) Scanning, and Proton Magnetic Resonance (PMR) Spectroscopy in Diagnosis of Nonalcoholic Fatty Liver Disease (NAFLD) and Nonalcoholic Steatohepatitis or Steatohepatopathy (NASH) in the Overweight Diabetes Mellitus (DM) Patient

Indications for Test

Imaging of the liver is indicated in overweight and obese T2DM patients with:

- Dyslipidemia, hypertension, cardiovascular disease, or liver disease with or without a history of drug or alcohol abuse
- Possible liver disease associated with the cardiometabolic syndrome including NAFLD, as well as to differentiate alcoholic hepatitis from NASH and NAFLD
- An abnormally low AST/ALT ratio

Procedure

1. Obtain an USG of the liver as a screening procedure.
2. Obtain a CT scan of the liver as a more quantitative, readily available procedure.

3. Obtain a PMR spectroscopic study of the liver and estimate the patient's fat fraction using a modification of the Dixon method in which a fast gradient echo sequence allows imaging within a single breath hold.

Interpretation

1. Estimates of fat in the liver from ultrasound studies are qualitative and may underestimate the amount of hepatic fat.
2. Regional areas throughout the liver can be isolated for study using CT imaging, and a determination of the density of these areas can be made in Hounsfield units, with lower values (<10) indicating more fat.
3. PMR spectroscopy signals quantitatively detect intrahepatocellular lipids (IHCLs) which are significantly greater in subjects with hepatic steatosis (Table 5.9).

Notes

1. Obesity, DM, and hypertriglyceridemia were more common by 5.3-, 4.0-, and 6.7-fold, respectively, in subjects with severe hepatic steatosis at ultrasonography, compared to controls.
2. NAFLD diagnosed based on liver ultrasound study of T2DM had a prevalence of almost 70%. Those with NAFLD had higher prevalences of coronary artery, cerebrovascular, and peripheral vascular disease by about 50% (Targher et al., 2007)
3. Findings of a fatty liver on fast scan MR imaging correlate better with liver enzyme abnormalities than do findings on liver ultrasonography.
4. Based on PMR imaging, the frequency of NAFLD varied significantly with ethnicity (Hispanics, 45%; Caucasians, 33%; African-Americans, 24%) and sex (Caucasian men, 42%; Caucasian women, 24%).
5. The higher prevalence of hepatic steatosis in Hispanics may be the result of a higher prevalence of obesity and insulin resistance in this ethnic group (Browning et al., 2004).
6. IHCLs by PMR were found to be significantly greater in patients with biopsy-proven hepatic steatosis (geometric mean [GM], 11.5; interquartile range [IQR], 7.0–39.0) than in normal volunteers (GM, 2.7; IQR, 0.7–9.3; $p = 0.02$). In overweight subjects (body mass index [BMI] > 25 kg/m^2, $n = 23$), the GM was 7.7 (IQR, 4.0–28.6); in lean subjects (BMI < 25 kg/m^2, $n = 11$), the GM was 1.3 (IQR, 0.3–3.6; $p = 0.004$) (Thomas et al., 2005).
7. Most individuals with NAFLD in its uncomplicated form (simple steatosis) are asymptomatic. Four fifths of those with hepatic steatosis by PMR imaging have normal levels of serum ALT (Browning et al., 2004).
8. Clinical trial data have shown improvements in NASH and reductions in liver fat content after treatment with insulin-sensitizing agents, including the thiazolidinediones and metformin. Other agents such as 3-hydroxy-3-methylglutaryl coenzyme A inhibitors, or orlistat, may also reduce fatty infiltrates in the liver.

ICD-9 Codes

Refer to Test 5.5.1 codes.

Suggested Reading

Browning JD, Szczepaniak LS, Dobbins R, Nuremberg P, Horton JD, Cohen JC, Grundy SM, Hobbs HH. Prevalence of hepatic steatosis in an urban population in the United States: impact of ethnicity. *Hepatology* 2004; 40:1387–95.

Targher G, Bertolini L, Padovani R, Rodella S, Tessari R, Zenari L, Day C, Arcaro G. Prevalence of nonalcoholic fatty liver disease and its association with cardiovascular disease among type 2 diabetic patients. *Diabetes Care*, 2007; 30:1212–18.

Thomas EL, Hamilton G, Patel N, O'Dwyer R, Doré CJ, Goldin RD, Bell JD, Taylor-Robinson SD. Hepatic triglyceride content and its relation to body adiposity: a magnetic resonance imaging and proton magnetic resonance spectroscopy study. *Gut* 2005; 54:122–7.

Chapter 6

Adrenal Gland

6.1 Glucocorticoid Reserve

6.1.1 Cosyntropin Stimulation Test (CST): Short or Rapid with High-Dose (250-mcg) Cosyntropin in Primary Adrenal Insufficiency (1°AD) and in Critical Illness

Indications for Test

The high-dose CST is indicated in:

- Screening for primary hypoadenalism when fatigue, hypotension, electrolyte imbalances (i.e., hyperkalemia), hyperpigmentation, muscle weakness, pediatric hypoglycemia, or weight loss is found (Algorithm 6.1)
- Patients who have recently been treated with short-term suppressive doses of glucocorticoids and have symptoms consistent with steroid withdrawal (e.g., orthostatic hypotension)
- Patients who received abdominal irradiation, usually years previously
- Patients with one or more autoimmune diseases (e.g., thyroid dysfunction, diabetes, lupus) sometimes associated with primary autoimmune Addison's disease (1°AD)
- Critical illness, when attempting to determine prognosis

Procedure

1. Withhold prednisone therapy for at least 24 hours before performing this test. If prednisone or other corticosteroids are in use, do *not* do this test.
2. Obtain informed consent before performing this invasive test.
3. Insert a peripheral vein catheter for blood sampling at least 20 minutes before the CST.
4. Obtain blood samples for cortisol levels at -15 (T_{-15}) and -10 (T_{-10}) before administration of cosyntropin (baseline, T_0).
5. Administer high-dose cosyntropin (250 mcg i.v. or i.m.).
6. Obtain blood samples for cortisol levels at 30 (T_{30}) and 60 (T_{60}) minutes after the cosyntropin injection is given at 0 minutes (T_0).
7. Choose as T_0 the higher of the two cortisol value at T_{-15} and T_{-10}.
8. Determine the change (Δ) in cortisol values between T_{30} and T_0 and between T_{60} and T_0:
   - $\Delta_{30} = ([\text{cortisol}] @ T_{30}) - ([\text{cortisol}] @ T_0)$.
   - $\Delta_{60} = ([\text{cortisol}] @ T_{60}) - ([\text{cortisol}] @ T_0)$.
   - Δ_{max} = highest change in [cortisol] (i.e., Δ_{30} or Δ_{60}, whichever is the greater value).

Typical Symptoms and Clinical Findings in Addison's Disease (AD)

1° AD	Both 1° and 2° AD	
Hyperpigmentation	Fatigue	Hyponatremia
Salt craving	Weakness	Hypoglycemia
Marked orthostatic hypotension	Listlessness	Hyperkalemia
	Orthostatic dizziness	Metabolic acidosis
Pallor	Weight loss	Eosinophilia
Thinning of axillary and pubic hair	Anorexia	Lymphocytosis
	Erectile dysfunction	Hypercalcemia
	Decreased libido	
	Amenorrhea	
	G.I. complaints	

```
                        Obtain AM Cortisol

      ┌──────────────────────┼──────────────────────┐

   ≤3 µg/dL        >3 µg/dL but <19 µg/dL         ≥19 µg/dL

  Clinical AD                                      AD ruled out
  ruled out            ACTH ≥100 pg/mL

                  ┌────────────┴────────────┐
                Yes                        No

               1° AD              Cortrosyn Stimulation Test
                                      (1 µg or 250 µg)

                                  60' Cortisol ≥18 µg/dL (1 µg test)
                                  60' Cortisol ≥20 µg/dL (250 µg test)

                               ┌────────────┴────────────┐
                             Yes                        No

                          AD unlikely             AD likely; search
                                                     for cause
```

Algorithm 6.1

Identification of patients with typical symptoms or clinical findings suggestive of primary (1°) or secondary (2°) Addison's disease (AD).

Interpretation

1. Peak serum [cortisol] within 60 minutes after administration of 250 mcg cosyntropin:
   - ≥20 mcg/dL (552 nmol/L) is consistent with adrenal sufficiency.
   - <9 mcg/dL is suggestive of primary adrenal failure or adrenal glands inadequately stimulated by pituitary adrenocorticotropic hormone (ACTH).
2. In critically ill (e.g., septic shock) patients, if:
   - Baseline [cortisol] ≤ 34 mcg/dL and Δ_{max} > 9 mcg/dL, then 28-day patient mortality is 26%.
   - Baseline [cortisol] > 34 mcg/dL and Δ_{max} ≤ 9 mcg/dL, then 28-day patient mortality is 82%.

Notes

1. Cosyntropin is a 24-amino-acid-containing polypeptide homologous with the *N*-terminal first 24 amino acids of native adrenocorticotropic hormone, a 39-amino-acid-containing peptide. Its commercial name is Cortrosyn®.
2. Repeating the high-dose CST in the same patient does not improve diagnostic accuracy.
3. Up to 10% of high-dose CST results represent false positives even in cases of primary hypoadrenalism. False-negative results occur even more frequently in secondary adrenal insufficiency secondary to hypothalamic or pituitary dysfunction; therefore, when evaluating patients for hypothalamic–pituitary–adrenal insufficiency, the low-dose (1-mcg) CST (see Test 2.1.2) is preferred over the high-dose (250-mcg) CST
4. Because of its greater sensitivity and accuracy, the low-dose 1-mcg CST has been proposed as a replacement for the high-dose 250-mcg CST in screening for 1°AD.
5. Because cortisol assays are influenced by prednisone, but not dexamethasone, it is inappropriate to attempt measurement of cortisol levels in patients on prednisone therapy.
6. 1°AD may be present even in patients whose [ACTH] is not clearly elevated and no other hypothalamic/pituitary disturbance is present.
7. [ACTH] may be normal (i.e., not suppressed) even if the adrenal glands are chronically suppressed by glucocorticoids.

ICD-9 Codes

Conditions that may justify this test *include but are not limited to*:

255.4	corticoadrenal insufficiency
276.7	hyperkalemia
458.9	hypotension, unspecified
728.87	muscle weakness (generalized)
780.7	malaise and fatigue
783.21	weight loss

Suggested Reading

Annane D, Sebille V, Troche G, Raphael JC, Gajdos P, Bellissant E. A 3-level prognostic classification in septic shock based on cortisol levels and cortisol response to corticotropin. *JAMA* 2000; 283:1038–1045; 284:308–9.

Beishuizen A, van Lijf JH, Lekkerkerker JF, Vermes I. The low-dose (1 microg) ACTH stimulation test for assessment of the hypothalamo–pituitary–adrenal axis. *Neth J Med* 2000; 56:91–9.

May ME, Carey RM. Rapid adrenocorticotropic hormone test in practice: retrospective review. *Am J Med* 1985; 79:679–84.

Tordjman K, Jaffe A, Trostanetsky Y, Greenman Y, Limor R, Stern N. Low-dose (1 microgram) adrenocorticotrophin (ACTH) stimulation as a screening test for impaired hypothalamo–pituitary–adrenal axis function: sensitivity, specificity, and accuracy in comparison with the high-dose (250 microgram) test. *Clin Endocrinol (Oxf)* 2000; 52:633–40.

6.1.2 Cortisol and ACTH Testing: Early-Morning (9 a.m.) Test Panel for Primary Adrenal Insufficiency (1°AD)

Indications for Test

A morning cortisol test or simultaneously obtained tests for morning cortisol and ACTH is indicated:

- As a screening test for 1°AD

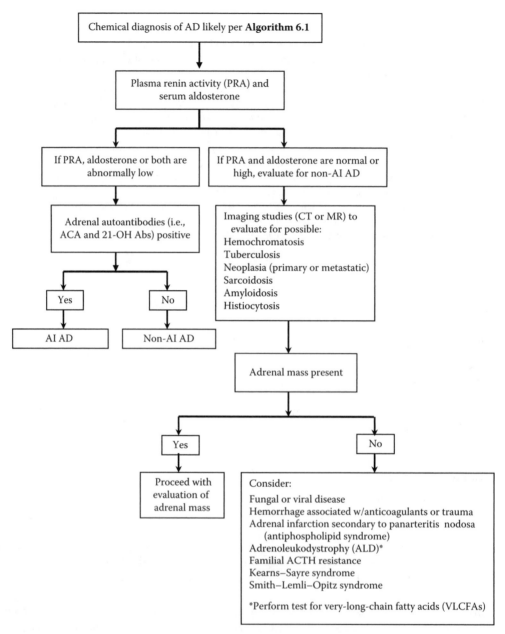

Algorithm 6.2
Testing to determine the cause of Addison's disease (AD) in patients with a positive chemical diagnosis of AD. AI, autoimmune.

Procedure

1. Refer to Algorithm 6.2.
2. Obtain an early-morning (e.g., 9 a.m.) blood sample for cortisol testing as a confirmatory test in patients suspected of having overt hypoadrenalism.
3. In patients in whom adrenal status is less certain, obtain early-morning simultaneous blood samples for baseline cortisol and ACTH testing.
4. In patients for whom the baseline cortisol and ACTH baseline results are equivocal, discordant, or not consistent with the clinical presentation of the patient, proceed to a high-dose CST (see Test 6.1.1).

Interpretation

1. If the morning [cortisol] < 10 mcg/dL *and* a simultaneous [ACTH] > 250 pg/dL, then the diagnosis of 1°AD is established and no further tests of the hypothalamic–pituitary–adrenal (HPA) axis are indicated.
2. If the morning [cortisol] or [ACTH] does not meet the criteria indicated above, a Cortrosyn® stimulation test (high or low dose, depending on the patient's history of pituitary or primary adrenal abnormality) is warranted to localize the defect in the HPA axis.

Notes

1. 1°AD is usually sporadic, caused by autoimmune processes affecting the adrenal cortices or the result of infections such as tuberculosis that impair the function of both adrenal glands.
2. Radiation therapy to the abdomen or pelvis may result in primary hypoadrenalism many years later.
3. The differential diagnosis of adrenal insufficiency includes unusual congenital conditions such as X-linked adrenoleukodystrophy (ALD), autoimmune polyendocrinopathy–mucocutaneous candidiasis–ectodermal dystrophy (APECED) syndrome, autoimmune polyglandular syndrome type 2 (APS-2), and achalasia cardia, alacrima, and 1°AD (triple A syndrome).
4. There is no absolute morning [cortisol] that distinguishes an adequate from an insufficient adrenal response in patients with septic shock; however, a random [cortisol] of <25 mcg/dL in such highly stressed patients may be a useful in the preliminary diagnosis of adrenal insufficiency (Marik and Zaloga, 2003).
5. Nearly 40% of critically ill patients with hypoalbuminemia have subnormal serum *total* [cortisol]. Request measurement of the biologically active *free* [cortisol] in the hypoalbuminemic patient and avoid unnecessary glucocorticoids therapy in euadrenal patients (Hamrahian et al., 2004).

ICD-9 Codes

Refer to Test 6.1.1 codes.

Suggested Reading

Hamrahian AH, Oseni TS, Arafah BM. Measurement of serum free cortisol in critically ill patients. *N Engl J Med* 2004; 350:1629–38.

Loriaux DL, McDonald, WJ. Adrenal insufficiency. In: *Endocrinology*, 3rd ed., DeGroot et al. (Eds). 1995. WB Saunders, pp. 1731–40.

Marik PE, Zaloga GP. Adrenal insufficiency during septic shock. *Crit Care Med* 2003; 31:141–5.

6.1.3   24-Hour Urinary Free Cortisol (UFC) Excretion Test to Monitor the Treatment of Hypoadrenalism

Indications for Test

A UFC excretion test is indicated when:

- Making adjustments in corticosteroid doses, usually for the purpose of avoiding overdose in patients on hydrocortisone or prednisone therapy

Procedure

1. Have the patient collect a 24-hour urine specimen (Specimen Collection Protocol P1 in Appendix 2) for UFC and urine creatinine determinations.
2. Specifically request that UFC testing be performed by high-performance liquid chromatography (HPLC) or tandem mass spectrometry (MS/MS).

Interpretation

1. A UFC excretion rate near or above the upper limit of the reference interval may be an indication to reduce the replacement dose of hydrocortisone therapy in patients with adrenal insufficiency.

Notes

1. A high proportion of patients on conventional corticosteroid replacement therapy (e.g., 20 mg hydrocortisone p.o. in the morning and 10 mg p.o. in the evening) are overtreated or on inappropriately excessive catabolic steroid "replacement" regimens and suffer the long-term risks of osteoporosis, diabetes, and infections.
2. The 24-hour UFC excretion rate progressively decreases with increasing division of the same dose of hydrocortisone given once to twice to five times a day.
3. HPLC/MS/MS methods for determination of UFC are not affected by interferents that affect immunoassay methods, with or without a specimen prepurification step prior to immunoassay testing. Moreover, non-immunoassay-based methods have higher analytical precision (i.e., lower interassay coefficient of variation [CV] than immunoassay methods) and can detect the presence of surreptitious corticosteroid use.

ICD-9 Codes

Refer to Test 6.1.1 codes.

Suggested Reading

Bliesener N, Steckelbroeck S, Redel L, Klingmuller D. Dose distribution in hydrocortisone replacement therapy has a significant influence on urine free cortisol excretion. *Exp Clin Endocrinol Diabetes* 2003; 111:443–6.

Peacey SR, Guo CY, Robinson AM, Price A, Giles MA, Eastell R, Weetman AP. Glucocorticoid replacement therapy: are patients over treated and does it matter? *Clin Endocrinol (Oxf)* 1997; 46:255–61; 269–70 [comment].

6.1.4 Very-Long-Chain Fatty Acids (VLCFAs) Testing in Adrenoleukodystrophy (ALD)

Indications for Test

Measurement of VLCFAs (C22 to C27) is indicated:

- When ALD is suspected as a cause of primary Addison's disease (1°AD) in the context of normal-sized adrenals
- When a patient has a family history of ALD and associated clinical findings such as auditory processing difficulties
- To screen neonates suspected of adrenal disease secondary to ALD as per "Features and Clinical Criteria for Diagnosis of Adrenoleukodystrophy (ALD)" (Table 6.1)

Procedure

1. Obtain a randomly obtained blood sample for VLCFA testing, including docosanoic acid (C22:0), tetracosanoic acid (C24:0), and hexacosanoic acid (C26:0), and the ratios of C26:0/C24:0, C24:0/C22:0, and C26:0/C22:0.

TABLE 6.1
Features and Clinical Criteria for Diagnosis of Adrenoleukodystrophy (ALD)

Forms	Non X-linked neonatal onset (severe)
	X-linked childhood onset (classic form associated with severe signs and symptoms)
	X-linked adult onset[a] (less severe than classic form)
Primary features	Loss of myelin in brain and peripheral nerves and progressive dysfunction of the adrenal gland (i.e., adrenal failure)
Neonatal signs and symptoms	Both sexes affected by mental retardation, facial abnormalities, seizures, retinal degeneration, weak muscle tone, enlarged liver, adrenal dysfunction
Classic signs of primary adrenal and neurologic dysfunction in childhood	Increased skin pigmentation, intermittent vomiting, hyperkalemia, disturbances of gait and coordination, seizures, poorly articulated speech, difficulty swallowing, deafness
Classic symptoms in childhood	Fatigue, learning disabilities, poor memory and school performance, visual loss, progressive dementia, behavioral changes (abnormal withdrawal or aggression)
Adult male onset signs and symptoms	Progressive stiffness, weakness or paralysis of the lower limbs, and ataxia with slowly progressive deterioration of brain function

[a] Female carriers may have progressive stiffness, weakness of the lower limbs, ataxia, excessive muscle tone, mild peripheral neuropathy, and bladder control problems. Adrenal dysfunction may be subclinical.

Interpretation

1. Ordinarily, elevated C24:0/C22:0 and C26:0/C22:0 ratios identify patients with ALD; however, the finding of nondiagnostic concentrations of VLCFAs may still be associated with all of the clinical features of X-linked ALD.
2. Increased VLCFA levels may be found in patients homozygous for adult and neonatal ALD, Zellweger syndrome, and infantile Refsum's disease and in patients with deficiencies of peroxisomal acyl–coenzyme A (CoA) oxidase, bifunctional enzyme, and 3-oxoacyl-coenzyme A thiolase.

Notes

1. X-linked ALD is a neurometabolic disorder associated with elevated levels of saturated unbranched VLCFA of chain length >22:0 in plasma and tissues, with reduced VLCFA beta-oxidation in fibroblasts, white blood cells, and amniocytes.
2. Clinically, three distinct forms of X-linked ALD are recognized:
   • A classic childhood cerebral form characterized by demyelinization, neurologic signs, and adrenal insufficiency affects 35 to 40% of patients
   • An adult-onset form (adrenomyeloneuropathy) having a slow progression and seldom involving the central nervous system affects 45 to 50% of patients
   • A milder form, with adrenal insufficiency only, affecting 5 to 7% of all ALD patients
3. VLCFA levels are no different in individuals with the childhood and adult forms of ALD and do not change with age. Because VLCFA levels are elevated on the day of birth, there is the potential for neonatal mass screening for this disorder.
4. Impaired adrenal function was found in 39/49 (80%) patients (age, 4.5 ± 3.5 years) identified with ALD based on screening with VLCFA testing. Of these, ACTH levels were elevated in 34 (69%), while the high-dose CST was abnormal in only 21 (43%). Adrenal steroid replacement therapy is effective for the treatment of the adrenal deficiency of ALD but does not affect the neurodegenerative component of ALD.
5. Multifunctional protein 2 (MFP-2) catalyzes the breakdown of VLCFAs, while normal metabolism of long, straight-chain fatty acids, such as palmitic acid (C16), is accomplished by the enzymes of the classical peroxisomal beta-oxidation pathway (Baes et al., 2000).

6. In humans, ALD might not be caused by peroxisomal dysfunction directly but rather by a primary defect in mitochondrial processing of long-chain fatty acids (LCFAs), plus a reduced capacity of fatty acyl–CoA synthetase to metabolize VLCFAs.
7. Saturated VLCFAs accumulate in X-linked ALD, appear to disrupt membrane structure, and may play a role in the pathogenesis of the inflammatory responses in the brain observed in some patients with ALD.
8. Dietary therapy with 4:1 mixture of glyceryl trioleate and glyceryl trierucate (Lorenzo's oil) initiated as soon as possible when ALD patients are still asymptomatic appears to reduce the risk of developing brain MR abnormalities by significantly reducing C26:0 concentrations (Moser et al., 2005).
9. Up to 35% of persons surveyed with idiopathic non-autoimmune adrenal insufficiency had biochemical or genetic evidence for affliction with ALD (Dubey et al., 2005).

ICD-9 Codes

Conditions that may justify this test *include but are not limited to*:

277.86   peroxisomal disorders

Refer also to Test 6.1.1 codes.

Suggested Reading

Baes M, Huyghe S, Carmeliet P, Declercq PE, Collen D, Mannaerts GP, Van Veldhoven PP. Inactivation of the peroxisomal multifunctional protein-2 in mice impedes the degradation of not only 2-methyl-branched fatty acids and bile acid intermediates but also of very long chain fatty acids. *J Biol Chem* 2000; 275:16329–36.

Bamiou DE, Davies R, Jones S, Musiek FE, Rudge P, Stevens J, Luxon LM. An unusual case of X-linked adrenoleukodystrophy with auditory processing difficulties as the first and sole clinical manifestation. *J Am Acad Audiol* 2004; 15:152–60.

Dubey P, Raymond GV, Moser AB, Kharkar S, Bezman L, Moser HW. Adrenal insufficiency in asymptomatic adrenoleukodystrophy patients identified by very long-chain fatty acid screening. *J Pediatr* 2005; 146:528–32.

Johnson DW, Trinh MU, Oe T. Measurement of plasma pristanic, phytanic and VLCFA by liquid chromatography-electrospray tandem mass spectrometry for the diagnosis of peroxisomal disorders. *J Chromatogr B Analyt Technol Biomed Life Sci* 2003; 798:159–62.

Kennedy CR, Allen JT, Fensom AH, Steinberg SJ, Wilson R. X-linked adrenoleukodystrophy with non-diagnostic plasma very long chain fatty acids. *J Neurol Neurosurg Psychiatry* 1994; 57:759–61.

Mo YH, Chen YF, Liu HM. Adrenomyeloneuropathy, a dynamic progressive disorder: brain magnetic resonance imaging of two cases. *Neuroradiology* 2004; 46:296–300.

Moser AB, Kreiter N, Bezman L, Lu S, Raymond GV, Naidu S, Moser HW. Plasma very long chain fatty acids in 3000 peroxisome disease patients and 29,000 controls. *Ann Neurol* 1999; 45:100–10.

Moser HW, Moser AB. Very long-chain fatty acids in diagnosis, pathogenesis, and therapy of peroxisomal disorders. *Lipids* 1996; 31(Suppl):S141–4.

Moser HW, Dubey P. Fatemi A. Progress in X-linked adrenoleukodystrophy. *Curr Opin Neurol* 2004; 17:263–9.

Moser HW, Raymond GV, Lu SE, Muenz LR, Moser AB, Xu J, Jones RO, Loes DJ, Melhem ER, Dubey P, Bezman L, Brereton NH, Odone A. Follow-up of 89 asymptomatic patients with adrenoleukodystrophy treated with Lorenzo's oil. *Arch Neurol* 2005; 62:1073–80.

Paik MJ, Kim KR, Yoon HR, Kim HJ. Diagnostic patterns of very-long-chain fatty acids in plasma of patients with X-linked adrenoleukodystrophy. *J Chromatogr B Biomed Sci Appl* 2001; 760:149–57.

Smith KD. X-linked adrenoleukodystrophy: genetic causes of adrenal insufficiency. In: *Program and Abstracts of the 82nd Annual Meeting of the Endocrine Society*, June 21–24, 2000, Toronto, Ontario, Canada.

Wanders RJ, van Roermund CW, Lageweg W, Jakobs BS, Schutgens RB, Nijenhuis AA, Tager JM. X-linked adrenoleukodystrophy: biochemical diagnosis and enzyme defect. *J Inherit Metab Dis* 1992; 15:634–44.

TABLE 6.2
Features and Clinical Criteria for Diagnosis of
Primary Hypoadrenalism (Addison's Disease)

Organ System	Type of Syndrome	
	Acute	Chronic
General	Muscle and joint pain, fever, hyperkalemia, hyponatremia, eosinophilia, lymphocytosis, metabolic acidosis	Weakness, fatigue, listlessness, loss of libido, amenorrhea, weight loss, salt craving, hyperkalemia, hyponatremia, hypoglycemia
Gastrointestinal	Vomiting, occasionally associated with hypokalemia	Anorexia, nausea, diarrhea, abdominal pain
Cardiovascular	Hypotension	Orthostasis, syncope
Skin and mucous membranes[a]	No acute skin changes; dehydration	Hyperpigmentation, pallor, vitiligo, alopecia, thinning of axillary and pubic hair
Neurologic	Clouded sensorium	Apathy, erectile dysfunction

[a] In central hypoadrenalism, the renin–angiotensin system is usually intact; also, there is no hypoaldosteronism, no hyperpigmentation, and a tendency toward increased vasopressin secretion leading to hemodilution.

6.1.5 Adrenocortical Antibody (ACA) Screening for Autoimmune Adrenal Disease with Reflexive Immunoprecipitation Assay (IPA) for Antibodies to the 21-Hydroxylase Enzyme (21OH-Abs)

Indications for Test

Measurement of ACA is indicated:

- For the diagnosis of autoimmune adrenalitis as a cause for primary Addison's disease (1°AD)
- If clinical findings suggestive of 1°AD are present as per "Features and Clinical Criteria for Diagnosis of Primary Hypoadrenalism" (Table 6.2)

Procedure

1. Obtain a random blood sample for qualitative determination of ACA.
2. If the ACA screen is negative, no further testing is required.
3. If the ACA screen is positive, order quantitative testing by IPA for antibodies to the 21-hydroxylase enzyme (21OH-Abs).

Interpretation

1. Reference intervals in healthy individuals:
   - ACA titer, <1:64
   - 21OH-Ab, 2.6 to 1311 U/mL

Notes

1. ACA are primarily autoantibodies to the 21-hydroxylase enzyme, although other adrenal cortical enzymes may act as antigens. They are usually detected using an indirect immunofluorescence assay (IFA).
2. Widely variable positive titers for ACA have been reported. A positive correlation exists between 21OH-Abs (range, 2.6 to 1311 U/mL) and ACA titers (range, 1:16 to 1:64).
3. Overt 1°AD was found to develop in 21% and subclinical hypoadrenalism in 29% of ACA-positive adults. Overt 1°AD developed in 9 of 10 (90%) ACA and 21OH-Ab-positive children after 3 to 121 months, with the remaining child developing subclinical hypoadrenalism (Betterle et al., 1997a,b).

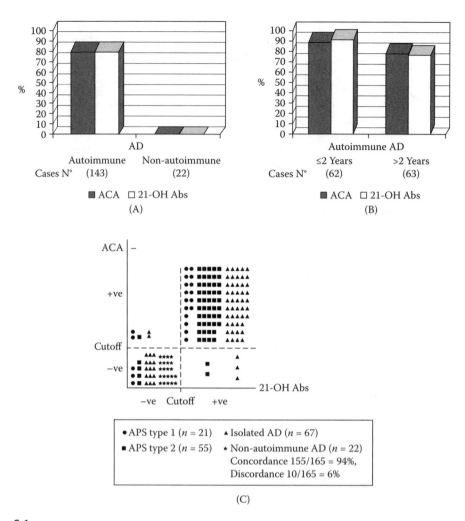

Figure 6.1
Frequency of adrenal cortex autoantibodies (ACAs) and 21-hydroxylase antibodies (21OH-Abs) in patients with (A) autoimmune (AI) and non-AI Addison's disease (AD), (B) AD of ≤2 years' or >2 years' duration, or (C) autoimmune polyglandular syndromes (APSs) or AD. +, positive; –, negative. (From Betterle, C. et al., *Endocr. Rev.*, 23, 327–364, 2002. With permission.)

4. Early in the course of autoimmune adrenalitis associated with autoimmune polyglandular syndromes, ACA is positive in 100% of cases. Over time, this percentage drops to about 80%.

5. Isolated autoimmune 1°AD is positive for ACA in about 80% of cases.

6. As of 2008, standardization of ACA and 21OH-Ab assays available in different laboratories had not been achieved.

7. 21OH-Abs were found to be positive in 91% of ACA-positive sera (Betterle et al., 1997a,b). A concordance of 84% and discordance of 6% was found between these tests (Figure 6.1).

8. In addition to IPA, 21OH-Abs can be reliably determined by western blotting or by using radiolabeled 21OH enzyme expressed in an *in vitro* transcription–translation system.

9. The disappearance of both 21OH-Ab and ACA and their prolonged absence following relatively short-term (6 months) corticosteroid treatment suggests that such treatment can induce long-term remission of subclinical adrenal insufficiency and prevent the onset of the clinical phase of 1°AD (De Bellis et al., 2001).

10. Other endocrine syndromes such as T1DM, autoimmune thyroid disease, and other autoimmune polyglandular syndromes (APS 1 to 4) are frequently found in patients with established 1°AD.

ICD-9 Codes

Conditions that may justify this test *include but are not limited to*:

036.3   Waterhouse–Friderichsen syndrome, meningococcal

Refer also to Test 6.1.1 codes.

Suggested Reading

Betterle C, Volpato M, Rees Smith B, Furmaniak J, Chen S, Greggio NA, Sanzari M, Tedesco F, Pedini B, Boscaro M, Presotto F. Adrenal cortex and steroid 21-hydroxylase autoantibodies in adult patients with organ-specific autoimmune diseases: markers of low progression to clinical Addison's disease, Part I. *J Clin Endocrinol Metab* 1997a; 82:932–4.

Betterle C, Volpato M, Rees Smith B, Furmaniak J, Chen S, Zanchetta R, Greggio NA, Pedini B, Boscaro M, Presotto F. Adrenal cortex and steroid 21-hydroxylase autoantibodies in children with organ-specific autoimmune diseases: markers of high progression to clinical Addison's disease, Part II. *J Clin Endocrinol Metab* 1997b; 82:939–42.

Betterle C, Dal Pra C, Mantero F, Zanchetta R. Autoimmune adrenal insufficiency and autoimmune polyendocrine syndromes: autoantibodies, autoantigens, and their applicability in diagnosis and disease prediction. *Endocr Rev* 2002; 23:327–64.

Dal Pra C, Chen S, Furmaniak J, Smith BR, Pedini B, Moscon A, Zanchetta R, Betterle C. Autoantibodies to steroidogenic enzymes in patients with premature ovarian failure with and without Addison's disease. *Eur J Endocrinol* 2003; 148:565–70.

De Bellis A, Falorni A, Laureti S, Perrino S, Coronella C, Forini F, Bizzarro E, Bizzarro A, Abbate G, Bellastella A. Time course of 21-hydroxylase antibodies and long-term remission of subclinical autoimmune adrenalitis after corticosteroid therapy: case report. *J Clin Endocrinol Metab* 2001; 86:675–8.

6.1.6 Steroid-Producing Cell Autoantibodies (StCAs) in Evaluation of Risk of Primary Hypoadrenalism or Addison's Disease (1°AD)

Indications for Test

Determination of StCA is indicated:

- To establish the diagnosis of autoimmune adrenalitis as a cause for established 1°AD when adrenocortical 21-hydroxylase antibodies are negative
- In cases of premature ovarian failure (POF) as suggested by elevated FSH (see Test 9.2.1)
- If clinical findings suggestive of 1°AD are present (Table 6.2)

Procedure

1. Obtain a blood sample for steroid-producing cell autoantibodies reactive with steroid 17α-hydroxylase or P450 side-chain cleavage enzymes.

Interpretation

1. In various reports, a positive StCA test may be found in 0.08 to 1.8% of patients with various non-adrenal autoimmune diseases.
2. A positive StCA test is highly predictive for the development of adrenal insufficiency within 10 years.

3. The standard high-dose CST (see Test 6.1.1) should be reserved to confirm adrenal insufficiency in women with positive StCA associated with POF or in those with signs and symptoms of adrenal insufficiency. Expect a higher incidence of false-positive CST results in POF patients who are StCA negative.

Notes

1. In 123 women with POF, a new diagnosis of adrenal insufficiency was found in four (3.2%; 95% confidence interval [CI], 0.2 to 6.4%) (Bakalov et al., 2002). In 8840 adults with organ-specific autoimmune diseases, 67 (0.8%) were ACA positive, with the highest prevalence in those with POF (8.9%) (Betterle et al., 1997a,b).
2. 1°AD caused by autoimmune adrenalitis and POF caused by lymphocytic oophoritis are almost always associated with the presence of ACA. In one study, 73% of patients with POF and adrenal insufficiency were positive for StCA, 93% for ACA, and 100% for 21OH-Abs (Dal Pra et al., 2003).

ICD-9 Codes

Conditions that may justify this test *include but are not limited to*:

256.3   ovarian failure

Refer also to Test 6.1.1 codes.

Suggested Reading

Bakalov VK, Vanderhoof VH, Bondy CA, Nelson LM. Adrenal antibodies detect asymptomatic auto-immune adrenal insufficiency in young women with spontaneous premature ovarian failure. *Hum Reprod* 2002; 17:2096–100; *Hum Reprod* 2003; 18:1132–3 [comment].

Betterle C, Volpato M, Rees Smith B, Furmaniak J, Chen S, Greggio NA, Sanzari M, Tedesco F, Pedini B, Boscaro M, Presotto F. Adrenal cortex and steroid 21-hydroxylase autoantibodies in adult patients with organ-specific autoimmune diseases: markers of low progression to clinical Addison's disease, Part I. *J Clin Endocrinol Metab* 1997a; 82:932–8.

Betterle C, Volpato M, Rees Smith B, Furmaniak J, Chen S, Zanchetta R, Greggio NA, Pedini B, Boscaro M, Presotto F. Adrenal cortex and steroid 21-hydroxylase autoantibodies in children with organ-specific autoimmune diseases: markers of high progression to clinical Addison's disease, Part II. *J Clin Endocrinol Metab* 1997b; 82:939–42.

Dal Pra C, Chen S, Furmaniak J, Smith BR, Pedini B, Moscon A, Zanchetta R, Betterle C. Autoantibodies to steroidogenic enzymes in patients with premature ovarian failure with and without Addison's disease. *Eur J Endocrinol* 2003; 148:565–70.

6.2   Glucocorticoid Excess in Cushing's Syndrome

6.2.1   Cortisol Testing: Diurnal Variation of Salivary vs. Serum Cortisol Concentration in Cushing's Syndrome (CS)

Indications for Test

Testing for the diurnal variation of serum or salivary cortisol concentration is indicated in:

*   Screening for CS
*   Patients with central nervous system disorders in which the normal hypothalamic–pituitary-driven diurnal rhythm of ACTH secretion is lost
*   Postoperative assessment of adrenal status in patients who have undergone pituitary surgery for Cushing's disease

Procedure

1. Select patients for study who have a normal sleep–wake pattern (do not work an evening or night shift).
2. Do not pursue diurnal cortisol determinations in patients with erratic sleep–wake cycles or chronic gum or mouth bleeding conditions. Blood in the saliva will invalidate the test.
3. If saliva collection (see Specimen Collection Protocol P2 in Appendix 2) for determination of cortisol is unavailable or impractical, obtain a blood sample for serum cortisol testing at both 0800 hours and 2300 hours. Make special arrangements with late-night clinic, hospital laboratory, or emergency department personnel to obtain the 2300-hour (11 p.m.) specimen.
4. Use a lab with a sensitive assay for cortisol testing. Be sure that such an assay is available at the lab chosen to do the test.

Interpretation

1. Blood [cortisol] at 2300 hours is normally <50% of the concentration at 0800 hours.
2. Salivary [cortisol] ranges from 0.18 to 0.95 mcg/dL at 0800 hours to 0.05 to 0.17 mcg/dL at 2300 hours.
3. CS patients show loss of diurnal variation, with elevated 2300-hour [ACTH] and [cortisol].
4. If the 2300-hour [cortisol] is >50% of the 0800-hour level, the normal diurnal rhythm of ACTH secretion has been lost and CS is possible, particularly when the [cortisol] at both time points are in the high normal or elevated range.
5. If cortisol concentrations at both time points are in the low range, consistent with lack of normal diurnal variation in [cortisol], central hypoadrenalism is a likely possibility.
6. Measurement of late-night salivary [cortisol] is as sensitive as and more convenient than the blood test for [cortisol] and eliminates the stress of venipuncture. A cutoff value of 0.27 mcg/dL (7.5 nmol/L) for the salivary test offers a diagnostic accuracy of 93% (Gafni et al., 2000) for CS, which is similar to that of a midnight serum [cortisol] (95.7%) and a urinary free [cortisol] (95.3%) (Test 6.2.2) (Putignano et al., 2003).
7. Extra-adrenal as well as adrenal sources of hormones that suppress ACTH, including surreptitious ingestion or cutaneous application of glucocorticoids or hypersecretion of dehydroepiandrosterone (DHEA) from the adrenal or ovary, may lower [cortisol] and interfere with the diurnal variation of cortisol secretion.

Notes

1. Pulses of ACTH are normally released in a diurnal pattern, more frequently and of higher amplitude in the early morning hours and less frequently and of lower amplitude in the evenings.
2. Serum and salivary cortisol concentrations follow the same time-dependent concentration pattern, because saliva is in equilibrium with free plasma cortisol, independent of the rate of saliva production.
3. Patients having pseudo-Cushing's syndrome associated with alcoholism, obesity, and stress usually, but not always, maintain a normal diurnal pattern of cortisol secretion.
4. Blood sampling times of 0800 and 2300 hours to test for diurnal variation in cortisol secretion are unwieldy and impractical in an outpatient setting.
5. Prednisone cross-reacts in the immunoassay for cortisol.
6. A laboratory that requires over 1 mL of saliva for measurement of [cortisol] does not have the sensitive assay required for doing this test. Send blood samples, not saliva, for cortisol testing to such a laboratory.

ICD-9 Codes

Conditions that may justify this test *include but are not limited to*:

255.X	Cushing's syndrome
278.1	localized adiposity/supraclavicular fat pad/abdomen
701.3	abdominal striae/stretch marks (atrophic)
733	osteoporosis
754.0	facial edema/moon facies (puffiness)

Suggested Reading

Castro M, Elias PC, Quidute AR, Halah FP, Moreira AC. Outpatient screening for Cushing's syndrome: the sensitivity of the combination of circadian rhythm and overnight dexamethasone suppression salivary cortisol tests. *J Clin Endocrinol Metab* 1999; 84:878–82.

Gafni RI, Papanicolaou DA, Nieman LK. Nighttime salivary cortisol measurement as a simple, noninvasive, outpatient screening test for Cushing's syndrome in children and adolescents. *J Pediatr* 2000; 137:30–5.

Putignano P, Toja P, Dubini A et al. Midnight salivary cortisol versus urinary free and midnight serum cortisol as screening tests for Cushing's syndrome. *J Clin Endocrinol Metab* 2003; 88:4153–57.

Viardot A, Huber P, Puder JJ, Zulewski H, Keller U, Muller B. Reproducibility of nighttime salivary cortisol and its use in the diagnosis of hypercortisolism compared with urinary free cortisol and overnight dexamethasone suppression test. *J Clin Endocrinol Metab* 2005; 90:5730–36.

6.2.2 Urinary Free Cortisol (UFC) 24-Hour Excretion Rate and UFC/Urine Creatinine (UCr) Ratio in Cushing's Syndrome (CS)

Indications for Test

Determination of UFC secretion rate and the UFC/UCr ratio is indicated when:

- Patients present with signs and symptoms of overt CS, including unexplained weight gain, diabetes, hypertension, appearance of moon facies, or dark or purplish abdominal striae appear, usually in the context of an adrenal or pituitary mass.
- The [ACTH] is low or in the lower quartile of the reference range, the evening [cortisol] is elevated, or loss of diurnal variation in cortisol secretion has been documented.

Procedure

1. Instruct the patient to collect a 24-hour urine specimen (see Specimen Collection Protocol P1 in Appendix 2). Be sure the laboratory has a request for UFC (mcg/dL) and UCr (mg/dL) testing on the patient's 24-hour urine specimen so the UFC excretion rate and the UFC/UCr ratio can be calculated.
2. Specifically request that UFC testing be performed by (high-performance) liquid chromatography (LC), tandem mass spectrometry (MS/MS), or LC-MS/MS, as these methods are not affected by interferents that affect immunoassay methods, with or without a specimen prepurification step prior to immunoassay testing. The MS-MS method is preferred for its precision, which is better than that of LC.
3. Because 24-hour urine collections are cumbersome, potentially inadequate, and inherently inaccurate, obtain an overnight 12-hour urine collection for UFC testing as a screening test for Cushing's syndrome (referred to as the overnight UFC test).
4. To collect an overnight 12-hour urine, instruct the patient to empty his or her bladder prior to 2000 hours (8 p.m.) and then collect all urine voided between 2000 hours (8 p.m.) and 0800 hours (8 a.m.).
5. Calculations:
   - For a 12-hour urine collection:

$$\text{UFC (mcg/12 hr)} = \frac{\text{UFC (mcg/dL)} \times U_{vol} \text{ (mL/12 hr)}}{100 \text{ mL/dL}}$$

$$\text{UFC/UCr ratio (mcg/g)} = \frac{\text{UFC (mcg/12 hr)}}{\text{UCr (g/12 hr)}}$$

   - For a 24-hour urine collection:

$$\text{UFC (mcg/24 hr)} = \frac{\text{UFC (mcg/dL)} \times U_{vol} \text{ (mL/24 hr)}}{100 \text{ mL/dL}}$$

$$\text{UFC/UCr ratio (mcg/g)} = \frac{\text{UFC (mcg/24 hr)}}{\text{UCr (g/24 hr)}}$$

Examples
- 12-hour urine collection:

$$U_{vol} = 825 \text{ (mL/12 hr)}$$

$$UFC = 6.23 \text{ (mcg/dL)}$$

$$UCr = 50.4 \text{ mg/dL}$$

$$UFC \text{ excretion rate } = \frac{6.23 \text{ mcg/dL} \times 825 \text{ mL/12 hr)}}{100 \text{ mL/dL}} = 51.4 \text{ mcg/12 hr}$$

$$UCr = \frac{50.4 \text{ mg/dL} \times 8.25 \text{ mL/12 hr}}{100 \text{ mL/dL}} = 415.8 \text{ mg/12 hr} = 0.416 \text{ g/12 hr}$$

$$UFC/UCr \text{ ratio} = \frac{51.4 \text{ mcg/12 hr}}{0.416 \text{ g/12 hr}} = 123.6 \text{ mcg/g}$$

- 24-hour urine collection:

$$U_{vol} = 750 \text{ (mL/24 hr)}$$

$$UFC = 1.28 \text{ (mcg/dL)}$$

$$UCr = 149 \text{ mg/dL}$$

$$UFC \text{ excretion rate } = \frac{1.28 \text{ mcg/dL} \times 750 \text{ mL/24 hr)}}{100 \text{ mL/dL}} = 9.6 \text{ mcg/24 hr}$$

$$UCr = \frac{149 \text{ mg/dL} \times 750 \text{ mL/24 hr}}{100 \text{ mL/dL}} = 1117 \text{ mg/24 hr} = 1.12 \text{ g/24 hr}$$

$$UFC/UCr \text{ ratio} = \frac{9.6 \text{ mcg/24 hr}}{1.12 \text{ g/24 hr}} = 8.6 \text{ mcg/g}$$

Interpretation

1. Refer to "Interpretation of UFC and UFC/UCr Tests" (Table 6.3).

Notes

1. If sensitivity is set at 100%, a specificity of 97% for the timed overnight (i.e., 12-hour) UFC/UCr ratio vs. 87% for the 24-hour UFC/UCr ratio may be found.
2. The UFC excretion rate and UFC/UCr ratio are not sensitive enough for diagnosis of less than overt CS.
3. The definitive diagnosis of CS is best established by combining information provided by measurement of the UFC excretion rate, morning and evening serum or salivary [cortisol], and [ACTH] measured in the morning. One-hour spot urine collections for measurement of cortisol (ng)/creatinine (mg) from 0700 to 0800 hours and from 2200 to 2400 hours can detect the same loss of diurnal variation in cortisol excretion in CS as can be seen in salivary and serum cortisol testing.
4. In 14 patients with CS, the morning spot urinary cortisol was elevated (207 ± 176 ng/mg creatinine), but there was overlap with values in normal subjects. In contrast, evening values in CS (248 ± 208 ng/mg creatinine) were elevated, with no diurnal variation and no overlap with normal subjects (Contreras et al., 1986).

TABLE 6.3

Interpretation of UFC and UFC/UCr Tests

Test	Patient's Result	Interpretation
24-hour UFC (mcg/24 hr)	≤45[a]	Healthy individuals[b]
	≥90	Cushing's syndrome
24-hour UFC/UCr ratio (mcg/g)	0.2–16.6	Healthy individuals[c]
	≥14	Cushing's syndrome
12-hour overnight[d] UFC/UCr ratio (mcg/g)	≤12.5	Healthy individuals[c]
	≥16.2	Cushing's syndrome

[a] When UFC is measured by liquid chromatography–tandem mass spectrometry (LC-MS/MS).

[b] False positives may occur in patients with alcoholism, major stress, or exogenous glucocorticoid administration but are not seen with obesity alone.

[c] Patients with renal failure may have UFC/UCr ratios that overlap with normal (8.4 ± 4.1 mcg/g; range, 1.5–21.2). A false-negative rate of 5 to 10% may be expected.

[d] Urine collection between bedtime and 12 hours later; for example, the patient goes to bed at 10:00 p.m. and the urine void at bedtime is not collected (discarded), but 12 hours later (10:00 a.m.), the patient must urinate and collect the entire void volume at that time.

5. The 2-mg low-dose and 8-mg high-dose dexamethasone suppression tests done in an attempt to differentiate a cortisol- from an ACTH-secreting tumor may be misleading and are obsolete if high-quality imaging modalities of the pituitary and adrenal glands are available.

6. Magnetic resonance (MR) imaging has greater sensitivity for detecting ACTH-producing pituitary adenomas than computed tomography (CT).

7. Because non-immunoassay-based methods (e.g., LC-MS/MS) have higher analytical precision (i.e., lower interassay coefficient of variation) than immunoassay methods, these methods can detect the presence of surreptitious corticosteroid use.

ICD-9 Codes

Conditions that may justify this test *include but are not limited to*:

250.X	diabetes mellitus
255.0	Cushing's syndrome
401.1	hypertension, benign
754.0	facial edema (puffiness)
783.1	weight gain, abnormal

Suggested Reading

Contreras LN, Hane S, Tyrrell JB. Urinary cortisol in the assessment of pituitary–adrenal function: utility of 24-hour and spot determinations. *J Clin Endocrinol Metab* 1986; 62:965–9.

Corcuff JB, Tabarin A, Rashedi M, Duclos M, Roger P, Ducassou D. Overnight urinary free cortisol determination: a screening test for the diagnosis of Cushing's syndrome. *Clin Endocrinol (Oxf)* 1998; 48:503–8.

Kaye TB, Crapo L. The Cushing syndrome: an update on diagnostic tests. *Ann Intern Med* 1990; 112:434–44.

Terzolo M, Osella G, Ali A, Borretta G, Cesario F, Paccotti P, Angeli A. Subclinical Cushing's syndrome in adrenal incidentaloma. *Clin Endocrinol (Oxf)* 1998; 48:89–97.

TABLE 6.4
Interpretation of Plasma ACTH Values in
Early-Morning EDTA Whole-Blood Specimens

Demographics	[ACTH] (pg/mL)[a]	Consistent with ...
All ages, either gender	Undetectable (<LLD)	ACTH-independent, primary adrenal hypercortisolism
	<15	Subclinical autonomous hypersecretion of adrenal cortisol (e.g., adrenal incidentaloma)
	Between LLD and ULN	Indeterminate
Adult men	>50	Cushing's disease
Adult women	>30	Cushing's disease
Children (pubertal)	>50	Cushing's disease
All ages, either gender	>200 but <1000	Possible ectopic[b] hypersecretion of ACTH
	>1000	Overt ectopic[b] hypersecretion of ACTH

[a] Considerable overlap occurs in the ACTH range (5–500 pg/mL) in pituitary conditions or ectopic sources of ACTH.

[b] Nonpituitary, nonadrenal tumor.

Note: ACTH, adrenocorticotropic hormone; LLD, lower limit of assay detection (i.e., the analytical sensitivity of the ACTH assay); ULN, upper limit of normal (reference).

6.3   Determining the Etiology of Cushing's Syndrome

6.3.1   ACTH Testing in the Differential Diagnosis of Causes for Cushing's Syndrome (CS)

Indications for Test

ACTH measurement is indicated:

- When determining the etiology of established or overt CS as ACTH-dependent Cushing's disease (CD) vs. ACTH-independent adrenal hypercortisolism or ectopic (i.e., nonpituitary, nonadrenal tumor) hypersecretion of ACTH.

Procedure

1. Collect an early-morning EDTA whole-blood sample between 0700 hours (7 a.m.) and 0900 hours (9 a.m.) from a patient with a regular sleep–wake cycle.
2. Request an ACTH determination on this sample.

Interpretation

1. Refer to "Interpretation of Plasma ACTH Values in Early-Morning EDTA Whole-Blood Specimens" (Table 6.4).

Notes

1. Use of the [ACTH] to help classify a hypercortisolemic state as ACTH-dependent or ACTH-independent is limited when faced with the problem of subclinical glucocorticoid hypersecretion in patients with an adrenal "incidentaloma."
2. Ectopic ACTH-secreting tumors may produce abnormal, but bioactive, ACTH fragments that may not be detected with a two-site immunometricassay for intact ACTH.
3. Elevated [ACTH] have not been found to be very helpful in distinguishing between pituitary overproduction and ectopic ACTH secretion except when extremely high concentrations are found and a potentially ectopic source of ACTH is found (e.g., lung mass).

4. A 50% increase in [ACTH] and [cortisol] after administration of ovine corticotropin-releasing hormone (oCRH) yields a diagnostic accuracy of 86% and 61%, respectively, in the differential diagnosis of ACTH-dependent CS.
5. The insulin tolerance test in differential diagnosis of CS—in which a cortisol increment over baseline of <5 mcg/dL is consistent with CD while a cortisol increment over baseline of ≥5 mcg/dL is consistent with chronic severe alcoholism or severe depression—fails to clearly separate ACTH-dependent from ACTH-independent etiologies of CS and has an 18% false-negative rate (Crapo, 1979).
6. Out of 426 patients with CS, 288 (68%) had CD, 80 (19%) had an adrenal adenoma, 24 (5.6%) had an adrenal carcinoma, 25 (5.9%) had ectopic ACTH and/or CRH secretion, and 9 (2.1%) had ACTH-independent nodular adrenal hyperplasia (Invitti et al., 1999).
7. In CD, an overall relapse rate after pituitary surgery of 17% has been reported. The probability of relapse-free survival, as assessed by Kaplan–Meier analysis, was 95% at 12 months, 84% at 2 years, and 80% at 3 years (Invitti et al., 1999).

ICD-9 Codes

Conditions that may justify this test *include but are not limited to*:

255.0   Cushing's syndrome

Suggested Reading

Bornstein SR, Chrousos GP. Clinical review 104: adrenocorticotropin (ACTH)- and non-ACTH-mediated regulation of the adrenal cortex—neural and immune inputs. *J Clin Endocrinol Metab* 1999; 84:1729–36.

Crapo L. Cushing's syndrome: a review of diagnostic tests. *Metabolism* 1979; 28:955–77.

Findling JW, Doppman JL. Biochemical and radiologic diagnosis of Cushing's syndrome. *Endocrinol Metab Clin North Am* 1994; 23:521–37.

Invitti C, Giraldi FP, de Martin M, Cavagnini F. Diagnosis and management of Cushing's syndrome: results of an Italian multicentre study—Study Group of the Italian Society of Endocrinology on the Pathophysiology of the Hypothalamic–Pituitary–Adrenal Axis. *J Clin Endocrinol Metab* 1999; 84:440–8.

6.3.2   Bilateral Inferior Petrosal Sinus Sampling (BIPSS) after Administration of Ovine Corticotropin-Releasing Hormone (oCRH) to Exclude Ectopic ACTH Syndrome

Indications for Test

BIPSS with oCRH stimulation to exclude ectopic ACTH syndrome is indicated:

- In patients with equivocal imaging studies for an ACTH-secreting pituitary adenoma, but clear biochemical evidence (e.g., elevated ACTH and cortisol levels) of Cushing's syndrome (CS)

Procedure

1. Refer to Specimen Collection Protocol P5 in Appendix 2.

Interpretation

1. If the left petrosal sinus (LPS)/right petrosal sinus (RPS) or RPS/LPS [ACTH] ratio is >1.4 and the inferior petrosal sinus (IPS)/peripheral vein (PV) ACTH ratio (i.e., $[ACTH]_{RPS}/[ACTH]_{PV}$ or $[ACTH]_{LPS}/[ACTH]_{PV}$) is ≥3 in any two or more samples, the presence of a pituitary ACTH-producing adenoma is diagnosed with a high degree of specificity and sensitivity (Oldfield et al., 1991).
2. Refer to "Interpretation of Bilateral Inferior Petrosal Sinus Sampling (BIPSS) ACTH Results" (Table 6.5) and Figure 6.2.

TABLE 6.5
Interpretation of Bilateral Inferior Petrosal Sinus Sampling (BIPSS) ACTH Results

ACTH Ratio	ACTH Ratio in Any Two or More Samples	Most Likely Interpretation
$[ACTH]_{RPS}/[ACTH]_{LPS}$ _or_ $[ACTH]_{LPS}/[ACTH]_{RPS}$	1.4	Unsuccessful cannulation of right and/or left petrosal sinus or absence of an ACTH-secreting pituitary lesion
$[ACTH]_{RPS}/[ACTH]_{LPS}$ _or_ $[ACTH]_{LPS}/[ACTH]_{RPS}$	>1.4	Successful cannulation of right and/or left petrosal sinus; review RPS/PV and LPS/PV ACTH ratios
and		
$ACTH]_{RPS}/[ACTH]_{PV}$ _or_ $[ACTH]_{LPS}/[ACTH]_{PV}$	<2.0	Primary adrenal tumor or the ectopic ACTH syndrome[a]
	2.0–2.9	Abnormal pituitary ACTH secretion[b]
	≥3.0	Pituitary source of ACTH

[a] An IPS/PV ACTH ratio > 2.0 has a sensitivity of 95% but a specificity of 100%; lower ratios (≤2.0) suggest ectopic ACTH syndrome or failed cannulation of the petrosal sinus.

[b] Or primary adrenal tumor, ectopic ACTH syndrome, pituitary ACTH-producing adenoma. An IPS/PV ACTH ratio < 3.0 was found in all patients with ectopic ACTH secretion; however, this ratio was found to occur in 11/76 (15%) patients with eventually proven Cushing's disease (i.e., the BIPSS results in these 11 patients were falsely negative). The rare case of ectopic CRH production as the cause of Cushing's syndrome would be expected to result in an elevated IPS/PV ACTH ratio, falsely suggesting a pituitary lesion.

Important: If there is a consistent left- or right-side ACTH gradient (i.e., $[ACTH]_{RPS}/[ACTH]_{LPS}$ or $[ACTH]_{LPS}/[ACTH]_{RPS}$ > 1.4) between the two petrosal sinus samples, the pituitary adenoma may be located on the same side as the elevated ACTH level; however, BIPSS data cannot be used to unequivocally indicate laterality of an ACTH-secreting adenoma. Therefore, the results of BIPSS testing may or may not be useful in providing preoperative guidance to identify which side of the pituitary contains an ACTH-secreting pituitary tumor in patients with biochemically confirmed hypercortisolism or in patients with low basal ACTH levels.

Notes

1. Pituitary imaging alone may be expected to identify an adenoma in more than 60% of patients with CD. An even higher success rate for imaging of tumors in patients with CD might be expected with the advent of multidetector technology in imaging modalities such as MR and 64-slice computed tomography.

2. The principal use of BIPSS testing is to exclude the ectopic ACTH syndrome. BIPSS can be used to identify the source of ACTH as pituitary or extrapituitary, reliably distinguishing between pituitary Cushing's disease and Cushing's syndrome caused by ectopic sources of ACTH such as bronchial carcinoid tumors.

3. BIPSS testing does not distinguish between healthy patients and those with pseudo-Cushing's states or pituitary Cushing's disease, as the pituitary is the source of ACTH in all such cases.

4. Injection of 10 mcg of desmopressin acetate (DDAVP®) may be used as an alternative to injection of oCRH when performing the BIPSS test (Salgado et al., 1997).

5. Approximately 85% of patients with Cushing's disease respond to a single injection of oCRH with an increase in peripheral vein [ACTH] and [cortisol]. Only 5% of patients with ectopic ACTH-secreting tumors have an increase in these analytes after oCRH.

6. A rise in peripheral vein [ACTH] > 35% over baseline measured at 15 and 30 minutes after oCRH yields a 100% rate of specificity and a 93% rate of sensitivity in the diagnosis of CD.

7. An increase of at least 20% in peripheral vein [cortisol] over baseline measured 30 and 45 minutes after oCRH yields a specificity of 88% and a sensitivity of 91% in the diagnosis of CD.

8. When the oCRH stimulation test is performed in conjunction with a dexamethasone suppression test, nondiagnostic results from both tests rule out a diagnosis of CD with a diagnostic accuracy > 98%.

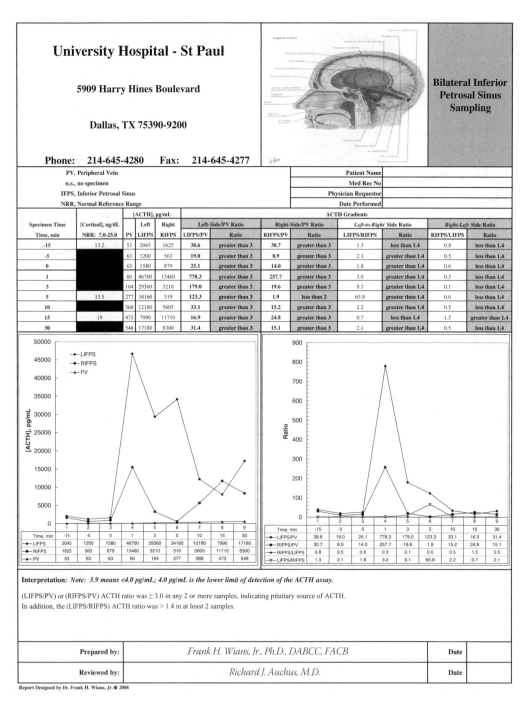

Figure 6.2

Sample bilateral inferior petrosal sinus sampling report. (Figure courtesy of Dr. Frank H. Wians, Jr.)

ICD-9 Codes

Conditions that may justify this test *include but are not limited to*:

195	malignant neoplasm of other and ill-defined sites
227.3	pituitary adenoma
255.0	Cushing's syndrome

Suggested Reading

Auchus RJ, Michaelis C, Wians FH Jr et al. Rapid cortisol assays improve the success rate of adrenal vein sampling for primary aldosteronism. *Ann Surg* 2009; 249:1–4.

Oldfield EH, Doppman JL, Nieman LK et al. Petrosal sinus sampling with and without corticotropin-releasing hormone for the differential diagnosis of Cushing's syndrome. *N Engl J Med* 1991; 325:897–905.

Salgado LR, Mendoca BB, Pereira MAA et al. Use of desmopressin in bilateral and simultaneous inferior petrosal sinus sampling for differential diagnosis of ACTH-dependent Cushing's syndrome. *Endocrinologist* 1997; 7:135–40.

Wians FH Jr. Use of Excel spreadsheets to create interpretive reports for laboratory tests requiring complex calculations. *Lab Med* 2009; 40:5–12.

6.4 Mineralocorticoid Reserve

6.4.1 Potassium Testing: Screening Test for Hypoaldosteronism

Indications for Test

Testing for potassium concentration [K^+] in blood as a screening test for hypoaldosteronism is indicated in patients who:

- Are suspected of having a disorder of aldosterone secretion or primary adrenal disease (e.g., Addison's disease) and are not in Stage 4 or 5 chronic kidney disease (CKD)
- Have symptoms of hypoadrenalism such as hypotension, fatigue, asthenia, etc. (see Table 6.2)
- Have abnormal EKG with peaked T-waves

Procedure

1. To ensure that the patient is not in Stage 4 or 5 CKD, obtain a blood sample for serum creatinine testing and estimation of glomerular filtration rate (GFR) (see Test 5.2.1).
2. Obtain a recumbent and standing blood pressure (BP). If there is a drop in mean arterial BP on standing, orthostatic hypotension consistent with primary hypoadrenalism is possible.
3. Use Specimen Collection Protocol P6 in Appendix 2 to collect a blood sample for [K^+].
4. If the [K^+] is elevated or >5.0 mEq/L, obtain a blood sample for serum potassium, sodium, and aldosterone determination.

Interpretation

1. [K^+] > 5.3 mEq/L is consistent with hypoaldosteronism and should prompt testing for serum [aldosterone] as well as special testing (e.g., adrenal imaging studies, blood sampling for cortisol, sodium, adrenal autoantibodies) to determine the cause of the aldosterone deficiency.
2. Hyperkalemia occurs in about two thirds of primary adrenal insufficiency patients with hypoaldosterone, resulting in water retention and hyponatremia in up to 90% of cases.

Notes

1. Aldosterone is the primary regulator of [K^+].
2. The serum [K^+] is a reliable reflection of aldosterone secretion in patients not taking medications known to affect serum [K^+] (e.g., diuretics, K^+ binding resins, beta blockers, NSAIDs, ACEIs).
3. Renal insufficiency may cause an increase in [K^+].
4. Hyperkalemia with hypoaldosteronism may result from:
   - Defects in adrenal synthesis of aldosterone from almost complete destruction of the adrenal glands usually from autoimmune adrenalitis, as well as infections, amyloid infiltration, hemorrhage, metastatic cancers (breast > lung > kidney), or adrenoleukodystrophy
   - Inadequate stimulation of aldosterone secretion (hyporeninemic hypoaldosteronism)

- Resistance to the ion transport effects of aldosterone, such as are seen in patients with pseudo-hypoaldosteronism type I related to mutations involving the amiloride-sensitive epithelial sodium channel
- Gordon's syndrome (pseudohypoaldosteronism type II) resulting from resistance to the kaliuretic, but not sodium reabsorptive, effects of aldosterone for which the genetic basis is still unknown (Gordon, 1986)

5. Patients suspected of having Addison's disease should receive the cosyntropin test (Test 6.1.1).

ICD-9 Codes

Conditions that may justify this test *include but are not limited to*:

80.79	fatigue/malaise
255.1	hyperaldosteronism
255.4	corticoadrenal insufficiency
276.7	hyperpotassemia
585.X	chronic kidney disease (CKD), Stages 3 to 5

Suggested Reading

De Fronzo RA. Hyperkalemia and hyporeninemic hypoaldosteronism. *Kidney Int* 1980; 17:118–34.

Fuller PJ. Aldosterone and its mechanism of action: more questions than answers. *Aust NZ J Med* 1995; 25:800–7.

Gordon RD. Syndrome of hypertension and hyperkalemia with normal glomerular filtration rate. *Hypertension* 1986; 8:93–102.

Torpy DJ, Stratakis CA, Chrousos GP. Hyper- and hypoaldosteronism. *Vitam Horm* 1999; 57:177–216.

6.5 Mineralocorticoid Excess

6.5.1 Potassium Testing: Screening Test for Hyperaldosteronism

Indications for Test

Testing for potassium concentration [K$^+$] in serum as a screening test for hyperaldosteronism is indicated in patients with:

- Moderate to severe hypertension resistant to therapy, usually of onset at less than 50 years of age, palpitations, abnormal EKG with U-waves or symptomatic orthostasis without CKD in whom an endocrine cause for hypertension is suspected

Procedure

1. Obtain a blood sample for serum creatinine and estimation of GFR.
2. Obtain a recumbent and standing blood pressure.
3. Use Specimen Collection Protocol P6 in Appendix 2 in collection of a blood sample for [K$^+$].
4. If the [K$^+$] is low (<3.5 mEq/L), obtain another blood sample for both serum potassium and aldosterone determination.

Interpretation

1. In the context of Stage 2 hypertension (see Table 4.2), normal diet, normal fluid intake, and the absence of diuretic use, a random [K$^+$] < 3.5 mEq/L suggests hyperaldosteronism.
2. Once the combination of hypokalemia and hyperaldosteronism (see Test 6.5.2) has been well established, CT imaging of the adrenal glands (see Test 6.5.6) is indicated.

Notes

1. Autonomous aldosterone hypersecretion leads to hypertension and hypokalemia.
2. A $[K^+] < 3.5$ mEq/L occurs in ~72% of proven cases of hyperaldosteronism and should prompt testing for serum aldosterone concentration as well as evaluation and testing for the cause of aldosterone excess (e.g., history of licorice ingestion, genetic analyses, adrenal imaging studies, surreptitious use of diuretics).
3. The use of serum $[K^+]$ alone as a screening test for hyperaldosteronism would miss about a third of cases as well as result in finding a high percentage of false positives given the multiplicity of reasons for hypokalemia other than hyperaldosteronism.
4. About 10% of hypertensives may have underlying hyperaldosteronism.
5. The frequency of primary aldosteronism among adults with hypertension is uncertain. Normokalemic cases of hyperaldosteronism may be more frequent than was thought in the years prior to 2003.
6. Aldosterone is a key mediator of blood volume and serum potassium homeostasis, which in turn is regulated by aldosterone secretion by the zona glomerulosa of the adrenal cortex.
7. Advanced renal insufficiency (CKD Stage 4 or 5) may result in an increase in serum $[K^+]$ and obscure the diagnosis of hyperaldosteronism otherwise associated with hypokalemia.
8. Malignant hypertension and hypokalemia may be associated with glucocorticoid remediable aldosteronism (GRA).
9. The genetic basis of GRA is known, but the basis for the syndrome of familial hyperaldosteronism type II is not, and it appears to be unrelated to mutations in the genes for aldosterone synthase or the angiotensin II receptor.
10. Hypokalemia may be due to non-aldosterone-mediated mineralocorticoid excess, including conditions such as:
    - Congenital adrenal hyperplasia secondary to 11β-hydroxylase or 17α-hydroxylase deficiency
    - Syndrome of apparent mineralocorticoid excess secondary to 11β-hydroxysteroid dehydrogenase deficiency
    - Primary glucocorticoid resistance
    - Liddle's syndrome caused by activating mutations of the renal epithelial sodium channel
    - Exogenous sources of mineralocorticoids, such as licorice or drugs such as carbenoxolone
11. Causes of secondary hyperaldosteronism with hypokalemia include renovascular hypertension related to:
    - Use of diuretics
    - Renin-secreting tumor
    - Cushing's syndrome
    - Bartter's syndrome
    - Pheochromocytoma

ICD-9 Codes

Conditions that may justify this test *include but are not limited to*:

255.1   primary hyperaldosteronism
255.10 hyperaldosteronism

Suggested Reading

Nadar S, Lip GY, Beevers DG. Primary hyperaldosteronism. *Ann Clin Biochem* 2003; 40(Pt 5):439–52.
Torpy DJ, Stratakis CA, Chrousos GP. Hyper- and hypoaldosteronism. *Vitam Horm* 1999; 57:177–216.

6.5.2 Plasma Aldosterone Concentration (PAC)/Renin (R) Ratio as a Screening Test for Primary Hyperaldosteronism (PHA)

Indications for Test

Determination of the PAC/R ratio, where R is plasma renin activity (PRA) or plasma renin concentration (PRC), is indicated when:

- Screening patients with hypokalemia ($[K^+]$ usually < 3.5 mEq/L) for possible PHA
- Screening patients in whom risk factors and family history of hypertension are negative, and new-onset, severe, or malignant hypertension resistant to treatment is found
- An incidental adrenal tumor is found in a hypertensive individual

Procedure

1. Withdraw all antihypertensive agents except alpha blockers from the patient for at least 2 weeks. This includes calcium channel blockers, beta blockers, nonsteroidal antiinflammatory drugs (NSAIDs), ACE inhibitors, and angiotensin receptor blockers (ARBs). Withdraw the use of all diuretics and licorice for 4 weeks. Use hydralazine, prazosin, or doxazocin to control the hypertension, if necessary.
2. Liberalize salt intake before testing. Allow potassium supplementation but not high-dose loading.
3. After the patient has been upright for 2 hours and seated for less than 10 minutes, obtain a blood sample for plasma aldosterone concentration and either plasma renin activity (ng/mL/hr) or plasma renin concentration using a direct renin immunoassay.
4. Calculate the PAC/R ratio, where R is either PRA or PRC.

Interpretation

1. Upright healthy individuals:
   - PAC, <30 ng/dL (0.832 nmol/L)
   - PRA, 0.4 to 9.0 ng/mL/hr
   - PRC, 2.1 to 26.0 pg/mL
   - PAC/PRA ratio, 3 to 75
   - PAC/PRC ratio, 1.2 to 14.3
2. Patients with PAC > 20 ng/dL (0.56 nmol/L) may have an adrenal adenomas.
3. A PRA (upright) < 0.5 ng/mL/hr is suggestive of PHA or essential hypertension.
4. A PRC (upright) < 10 pg/mL is suggestive of PHA.
5. A PAC/PRA ratio > 250 constitutes a positive screening test for the diagnosis of PHA, while a ratio > 500 is virtually diagnostic of PHA with a 90 to 100% sensitivity and specificity (Ferrari et al., 2004).
6. A PAC/PRC ratio ≥ 62 in patients with PAC > 20 ng/dL reliably identifies PHA in patients with an aldosterone-secreting adrenal adenoma (Unger et al., 2004).

Notes

1. Commercially available immunoassays for the measurement of direct renin (i.e., PRC) became limited after the purchase of Nichols Institute Diagnostics, the manufacturer of the first direct renin immunoassay, by Quest Diagnostics in 2006.
2. Typical findings for *supine* PRA and PRC were 0.08 ± 0.03 ng/mL/hr and 2.6 ± 0.5 pg/mL, respectively, in patients with PHA, and 7.2 ± 2.5 ng/mL/hr and 138 ± 51 pg/mL, respectively, in patients with hepatic cirrhosis (Morganti et al., 1995).
3. Tests for diurnal variations in PAC (i.e., 8 a.m. and 11 p.m.) may help to demonstrate autonomy of aldosterone secretion in hyperaldosteronism.
4. PRA and PRC are highly correlated (Ferrari et al., 2004), while the inter- and intralaboratory coefficients of variation (CVs) for PRC measurements are lower than those for PRA (Morganti et al., 1995; Tanabe et al., 2003).
5. A two- to threefold elevated PRA is associated with myocardial infarction (MI) in the emergency room setting and increased risk of future MI in ambulatory hypertensive patients. In contrast, an increased PAC with decreased PRA is the abnormal endocrine finding in patients with PHA.
6. PHA may be the result of aldosterone-secreting tumors or bilateral adrenal hyperplasia but rarely results from adrenal carcinoma or genetic causes. In addition, 5 out of 17 patients with CKD Stage 4 or 5 were reported to have elevated PAC/PRA ratios but did not have PHA (McKenna et al., 1991).
7. The PAC/R ratio is only a screening test. Other testing is required to arrive at a precise diagnosis of the cause for hyperaldosteronism and hypertension. This is particularly true in renovascular hypertension in which renins are usually normal to upper quartile of normal.

TABLE 6.6
Results of Workup in 154 Cases of Primary Hyperaldosteronism to Determine Underlying Causes of Elevated Aldosterone

No. of Cases (% of Total)	Results of Workup
12 (7.8)	These patients did not fulfill established criteria for either an aldosterone-producing adenoma (APA) or idiopathic hyperaldosteronism (IHA).
11 (7.1)	Unilateral adrenalectomy was performed due to an APA.
8 (5.2)	Bilateral nodular adrenocortical hyperplasia was found with plasma and urinary aldosterone elevated and responses to stimulatory and suppressive maneuvers demonstrating the same autonomy seen in patients with an APA. This subset of patients was designated as primary adrenal hyperplasia (PAH). Surgery was not performed.
4 (2.6)	Elevated aldosterone levels were responsive to stimulatory and suppressive maneuvers, similar to IHA, but patient harbored a unilateral adrenal tumor. This subset of patients was designated aldosterone-producing, renin-responsive adenoma.
1 (0.6)	No clear diagnosis. Patient preferred to undergo prolonged spironolactone therapy, which resulted in a sustained cure or amelioration of hypertension, hypokalemia, and normalization of aldosterone production.

Source: Irony, I. et al., *Am. J. Hypertens.*, 3, 576–382, 1990. With permission.

8. Only in severe renal artery stenosis will high levels of renin be associated with hyperaldosteronism and hypokalemia.
9. Hyperaldosteronism is one of the few treatable causes of hypertension. A systematic approach is required to ensure that those patients with an aldosterone-secreting adrenal adenoma are properly identified and referred for adrenalectomy. Refer to "Results of Workup in 154 Cases of Primary Hyperaldosteronism to Determine Underlying Causes of Elevated Aldosterone" (Table 6.6).

ICD-9 Codes

Conditions that may justify this test *include but are not limited to*:

255.10   primary hyperaldosteronism

Suggested Reading

Blumenfeld JD, Sealey JE, Alderman MH, Cohen H, Lappin R, Catanzaro DF, Laragh JH. Plasma renin activity in the emergency department and its independent association with acute myocardial infarction. *Am J Hypertens* 2000; 13:855–63.

Brunner HR, Laragh JH, Baer L et al. Essential hypertension: renin and aldosterone, heart attack and stroke. *N Engl J Med* 1972; 286:441–9.

Chobanian AV, Bakris GL, Black HR et al. The Seventh Report of the Joint National Committee on Prevention, Detection, Evaluation, and Treatment of High Blood Pressure: the JNC7 report. *JAMA* 2003; 289:2560–72.

Ferrari P, Shaw SG, Nicod J, Saner E, Nussberger J. Active renin versus plasma renin activity to define aldosterone-to-renin ratio for primary aldosteronism. *J Hypertens* 2004; 22:377–81.

Irony I, Kater CE, Biglieri EG, Shackleton CH. Correctable subsets of primary aldosteronism: primary adrenal hyperplasia and renin responsive adenoma. *Am J Hypertens* 1990; 3:576–82.

Laragh JH. Vasoconstriction-volume analysis for understanding and treating hypertension: the use of renin and aldosterone profiles. *Am J Med* 1973; 55:261–74.

Laragh JH. The renin system and future trends in management of high blood pressure [review]. *Clin Exp Hypertens* 1980; 2:525–52.

Laragh JH, Sealey JE. Relevance of the plasma renin hormonal control system that regulates blood pressure and sodium balance for correctly treating hypertension and for evaluating ALLHAT. *Am J Hypertens* 2003; 16:407–415.

McKenna TJ, Sequeira SJ, Heffernan A, Chambers J, Cunningham S. Diagnosis under random conditions of all disorders of the renin–angiotensin–aldosterone axis, including primary hyperaldosteronism. *J Clin Endocrinol Metab* 1991; 73:952–7.

Morganti A, Pelizzola D, Mantero F, Gazzano G, Opocher G, Piffanelli A. Immunoradiometric versus enzymatic renin assay: results of the Italian Multicenter Comparative Study. Italian Multicenter Study for Standardization of Renin Measurement. *J Hypertens* 1995; 13:19–26.

Moser M, Izzo JL. Plasma renin measurement in the management of hypertension: the V and R hypothesis. *J Clin Hypertens (Greenwich)* 2003; 5:373–6.

Nadar S, Lip GY, Beevers DG. Primary hyperaldosteronism. *Ann Clin Biochem* 2003; 40(Pt 5):439–52.

Tanabe A, Naruse M, Takagi S, Tsuchiya K, Imaki T, Takano K. Variability in the renin/aldosterone profile under random and standardized sampling conditions in primary aldosteronism. *J Clin Endocrinol Metab* 2003; 88:2489–94.

Unger N, Lopez Schmidt I, Pitt C, Walz MK, Phillipp T, Mann K, Petersenn S. Comparison of active renin concentration and plasma renin activity for the diagnosis of primary hyperaldosteronism in patients with an adrenal mass. *Eur J Endo* 2004; 150:517–23.

Weber MA, Drayer JI, Rev A et al. Disparate patterns of aldosterone response during diuretic treatment of hypertension. *Ann Intern Med* 1977; 87:558–63.

6.5.3   18-Hydroxycorticosterone (18OH-B) Testing to Differentiate Idiopathic Hyperaldosteronism (IHA) from Aldosterone-Producing Adenoma (APA)

Indications for Test

The 18OH-B test is indicated:

- In patients with documented primary hyperaldosteronism (PHA) and an equivocal or unavailable imaging study of the adrenals or lack of adrenal venous sampling capability

Procedure

1. At 0800 hours, after at least 12 hours of recumbence, obtain a blood sample for 18OH-B testing.

Interpretation

1. 18OH-B < 50 ng/dL is consistent with IHA or normal adrenal status, whereas 18OH-B ≥ 50 ng/dL strongly suggests APA
2. Severe, persistent hypokalemia, increased plasma 18OH-B, and an anomalous postural decrease in the plasma aldosterone concentration, when present, provide the best indicators of the presence of an adrenal aldosterone-secreting adenoma.

Notes

1. This test helps in the differential diagnosis of APA vs. IHA and is a useful predictor of the etiology of primary aldosteronism.
2. Adrenal venous sampling for plasma aldosterone concentration remains the most precise technique for identification and localization of aldosterone-secreting tumors of the adrenal.

3. In a study of 44 individuals—9 healthy subjects, 10 with essential hypertension, 23 with an aldosterone-producing adenoma (APA), and 2 with unusual macro- and micronodular hyperplasia of the adrenal glands—except for 1 patient with an APA all of the individuals had an [18OH-B] > 100 ng/dL. By contrast, 9 patients with idiopathic adrenal hyperplasia out of 34 patients with documented primary aldosteronism had an [18OH-B] < 100 ng/dL (Kem et al., 1985).
4. All 9 patients with idiopathic adrenal hyperplasia out of 34 patients with documented primary aldosteronism had an [18OH-B] < 100 ng/dL (Kem et al., 1985).
5. In PHA, refractory hypertension, hyperkinetic circulation, and hypovolemia are frequent occurrences, while use of serum potassium concentration and plasma renin activity as screening tests for disease has led to high rates of false-positive and false-negative results.

ICD-9 Codes

Conditions that may justify this test *include but are not limited to*:

255.10   primary hyperaldosteronism

Suggested Reading

Auchus RJ, Chandler DW, Singeetham S, Chokshi N, Nwariaku FE, Dolmatch BL, Holt SA, Wians FH Jr, Josephs SC, Trimmer CK, Lopera J, Vongpatanasin W, Nesbitt SD, Leonard D, Victor RG. Measurement of 18-hydroxycorticosterone during adrenal vein sampling for primary aldosteronism. *J Clin Endocrinol Metab* 2007; 92:2648–51.

Bravo EL, Tarazi RC, Dustan HP, Fouad FM, Textor SC, Gifford RW, Vidt DG. The changing clinical spectrum of primary aldosteronism. *Am J Med* 1983; 74:641–51.

Kem DC, Tang K, Hanson CS, Brown RD, Painton R, Weinberger MH, Hollifield JW. The prediction of anatomical morphology of primary aldosteronism using serum 18-hydroxycorticosterone levels. *J Clin Endocrinol Metab* 1985; 60:67–73.

6.5.4   18-Hydroxycorticosterone/Cortisol Ratio after Saline Infusion in Hyperaldosteronism

Indications for Test

Determination of the 18-hydroxycorticosterone (18OH-B)/cortisol (F) ratio after intravenous saline is indicated:

• To differentiate IHA from APA in patients with documented hyperaldosteronism

Procedure

1. Provide patient with a carbohydrate- and protein-balanced low-sodium diet containing <100 mmol/day (approximately 2 g/day) sodium for a week before testing.
2. Obtain blood samples for 18OH-B and cortisol (F) testing before and after a 1-liter infusion of normal saline given over 2 hours (8 a.m. to 10 a.m.) after recumbence overnight.
3. Calculate the 18OH-B (ng/dL)/F (mcg/dL) ratio.

Interpretation

1. 18OH-B:F ratio:
   • ≤3 suggests IHA.
   • >3 suggests a hyperaldosterone state (e.g., APA) that may be surgically correctable.

Notes

1. Angiotensin II activity is a major mediator of mineralocorticoid hormone synthesis in patients with idiopathic hyperaldosteronism but has little or no effect in those with an aldosterone-producing adenoma.
2. After saline infusion, plasma [cortisol] decreases in both IHA and APA patients, but plasma renin concentration decreases only in those patients with IHA ($p < 0.05$).
3. An aldosterone/cortisol ratio of <2.2 after saline infusion is consistent with IHA.
4. The saline infusion test with measurement of aldosterone has low sensitivity and specificity in diagnosis of normokalemic hyperaldosteronism (Schirpenbach et al., 2006).

ICD-9 Codes

Conditions that may justify this test *include but are not limited to*:

227 benign neoplasm of other endocrine glands and related structures
255.10 hyperaldosteronism

Suggested Reading

Arteaga E, Klein R, Biglieri EG. Use of the saline infusion test to diagnose the cause of primary aldosteronism. *Am J Med* 1985; 79:722–8.

Schirpenbach C, Reincke M. Screening for primary aldosteronism. *Best Prac Res Clin Endocrinol Metab* 2006; 20:369–84.

6.5.5 Adrenal Vein Sampling (AVS) Protocol for Localization of Adrenal Source of an Elevated Aldosterone Concentration

Indications for Test

Performing AVS is indicated in patients with:

- Confirmed hyperaldosteronism based on clinical and laboratory findings including fatigue, muscle weakness, hypertension, alkalosis, hypokalemia, and elevated aldosterone levels but negative or equivocal imaging studies of the adrenals in localizing an adrenal tumor

Procedure

1. Refer to Specimen Collection Protocol P7 in Appendix 2 for assembly of materials required in advance.
2. On the day of the scheduled AVS procedure:
    - Contact the angiography suite and confirm that the patient has arrived and the procedure is on schedule.
    - Imprint patient information on the appropriate lab requisition forms. Note that a separate lab requisition is usually required for each blood specimen obtained during the procedure.
    - Ask the radiologist to contact you when the patient is ready for the procedure so you can be there to ensure the collection of a baseline sample prior to the infusion of ACTH and to ensure proper infusion of the ACTH.
    - Upon notification from radiology service, take the labels and test tube tray in an insulated cooler with ice, a copy of the protocol, and a pen and go to the angiography suite. Bring extra tubes in case one breaks.
    - Review the AVS protocol briefly with the radiologist.
    - Prepare five patient labels with specimen site, date, time, and your initials as collector. Affix the labels to five serum tubes. Prepare a duplicate set of labels for the lab to use for their pour-off tubes. Be sure that the actual time of sampling is indicated to the minute (e.g., 11:18 a.m.) on each tube, as well as the identifiers RAV, LAV, IVC, PV, and PV-B.

3. Prior to the infusion of ACTH, a peripheral vein i.v. line should be inserted and a 5-mL blood sample obtained for cortisol testing. This is the baseline specimen (PV-B).
4. Dissolve 0.25 mg of cosyntropin in 250 mL of D5W and infuse this solution through the peripheral vein i.v. line at a rate of 50 mL/hr. This provides a sustained stimulation of the adrenal glands and minimizes fluctuations in cortisol concentration.
5. Have the interventional radiologist insert catheters simultaneously into the right (RAV) and left (LAV) adrenal veins via the femoral veins.
6. After the start of the ACTH infusion, the radiologist will engage the adrenal vein catheters and obtain four more blood samples, one each from both adrenal veins (RAV and LAV), the inferior vena cava (IVC) below the renal veins, and the PV during the ACTH infusion and hand them to the assistant.
7. Refer to Specimen Collection Protocol P7 in Appendix 2 for details regarding sample collection and transfer at this point.
8. If the RAV cannot be successfully catheterized, use the aldosterone and cortisol results from the blood sample obtained from the IVC to help determine lateralization.
9. Be sure that all blood samples are placed in the correctly labeled SST. On each tube, note the actual clock time that these samples were obtained. Stop the cosyntropin infusion after the radiologist is satisfied that the catheters have been appropriately engaged in the RAV and LAV and blood specimen collection is complete.
10. Transport all blood specimens to the lab for measurement of [aldosterone] and [cortisol] in all five blood samples (i.e., PV-B, RAV, LAV, IVC, and PV). Leave the second set of labels with these samples for use by laboratory personnel.
11. Determine the serum aldosterone (A) concentration (ng/dL) and cortisol (C) concentration (mcg/dL) in the RAV, LAV, IVC, and PV blood specimens obtained during ACTH infusion. Identify aldosterone results as A_{RAV}, A_{LAV}, A_{IVC}, and A_{PV} and cortisol results as C_{RAV}, C_{LAV}, C_{IVC}, and C_{PV}.
12. Calculate the A/C ratio for each site from which the blood specimens were obtained: $(A/C)_{RAV}$, $(A/C)_{LAV}$, $(A/C)_{IVC}$, and $(A/C)_{IVC}$.

Interpretation

1. To simplify the interpretation of the results of the AVS procedure, use a form similar to that shown in Figure 6.3. The information appearing on this form includes:
   - A/C value for each sample—that is, four A/C values, one for each site (RAV, LAV, IVC, PV) from which a blood specimen was obtained:

     $(A/C)_{RAV}$ $(A/C)_{LAV}$
     $(A/C)_{IVC}$ $(A/C)_{PV}$
   - $(A/C)_D$, the dominant A/C value (the *highest* A/C value among all four A/C values), and $(A/C)_{ND}$, the nondominant value (the *lowest* A/C value among all four A/C values)
   - Values for $(A/C)_D/(A/C)_{IVC}$ (D/IVC) and $(A/C)_{ND}/(A/C)_{IVC}$ (ND/IVC)
   - Value obtained for $(A/C)_D/(A/C)_{ND}$
   - The answers (Yes or No) to the following questions:
     a. Overall, was AVS successful (i.e., $C_{RAV} > 3C_{IVC}$ *and* $C_{LAV} > 3 C_{IVC}$)?
     b. Was the $C_{PV} > 20$ mcg/dL?
     c. Were criteria met for APA or PAH; that is, $(A/C)_D/(A/C)_{ND} \geq 4$ *and* $(A/C)_{ND} < (A/C)_{IVC}$?
     d. Were criteria met for BAH; that is, $(A/C)_D/(A/C)_{ND} < 4$ *and* both $(A/C)_D$ and $(A/C)_{ND} > (A/C)_{IVC}$?

   Abbreviations: APA, aldosterone-producing adenoma; AVS, adrenal vein sampling; BAH, bilateral adrenal hyperplasia; C_{PV}, cortisol concentration in the peripheral vein (PV) sample after ACTH infusion; PAH, primary adrenal hyperplasia
2. Based on the answers to the questions above and using the criteria included in the Interpretive Notes section of the form, the form will also indicate, automatically, whether the laterality of the aldosterone-producing adenoma is the right (RAGA) or left (LAGA) adrenal gland, or that bilateral adrenal hyperplasia is the most likely interpretation of all data.
3. Provide a copy of the report form to the appropriate pathologist/physician for review and signature.
4. Overall AVS was successful if both the C_{RAV} and C_{LAV} were >$3C_{IVC}$.
5. $C_{PV} > 20$ mcg/dL is the expected result following cosyntropin infusion.

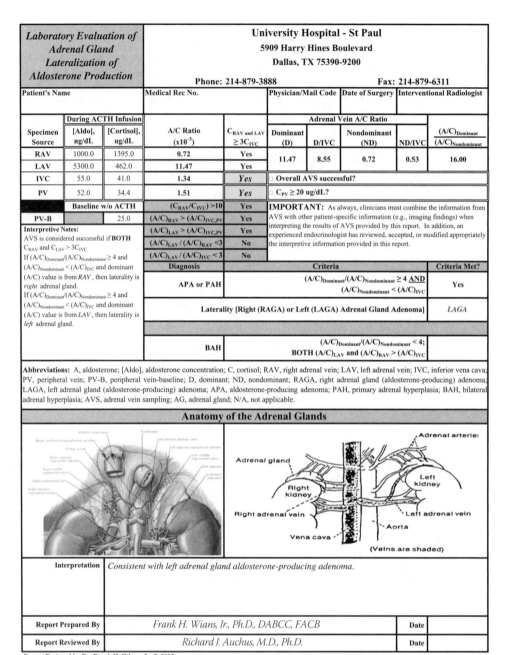

Figure 6.3

Sample adrenal vein sampling (AVS) report from the laboratory of one of the authors, illustrating the calculations, interpretive criteria, and interpretation of aldosterone and cortisol data from an AVS. (Figure courtesy of Dr. Frank H. Wians, Jr.)

6. If $(A/C)_D/(A/C)_{ND} \geq 4$ and $(A/C)_{ND} < (A/C)_{IVC}$, then laterality is to the side (right or left adrenal gland) from which the dominant A/C value was obtained.

7. Thus, when both the RAV and LAV are successfully catheterized:
   - If $(A/C)_{RAV} \geq 4(A/C)_{LAV}$ and $(A/C)_{LAV} < (A/C)_{IVC}$, laterality is to the right adrenal gland.
   - If $(A/C)_{LAV} \geq 4(A/C)_{RAV}$ and $(A/C)_{RAV} < (A/C)_{IVC}$, laterality is to the left adrenal gland.
   - If $(A/C)_D/(A/C)_{ND} < 4$ and both $(A/C)_{RAV}$ and $(A/C)_{LAV}$ are $\geq (A/C)_{IVC}$, consider bilateral adrenal hyperplasia or idiopathic hyperaldosteronism as the cause of these findings.

Notes

1. AVS testing supplements the use of CT and adrenal scintigraphy in differentiating between hyper-aldosteronism due to bilateral adrenal gland hyperplasia (BAH) vs. primary adrenal hyperplasia (PAH) due to an aldosterone-producing adenoma (APA) and to provide additional information regarding which adrenal gland, right or left, contains the adenoma.

2. This test helps to identify which adrenal gland (right or left) contains a functional adenoma and is particularly useful if no nodule or bilateral nodules are seen on imaging studies in cases when the suspicion of an APA is high.

3. Continuous ACTH infusion during the AVS test minimizes fluctuations in cortisol concentration by stimulating the adrenal glands in a sustained fashion.

4. A unilateral source of aldosterone excess may be found in up to 40% of patients whose adrenal glands appear normal or show only a minimally expanded adrenal limb on CT scan.

5. AVS findings are crucial in determining the source of aldosterone excess in patients with bilateral adrenal masses or atypical-appearing macroadenomas by CT.

6. A concordance of only 54% (22 out of 41 patients) was observed between the CT finding of an adrenal tumor and adrenal vein sampling in the localization of a source of abnormal aldosterone production (Nawariaku et al., 2006).

7. Despite the difficulty in successfully catheterizing the RAV, compared to the LAV, expect successful adrenal vein catheterization for both the RAV and LAV by an experienced interventional radiologist in more than 95% of patients (Young et al., 1996).

8. An Excel® (Microsoft; Seattle, WA) spreadsheet program for calculating all A/C values and interpreting the results of AVS testing per the criteria indicated above is available from Department of Pathology University of Texas Southwestern Medical Center in Dallas, TX (c/o Dr. Frank H. Wians, Jr.).

ICD-9 Codes

Conditions that may justify this test *include but are not limited to*:

Electrolytes
276.3   alkalosis
276.8   hypopotassemia

Other
255.1   hyperaldosteronism
401.1   hypertension, benign
728.87   muscle weakness (generalized)
780.7   malaise and fatigue

Suggested Reading

Doppman JL, Gill JR Jr. Hyperaldosteronism: sampling the adrenal veins. *Radiology* 1996; 198:309–12.

Iwaoka T, Umeda T, Naomi S, Miura F, Inoue J, Sasaki M, Hamasaki S, Sato T. Localization of aldosterone-producing adenoma: venous sampling in primary aldosteronism. *Endocrinol Jpn* 1990; 37:151–7.

Magill SB, Raff H, Shaker JL, Brickner RC, Knechtges TE, Kehoe ME, Findling JW. Comparison of adrenal vein sampling and computed tomography in the differentiation of primary aldosteronism. *J Clin Endocrinol Metab* 2001; 86:1066–71.

Nwariaku FE, Miller BS, Auchus R, Holt S, Watumull L, Dolmatch B, Nesbitt S, Vongpatanasin W, Victor R, Wians FH Jr, Livingston E, Snyder WH III. Primary hyperaldosteronism: effect of adrenal vein sampling on surgical outcome. *Arch Surg* 2006; 141:497–503.

Wians FH Jr, Stein DT, Keffer JH. The case for adrenal venous sampling in a patient with primary hyperaldosteronism. *J Diagn Endocrinol Immunol Metab* 1999; 17:289–94.

Young WF Jr, Stanson AW, Grant CS, Thompson GB, van Heerden JA. Primary aldosteronism: adrenal venous sampling. *Surgery* 1996; 120:913–20.

TABLE 6.7
Characteristics and Differential Diagnosis of
Adrenocortical Nodules Based on CT Imaging

Nodule Characteristic	Conditions Among Differential Diagnosis			
	Adenoma	Adrenal Carcinoma	Pheochromo-cytoma	Metastasis to Adrenal Glands
Size (diameter) (cm)	≤3	>4	>3	Variable
Shape	Round, oval, smooth	Irregular, indistinct margins	Round, oval, distinct margins	Oval, irregular, indistinct margins
Texture	Homogenous	Heterogeneous	Heterogeneous	Heterogeneous
Laterality	Solitary, unilateral	Solitary, unilateral	Solitary, unilateral	Frequently bilateral
Density (Hounsfield units)	≤10	>10[a]	>10[a]	>10[a]
Vascularity	Not very vascular	Usually vascular	Usually vascular	Usually vascular
Contrast washout[b] (%)	≥50	<50	<50	<50

[a] Typically, >25 HU.

[b] At 10 minutes.

Source: Adapted from Young, Jr., W.F. et al., *Endocrinol. Metab. Clin. North Am.*, 29, 159–185, 2000.

6.5.6 Imaging of Adrenal Glands Using Cholesterol Analogs (NP-59, [131]I- or [75]Se-6β-Methyl-19-Norcholesterol) as an Adjunct to Multislice Computerized Tomography (M-CT)

Indications for Test

Imaging of the adrenal glands using the isotope-labeled cholesterol analogs NP-59 or radioisotope labeled 6β-methyl-19-norcholesterol as an adjunct to M-CT is indicated in patients with:

- Hyperaldosteronism to discriminate between a unilateral adrenal adenoma vs. bilateral adrenal hyperplasia as the cause of their hyperaldosteronism
- ACTH-independent Cushing's syndrome (CS) to discriminate between a space-occupying lesion (i.e., a hyperfunctional adenoma) in the adrenal glands, adrenal carcinoma, or adrenal nodular hyperplasia
- Virilizing hyperandrogenism and an anatomic abnormality, with unknown functionality, of the ovaries or adrenal glands

Procedure

1. Perform M-CT imaging of the adrenals. Obtain results of the noncontrast CT attenuation coefficient in Hounsfield units (HU) of the space-occupying lesions present in the adrenals.
2. Quantitatively identify the size, shape, laterality, density, and vascularity of a lesion on CT imaging with and without contrast enhancement and appearance on MR if available. Refer to "Characteristics and Differential Diagnosis of Adrenocortical Nodules Based on CT Imaging" (Table 6.7).
3. Do not perform this test in patients with ACTH-dependent (i.e., pituitary-related) Cushing's disease (CD). Note that CD accounts for more than 90% of all cases of CS.
4. Establish whether or not the patient's ambient diet is relatively low in iodine and salt intake. If not, place the patient on a low-iodine, low-salt (i.e., <1.5 g Na) diet for 2 weeks prior to imaging studies using isotope-labeled cholesterol.
5. Block thyroidal ^{131}I uptake with unlabeled iodide treatment (e.g., saturated solution of potassium iodide [SSKI]; one drop p.o. t.i.d.) or administration of T_3 (20 mcg q 8h p.o.) for 2 days prior to and 14 days after isotope-labeled cholesterol dosing.

6. Decrease colonic retention of excreted isotope-labeled cholesterol by administering, 2 hours prior to the first adrenal imaging scan, a mild laxative (e.g., bisacodyl) that does not interfere with the enterohepatic circulation of isotope-labeled cholesterol.

7. If a dexamethasone suppression test (DST) is not being performed prior to imaging studies using isotope-labeled cholesterol analogs, administer a precisely calculated and verified dose (37 MBq [1.0 mCi]/1.73 m^2 BSA) of isotope-labeled cholesterol i.v. over 3 to 4 minutes. *Important:* Extravasation of this highly caustic agent must be scrupulously avoided.

8. Using a gamma camera collimator system, obtain posterior, anterior, and lateral abdominal isotope marker images after i.v. administration and at 5 and 7 days after administration of isotope-labeled cholesterol.

9. If a DST (1 mg dexamethasone p.o. q.i.d. for 7 days prior to isotope-labeled cholesterol imaging study) is performed, obtain posterior, anterior, and lateral abdominal marker images after i.v. administration and at 3 and 5 days after administration of isotope-labeled cholesterol.

Interpretation

1. In healthy individuals, isotope-labeled cholesterol (marker) uptake $\leq 0.26\%$ of baseline marker level (mean, 0.16%; range, 0.07–0.26%).

2. A noncontrast CT attenuation coefficient of ≤ 10 HU makes diagnosis of adrenal metastases or adrenal carcinoma very unlikely.

3. In patients with CS for whom a DST was performed prior to imaging studies, marker uptake is characterized by:
   - 50% lower uptake compared to those in whom no DST was performed
   - Early visualization of both adrenals in patients with adrenal hyperplasia
   - Early visualization of a hyperfunctioning adenoma in an affected adrenal gland
   - Visualization (using a deeper view of the pelvis) of the ovary that is the source of androgen secretion in a hyperandrogenic patient

4. In patients with CS for whom a DST was not performed prior to marker imaging studies, imaging helps to identify patients with:
   - A single adenoma with ipsilaterally increased marker uptake and contralateral suppression
   - A carcinoma in which there is bilateral nonvisualization of marker uptake usually with HU > 10
   - Adrenal hyperplasia in which there is typically bilaterally increased symmetric uptake of marker

5. Surgical excision of lesions ≥ 6 cm is usually indicated depending on radiologic characteristics of the mass.

Notes

1. The vehicle for isotope-labeled cholesterol is caustic; thus, extravasation must be stringently avoided.

2. This imaging modality exploits the ability of isotope-labeled cholesterol analogs to be incorporated into low-density lipoprotein cholesterol (LDL-C) particles and undergo esterification but not be further metabolized.

3. High-resolution M-CT scan imaging and MR scanning with pulse sequencing of the adrenal glands have high diagnostic sensitivity for detection of anatomic abnormalities but low specificity for adrenal gland disease so the frequency of unnecessary follow-up procedures (e.g., surgery) can be unacceptably high.

4. Unless a clinical or biochemical adrenal gland abnormality is present, avoid advanced imaging studies of the adrenal glands.

5. Lesions seen on M-CT with HU ≤ 10 tend to have higher lipid content. Lesions with HU > 10 tend to be pheochromocytomas, adrenocortical carcinomas, or metastases to the adrenal (Boland et al., 1998).

6. Isotope-labeled cholesterol imaging studies can visualize functional abnormalities in all three layers of the adrenal cortex.

ICD-9 Codes

Conditions that may justify this test *include but are not limited to*:

255.0   Cushing's syndrome
255.1   hyperaldosteronism
255.9   unspecified disorder of adrenal glands

Suggested Reading

Boland GW, Lee MJ, Gazelle GS, Halpern EF, McNicholas MM, Mueller PR. Characterization of adrenal masses using unenhanced CT: an analysis of the CT literature. *AJR Am J Roentgenol* 1998; 171:201–4.

Young WF Jr. Management approaches to adrenal incidentalomas: a view from Rochester, Minnesota. *Endocrinol Metab Clin North Am* 2000; 29:159–85.

6.6   Catecholamine Excess (Pheochromocytoma)

6.6.1   Free Metanephrines (FMNs): Screening Test in Plasma (PFMN) and Random Urine Samples

Indications for Test

Testing for free or total MNs (i.e., normetanephrine + metanephrine) or individual catecholamines is indicated in patients with:

- Signs and symptoms of pheochromocytoma, such as flushing or episodic hypertension, with or without an associated medullary carcinoma
- von Hippel–Lindau disease or multiple endocrine neoplasia type 2 (MEN 2) with labial neuromas (i.e., "bumpy lips")

Procedure

1. Prior to the collection of a blood sample for plasma free metanephrine (PFMN) testing or random urine sample for MNs, be sure that the patient has abstained from caffeinated and decaffeinated beverages, beta blockers, tobacco, alcohol for 4 hours, acetaminophen for 5 days, and epinephrine or epinephrine-like drugs (e.g., alpha-methyldopa) for 1 week.
2. With the patient in a supine position for at least 20 minutes, obtain a blood sample for PFMN testing by high-performance liquid chromatography (HPLC).
3. Repeat testing for PFMNs if clinical suspicion for pheochromocytoma is high or the initial test is negative or only slightly above reference range, as tumor secretion of hormone may be intermittent.
4. Before collection of random urine sample for MNs, have patient abstain from the substances noted above for at least 3 days.

Interpretation

1. In healthy individuals, the reference intervals for PFMNs by HPLC are:
   - Normetanephrine, <610 pmol/L (111 pg/mL)
   - Metanephrine, <310 pmol/L (61 pg/mL)
2. Interpretive data for PFMNs by HPLC to clearly discriminate between healthy individuals and individuals with signs and symptoms of pheochromocytoma given the following levels in plasma are:
   - Normetanephrine, >2190 pmol/L (401 pg/mL)
   - Metanephrine, >1200 pmol/L (239 pg/mL)

   These levels have a 100% diagnostic specificity and 79% sensitivity in identifying patients with pheochromocytoma (Lenders et al., 2002a,b).
3. PFMN reference intervals by LC-MS/MS (Quest Diagnostics):
   - Normetanephrine—adult male, 207 to 956 pmol/L (38–175 pg/mL); adult female, 80 to 710 pmol/L (33–130 pg/mL)
   - Metanephrine—adult male, 416 pmol/L (≤82 pg/mL); adult female, 487 pmol/L (≤96 pg/mL)
   - Total metanephrines—adult male, 321 to 1294 pmol/L (61–246 pg/mL); adult female, 263 to 1052 pmol/L (50–200 pg/mL)

4. Fractionated urine metanephrines reference intervals by LC-MS/MS random urine screen (Quest Diagnostics):
   - Normetanephrine (>40-year-old adult), 85 to 524 mcg/g creatinine
   - Metanephrine (>40-year-old adult), 21 to 192 mcg/g creatinine
   - Total metanephrines (>40-year-old adult), 149 to 608 mcg/g creatinine

Notes

1. In addition to pheochromocytoma, the differential diagnosis of flushing includes carcinoid syndrome, niacin reactions, alcohol dehydrogenase gene polymorphism in Orientals who have alcohol intolerance and ingest alcohol, perimenopause, menopause rosacea, and septic infections.
2. Catecholamines are metabolized to free metanephrines within pheochromocytoma tumor cells independent of episodic catecholamine release.
3. False-negative test results may occur in asymptomatic patients tested because of an adrenal incidentaloma or familial pheochromocytoma or when sampling is carried out between episodes of paroxysmal hypertension.
4. Measurements of urinary total metanephrines or vanillylmandelic acid (VMA), although highly specific (93–95%), are of lesser value in initial screening for pheochromocytoma given their lower sensitivity compared to PFMNs.
5. The sensitivity of measurements of PFMNs for the detection of pheochromocytoma was 97% in patients with a familial predisposition to these tumors (Eisenhofer et al., 1999).
6. The diagnostic sensitivity and specificity for detection or exclusion of a pheochromocytoma are highest for PFMNs when compared to total catecholamines in plasma and urine, total and fractionated urine metanephrines, and VMA in urine.

ICD-9 Codes

Conditions that may justify this test *include but are not limited to*:

255.6	medulloadrenal hyperfunction
401.1	hypertension, benign
528.5	diseases of lips
759.6	hamartoses
782.62	flushing

Suggested Reading

Eisenhofer G, Lenders JW, Linehan WM, Walther MM, Goldstein DS, Keiser HR. Plasma normetanephrine and metanephrine for detecting pheochromocytoma in von Hippel–Lindau disease and multiple endocrine neoplasia type 2. *N Engl J Med* 1999; 340:1872–9.

Lenders JW, Pacak K, Eisenhofer G. New advances in the biochemical diagnosis of pheochromocytoma: moving beyond catecholamines. *Ann NY Acad Sci* 2002a; 970:29–40.

Lenders JW, Pacak K, Walther MM, Linehan WM, Mannelli M, Friberg P, Keiser HR, Goldstein DS, Eisenhofer G. Biochemical diagnosis of pheochromocytoma: which test is best? *JAMA* 2002b; 287:1427–34.

6.6.2 Metaiodobenzylguanidine (MIBG) Scan as an Adjunct to Magnetic Resonance (MR) or Multislice Computerized Tomography (M-CT) Imaging for Neuroendocrine Tissue

Indications for Test

MIBG scanning is indicated:

- For localization of sympatomedullary adrenal lesions when extraadrenal or metastatic neoplasms are suspected in patients with known pheochromocytomas or neuroblastomas especially when adrenal imaging with M-CT or MR is negative or equivocal and surgical exploration is planned

Procedure

1. If an initial MR or M-CT scan fails to identify a neuroendocrine lesion (see Table 6.7), obtain a repeat MR of the abdomen prior to MIBG scan as it may be as sensitive for detection of similar size metastatic neuroendocrine tumors (up to 100%) and does not involve radioisotopes.
2. Confine imaging of a pregnant patient's abdomen to MR and ultrasound, as MIBG scanning is contraindicated in pregnant patients.
3. If MR is unavailable or nonspecific and MIBG scan is needed to confirm or better localize the lesion, instruct the patient to discontinue use of the drugs guanethidine, catecholamine agonists, antipsychotics, calcium channel blockers, and cocaine, which may interfere with MIBG uptake.
4. Pretreat the patient with oral potassium iodide for 2 days to block thyroidal uptake of radioiodine-labeled MIBG.
5. Administer a 0.5- to 1.0-mCi i.v. dose of [131]I- or ^{123}I-MIBG.
6. Continue iodide therapy for 6 days after dosing if ^{131}I-MIBG is used and for 4 days after dosing if ^{123}I-MIBG is used.
7. Image the whole body with attention to the abdomen 24, 48, and 72 hours after dosing with ^{131}I-MIBG or 2 hours and 24 hours (rarely 48 hours) with use of ^{123}I-MIBG.

Interpretation

1. The radiologist bases assessment of the appearance of abnormal or increased uptake on the scan of abdomen and spine.

Notes

1. MIBG is an analog of the endogenous neurotransmitter norepinephrine.
2. MIBG uptake is 77 to 88% sensitive, 88 to 100% specific, and 86 to 94% accurate for localization of a pheochromocytoma (Witteles et al., 2000).
3. MR scan with gadolinium enhancement is highly sensitive but less specific for identifying extra-adrenal pheochromocytomas.
4. Other neuroendocrine tumors, such as those associated with amine precursor uptake and decaboxylation (APUD) cells in the gastrointestinal tract, may be imaged with MIBG, but not as well as pheochromocytomas or neuroblastomas.
5. MIBG scanning is expensive and not widely available and should be reserved for circumstances when the M-CT scan is negative or equivocal in the clinical setting of a positive biochemical test for pheochromocytoma and suspected neuroendocrine tumor.
6. The diagnostic sensitivity of single detector spiral CT scan for pheochromocytoma masses is only 88%; hence, M-CT is the preferred imaging technique.

ICD-9 Codes

Conditions that may justify this test *include but are not limited to*:

194.0X   malignant neoplasm of the adrenal glands

Suggested Reading

Boersma HH, Wensing JW, Kho TL, de Brauw LM, Liem IH, van Kroonenburgh MJ. Transient enhanced uptake of ^{123}I-metaiodobenzylguanidine in the contralateral adrenal region after resection of an adrenal pheochromocytoma. *N Engl J Med* 2000; 342:1450–1.

Boyd CA. Amine uptake and peptide hormone secretion: APUD cells in a new landscape. *J Physiol*. 2001; 531:581.

Witteles RM, Kaplan EL, Roizen MF. Sensitivity of diagnostic and localization tests for pheochromocytoma in clinical practice. *Arch Intern Med* 2000; 160:2521–4.

6.7 Virilizing Gonadal Steroids of Adrenal Origin in Females

6.7.1 Cosyntropin Stimulation Test (CST) in Hirsutism and Possible Late-Onset Congenital Adrenal Hyperplasia (CAH)

Indications for Test

The CST is indicated in female patients with:

- Idiopathic precocious pubarche or adult-onset polycystic ovary syndrome, moderate to severe hirsutism, virilization, oligomenorrhea, or infertility
- Elevations of baseline levels of serum total testosterone, [TT]; 17-hydroxyprogesterone, [17OH-Prog]; 17-hydroxypregnenolone, [17OH-Preg]; and/or 11-deoxycortisol, [11-DC], suggestive of late-onset CAH
- Elevations of baseline serum total testosterone [TT]; free testosterone, [FT]; androstenedione, [A]; dehydroepiandrosterone, [DHEA]; and sulfated DHEA, [DHEAS] (see Test 9.4.1)

Procedure

1. Obtain a blood sample for baseline determination of serum [cortisol], [TT], [DHEAS], [17OH-Prog], [17OH-Preg], and [11-DC].
2. If values for all baseline tests indicated above are in the upper reference range or only slightly elevated, be sure that the patient discontinues use of any oral contraceptives or other gonadal or adrenal hormone-related therapy (i.e., glucocorticoids, estrogens).
3. Obtain repeat blood samples for confirmation of elevated test results. Terminate glucocorticoid and androgenic steroid therapy prior to any further testing as such treatment interferes with the measurement of these analytes.
4. Schedule the CST during the follicular phase of the menstrual cycle if the patient is cycling.
5. Administer high-dose (250 mcg i.v.) cosyntropin and obtain blood samples at 0, 30, and 60 minutes after injection for [cortisol], [TT], [17OH-Prog], [17OH-Preg], and [11-DC] during the follicular phase of the menstrual cycle if present.

Interpretation

1. Refer to "Interpretation of Steroid Hormone Test Results Following Cosyntropin Stimulation Test" (Table 6.8), "Concentration of Baseline 17OH-Preg and the 17-P/F Ratio in Individuals with Type II 3β-HSD Deficiency and Proven Genotypic Abnormalities (GAs) Compared to a Healthy Reference Population" (Table 6.9), and "Cosyntropin (ACTH) Stimulation of Steroids in Normal Adults" (Table 6.10) for reference ranges and changes in steroid hormones consistent with various clinical diagnoses.
2. A baseline [TT] ≥ 200 ng/dL (6.94 nmol/L) prompts a search for palpable adnexal masses in the pelvis, imaging of the pelvis if no mass is felt, and, eventually, surgical exploration.
3. A baseline [17OH-Prog] ≥ 200 ng/dL (6 nmol/L) prompts performance of the CST.
4. A CST-stimulated [17OH-Prog] ≥ 1000 ng/dL (2.2 nmol/L) establishes the diagnosis of CAH, usually of late-onset adrenal hyperplasia.

Notes

1. Classic CAH, characterized by genital ambiguity, occurs in 1 in 15,000 live births worldwide, 75% of whom have mineralocorticoid deficiency, which can be fatal if not discovered promptly. Salt-wasting and genital ambiguity in neonatal or prenatal females are strong evidence for the diagnosis of classic CAH. A high index of suspicion is required in less overt cases.
2. In the United States, 21-hydroxylase-deficient late-onset or nonclassical adrenal hyperplasia (ncAH) appears to affect only 0.1% to rarely more than 3% of hyperandrogenic women and is even less common in cases of children with premature adrenarche.

TABLE 6.8
Interpretation of Steroid Hormone Test Results
Following Cosyntropin Stimulation Test

Test and Subject	[Steroid] Range after Administration of High-Dose (250-mcg) Cosyntropin		Interpretation
	0 (Baseline)	30 or 60 Min	
11-Deoxycortisol (ng/dL)			
Term infants through prepuberty (age 12 years)	7–210	<400	Normal response[a]
Puberty through adult	10–150	<300	Normal response
17OH-Preg (ng/dL)			
Healthy adults	20–190	<1000[b]	Normal response
Hirsute women	—	<1500	Normal response
Children with premature pubarche without 3β-HSD deficiency	—	≤2000	Normal response
Neonatal infants with ambiguous genitalia	—	≥12000	3β-HSD deficiency
Tanner stage 1 children with ambiguous genitalia	—	≥5500	3β-HSD deficiency
Tanner stage 2 children with premature pubarche	—	≥9500	3β-HSD deficiency
17OH-Prog (ng/dL)			
Pubertal and prepubertal subjects	<180	<500	Normal response
	>200	—	Nonclassical 21-hydroxylase CAH
	—	>1000	Partial 21-hydroxylase deficiency
	>500	—	Classic 21-hydroxylase CAH
17OH-Preg (ng/dL); 17OH-Preg/17OH-Prog ratio	—	≥1354; >10.4	Partial 3β-HSD deficiency in healthy pubertal children
17OH-Preg (ng/dL)/cortisol (mcg/dL) ratio			
Hirsute women	—	≤151	Normal response
Children with premature pubarche without 3β-HSD deficiency	—	≤67	Normal response

Gender	F	M	F and M	
17OH-Prog (ng/dL)	9–300	36–364	<909	Normal response
17OH-Preg (ng/dL)	91–273	91–636	<2000[c]	Normal response
11-Deoxycortisol (ng/dL)	45–218	45–218	<340	Normal response

[a] Nonclassical or partial 11β-hydroxylase deficiency is unlikely; convert ng/dL of Prog, Preg, or deoxycortisol to nmol/L by multiplying ng/dL by 0.0330.

[b] Quest Diagnostics data (2007) from 10 males and 10 females.

[c] Data from Eldar-Geva (1990), New (1983), and Azziz (1989) for populations with a high prevalence of CAH.

Note: 17OH-Preg, 17-hydroxypregnenolone; 17OH-Prog, 17-hydroxyprogesterone; 3β-HSD, 3β-hydroxysteroid dehydrogenase; CAH, congenital adrenal hyperplasia; M, males; F, females.

3. About one third of Israeli Jewish women with hirsutism, menstrual disorders, or unexplained infertility appeared to have ncAH and an excess of intraadrenal androgen production associated with one or more of three adrenal enzyme deficiencies (Eldar-Geva et al., 1990).

4. The CST, but not basal steroid levels, uncovers mild defects in adrenal steroidogenesis in a substantial proportion of hirsute women.

TABLE 6.9
Concentration of Baseline 17OH-Preg and the 17-P/F Ratio in Individuals with Type II 3β-HSD Deficiency and Proven Genotypic Abnormalities (GAs) Compared to a Healthy Reference Population

| | [17OH-Preg] (ng/dL) in Individuals Who Are | | 17-P/F Ratio in Individuals Who Are | |
Patient	Healthy	GA+	Healthy	GA+
Neonate	42–148	>378	34–142	>434
Tanner stage 1	7.7–16.3	>165	3–21	>216
Premature pubarche	12–22[a]	>294	11–29	>363
Adults	13–37	>289	11–47	>4010

[a] Individuals greater than Tanner stage 2 sexual development.

Note: 17OH-Preg, 17-hydroxypregnenolone; 17-P/F, 17-hydroxypregnenolone/cortisol ratio (ng/dL/mcg/dL); 3β-HSD, 3β-hydroxysteroid dehydrogenase; GA+, individuals positive for a genotypic abnormality associated with type II 3β-HSD deficiency.

TABLE 6.10
Cosyntropin (ACTH) Stimulation Results in Normal Adults[a]

| | Range of Blood Levels (ng/dL) | | | | |
| | Time after ACTH Administration (minutes) | | | | |
Steroid	0	15	30	60	90
Glucocorticoids					
17-Hydroxpregnenolone	29–189	267–856	305–847	293–913	252–846
17-Hydroxyprogesterone	27–122	63–183	74–211	72–187	58–180
Dehydroepiandrosterone	230–955	472–1681	545–1785	545–1846	545–1855
Androstenedione	56–134	69–206	88–271	72–288	98–292
11-Deoxycortisol	21–133	82–232	82–292	82–261	82–246
Cortisol (mcg/dL)	5.8–19	13–30	14–36	14–41	18–43
Mineralocorticoids					
Progesterone	5–33	—	11–44	21–44	21–69
Deoxycorticosterone	3–10	—	11–35	14—33	13–30
18-Hydroxycorticosterone	11–46	43–131	45–151	54–161	61–233
Aldosterone	2–9	6–33	5–27	5–20	4–15

[a] 250 mcg; data from normal volunteers (10 females and 10 males), courtesy of Quest Diagnostics.

5. An abnormal CST is sufficient in the making of a preliminary diagnosis of CAH and, specifically, deficiencies of the enzymes 21-hydroxylase (21OH), 3β-hydroxysteroid dehydrogenase (3β-HSD) or 11β-hydroxylase (11β-OHase).

6. CAH secondary to 21OHase deficiency is a disorder of cortisol and aldosterone biosynthesis that results from mutations in the CYP21 gene encoding the adrenal 21OHase P-450c21 enzyme.

7. Although patients with salt-wasting 21OHase deficiency have functionally equivalent mutations in their CYP21 genes, they may vary from one another and, over time, in their ability to produce mineralocorticoids. This variation may be attributable to other adrenal enzymes with 21OHase activity.

8. Baseline and ACTH-stimulated [DHEA] and 17OH-Preg/17OH-Prog and DHEA/A ratios in genotype-proven 3β-HSD deficiency patients overlap with those of genotype-normal patients or control subjects (Lutfallah et al., 2002).

9. [Epinephrine] and [metanephrine] were 40 to 80% lower in the patients with more advanced CAH than in normal subjects ($p < 0.05$) and were lowest in the patients with the most severe deficits in cortisol production, consistent with compromises in both the development and functioning of the adrenomedullary system.
10. A consensus statement on the diagnosis, and treatment of CAH has been presented (LWPES/ESPE, 2002).

ICD-9 Codes

Conditions that may justify this test *include but are not limited to*:

255.2	adrenogenital disorders
259.1	precocious sexual development and puberty, not elsewhere classified
620.2	ovarian cyst, other and unspecified
626.1	scanty or infrequent menstruation
704.1	hirsutism

Suggested Reading

Azziz R, Zacur HA. 21-Hydroxylase deficiency in female hyperandrogenism: screening and diagnosis. *J Clin Endocrinol Metab* 1989; 69:577–84.

Azziz R, Dewailly D, Owerbach D. Clinical review 56: nonclassic adrenal hyperplasia: current concepts. *J Clin Endocrinol Metab* 1994; 78:810–5.

Dewailly D. ACTH stimulation tests in women with hirsutism. *N Engl J Med* 1991 21; 324:564–5.

Eldar-Geva T, Hurwitz A, Vecsei P, Palti Z, Milwidsky A, Rosler A. Secondary biosynthetic defects in women with late-onset congenital adrenal hyperplasia. *N Engl J Med* 1990; 323:855–63.

Erel CT, Senturk LM, Oral E, Mutlu H, Colgar U, Seyisoglu H, Ertungealp E. Results of the ACTH stimulation test in hirsute women. *J Reprod Med* 1999; 44:247–52.

LWPES/ESPE (Joint Lawson Wilkins Pediatric Endocrine Society and the European Society for Paediatric Endocrinology). Congenital Adrenal Hyperplasia Working Group consensus statement on 21-hydroxylase deficiency from the LWPES/ESPE. *J Clin Endocrinol Metab* 2002; 87:4048–53.

Lutfallah C, Wang W, Mason JI, Chang YT, Haider A, Rich B, Castro-Magana M, Copeland KC, David R, Pang S. Newly proposed hormonal criteria via genotypic proof for type II 3beta-hydroxysteroid dehydrogenase deficiency. *J Clin Endocrinol Metab* 2002; 87:2611–22.

Merke DP, Chrousos GP, Eisenhofer G, Weise M, Keil MF, Rogol AD, Van Wyk JJ, Bornstein SR. Adrenomedullary dysplasia and hypofunction in patients with classic 21-hydroxylase deficiency. *N Engl J Med* 2000; 343:1362–8.

Morris AH, Reiter EO, Geffner ME, Lippe BM, Itami RM, Mayes DM. Absence of nonclassical congenital adrenal hyperplasia in patients with precocious adrenarche. *J Clin Endocrinol Metab* 1989; 69:709–15.

New MI, Lorenzen F, Lerner AJ, Kohn B, Oberfield SE, Pollack MS, Dupont B, Stoner E, Levy DJ, Pang S, Levine LS. Genotyping steroid 21-hydroxylase deficiency: hormonal reference data. *J Clin Endocrinol Metab* 1983; 57:320–6.

Siegel SF, Finegold DN, Lanes R, Lee PA. ACTH stimulation tests and plasma dehydroepiandrosterone sulfate levels in women with hirsutism. *N Engl J Med* 1990; 323:849–54.

Speiser, PW, Brenner D. Congenital adrenal hyperplasia resulting from 21-hydroxylase deficiency. *Endocrinologist* 2003; 13:334–40.

Speiser PW, White PC. Congenital adrenal hyperplasia. *N Engl J Med* 2003; 349:776–88.

Speiser PW, Agdere L, Ueshiba H, White PC, New MI. Aldosterone synthesis in salt-wasting congenital adrenal hyperplasia with complete absence of adrenal 21-hydroxylase. *N Engl J Med* 1991; 324:145–9.

Chapter 7

Bone and Parathyroid Glands

7.1 Assessment of the Axial Skeleton (Spine and Hip)

7.1.1 Measurement of Bone Mineral Density (BMD) by Dual-Energy X-Ray Absorptiometry (DXA) of Hip and Lumbar Spine

Indications for Test

Measurement of BMD by DXA is indicated for:

- Populations at higher risk for osteoporosis as per "Risk Factors for Osteoporosis in Patients Identified as Having Limited Weight-Bearing Exercise/Exercise Intolerance, Impaired Corrected Vision, Dementia, History of Falls, Positive Rhomberg Sign, Vitamin D Deficiency, and/or Kyphosis/Bone and Joint Deformity" (Table 7.1)
- The assessment of the degree of bone loss in patients with hyperparathyroidism
- Monitoring patients with possible osteopenia (lower than average BMD) or osteoporosis (higher risk for fracture) and the effectiveness of intervention for these conditions

Procedure

1. To avoid the necessity for disrobing at the test site and reduce time needed for test preparation, advise the patient to come for testing clothed in woolen or cotton outfits without belts, zippers, or buttons if only the hip and spine are to be scanned.
2. Permanent or semipermanent appliances in the field of the scan, such as breast implants, metal sutures, prostheses, or body jewelry, should be noted and removed if possible.
3. The patient must not be wearing shoes when tested.
4. Record gender, age, height, and weight as this information is needed for optimum interpretation of scan.
5. Obtain scan of both hips or of the dominant, weight-bearing hip of ambulatory patient using a phantom-calibrated DXA device.
6. Scan contralateral hip if the T-score < -1.0 but > -2.0 (i.e., in the osteopenic range).
7. Obtain scans of both anterior–posterior (AP) and lateral lumbar spine, if possible based on clinical status (e.g., evidence of kyphosis, fractures, or history of back surgery).

TABLE 7.1
Risk Factors for Osteoporosis in Patients Identified as Having Limited Weight-Bearing Exercise/Exercise Intolerance, Impaired Corrected Vision, Dementia, History of Falls, Positive Rhomberg Sign, Vitamin D Deficiency, and/or Kyphosis/Bone and Joint Deformity

Characteristic	Clinical Feature
Increased risk of fractures	Male hypogonadism with low testosterone levels
	Premenopause status with hypothalamic amenorrhea
	Onset menopause or postmenopause status with estrogen deprivation or proscription
	Thin body habitus in older white females
	Disordered nutrition, including celiac disease, bulimia, anorexia, chronic kidney disease (CKD)
Drug intake and medication therapy that promote osteoporosis	Inadequate (i.e., low dose) calcium and/or vitamin D intake or supplementation
	Long-term parenteral nutrition
	Excessive alcohol intake and smoking in general
	Glucocorticoids taken orally or inhaled
	Anticonvulsant, cytotoxic, or immunosuppressant chemotherapy; treatment with lithium, depoprogesterone, or tamoxifen premenopause
Risk factors for vitamin D deficiency[a]	Older age; malabsorption; abnormally low or high body mass index (BMI)
	Overexercise
	Low sun exposure in Caucasians; use of higher strength sun blockers
	Chronic kidney disease (CKD) Stages 2 to 5
	High-dose glucocorticoid therapy; lack of education and counseling about nutrition and vitamin D therapy

[a] [25-hydroxyvitamin D] < 32 ng/mL.

Interpretation

1. The reference interval for adult BMD is a function of age, ethnicity, and gender of patient. Z-scores identify bone density in relationship to the patient's age-related peer group.

2. T-scores are based on the standard deviation from the gender-specific mean BMD of young adults. The designation of osteoporosis vs. osteopenia based on T-scores of < –2.5 or –2.0 vs. –1.0 to –2.0 (or –2.5) remains controversial, but T-scores are practical indices of bone mass.

3. Prepubertal BMD is independent of ethnicity.

4. The magnitude of increase in BMD at puberty up to age 20 is approximately threefold greater in African-American than in Caucasian subjects.

5. Fracture thresholds based on BMD have been proposed but have yet to be adequately verified. BMD does not necessarily correlate with risk of fracture as shown for fluoride-treated patients.

6. The WHO criteria (Kanis, 1994) for the diagnosis of osteoporosis and osteopenia are based on standard deviations from the mean T-score of a single population and remain controversial.

7. A marked increase in BMD suggests bone dysplasia or myelophthistic disease.

8. An increase in spinal BMD coupled with a decrease in vertebral height is usually consistent with partial collapse of a vertebra.

9. Patients with an increase in BMD in response to bisphosphonate treatment have a lower fracture risk than patients with a decrease in BMD; however, greater increases in BMD do not reliably predict greater decreases in fracture risk.

10. During growth hormone therapy in adults, a 2 to 3% decrease in BMD may occur at 6 months; however, this transient decline is followed by a 4 to 6% rise in density at 18 to 36 months.

11. An above-average BMD makes a contribution of <30% to any reduction in fracture risk, with other factors (i.e., degree of bone damage accumulation, turnover, microarchitectural change) making larger contributions.

12. A total body scan by DXA gives a useful estimate of body composition (i.e., percent body fat, water, and lean body mass) but is not of proven use as a test of BMD. In body composition assessment, all jewelry, hairpins, dentures, or partials are to be removed before a total body scan is done.

Notes

1. BMD testing is not meant for the diagnosis of advanced osteoporosis, which is primarily a clinical determination supported by conventional radiographs, history of fractures, age of patient, evidence of kyphosis, and combinations of risk factors (i.e., hypogonadism, premature menopause, diet deficient in calcium or vitamin D, a measured vitamin D deficiency, lack of sun exposure, immobilization, contact with bone toxins, past illnesses such as hyperthyroidism).

2. The marked difference between African-American and Caucasian females in cancellous vertebral bone density occurs during a relatively brief period late in puberty.

3. DXA is used to measure bone density of the entire skeleton and a variety of selected areas and involves very low radiation exposure.

4. Reproducibility of total bone density measurements by DXA is within ~1%. Single-photon absorptiometry and quantitative computerized tomographic (QCT) bone densitometry are much less reproducible.

5. Phantoms made up of bone-like materials (epoxy resins or polyvinylchloride with marrow-mimicking material) should be used to calibrate devices used for measuring BMD including DXA, QCT, and heel ultrasound.

6. Scanning devices that utilize higher doses of radiation yield results within a shorter test time. Radiation exposure, however, is still less than that from a routine chest radiograph.

7. The antifracture efficacy of antiresorptive therapies (i.e., bisphosphonates) is only partially explained by increases in BMD.

8. Although DXA is the preferred method of diagnosing bone loss, it is two to three times more expensive than measurement of bone turnover markers (see Tests 7.2.1 and 7.2.2); therefore, bone turnover markers are more cost effective for monitoring response to various therapies for osteoporosis.

9. In the patient with anorexia nervosa, low BMD, especially in the lumbar spine, correlates with both a longer duration of amenorrhea and anorexia. Strikingly, even a short period of anorexia and amenorrhea may result in rapid bone loss with development of osteopenia.

10. The prevalence of celiac disease is 3.4% among osteoporotic patients compared to 0.2% among nonosteoporotics, prompting a recommendation to screen all osteoporotics for celiac disease (Stenson et al., 2005).

11. In older patient populations, apparent increases in BMD of the spine over several years time usually reflect either partial collapse of vertebrae or an increase in aortic calcification.

12. Use of inhaled corticosteroids is associated with independent dose-related increased fracture risk in older adults with airflow obstruction. The rate ratio of those with a mean daily dose of more than 600 mcg was 2.53 (95% confidence interval [CI], 1.65–3.89; p for trend < .0001).

13. Other risk factors for fractures in older adults with airflow obstruction are previous fracture (rate ratio, 2.24; 95% CI, 1.63–3.06) and oral, but not injected, corticosteroid use.

ICD-9 Codes

Conditions that may justify this test *include but are not limited to*:

BMD disorders
733.0   osteoporosis, unspecified, osteopenia (lower than average BMD)
733.01   osteoporosis, senile postmenopausal

Nutrition disorders
268   vitamin D deficiency
783.0   anorexia
783.22   underweight

Alcohol problems
303   dependence syndrome (V11.3)
303.9   alcoholism

Lifestyle issues
V69   problems related to lifestyle, smoking, lack of physical exercise
V69.1   inappropriate diet and eating habits

Other

252.0	hyperparathyroidism
257.2	testicular hypofunction
290.0	senile dementia
369.9	visual loss, unspecified
579.0	celiac disease
626.0	absence of menstruation
627.X	menopausal and postmenopausal disorders

Suggested Reading

Eastell, R. Treatment of postmenopausal osteoporosis. *N Engl J Med* 1998; 338:736–46.

Gilsanz V, Roe TF, Mora S, Costin G, Goodman WG. Changes in vertebral bone density in black girls and white girls during childhood and puberty. *N Engl J Med* 1991; 325:1597–600.

Johnston CC, Slemenda CW, Melton LJ. Clinical use of bone densitometry. *N Engl J Med* 1991; 324:1105–9.

Kanis JA. Assessment of fracture risk and its application to screening for postmenopausal osteoporosis: synopsis of a WHO report: WHO Study Group. *Osteoporos Int* 1994; 4:368–81.

Stenson WF, Newberry R, Lorenz R, Baldus C, Civitelli R. Increased prevalence of celiac disease and need for routine screening among patients with osteoporosis. *Arch Intern Med* 2005; 165:393–9.

Watts NB, Cooper C, Lindsay R, Eastell R, Manhart MD, Barton IP, van Staa TP, Adachi JD. Relationship between changes in bone mineral density and vertebral fracture risk associated with risedronate: greater increases in bone mineral density do not relate to greater decreases in fracture risk. *J Clin Densitom* 2004; 7:255–61.

7.1.2   Imaging of the Vertebral Spine to Determine Presence of and Risk for Fractures

Indications for Test

Imaging of the vertebral spine is indicated in cases of:

- Back pain, whether acute or chronic
- Fall-prone elderly, particularly those with back pain unresponsive to conservative treatment
- Subtle or more overt kyphosis or kyphoscoliosis of the spine, particularly if bone density in the lower spine or hip is within the age-specific reference range (see Test 7.1.1)
- Individuals with previous fractures of the spine or other bones

Procedure

1. Obtain the patient's history of back pain, injury, falls, and any prior imaging studies.
2. Examine the patient for evidence of kyphosis and scoliosis and obtain the patient's height by stadiometer.
3. Make a semiquantitative assessment of vertebral deformity (Genant et al., 1993) from plain radiograph or a fan-beam dual-energy x-ray absorptiometry (DXA) high-resolution lateral spine image.
4. Use conventional computed tomography (CT) scan imaging with T1 and T2 images to evaluate spine fractures.
5. Observe for wedging of vertebrae and measure vertebral height from CT image.
6. Use higher resolution (<20 µm) multidetector computed tomography (M-CT) scanning and automated voxel-based finite-element modeling techniques to make a noninvasive assessment of bone strength from three-dimensional, computer-generated images of bone trabeculae.

7. Use nonenhanced T1-weighted spin-echo and short inversion time inversion–recovery sequence or single-shot fast spin echo diffusion-weighted magnetic resonance (MR) imaging to define fracture-related abnormalities in the spine in the absence of overt fractures visible on conventional radiographs.

8. Analyze the MR image for the presence of a *fluid sign* with a signal intensity isodense to that of cerebrospinal fluid adjacent to a fractured end plate as well as to detect edema in the marrow of acutely fractured bone. Characterize the MR fluid sign as linear, triangular, or focal.

Interpretation

1. Genant's semiquantitative morphometric technique for assessment of conventional radiographs is a more reproducible approach to objectively determining the nature of a vertebral fracture than other techniques (e.g., Melton, Eastell, Singh, McCloskey-Kanis) and is considered the gold standard method of radiographic analysis of vertebral bone status (Genant et al., 1993).

2. Vertebral fractures are detected on the basis of the presence of endplate deformities, the lack of parallelism of the endplates, and a generally altered appearance compared with neighboring vertebrae.

3. Fractures are classified as wedge, crush, or biconcave ("fish mouth"), depending on whether the anterior, posterior, or middle portion of the vertebral body is most diminished in height exceeding 20%.

4. Difficulties in the diagnosis of vertebral fractures may result from poor technique in which the lateral projection is really an oblique projection leading to the appearance of a fracture. Pseudofractures may be detected on lateral projection views in patients with scoliosis.

5. Vertebral deformities may mimic a fracture. These include Schmorl nodes (vertebral osteochondrosis or Scheuermann's disease), H-shaped vertebrae (sickle cell disease or Gaucher's disease), and Cupid's bow and limbus vertebra, which are developmental variants.

6. Not every deformed vertebra is a vertebral fracture caused by osteoporosis.

7. Postmenopausal women (mean age, 74 years) who develop a vertebral fracture are at substantial risk for additional fracture within the next year (incidence, 19.2%) and have a relative risk (RR) of a fracture fivefold greater than in subjects without previous vertebral fractures.

8. Less than 10% of all vertebral fractures are neoplastic in origin, and less than 10% of these are associated with a fluid sign on MR.

9. High or intermediate signal intensity on single-shot fast spin echo diffusion-weighted MR is highly specific for the diagnosis of metastatic tumor infiltration of the spine.

10. MR can be used to accurately differentiate between benign and malignant causes of vertebral collapse. Further differentiation of an osteoporotic, a traumatic, or an infective cause can be done with the help of clinical history and evaluation of end-plate integrity.

11. On MR, a linear fluid sign is most commonly associated with an acute vertebral fracture. In less than a third of acute fractures, the fluid sign is triangular or focal.

12. On MR of a vertebral deformity, a normal, nonedematous bone marrow signal indicates an old fracture.

13. With high-resolution lateral spine image using DXA, qualitative visual identification of vertebral fractures is superior to quantitative assessment of the DXA image and may obviate conventional radiography of the spine for diagnosis of vertebral fractures.

Notes

1. The correlation between BMD determined by DXA and vertebral strength or resistance to fracture is not based on mechanical principles and does not reflect the effects of subtle geometric features and densitometric inhomogeneities in trabeculae that may significantly alter fracture risk.

2. A voxel or volume element can be represented as an eight-node cuboidal or four-node tetrahedral area making up part of trabecular bone. In finite-element volume meshing, trabecular bone is represented by an array of voxels from which three-dimensional images can be generated and the mechanical properties (strength) of bone tissue predicted.

3. In studies of cadaver spines, highly automated voxel finite-element models are superior to correlation-based QCT bone density methods in predicting vertebral compressive strength.

4. Using positron emission tomography (PET) scans, acute vertebral fractures that originated from osteoporosis or preclinical osteoporosis tend to have no pathologically increased 18-fluorodeoxyglucose isotope uptake.

5. Unlike MR, CT combined with radionuclide bone scans does not reliably distinguish between benign and malignant causes of vertebral collapse; hence, MR is preferred over CT imaging in the assessment of suspected spine malignancy.

6. Non-trauma-related chronic low back pain without radiculopathy can usually be evaluated clinically without imaging unless unresponsive to conservative treatment.

7. The pain of a vertebral fracture may occur in one of two ways: (a) acute and severe, improving gradually but persisting for in excess of 4 to 8 weeks, or (b) less severe and of short duration (1 to 2 weeks) but after 6 to 16 weeks a new attack of acute pain presents and recurs over 6 to 18 months' time. With acute and severe pain, the vertebral wedging is obvious from the beginning and remains unchanged. In cases of less severe pain, an obvious collapse of vertebrae occurs over a more extended period.

8. Wedge deformities are the most frequent and tend to cluster at the mid-thoracic and thoracolumbar regions of the spine in both men and women.

9. In both sexes, the frequency of biconcave deformities is higher in the lumbar than the thoracic spine and, unlike crush and wedge types, do not decline in frequency at lower lumbar vertebral levels.

10. Nonvertebral fractures account for approximately 55% of all osteoporotic fractures.

ICD-9 Codes

Conditions that may justify this test *include but are not limited to*:

Back deformity
737.3   kyphoscoliosis and scoliosis
737.41   kyphosis

Other
724.5   backache, unspecified

Suggested Reading

Baur A, Stabler A, Arbogast S, Duerr HR, Bartl R, Reiser M. Acute osteoporotic and neoplastic vertebral compression fractures: fluid sign at MR imaging. *Radiology* 2002; 225:730–5.

Crawford RP, Cann CE, Keaveny TM. Finite element models predict *in vitro* vertebral body compressive strength better than quantitative computed tomography. *Bone* 2003a; 33:744–50.

Crawford RP, Rosenberg WS, Keaveny TM. Quantitative computed tomography-based finite element models of the human lumbar vertebral body: effect of element size on stiffness, damage, and fracture strength predictions. *J Biomech Eng* 2003b; 125:434–8.

Ferrar L, Jiang G, Eastell R, Peel NF. Visual identification of vertebral fractures in osteoporosis using morphometric x-ray absorptiometry. *J Bone Miner Res* 2003; 18:933–8.

Genant HK, Wu CY, van Kuijk C, Nevitt MC. Vertebral fracture assessment using a semiquantitative technique. *J Bone Miner Res* 1993; 8:1137–48.

Guermazi A, Mohr A, Grigorian M, Taouli B, Genant HK. Identification of vertebral fractures in osteoporosis. *Semin Musculoskelet Radiol* 2002; 6:241–52.

Ismail AA, Cooper C, Felsenberg D, Varlow J, Kanis JA, Silman AJ, O'Neill TW. Number and type of vertebral deformities: epidemiological characteristics and relation to back pain and height loss. European Vertebral Osteoporosis Study Group. *Osteoporos Int* 1999; 9:206–13.

Lindsay R, Silverman SL, Cooper C, Hanley DA, Barton I, Broy SB, Licata A, Benhamou L, Geusens P, Flowers K, Stracke H, Seeman E. Risk of new vertebral fracture in the year following a fracture. *JAMA* 2001; 285:320–3.

Lenchik L, Rogers LF, Delmas PD, Genant HK Diagnosis of osteoporotic vertebral fractures: importance of recognition and description by radiologists. *Am J Roentgenol* 2004; 183:949–58.

Lyritis GP, Mayasis B, Tsakalakos N, Lambropoulos A, Gazi S, Karachalios T, Tsekoura M, Yiatzides A. The natural history of the osteoporotic vertebral fracture. *Clin Rheumatol* 1989; 8(Suppl 2):66–9.

McNally EG, Wilson DJ, Ostlere SJ. Limited magnetic resonance imaging in low back pain instead of plain radiographs: experience with first 1000 cases. *Clin Radiol* 2001; 56:922–5.

Park SW, Lee JH, Ehara S, Park YB, Sung SO, Choi JA, Joo YE. Single shot fast spin echo diffusion-weighted MR imaging of the spine: is it useful in differentiating malignant metastatic tumor infiltration from benign fracture edema? *Clin Imaging* 2004; 28:102–8.

Ulrich D, Reitbergen B, Laib A, Ruegsegger P. Mechanical analysis of bone and its microarchitecture based on *in vivo* voxel images. *Technol Health Care* 1998; 6:421–7.

Wu CY, Li J, Jergas M, Genant HK. Comparison of semiquantitative and quantitative techniques for the assessment of prevalent and incident vertebral fractures. *Osteoporos Int* 1995; 5:354–70.

7.1.3   Stadiometry for Assessment of Height Loss and Knemometry for Assessment of Height Velocity

Indications for Test

Stadiometry is indicated:

- For the routine evaluation of all endocrinology patients
- To assess for height loss secondary to osteoporosis, hypogonadism, and vitamin D deficiency

Knemometry is indicated:

- For assessment of height velocity in children with possible growth delay
- For assessment of the effects of drug therapy, particularly steroids and growth hormone, on growth in children

Procedure

1. Use a knemometer in children or mini-knemometer or electronic caliper in premature infants to measure the heel to 90°-flexed knee length of a sitting child with accuracy (technical error) of 90 to 160 µm.
2. Allow a learning period of at least 3 weeks for a new operator to achieve reliable and reproducible knemometry results.
3. To maximize reduction in technical measurement errors to <100 µm, allow 4 months of experience. After only 1 week of training, expect technical errors up to 150 µm.
4. For accurate growth velocity assessment, repeat knemometer measures as often as every day or every 7 to 10 days.
5. For an adult subject, repeat the stadiometer height measurement 3 times to allow stabilization in height (maximum height decrement typically occurs after 3.0 to 3.5 minutes of standing) and to reduce random variation due to posture changes that occur when the subject repeatedly steps in and out of the stadiometer device.

Interpretation

1. Figure 7.1 shows the normal growth curve of height velocity in cm/year.
2. In 2- to 3-year-olds, changing measurements from length to height results in an upward shift of calculated body mass index (BMI).

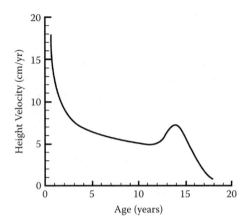

Figure 7.1
Normal height velocity growth curve in children and adolescents.

3. In healthy children ages 8 to 12 years, the mean short-term velocity of lower leg growth was 400 μm per week (SD, ±120 μm/week) when knemometric measurements were made at an interval of one year.
4. Height loss in aging adults of ≥1 inch over historically maximum height is significant when measured by stadiometer and consistent with vertebral bone loss.
5. Height loss of 4 cm or more is practically definitive evidence for spinal osteoporosis.

Notes

1. No significant differences were found between use of an ultrasound-based knemometer (G-100) and the Harpenden stadiometer to measure height and growth velocity in children after 3 months of growth hormone therapy.
2. The technical handling error, calculated as the mean, was 0.360 μm for the simple electronic caliper and 590 μm for the mini-knemometer (Ahmed et al., 1995).
3. Knemometry has shown that the initial growth-inhibiting effect on the epiphyses of oral ethinylestradiol (EE2) therapy in the treatment of boys growing too rapidly is overridden by the effects of long-term, high-dose testosterone enanthate injections which, when added to EE2 therapy, have a more potent inhibitory effect on growth velocity (Decker et al., 2002).
4. Knemometry failed to show any change in growth rates (i.e., 280 ± 40 μm/week with or without oral administration of relatively low-dose vitamin D_3) in healthy children during winter in Denmark (Schou et al., 2003).
5. Knemometry was found to be sensitive enough to detect the growth delay-sparing effect of alternate-day steroid medication in children with Crohn's disease.
6. Stadiometry is precise and reproducible, and can detect true changes in height over periods of 1 month in mid-childhood.

ICD-9 Codes

Conditions that may justify this test *include but are not limited to*

Bone loss
733.0   osteoporosis, unspecified
733.01   osteoporosis, senile postmenopausal

Other
783.43   short stature

Suggested Reading

Ahmed SF, Wallace WH, Kelnar CJ. Knemometry in childhood: a study to compare the precision of two different techniques *Ann Hum Biol* 1995; 22:247–52.

Buyken AE, Hahn S, Kroke A. Differences between recumbent length and stature measurement in groups of 2- and 3-y-old children and its relevance for the use of European body mass index references. *Int J Obes Relat Metab Disord* 2005; 29:24–8.

Coles RJ, Clements DG, Evans WD. Measurement of height: practical considerations for the study of osteoporosis. *Osteoporos Int* 1994; 4:353–6.

Decker R, Partsch CJ, Sippell WG. Combined treatment with testosterone (T) and ethinylestradiol (EE2) in constitutionally tall boys: is treatment with T plus EE2 more effective in reducing final height in tall boys than T alone? *J Clin Endocrinol Metab* 2002; 87:1634–39.

Engstrom E, Wallgren K, Hellstrom A, Niklasson A. Knee–heel length measurements in preterm infants: evaluation of a simple electronically equipped instrument. *Acta Paediatr* 2003; 92:211–5.

Glock M, Hermanussen M, Keller E, Hartmann KKP. Gulliver G-100: a new device to evaluate daily growth measurement in comparison with Harpenden stadiometer. *Hormone Res* 1999; 52:287–9.

Hermanussen M. Knemometry, a new tool for the investigation of growth: a review. *Eur J Pediatr* 1988; 147:350–5.

Hermanussen M, Gausche R, Keller A, Kiess W, Brabec M, Keller E. Short-term growth response to GH treatment and considerations upon the limits of short-term growth predictions. *Horm Res* 2002; 58:71–7.

Keller A, Keller E, Hermanussen M, Vogtmann C. Short-term growth of premature infants treated with dexamethasone assessed by mini-knemometry. *Ann Hum Biol* 2004; 31:389–97.

Schou AJ, Heuck C, Wolthers OD. A randomized, controlled lower leg growth study of vitamin D supplementation to healthy children during the winter season. *Ann Hum Biol* 2003; 30:214–9.

Stothart JP, McGill SM. Stadiometry: on measurement technique to reduce variability in spine shrinkage measurement. *Clin Biomech (Bristol, Avon)* 2000; 15:546–8.

Vogt TM, Ross PD, Palermo L et al. Vertebral fracture prevalence among women screened for the fracture intervention trial and a simple clinical tool to screen for undiagnosed vertebral fractures. *Mayo Clin Proc* 2000; 75:888–96.

Wales JK, Milner RD. Variation in lower leg growth with alternate day steroid treatment. *Arch Dis Child* 1988; 63:981–3.

Watt V, Pickering M, Wales JKH. A comparison of ultrasonic and mechanical stadiometry. *Arch Dis Child* 1998; 78:269–70.

Wolthers OD, Konstantin-Hansen K, Pedersen S, Petersen KE. Knemometry in the assessment of short-term linear growth in a population of healthy school children. *Horm Res* 1992; 37:156–9.

7.2 Bone Turnover Assessment

7.2.1 Cross-Linked C-Terminal (CTx) and N-Terminal (NTx) Telopeptides of Type 1 Bone Collagen as Markers of Accelerated Bone Turnover

Indications for Test

Testing for telopeptides may be indicated:

- When distinguishing among high bone turnover states associated with active Paget's disease, hyper-parathyroidism, teriparatide therapy, thyrotoxicosis, or postmenopausal type I osteoporosis and low bone turnover states characteristic of glucocorticoid excess or senile type II osteoporosis

TABLE 7.2
Age-Stratified Reference Interval for Serum C-Terminal Telopeptide of Type I Collagen (CTx-I) in Men and Women

Age Group	CTx-I (pg/mL)	Women	
(years)	Men	Premenopausal	Postmenopausal
18–29	87–1200	64–640	Not applicable
30–39	70–780	60–650	Not applicable
40–49	60–700	40–465	104–1008
50–69	87–345	Not applicable	104–1008
≥70	80–1050	Not applicable	104–1008

- As a marker of effective treatment for osteoporosis with parathyroid hormone (e.g., Forteo®)
- As a marker of the dissipation of effect of single-dose, long-acting intravenous bisphosphonate therapy (e.g., Reclast®)

Procedure

1. Obtain a blood sample for CTx or NTx measurement. Alternatively, obtain a second void morning urine collection for NTx and creatinine measurement.
2. Because CTx and NTx levels in blood fluctuate in a diurnal cycle, specimen collection should occur at the same time of day when performing serial testing for CTx or NTx.

Interpretation

1. Biochemical markers of bone turnover have not been established for predicting fracture risk or selecting patients for various forms of antiresorptive therapy.
2. Bone turnover markers are, at best, only weakly associated with bone mass.
3. Failure of CTx or NTx to increase, usually more than twice the baseline, after initiation of treatment with Forteo® suggests a lack of adherence to therapy or its lack of effectiveness in an individual metabolic bone disease patient, possibly related to the inappropriate or concurrent therapy with bisphosphonates.
4. Sample serum CTx reference intervals to supplement information provided in Table 7.2:
   - Men, ages 18 to 70 years: 35–873 pg/mL (0.035–0.873 µg/L)
   - Premenopausal healthy adult women: 25–573 pg/mL (0.025–0.573 µg/L)
   - Postmenopausal healthy adult women: 104–1008 pg/mL (0.104–1.008 µg/L)

   Note: It is not possible to provide a conversion factor for converting pg/mL to pmol/L because immunoassays using antibodies to the 8-amino-acid sequence specific to part of the CTx molecule quantify several differing molecular weight species of type I collagen breakdown products (Garnero et al., 2001).
5. Urine NTx (Osteomark®; Ostex International, Seattle, WA) reference intervals (nmol bone collagen equivalents [BCE] per mmol creatinine):
   - Males, 21–66
   - Females, 19–63
   - Females, premenopausal, <80

Notes

1. In general, serum-based assays for telopeptides have replaced urine-based assays. CTx measurement in blood has superceded NTx measurement in urine as a marker of bone turnover.
2. Patients with type I postmenopausal osteoporosis and high bone turnover have significantly elevated [NTx] (>2× upper limit of reference) compared to premenopausal women. [CTx] may be similarly elevated in these patients.
3. Markers of bone resorption may be used to predict the patient's response to antiresorptive therapy in states of increased bone turnover characterized by a net increase in bone resorption.
4. Greater decreases in CTx and NTx in response to bisphosphonate therapy were associated with greater decreases in vertebral and nonvertebral fractures.

5. As reflected by [CTx], there may be a level of bone resorption reduction below which there is no further decrease in fractures in patients on bisphosphonate therapy.

6. Reclast® treatment reduces [CTx] more than 8-fold into the lower end of the premenopausal reference interval within 7 days (Saag et al., 2007). After Reclast® therapy, the [CTx] rises slowly, from about 100 pg/mL to 200 pg/mL over 36 months, prompting retreatment within that interval (Black et al., 2007). In contrast, the [BSAP] fell more gradually into the premenopausal range within 12 weeks.

7. Profoundly low levels of bone turnover markers may mark patients at risk for osteonecrosis in the event of serious dental infections in which increased bone angiogenesis is required but has been inhibited by bisphosphonate treatment.

8. Bone stimulants, such as teriparatide (Forteo®), and conditions associated with increased bone turnover (e.g., fractures) result in high [CTx] and [NTx] with a net increase in bone formation.

9. When [NTx] and [CTx] are corrected for whole-body bone mineral content (BMC), the fraction of bone resorbed per day does not correlate with age in postmenopausal women and remains elevated for up to 40 years after menopause.

10. Reductions in urinary CTx of 60% and in NTx of 51% after 3 to 6 months of bisphosphonate therapy were associated ($p < 0.05$) with a reduction in vertebral fracture risk of 75% at 1 year and 50% over 3 years (Worsfold et al., 2004).

11. With bisphosphonate treatment, resorption markers (serum CTx, urine NTx, and urinary free pyridinoline) decreased earlier than markers of bone formation (osteocalcin, bone-specific alkaline phosphatase, and carboxy-terminal propeptide of type I collagen), consistent with a direct action of the drug in inhibiting osteoclastic bone resorption.

12. A rapid decline in [NTx] occurs when bone resorption rates fall in response to bisphosphonate therapy as the capacity to degrade NTx matches and exceeds its rate of production.

13. Alendronate and ibandronate produce a decrease in urinary CTx-II levels of about 50% of baseline levels, suggesting that bisphosphonates may have chondroprotective effects.

14. The effects of antiresorptive agents (bisphosphonates and estrogen) on bone turnover in postmenopausal osteoporosis are restricted to cancellous bone, which comprises only 20% of bone mass.

15. Although the BMD response to bisphosphonates in the hip is less than that of spine, the log NTx value indicates a response in 78% of treated osteoprotics, regardless of BMD, but less often to estrogen therapy (67%). The log urinary deoxypyridinoline value indicates response to bisphosphonate therapy less frequently (31% at 12 months) than log NTx values.

16. In 11 women (mean age, 77 years), markers of bone resorption were shown to decrease during estrogen replacement therapy (ERT) and return to baseline after ERT was discontinued. Markers of bone formation were found to decline less than markers of bone resorption during ERT but continued to decline after ERT was stopped (Prestwood et al., 1994).

17. Intranasal and subcutaneous administration of salmon calcitonin results in a dose-dependent reduction in plasma [CTx-I] by 50 to 60% as early as 1 hour after calcitonin administration (Zikan and Stepan, 2002).

18. CTx and NTx assays detect fragments arising from the telopeptide region of the collagen molecule. A degradation fragment originating from the helical part of type I collagen consists of the 620- to 633-amino-acid sequence of the alpha-1 chain and is called *urinary helical peptide*. This peptide can be measured using an enzyme-linked immunosorbent assay (ELISA) (Metra™ Helical Peptide; Quidel Corporation; San Diego, CA). Urinary helical peptide has been shown to be highly correlated with urinary [CTx] (Garnero and Delmas, 2003).

19. Measures of bone resorption with lesser diagnostic specificity and sensitivity for disease process classification include: (a) pyridinoline in urine, (b) deoxypyridinoline cross-links, and (c) free and total hydroxyproline.

ICD-9 Codes

Conditions that may justify this test *include but are not limited to*:

Bone disease

731.0	osteitis deformans without mention of bone tumor
733.0	osteoporosis
733.01	osteoporosis, senile postmenopausal
733.9	disorder of bone and cartilage, unspecified
588.0	renal osteodystrophy

Other

242	thyrotoxicosis with or without goiter
252.0	hyperparathyroidism
255.0	Cushing's syndrome
627	menopausal and postmenopausal disorders

Suggested Reading

Black DM, HORIZON Pivitol Fracture Investigators. Once-yearly zoledronic acid for treatment of postmenopausal osteoporosis. *N Engl J Med* 2007; 356:1809–22.

Eastell R, Barton I, Hannon RA, Chines A, Garnero P, Delmas PD. Relationship of early changes in bone resorption to the reduction in fracture risk with risedronate. *J Bone Miner Res* 2003; 18:1051–56.

Garnero P, Delmas PD. An immunoassay for type I collagen alpha 1 helicoidal peptide 620–633, a new marker of bone resorption in osteoporosis. *Bone* 2003; 32:20–6.

Garnero P, Borel O, Delma PD. Evaluation of a fully automated serum assay for C-terminal cross-linking telopeptide of type I collagen in osteoporosis. *Clin Chem* 2001; 47:694–702.

Garnero P, Shih WJ, Gineyts E, Karpf DB, Delmas PD. Comparison of new biochemical markers of bone turnover in late postmenopausal osteoporotic women in response to alendronate treatment. *J Clin Endocrinol Metab* 1994; 79:1693–1700.

Hsin-Shan JJ, Leung S, Brown B, Stringer MA, Leigh S, Scherrer C, Shepard K, Jenkins D, Knudsen J, Cannon R. Comparison of analytical performance and biological variability of three bone resorption assays. *Clin Chem* 1997; 43:1570–76.

Lehmann HJ, Mouritzen U, Christgau S, Cloos PA, Christiansen C. Effect of bisphosphonates on cartilage turnover assessed with a newly developed assay for collagen type II degradation products. *Ann Rheum Dis* 2002; 61:530–3.

Prestwood KM, Pilbeam CC, Burleson JA, Woodiel FN, Delmas PD, Deftos LJ, Raisz LG. The short-term effects of conjugated estrogen on bone turnover in older women. *J Clin Endocrinol Metab* 1994; 79:366–71.

Qvist P, Munk M, Hoyle N, Christiansen C. Serum and plasma fragments of C-telopeptides of type I collagen (CTx) are stable during storage at low temperatures for 3 years. *Clin Chim Acta* 2004; 350:167–73.

Rosenquist C, Fledelius C, Christgau S, Pedersen BJ, Bonde M, Qvist P, Christiansen C. Serum CrossLaps™ one-step ELISA: first application of monoclonal antibodies for measurement in serum of bone-related degradation products from C-terminal telopeptides of type I collagen. *Clin Chem* 1998; 44:2281–89.

Saag K, Lindsay R, Kriegman A, Beamer E, Zhou W. A single zoledronic acid infusion reduces bone resorption markers more rapidly than weekly oral alendronate in postmenopausal women with low bone mineral density. *Bone* 2007; 40:1238–43.

Worsfold M, Powell DE, Jones TJW, Davie MWJ. Assessment of urinary bone markers for monitoring treatment of osteoporosis. *Clin Chem* 2004; 50:2263–70.

Zikan V, Stepan JJ. Plasma type 1 collagen cross-linked C-telopeptide: a sensitive marker of acute effects of salmon calcitonin on bone resorption. *Clin Chim Acta* 2002; 316:63–9.

7.2.2 Alkaline Phosphatase (AP), Bone-Specific Alkaline Phosphatase (BSAP), and Osteocalcin as Serum Markers of Bone Formation in Conditions Associated with Accelerated Bone Turnover

Indications for Test

Measurement of AP, BSAP, and osteocalcin are indicated as measures of bone formation in:

- States of elevated bone turnover such as Paget's disease, hyperparathyroidism, thyrotoxicosis, and postmenopausal type I osteoporosis
- The follow-up of patients who have undergone parathyroid surgery
- Monitoring effectiveness of treatments for osteoporosis (i.e., Forteo®, bisphosphonates) and its prevention (i.e., estrogens, calcium, and vitamin D)

Procedure

1. Obtain blood samples for the measurement of AP, BSAP, and osteocalcin.
2. Because blood [osteocalcin] fluctuates in a diurnal cycle, obtain specimens at the same time of day when performing serial testing for osteocalcin.

Interpretation

1. Biochemical markers of bone turnover have not been established in predicting fracture risk or select patients for various forms of antiresorptive therapy.
2. Bone turnover markers are, at best, only weakly associated with bone mass.
3. Reference intervals:
   - [AP] (adults), 25–125 IU/L
   - [BSAP] (males), ≤20 mcg/L
   - [BSAP] (females, premenopausal) ≤14 mcg/L
   - [BSAP] (females, postmenopausal) ≤22 mcg/L
   - [Osteocalcin], 1.7 mcg/L (evening) to 18.1 mcg/L (morning) (diurnal variation affects interval)

Notes

1. Children who are gaining length or height have higher AP and BSAP reference intervals than adults.
2. [Osteocalcin] is generally higher in children than in adults, with the highest levels being reached during the puberty growth spurt at about 12 years of age in girls and 14 years of age in boys, thereafter rapidly declining toward adult levels.
3. An elevated [AP] or [BSAP] in patients who have undergone successful parathyroid surgery is predictive of a slower recovery from the "hungry bones" syndrome, usually characterized by symptomatic hypocalcemia.
4. Markers of bone formation are most useful in Paget's disease patients, particularly those with asymptomatic monostotic disease, as a way to monitor response to treatment in these patients.
5. Compared to [CTx], a baseline [osteocalcin] discriminates poorly between different levels of Paget's disease activity.
6. A decrease in [NTx] was found to provide the greatest sensitivity and specificity in monitoring the response to estrogen hormone replacement therapy in early postmenopausal women ($p = 0.0001$). [BSAP] was next most effective ($p = 0.001$), and [osteocalcin] was the least effective ($p = 0.3$) (Rosen et al., 1996).
7. The overall rates of both bone formation and bone resorption remain high in elderly women.

ICD-9 Codes

Conditions that may justify this test *include but are not limited to*:

Bone disease

731.0	osteitis deformans without mention of bone tumor
733.0	osteoporosis
733.01	osteoporosis, senile postmenopausal

Other

242	thyrotoxicosis with or without goiter
252.0	hyperparathyroidism
790.5	nonspecific abnormal serum enzyme levels (i.e., alkaline phosphatase)

Suggested Reading

Bauer DC, Black DM, Garnero P, Hochberg M, Ott S, Orloff J, Thompson DE, Ewing SK, Delmas PD. Change in bone turnover and hip, non-spine, and vertebral fracture in alendronate-treated women: the fracture intervention trial. *J Bone Miner Res* 2004; 19:1250–58.

Panigrahi K, Delmas PD, Singer F, Ryan W, Reiss O, Fisher R, Miller PD, Mizrahi I, Darte C, Kress BC et al. Characteristics of a two-site immunoradiometric assay for human skeletal alkaline phosphatase in serum. *Clin Chem* 1994; 40:822–8.

Randall AG, Kent GN, Garcia-Webb P, Bhagat CI, Pearce DJ, Gutteridge DH, Prince RL, Stewart G, Stuckey B, Will RK, Retallack RW, Price RI, Ward L. Comparison of biochemical markers of bone turnover in Paget disease treated with pamidronate and a proposed model for the relationships between measurements of the different forms of pyridinoline cross-links. *J Bone Miner Res* 1996; 11:1176–84.

Rosen C, Mallinak N, Cain D, Flessland K, Chesnut C. A comparison of biochemical markers in monitoring skeletal response to hormone replacement therapy in early postmenopausal women. *J Bone Miner Res* 1996; 11(Suppl 1): pS10, pS119.

Seydewitz HH, Henschen M, Kuhnel W, Brandis M. Pediatric reference ranges for osteocalcin measured by the Immulite analyzer. *Clin Chem Lab Med* 2001; 39:980–2.

7.3   Calcium Metabolism Assessment

7.3.1   Calcium, Phosphorus, and Albumin Test Panel

Indications for Test

Measurement of serum calcium, phosphorus, and albumin is indicated for:

- Screening for abnormal parathyroid gland function, particularly before thyroid surgery, for conditions other than known hyperparathyroidism
- Evaluation of patients with acute or chronic kidney disease (CKD), malnutrition, chronic diarrhea, alcoholism, uncontrolled diabetes mellitus (DM), and neoplasms that cause hypercalcemia, including nephromas, hepatomas, multiple myeloma, and lung, bone, and skin cancer

Procedure

1. Obtain a blood sample for total calcium [Ca], phosphorus [P], and albumin [Alb] concentration.
2. Correct the measured [Ca] for [Alb] (see Test 7.3.2).
3. In cases of hypocalcemia or hypophosphatemia, obtain a blood sample for measurement of magnesium and 25-hydroxyvitamin D.

Interpretation

1. Refer to "Correlation of Serum Total Calcium and Phosphorus Levels with Various Clinical Conditions" (Table 7.3).
2. Reference intervals:
   - Total calcium, 8.4 to 10.4 mg/dL
   - Phosphorus, 3.0 to 4.5 mg/dL
   - Albumin, 3.5 to 5.5 g/dL
3. If the albumin-corrected total [Ca], intact parathyroid hormone (iPTH), and [P] are within normal limits, the patient is probably, but not unequivocally, euparathyroid.
4. Low to undetectable levels of iPTH with elevated calcium levels are characteristic of a variety of neoplastic conditions as noted above.

Notes

1. In hypophosphatemia, pseudofractures (Looser's transformation zones) may appear in both oncogenic osteomalacia and X-linked hypophosphatemic rickets with reduced calcitriol formation by the kidney.
2. Animal protein is a major source of dietary phosphorus, and its excess ingestion should be avoided in patients with CKD. Overtreatment with calcium supplements or phosphate supplements (e.g., Neutra-Phos®) may cause either hypo- or hyperphosphatemia, respectively.

TABLE 7.3
Correlation of Serum Total Calcium and Phosphorus Levels with Various Clinical Conditions

[Total Calcium]	[Phosphorus]	Possible Etiology
Variable	↓↓↓	Hyperphosphaturia due to alcohol withdrawal, diabetic ketoacidosis, respiratory alkalosis, recovery from thermal burn injury, renal tubular defects, intestinal absorption defects, starvation
↓	↓	Vitamin D deficiency or resistance in patients with normal renal function
↓	↑	Pseudohypoparathyroidism
↓	↑↑	Severe recent-onset end-stage renal disease (ESRD), hypoparathyroidism, acute-onset rhabdomyolysis, tumor lysis, overdose with phosphate-containing laxatives, or overuse of phosphate-containing purgatives
↓ or ↔	↓ or ↔	Other than parathyroid hormone (PTH) deficiency (e.g., malnutrition or vitamin D deficiency)
↑	↓	Hyperparathyroidism
↑	↑	Vitamin D intoxication, milk-alkali syndrome, multiple myeloma, neoplasms metastatic to bone
High ↔ or ↑	Mod ↓ or low ↔	Hyperparathyroidism, alcohol abuse, uncontrolled diabetes mellitus, oncogenic osteomalacia, antacid abuse, recovery from hypothermia, X-linked hypophosphatemic rickets

Note: ↓, decreased; ↑, increased; ↔, within normal reference limits; low ↔, low-normal; high ↔, high-normal; mod, moderately.

3. In hypophosphatemia, erythrocyte 2,3-diphosphoglycerate levels increase, resulting in increased affinity of hemoglobin for oxygen and reduced oxygen release at the tissue level along with a decreased intracellular adenosine triphosphate (ATP) levels and leukocyte dysfunction (impaired phagocytosis and decreased granulocyte chemotaxis).

4. In a prospective study, 9 of 748 patients (1.2%) screened prior to thyroid surgery were found to meet criteria for hyperparathyroidism based on elevated random [Ca] and follow-up [iPTH]. All were found to have a parathyroid adenoma. Three of the parathyroid glands were not obviously enlarged. In the 739 patients with a normal [Ca] on screening, 12 had the appearance of an incidental thyroid nodule, two of which were actually parathyroid adenomas (Denizot et al., 2002).

5. Hypercalcemia and hyperparathyroidism related to CKD may occur in patients with chronic hypertension or diabetes or who are on long-term lithium therapy even when lithium therapy is interrupted for several weeks.

6. Hypercalcemia and elevated iPTH, independent of the appearance of an obvious parathyroid adenoma, is seen more frequently in lithium-treated patients than in non-lithium-treated patients (Bendz et al., 1996).

7. Most cases of hyperparathyroidism in lithium therapy patients are related to the presence of one or more hyperfunctioning adenomas rather than four-gland hyperplasia. Recurrence of the adenoma may occur if lithium therapy is continued (Awad et al., 2003).

8. Hypocalcemia secondary to hypomagnesemia, particularly associated with seizures, can occur at a serum magnesium concentration of <1.0 mg/dL.

9. Autoimmune polyendocrine syndrome type 1 (APS-1) is clinically defined as the presence of two components out of the triad of hypoparathyroidism, adrenal insufficiency, and mucocutaneous candidiasis. Reactivity to the tissue-specific autoantigen NACHT leucine-rich-repeat protein 5 (NALP5) may result in the autoantibodies responsible for hypoparathyroidism in APS-1 (Alimohammadi et al., 2008).

10. Resistance to parathyroid hormone or pseudohypoparathyroidism is an uncommon condition in which the iPTH is elevated while the [Ca] is below or in the lowest quintile of the reference range. Its diagnosis is usually made based on clinical presentation and genetic testing. Testing for the response to PTH infusion in these patients (the Ellsworth–Howard test) is obsolete.

ICD-9 Codes

Conditions that may justify this test *include but are not limited to*:

Parathyroid
252.0   hyperparathyroidism
252.8   other specified disorders of parathyroid gland

Calcium
275.41   hypocalcemia
275.42   hypercalcemia

Other
195   malignant neoplasm of other and ill-defined sites
203.0   multiple myeloma
250.X   diabetes mellitus (DM)
263.9   malnutrition
303   alcohol dependence syndrome
585   chronic kidney disease (CKD)
787.91   diarrhea

Suggested Reading

Alimohammadi M et. al. Autoimmune polyendocrine syndrome type 1 and NALP5, a parathyroid autoantigen. *N Engl J Med* 2008; 358:1018–28.

Aurbach GD, Marx SJ, Spiegel, AM. Parathyroid hormone, calcitonin, and the calciferols. In: *Williams Textbook of Endocrinology*, 8th ed., Wilson JD and Foster DW (Eds). 1992. WB Saunders, pp. 1397–1476.

Awad SS, Miskulin J, Thompson N. Parathyroid adenomas versus four-gland hyperplasia as the cause of primary hyperparathyroidism in patients with prolonged lithium therapy. *World J Surg* 2003; 27:486–8.

Bendz H, Sjodin I, Toss G, Berglund K. Hyperparathyroidism and long-term lithium therapy: a cross-sectional study and the effect of lithium withdrawal. *J Intern Med* 1996; 240:357–65.

Denizot A, Dadoun F, Meyer-Dutour A, Alliot P, Argeme M. Screening for primary hyperparathyroidism before thyroid surgery: a prospective study. *Surgery* 2002; 131:264–9.

Shaw NJ, Wheeldon J, Brocklebank JT. Indices of intact serum parathyroid hormone and renal excretion of calcium, phosphate, and magnesium. *Arch Dis Child* 1990; 65:1208–11.

7.3.2   Calcium Concentration Corrected for [Albumin] in Serum

Indications for Test

Total calcium corrected for albumin concentration is indicated:

- When the total calcium level and clinical findings of hypo- or hypercalcemia are discordant
- In patients who have just received multiple transfusions

Procedure

1. Obtain a blood sample for serum total calcium and albumin measurement from patients who are not dehydrated or critically ill.
2. Calculate the albumin-corrected total calcium concentration as follows:

$$[\text{Total calcium}]_{\text{corrected}} = \left[0.8 \times (3.5 - [\text{albumin}])\right] + [\text{total calcium}]_{\text{measured}}$$

3. Use a correction factor of 4.0 instead of 3.5 in making this calculation if the patient is dehydrated.

Interpretation

1. Reference interval for [albumin] is 3.5 to 5.5 g/dL.
2. Because calcium binds to albumin, for every 1.0 g/dL of albumin lower or higher than 3.5 g/dL, the total serum calcium concentration, [Ca], is lower or higher, respectively, by ~0.8 mg/dL.

Notes

1. A routine serum total calcium level measures both protein-bound and free (or ionized) calcium (Ca^{2+}) components.
2. At normal levels of albumin and calcium, approximately half the total serum calcium is bound to albumin, while the other half is free in the serum.
3. A linear decrease in [Ca] and [albumin] occurs with increasing numbers of units of blood transfused during treatment for trauma, as the anticoagulant citrate binds calcium in blood.
4. In critically ill multiple-trauma patients, methods for detecting hypocalcemia lack sensitivity, thus yielding an inordinately high rate (mean, 75%; SD, 32%) of false-negative [Ca] levels. Direct measurement of ionized calcium [Ca^{2+}] is preferred in this population of patients.

ICD-9 Codes

Conditions that may justify this test *include but are not limited to*:

Calcium
275.41   hypocalcemia
275.42   hypercalcemia

Suggested Reading

Dickerson RN, Alexander KH, Minard G, Croce MA, Brown RO. Accuracy of methods to estimate ionized and "corrected" serum calcium concentrations in critically ill multiple trauma patients receiving specialized nutrition support. *J Parenter Enteral Nutr* 2004; 28:133–41.

Frolich A. Combined reports on serum calcium and discriminant functions increase the diagnostic rate of hypercalcaemia. *Scand J Clin Lab Invest* 1997; 57:725–9.

Ward RT, Colton DM, Meade PC, Henry JC, Contreras LM, Wilson OM, Fleming AW. Serum levels of calcium and albumin in survivors versus nonsurvivors after critical injury. *J Crit Care* 2004; 19:54–64.

7.3.3   Calcium Concentration in 24-Hour Urine Collections

Indications for Test

Measurement of 24-hour urine calcium excretion (U_{Ca}) is indicated when:

- The serum calcium concentration [Ca] is low or normal in patients with nephrolithiasis, sarcoidosis, renal tubular acidosis, hyperthyroidism, hyperparathyroidism, malignancy, Paget's disease, primary aldosteronism, Cushing's syndrome, or evidence of surreptitious or chronic use of glucocorticoids or loop diuretics such as furosemide
- Making an assessment of the severity of hyperparathyroidism and other causes of hypercalcemia
- Searching for decreased calcium excretion in patients with malabsorption or vitamin D deficiency-related hyperparathyroidism
- Deciding whether to perform a parathyroidectomy in an asymptomatic hyperparathyroid patient as per "A Comparison of 2002 vs. 1990 Guidelines for Parathyroid Surgery in Patients with Asymptomatic Primary Hyperparathyroidism" (Table 7.4)

TABLE 7.4

A Comparison of 2002 vs. 1990 Guidelines for Parathyroid Surgery[a] in Patients with Asymptomatic Primary Hyperparathyroidism

Parameter	Guidelines	
	2002	1990
Age (years)	50	50
Serum calcium, mg/dL above the upper limit of normal (ULN)	1.0	1.0–1.6
Urinary calcium concentration (mg/24 hr)	>400	>400
% Reduction in creatinine clearance	>30	>30
Bone mineral density	T-score –2.5 at any site	Z-score –2.0 (forearm)

[a] Surgery is also indicated in patients for whom medical surveillance is neither desired nor possible.

Source: Bilezikian, J.P. et al., *J. Clin. Endocrinol. Metab.*, 87, 5353–5361, 2002. With permission.

Procedure

1. An evaluation for hypercalciuria should begin with an assessment of [Ca] in serum.
2. Prior to the patient's collection of a 24-hour urine specimen, inform patients: (a) not to ingest calcium immediately prior to or during urine collection unless it is prescribed or taken routinely, and (b) to maintain their typical dietary intake of calories and proportion of protein, fat, and carbohydrates.
3. Instruct the patient on the proper collection of a 24-hour urine specimen as per Specimen Collection Protocol P1 in Appendix 2.
4. If the adequacy of the 24-hour urine collection is questionable (e.g., 24-hour urine total volume is lower than expected, given the patient's BMI, renal function, and usual hydration status), consider repeat collection after discussing with the patient the details of proper collection of a 24-hour urine specimen.
5. Order urine total calcium and creatinine tests so the ratio of total mg of calcium per total mg creatinine (U_{Ca}/U_{Creat}) in the 24-hour urine collection can be determined.
6. If U_{Ca} excretion < 50 mg/24 hr, obtain a blood sample for 25-hydroxyvitamin D level.
7. In patients suspected of malabsorption or deficient calcium intake, recheck the 24-hour U_{Ca} excretion 4 to 6 weeks after oral loading with calcium.

Interpretation

1. Estimated reference interval for U_{Ca}:
    *   Males, 100 to 250 mg/24 hr
    *   Females, 100 to 200 mg/24 hr
2. Hypocalciuria is defined as:
    * U_{Ca} < 50 mg/24 hr
3. Hypercalciuria is defined as:
    * U_{Ca} > 4 mg/kg of body weight
    * U_{Ca} > 300 mg/24 hr in men
    * U_{Ca} > 250 mg/24 hr in women
    * U_{Ca}/U_{Creat} ratio > 0.14
4. U_{Ca} < 1 mg/kg is typical of patients with familial hypocalciuric hypercalcemia (FHH).
5. Hypercalciuria can be divided into three categories:
    *   Absorptive hypercalciuria due to excessive gut absorption of calcium
    *   Renal leak hypercalciuria due to renal tubular disease or hypoparathyroidism
    *   Resorptive hypercalciuria due to bone resorption (e.g., in Paget's disease, hyperthyroidism, skeletal metastases, hyperparathyroidism, or immobility)

Notes

1. Conditions associated with hypocalciuria include osteomalacia, malnutrition, and FHH. In patients with FHH or those being treated with thiazide diuretics, the serum calcium concentration may be high normal or elevated.
2. The majority of patients with eucalcemic hypercalciuria will not have an identifiable cause for this abnormality.
3. Prior to 2000, the majority of studies reporting reference intervals for 24-hour urine calcium excretion were limited by the inclusion of recurrent stone formers and poorly defined controls.
4. Long-term potassium (K) citrate therapy increases forearm BMD in idiopathic Ca stone formers but does not alter U_{Ca}.
5. Diets rich in calcium (800–1000 mg/day) reduce the risk of calcium oxalate kidney stones in patients with idiopathic nephrolithiasis, while low-calcium diets in this condition may result in reduced bone mineral density.
6. A high intake of calcium taken in pill form as a drug supplement does not reduce the urinary excretion of oxalate and formation of kidney stones.
7. The relative risk of kidney stones for men in the highest as compared with the lowest quintile for calcium intake was shown to be 0.56 (95% CI, 0.43–0.73; p for trend < 0.001).
8. Obesity and weight gain increase the risk of kidney stone formation. The magnitude of the increased risk appears to be greater in women than in men.
9. Potential risk factors promoting kidney stone formation include advancing age, untreated urinary tract infection, excess consumption of sodium, sucrose, grapefruit juice, alcohol, or animal protein, particularly in the context of elevated uric acid excretion.
10. Animal protein intake was directly associated with a relative risk of stone formation in men of 1.33 (95% CI, 1.00–1.77) for those with the highest protein intake compared to the lowest.
11. Independent of urinary calcium excretion, idiopathic calcium nephrolithiasis may be prevented by treatment with thiazides, allopurinol, potassium or potassium–magnesium citrate, and possibly neutral potassium phosphate.
12. Potassium and fluid intake are inversely related to the risk of kidney stones. For high potassium intake, the relative risk of stones is 0.49 (95% CI, 0.35–0.68). For high fluid intake, the relative risk is 0.71 (95% CI, 0.52–0.97).
13. In health professionals, the relative risk of stone formation among those with a U_{Ca} of ≥200 mg/L (mean age of 60 years) was 4.3 compared to those with a U_{Ca} of <75 mg/L. For those of mean age of 42 years, the relative risk was 51.09 (4.27–611.1). Urine [oxalate] and [citrate] did not differ between groups, but urine volumes were lower in stone formers (Curhan et al., 2001).
14. Vitamin C supplementation (1–2 g/day) appears to increase urinary oxalate excretion and the risk of calcium oxalate crystallization in calcium stone-forming patients without altering urinary pH.
15. In Bartter's syndrome, hypercalciuria occurs associated with a primary defect in NaCl reabsorption in the medullary thick ascending limb of the loop of Henle and hypokalemic metabolic alkalosis.
16. In Gitelman's syndrome, hypocalciuria occurs as the result of a distal tubular defect with normal urinary concentrating ability.
17. Bartter's and Gitelman's syndromes can be diagnosed only if diuretics are not in use.

ICD-9 Codes

Conditions that may justify this test *include but are not limited*

Parathyroid
252.0   hyperparathyroidism
252.1   hypoparathyroidism
252.8   other specified disorders of parathyroid gland

Adrenal
255.0   Cushing's syndrome
255.10   primary aldosteronism

Other

242.9	thyrotoxicosis without mention of goiter or other cause
268	vitamin D deficiency
275.42	hypercalcemia
276.2	acidosis
321.4	meningitis in sarcoidosis
592.0	calculus of kidney (nephrolithiasis)
731.0	osteitis deformans without mention of bone

Suggested Reading

Baxmann AC, de OG Mendonca C, Heilberg IP. Effect of vitamin C supplements on urinary oxalate and pH in calcium stone-forming patients. *Kidney Int* 2003; 63:1066–71.

Bilezikian JP, Potts JT Jr, Fuleihan GE-H, Kleerekoper M, Neer R, Peacock M, Rastad J, Silverberg SJ, Udelsman R, Wells SA. Summary statement from a workshop on asymptomatic primary hyperparathyroidism: a perspective for the 21st century. *J Clin Endocrinol Metab* 2002; 87:5353–61.

Colussi G, De Ferrari ME, Brunati C, Civati G. Medical prevention and treatment of urinary stones. *J Nephrol* 2000; 13(Suppl 3):S65–70.

Curhan GC, Willett WC, Rimm EB, Stampfer MJ. A prospective study of dietary calcium and other nutrients and the risk of symptomatic kidney stones. *N Engl J Med* 1993; 328:833–8.

Curhan GC, Willett WC, Speizer FE, Stampfer MJ. Twenty-four-hour urine chemistries and the risk of kidney stones among women and men. *Kidney Int* 2001; 59:2290–98.

Prien EL Jr, Curhan GC. Calcium nephrolithiasis. In: *Endocrinology*, 3rd ed., DeGroot LJ (Ed). WB Saunders, pp. 1179–89.

Vescini F, Buffa A, La Manna G, Ciavatti A, Rizzoli E, Bottura A, Stefoni S, Caudarella R. Twenty-four-hour urine chemistries and the risk of kidney stones among women and men: long-term potassium citrate therapy and bone mineral density in idiopathic calcium stone formers. *J Endocrinol Invest* 2005; 28:218–22.

7.3.4   Fractional Excretion of Calcium (F_ECa)

Indications for Test

The F_ECa test is indicated:

- When attempting to discriminate between hyperparathyroidism and familial hypocalciuric hypercalcemia (FHH)
- In the diagnostic work-up of patients with primary aldosteronism or glucocorticoids excess
- As a screening test preliminary to molecular genetic studies of patients suspected of having FHH

Procedure

1. Instruct the patient on the proper collection of a 24-hour urine specimen as per Specimen Collection Protocol P1 in Appendix 2. *Important:* Remind patients to have a blood sample obtained when they deliver their 24-hour urine specimen to the laboratory.
2. If the adequacy of the 24-hour urine collection is questionable (e.g., 24-hour urine total volume is lower than expected, given the patient's BMI, renal function, and usual hydration status), consider repeat collection after discussing with the patient the details of proper collection of a 24-hour urine specimen.
3. Order calcium and creatinine testing on *both* the patient's 24-hour urine and blood samples.
4. Calculate the F_ECa as follows:

$$F_E Ca\ (\%) = \frac{[Ca]_U / [Ca]_S}{[Creat]_U / [Creat]_S} \times 100$$

where:

[Ca]$_U$ = urine calcium concentration (mg/dL).
[Ca]$_S$ = serum calcium concentration (mg/dL).
[Creat]$_U$ = urine creatinine concentration (mg/dL).
[Creat]$_S$ = serum creatinine concentration (mg/dL).

Interpretation

1. Estimated reference interval for F$_E$Ca is 0.9 to 1.4%.
2. Patients with FHH typically have an F$_E$Ca well below 1%.
3. When the serum calcium concentration is elevated, a F$_E$Ca > 2% is consistent with hyperparathyroidism. Less than 1% of patients with these findings will have FHH.
4. When the serum calcium concentration is decreased, an increased F$_E$Ca is typical of patients with primary aldosteronism and Cushing's syndrome or those being treated with glucocorticoids.

Notes

1. Based on lower tubular reabsorption and decrease in serum calcium following inhibition of 11β-hydroxysteroid dehydrogenase type 2 activity by glycyrrhetinic acid (licorice), the most likely site of steroid-regulated renal Ca^{2+} handling appears to be the distal renal tubule.
2. A hot spot for calcium-sensing receptor (CaSR) gene mutations in patients with FHH has been found in exon 4 encoding for the extracellular domain of the CaSR.
3. Molecular abnormalities of the CaSR are responsible for three clinical disorders—FHH, neonatal severe hyperparathyroidism, and autosomal dominant hypocalcemia with hypercalciuria—and can be detected by clinically available genetics testing.

ICD-9 Codes

Conditions that may justify this test *include but are not limited to*:

Adrenal disorders
255.0   Cushing's syndrome
255.10   primary aldosteronism

Other
252.0   hyperparathyroidism
275.4   disorders of calcium metabolism/familial hypocalciuric hypercalcemia (FHH)

Suggested Reading

Felderbauer P, Hoffmann P, Klein W, Bulut K, Ansorge N, Epplen JT, Schmitz F, Schmidt WE. Identification of a novel calcium-sensing receptor gene mutation causing familial hypocalciuric hypercalcemia by single-strand conformation polymorphism analysis. *Exp Clin Endocrinol Diabetes* 2005; 113:31–4.
Ferrari P, Bianchetti MG, Sansonnens A, Frey FJ. Modulation of renal calcium handling by 11 beta-hydroxysteroid dehydrogenase type 2. *J Am Soc Nephrol* 2002; 13:2540–46.

7.3.5   Calcium Load Test for Suspected or Mild Hyperparathyroidism

Indications for Test

The calcium load test is indicated to:

- Confirm suspected mild primary hyperparathyroidism in patients who may have symptoms of hyperparathyroidism but have minimal, intermittent, or no elevation in their levels of total calcium and/or intact PTH
- Distinguish mild hyperparathyroidism after parathyroid surgery from persistent primary hyperparathyroidism

Procedure

1. Be sure that the patient does not have severe chronic kidney disease (CKD) Stage 4 or 5, as this test should not be performed in such patients.
2. Instruct the patient to ingest a diet restricted to 400 mg/day of calcium for 7 days prior to the calcium load test. To ensure the adequacy of the patient's compliance with the calcium-restricted diet, review the patient's food log record with a certified nutritionist.
3. Instruct the patient to consume a calcium load of 25 mg/kg in 240 mL of milk or other compatible fluid after an 8-hour fast. Neo-Calglucon® (calcium glubionate) containing 23 mg Ca^{2+}/mL is a palatable and readily absorbed calcium source but may be difficult to obtain in adult human dose quantities.
4. Obtain blood samples at baseline, 30, 60, and 120 minutes after the calcium load for total [Ca] and intact PTH testing.

Interpretation

1. Healthy individuals will suppress their intact PTH levels to a nadir of <10 pg/mL at any time point during a calcium load.
2. Hyperparathyroid patients will exhibit an intact PTH > 20 pg/mL at all time points during the calcium load.
3. Failure of the calcium load test to suppress a persistently elevated intact PTH, even after apparently successful parathyroidectomy, suggests the diagnosis of secondary hyperparathyroidism usually secondary to CKD Stage 3 in which both an elevated intact PTH and low to normal [Ca] are present.

Notes

1. This test relies on the principle that normal parathyroid tissue suppresses its secretion of PTH upon the administration of a calcium load, while adenomatous parathyroid tissue is less susceptible to suppressed production of PTH after calcium loading.
2. Expect a rise in serum calcium of 0.3 to 0.6 mg/dL during this test, with higher values to be found in hyperparathyroid subjects.
3. The intravenous infusion of as little as 1000 mg calcium suppresses the level of PTH to only 83.7% of baseline in subjects with hyperparathyroidism; in control subjects, PTH falls to a much lower level (58.8% of baseline) (Monchik et al., 1992).

ICD-9 Codes

Conditions that may justify this test *include but are not limited to*:

    252.0   hyperparathyroidism

Suggested Reading

McHenry CR, Rosen IB, Walfish PG, Pollard A. Oral calcium load test: diagnostic and physiologic implications in hyperparathyroidism. *Surgery* 1990; 108:1026–32.

Monchik JM, Lamberton RP, Roth U. Role of the oral calcium-loading test with measurement of intact parathyroid hormone in the diagnosis of symptomatic subtle primary hyperparathyroidism. *Surgery* 1992; 112:1103–10.

Tohme JF, Bilezikian JP, Clemens TL, Silverberg SJ, Shane E, Lindsay R. Suppression of parathyroid hormone secretion with oral calcium in normal subjects and patients with primary hyperparathyroidism. *J Clin Endocrinol Metab* 1990; 70:951–6.

TABLE 7.5
Differential Diagnosis of Disorders of Calcium Balance

Calcium Concentration (mg/dL)		Intact Parathyroid Hormone (iPTH) Concentration (ng/mL)		
Total[a]	Ionized	<10	10–65	>65
<8.0	<4.48	Hypoparathyroidism	Hypoparathyroidism	Pseudohypoparathyroidism, renal failure, or vitamin D deficiency
8.0–10.6	4.48–5.20	Increased calcium absorption or resorption by mechanisms other than PTH	Normal	2° Hyperparathyroidism[b]
>10.6	>5.2	Hypercalcemia due to causes other than elevated PTH	1° or 2° Hyperparathyroidism	1° Hyperparathyroidism

[a] Corrected for serum albumin concentration.

[b] For example, due to renal failure.

7.3.6 Intact* Parathyroid Hormone (PTH) Testing

Indications for Test

Intact PTH testing using an intact N-terminal-specific PTH assay or an assay that measures whole-molecule (or bio-intact) PTH (PTH_{1-84}) and a simultaneously obtained serum total calcium, $[S_{Ca}]$, or ionized calcium, $[Ca^{2+}]$, test is indicated in:

- The diagnosis of hyperparathyroidism (HPT) causing hypercalcemia
- Cases of abnormal $[S_{Ca}]$ and chronic kidney disease (CKD) in the diagnosis of secondary or tertiary hyperparathyroidism
- Periodically assessing [PTH] status in patients with CKD Stages 3 to 5 (renal dialysis)
- Cases of hypocalcemia

Procedure

1. Obtain a blood sample for intact or whole-molecule PTH ($[PTH_{1-84}]$) and $[S_{Ca}]$ or $[Ca^{2+}]$ 1 or 2 hours after the patient arises from sleep, preferably after an overnight fast.
2. If possible, instruct the phlebotomist obtaining the blood sample for PTH and calcium to not leave the tourniquet on too long before performing the phlebotomy, as the pH of the blood will fall and falsely increase the $[S_{Ca}]$.

Interpretation

1. Always interpret the [PTH] in the context of the total, albumin-corrected, or ionized calcium value as per "Differential Diagnosis of Disorders of Calcium Balance" (Table 7.5) and the serum creatinine concentration.
2. In patients with normal renal function not on thiazides and with a normal total (or albumin-corrected) serum calcium concentration (i.e., $[S_{Ca}]$ = 8.4–10.4 mg/dL) or normal ionized calcium concentration (i.e., $[Ca^{2+}]$ = 5.0–5.2 mg/dL), the reference interval for intact [PTH] is typically 12 to 65 pg/mL.

---

* *Note:* PTH assay nomenclature is confusing because prior to the advent in 2006 of PTH assays that unequivocally quantify only whole molecule (or bio-intact) PTH (i.e., PTH_{1-84}), PTH assays referred to as "intact" PTH assays quantified both whole-molecule PTH and N-terminal PTH fragments. Because of the greater analytical specificity of bio-intact PTH assays in measuring only PTH_{1-84}, the upper limit of normal (ULN) of the reference intervals for these assays (e.g., 55 pg/mL) is lower than for so-called "intact" PTH assays (ULN = 65 pg/mL).

TABLE 7.6
Stages of Chronic Kidney Disease (CKD), Estimated Glomerular Filtration Rate (GFRe), and Target [PTH] in Response to Therapy

CKD Stage	GFRe[a] (mL/min/ 1.73 m^2)	Target[b] [PTH] in Vitamin D-Replete Individuals	Interpretation of CKD Stage
1	≥90	<65[c]	Normal GFR with evidence of kidney damage for ≥3 months
2	60–89	<65[c]	Impaired GFR for ≥3 months [mild ↓ GFR]
3	30–59	35–70[c] or 17–35[d]	Onset of azotemia [moderate ↓ GFR]
4	15–29	70–110[c] or 35–55[d]	Severe ↓ GFR
5	<15	150–300[c]	End-stage renal disease usually requiring dialysis

[a] Based on the Modification of Diet in Renal Disease (MDRD) equation for calculating GFRe.

[b] Based on current interventions to reduce or suppress [PTH].

[c] Intact parathyroid hormone level (pg/mL).

[d] PTH_{1-84} (pg/mL).

Source: Adapted from National Kidney Foundation, *Am. J. Kidney Dis.*, 42, 1–201, 2003.

3. Patients with progressively worsening renal failure or insufficiency tend to have progressively higher PTH levels associated with more advanced stages of secondary HPT.
4. Patients treated with thiazides tend to have slightly higher [S_{Ca}] and lower [PTH].
5. Refer to "Stages of Chronic Kidney Disease (CKD), GFRe, and Target [PTH] in Response to Therapy" (Table 7.6) for the diagnosis of CKD and guidelines with regard to achievable intact PTH.

Notes

1. PTH is normally secreted in response to hypocalcemia and is suppressed by hypercalcemia.
2. Severe primary HPT is associated with an advanced bone disorder known as osteitis fibrosa cystica in which brown tumors of the bone may develop.
3. Although higher S_{Ca} concentrations in patients with primary HPT are associated with increased fracture risk and parathyroid surgery may have a protective effect, only increasing age (relative hazard [RH] per 10-year increase, 1.6; 95% CI, 1.4–1.9) and female gender (RH, 2.3; 95% CI, 1.2–4.1) are significant independent predictors of fracture risk.
4. Among unselected patients, most with mild primary HPT (e.g., total [S_{Ca}] = 10.9 ± 0.6 mg/dL), there is still a significant increase in the risk of vertebral, wrist, rib, and pelvic fractures (Khosla et al., 1999).
5. One out of 20 patients with hypercalcemia of malignancy had a [PTH] within the normal range (Kao et al., 1992).
6. Native or whole-molecule PTH (PTH_{1-84}) is biologically active and has a half-life measured in minutes, whereas the carboxy-terminal PTH (C-PTH) and mid-molecule fragments of PTH (MM-PTH) are biologically inactive and have half-lives 10- to 20-fold longer.
7. High concentrations of biologically inactive PTH fragments build up in the blood of patients with impaired renal function and may obscure the diagnosis of their parathyroid status.
8. Cinacalcet therapy lowers [PTH] as well as calcium and phosphorus levels.

ICD-9 Codes

Conditions that may justify this test *include but are not limited to:*

Hyperparathyroidism
252.0   primary
252.8   tertiary

Other
275.42   hypercalcemia
585   chronic kidney disease (CKD)

Suggested Reading

Endres DB, Villanueva R, Sharp CF, Singer FR. Immunochemiluminometric and immunoradiometric determinations of intact and total immunoreactive parathyrin: performance in the differential diagnosis of hypercalcemia and hypoparathyroidism. *Clin Chem* 1991; 37:162–8.

Kao PC, van Hearden J, Grant CS, Klee GG, Khosla S. Clinical performance of parathyroid hormone immunometric assays. *Mayo Clin Proc* 1992; 67:637–45.

Khosla S, Melton LJ, Wermers RA, Crowson CS, O'Fallon W, Riggs B. Primary hyperparathyroidism and the risk of fracture: a population-based study. *J Bone Miner Res* 1999; 14:1700–7.

Levey AS, Coresh J, Balk E, Kausz AT, Levin A, Steffes MW, Hogg RJ, Perrone RD, Lau J, Eknoyan G. National Kidney Foundation practice guidelines for chronic kidney disease: evaluation, classification, and stratification. *Ann Intern Med* 2003; 139:137–47.

7.3.7 Parathyroid Hormone-Related Protein (PTHrP) Testing

Indications for Test

PTHrP testing is indicated in:

- Patients with hypercalcemia of malignancy and a low-normal intact or whole-molecule PTH level
- The monitoring of successful treatment of a PTHrP-producing tumor in patients with hypercalcemia of malignancy

Procedure

1. Restrict PTHrP testing to patients with unexplained hypercalcemia after thorough evaluation or those with known tumor-related hypercalcemia.
2. In any hypercalcemic patient, always obtain a blood sample for PTH testing before requesting PTHrP testing.
3. Obtain a random blood sample for PTHrP testing if the hypercalcemic patient has a [PTH] in the reference range.

Interpretation

1. Expected [PTHrP] in healthy individuals < 3 pmol/L.
2. If treatment of a PTHrP-producing tumor is successful, [PTHrP] declines and hypercalcemia improves.
3. Compared to patients with a [PTHrP] < 150 pmol/L, patients with hypercalcemia of malignancy and a [PTHrP] ≥ 150 pmol/L have an increased rate of bone metastases and decreased survival (Hiraki et al., 2002).

Notes

1. Solid tumors may produce a peptide hormone (PTHrP) that shares significant homology with the first 13 amino acids of the N-terminal region of native PTH (PTH$_{1-84}$) and exerts PTH-like biological activity.
2. Measures of total calcium are less expensive and easier to perform than ionized calcium or PTHrP for monitoring treatment of patients with humoral hypercalcemia of malignancy.
3. PTHrP is of little use in the differential diagnosis of hypercalcemia.
4. Elevated PTHrP levels are a common finding in cancer patients without bone metastases.
5. Biointact PTH should always be measured in patients with apparent hypercalcemia of malignancy because concurrent primary hyperparathyroidism is not rare in these patients.
6. Primary hyperparathyroidism accounts for hypercalcemia in the 90% of cancer-free patients whose PTHrP levels are found to be elevated.

ICD-9 Codes

Conditions that may justify this test *include but are not limited to*:

275.42   hypercalcemia

Suggested Reading

Casez J, Pfammatter R, Nguyen Q, Lippuner K, Jaeger P. Diagnostic approach to hypercalcemia: relevance of parathyroid hormone and parathyroid hormone-related protein measurements. *Eur J Intern Med* 2001; 12:344–9.

Hiraki A, Ueoka H, Bessho A, Segawa Y, Takigawa N, Kiura K, Eguchi K, Yoneda T, Tanimoto M, Harada M. Parathyroid hormone-related protein measured at the time of first visit is an indicator of bone metastases and survival in lung carcinoma patients with hypercalcemia. *Cancer* 2002; 95:1706–13.

7.3.8   Aluminum, [Al], and Parathyroid Hormone, [PTH]: Test Panel in Renal Osteodystrophy

Indications for Test

Measurement of [PTH] and [Al] may be indicated in:

- Renal dialysis (CKD Stage 5) patients with low BMD in whom aluminum exposure has not been prospectively and scrupulously avoided
- Patients with symptoms of encephalopathy including seizures, incoordination, dysarthria, postural tremor, and fluctuating short-term memory (Alfrey et al., 1976; Becaria et al., 2002)

Procedure

1. Obtain the CKD Stage 5 patient's history of intake of aluminum-based antacids and exposure to dialysis solutions not specifically depleted of aluminum. Obtain a history of exposure to ingestion or infusions of aluminum related to aluminum cooking utensils, intravenous feeding, or intravesical alum irrigation for hemorrhagic cystitis.
2. Assess patient's BMD using DXA.
3. If BMD is in the osteopenic or osteoporotic range, obtain a random blood sample for [Al] and biointact [PTH]. Ensure that special, aluminum-free tubes and phlebotomy equipment are used.

Interpretation

1. A baseline serum [Al] > 100 mcg/L and a normal or low-normal [PTH] have an 88% specificity, 80% sensitivity, and 80% positive predictive value for aluminum osteodystrophy.
2. Short-duration exposure to high to high-normal [PTH] may be protective against aluminum-induced renal osteodystrophy.
3. Extended exposure to an excessively high [PTH] may result in other serious complications including hypercalcemia, ectopic calcification, calciphylaxis, or skeletal pain in addition to aluminum-induced osteodystrophy.

Notes

1. Aluminum suppresses the pulsatile secretion of PTH and may reduce PTH synthesis.
2. Patients with CKD Stage 5 are at risk for aluminum-related aplastic bone disease and renal osteodystrophy if they are in a long-term dialysis program using dialysis fluids containing more than 120 mcg/L of aluminum.

3. Aluminum is known to have a causal role in dialysis-related encephalopathy, microcytic anemia, adynamic bone disease, and osteomalacia and may play a role in the pathogenesis of Alzheimer's disease.
4. Aluminum-based hypophosphatemics are no longer used in CKD patients, resulting in the near disappearance of aluminum toxicity in these patients.

ICD-9 Codes

Conditions that may justify this test *include but are not limited to*:

Neurologic
333.1   essential and other specified forms of tremor
348.31   metabolic encephalopathy
780.39   convulsions
780.93   memory loss
781.3   lack of coordination
784.5   speech disturbance

Other
985   toxic effect of metals

Suggested Reading

Alfrey AC, LeGendre GR, Kaehny WD. The dialysis encephalopathy syndrome: possible aluminum intoxication. *N Engl J Med* 1976; 294:184–8.
Becaria A, Campbell A, Bondy SC. Aluminum as a toxicant. *Toxicol Ind Health* 2002; 18:309–20.
Diaz-Corte C, Fernandez-Martin JL, Barreto S, Gomez C, Fernandez-Coto T, Braga S, Cannata JB. Effect of aluminum load on parathyroid hormone synthesis. *Nephrol Dial Transplant* 2001; 16:742–5.
Favus MJ. Factors affecting calcium metabolism in disorders of the kidney. *Ann Clin Lab Sci* 1981; 11:327–32.
Favus MJ (Ed). *Primer on the Metabolic Bone Diseases and Disorders of Mineral Metabolism*, 2nd ed. 1993. American Society for Bone and Mineral Research, pp. 113, 318.

7.4   Parathyroid Adenoma Localization

7.4.1   Dual-Phase ^{99m}Tc-Sestamibi (^{99m}Tc-MIBI) Scan for Parathyroid Adenoma Localization

Indications for Test

Dual-phase ^{99m}Tc-MIBI scanning of the neck is indicated:

- After diagnosis of hyperfunctioning parathyroid tissue is made based on calcium and PTH testing
- To localize and differentiate between solitary and multiglandular parathyroid disease in known hyperparathyroid patients

Procedure

1. Prior to imaging, establish that the [PTH] is clearly elevated for level of serum calcium and that the vitamin D level is normal (i.e., 25-hydroxyvitamin D > 30 ng/mL). Correction of vitamin D deficiency is recommended before parathyroid imaging.
2. Discontinue any thiazide diuretics and calcium supplements for 2 or more weeks before MIBI scanning, as these agents tend to raise serum calcium concentration and interfere with the uptake of ^{99m}Tc-hexakis-2-methoxyisobutyl isonitrile (^{99m}Tc-MIBI) by parathyroid adenomas.

3. In selected patients, especially those with larger, multinodular goiters, proceed with the triiodothy-ronine suppression protocol as follows:
    *   Before MIBI scan, administer triiodothyronine (Cytomel®; 25 mcg p.o. b.i.d. for 5 days).
    *   Administer a dose of 20 to 25 mCi ^{99m}Tc-MIBI intravenously.
4. With the patient in the supine position, place a roll under the shoulders and extend the neck. The neck is kept midline for all studies. Closely apply detector to the skin.
5. Acquire images using a gamma camera with a low-energy all-purpose or low-energy high-resolution collimator with a 20% window and 1.6× magnification. Computer-enhanced imaging is desirable but not mandatory.
6. Obtain early (first-phase) planar imaging of the anterior neck and upper mediastinum within 5 to 15 minutes after injection of ^{99m}Tc-MIBI and delayed (second-phase) washout images at 1.2 to 2.5 hours. Rapid image acquisition is possible with modern (later than 1990s) imaging devices.
7. Obtain frontal, right, and left anterior oblique (RAO and LAO) views of the neck in all patients. Oblique views are to be obtained at about 30° from vertical by moving the camera, not the patient's head. This is facilitated by the availability of a dual- or triple-head camera.
8. If indicated, the mediastinal view should include the top half of the heart. Views of the neck should not show more than a small sliver of the heart.
9. If atypically located parathyroid tissue is suspected or standard images fail to clearly identify the location of a biochemically established parathyroid adenoma, acquire images of the lower neck and chest using single-photon emission computed tomography (SPECT) followed by neck and/or chest CT/MR imaging.
10. Obtain a combined ^{201}thallium and pertechnetate subtraction scan to confirm the presence of multiple parathyroid adenomas if multiple abnormalities are seen on dual-phase MIBI scintigraphy and a multinodular goiter (MNG) is not evident. Note that multiple parathyroid adenomas are exceedingly rare (<1%).
11. As an alternative to combined ^{201}thallium and pertechnetate subtraction scanning, obtain a second dual-phase ^{99m}Tc-MIBI scan after Cytomel® suppression as per above protocol.

Interpretation

1. Using scanning technology available in the 1980s, a clear pattern of delayed washout was seen in only 38.5% of patients with parathyroid adenomas. Subtle abnormalities were seen in another 38.5% of patients in whom smaller lesions were observed, thus raising the diagnostic sensitivity of ^{99m}Tc-MIBI scan in identifying patients with parathyroid gland adenomas to little more than 75% (Taillifer et al., 1992).
2. In a metaanalysis of the diagnostic accuracy of MIBI scanning, the average sensitivity and specificity of scanning for helping to identify patients with parathyroid gland adenomas were reported to be 90.7% and 98.8%, respectively (Denham and Norman, 1998).
3. SPECT or ^{201}thallium and pertechnetate thyroid subtraction scanning with computer-enhanced graphic analysis is rarely required and may be even less effective at localizing adenomas than RAO and LAO images on dual-phase MIBI scanning, even in patients with smaller (<500 g) parathyroid glands.
4. MIBI scan identifies parathyroid adenomas weighing >500 mg with 91% sensitivity and <500 mg with 80% sensitivity. Neither intact PTH nor calcium levels are predictive of a positive scan.
5. In 20 to 25% of confirmed parathyroid adenomas, multiple uptake abnormalities related to large multinodular goiters may occur and interfere with the localization of adenomas. Cytomel® suppression may help to improve localization in cases of large MNGs.
6. In patients who appear to have multiple abnormalities, a pinhole collimation or subtraction scan may provide adjunctive information helpful for localizing a significant lesion. Used alone, ^{99m}Tc-MIBI scans do not reliably predict multigland hyperplasia.
7. Patients with incorrectly interpreted or nondiagnostic scintigrams tend to have a higher percentage of upper pole adenomas than patients with diagnostic scans. There is an even distribution of upper vs. lower parathyroid adenomas.
8. The appearance of a lower pole adenoma on scan may occur when the lesion seen is actually present in the mediastinum. About 4 to 5% of all parathyroid adenomas occur in the chest.
9. In high-resolution MIBI scans not requiring pinhole technology, first-phase images identify most parathyroid adenomas better than the delayed second-phase images.

Notes

1. Multiple comparative studies suggest that the diagnostic utility of sestamibi protocols equals or exceeds other noninvasive, nonscintigraphic imaging strategies, including high-resolution US, CT, and MR imaging (see Test 7.4.2).
2. The dual-phase MIBI scan is preferred for its sensitivity, specificity, and simplicity over dual-isotope scans and CT or MR and is much less expensive than SPECT.
3. SPECT followed by CT scanning is the procedure of choice for localizing an adenoma in the anterior mediastinum.
4. The superior soft-tissue contrast of MR over CT may allow differentiation of thyroid nodules, thyroid cysts, and parathyroid tumors from normal thyroid tissue.
5. Older gamma cameras (e.g., Ohio-Nuclear Sigma 410, *circa* 1979) require the use of a pinhole collimator to resolve an image and will yield suboptimal oblique views. More current devices (e.g., Siemens ZLC 7500 Orbitor camera and Picker Odyssey 750 computer) do not require a pinhole collimator to get low-energy, high-resolution images.
6. The sensitivity and specificity of current dual-phase MIBI scanning techniques for parathyroid adenoma localization are typically 85 to 100% but are consistently better than 95% when technical factors are optimized using 21st-century devices.
7. Although the sensitivity of adenoma detection may increase to >95% using SPECT and three-dimensional image display (volume-rendered reprojection for visualization), the cost of SPECT is usually prohibitive. Computer-generated image enhancement is a standard feature of modern scans.
8. With high-resolution MIBI scanning, minimally invasive surgery with a better cosmetic result and avoidance of the complications of bilateral neck exploration are routinely achieved.
9. Differential uptake and washout of isotope from parathyroid adenomas and hyperplastic glands occurs compared to normal parathyroid and thyroid tissue. The high density of mitochondria in both adenomatous and hyperplastic parathyroids permits imaging on delayed MIBI scan images, but adenomas are usually much better seen.
10. In patients who require diuretics for antihypertensive therapy, torsemide is an alternative to a thiazide prior to scanning. Unlike thiazide therapy, torsemide does not block renal excretion of calcium.
11. Giving lemon juice 20 minutes before imaging (20% juice/80% water) does not significantly decrease isotope uptake by the salivary glands.
12. Long-standing vitamin D deficiency may promote the appearance and recurrence of parathyroid hyperplasia in either a single or multiple parathyroid glands.

ICD-9 Codes

Conditions that may justify this test *include but are not limited to*:

Parathyroid
252   hyperparathyroidism
252.8   parathyroid adenoma

Suggested Readings

Denham DW, Norman J. Cost-effectiveness of preoperative sestamibi scan for primary hyperparathyroidism is dependent solely upon the surgeon's choice of operative procedure. *J Am Coll Surg* 1998; 186:293–305.

McBiles M, Lambert AT, Cote MG, Kim SY. Sestamibi parathyroid imaging. *Semin Nucl Med* 1995; 25:221–34.

Moka D, Voth E, Dietlein M, Larena-Avellaneda A, Schicha H. Technetium 99m-MIBI-SPECT: a highly sensitive diagnostic tool for localization of parathyroid adenomas. *Surgery* 2000; 128:29–35.

Norman JG, Chheda H. Minimally invasive parathyroidectomy facilitated by intraoperative nuclear mapping. *Surgery* 1997; 122:998–1004.

Norman JG, Jaffray CE, Chheda H. The false-positive parathyroid sestamibi: a real or perceived problem and a case for radioguided parathyroidectomy. *Ann Surg* 2000; 231:31–7.

Norton KS, Johnson LW, Griffen FD, Burke J, Kennedy S, Aultman D, Li BD, Zibari G. The sestamibi scan as a preoperative screening tool. *Am Surg* 2002; 68:812–5.

Schneider PB. Parathyroid scanning accuracies. *J Nucl Med* 1999; 40:361.

Staudenherz A, Telfeyan D, Steiner E, Niederle B, Leitha T, Kletter K. Scintigraphic pitfalls in giant parathyroid glands. *J Nucl Med* 1995; 36:467–9.

Taillefer R, Boucher Y, Potvin C, Lambert R. Detection and localization of parathyroid adenomas in patients with hyperparathyroidism using a single radionuclide imaging procedure with technetium-99m-sestamibi (double-phase study). *J Nucl Med* 1992; 33:1801–9.

7.4.2 Parathyroid Imaging Techniques: Multislice Computed Tomography (M-CT), Magnetic Resonance (MR), and Ultrasonography (USG) Imaging as Adjuncts to Dual-Phase Sestamibi (^{99m}Tc-MIBI) Scanning

Indications for Test

Imaging of the neck and chest by M-CT, MR, and/or USG imaging techniques may be indicated if:

- High-resolution, dual-phase ^{99m}Tc-MIBI scan is unavailable or yields an equivocal result, the patient is unable to remain still for the time required for scanning, or venous access for injection of isotope is limited.
- The presence of a large parathyroid adenoma is suspected based on major elevations in calcium and PTH levels, and rapid screening for location of the parathyroid adenoma is desired.

Procedure

1. Obtain a baseline and 2.5- to 3-hour delayed ^{99m}Tc- MIBI image of the neck as described in Test 7.4.1 and determine whether uptake in one or more areas is increased.
2. If the dual-phase ^{99m}Tc-MIBI image fails to identify an abnormality, obtain a USG image of the neck prior to M-CT or MR of the neck and mediastinum.
3. If dual-phase ^{99m}Tc-MIBI and USG scans are negative in the face of biochemically documented hyperparathyroidism, obtain SPECT or MR scanning of the retrothyroid space or mediastinum.

Interpretation

1. Areas of increased uptake on ^{99m}Tc-MIBI imaging suggest the presence of parathyroid adenomas with a high degree of specificity but lower sensitivity, dependent on technical factors related to the scanning technique used to localize the adenomas (Figure 7.2).
2. A lucent area in the retrothyroid space consistent with a very large parathyroid adenoma may appear on USG.
3. Rounded masses seen on M-CT or MR of the neck and anterior mediastinum may be parathyroid adenomas, thyroid nodules, or lymph nodes.

Notes

1. MR and M-CT scans of the neck have a sensitivity of about 75% with a specificity of 85% in imaging of abnormal parathyroid tissue.
2. The sensitivity and specificity of modern-day dual-phase ^{99m}Tc-MIBI imaging for parathyroid adenomas is 85 to 100%, which is much better than results from older, single-detector CT scans (Doppman et al., 1977) or from MR scans obtained prior to 1984 (Stark et al., 1984).

ICD-9 Codes

Refer to Test 7.4.1 codes.

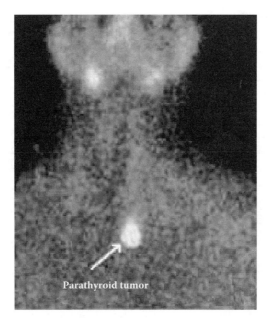

Figure 7.2
Sestamibi imaging illustrating the presence of a mediastinal parathyroid tumor.

Suggested Reading

Doppman JL, Brennan MF, Koehler JO, Marx SJ. Computed tomography for parathyroid localization. *J Comput Assist Tomogr* 1977; 1:30–6.

Stark DD, Clark OH, Moss AA. Magnetic resonance imaging of the thyroid, thymus, and parathyroid glands. *Surgery* 1984; 96:1083–91.

7.4.3   Selective Venous Sampling for Localization of Hyperfunctioning Parathyroid Glands

Indications for Test

Selective venous sampling for PTH is only indicated in patients:

- With persistent hyperparathyroidism, usually following unsuccessful neck exploration for removal of hyperfunctioning parathyroid tissue, or upon failure to clearly localize a PTH-secreting adenoma by advanced imaging techniques such as SPECT of the chest (see Test 7.4.1)

Procedure

1. Be sure that the patient is not pregnant and that a thorough review of previous imaging studies has been accomplished by an endocrinology specialist.
2. Refer to Specimen Collection Protocol P8 in Appendix 2.
3. If the laboratory does not provide a report indicating all PTH values, sorted from lowest to highest PTH values and their corresponding anatomic site, the endocrinologist must prepare such a report so the PTH gradient leading to an anatomic site where the PTH level is highest can be appreciated (Figure 7.3).
4. Discuss the report with the endocrinology surgeon in terms of the most likely anatomic location of a PTH-secreting adenoma for surgical resection.

University Hospital - St Paul 5909 Harry Hines Boulevard Dallas, TX 75390-9200 Phone: 214-879-3888; Fax: 214-879-6311		**Selective Venous Sampling (SVS) for PTH**		
Patient		Medical Record No. (MRN)		
Date of Surgery		Report Date		Lowest PTH Value, pg/mL
Requestor		Mail Code		67.2
Sample#	Anatomic Site	Time	[PTH], pg/mL	"Fold" Increase over *Lowest PTH Value*
5	Right Thyrodial	10:10	67.2	0.0
18	Right Middle Vertebral	10:56	73.8	0.1
21	Azygous	11:02	76.3	0.1
14	Left High Internal Jugular	10:43	84.7	0.3
12	Left Middle Vertebral	10:39	86.3	0.3
1	Right Subclavian	9:55	90.1	0.3
19	Right Low Vertebral	10:58	91.1	0.4
17	Left Thyroidal	10:51	92.8	0.4
6	Right Low Thyrodial	10:17	93.5	0.4
8	Right Brachiocephalic	10:29	97.7	0.5
7	Right Lowest Thyrodial	10:18	99.6	0.5
4	Right Low Internal Jugular	10:01	103.1	0.5
2	Right High Internal Jugular	9:59	104.4	0.6
15	Left Middle Internal Jugular	10:44	104.4	0.6
11	Left Subclavian	10:33	104.5	0.6
16	Left Low Internal Jugular	10:46	104.5	0.6
3	Right Middle Internal Jugular	10:00	110.0	0.6
10	Right Atruim	10:32	110.5	0.6
13	Left Low Vertebral	10:40	117.0	0.7
20	Left Brachiocephalic	10:59	126.1	0.9
9	SVC	10:31	169.4	1.5
22	Thymic	11:05	2999.3	43.6
Anatomic Map of SVS for PTH Samples				

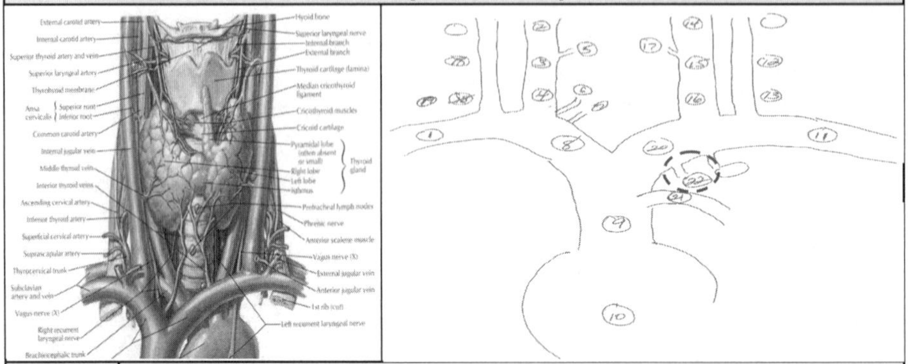

Interpretation	Possible PTH-secreting adenoma in the region of the thymic vein.		
Prepared by	*Frank H. Wians, Jr., Ph.D.*	Date	
Reviewed by	*Richard J. Auchus, M.D.*	Date	

Report Designed by Dr. Frank H. Wians, Jr. © 2008

Figure 7.3
Example of Excel spreadsheet-based report for the results of selective venous sampling (SVS) for PTH procedures (dotted oval identifies the anatomic location of the blood samples with the highest fold-increase over the PTH concentration in the blood sample with the lowest PTH concentration.

Interpretation

1. The anatomic site associated with the highest PTH concentration is the most likely site of a PTH-secreting adenoma; however, the magnitude of this PTH value and the characteristics of the PTH gradient leading to and from this site require evaluation.

2. Typically, a PTH-secreting adenoma in a particular anatomic site is associated with a significantly increased PTH concentration in the blood sample obtained from this site, compared to the PTH values in samples taken from other anatomical sites that do not harbor a PTH-secreting adenoma.

3. If the magnitude of PTH values and the PTH gradient are not noticeably different throughout the anatomic tree from which blood samples were obtained, an ectopic site, such as high in the neck toward an ear or deep in the thorax toward the heart, may be involved.

Notes

1. Significant anatomic variation exists in the venous drainage system from the four parathyroid glands between individuals.
2. Heparin used to prevent clotting in the catheters will not affect the PTH assay.

ICD-9 Codes

Refer to Test 7.4.1 codes.

Suggested Reading

Udelsman R, Aruny JE, Donovan PI, Sokoll LJ, Santos F, Donabedian R, Venbrux AC. Rapid parathyroid hormone analysis during venous localization. *Ann Surg* 2003; 237:714–9.

Wians FH Jr. Use of Excel spreadsheet to create interpretive reports for laboratory tests requiring complex calculations. *Lab Med* 2009; 40:5–12.

7.4.4   Rapid Parathyroid Hormone (PTH) Testing and Use of the Radioisotopic Gamma Probe (GP) in the Intraoperative Localization and Removal of Parathyroid Adenomas

Indications for Test

Rapid PTH testing intraoperatively or in a central laboratory for localization of parathyroid adenomas may be indicated to:

- Shorten anesthesia time in patients undergoing bilateral neck exploration for suspected parathyroid gland adenomas
- Assist the neck surgeon in successful removal of all offending parathyroid glands responsible for the patient's hypercalcemia, including those in ectopic areas such as the thymus and mediastinum

Procedure

Dual-Phase ⁹⁹ᵐTc-Sestamibi (⁹⁹ᵐTc-MIBI) Scanning

1. Obtain a ⁹⁹ᵐTc-MIBI scintigraphy scan of the patient's neck and upper chest (see Test 7.4.1).

Radioisotopic Gamma Probe (GP) Localization

1. At 2.5 to 3.0 hours prior to the time of the scheduled parathyroidectomy, inject a dose (20–25 mCi) of ⁹⁹ᵐTc-MIBI.
2. With a sterile-covered gamma probe (GP) passed over neck, thymus, and mediastinum, identify the parathyroid lesion intraoperatively to help direct its surgical excision. Use of the probe 2.5 to 3 hours after dosing with ⁹⁹ᵐTc-MIBI is critical. *Note:* The criticality of this timing limits the clinical usefulness of this technique.

Intraoperative PTH Testing

1. Be sure that the laboratory or operating room-based assay system uses rapid (10- to 12-minute) test turnaround times after blood specimen collection for PTH assay during intraoperative PTH (ioPTH) testing done either directly in the surgical suite or in a central laboratory after rapid delivery of blood specimens collected intraoperatively.
2. Have the neck surgeon review the ⁹⁹ᵐTc-MIBI scan of the neck and upper chest to determine the putative location of PTH-secreting parathyroid adenomas prior to surgery.

3. Have a member of the surgical team obtain peripheral blood specimens according to the established sequence below (Irvin et al., 1994):
   - Specimen 1—After induction of anesthesia, but before neck incision (0 minutes or baseline)
   - Specimen 2—During manipulation of a suspected hyperfunctioning parathyroid gland
   - Specimen 3—5 minutes after gland excision
   - Specimen 4—10 minutes after gland excision
   - Specimen 5—20 minutes (occasionally) after gland excision
4. Have the neck surgeon review results of intraoperative PTH testing, after excision of an adenoma, to aid in the assessment of complete removal of all PTH hypersecreting tissue.
5. Failure of ioPTH levels in these timed blood specimens to meet the criterion (see Interpretation point 4 below) for successful excision of all hyperfunctioning parathyroid gland or tissue will prompt additional exploration of the neck and mediastinal thorax for the site of additional adenomas and repeat collection of specimens 2 through 5 noted above.

Interpretation

1. Intraoperative radioactivity detected by GP will increase over the site of an adenoma at about 3 hours ± 15 minutes from the time the isotope (^{99m}Tc-MIBI) was administered prior to neck exploration.
2. If the GP is applied too early (1 to 2 hours after administration of the isotope) or too late (4 hours after administration of the isotope), radioactivity over the site of an adenoma may not be different from background.
3. Increased isotope uptake on GP testing is not specific for parathyroid tissue and may falsely identify thyroid adenomas.
4. For ioPTH testing, the criterion used to indicate that the offending parathyroid glands have been excised and the remaining parathyroid glands are not hyperfunctioning is a >50% decline in the [ioPTH] of the 5-minute, post-excision blood specimen over the 0-minute [ioPTH] (Figure 7.4).
5. Normocalcemia occurs postoperatively in >95% of patients who undergo successful ioPTH testing, with even higher postoperative success rates occurring in patients with only uniglandular disease.
6. Failure of a post-excision ioPTH level to decrease by greater than 50%, relative to the highest baseline ioPTH value can occur in up to half of all patients with multiglandular parathyroid disease.

Notes

1. GP localization of a parathyroid gland lesion has a lower sensitivity and specificity for identification of adenomas than ioPTH testing and gives no better information than preoperative ^{99m}Tc-MIBI scanning in patients with clearly positive scans.
2. Two out of six patients who required repeat surgery for persistent hyperparathyroidism also required both GP and ioPTH testing for successful localization of the offending glands (Jaskowiak et al., 2002; Sullivan et al., 2001).
3. Formerly, the monitoring of parathyroid function/hyperfunction intraoperatively was accomplished by obtaining serial samples of urine for measurement of cAMP levels. The ioPTH testing has superceded this method.
4. In 50 cases of primary hyperparathyroidism with a solitary adenoma, 45 had a ≥50% reduction in ioPTH 10 minutes after excision, providing diagnostic sensitivity, specificity, accuracy, and positive predictive value (PPV) of 88%, 50%, 85%, and 97%, respectively. With preoperative ^{99m}Tc-MIBI scanning alone, values for diagnostic sensitivity, specificity, accuracy, and PPV were 90%, 50%, 87%, and 96%, respectively. Combining ^{99m}Tc-MIBI scintigraphy with ioPTH testing resulted in values for diagnostic sensitivity and PPV of 100% and 97.5%, respectively, for localization of the parathyroid adenoma (Miller et al., 2003).
5. When used alone, neither preoperative imaging tests nor ioPTH testing has been shown to adequately identify the 5 to 10% of patients with primary hyperparathyroidism due to multiglandular disease.
6. The accuracy of combining neck ultrasonography, ^{99m}Tc-MIBI scintigraphy, and ioPTH testing in predicting multiglandular disease has not been shown to exceed 80%.
7. In a study of 103 patients, 12 of whom had multiglandular disease, ioPTH testing identified 8 of the 9 patients with a positive ^{99m}Tc-MIBI scan (a solitary focus) and multiglandular disease (Haciyanli et al., 2003).

RESULTS OF STAT INTRA-OPERATIVE PTH TESTING			Patient's Name			Physician	
Aston Pathology Lab	Dallas, TX 75235-9072	Phone: 214-648-9152	Med Rec No.			Date of Surgery	
5323 Harry Hines Blvd	CLIA I.D. No. 45D0659587	Fax: 214-648-3972	Weight of PTH Gland			Surgery Performed @	
Sestamibi Scan Findings			Date Scan Performed			Initial/Repeat Surgery	
Sample Number	Specimen Type	Specimen Time	(A) Q-[PTH], pg/mL	(B) Q-[PTH] - Q-[PTH]$_{Baseline}$	(B)/(A) x 100 % Change in Q-[PTH]	Comment(s)	
0	Pre-Op (Baseline)		1256.0				
1	5 min PE		449.0	-807.0	-64.3		
2	10 min PE		388.0	-868.0	-69.1		
3	20 min PE		281.0	-975.0	-77.6		
5							
6							
7							
8							
9							
10							

NOTE: Lower limit of detection (LLD) of Q-PTH assay is 4 pg/mL. Q-PTH assay is linear up to approx. 1400-1600 pg/mL.

Interpretive Note: A decrease in Q-PTH level of 50% or more in the **10-min** post-PTG excision sample is the criterion used typically to indicate that the offending parathyroid gland(s) has(have) been excised and the remaining parathyroids are not hyperfunctioning, thus predicting a post-operative return to normal serum calcium levels.

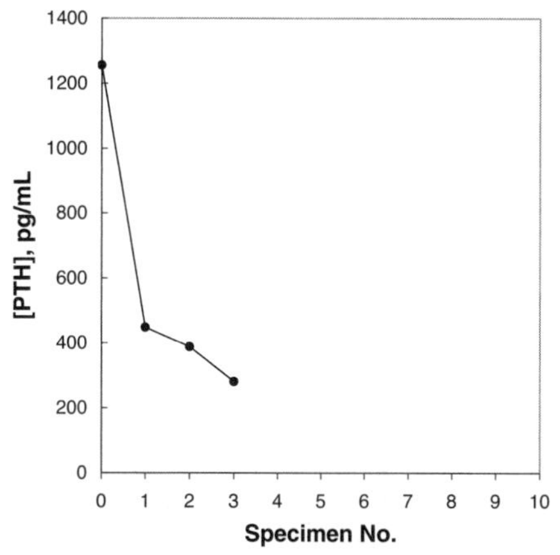

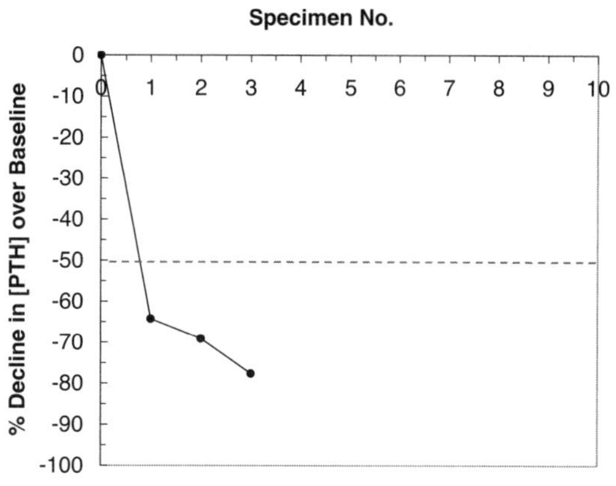

Interpretation: All post-excision specimens demonstrated an ioPTH decline of >50% consistent with successful removal of all hyperfunctioning PTGs.

Technologist		Reviewed By		Date	

Figure 7.4

Example of Excel spreadsheet-based report for the results of intra-operative parathyroid hormone (ioPTH) testing (dotted line identifies 50% decline in PTH concentration over baseline).

8. The incidental discovery of a large parathyroid adenoma at the time of thyroid surgery for other conditions, even in the absence of hypercalcemia, is an uncommon, but well described, phenomenon that should prompt excision of the adenoma and evaluation of the remaining parathyroid glands.

ICD-9 Codes

Refer to Test 7.4.1 codes.

Suggested Reading

Denizot A, Dadoun F, Meyer-Dutour A, Alliot P, Argeme M. Screening for primary hyperparathyroidism before thyroid surgery: a prospective study. *Surgery* 2002; 131:264–9.

Haciyanli M, Lal G, Morita E, Duh QY, Kebebew E, Clark OH. Accuracy of preoperative localization studies and intraoperative parathyroid hormone assay in patients with primary hyperparathyroidism and double adenoma. *J Am Coll Surg* 2003; 197:739–46.

Irvin, GL III, Prudhomme DL, Deriso GT et al. A new approach to parathyroidectomy. *Ann Surg* 1994; 219:574–9.

Jaskowiak NT, Sugg SL, Helke J, Koka MR, Kaplan EL. Pitfalls of intraoperative quick parathyroid hormone monitoring and gamma probe localization in surgery for primary hyperparathyroidism. *Arch Surg* 2002; 137:659–69.

Merritt RM, Anarayana S, Wei JP. Asymptomatic and normocalcemic parathyroid adenoma: detection by preoperative ultrasound. *Am Surg* 1993; 59:232–4.

Miller P, Kindred A, Kosoy D, Davidson D, Lang H, Waxman K, Dunn J, Latimer RG. Preoperative sestamibi localization combined with intraoperative PTH assay predicts successful focused unilateral neck exploration during surgery for primary hyperparathyroidism. *Am Surg* 2003; 69:82–5.

Reddy V, Khan AI, Remaley AT, Wians FH Jr. Update on point-of-surgery testing. *Lab Med* 2006; 37:754–756.

Sokoll LJ, Wians FH Jr, Remaley AT. Rapid intraoperative immunoassay of parathyroid hormone and other hormones: a new paradigm for point-of-care testing. *Clin Chem* 2004; 50:1126–35.

Sullivan DP, Scharf SC, Komisar A. Intraoperative gamma probe localization of parathyroid adenomas. *Laryngoscope* 2001; 111:912–7.

Westerdahl J, Bergenfelz A. Sestamibi scan-directed parathyroid surgery: potentially high failure rate without measurement of intraoperative parathyroid hormone. *World J Surg* 2004; 28:1132–8.

Wians FH Jr, Balko JA, Hus RM, Byrd W, Snyder WH. Intraoperative vs. central laboratory PTH testing during parathyroidectomy surgery. *Lab Med* 2000; 31:616–21.

7.5   Discriminating between Osteoporosis and Osteomalacia

7.5.1   Bone Biopsy Labeling Technique

Indications for Test

Bone biopsy labeling and microscopic examination of bone is indicated in:

- The differential diagnosis of osteopenia or osteoporosis found on bone mineral densitometry
- Patients with suspected osteomalacia or bone pain or hypophosphatemia of unknown etiology
- Patients with rapidly progressive bone loss
- The presence of elevated markers of bone turnover (see Test 7.2.1 and/or 7.2.2) for which there is no clear etiology
- Cases of pathologic fractures

Procedure

1. Be sure that the patient undergoing this test is not pregnant, does not have any acute illness, and is not taking any medication that affects bone metabolism. *Caution:* Pregnant patients should not receive tetracycline.

2. To help identify secondary causes of osteoporosis, review results from a standard laboratory work-up including renal and hepatic function tests, electrolyte levels, complete blood count, 24-hour urine calcium excretion, [TSH] if patient is treated with thyroid hormones, and testosterone and 25-hydroxyvitamin D levels. Correct abnormalities before proceeding with bone biopsy.

3. Inform the patient that both short- and long-term discomfort (i.e., pain at bone biopsy site) may result from this procedure.

4. To obtain the most information from this invasive procedure, carefully perform time labeling with tetracycline as follows:
   - Administer oxytetracycline (250 mg p.o. t.i.d.) or demeclocycline (150 mg p.o. t.i.d.) for 3 days (i.e., 9 doses).
   - Ensure that no tetracycline antibiotics are given to the patient on days 4 through 15.
   - For 3 days after day 15, administer a higher dose course of tetracycline—oxytetracycline (250 mg p.o. q.i.d.) or demeclocycline (300 mg p.o. b.i.d.).

5. Perform a bone biopsy on day 20 or 21. The biopsy should be taken 3 cm inferior and posterior to the anterior iliac spine, and both cortices should be visible on the biopsy. The cortices should be roughly parallel. The biopsy should be 5 to 8 mm in diameter and obtained with an 8-gauge Jam-Shidi needle.

6. Handle the biopsy specimen as follows:
   - Fix the biopsy specimen in absolute methanol cooled to the temperature of dry ice for 2 hours.
   - Process tissue, without decalcification, into glycol or methyl methacrylate.
   - Section the trabecular portion of the biopsy specimen into sections 8 μm and 4 μm thick.
7. If the biopsy specimen cannot be delivered to the histopathology laboratory on the same day that it was obtained, place the specimen immediately in 70% ethanol, store for 48 hours or longer, and send to a referral laboratory, such as the Armed Forces Institute of Pathology (AFIP), Division of Orthopedic Pathology, with a copy of the patient's bone x-rays and clinical summary.
8. Have the histopathologist prepare 8-μm sections of the biopsy specimen for tetracycline analysis using a fluorescent microscope to visualize areas of increased tetracycline uptake. The 4-μm sections are to be stained for:
   - Calcified bone and osteoid by the von Kossa method
   - Histology by the hematoxylin and eosin (H&E) stain
   - Osteocytes and aluminum by tartrate-resistant acid phosphatase
   Insist that the anatomic pathology report indicate the results of histomorphometric analyses using these various staining techniques.

Interpretation

1. Review the anatomic pathology report for the results of histomorphometric analyses and their interpretation with regard to:
   - Resorptive activity
   - Rate of bone formation and mineralization
   - Osteoblast and osteoclast number
   - Trabecular bone volume
   - Quantity of osteoid

Notes

1. Conventional radiography may reveal abnormalities of long bones, skull, or pelvis that may prompt directed bone biopsy to diagnose bone dysplasia.
2. Bone biopsy is rarely needed in the diagnosis of Paget's disease, which is usually confirmed based on signs and symptoms, plain radiographs, and serum alkaline phosphatase test results.
3. Based on bone biopsy results, bone remodeling rates (activation frequencies) have been shown to go from 0.13/year to 0.24/year at menopause, triple from baseline 13 years later, and remain elevated in osteoporosis (Recker et al., 2004).

ICD-9 Codes

Conditions that may justify this test *include but are not limited to*:

Bone disorders
733   osteoporosis
733.1   pathologic fracture
733.9   osteopenia

Other
275.3   disorders of phosphorus metabolism

Suggested Reading

Eriksen EF, Axelrod DW, Melsen F. *Bone Histomorphometry*. 1994. Raven Press.

Recker R, Lappe J, Davies KM, Heaney R. Bone remodeling increases substantially in the years after menopause and remains increased in older osteoporosis patients. *J Bone Miner Res* 2004; 19:1628–33.

Vigorita VJ. The tissue pathologic features of metabolic bone disease. *Orthop Clin North Am* 1984a; 15:613–29.

Vigorita VJ. The bone biopsy protocol for evaluating osteoporosis and osteomalacia. *Am J Surg Pathol* 1984b; 8:925–30.

7.5.2   Vitamin D Testing

Indications for Test

Measurement of total 25-hydroxyvitamin D, [25(OH)D], which is the sum of 25-hydroxyvitamin D_2 and D_3, is indicated when:

- An overall assessment of a patient's nutritional condition is required.
- A gastrointestinal malabsorption syndrome is diagnosed (e.g., postmalabsorptive gastrointestinal bypass surgery).
- Hypercalcemia is present and excessive or surreptitious vitamin D ingestion is suspected.
- The bone mineral density (BMD) is low.
- Hypocalcemia is documented.
- Vitamin D inadequacy is suspected based on the following findings: older age, low sun exposure in Caucasians, use of higher strength sun blockers, abnormally low or high BMI, chronic kidney disease (CKD), use of medications known to affect vitamin D metabolism (e.g., high-dose glucocorticoids), inadequate vitamin D supplementation, overexercise, or prior lack of education and counseling about nutrition and vitamin D therapy.

Measurement of 1,25-dihydroxyvitamin D_3, [1,25(OH)$_2$D$_3$], is indicated:

- In the differential diagnosis of causes for hypercalcemia associated with granulomatous disease

Procedure

1. Obtain a random blood sample for total 25(OH)D testing. The blood sample should be protected from ambient ultraviolet light (i.e., wrap specimen tube in tin foil).
2. Use an assay (i.e., LC-MS/MS) for total 25(OH)D testing that does not underestimate (under-recover) vitamin D_2 especially in patients treated with vitamin D_2 (i.e., Drisdol®, ergocalciferol).
3. Be prepared to specifically instruct laboratory personnel to send the patient's blood sample for total vitamin D testing to a referral lab that uses LC-MS/MS for the quantitative measurement of the 25-hydroxy metabolites of both vitamins D_2 and D_3, as outmoded radioimmunoassays for vitamin D may still be in use.
4. Obtain a random blood sample for 1,25(OH)$_2$D$_3$ testing only if hypercalcemia is unexplained in patients without known or suspected renal or parathyroid disease and granulomatous disease is suspected.

Interpretation

1. The reference interval for total [25(OH)D] (i.e., combined vitamins D_2 and D_3) is typically >30 ng/mL with no toxicity known up to 100 ng/mL.
2. The reference interval for 1,25(OH)$_2$D$_3$ in Caucasians is 20 to 76 pg/mL.
3. A total [25(OH)D] $\leq$ 30 ng/mL is strongly suggestive of vitamin D deficiency in the general population.
4. A broad range of expert opinions suggests that the minimum level of total 25(OH)D for fracture prevention varies between 20 and 32 ng/mL (50 and 80 nmol/L), with higher levels preferred (i.e., up to 100 ng/mL) (Dawson-Hughes et al., 2005).
5. In 2005, the product insert for the total 25(OH)D radioimmunoassay from DiaSorin (Stillwater, MN) indicated 20 ng/mL (50 nmol/L) as the cutoff value for identifying patients with vitamin D deficiency, whereas the product insert for the LIAISON 25OHD assay indicated that the 3rd to the 97th percentile reference interval is 5 to 60 ng/mL (13–151 nmol/L).
6. Healthy African-American or more pigmented populations, when compared to healthy Caucasians, have lower [25(OH)D$_3$] and urinary calcium levels while their PTH, 1,25(OH)$_2$D$_3$, and urinary cyclic adenosine 3′,5′-monophosphate levels are higher.
7. Vitamin D insufficiency or deficiency exacerbates the effects of mild primary hyperparathyroidism on biochemical, densitometric, and bone histomorphometric indices.

8. Normal or decreased [25(OH)D] may be seen in patients with granulomatous disease and hypercalcemia, and elevated [1,25(OH)$_2$D$_3$] may occur in patients with hypophosphatemia, hyperparathyroidism, and granulomatous disease, including sarcoidosis, tuberculosis, and lymphomas.

9. A low [25(OH)D] may indicate malabsorption, decreased vitamin D intake, decreased sun exposure, or decreased vitamin D binding globulin. Any of these may result in rickets in children or osteomalacia in adults.

10. Elevated [25(OH)D] is seen with excessive ingestion of the more active forms of vitamin D (i.e., >2000 units of vitamin D$_3$/day).

11. Elevated [PTH] in patients with CKD Stage 3 or 4 will increase the efficiency with which the failing kidney converts 25(OH)D$_2$ or D$_3$ to 1,25(OH)$_2$D$_3$, resulting in normal [1,25(OH)$_2$D$_3$] even when the patient is vitamin D deficient based on the total [25(OH)D].

12. When excessive vitamin D$_2$ is ingested, the [1,25(OH)$_2$D$_3$] usually equilibrates into the normal range.

13. Low [1,25(OH)$_2$D$_3$] is seen in patients with hypoparathyroidism, CKD Stage 4 or 5, or malabsorption.

14. Patients with primary hyperparathyroidism and normal [1,25(OH)$_2$D$_3$] can liberalize their calcium intake without adverse consequences; however, those with elevated [1,25(OH)$_2$D$_3$] are prone to hypercalciuria at higher calcium intake.

Notes

1. Vitamin D$_2$ is an orally ingested, therapeutic form of vitamin D, whereas vitamin D$_3$ is a more active, endogenous form produced in the body. The LC-MS/MS assay measures each type of vitamin D.

2. 1,25(OH)$_2$D$_3$ is the active form of vitamin D$_3$ produced in the kidney via 1α-hydroxylase activity.

3. 1α-Hydroxylation of 25(OH)D$_3$ in the kidney is the rate-limiting step in the production of 1,25(OH)$_2$D$_3$ and is regulated by PTH, 25(OH)D, and serum phosphorus levels.

4. Normally, elevated [PTH] and low phosphorus levels stimulate 1α-hydroxylase activity.

5. Vitamin D$_3$ is produced from 7-dehydrocholesterol by skin exposure to ultraviolet radiation or is ingested in the diet. Vitamin D$_3$ from these sources is hydroxylated at the 25 position in the liver to form 25(OH)D$_3$.

6. The average older man and woman need intakes of at least 20 to 25 mcg (800 to 1000 IU) per day of vitamin D$_3$ to reach a serum [25(OH)D$_3$] of 30 ng/mL (75 nmol/L).

7. Good, nondairy sources of vitamin D include dark green leafy vegetables and fish (e.g., herring, 1300 IU/3 oz; pink salmon, 530 IU/3oz; cod liver oil, 450 IU/tsp).

8. Pharmacologic doses of vitamin D$_2$ (i.e., 50,000 units) may significantly increase serum [25(OH)D$_3$] and urinary [calcium] and reduce [1,25(OH)$_2$D$_3$] and urinary cyclic adenosine 3′,5′-monophosphate (cAMP) levels, which are indices of PTH function.

9. Endogenous [1,25(OH)$_2$D$_3$] can be determined within 5 hours after: (a) acetonitrile extraction, (b) treatment of the crude extract supernatant fluid with sodium periodate, (c) extraction, (d) purification by solid-phase chromatography, and (e) quantification by RIA. This time-consuming and expensive procedure was an improvement over that of the radioreceptor assay used prior to 2000.

10. Competitive binding techniques are unable to distinguish between 25(OH)D$_2$ and D$_3$ and suffer from interference from other hydroxy metabolites of vitamin D.

11. Under-recovery of vitamin D$_2$ occurs with the use of the LIASON and IDS Ltd. (Tyne and Wear, U.K.) 25(OH)D assays. Thus, there is a significant problem with the reliability of [25(OH)D] using these assays when 25(OH)D$_2$ constitutes an appreciable part of the total circulating [25(OH)D], usually as a result of vitamin D$_2$ therapy.

12. The mean [1,25(OH)$_2$D$_3$] was significantly lower in patients with gout compared with control subjects (38 ± 12 vs. 44 ± 11 pg/mL; $p <$ .005), suggesting that uric acid inhibits 1α-hydroxylase (Takahashi et al., 1998).

13. More than half of North American women receiving therapy to treat or prevent osteoporosis have an inadequate [25(OH)D$_2$] (i.e., < 30 ng/mL) with <20 ng/mL found in 18% of these women (Holick et al., 2005).

14. In a well-nourished Swedish population, [1,25(OH)$_2$D$_3$] (Landin-Wilhelmsen et al., 1995):
   • Declined with age and was inversely related to height in men
   • Was positively correlated with intact PTH and osteocalcin levels
   • Did not correlate with sun exposure, although 25(OH)D$_3$ levels were affected by sunlight

ICD-9 Codes

Conditions that may justify this test *include but are not limited to*:

Bone loss
733.0   osteoporosis
733.9   osteopenia

Other
E933.54 vitamin D toxicity
135   sarcoidosis
268.9   vitamin D deficiency
275.42   hypercalcemia
579.9   intestinal malabsorption
585   chronic kidney disease (CKD)

Suggested Reading

Bell NH. 25-Hydroxyvitamin D_3 reverses alteration of the vitamin D–endocrine system in blacks. *Am J Med* 1995; 99:597–9.

Carter GD, Carter R, Jones J, Berry J. How accurate are assays for 25-hydroxyvitamin D? Data from the international vitamin D external quality assessment scheme. *Clin Chem* 2004; 50:2195–7.

Dawson-Hughes B, Heaney RP, Holick MF, Lips P, Meunier PJ, Vieth R. Estimates of optimal vitamin D status. *Osteoporos Int* 2005; 16:713–6.

Holick MF, Siris ES, Binkley N, Beard MK, Khan A, Katzer JT, Petruschke RA, Chen E, de Papp AE. Prevalence of Vitamin D inadequacy among postmenopausal North American women receiving osteoporosis therapy. *J Clin Endocrinol Metab* 2005; 90:3215–24.

Hollis BW, Kamerud JQ, Kurkowski A, Beaulieu J, Napoli JL. Quantification of circulating 1,25-dihydroxy-vitamin D by radioimmunoassay with ^{125}I-labeled tracer. *Clin Chem* 1996; 42:586–92.

Landin-Wilhelmsen K, Wilhelmsen L, Wilske J et al. Sunlight increases serum 25(OH) vitamin D concentration whereas $1,25(OH)_2D_3$ is unaffected: results from a general population study in Goteborg, Sweden (the WHO MONICA Project). *Eur J Clin Nutr* 1995; 49:400–7.

Locker FG, Silverberg SJ, Bilezikian JP. Optimal dietary calcium intake in primary hyperparathyroidism. *Am J Med* 1997; 102:543–50.

Silverberg SJ, Shane E, Dempster DW, Bilezikian JP. The effects of vitamin D insufficiency in patients with primary hyperparathyroidism. *Am J Med* 1999; 107:561–7.

Takahashi S, Yamamoto T, Moriwaki Y, Tsutsumi Z, Yamakita J, Higashino K. Decreased serum concentrations of $1,25(OH)_2$-vitamin D_3 in patients with gout. *Metabolism* 1998; 47:336–8.

Vogeser M, Kyriatsoulis A, Huber E, Kobold U. Candidate reference method for the quantification of circulating 25-hydroxyvitamin D_3 by liquid chromatography–tandem mass spectrometry. *Clin Chem* 2004; 50:1415–7.

7.5.3   Anion Gap (AG), Serum Anion Gap (S_{AG}), and Urine Anion Gap (U_{AG}): Test for Differential Diagnosis of Renal Tubular Acidosis (RTA) and Prediction of Bone Disease of Chronic RTA

Indications for Test

Measurement of the urine anion gap is used to:

- Discriminate between proximal and distal RTA in patients with chronic non-anion-gap or hyperchloremic metabolic acidosis.
- Identify causes for calcium phosphate-type nephrolithiasis, nephrocalcinosis, and low bone mineral density (BMD).

Procedure

1. Obtain history of ketosis-prone diabetes mellitus, urinary tract symptoms, chronic obstructive pulmonary disease, vomiting and diarrhea, if any.
2. Instruct the patient on the proper collection of a random urine specimen as per Specimen Collection Protocol P1 in Appendix 2 and measure urine pH.
3. Measure electrolyte concentrations in this urine: sodium (Na^+), potassium (K^+), and chloride (Cl^-). For monovalent ions, mEq/L = mmol/L.
4. Calculate the U_{AG} using the formula:

$$U_{AG} = \left(U_{Na^+} + U_{K^+}\right) - U_{Cl^-}$$

5. Estimate the urine ammonium concentration, $[U_{NH4}]$, using the formula:

$$[U_{NH_4}] = U_{Cl^-} - \left(U_{Na^+} + U_{K^+}\right) + 80 \text{ mmol/L}$$

6. Obtain a blood sample for sodium (Na^+), chloride (Cl^-), and bicarbonate.
7. To calculate the serum or plasma anion gap (S_{AG}), obtain a blood sample for both serum albumin and electrolytes: sodium (Na^+), bicarbonate (HCO_3^-) or CO_2, and chloride (Cl^-).
8. To calculate the S_{AG} in a normoalbuminemic (4.0 ± 0.4 g/dL) patient, use the formula:

$$S_{AG} = S_{Na^+} - \left(S_{HCO_3^-} + S_{Cl^-}\right)$$

Example—Consider a patient with a normal S_{AG} metabolic acidosis (e.g., serum $[HCO_3^-] = 10$ mEq/L), hypokalemia, and a more alkaline urine pH (>6.0) who might be losing K^+ and HCO_3^- in diarrheal stool or may actually have type I RTA. If the urine showed:

$$[Na^+] = 50 \text{ mEq/L}$$

$$[K^+] = 28 \text{ mEq/L}$$

$$[Cl^-] = 55 \text{ mEq/L}$$

Then,

$$U_{AG} = \left(U_{Na^+} + U_{K^+}\right) - U_{Cl^-}$$

$$= (50 + 28) - 55 = 78 - 55 = 23 \text{ mEq/L}$$

which is a value that supports a diagnosis of type I RTA.

Interpretation

1. The reference interval for S_{AG} is 3 to 13 mEq/L (mean, ~10 mEq/L). It is accounted for by negatively charged serum albumin. The anion gap falls by ~2.5 mEq/L for every 1-g/dL reduction in serum albumin level.
2. In metabolic acidosis, the urinary osmolal (U_{osm}) gap:

$$\text{Measured } U_{osm} - \text{calculated } U_{osm}$$

where

$$\text{Calculated } U_{osm} = 2(Na + K) + \text{urea} + \text{glucose}$$

is usually significantly increased over the U_{osm} reference interval of 80 to 100 mmol/kg H_2O. A normal or low U_{osm} in an individual with metabolic acidemia is the result of impaired NH_4^+ and Cl^- excretion and is consistent with the diagnosis of type I RTA.
3. The reference interval for urine pH is from 4.6 to 8.0.

TABLE 7.7
Differentiation of Type I, II, and IV Renal Tubular Acidosis (RTA)

Feature	RTA Type		
	I[a]	II[b]	IV
Non-anion-gap acidosis present[c]	Yes	Yes	Yes
Minimum urine pH	>5.5	<5.5	<5.5
Serum potassium concentration	Low or high	Low	High
Fanconi[d] syndrome present	No	Yes	No
Daily acid excretion in urine	Low	Normal	Low
Urine ammonium (NH_4^+) excretion	Low	Normal[e]	Low
Correction of acidosis with mineralocorticoid therapy	No	No	Yes

[a] Associated with distal renal tubule.

[b] Associated with proximal renal tubule.

[c] Serum anion gap = $[Na^+] - ([Cl^-] + [HCO_3^-])$; reference interval, 10–14.

[d] Fanconi syndrome, familial or acquired, is defined as the excessive loss of urinary amino acids, glucose, phosphate, and bicarbonate in the absence of high plasma concentrations of these substances.

[e] 80–100 mEq/L.

4. Urine tends to be acidic (pH < 5.3) in patients with diarrhea and tends to be alkaline (pH > 5.3) in patients with type I RTA. In patients with diarrhea, an alkaline urine, and hypokalemia, ammonia production may have increased to the point where it becomes a urinary buffer neutralizing excess hydrogen ion secretion.

5. If U_{AG} is a negative number, then proximal RTA is diagnosed and distal RTA is excluded, but type II RTA is not differentiated from type IV RTA as per "Differentiation of Type I, II, and IV Renal Tubular Acidosis (RTA)" (Table 7.7).

6. A negative-value U_{AG} may occur with gastrointestinal loss of bicarbonate, and in patients with diarrhea and proximal RTA the calculated $[U_{NH4}]$ should be high (>100 mmol/L).

7. If U_{AG} is a positive number, then classic distal RTA is probably present. Alternatively, CKD with hyperkalemia and decreased ammoniagenesis may be responsible for the positive U_{AG}.

8. The anions β-hydroxybutyrate and acetoacetic acid can contribute to a positive U_{AG}.

9. Patients with type I (distal) RTA will have a low calculated $[U_{NH4}]$ (<80 mmol/L), in spite of acidosis.

10. Recognize that the U_{AG} is not valid if:
    - HCO_3^- is present in the urine (usually suggested by a more alkaline urine).
    - High levels of unmeasured anions (e.g., ketoacids, lactic acid, salicylates, or carbenicillin) are present.
    - Excretion of uncommon cations (e.g., lithium) is occurring in high concentrations.

11. Metabolic acidosis secondary to a fall in the plasma $[HCO_3^-]$ is associated with one of two phenomena:
    - An anion gap increase
    - No significant change in the anion gap as the plasma $[Cl^-]$ increases, and a hyperchloremic metabolic acidosis develops

12. In CKD Stage 3 or early Stage 4, the plasma $[HCO_3^-]$ rarely falls below <12 mEq/L, and the S_{AG} is usually <20 mEq/L.

13. An increased S_{AG} may be caused by:
    - Lactic acidosis, the most common cause of metabolic acidosis in hospitalized patients with appearance of lactate ($CH_3CHOHCOO^-$)
    - Advanced renal failure (CKD Stages 4 and 5), the most common cause of metabolic acidosis in outpatients with the appearance of high concentrations of sulfate and phosphate anions
    - Ketoacidosis, with the appearance of acetoacetate and β-hydroxybutyrate anions
    - Rhabdomyolysis, massive muscle breakdown resulting in the release of phosphate anions with metabolic acidosis occurring when renal failure develops and phosphorus is retained
    - Toxins, including salicylates, ethylene glycol (found in antifreeze and solvents), and methanol (found in sterno, shellac, varnish, deicing solutions)

Notes

1. Extrarenal etiologies of acidosis include hypoaldosteronism, established ureterosigmoid fistula, and chronic diarrhea.
2. The U_{AG} may be used in combination with furosemide infusion (1 mg/kg) and measurement of urine pH 2 and 3 hours later to distinguish type I from type II or IV RTA (Dons et al., 1997).
3. Direct measurement of urine $[U_{NH4}]$ is not available in most clinical laboratories but may be more useful in establishing the differential diagnosis of hyperchloremic metabolic acidosis than the calculated $[U_{NH4}]$ obtained as described above.
4. Low BMD is a common occurrence in the chronic, systemic acidosis found in patients with CKD.
5. Measurement of the $[Na^+]$ in a random urine specimen may be used to distinguish between volume depletion (low urinary $[Na^+]$ usually < 20 mEq/L) and euvolemia (urinary $[Na^+]$ > 40 mEq/L); however, in volume depletion coexistent with metabolic alkalosis, the urinary $[Na^+]$ may not be low, as the kidney has the capacity to retain sodium. In such cases, the presence of underlying hypovolemia can be detected by the finding of a urinary $[Cl^-]$ < 25 mEq/L.
6. In short-bowel syndrome, an increased local bacterial overgrowth occurs with the metabolism of carbohydrates to D-lactic acid. If D-lactic acidosis develops, the serum anion gap rises initially but may fall over time because renal tubular reabsorption of L-lactate exceeds that of D-lactate.
7. Drugs that increase urine pH include acetazolamide, potassium citrate, and sodium bicarbonate as well as a diet high in citrus fruits, vegetables, or dairy products. The urine pH decreases in response to treatment with ammonium chloride, chlorothiazide diuretics, or methenamine mandelate, as well as with a diet high in meat products or cranberries.
8. Acidic urine is associated with xanthine, cystine, uric acid, and calcium oxalate kidney stones, while alkaline urine is associated with calcium carbonate, calcium phosphate, and magnesium phosphate stones.
9. Urine pH can be affected by leaving the urine standing in an uncovered container. Bacteria, if present, increase the pH as they break down urea in the urine to ammonia.
10. The *acid loading test*, with administration of ammonium chloride and measurement of urine pH, is an adjunctive test for diagnosis of RTA.

ICD-9 Codes

Conditions that may justify this test *include but are not limited to*:

Bone loss
733.0   osteoporosis
733.9   osteopenia

Other
275.49   disorders of calcium metabolism/nephrocalcinosis
588.89   renal tubular acidosis
592.0   nephrolithiasis

Suggested Reading

Batlle DC, Hizon M, Cohen E, Gutterman C, Gupta R. The use of the urinary anion gap in the diagnosis of hyperchloremic metabolic acidosis. *N Engl J Med* 1988a; 318:594–9.

Batlle DC, Hizon M, Cohen E, Gutterman C, Gupta R. Urinary anion gap in hyperchloremic metabolic acidosis. *N Engl J Med* 1988b; 319:585–7.

Black, RM. Disorders of acid–base and potassium balance. In: *ACP Medicine*. Vol. 10. *Nephrology*, Dale DC, Federman DD (Eds). 2000. American College of Physicians.

Dons RF. *Endocrine and Metabolic Testing Manual*, 3rd ed. 1997. CRC Press, pp. AII-4–AII-5.

Halperin ML, Richardson RM, Bear RA, Magner PO, Kamel K, Ethier J. Urine ammonium: the key to the diagnosis of distal renal tubular acidosis. *Nephron* 1988; 50:1–4

7.5.4 Citrate: 24-Hour Urine in Nephrolithiasis and Renal Tubular Acidosis (RTA)

Indications for Test

Measurement of the citrate concentration in a 24-hour urine collection may be indicated in:

- Cases of recurrent nephrolithiasis, particularly in females with suspected low urine volume
- Patients with type 1a glycogen storage disease
- Patients with RTA

Procedure

1. Obtain the patient's history of fluid, vitamin, and mineral intake, especially the type of calcium supplements (carbonate or citrate) and the amount of vitamin C intake.
2. Instruct the patient to obtain a spot urine specimen and order pH testing on this specimen.
3. Have the patient collect a 24-hour urine specimen (see Specimen Collection Protocol P1 in Appendix 2). *Important:* Remind patients to have a blood sample obtained for creatinine testing when they deliver their 24-hour urine specimens to the laboratory.
4. If the adequacy of the 24-hour urine collection is questionable (e.g., 24-hour urine total volume is lower than expected based on the patient's BMI, renal function, and usual hydration status), repeat the 24-hour collection after discussing with the patient the technique of proper collection.
5. Obtain total citrate, free (or ionized) citrate (F-Cit), calcium, phosphorus, oxalate, uric acid, and creatinine clearance tests on the patient's 24-hour urine specimen. Obtain a serum creatinine on a blood sample obtained at the time of delivery of the 24-hour urine specimen to calculate the creatinine clearance.

Interpretation

1. Because urine citrate excretion increases with increasing age, reference values for urine citrate excretion are age dependent; therefore, review the reference interval provided with the patient's laboratory report in the context of the age of the patient.
2. 24-hour urine citrate, uric acid, calcium, and oxalate excretion may be useful in predicting the risk of stone formation, but the magnitude and significance of the association between these urine analytes and risk of stone formation differ by age and gender.
3. F-Cit, as opposed to total or complexed citrate, is lower in patients with idiopathic recurrent urolithiasis and rises in a pH-dependent manner. The ratio of F-Cit at the original urine pH to F-Cit at a urine pH adjusted to 6.0 correlates inversely with the nucleation index (i.e., total oxalate/total calcium ratio).
4. Alkalinization of the urine increases the excretion of citrate. The pH of a 24-hour hour urine collection may be higher (>6.0) than a spot urine from the same patient.
5. Oral supplements of calcium citrate increase urinary calcium and citrate excretion but decrease urinary oxalate and phosphate excretion. Oral potassium citrate decreases urinary calcium excretion and increases urinary citrate excretion and urine pH.
6. A pH < 5.5 in a spot urine sample, but not in a 24-hour urine sample, is associated with an increased tendency to form uric acid kidney stones.
7. Patients with type 1a glycogen storage disease tend to have low levels of citrate excretion (2.4 ± 1.8 mg/kg/24 hr or 129 ± 21 mg citrate per g creatinine) (Weinstein et al., 2001).
8. The reference interval for urinary citrate concentration in term babies and in those born premature who subsequently achieve a full-term age was reported to vary widely between 0.025 and 2.97 mmol/L (mean, 1.03) (White et al., 2005).
9. Prior to 2000, most reference intervals for 24-hour urine chemistry tests were limited by poorly defined controls and the inclusion of recurrent stone formers in the reference population.

Notes

1. Citrate inhibits calcium stone formation by complexing calcium in a soluble form that is excreted in the urine and thus prevents the formation of kidney stones.
2. Solution, surface, and interface chemistry factors interact in a complex manner in physiologic urinary environments to either inhibit, as in the case of the protein mucin, or promote, as in the case of oxalate ion, crystal aggregate and kidney stone formation.
3. A low urine volume and high urate excretion rate are typical, but not necessary, in urate stone formers. The major function of the renal medullary ammonia shunt pathway appears to be maintenance of a urine pH close to 6.0 throughout the day to minimize formation of uric acid kidney stones.
4. While an increase in urinary pH and citrate and magnesium excretion and a decline in calcium excretion occur in patients on a vegetarian diet, urinary oxalate excretion increases, on average, by about a third, prompting the recommendation to supplement the diet of these patients with calcium to avoid supersaturated oxaluria.
5. Use of oral supplements of calcium carbonate, magnesium oxide, and sodium citrate–bicarbonate results in three different synergistic effects:
   - Magnesium and citrate lower the relative supersaturation of brushite.
   - Magnesium and citrate raise the urine pH.
   - Calcium and citrate lower the relative supersaturation of uric acid.
6. In the Nurses' and Health Professionals' Follow-Up Studies, mean 24-hour urine oxalate and citrate excretion rates of kidney stone formers did not differ from controls (Curhan et al., 2001).
7. Metabolically compensated patients with type 1a glycogen storage disease may develop hypocitraturia that worsens with age. When combined with hypercalciuria, low citrate excretion appears to promote nephrocalcinosis and nephrolithiasis in these patients.

ICD-9 Codes

Conditions that may justify this test *include but are not limited to*:

271.0	glycogenosis
588.89	renal tubular acidosis
592.0	nephrolithiasis

Suggested Reading

Allie S, Rodgers A. Effects of calcium carbonate, magnesium oxide and sodium citrate bicarbonate health supplements on the urinary risk factors for kidney stone formation. *Clin Chem Lab Med* 2003; 41:39–45.

Christmas KG, Gower LB, Khan SR, El-Shall H. Aggregation and dispersion characteristics of calcium oxalate monohydrate: effect of urinary species. *J Colloid Interface Sci* 2002; 256:168–74.

Curhan GC, Willett WC, Speizer FE, Stampfer MJ. Twenty-four-hour urine chemistries and the risk of kidney stones among women and men. *Kidney Int* 2001; 59:2290–8.

Kamel KS, Cheema-Dhadli S, Shafiee MA, Davids MR, Halperin ML. Recurrent uric acid stones. *QJM* 2005; 98:57–68.

Sakhaee K, Poindexter JR, Griffith CS, Pak CY. Stone forming risk of calcium citrate supplementation in healthy postmenopausal women. *J Urol* 2004; 172:958–61.

Schwille PO, Schmiedl A, Manoharan M. Is calcium oxalate nucleation in postprandial urine of males with idiopathic recurrent calcium urolithiasis related to calcium phosphate nucleation and the intensity of stone formation? Studies allowing insight into a possible role of urinary free citrate and protein. *Clin Chem Lab Med* 2004; 42:283–93.

Siener R, Hesse A. The effect of different diets on urine composition and the risk of calcium oxalate crystallisation in healthy subjects. *Eur Urol* 2002; 42:289–96.

Weinstein DA, Somers MJ, Wolfsdorf JI. Decreased urinary citrate excretion in type 1a glycogen storage disease. *J Pediatr* 2001; 138:378–82.

White MP, Aladangady N, Rolton HA, McColl JH, Beattie J. Urinary citrate in preterm and term babies. *Early Hum Dev* 2005; 81:319–23.

Chapter 8

Male Reproductive Organ and Sexual Differentiation Testing

8.1. Testes

8.1.1 Orchidometry in the Assessment of Hypogonadism and Pubertal Development in Males

Indications for Test

Orchidometry, including measurement of size, volume, and turgor of the descended testicle, is indicated for:

- The assessment of normal, precocious, or delayed pubertal development
- Estimation of the mass of Sertoli cells or seminiferous tubuli in children with abnormal sexual development or in infertile adult males
- Patients with Klinefelter's syndrome, hypogonadotropic hypogonadism (Kallmann's syndrome), or hypopituitarism
- The assessment of patient status after mumps orchitis, hypophysectomy with hypopituitarism, repair of cryptorchidism, or radio- or chemotherapy

Procedure

1. Obtain a string of wooden or plastic beads consisting of solid ellipsoids with volumes (mL) of 1, 2, 3, 4, 5, 6, 8, 10, 12, 15, 20, 25 and corresponding longitudinal dimensions (mm) of 15, 19, 23, 25, 27, 29, 32, 35, 37, 39, 43, 45, respectively.
2. Select the bead (model testis) that is closest in size to the testicle as determined by palpation.
3. Record the volume of the right and left testes and age of the patient.
4. Obtain serial measurements of testicular volume to estimate the progress of pubertal development or effects of interventional therapies.
5. If a ruler or caliper type of measuring device is used to quantify testicular dimensions instead of an ellipsoidal bead, calculate the volume of an ellipsoid from the width (W), length (L), and height (H) axes. A variety of formulae may be used to calculate a testicular volume (V) using these linear measurements
   - Prolate ellipsoid:

$$V\ (cm^3;\ mL) = 4/3 \times \left[(L/2) \times (W/2) \times (H/2)\right] \tag{8.1}$$

$$V\ (cm^3;\ mL) = 0.52 \times \left[(L) \times (W) \times (H)\right] \tag{8.2}$$

- Prolate spheroid:

$$V \text{ (cm}^3\text{; mL)} = 0.52 \times \left[(L) \times (W)^2 \right] \tag{8.3}$$

- Empiric formula of Lambert (Lambert, 1951):*

$$V \text{ (cm}^3\text{; mL)} = 0.71 \times \left[(L)^2 \times (W) \right] \tag{8.4}$$

Example—If L = 3.7 cm, W = 2.5 cm, H = 2.5 cm, then:

By Equation 8.1,

$$V \text{ (mL)} = \left[(4/3) \times (3.1417) \right] \times \left[(3.7/2) \times (2.5/2) \times (2.5/2) \right]$$

$$4.189 \times [1.85 \times 1.25 \times 1.25] = 4.189 \times [2.891] = 12.11 \text{ cm}^3 = 12.1 \text{ mL}$$

By Equation 8.2,

$$V \text{ (mL)} = 0.52 \times \left[(3.7) \times (2.5) \times (2.5) \right] = 12.0 \text{ mL}$$

By Equation 8.3,

$$V \text{ (mL)} = 0.52 \times \left[(3.7) \times (2.5)^2 \right] = 12.0 \text{ mL}$$

By Equation 8.4,

$$V \text{ (mL)} = 0.71 \times \left[(3.7)2 \times (2.5) \right] = 16.4 \text{ mL}$$

6. Qualitatively estimate testicular turgor as either "firm" or "soft," or use a testicular tonometer to obtain a quantitative measurement.

Interpretation

1. The testes are considered to be prepubertal if the estimated volume is less than 4 mL.
2. The average adult testicular volume is 20 to 30 mL (cm³). Testicular size enlarges from prepubertal (age <9 years) volume of 1.5 mL to normal adult size over the next 5 to 6 years (late puberty).
3. A testicular length ≤ 3.7 cm (volume, ≤12 mL) is subnormal, and a qualitative estimate of the testicles as "soft" is abnormal for an adult male.
4. There is a weak positive correlation ($r > 0.5$) between body surface area (BSA) and testicular size.
5. Nonobese individuals with a body surface area > 1.73 m² may be expected to have a larger length and volume of the testes, usually >25 mL.

Notes

1. Testicular size and volume are most conveniently estimated clinically using ellipsoid models (Prader beads) in contrast to ultrasound and caliper measurement. From birth to age 9, testicular size increases from 0.5 to 1.5 mL, a change not detectable clinically.
2. Of the formulas noted above, the empiric formula of Lambert gave the most accurate estimate of testicular volume (Sakamoto et al., 2007a). Good correlation between Prader orchiometry and ultrasonographic measurement ($r > 0.7$) was found, but the orchiometer overestimated the testicular volume, especially in small testes (Sakamoto et al., 2007b).

---

* Equation 8.4, with ultrasound measurements of testicular length (L), width (W), and height (H), provides a superior estimate of testicular volume and should be used in clinical practice (Paltiel et al., 2008).

3. Tonometry measures testicular turgor quantitatively and reflects tubular function and spermatozoa production.

4. Cryptorchidism is the most common malformation in newborn boys. Risk factors for cryptorchidism include prematurity, low birth weight for gestational age, and abnormal maternal glucose metabolism. Testicular descent appears to be influenced by the balance between androgens and estrogens.

ICD-9 Codes

Conditions that may justify this test *include but are not limited to*:

Testes
604   orchitis and epididymitis
606   male infertility
752.5   undescended and retractile testicle/cryptorchism

Pituitary
253.2   panhypopituitarism
253.7   iatrogenic pituitary disorders (i.e., neurosurgery)
257.2   hypogonadotropism

Puberty
259.1   precocious development

Other
072   mumps
259.0   delay in sexual development
758.7   Klinefelter's (XXY) syndrome
E926.3   exposure to x-rays and other electromagnetic ionizing radiation

Suggested Reading

Lambert B. The frequency of mumps and of mumps orchitis and the consequences on sexuality and fertility. *Acta Genet Stat Med* 1951; 2(Suppl 1):1–166.

Paltiel HJ, Diamond DA, Di Canzio J, Zurakowski D, Borer JG, Atala A. Testicular volume: comparison of orchidometer and US measurements in dogs. *Radiology* 2002; 222:114–9.

Sakamoto H, Saito K, Ogawa Y, Yoshi H. Testicular volume measurements using Prader orchidometer versus ultrasonography in patients with infertility. *Urology* 2007a; 69:158–62.

Sakamoto H, Saito K, Oohta M, Inoue K, Ogawa Y, Yoshida H. Testicular volume measurement: comparison of ultrasonography, orchidometry, and water displacement. *Urology* 2007b; 69:152–7.

Steeno OP. Clinical and physical evaluation of the infertile male: testicular measurement or orchidometry. *Andrologia* 1989; 21:103–12.

Tanner JM, Whitehouse RH. Clinical longitudinal standards for height, weight, height velocity, weight velocity, and stages of puberty. *Arch Dis Child* 1976; 51:170–9.

8.1.2   Human Chorionic Gonadotropin (hCG): Single-Injection hCG Stimulation Test for Hypothalamic Hypogonadism (Kallmann's Syndrome)

Indications for Test

hCG stimulation of total testosterone (TT) secretion is indicated:

- In the evaluation of testicular Leydig cell function and reserves in patients suspected of having hypothalamic hypogonadism (Kallmann's syndrome)

TABLE 8.1
Testicular Response to Stimulation with a Single Dose of Human Chorionic Gonadotropin (hCG) in Males

Testicular Status	Testosterone Concentration, [T] (ng/dL)	
	Baseline	Stimulated
Healthy prepubertal male	Variable	>150 over baseline [T]
Healthy adult	Variable, typically >200	>2× baseline [T]
Adequate Leydig cell reserve in Kallmann's syndrome	<100	> 400
Leydig cell insufficiency[a]	Variable, typically <200	<2× baseline [T]
Partial suppression of Leydig cells by sex steroids	Variable	>400
Functionally or anatomically anorchid	Variable, typically <100	No change from baseline

[a] Hypogonadism, Klinefelter's syndrome.

Procedure

1. Obtain three separate baseline blood specimens 15 minutes apart. Order a pooled TT test on these three baseline specimens (i.e., laboratory personnel will separate the serum from all three specimens, pool them together, and perform TT testing on the pooled serum specimen).
2. Administer a single intramuscular (i.m.) injection of 5000 units of hCG per 1.7 m² body surface area (BSA).
3. At 72 hours (beginning of 4th day) from i.m. injection of hCG, obtain three separate blood specimens 15 minutes apart and order a pooled TT test on these three stimulated specimens.

Interpretation

1. The testicular response to a single dose of hCG is shown in Table 8.1.
2. A low [TT], readily stimulated by hCG, is definitive for hypothalamic hypogonadism (i.e., Kallmann's syndrome).
3. The hCG stimulation test may not clearly distinguish between patients with loss of Leydig cell reserve (e.g., older patients with Kallmann's syndrome) and those with primary hypogonadism.

Notes

1. In patients with micropenis, cryptorchidism, hermaphroditism, male pseudohermaphroditism, hypospadias, or sex chromosome anomalies, considerable variation in hCG-stimulated [TT] may be observed.
2. In normal prepubertal boys, stimulated TT concentrations were elevated 22- to 29-fold from baseline after a 100-IU/kg total-body-weight-adjusted dose of hCG vs. a 34- to 35-fold increase from baseline after a BSA-adjusted dose of hCG (Kolon et al., 2001).
3. Using multiple injection of hCG (see Test 12.8.1) may assist in the determination of Leydig cell reserve in the cryptorchid patient.

ICD-9 Codes

Conditions that may justify this test *include but are not limited to*:

257.2   hypogonadotropism/testicular hypofunction

Suggested Reading

Dunkel L, Perheentupa J, Sorva R. Single versus repeated dose human chorionic gonadotropin stimulation in the differential diagnosis of hypogonadotropic hypogonadism. *J Clin Endocrinol Metab* 1985; 60:333–7.

Grant DB, Laurance BM, Atherden SM, Ryness J. HCG stimulation test in children with abnormal sexual development. *Arch Dis Child* 1976; 51:596–601.

Kolon TF, Miller OF. Comparison of single versus multiple dose regimens for the human chorionic gonadotropin stimulatory test. *J Urol* 2001; 166:1451–4.

8.1.3   Sperm Count in Evaluation of the Testes

Indications for Test

A sperm count is indicated in the assessment of:

- The integrity of the hypothalamic–pituitary–testicular (HPT) axis in the male member of an infertile couple
- Exposure of the testes to toxins, such as irradiation, or after hemiorchiectomy
- The degree of success achieved after stimulating spermatogenesis using hCG or gonadotropin preparations
- Patients with varicoceles

Procedure

1. The most reliable and accurate method of obtaining a semen specimen (seminal fluid) is by masturbation and collection of the ejaculate in a glass container (see Specimen Collection Protocol P3 in Appendix 2). Present this as the preferred method of collection to all patients. If this method is unacceptable, as may be the case for some devout Catholics and Orthodox Jews, offer the use of a nonrubber, nonplastic perforated pouch to Catholics and an unperforated pouch to Jewish patients during intercourse.
2. Obtain a computer-assisted semen analysis (CASA) in preference to a manual sperm count, as this method greatly simplifies the assessment of both numbers of sperm as well as other seminal fluid parameters (see Test 8.2.1).
3. Because of the marked day-to-day intraindividual variability in seminal fluid parameters, at least two sperm counts, a minimum of 2 weeks apart, should be performed as part of any infertility evaluation.

Interpretation

1. Sperm counts of >20 million/mL (or >40 million/ejaculate) are normal (reference range, 20–250 million/mL; ejaculate volume, 1.5–5.5 mL).
2. Sperm counts of <10 million/mL are highly abnormal. *In vitro* fertilization is to be recommended if only 2 to 4 million sperm/ejaculate are present.
3. The absolute number of sperm required for fertility is unknown; males with sperm counts < 1 million/mL have fathered children, but the probability of conception decreases as the sperm concentration decreases below 20 million/mL.
4. Many viral illnesses, some severe illnesses, and exogenous androgen exposure inhibit sperm production. Because spermatogenesis requires 72 to 75 days, it may take 2 to 3 months before the sperm count returns to normal after recovery from the effects of these conditions or exposures.
5. A body mass index (see Test 11.1.3) < 20 kg/m^2 or > 25 kg/m^2 has been associated with reduced sperm counts of 20 to 30%.
6. Infertile men with varicoceles tend to have higher levels of follicle-stimulating hormone (FSH), smaller testes, and lower sperm concentration and motility compared with fertile controls with or without varicoceles.
7. No statistical differences have been found in testicular size, sperm concentration, or sperm motility among fertile men with or without incidental varicoceles detected at physical examination (Pasqualotto et al., 2005).

TABLE 8.2
Gonadotoxic Agents That Adversely Affect
the Male Reproductive System

Type of Gonadotoxic Agent	Abnormal Sperm[a]	Lower Pregnancy Rates
Chemotherapy drugs		
Alkylating agents	+	+
Vinca alkaloids	+	+
Antibiotics		
Erythromycin	+	+
Gentamycin	+	+
Nitrofurantoin	+	+
Recreational/illicit toxins		
Tobacco smoke	+	+
Alcohol	+	+
Cannabis smoke (marijuana)	+	+
Cocaine	+	+
Environmental toxins		
Heat exposure	+	+
Lead	+	+
Ethylene glycol ethers	+	+
Polychlorinated biphenyls (PCBs)	+	+
Perchloroethelyne as used in dry cleaning	+	+
Miscellaneous		
Sulfasalazine	+	+
Bromine vapor	+	+
Ethylene dibromide	+	+
Phthalate esters	+	+

[a] Abnormal forms and/or numbers of sperm.

Note: +, positive effect

Source: Nudell, D.M. et al., *Urol. Clin. North Am.*, 29, 965–973, 2002. With permission.

8. Intake of recreational or illicit drugs, chemotherapeutic agents, or antibiotics may damage the testes with resultant inhibition of spermatogenesis and lower sperm counts (Table 8.2).

9. The adverse effects on testes of a variety of agents, including anabolic steroids, cimetidine, spironolacone, tetracycline, and cyclosporine, may alter the hypothalamic–pituitary–gonadal axis, thus reducing sperm counts and male fertility.

Notes

1. Advanced techniques (e.g., intracytoplasmic sperm injection) with or without testicular sperm extraction may facilitate fertility in patients with <10,000 sperm/ejaculate.

2. Sperm count and total motile sperm increased after a 14-day vs. 4-day abstinence period, with no change in motility in patients with nonobstructive azoospermia. There were no advantages to the longer abstinence if 10 motile sperm could be obtained and used for *in vitro* fertilization (Raziel et al., 2001).

3. Penicillin, cephalosporin, and sulfonamide antibiotics do not impair spermatogenesis.

4. Seasonal variation in some semen characteristics were observed in men living in a Mediterranean climate, including an increase in March and a decrease in September in the adjusted mean ($p <$ 0.0005), total ($p < 0.0005$), and motile ($p = 0.01$) and normal ($p = 0.002$) sperm counts, but no variations in semen volume (Andolz et al., 2001).

5. Histologic changes in the seminiferous tubules consistent with excretory duct obstruction include ectasia, indented outline of the seminiferous epithelium, intratesticular spermatocele, apical cytoplasmic vacuolation of Sertoli cells, and mosaic distribution of testicular lesions.

6. Because the testes of many azoospermic men with excretory duct obstruction produce spermatozoa, testicular biopsy may be required in the evaluation of male infertility.

7. Histologically, atrophic testicles from patients with mumps or Klinefelter's syndrome are indistinguishable.

8. Using an infusate of 3 rather than 7 million sperm did not impair pregnancy success after fallopian tube sperm perfusion, but patients will still be at high risk of multiple pregnancies if an ovarian hyperstimulation and ovulation induction procedure is used (Strandell et al., 2003).

ICD-9 Codes

Conditions that may justify this test *include but are not limited to*:

257.2	hypopituitarism/testicular hypofunction
456.4	scrotal varices
606	infertility, male

Suggested Reading

Andolz P, Bielsa MA, Andolz A. Circannual variation in human semen parameters. *Int J Androl* 2001; 24:266–71.

Jensen TK, Andersson AM, Jorgensen N, Andersen AG, Carlsen E, Petersen JH, Skakkebaek NE. Body mass index in relation to semen quality and reproductive hormones among 1558 Danish men. *Fertil Steril* 2004; 82:863–70.

Nistal M, Riestra ML, Paniagua R. Correlation between testicular biopsies (prepubertal and postpubertal) and spermiogram in cryptorchid men. *Hum Pathol* 2000; 31:1022–30.

Pasqualotto FF, Lucon AM, de Goes PM, Sobreiro BP, Hallak J, Pasqualotto EB, Arap S. Semen profile, testicular volume, and hormonal levels in infertile patients with varicoceles compared with fertile men with and without varicoceles. *Fertil Steril* 2005; 83:74–7.

Raziel A, Friedler S, Schachter M, Kaufman S, Omanski A, Soffer Y, Ron-El R. Influence of a short or long abstinence period on semen parameters in the ejaculate of patients with nonobstructive azoospermia. *Fertil Steril* 2001; 76:485–90.

Strandell A, Bergh C Soderlund B, Lundin K, Nilsson L. Fallopian tube sperm perfusion: the impact of sperm count and morphology on pregnancy rates. *Acta Obstet Gynecol Scand* 2003; 82:1023–29.

8.1.4 Testosterone: Total, Free, Weakly Bound, and Bioavailable Testosterone Testing for Male Hypogonadism

Indications for Test

Measurement of testosterone as total [TT], free [FT], and bioavailable [BioT] are indicated in patients with:

- Signs and symptoms of primary hypogonadism, including erectile dysfunction, diminished libido, decreased energy or motivation, depressed mood, poor concentration and memory, increased body fat, body mass index >27, reduced muscle bulk and strength, low bone mineral density, reduced shaving, reduced testicular volume, or anemia consistent with hypogonadism as a primary condition

- Risk factors for secondary or primary hypogonadism, including erectile dysfunction with or without headaches, toxin exposure, or pituitary or testicular surgery or trauma
- Confirmed hypogonadism and treatment with androgens who are undergoing monitoring of therapy

Procedure

1. To screen for androgen deficiency, obtain a single random blood specimen for determination of [TT] alone. Beware of laboratory sampling bias in the establishment of reference intervals for androgen levels. Ask your lab which type of testosterone assay is used and the origin of samples used to establish its reference interval.
2. In patients who are suspected of having abnormalities of androgen-binding proteins (i.e., sex hormone binding globulin, or SHBG), obtain a random blood sample for [TT] (ng/dL) and [SHBG] (nmol/L) (see Test 8.1.5) and calculate the free androgen index (FAI) according to the formula (calculator available at www.issam.ch):

$$FAI = (TT/SHBG) \times 100$$

3. Alternatively, because the free plus weakly bound (non-SHBG bound) testosterone or BioT is a better test of bioactive testosterone concentration than FAI or FT by analog assay, test the patient's blood specimen for [BioT] first rather than for TT, SHBG, or FT levels.
4. In cases with a borderline low [TT] on a single screening sample, collect three blood specimens 20 minutes apart (see Specimen Collection Protocol P9 in Appendix 2). For screening purposes, repeated sampling is not necessary.
5. To differentiate between patients with primary or secondary hypogonadism and to assess the impact of low or high [BioT] in these patients following androgen therapy, obtain additional blood specimens for FSH, luteinizing hormone (LH), prostate-specific antigen (PSA), and complete blood count (CBC) testing. Once the baseline [TT] has been established and androgen therapy initiated, monitor [TT] as a more cost-effective option than monitoring [FT] or [BioT].

Interpretation

1. Normal adult (age ≥18 years; Tanner stage 5) [TT] is shown in Table 8.3. The typical lower limit for the reference interval of [TT] is 250 to 500 ng/dL (8.7–17.3 nmol/L).
2. In younger adults, an early morning peak (0600 to 0900 hours) in [TT] may occur. No clear diurnal pattern of testosterone secretion has been demonstrated in older individuals. A nadir in [TT] tends to occur between 1600 and 2000 hours.
3. Normative data for [TT] in infants are difficult to obtain, so historical data are used (Forest et al., 1973, 1974).

TABLE 8.3
Relationship Between Tanner Stage in Healthy Males vs. Range of Total Testosterone Concentration, [TT]

Tanner Stage	[TT] (ng/dL)
1	2–23
2	5–70
3	15–280
4	105–545
5	260–1000

TABLE 8.4
Effect of Weight Loss on TT and FT Concentration in Massively Obese[a] Male Subjects

Type of Testosterone	Before Weight Loss	After Weight Loss	Mean Absolute Change	Mean % Change
Total (ng/dL)	240 ± 116	377 ± 113	137	50.7
Free (ng/dL)	9.5 ± 5.0	13.4 ± 4.3	3.9	41.0

[a] 100 to 300% above ideal body weight or a body mass index (BMI) > 40 kg/m^2.

Note: To convert [TT] in ng/dL to [TT] in nmol/L, multiply ng/dL $\times$ 0.0354; to convert [FT] in ng/dL to [FT] in pmol/L, multiply ng/dL $\times$ 34.6316. [TT], total testosterone concentration; [FT], free testosterone concentration.

4. [TT] > 1200 ng/dL in males should raise suspicion of the presence of a virilizing tumor or exogenous intake of androgens.
5. If the [SHBG] changes, the [TT] will change in a similar direction, but the [BioT] will tend to remain the same.
6. [TT] > 320 ng/dL (11.1 nmol/dL) is decidedly normal. [TT] < 200 ng/dL (6.9 nmol/L) is diagnostic of hypogonadism, but a [TT] 200 to 320 ng/dL (6.9–11.1 nmol/L) is equivocal. An [FT] of 6.5 ng/dL (0.23 nmol/L) and a [BioT] of 150 ng/dL (5.2 nmol/L) mark the lower quintile of normal (Vermeulen et al., 2005).
7. Low [TT], [FT], or [BioT] with elevated [LH] and [FSH] indicates primary hypogonadism.
8. Low [TT] with low to low-normal [FSH] and [LH] suggest secondary hypogonadism (pituitary or hypothalamic deficiency), excess estrogen production, usually from adipose tissue conversion of testosterone to estrogen, exposure to exogenous estrogen, androgen steroid abuse occurring up to a year previously, or hyperprolactinemia.
9. Significant increases in [TT] and [FT] occur with weight loss of 26 to 129 kg in the massively obese, defined as 100 to 300% above ideal body weight or with a body mass index (BMI) > 40 kg/m^2 (Table 8.4) (Strain et al., 1988).
10. In obese (BMI $\geq 33.4 \pm 0.8$ kg/m^2) type 2 diabetes mellitus (T2DM) patients, there was a significant inverse correlation of BMI with [FT] ($r = -0.382$; $p < 0.01$) and [TT] ($r = -0.327$; $p < 0.01$), with up to one third of these patients being hypogonadal with a low [FT] (Dhindsa et al., 2004).
11. Metaanalysis of studies in T2DM patients showed that 30 to 40% have low testosterone. Many men with a variety of chronic illnesses also have a low [TT]. Thus, whereas about a third of all diabetic men have low [TT], a third of those with hypertension and two thirds of men with obstructive sleep apnea also are hypogonadal (Guay and Seftel, 2008).

Notes

1. The testes secrete large amounts of testosterone during the first year of life, but gonadal steroidogenesis is very low in both boys and girls thereafter until the start of puberty.
2. Low [TT] does not distinguish between individuals with primary or secondary hypogonadism.
3. FSH and LH are secreted in a pulsatile fashion with the LH pulse having wide variability in terms of both amplitude and frequency during the day contributing to considerable variability in male [TT] on an hourly and daily basis.
4. Pulses of LH stimulate testosterone secretion. LH pulses are stimulated by pulses of gonadotropin-releasing hormone (GnRH), which are suppressed by circulating testosterone or estrogen.
5. Androgen levels may fall into the low-normal to below-normal reference interval just prior to an LH pulse which may occur at intervals up to ≥ 4 hours.
6. [TT] and [FT] reference intervals must specify gender, age, and Tanner stage (Kushnir et al., 2006).
7. Avoid determination of [FT] by direct analog assay methods, as they may underestimate the true [FT] by as much as 100%.
8. Testosterone is firmly bound to SHBG but only weakly bound to albumin, which enables albumin-bound testosterone to be available to the tissues as BioT. FT comprises, at most, only 2 to 3% of the [TT].

9. The FT concentration by dialysis is an acceptable, but costly and labor intensive, method for determining [FT]. Charges for testosterone testing typically decreases in the order of TT < FT by analog assay << BioT << FT by dialysis.

10. The FAI/FT by dialysis ratio varies as a function of [SHBG]. Thus, neither [FT] by analog assay or the FAI alone is a reliable index of [BioT].

11. Although transdermal dihydrotestosterone (DHT) treatment has no adverse effects on prostate volume, [PSA], vascular endothelium, or lipids, treatment with 1% testosterone gel over a 42-month period showed that 18% of hypogonadal men had an increase in [PSA] over baseline (Ly et al., 2001).

12. Testosterone supplements had no effect on strength, mobility, or cognitive ability for Dutch men between 60 and 80 years of age with [TT] < 395 ng/dL (< 13.7 nmol/L), but decreased body fat and increased lean body mass were observed. These men also had lower total and high-density lipoprotein (HDL) cholesterol levels, lower glucose levels, and increased insulin sensitivity after 6 months of testosterone therapy (Emmelot-Vonk et al., 2008).

ICD-9 Codes

Conditions that may justify this test *include but are not limited to*:

257.2	hypogonadotropism/testicular hypofunction
302.70	psychosexual dysfunction, unspecified
E879	pituitary or testicular surgery

Suggested Reading

Dhindsa S, Prabhakar S, Sethi M, Bandyopadhyay A, Chaudhuri A, Dandona P. Frequent occurrence of hypogonadotropic hypogonadism in type 2 diabetes. *J Clin Endocrinol Metab* 2004; 89:5462–68.

Emmelot-Vonk MH, Verhaar HJJ, Nakhai Pour HR, Aleman A, Lock TMTW, Ruud Bosch JLH, Grobbee DE, van der Schouw YT. Effect of testosterone supplementation on functional mobility, cognition, and other parameters in older men. *JAMA* 2008; 299:39–52.

Forest MG, Cathiard AM, Bertrand JA. Total and unbound testosterone levels in the newborn and in normal and hypogonadal children: use of a sensitive radioimmunoassay for testosterone. *J Clin Endocrinol Metab* 1973; 36:1132–42.

Forest MG, Sizonenko PC, Cathiard AM, Bertrand J. Hypophyso-gonadal function in humans during the first year of life. 1. Evidence for testicular activity in early infancy. *J Clin Invest* 1974; 53:819–28.

Guay AT, Seftel AD. Men with erectile dysfunction have hypogonadism due to varied chronic illnesses. In: *Proc. of the 6th World Congress on the Aging Male*, Tampa, FL, February, 2008.

Kushnir MM, Rockwood AL, Roberts WL, Pattison EG, Bunker AM, Fitzgerald RL, Meikle AW. Performance characteristics of a novel tandem mass spectrometry assay for serum testosterone. *Clin Chem* 2006; 52:120–8.

Ly LP, Jimenez M, Zhuang TN, Celermajer DS, Conway AJ, Handelsman DJ. A double-blind, placebo-controlled, randomized clinical trial of transdermal dihydrotestosterone gel on muscular strength, mobility, and quality of life in older men with partial androgen deficiency. *J Clin Endocrinol Metab* 2001; 86:4078–88.

Rosner W, Auchus RJ, Azziz R, Sluss PM, Raff H. Utility, limitations, and pitfalls in measuring testosterone: an Endocrine Society position statement. *J Clin Endocrinol Metab* 2007; 92: 405–13.

Strain GW, Zumoff B, Miller LK, Rosner W, Levit C Kalin M, Hershcopf RJ, Rosenfeld RS. Effect of massive weight loss on hypothalamic–pituitary–gonadal function in obese men. *J Clin Endocrinol Metab* 1988; 66:1019–23.

Vermeulen A. Hormonal cut-offs of partial androgen deficiency: a survey of androgen assays. *J Endocrinol Invest* 2005; 28:28–31.

Vermeulen A, Verdonck L, Kaufman JM. A critical evaluation of simple methods for the estimation of free testosterone in serum. *J Clin Endocrinol Metab* 1999; 84:3666–72.

8.1.5   Sex Hormone Binding Globulin (SHBG) and Androgen Testing, Including Dihydrotestosterone (DHT), in Males

Indications for Test

Androgen and SHBG testing are indicated when:

- A screening total testosterone concentration, [TT], fails to correlate with symptoms of erectile dysfunction or hypogonadism.
- Liver disease, obesity, or adrenal factors that might dynamically alter the levels of gonadal steroids or their binding proteins are present.

Measurement of dihydrotestosterone concentration, [DHT], is indicated in:

- Undervirilized newborn males to help distinguish rare syndromes of 5α-reductase deficiency from partial androgen insensitivity syndrome (AIS) or 17β-hydroxysteroid dehydrogenase type 3 deficiency

Procedure

1. Identify factors that are associated with decreased [SHBG], including moderate to severe obesity, nephrotic syndrome, hypothyroidism, use of glucocorticoids, progestins, and androgenic steroids.
2. Identify factors that are associated with increased [SHBG], including aging, hepatic cirrhosis, hyperthyroidism, use of anticonvulsants, use of estrogens, and HIV infection.
3. Determine the patient's body mass index (BMI) and obtain a blood or saliva sample (see Specimen Collection Protocol P2 in Appendix 2) for determination of [TT].
4. If the [TT] is in the range of 250 to 500 ng/dL, obtain a second blood sample for determination of [SHBG] and [TT] and calculation of the free testosterone concentration, [FT], or direct measurement of the bioavailable testosterone concentration, [BioT].
5. In men with documented hypogonadism whose response to testosterone therapy has been less than satisfactory, obtain a blood specimen for TSH and prolactin testing. In such men, search for factors (e.g., liver disease) that may alter the [SHBG].
6. In undervirilized newborn males, measure [DHT] as a screening test. As an alternative to serum DHT testing, assess the quantity of total 5α-reductase androgens and androgen precursors using a timed urine specimen for determination of the urinary steroid metabolite profile by liquid chromatography–tandem mass spectrometry (LC-MS/MS). Refer to Test 8.4.5.

Interpretation

1. Reference intervals for SHBG, TT, FT, and BioT in healthy individuals are shown in Table 8.5.
2. Salivary [TT] in male adolescents (ages 11 to 16 years) declines by about 50% from 0730 to 1630 hours. In the age range from 11 to 13 years, the 0730-hour [TT] is about 40 pg/mL and at age 14 to 16 the 0730-hour [TT] is about 71 pg/mL (Granger et al., 2003).
3. High TT/DHT ratios are consistent with 5α-reductase deficiency, whereas normal ratios in cases of ambiguous genitalia suggest partial androgen insensitivity.
4. Tissue concentrations of DHT derived from testosterone precursors, rather than circulating DHT levels, are more important in the diagnosis of milder disorders of sexual differentiation. Thus, the use of serum [DHT] will decline as tests for tissue levels of DHT become more widely used in the diagnosis of these disorders.
5. In both men and women, increased serum [SHBG] correlates with reduced adiposity and fewer risk factors for cardiovascular disease.
6. Serum [SHBG] in weight-stable men shows a negative linear correlation with BMI, decreasing 0.2 nmol/L per unit increase in BMI.
7. During weight loss, serum [SHBG] increases at an average slope of 0.43 nmol/L per unit decrease in BMI, which is much greater than the negative slope of 0.2 nmol/L per unit increase in BMI occurring in weight-stable men over a broad range of BMI values.

TABLE 8.5
Assay- and Age-Dependent Reference Intervals for the Serum Concentration of Sex Hormone Binding Globulin and Selected Androgens in Healthy Adult Males

Analyte	Type of Assay	Age (years)	Reference Interval
Sex hormone binding globulin (SHBG)	ICMA	18–49	7–49 nmol/L
		50–91	17–65 nmol/L
Total testosterone (TT)	ICMA (screening)	20–60	>300–827 ng/dL
	LC-MS/MS	18–69	250–1100 ng/dL
	LC-MS/MS	70–89	90–890 ng/dL
Free testosterone (FT)[a]	ED	18–89	35–155 pg/mL
	LC-MS/MS	18–69	46–224 pg/mL
		70–89	6–73 pg/mL
Bioavailable testosterone (BioT)[a]	LC-MS/MS	18–69	110–575 pg/mL
		70–89	15–150 pg/mL

[a] Calculated value.

Note: ED, equilibrium dialysis; ICMA, immunochemiluminometric assay; LC-MS/MS, liquid chromatography–tandem mass spectrometry.

8. Based on data from serial SHBG and TT testing on three or more occasions, when there is a change in [SHBG] or [TT] more than expected based on biological variation alone:
   - An increased [SHBG] with no change in [TT] suggests a decrease in [BioT] or [FT].
   - A decrease in [SHBG] associated with no change in [TT] suggests an increase in [BioT] or [FT] in nonobese individuals.
   - A similar change in [SHBG] and [TT] suggests that the [BioT] will be unchanged.

Notes

1. Testosterone is firmly bound to SHBG in serum.
2. Direct measurement of serum [FT] by equilibrium dialysis or of [BioT] by direct assay is more precise and accurate than a calculated [FT], but both of these tests are more expensive than the calculated or measured [FT] by analog assay.
3. Significant genetic and ethnic variations exist in the reference interval for [SHBG].
4. The [SHBG] varies from low levels in hirsute women to extremely high levels in individuals with hyperthyroidism regardless of gender.
5. Male obesity and excess abdominal fat are associated with reductions in circulating androgen derived from the testes as increased conversion to estrogens takes place in adipose tissue.
6. The inverse relationship between serum insulin and [SHBG] indicates that insulin has control over SHBG synthesis. [BioT] is correlated inversely with BMI.
7. [SHBG] rises with massive weight loss in obese individuals (e.g., from 9.2 ± 3.2 to 12.9 ± 5.4 nmol/L; $p < 0.005$) (Strain et al., 1994), with an expected increase in both [TT] and [FT] (Table 8.5).
8. Approximately 30% of men 60 to 70 years of age and 70% of men 70 to 80 years of age had low serum [FT] measured by analog assay (Hijazi and Cunningham, 2005).
9. Symptoms and findings of testosterone deficiency are similar to those associated with normal aging during which a decline in [TT] of about 1 to 2% per year starting at age 30 years results in biochemical hypogonadism in at least 20% of men 60 to 80 years of age.
10. Significant increases in serum [TT] and [FT] were observed in adolescents with type 1 diabetes mellitus (T1DM) when compared to healthy sex- and pubertal-stage-matched controls in late puberty, but no differences were observed in concentrations of dehydroepiandrosterone sulfate (DHEAS), SHBG, DHT, and 3α-androstanediol (Meyer et al., 2000).
11. In 116 men with T2DM and 630 healthy nondiabetic men (control group), [FT] was low in 46% of those with T2DM vs. 24% of the men in the control group, while the [TT] was low in 34% of the men with T2DM and 23% of the men in the control group.
12. In T2DM patients, most of whom were obese, serum [SHBG] correlated inversely with BMI ($r = -0.27$; $p < 0.05$) but positively with age ($r = 0.54$; $p < 0.001$) and [TT] ($r = 0.57$; $p < 0.001$).

13. When analyzed separately, low [FT], low [TT], BMI > 25 kg/m^2, and increased waist/hip ratios were independently associated with T2DM in men. After adjusting for waist/hip ratios, T2DM men were three times as likely as nondiabetic men to have low [FT] and twice as likely to have low [TT]. Thus, subnormal [FT] and T2DM are directly associated, whereas [TT] appears to be more strongly related to obesity and central adiposity than to T2DM (Rhoden et al., 2005).

14. For adolescent males, lower overall [TT] and [TT] that decrease more slowly across the day are associated with higher levels of anxiety/depression and attention problems. These associations were not moderated by pubertal development (i.e., change in age from 12 to 15 years) (Granger et al., 2003).

ICD-9 Codes

Conditions that may justify this test *include but are not limited to*:

257.2	hypopituitarism/testicular hypofunction
278.00	obesity
302.70	psychosexual dysfunction, unspecified
573	liver disease

Suggested Reading

Cunningham GR, Swerdloff RS. *Summary from the Second Annual Andropause Consensus Meeting*, Chevy Chase, MD. 2001. Endocrine Society.

Granger DA, Shirtcliff EA, Zahn-Waxler C, Usher B, Klimes-Dougan B, Hastings P. Salivary testosterone diurnal variation and psychopathology in adolescent males and females: individual differences and developmental effects. *Dev Psychopathol* 2003; 15:431–49.

Hijazi RA, Cunningham GR. ANDROPAUSE: is androgen replacement therapy indicated for the aging male? *Annu Rev Med* 2005; 56:117–37.

Mazen I, Hafez M, Mamdouh M, Sultan C Lumbroso S. A novel mutation of the 5 alpha-reductase type 2 gene in two unrelated Egyptian children with ambiguous genitalia. *J Pediatr Endocrinol Metab* 2003; 16:219–24.

Meyer K, Deutscher J, Anil M, Berthold A, Bartsch M, Kiess W. Serum androgen levels in adolescents with type 1 diabetes: relationship to pubertal stage and metabolic control. *J Endocrinol Invest* 2000; 23:362–8.

Rhoden EL, Ribeiro EP, Teloken C, Souto CA. Diabetes mellitus is associated with subnormal serum levels of free testosterone in men. *BJU Int* 2005; 98:867–70.

Strain G, Zumoff B, Rosner W, Pi-Sunyer X. The relationship between serum levels of insulin and sex hormone-binding globulin in men: the effect of weight loss. *J Clin Endocrinol Metab* 1994; 79:1173–6.

Tchernof A, Despres JP. Sex steroid hormones, sex hormone-binding globulin, and obesity in men and women. *Horm Metab Res* 2000; 32:526–36.

Zumoff B, Strain GW. A perspective on the hormonal abnormalities of obesity: are they cause or effect? *Obes Res* 1994; 2:56–67.

Zumoff B, Strain GW, Miller LK, Rosner W, Senie R, Seres DS, Rosenfeld RS. Plasma free and non-sex-hormone-binding-globulin-bound testosterone are decreased in obese men in proportion to their degree of obesity. *J Clin Endocrinol Metab* 1990; 71:929–31.

8.1.6 Hypogonadism in Potentially Androgen-Deficient Aging Males (ADAM): Screening Questionnaires (ADAM and Daily Assessment of Mood Score)

Indications for Test

Administration of the androgen deficiency symptom questionnaire (Q8.1 in Appendix 1) and assessment of mood questionnaire (Q8.2 in Appendix 1) is indicated:

- If ADAM is suspected based on chronic complaints of dysphoria, negativism, bad mood, fatigue, depression, decreased libido, and erectile dysfunction
- In the assessment of response to testosterone therapy in patients with established hypogonadism

Procedure

1. Screen for severe depression, which may interfere with interpretation of mood scores, by noting the presence of any of the following three physiologic, or "vegetative," symptoms:
   - A markedly diminished interest or pleasure in almost all activities nearly every day
   - Significant unintentional weight loss or weight gain or marked change in appetite
   - Insomnia or hypersomnia nearly every day
2. Proceed with administration of the Beck Depression Inventory assessment as appropriate. If vegetative symptoms of depression are identified, refer the patient for antidepressant therapy before proceeding with further endocrine evaluation of the gonads.
3. Administer the 10-item ADAM questionnaire (Q8.1 in Appendix 1) to screen for hypogonadism.
4. Provide the patient with the Daily Assessment of Mood Score questionnaire, a list of true–false questions regarding positive and negative moods (Q8.2 in Appendix 1).
5. To identify the patient's mood on a daily basis, have the patient record and score parameters for sexual desire, sexual enjoyment, sexual performance, and sexual activity at the end of the day for 7 days before a clinic visit.
6. Use the Q8.2 questionnaire and a one-week self-report diary to monitor on a daily basis the patient's overall mood as follows:
   - A positive mood is defined as alert, full of pep or energy, and friendly with a good sense of well being or vigor.
   - A negative mood is defined as angry, irritable, tired, or fatigued and tense or nervous with feelings of being sad, blue, or depressed.
7. Have the patient score each mood question on a rating scale from 0 to 7 (0, not true at all; 3, slightly false; 4, slightly true; 7, very true).
8. Repeat scoring of mood questions after 30 to 60 days of testosterone replacement therapy.

Interpretation

1. The patient's responses on the Q8.1 questionnaire are scored as noted. An abnormal score on the Q8.1 questionnaire should prompt tests for androgen levels.
2. In individuals with a positive score on the Q8.1 questionnaire, conditions other than hypogonadism, including depression, hemochromatosis, micro- or macroprolactinoma of the pituitary, hypothyroidism, benign prostatic hyperplasia (BPH), or prostate carcinoma, should be considered.
3. Mood parameters may not clearly differentiate eugonadal from hypogonadal men before therapy with testosterone.
4. In the Q 8.2 questionnaire, scores reflect the degree of positive and negative moods on a daily basis using a scale from −7 to +7.
5. Within 30 days of testosterone treatment in men with hypogonadism, positive mood parameters tend to appear and negative mood parameters improve.
6. Neither the ADAM nor Mood Score questionnaires can clearly distinguish hypogonadism from nonorganically based depression.
7. If vegetative symptoms of depression, as noted above, are identified, consider referral for antidepressant therapy before proceeding with further endocrine evaluation of the gonads.

Notes

1. The Q8.1 questionnaire has a high sensitivity (81%) in identifying aging males with low free testosterone concentrations measured by equilibrium dialysis; however, it cannot be used as a surrogate for serum FT testing because of its low specificity (21.6%).
2. In 316 Canadian physicians ages 40 to 62 years, low bioavailable [T] were found in 25% of them, but none had elevated [LH]. The Q8.1 questionnaire had 88% sensitivity and 60% specificity for diagnosis of hypogonadism in this population (Morley et al., 2002).

3. A simple self-report diary may be useful in assessing the sexual function and mood profile of hypogonadal subjects. A good correlation has been shown between mood parameters assessed by a diary with those assessed by the Q8.2 questionnaire.

4. ADAM is a clinical entity characterized biochemically by a decrease not only in serum androgen concentrations but also in growth hormone and melatonin concentrations.

5. The onset of ADAM is unpredictable and its manifestations are subtle and variable but may include any combination of the following symptoms: fatigue, depression, decreased libido, alterations in mood, cognition sexual dysfunction, decrease in muscle mass and strength, or increase in fat mass.

6. An inability to control one's behavior (i.e., impulsivity), when such control is required, has been found to significantly predict levels of aggression over and above one's age and [T].

7. Achievement of supraphysiological [T] by administering 200 mg i.m. on a weekly basis in men with hypogonadism was not found to lead to an increase in self- or partner-reported aggression or mood disturbances (O'Conner et al., 2002).

8. In the Massachusetts Male Aging Study (MMAS), potential risk factors for testosterone deficiency, including age, obesity, chronic diseases, health behaviors, results on the Jackson dominance scale, and symptoms of stress, were recorded and assessed. The prevalence of testosterone deficiency was 20.4% in the MMAS, with results from a screening questionnaire being a sensitive predictor of testosterone deficiency (area under the receiver–operator characteristic [ROC] curve [AUC] = 0.66) (Morley et al., 2000).

ICD-9 Codes

Conditions that may justify this test *include but are not limited to*:

257.2	hypopituitarism/testicular hypofunction
296.90	unspecified episodic mood disorder/dysphoria, bad mood, negativism,
302.70	psychosexual dysfunction, unspecified
780.7	malaise and fatigue
799.81	decreased libido
V79.0	depression

Suggested Reading

Lee KK, Berman N, Alexander GM, Hull L, Swerdloff RS, Wang C. A simple self-report diary for assessing psychosexual function in hypogonadal men. *J Androl* 2003; 24:688–98.

Morales A, Heaton JP, Carson CC. Andropause: a misnomer for a true clinical entity. *J Urol* 2000; 163:705–12.

Morley JE, Charlton E, Patrick P, Kaiser FE, Cadeau P, McCready D, Perry HM. Validation of a screening questionnaire for androgen deficiency in aging males. *Metabolism* 2000; 49:1239–42.

O'Connor DB, Archer J, Hair WM, Wu FC. Exogenous testosterone, aggression, and mood in eugonadal and hypogonadal men. *Physiol Behav* 2002; 75:557–66.

Smith KW, Feldman HA, McKinlay JB. Construction and field validation of a self-administered screener for testosterone deficiency (hypogonadism) in ageing men. *Clin Endocrinol (Oxf)* 2000; 53:703–11.

Tancredi A, Reginster JY, Schleich F, Pire G, Maassen P, Luyckx F, Legros JJ. Interest of the androgen deficiency in aging males (ADAM) questionnaire for the identification of hypogonadism in elderly community-dwelling male volunteers. *Eur J Endocrinol* 2004; 151:355–60.

T'Sjoen G, Feyen E, De Kuyper P, Comhaire F, Kaufman JM. Self-referred patients in an aging male clinic: much more than androgen deficiency alone. *Aging Male* 2003; 6:157–65.

Wang C Swedloff RS, Iranmanesh A, Dobs A, Snyder PJ, Cunningham G, Matsumoto AM, Weber T, Berman N. Transdermal testosterone gel improves sexual function, mood, muscle strength, and body composition parameters in hypogonadal men: Testosterone Gel Study Group. *J Clin Endocrinol Metab* 2000; 85:2839–53.

8.2   Male Infertility

8.2.1   Semen Analysis: Complete Analysis in Male Infertility

Indications for Test

Complete semen analysis is indicated in the evaluation of:

- All infertile couples to identify male factors contributing to infertility
- Men with gynecomastia, small or atrophic testes, or a history of mumps or cryptorchidism
- Status after radiotherapy to the pelvic region or chemotherapy for cancer

Procedure

1. Obtain history of childhood mumps and coital practices, and assess body mass index (BMI).
2. From the male patient, collect seminal fluid as described in Specimen Collection Protocol P3 (Appendix 2).
3. Use computer-assisted sperm motility analysis (CASMA) with a multiple-exposure photography (MEP) system to measure sperm motility, straightline velocity (linearity of travel), and head displacement (lateral/side-to-side motion of sperm head).
4. Calculate the motility index as the product of motility and velocity.
5. Analyze sperm for motility using CASMA or a microscope with 10× and 40× objectives.
6. In the postcoital female infertility patient, use fluid obtained from a subcervical vaginal pool for analysis of sperm motility, velocity, motility index, and morphology.

Interpretation

1. Reference intervals for male semen analysis parameters are shown in Table 8.6.
2. If the patient has azoospermia and:
   - The semen fructose concentration is undetectable, these findings indicate a congenital absence or obstruction of seminal vesicles and distal vas deferens.
   - The semen fructose concentration is detectable, these findings indicate germinal, sperm production problems, or epididymal or proximal obstruction of the vas deferens.
3. Teratospermia is defined as >40% abnormal forms.
4. If red blood cells (RBCs) are found in seminal fluid, the patient has hematospermia.
5. Asthenozoospermia (hypomotile sperm) is characterized by sperm velocity of <20 and a motility index of <8 μm/sec.
6. Marked obesity or exposure to a variety of gonadotoxins (Table 8.2) may result in abnormal sperm parameters and lead to lower pregnancy rates.
7. Antidepressants may elevate prolactin levels, which can lead to significant but reversible suppression of spermatogenesis.

Notes

1. Workplace exposure to toxic substances may contribute to male infertility, but most of the studies prior to 2004 on this topic are limited to either case reports or epidemiological studies (population-based, case-control, or cohort analyses).
2. Sperm motility was found to be lower in infertile men with varicoceles (37% ± 24%) than in fertile men with varicoceles (54% ± 17%) or fertile men without varicoceles (59% ± 16%).
3. Surgery to correct a varicocele may improve sperm motility.
4. Viral orchitis followed by obstruction of the epididymis or vas deferens is the most common cause of male infertility.
5. There is a significant lack of standardization in the performance and reporting of results of semen analyses among laboratories. The large degree of variation and discordance between different laboratories may stem from their lack of routinely exercised quality control procedures (Keel, 2004).

TABLE 8.6
Reference Intervals for Sperm Parameters

Parameter	Interval
Ejaculate volume	1.5–5.5 mL
Sperm concentration	20–250 × 10⁶/mL
Motility by MEP	>40%
Velocity by MEP	>20 μm/sec
Motility index	>8 μm %/sec
Morphology	>60% normal forms
Liquefaction	10–30 minutes
Fructose concentration	120–450 μMol/ejaculate or ≥13 mg/dL

Note: MEP, multiple exposure photography.

6. In couples undergoing artificial insemination of an ovum via the intracytoplasmic sperm injection (ICSI) procedure, the occurrence of *de novo* prenatal chromosomal anomalies was 2.1% for sperm concentrations of <20 × 10⁶/mL vs. 0.24% for sperm concentrations >20 × 10⁶/mL. No statistical difference in the frequency of chromosomal anomalies has been observed within the range of lower threshold values of sperm concentration (<1 × 10⁶ to <15 × 10⁶/mL).

7. In couples undergoing ICSI, abnormal chromosomes in fetal cells obtained by amniocentesis or chorionic villus sampling (CVS) were more frequent if sperm motility was poor but morphology was <40% abnormal.

8. A positive association between beta-carotene intake and sperm concentration ($p = 0.06$) and progressive motility ($p = 0.06$) has been observed, while increased folate and zinc intake was not associated with improved semen quality (Eskenazi et al., 2005).

9. Agents that may impair fertility or function of sperm without altering standard semen analysis parameters include calcium channel blockers and exposures to higher concentrations of manganese.

10. Other than a small, but significant, reduction in semen volume in diabetic men (2.6 vs. 3.3 mL; $p < 0.05$), conventional semen parameters did not differ significantly from control subjects. Diabetic subjects had significantly higher mean nuclear DNA fragmentation (53 vs. 32%; $p < 0.0001$) and median number of mitochondrial DNA deletions (4 vs. 3; $p < 0.05$) compared with control subjects (Agbaje et al., 2007).

ICD-9 Codes

Conditions that may justify this test *include but are not limited to*:

072	mumps
257.2	testicular hypofunction
611.1	gynecomastia
752.51	undescended testis

Suggested Reading

Agbaje IM, Rogers DA, McVicar CM, McClure N, Atkinson AB, Mallidis C, Lewis SEM. Insulin dependant diabetes mellitus: implications for male reproductive function. *Hum Reprod* 2007; 5:1–7.

Bonduelle M, Van Assche E, Joris H, Keymolen K, Devroey P, Van Steirteghem A, Liebaers I. Prenatal testing in ICSI pregnancies: incidence of chromosomal anomalies in 1586 karyotypes and relation to sperm parameters. *Hum Reprod* 2002; 17:2600–14.

Claman P. Men at risk: occupation and male infertility. *Sex Reprod Menopause* 2004; 2:19–23.

Eskenazi B, Kidd SA, Marks AR, Sloter E, Block G, Wyrobek AJ. Antioxidant intake is associated with semen quality in healthy men. *Hum Reprod* 2005; 20:1006–12.

Keel BA. How reliable are results from the semen analysis? *Fertil Steril* 2004; 82:41–4.

Keel BA, Webster BW (Eds). *CRC Handbook of the Laboratory Diagnosis and Treatment of Infertility*. 1990, CRC Press, p. 34.

Nudell DM, Monoski MM, Lipshultz LI. Common medications and drugs: how they affect male fertility. *Urol Clin North Am* 2002; 29:965–73.

Pasqualotto FF, Lucon AM, de Goes PM, Sobreiro BP, Hallak J, Pasqualotto EB, Arap S. Semen profile, testicular volume, and hormonal levels in infertile patients with varicoceles compared with fertile men with and without varicoceles. *Fertil Steril* 2005; 83:74–7.

8.2.2   Luteinizing Hormone (LH) and Follicle-Stimulating Hormone (FSH) Testing in Male Infertility

Indications for Test

Measurement of LH and FSH is indicated in:

- The assessment of hypothalamic–pituitary–gonadal (HPG) axis integrity and its contribution to male fertility particularly in cases of gynecomastia, galactorrhea, or obesity

Procedure

1. When screening for male fertility status, obtain a random blood sample for LH and FSH testing, usually in the context of men who have had previous total and/or bioavailable testosterone (BioT) and sperm count testing.
2. In men with low testosterone ([TT] or [BioT]) and low-normal [LH] and [FSH] on screening, order prolactin (PRL) testing and assess body mass index (BMI).
3. If the serum [PRL] is elevated, obtain a pituitary imaging study to assess for the presence of a pituitary tumor (prolactinoma) as the cause of a deranged HPG axis leading to the low levels of the patient's gonadotropins.

Interpretation

1. Interpret values for LH, FSH, testosterone, and prolactin using Table 8.7.

Notes

1. Obesity enhances the conversion of normal testicular testosterone to estrogen, which feeds back to the pituitary, lowering FSH, LH, and testosterone secretion.
2. An elevated [PRL] and secondary hypogonadism may be caused by a pituitary tumor.
3. Testosterone therapy for acquired hypogonadotropic hypogonadism may suppress pituitary production of FSH and LH, but fertility may be unimpaired in more than a third of treated cases. In fact, 8 out of 15 such patients had persistent spermatogenesis, 4 had sperm concentrations $\geq$ 15 million/mL, and only 6 were azoospermic (Drincic et al., 2003).

ICD-9 Codes

Conditions that may justify this test *include but are not limited to*:

278.00	obesity, unspecified
611.1	hypertrophy of breast/gynecomastia
628.1	hypothalamic–pituitary–gonadal (HPG) axis disorder
676.6	galactorrhea

TABLE 8.7
Interpretation of LH, FSH, Testosterone, and Prolactin Concentrations in Men Being Screened for Infertility

Hormone	Hormone Level	Interpretation
LH	< 10 mIU/mL	Within normal limits
FSH	< 16 mIU/mL	Within normal limits
TT	Variable	—
BioT	Variable	—
Prolactin[a]	<15 ng/mL	Within normal limits
LH	↑	Suggestive of an LH-secreting tumor
TT	High-normal or ↑	
LH	↑	Consistent with primary hypogonadism
TT	↓	
FSH	↑	
LH	↓	Consistent with secondary hypogonadism (pituitary or hypothalamic
TT	↓	deficiency) or excess estrogen from endogenous production in adipose
FSH	↓	tissue, external exposure, or ingestion.

[a] Test 2.9.2.

Note: BioT, bioavailable testosterone; FSH, follicle-stimulating hormone; LH, luteinizing hormone; TT, total testosterone.

Suggested Reading

Dhindsa S, Prabhakar S, Sethi M, Bandyopadhyay A, Chaudhuri A, Dandona P. Frequent occurrence of hypogonadotropic hypogonadism in type 2 diabetes. *J Clin Endocrinol Metab* 2004; 89:5462–8.

Drincic A, Arseven OK, Sosa E, Mercado M, Kopp P, Molitch ME. Men with acquired hypogonadotropic hypogonadism treated with testosterone may be fertile. *Pituitary* 2003; 6:5–10.

8.2.3　Antisperm Antibody (ASA) Testing in Male Infertility

Indications for Test

ASA determinations in cases of male infertility are indicated:

- Early in the course of an evaluation for the cause of a low sperm count in an infertile male
- When a sperm motility index of <8 μm/sec, suggestive of asthenozoospermia, is noted (see Test 8.2.1)

Procedure

1. Obtain blood and semen specimens for ASA testing by immunobead binding tests (IBTs) using direct mixed antiglobulin reaction (MAR), indirect gelatin agglutination (GA), or indirect tray agglutination (TA) methods.
2. Perform semen analysis using flow cytometry in preference to manual methods as it provides better precision and accuracy in the determination of all semen parameters, including ASA.

Interpretation

1. Reference intervals for ASA are dependent on assay. In general, the following cutoffs apply:
   - No impairment in fertility, <20% of motile sperm bound
   - Impaired fertility, >50% of motile sperm bound
   - Nondiagnostic, 20 to 50% of motile sperm bound

2. Sperm-bound autoantibodies are more frequently associated with hypomotile sperm (asthenozoospermia) and may impair fertility in men with normozoospermia as well.
3. Cell-mediated antisperm autoimmunity may play a significant role in impairment of spermiogenesis independent of the presence or absence of ASA.

Notes

1. The mechanism by which ASA leads to antibody-mediated infertility in the male is poorly understood.
2. Cell-mediated antisperm autoimmunity, determined by a migration-inhibition test, may play a significant role in impairment of spermiogenesis and cause asthenozoospermia (Dimitrov et al., 1992).
3. Semen with ASA had a significantly lower sperm concentration, motility, and total motile fraction as well as a higher percentage of vibratory sperm and percentage of bound antisperm antibodies compared to 44 specimens without ASA (Madar et al., 2002).
4. In patients with ASA, sperm morphology, liquefaction time, semen volume, and white blood cell concentration were not found to differ from those of a reference population.
5. Elevated thyroid peroxidase (TPO) antibodies were significantly correlated with pathozoospermia ($p = 0.036$) and asthenozoospermia ($p = 0.049$); however, the presence of ASA was not determined (Trummer et al., 2001).
6. The antigens detected by ASA from men who have had a vasectomy are mostly related to changes that occur upon epididymal passage of the sperm.
7. Patients with high levels of IgA antibodies to their sperm also have high antibody titers in serum such that the locally produced IgA antibodies reach the sperm and occupy the binding sites before the main bulk of IgG antibodies reaches the seminal compartment.

ICD-9 Codes

Conditions that may justify this test *include but are not limited to*:

Infertility
606   male type
606.01   oligospermia
V26.21   low sperm count

Suggested Reading

Bohring C Krause W. Differences in the antigen pattern recognized by antisperm antibodies in patients with infertility and vasectomy. *J Urol* 2001; 166:1178–80.

Clayton R, Moore H. Experimental models to investigate the pathology of antisperm antibodies: approaches and problems. *Hum Reprod Update* 2001; 7:457–9.

Dimitrov DG, Sedlak R, Nouza K, Kinsky R. A quantitative objective method for the evaluation of anti-sperm cell-mediated immunity in humans. *J Immunol Methods (Netherlands)*. 1992; 154:147–53.

Hjort T. Do autoantibodies to sperm reduce fecundity? A mini-review in historical perspective. *Am J Reprod Immunol* 1998; 40:215–22.

Kipersztok S, Kim BD Morris L, Drury KC Williams RS, Rhoton-Vlasak A. Validity of a rapid assay for antisperm antibodies in semen. *Fertil Steril* 2003; 79:522–8.

Lenzi A, Gandini L, Lombardo F, Rago R, Paoli D Dondero F. Antisperm antibody detection. 2. Clinical, biological, and statistical correlation between methods. *Am J Reprod Immunol* 1997; 38:224–30.

Madar J, Urbanek V, Chaloupkova A, Nouza K, Kinsky R. Role of sperm antibodies and cellular autoimmunity to sperm in the pathogenesis of male infertility. *Ceska Gynekol* 2002; 67:3–7.

Shai S, Roudebush W, Powers D Dirnfeld M, Lamb DJ. A multicenter study evaluating the flowcytometric-based kit for semen analysis. *Fertil Steril* 2005; 83:1034–38.

Trummer H, Ramschak-Schwarzer S, Haas J, Habermann H, Pummer K, Leb G. Thyroid hormones and thyroid antibodies in infertile males. *Fertil Steril* 2001; 76:254–7.

8.3   Neoplasms of the Male Reproductive Tract

8.3.1   Imaging of Testicles, Epididymis, and Anatomic Abnormalities of the Male Reproductive Apparatus Using Scrotal (SUS) and Transrectal (TRUS) Ultrasound

Indications for Test

Imaging of the testicles and epididymis is indicated when:

- Male infertility is being evaluated.
- One or both testicles fail to descend.
- A testicular mass is present or pain is noted in either gonad or epididymis.

Procedure

1. Recognize that ultrasound is both an operator- and equipment-dependent procedure.
2. Use pulsed or color Doppler SUS to image blood flow, the epididymis, and masses within the testicles. Follow up the finding of a solid or cystic-solid mass with measurement of alpha-fetoprotein (AFP) and human chorionic gonadotropin (hCG) levels (see Test 8.3.2).
3. Use TRUS to visualize patency of the ejaculatory ducts, seminal vesicles, and vas deferens, particularly in male infertility cases.

Interpretation

1. Testicular masses may be nonmalignant lesions or malignant seminomatous or nonseminomatous tumors.
2. Infections of the epididymis may result in edema or the appearance of a fluid-filled abscess by color Doppler SUS examination, which is more sensitive than SUS alone for diagnosis of testicular inflammation.
3. Decreased blood flow will be seen in cases of testicular torsion, with infarction appearing as early as 4 to 6 hours after symptoms of pain.
4. Microlithiasis of the testes and a mottled appearance of the seminiferous tubules in cases of testicular sclerosis and atrophy are newly recognized phenomena observed on ultrasound, the significance of which requires further investigation (Ragheb et al., 2002).

Notes

1. Doppler ultrasound can assess blood flow within the prepubertal testicle, allowing assessment of viability in the undescended testis and testicular torsion in neonates.
2. Epididymitis often has a less acute onset than testicular torsion.
3. The right testicle was found to be smaller in infertile patients with varicoceles (19 ± 8 mL) than in fertile men with varicoceles (25 ± 13 mL) or in fertile men without varicoceles (25 ± 11 mL).
4. The left testicle was found to be smaller in infertile men with varicoceles (18 ± 9 mL) than in fertile men with varicoceles (22 ± 8 mL) or in fertile men without varicoceles (23 ± 8 mL).
5. In one SUS study, 10 of 11 patients with acute epididymitis/orchitis had increased epididymal blood flow, and 8 of 11 had increased testicular blood flow (Tarantino et al., 2001).
6. Scrotal hydroceles may be anechoic or multiseptate.

ICD-9 Codes

Conditions that may justify this test *include but are not limited to*:

Neoplasms of male genital organs (testis and epididymis)
187.5   malignant
222.3   benign

Other
236.4   testicular mass/pain
606.X   male infertility
752.51   undescended testis

Suggested Reading

Munden MM, Trautwein LM. Scrotal pathology in pediatrics with sonographic imaging. *Curr Probl Diagn Radiol* 2000; 29:185–205.

Older RA, Watson LR. Ultrasound anatomy of the normal male reproductive tract. *J Clin Ultrasound* 1996; 24:389–404.

Pasqualotto FF, Lucon AM, de Goes PM, Sobreiro BP, Hallak J, Pasqualotto EB, Arap S. Semen profile, testicular volume, and hormonal levels in infertile patients with varicoceles compared with fertile men with and without varicoceles. *Fertil Steril* 2005; 83:74–7.

Ragheb D, Higgins JL. Ultrasonography of the scrotum: technique, anatomy, and pathologic entities. *J Ultrasound Med* 2002; 21:171–85.

Ralls PW, Jensen MC, Lee KP, Mayekawa DS, Johnson MB, Halls JM. Color Doppler sonography in acute epididymitis and orchitis. *J Clin Ultrasound* 1990; 18:383–6.

Tarantino L, Giorgio A, de Stefano G, Farella N. Echo color Doppler findings in postpubertal mumps epididymo-orchitis. *J Ultrasound Med* 2001; 20:1189–95.

Zahalsky M, Nagler HM. Ultrasound and infertility: diagnostic and therapeutic uses. *Curr Urol Rep* 2001; 2:437–42.

8.3.2   Human Chorionic Gonadotropin (hCG) and Alpha-Fetoprotein (AFP) Testing in Testicular Tumor Patients

Indications for Test

Measurement of hCG and AFP levels are indicated in the evaluation of men with:

- Gynecomastia
- Testicular masses
- Documented testicular cancer being monitored for its recurrence after therapy

Procedure

1. Obtain a blood specimen for *quantitative* measurement of serum [hCG] or a concentrated, nonacid-ified first morning urine specimen for *qualitative* measurement of [hCG].
2. For faster turnaround time of hCG test results, qualitatively measure [hCG] in a random blood or urine specimen.
3. If the [hCG] in the blood or urine specimens is increased, obtain a blood specimen for quantitative AFP determination.
4. Measure [hCG] and [AFP] at frequent intervals (e.g., weekly) after treatment of a marker-positive testicular tumor to calculate marker half-life (MHL).
5. If only two consecutive values are obtained, use Kohn's formula to determine the apparent half-life of the marker:

$$MHL = \ln(0.5/G)$$

   where G is the gradient (slope) of the change in marker concentration vs. time
6. If three or more marker values, obtained at intervals of 3 or more days over 2 to 4 weeks, are available, use simple linear regression analysis for calculation of MHL in days.

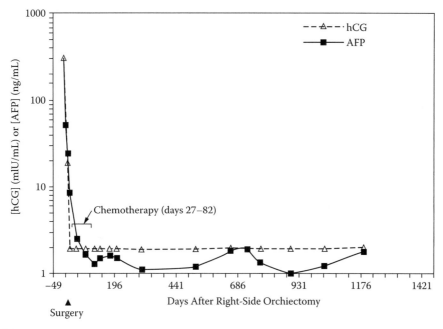

Figure 8.1
Changes in alpha-fetoprotein (AFP) and human chorionic gonadotropin (hCG) concentrations at 7 to 80 days after unilateral orchiectomy and chemotherapy from day 27 to 82 in a 23-year-old male with testicular cancer. Marker half-life (MHL) for hCG could not be assessed; however, normal value is ≤ 3.5 days. MHL for AFP, based on two consecutive values and Kohn's apparent half-life formula (MHL $= -0.693/M$, where M is the slope, was 3.2 days, a value considered normal (≤ 7 days). Marker decline is considered normal if both MHLs are within the previously mentioned limits or if one is within these limits and the other is not assessable.

Interpretation

1. HCG levels in healthy males are usually less than the lower limit of detection (LLD) of the hCG assay, typically <3 IU/L for screening tests and <0.5 IU/L in quantitative assays.
2. False positives may occur in patients with human anti-mouse antibodies (HAMAs), which result in markedly elevated hCG values.
3. If the [hCG] is $>10^5$ IU/L and on serial testing rises to $>3 \times 10^5$ IU/L, trophoblastic tumors (e.g., choriocarcinoma), lung cancer, hepatoma, or dysgerminoma in patients who have associated gynecomastia, precocious puberty, or elevated TSH (>1000 IU/L) may be present.
4. Do not rely on a qualitative screening hCG test for diagnosis of trophoblastic or nontrophoblastic lesions.
5. After therapy for testicular cancer, rising levels of AFP and hCG almost always mean recurrent disease.
6. The finding of a delayed rate of decline in a tumor marker after radio- or chemotherapy suggests that the malignancy is poorly responsive to the therapy used.
7. Tumor marker half-life is prolonged if it is >3.5 days for hCG and >7 days for AFP after surgical treatment (Figure 8.1). In such cases, a search for metastasis and chemotherapy are important considerations.

Notes

1. AFP and hCG measurements are useful in the management of nonseminomatous germ cell testicular tumors, but they should not be considered an adequate screen for the nature of a testicular mass.
2. In response to potentially curative therapy for nonseminomatous testicular tumors, a median marker half-life for AFP of 6.2 days (range, 2.6–65.4) and for hCG of 2.8 days (range, 0.7–16.7) was found. In another study, the median half-life for AFP was 3.9 days (range, 1.4–21.5) and 4.4 days (range, 1.4–21.0) for hCG (Inanc et al., 1999).

3. The histopathologically measured volume-weighted mean nuclear volume (MNV) was the only variable predicting lymph node metastasis in seminoma patients ($p = 0.0315$).

4. In Stage 1 (TxN0M0) patients with testicular seminomas, only the estimate of volume-weighted MNV was significantly correlated with progression-free survival ($p = 0.0118$).

5. Although the lactate dehydrogenase level and volume-weighted MNV were significantly greater in men with Stage 2 (TxN1-2M0) seminomas ($p = 0.001$), there were no significant differences between Stages 1 and 2 with respect to [β-hCG] ($p = 0.997$) and [AFP] ($p = 0.467$).

6. Testicular tumors are uncommon but occur most frequently in men between 15 and 35 years of age. Ninety percent of all testicular tumors originate from germ cells, the rest are of stromal origin.

7. A fourth of patients with persistently elevated AFP after orchiectomy have relapse of tumor.

8. Upon treatment with chemotherapy, about a quarter of patients with nonseminomatous testicular cancer will have a transient increase or surge in tumor markers. An hCG surge was of no prognostic importance for prediction of progression or survival, but an AFP surge was an adverse prognostic sign, independent of pretreatment characteristics (Oosterom et al., 1998).

9. Fluid from cystic germ cell teratomas of the testicle contain variably elevated levels of hCG and AFP which appear to be independent of serum [hCG] and [AFP] or tumor pathology.

10. Analysis of hCG and AFP in fluid from seminomatous and nonseminomatous hydroceles helped to classify patients as tumor marker positive or negative (Dorfinger et al., 1997).

ICD-9 Codes

Conditions that may justify this test *include but are not limited to*:

236.4	testicular mass/pain
611.1	gynecomastia

Suggested Reading

Bates SE, Longo DL. Tumor markers: value and limitations in the management of cancer patients. *Cancer Treatment Rev* 1985; 12:163–207.

Beck SD Patel MI, Sheinfeld J. Tumor marker levels in post-chemotherapy cystic masses: clinical implications for patients with germ cell tumors. *J Urol* 2004; 171:168–71.

de Wit R, Collette L, Sylvester R, de Mulder PH, Sleijfer DT, ten Bokkel Huinink WW, Kaye SB, van Oosterom AT, Boven E, Stoter G. Serum alpha-fetoprotein surge after the initiation of chemotherapy for non-seminomatous testicular cancer has an adverse prognostic significance. *Br J Cancer* 1998; 78:1350–5.

Dorfinger K, Kratzik C Madersbacher S, Dorfinger G, Berger P, Marberger M. Tumor markers in hydrocele fluids of patients with benign and malignant scrotal diseases. *J Urol* 1997; 158(3, Pt 1):851–5.

Fujikawa K, Matsui Y, Oka H, Fukuzawa S, Sasaki M, Takeuchi H. Prognosis of primary testicular seminoma: a report on 57 new cases. *Cancer Res* 2000; 60:2152–54.

Gerl A, Lamerz R, Clemm C Mann K, Hartenstein R, Wilmanns W. Does serum tumor marker half-life complement pretreatment risk stratification in metastatic nonseminomatous germ cell tumors? *Clin Cancer Res* 1996; 2:1565–70.

Gregory JJ, Finlay JL. Alpha-fetoprotein and beta-human chorionic gonadotropin: their clinical significance as tumour markers. *Drugs* 1999; 57:463–7.

Inanc SE, Meral R, Darendeliler E, Yasasever V, Onat H. Prognostic significance of marker half-life during chemotherapy in non-seminomatous germ cell testicular tumors. *Acta Oncol* 1999; 38:505–9.

Saxman SB, Nichols CR, Foster RS, Messemer JE, Donohue JP, Einhorn LH. The management of patients with clinical stage I nonseminomatous testicular tumors and persistently elevated serologic markers. *J Urol* 1996; 155:587–9.

Toner GC, Geller NL, Tan CC et al. Serum tumor marker half-life during chemotherapy allows early prediction of complete response and survival in nonseminomatous germ cell tumors. *Cancer Res* 1990; 50:5904–10.

8.3.3 Male Estrogens: Estrone (E1) and Total Estradiol (E2) Testing to Determine the Origin of Estrogens from Testicular Tissue, Nontesticular Tumors, or Exogenous Sources

Indications for Test

Measurement of [E1] and [E2] are indicated in males:

- Suspected of having increased estrogens from exogenous or endogenous sources
- With gynecomastia or feminization
- With liver cirrhosis or testicular tumors

Procedure

1. Obtain history of alcohol intake and any potential exposure to exogenous estrogens (e.g., intercourse with partner who uses vaginal estrogen cream).
2. Measure BMI (see Test 11.1.3) and obtain a random blood specimen for measurement of liver enzymes (i.e., ALT, AST), [E1], and [E2].
3. If prolonged (>2 weeks) exposure to exogenous estrogen is suspected, feminization is evident, or BMI is >30, obtain a blood specimen for FSH and LH testing, in addition to alanine aminotransferase (ALT), aspartate aminotransferase (AST), E1, and E2.
4. In patients with clearly elevated [E1] or [E2], obtain studies for assessment of liver diseases such as cirrhosis or hepatic steatosis or the presence of adrenal or testicular tumors (i.e., appropriate imaging, hCG levels).

Interpretation

1. Typically, reference intervals for estrogens in healthy males are:
    - E1, 10 to 50 pg/mL, with higher concentrations found in patients with adrenal tumors, obesity, or liver disease
    - E2, <70 pg/mL, with higher concentrations found in patients with hCG-producing gonadal tumors.
2. High [E1] or [E2] can occur with estrogen ingestion or cutaneous absorption, increased production in obesity (BMI > 30), or decreased clearance in liver disease and in patients with adrenal tumors, adrenal teratomas, or testicular tumors.

Notes

1. Estrogens are normally produced in fat cells by peripheral aromatization of androgens produced by the adrenal glands and testes.
2. Estrogens can feed back to the hypothalamus and inhibit secretion of gonadotropins, thereby decreasing androgen biosynthesis.
3. Epidemiologic studies suggest that men who develop breast cancer may have had elevated estradiol production or estrogen exposure.
4. There appears to be a survival advantage in men with established coronary artery disease (CAD) whose circulating concentrations of E1, dehydroisoandrosterone (DHA), dehydroisoandrosterone sulfate (DHAS), or androsterone glucuronide after a myocardial infarction (MI) are more elevated; however, men with a MI tend to have higher blood [E1], [DHA], and [DHAS] than men without such an event. Supplemental therapy with these hormones is not helpful in the treatment or prevention of CAD (Zumoff et al., 1982).
5. In a study of massive weight loss in hypogonadotropic hypogonad obese men, there was no significant change in plasma total [E2] (54–50 pg/mL), [free E2] (1.48–1.33 pg/mL), or [total E1] (75–82 pg/mL), although [SHBG] rose significantly from about 9 to 13 nmol/L (Strain et al., 1988).
6. Hypothalamic–pituitary function appears to change with weight loss in men, such that GnRH–gonadotropin secretion becomes less sensitive to suppression at any given level of estrogen.

ICD-9 Codes

Conditions that may justify this test *include but are not limited to*:

236.4	testicular mass/pain
571	chronic liver disease and cirrhosis
611.1	gynecomastia

Suggested Reading

Strain GW, Zumoff B, Miller LK, Rosner W, Levit C Kalin M, Hershcopf RJ, Rosenfeld RS. Effect of massive weight loss on hypothalamic–pituitary–gonadal function in obese men. *J Clin Endocrinol Metab* 1988; 66:1019–23.

Veldhuis JD Sowers JR, Rogol AD Klein FA, Miller N, Dufau ML. Pathophysiology of male hypogonadism associated with endogenous hyperestrogenism: evidence for dual defects in the gonadal axis. *N Engl J Med* 1985; 312:1371–5.

Zumoff B, Troxler RG, O'Connor J, Rosenfeld RS, Kream J, Levin J, Hickman JR, Sloan AM, Walker W, Cook RL, Fukushima DK. Abnormal hormone levels in men with coronary artery disease. *Arteriosclerosis* 1982; 2:58–67.

8.3.4   Prostate-Specific Antigen (PSA) as a Test to Screen for and Monitor Prostate Cancer (PCa)

Indications for Test

Determination of PSA concentration [PSA], in conjunction with a digital rectal exam (DRE), is indicated in the routine screening for PCa in:

- Caucasian men $\geq$ 50 years
- African-American men or Caucasian men $\geq$ 40 years with risk factors (e.g., a first-degree relative with or who has died of PCa)

Procedure

1. Obtain a blood sample for measurement of total [PSA] (and any of its forms or isoforms) in advance of the performance of a DRE or at least 3 days after the last DRE.
2. Because the ratio of free to total PSA (F/T) may improve the discrimination between benign prostatic hyperplasia (BPH) and PCa in men with total [PSA] of 4 to 10 ng/mL, obtain a follow-up blood sample for free PSA testing if the total [PSA] is in the 4- to 10-ng/mL range.
3. *Important:* Because of assay standardization issues that can lead to misleading results for F/T, measure both free and total [PSA] using assays from the same manufacturer.

Interpretation

1. The reference interval for [PSA] (and any of its forms or isoforms) in healthy individuals is assay dependent. Although a cutoff value of $\leq$4 ng/mL is used for many commercially available total PSA assays, a significant number of men with clinically significant PCa can have a total [PSA] of $\leq$4 ng/mL.
2. The prevalence of high-grade PCa increased from 12.5% of cancers associated with a [PSA] of $\leq$0.5 ng/mL to 25% of cancers associated with a [PSA] of 3.1 to 4.0 ng/mL (Thompson et al., 2004).
3. For many total PSA assays, values between 4 and 10 ng/mL are referred to as a diagnostic "gray zone" because 85% of men with total [PSA] in this range will have benign prostatic hyperplasia (BPH) and the remaining 15% can have clinically significant PCa.

4. Free/total PSA cutoff values of 25% or 27% have good diagnostic accuracy in discriminating between BPH and PCa (i.e., men with a total [PSA] of 4–10 ng/mL and an F/T value of ≥25%, or ≥27%) are more likely to have BPH than PCa. Those with an F/T value of <25%, or <27%, are more likely to have PCa than BPH.

5. Refer men with an abnormal DRE or PSA findings to an urologist for appropriate follow-up evaluation and testing (e.g., prostate biopsy or serial testing for [PSA]).

Notes

1. Out of 9459 men studied for 7 years, 2950 (31.2%) never had a [PSA] > 4 ng/mL or an abnormal mass on DRE, yet a prostate biopsy showed PCa in 449 out of the 2950 patients (15.2%) (Thompson et al., 2004).

2. Biopsy-detected PCa, including high-grade cancers (Gleason score ≥7), is not rare among men with [PSA] ≤ 4 ng/mL.

3. Typically, PCa begins in men in the second decade of life and steadily increases in incidence with age until the seventh decade, when about 80% of men may have some foci of PCa discovered on prostate biopsy.

4. Unfortunately, the PSA test, even in conjunction with a DRE, is not a very effective screening test for early-stage PCa, as it lacks both specificity and sensitivity for this disease. PSA testing has more value in monitoring PCa recurrence in men who have undergone a radical prostatectomy or have received chemo- or radiotherapy.

5. PCa with an elevated [PSA] is associated with high-fat, Western-style diets. Preventing and limiting the growth of prostate cancers, analogous to breast cancers, may be aided by the adoption of a low fat, plant-food-based diet.

ICD-9 Codes

Conditions that may justify these tests *include but are not limited to*:

599.60	obstructive uropathy
600.20	benign prostatic hyperplasia (BPH)
V10	personal and family history of malignant neoplasm

Suggested Reading

Park S, Cadeddu JA, Balko JA, Tortelli MW, Wians FH Jr. Persistently elevated prostate-specific antigen (PSA) level 4 months after successful laparoscopic radical prostatectomy in a 67-year-old man. *Lab Med* 2006; 37:474–7.

Roehrborn CG, Gregory A, McConnell JD, Sagalowsky AI, Wians FH Jr. Comparison of three assays for total serum prostate-specific antigen and percentage of free prostate-specific antigen in predicting prostate histology. *Urol* 1996; 48(6A):22–32.

Shariat SF, Roehrborn CG, Wians FH Jr. Update on prostate-specific antigen testing for the early diagnosis of prostate cancer. *ASCP Check Sample Clin Chem* 2003; 43:67–89.

Stamey TA, Caldwell M, McNeal JE, Nolley R, Hemenez M, Downs J. The prostate specific antigen (PSA) era in the United States is over for prostate cancer: what happened in the last 20 years. *J Urol*. 2004; 172:1297–301.

Thompson IM, Pauler DK, Goodman PJ, Tangen CM, Lucia MS, Parnes HL, Minasian LM, Ford LG, Lippman SM, Crawford ED, Crowley JJ, Coltman CA Jr. Prevalence of prostate cancer among men with a prostate-specific antigen level ≤4.0 ng/mL. *N Engl J Med* 2004; 350:2239–46.

Wians FH Jr. The "correct" PSA concentration. *Clin Chem* 1996; 42:1882–84.

Wians FH Jr. The role of prostate-specific antigen testing in the diagnosis, treatment, and follow-up of patients with adenocarcinoma of the prostate. *ASCP Check Sample Clin Chem* 1997; 37:77–103.

Wians FH Jr, Roehrborn CG. Human prostate gland: update on prostate cancer tests. *Dallas Med J* 1996; Dec:461–463.

Wians FH Jr, Cheli CD, Balko JA, Bruzek DJ, Chan DW, Sokoll LJ. Evaluation of the clinical performance of equimolar- and skewed-response total prostate-specific antigen assays versus complexed and free PSA assays and their ratios in discriminating between benign prostatic hyperplasia and prostate cancer. *Clin Chim Acta* 2002; 326:81–95.

Willis MS, Wians FH Jr. The role of nutrition in preventing prostate cancer: a review of the proposed mechanism of action of various dietary substances. *Clin Chim Acta* 2003; 328:1–27.

8.4   Sex Differentiation

8.4.1   Anogenital (AG) Ratio in the Evaluation of Ambiguous Genitalia of Neonates

Indications for Test

Calculation of the AG ratio is indicated:

- In the screening evaluation of female neonates with subtle anatomic abnormalities of the genitalia

Procedure

1. Apply this measurement to full-term and premature neonates (25 to 42 weeks' gestation).
2. Using a 0.1-cm calibrated, clear plastic ruler, measure the distance (AF) from the center of the anus (A) to the point of fusion of the labioscrotal folds (raphe) or fourchette (F) in mm.
3. Measure the distance (AC) from the center of the anus to the base of the clitoris (C) in mm.
4. Calculate the AG ratio = AF/AC.

Interpretation

1. The neonatal reference interval for the AG ratio is 0.23 to 0.51.
2. An AG ratio > 0.51 suggests an androgen-induced labioscrotal fusion abnormality in females.
3. Failure of normal labioscrotal fusion can occur in undervirilized males with hypoplastic penis. The AG ratio will usually be >0.5 in such individuals.
4. Abnormal fusion with an elevated AG ratio occurs in fetally masculinized females with clitoromegaly.

Notes

1. In pregnant adults, the reference interval for the AG ratio is 0.22 to 0.50.
2. Inhibition of aromatase activity caused by diesel exhaust inhalation resulted in significantly higher AG ratios in male and female rat fetuses. This finding suggests that toxicants in diesel exhaust may act on the feto–placental–ovarian unit to cause an accumulation of testosterone in the fetus.

ICD-9 Codes

Conditions that may justify this test *include but are not limited to*:

752   congenital anomalies of genital organs

Suggested Reading

Callegari C, Everett S, Ross M, Brasel JA. Anogenital ratio: measure of fetal virilization in premature and full-term newborn infants. *J Pediatr* 1987; 111:240–3.

Watanabe N, Kurita M. The masculinization of the fetus during pregnancy due to inhalation of diesel exhaust. *Environ Health Perspect* 2001; 109:111–9.

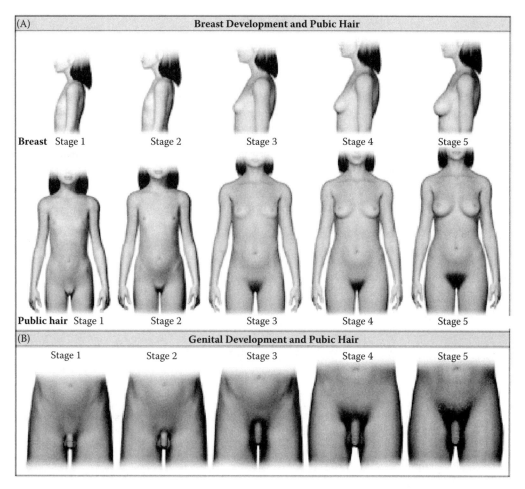

Figure 8.2
Pubertal rating according to Tanner stages, illustrating (A) breast and pubic hair development in girls (appearance of the breast bud marks the onset of pubertal development), and (B) genital development in boys (development stage 2 marks the onset of pubertal development, which is characterized by an enlargement of the scrotum and testes and by a change in the texture and a reddening of the scrotal skin). In normal boys, stage 2 pubic hair develops at an average of 12 to 20 months after stage 2 genital development. Pubic hair stage 2 marks the onset of pubic hair development in both sexes. (From Carel, J.-C. and Léger, J., *N. Engl. J. Med.*, 358, 2366–2377, 2008. With permission. Copyright © Massachusetts Medical Society.)

8.4.2 Determination of Secondary Sexual Characteristics and Pubertal (Tanner) Stage of Development in Boys and Girls

Indications for Test

The determination of pubertal stage in boys and girls is indicated for:

- Assessment of gonadal development and growth disorders (e.g., high growth velocity)
- Determination of precocious or delayed puberty

Procedure

1. Accurately identify and confirm the ethnicity and chronological age of the patient.
2. On clinical exam, observe and measure the following anatomic features (Figure 8.2):

TABLE 8.8
Causes of Precocious Puberty

Gonadotropin-Dependent (Both Sexes)	Gonadotropin-Independent	
	Boys	Girls
Idiopathic precocious puberty	*Testicular disorders (primary):*	*Ovarian disorders:*
Tumors of the central nervous system:	Familial male precocious puberty[a]	McCune–Albright syndrome
Craniopharyngioma	McCune–Albright syndrome	Granulosa or theca-cell tumors
Hypothalamic hamartoma	Leydig cell adenomas	Simple follicular cyst
Optic glioma, astrocytoma, and others	hCG-secreting tumors	Other estrogen-secreting tumors[c]
Nontumor central nervous system disorders:	Androgen-secreting teratomas	
Static encephalopathy[b]		
Low-dose cranial radiation		
Hydrocephalus		
Arachnoid cyst		
Septo-optic dysplasia		
Secondary causes of central precocious puberty:		
After prolonged delay in treatment of CAH		
Hypothyroidism		

[a] Testotoxicosis.

[b] Secondary to infection, hypoxia, trauma, etc.

[c] Teratomas, dysgerminomas.

Note: CAH, congenital adrenal hyperplasia; hCG, human chorionic gonadotropin.

- • Boys—size of the testes, scrotum, and penis, as well as the glans penis and coloration of the scrotum
- • Girls—size of the breasts, areola, and nipple papilla, as well as the contours and projections of each
- • Both sexes—the nature, distribution, texture, and curliness of pubic hairs

3. Pubic hair and genital or breast development stages are not necessarily synchronous and are to be scored separately.
4. Search for causes of gonadotropin-dependent and -independent precocious puberty as appropriate (Table 8.8). Check for central nervous system dysfunction including history of chronic headaches, visual impairment, seizures, and increased head circumference.
5. Search for causes of delayed puberty including delayed or impaired GnRH and/or gonadotropin secretion (e.g., CNS lesions, psychosocial dwarfism, hypothyroidism, Prader–Willi syndrome) and impaired gonadal function (e.g., toxin exposure, cryptorchidism, Klinefelter's syndrome).
6. Estimate effects of sex steroids on epiphyseal maturation by reference to atlas of bone age (e.g., Greulich and Pyle).

Interpretation

1. Tanner stages are identified in Table 8.9.
2. Onset of puberty is marked by Tanner stage 2 breast development in girls and testicular enlargement to volume of 4 mL or length of 25 mm in boys.
3. Age of onset of precocious puberty is controversial. The traditional pubertal threshold of 8 years for girls and 9 years, 6 months, for boys has been adjusted to as low as 6 years for African-American girls and 7 years for girls of other ethnicities, including Hispanics.

Notes

1. The peak of highest increment in height occurs at a mean age of 11.2 years in Europoid boys just after development of a Tanner stage 2 penis and scrotum at age 11.1 years.

TABLE 8.9
Identification of Tanner Stage Based on Secondary Sexual Characteristics in Boys and Girls

Tanner Stage	Description of Secondary Sexual Characteristics		
	Boys (Genitalia)	Girls (Breasts)	Both Sexes
1 (prepubertal)	Size and proportion of testes, scrotum, and penis same as in early childhood	Only papilla elevated	Vellus is not further developed over the pubes than the abdominal wall
2	Enlargement of the testes and scrotum; scrotum reddened and changed in texture	Breast bud stage; elevation of breast and papilla as a small mound; enlargement of the diameter of the areola	Sparse growth of long, slightly pigmented, downy hair, straight or curled, chiefly at the base of the penis or along the labia
3	Enlargement of the penis, at first mainly in length; further growth of the testes and scrotum	Further enlargement and elevation of the breasts and areola, with no separation of their contours	Considerably darker, coarser, and more curled hair; hair spreads sparsely over the junction of the pubes
4	Increased size of the penis with growth in the breadth and development of the glans; testes and scrotum larger; scrotal skin darkened	Projection of areola and papilla to form a secondary mound above the level of the breast	Pubic hair now adult in type but area covered by hair is still considerably smaller than in an adult; no spread of hair to the medial surface of the thighs
5	Genitalia adult in size and shape	Mature stage: projection of papilla only, related to recession of the areola to the general contour of the breast	Hair now adult in quantity and type with distribution of horizontal (or classically "feminine") pattern
6	—	—	Hair spreads up the linea alba (male-type pattern)

2. The peak of highest increment in height occurs at a mean age of 10.2 years in Europoid girls at onset of their progress toward sexual maturity before onset of Tanner stage 2 breast development at age 10.9 years.

3. Menarche occurs earlier in girls (mean age, 12.7 years) than does spermarche in boys (mean age, 13.8 years), marking the threshold for fertility in each sex.

4. Body habitus plays an important role in the onset of Tanner stage 2. Thin girls develop later (toward age 13 years), while thinner boys may develop slightly earlier.

5. African-American children tend to mature earlier than Mexican-American or non-Hispanic white children, but American children from these three ethnicities actually complete their sexual development at approximately the same ages (Sun et al., 2002).

6. Prevalence of precocious puberty is about 10 times as high in girls as in boys.

ICD-9 Codes

Conditions that may justify this test *include but are not limited to*:

Development and puberty
259.0   delayed sexual development
259.1   precocious sexual development
783.40   lack of normal physiological development, unspecified
783.43   short stature

Pituitary
253.0   acromegaly/gigantism
253.3   dwarfism
253.4   other anterior pituitary disorders

Suggested Reading

Carel J-C, Léger J. Precocious puberty. *N Engl J Med* 2008; 358:2366–77.

Greil H, Kahl H. Assessment of developmental age: cross-sectional analysis of secondary sexual characteristics. *Anthropol Anz* 2005; 63:63–75.

Sun SS, Schubert CM, Chumlea WC, Roche AF, Kulin HE, Lee PA, Himes JH, Ryan AS. National estimates of the timing of sexual maturation and racial differences among US children. *Pediatrics* 2002; 110:911–9.

Tanner JM, Whitehouse RH. Clinical longitudinal standards for height, weight, height velocity, weight velocity, and stages of puberty. *Arch Dis Child* 1976; 51:170–9.

8.4.3 Karyotype of Cells from Peripheral Circulation and Amniotic Fluid in Evaluation of Disorders of Sex Differentiation

Indications for Test

Determination of the gender-related karyotype is indicated:

- In the diagnosis of major alterations in chromosomes characteristic of a wide variety of congenital, gender-related endocrine syndromes involving the X and Y chromosomes

Procedure

1. In neonates and adults, obtain the appropriate peripheral blood specimen for the harvest of live leukocytes.
2. Obtain 5 mL of amniotic fluid and harvest live fetal cells for the prenatal diagnosis of genetic conditions if the locus of the variant is known or suspected based on kindred analysis.
3. Send these specimens to a cytogenetics lab for analysis. Typically, the analysis will involve:
   - Culturing of cells with phytohemagglutinin to stimulate cell division
   - Treating the culture with colchicine to arrest mitosis in metaphase
   - Subjecting the cells to hypotonic lysis to spread apart the chromosomes
   - Fixation and quinacrine or Giemsa staining and analysis of the banding and sizes of the chromosomes from at least 30 mitotic cells using 5 cells for screening and the rest for detailed analysis

Interpretation

1. The normal human karyotype is 46 XY (male) and 46 XX (female) or 22 pairs of autosomal and one pair of sex chromosomes.
2. Abnormalities that can be diagnosed with karyotyping include:
   - Klinefelter's syndrome (47, XXY and others)
   - Turner syndrome (45, X and others)
   - Androgen insensitivity (46, XY female)
   - Contiguous gene syndrome (X p21, adrenal hypoplasia, glycerol kinase deficiency)
3. Abnormalities caused by point mutations can have a normal karyotype.

Notes

1. Karyotyping from leukocytes or cells obtained from amniotic fluid is limited in that changes in an organ of interest (e.g., the gonad) may not be found in the cell sample studied.
2. Turner mosaicism with 45,X in skin fibroblasts was reported in a 55-year-old 46,XY female found to have an adnexal seminoma and an immature testis but previously normal menses. Her brother, with no impairment in fertility, had a karyotype—46,X,–Y,t(Y;15)(q12;p13)—in peripheral lymphocytes identical to that of his sister (Hoshi et al., 1998).

3. In cases of prenatal diagnosis of monozygotic female twins discordant for Turner syndrome, the phenotype of each twin is better predicted from karyotype analysis of cells from amniotic fluid than from fetal blood (Gilbert et al., 2002).
4. Genetic heterogeneity may be the cause of dual gonadal development in true hermaphroditism or, rarely, in 46,XX, Y-linked-gene-negative patients. Hidden mosaicism for the X and Y chromosomes, as a cause for intersexuality, can be detected by polymerase chain reaction (PCR) and fluorescence *in situ* hybridization (FISH) analysis of both peripheral blood cells and gonadal tissue (Nieto et al., 2004).
5. In Y-linked gene-positive hermaphrodites, an occult Y mosaicism, rather than an X–Y translocation, underlies the intersexuality disorder, while testis differentiation in Y-linked-gene-negative subjects may be caused by mutations of a gene on the X chromosome (Kojima et al., 1998).

ICD-9 Codes

Conditions that may justify this test *include but are not limited to*:

758.7 Klinefelter's syndrome (XXY)
795.2 nonspecific abnormal findings on chromosomal analysis

Suggested Reading

Gilbert B, Yardin C Briault S, Belin V, Lienhardt A, Aubard Y, Battin J, Servaud M, Philippe HJ, Lacombe D. Prenatal diagnosis of female monozygotic twins discordant for Turner syndrome: implications for prenatal genetic counselling. *Prenat Diagn* 2002; 22:697–702.

Hoshi N, Fujita M, Mikuni M, Fujino T, Okuyama K, Handa Y, Yamada H, Sagawa T, Hareyama H, Nakahori Y, Fujieda K, Kant JA, Nagashima K, Fujimoto S. Seminoma in a postmenopausal woman with a Y, 15 translocation in peripheral blood lymphocytes and a t(Y, 15)/45,X Turner mosaic pattern in skin fibroblasts. *J Med Genet* 1998; 35:852–6.

Kojima Y, Hayashi Y, Asai N, Maruyama T, Sasaki S, Kohri K. Detection of the sex-determining region of the Y chromosome in 46,XX true hermaphroditism. *Urol Int* 1998; 60:235–8.

Nieto K, Pena R, Palma I, Dorantes LM, Erana L, Alvarez R, Garcia-Cavazos R, Kofman-Alfaro S, Queipo G. 45,X/47,XXX/47,XX, del(Y)(p?)/46,XX mosaicism causing true hermaphroditism. *Am J Med Genet A* 2004; 130:311–4.

8.4.4 Imaging of Pelvic Structures with Ultrasonography (USG) in the Evaluation of Disorders of Sexual Differentiation

Indications for Test

Imaging of pelvic structures with USG scanning is indicated in individuals when:

- Grossly abnormal external genitalia are observed and/or gender assignment is uncertain.
- Subtle disturbances of the genitalia are seen.
- Fertility concerns are raised.
- Making a prenatal diagnosis of disorders of sexual differentiation.
- Family history of sexual differentiation disorders is obtained.

Procedure

1. Obtain an initial USG of the fetal pelvis at 13 to 15 weeks' gestational age (GA) and repeat USG at 22 to 24 weeks' GA.
2. Scan at a variety of angles to determine whether or not there is USG evidence for the presence of a penis or uterus.
3. In a nonpregnant female in whom a disorder of gonadal development is suspected, obtain a transvaginal ultrasound (see Test 9.5.2).

Interpretation

1. Ultrasound scan after approximately 19 weeks' GA enables detection of the uterus.
2. The presence of a uterus is diagnostic of a genetic female.
3. Repeat prenatal USG scans are a helpful tool in the prenatal diagnosis of sex differentiation disorders.
4. Both size and structural anomalies of the reproductive apparatus may evolve throughout fetal life, indicative of a developmental biological process rather than a single nonprogressive pathological event.

Notes

1. USG scanning is the modality of choice for screening patients with intersex disorders, but magnetic resonance (MR) is the preferred alternative test if USG is equivocal.
2. Females with severe, classic 21-hydroxylase (21OH) deficiency may be exposed to excess androgens prenatally and be born with virilized external genitalia, detectable with USG only in later stages of gestation, unless a mutation is detected early enough to permit prenatal treatment.
3. Out of approximately 10,000 gestations, 12 fetuses had abnormal USG imaging of the genitalia and 4 had a genotype–phenotype discrepancy. Of these 16 fetuses, 5 had female and 4 had male pseudohermaphroditism. Another 5 cases had chromosomal abnormalities, and 2 had 46,XX+ sex reversal found on analysis of the sex-determining region on the Y gene (Pinhas-Hamiel et al., 2002).
4. In Turner syndrome (i.e., gonadal failure associated with rudimentary gonads [95%] and pubertal delay), poor growth (98%), marked cubitus valgus (47%), and neck webbing (25%) are often observed on fetal USG.

ICD-9 Codes

Conditions that may justify this test *include but are not limited to*:

752.4	anomalies of cervix, vagina, and external female genitalia
752.49	other anomalies of cervix, vagina, and external female genitalia
758.6	gonadal dysgenesis

Suggested Reading

Biswas K, Kapoor A, Karak AK, Kriplani A, Gupta DK, Kucheria K, Ammini A. Imaging in intersex disorders. *J Pediatr Endocrinol Metab* 2004; 17:841–5.

Pinhas-Hamiel O, Zalel Y, Smith E, Mazkereth R, Aviram A, Lipitz S, Achiron R. Prenatal diagnosis of sex differentiation disorders: the role of fetal ultrasound. *J Clin Endocrinol Metab* 2002; 87:4547–53.

White PC Speiser PW. Congenital adrenal hyperplasia due to 21-hydroxylase deficiency. *Endocr Rev* 2000; 21:245–91.

8.4.5   Sex Hormone Binding Globulin (SHBG) Suppression Test in the Evaluation of Androgen Insensitivity Syndrome (AIS)

Indications for Test

SHBG suppression testing is indicated to:

*   Screen for male pseudohermaphrodites (XY karyotype) who may range in appearance from females born without a uterus to subfertile adult males.
*   Assess the *in vivo* biologic response to androgens in prepubertal patients with ambiguous genitalia, selecting those patients in whom it is worthwhile to perform second-level investigations to confirm a diagnosis of AIS.

TABLE 8.10
Classification of Androgen Insensitivity Syndrome (AIS) Based on Stanozolol Suppression of Sex Hormone Binding Globulin (SHBG)

AIS Type	Androgen Receptor Defect	Genitalia Characteristics	Reduction from Baseline [SHBG] (%)		
			Mean	Median	Range
I	None	Male/deformed	51.4	49.6	35.6–62.1
II	Mild	Predominately male	65.9	65.6	48.6–80.8
III	Moderate	Ambiguous	80.4	82.1	68.4–89.1
IV	Severe	Predominately female	83.8	83	81.3–87
V	Complete	Female *sans* uterus	102	97	92.4–129

Source: Sinnecker, G.H. et al., *Eur. J. Pediatr.*, 156, 7–14, 1997. With permission.

Procedure

1. Exclude from testing females with a uterus and infants who are <3 months of age whose blood SHBG level may be rising spontaneously.
2. To assess the *in vivo* biologic response to androgens in prepubertal patients with ambiguous genitalia, perform a screening hCG test as follows:
   - Collect blood specimen for SHBG testing prior to the administration of hCG (day 0).
   - Administer hCG i.m. (1500 IU for 3 consecutive days).
   - On day 5 after the administration of hCG, collect blood specimen for SHBG testing.
3. If the [SHBG] on day 5 compared to the [SHBG] on day 0 of the hCG test is essentially unchanged consistent with AIS, proceed with the *stanozolol suppression test*, as described below.
4. Obtain the androgenic steroid, stanozolol (17β-hydroxy-17α-methyl-5α-androstano-[3,2-*c*]pyrazol) (available from Sterling Research, Guildford, U.K.).
5. Obtain a blood specimen for SHBG testing (baseline, day 0).
6. Instruct the patient to administer the stanozolol (0.2 mg/kg/day) as a single oral dose at bedtime for 3 consecutive days.
7. Obtain blood specimens for SHBG testing on days 5, 6, 7, and 8 after administration of the first stanozolol dose on day 1.
8. Calculate the ratio of the lowest [SHBG] among the 4 blood samples collected on days 5, 6, 7, and 8 after the administration of stanozolol to the baseline [SHBG] as a percent:

$$\left([SHBG]_{lowest@day\,5-8} / [SHBG]_{day\,0}\right) \times 100$$

Interpretation

1. Refer to Table 8.10.
2. The degree of undermasculinization (AIS phenotype) and the SHBG response to stanozolol are closely correlated.

Notes

1. Girls with inguinal hernias should be karyotyped (see Test 8.4.3), as they are at high risk for AIS.
2. [SHBG] declines in response to administration of anabolic/androgenic steroids in a healthy reference population of individuals but not in individuals with AIS.
3. A reduction in SHBG fails to take place in AIS patients with androgen receptor (AR) mutations when they undergo treatment with androgens.
4. Different AR defects are associated with different degrees of reduction in [SHBG]. The SHBG suppression test helps to classify those with lesser or greater degrees of AIS who may be more or less responsive to androgen therapy.

5. The hCG test may be useful in deciding whom to select for definitive diagnostic testing for AIS. In 9 prepubertal patients with AIS, SHBG levels on day 5 after administration of hCG compared to day 0, prior to the administration of hCG, were essentially the same: 67.5 ± 18.6 nmol/L (day 5) and 66.2 ± 15.1 nmol/L (day 0); p = NS), representing a delta-variation of $1.7 \pm 12.7\%$ (Bertelloni et al., 1997).

6. In predominately female patients with partial AIS who underwent SHBG suppression testing, there was a lesser decrease in [SHBG] to about 80 to 87% (range, 92–129%) of baseline, indicating slight residual androgen responsiveness vs. complete AIS.

7. In individuals with ambiguous or predominately male AIS, the percent decline in SHBG levels over baseline after administration of stanozolol was 68 to 89% and 49 to 81%, respectively, while in control subjects the range was 32 to 65% (mean, 51%).

8. In 46,XY gonadal dysgenesis patients who were not on hormone replacement therapy (HRT), the mean nadir [SHBG] during the stanozolol suppression test was $52 \pm 9\%$, whereas in HRT control subjects it was significantly higher ($63 \pm 5\%$) (Krause et al., 2004).

9. Endogenous androgens and treatment with contraceptive medications interfere with the stanozolol-induced decrease in [SHBG].

ICD-9 Codes

Conditions that may justify this test *include but are not limited to*:

259.5   androgen insensitivity syndrome
752.7   indeterminate sex and pseudohermaphroditism screen for males
795.2   nonspecific abnormal findings on chromosomal analysis

Suggested Reading

Bertelloni S, Federico G, Baroncelli GI, Cavallo L, Corsello G, Liotta A, Rigon F, Saggese G. Biochemical selection of prepubertal patients with androgen insensitivity syndrome by sex hormone-binding globulin response to the human chorionic gonadotropin test. *Pediatr Res* 1997; 41:266–71.

Krause A, Sinnecker GH, Hiort O, Thamm B, Hoepffner W. Applicability of the SHBG androgen sensitivity test in the differential diagnosis of 46,XY gonadal dysgenesis, true hermaphroditism, and androgen insensitivity syndrome. *Exp Clin Endocrinol Diabetes* 2004; 112:236–40.

Sinnecker GH, Hiort O, Nitsche EM, Holterhus PM, Kruse K. Functional assessment and clinical classification of androgen sensitivity in patients with mutations of the androgen receptor gene: German Collaborative Intersex Study Group. *Eur J Pediatr* 1997; 156:7–14.

Chapter 9

Female Reproductive Organ Testing

9.1 Ovaries, Uterus, and Placenta

9.1.1 Progestin and Premarin®/Provera® (Conjugated Estrogen/Medroxyprogesterone) Challenge Tests for Amenorrhea

Indications for Test

The progestin and Premarin®/Provera® challenge tests are indicated in:

- Amenorrheic, nonpregnant females in whom a primary uterine disorder is suspected and in whom testing for remediable disorders (e.g., acromegaly, Cushing's syndrome, poorly controlled diabetes, hypothyroidism) is negative

Procedure

1. Review the patient's history for potential causes of secondary amenorrhea including anorexia nervosa, recent or severe weight loss, debilitating illness, psychological stress, or excessive exercise. Perform transvaginal ultrasonography (see Test 9.5.3) to estimate endometrial thickness if any of the above are noted.
2. If an endometrial thickness of ≤1.5 mm is found, the progestin challenge is not necessary because this finding alone predicts the absence of bleeding after a progestin challenge test with high diagnostic sensitivity (94%; 95% CI, 0.70–1.00), specificity (93%; 95% CI, 0.82–0.98), positive predictive value (79%), and negative predictive value (98%).
3. If Cushing's or Addison's disease is suspected as a cause for the patient's amenorrhea, instruct the patient to collect a 24-hour urine specimen for urine free cortisol determination (see Test 6.2.2).
4. In patients with an endometrial thickness > 1.5 mm, before initiating a progestin challenge test obtain a blood specimen for thyrotropin (TSH), free thyroxine (T_4), prolactin, insulin-like growth factor 1 (IGF-1), and human chorionic gonadotropin (hCG) testing. Be sure that the [hCG] is negative (<5 mIU/mL) before initiating the progestin challenge test.
5. When performing the progestin challenge test, administer 10 mg medroxyprogesterone per day orally for 5 days.

6. Perform the Premarin®/Provera® test only if a progestin challenge test was abnormal (i.e., no withdrawal vaginal bleeding is reported within 5 days of the last dose of medroxyprogesterone, indicating that the progesterone had no effect on the endometrium and no uterine lining sloughing occurred) and primary uterine failure is suspected.
7. Perform the Premarin®/Provera® challenge test by administering conjugated equine estrogens (2.5 mg orally for 21 days) followed by medroxyprogesterone (10 mg orally for 5 days) and instructing the patient to note the day of onset of withdrawal vaginal bleeding. As an alternative to oral progesterone administration, give a dose of 200 mg of progesterone in oil (not the contraceptive "depo" form) intramuscularly.

Interpretation

1. Withdrawal uterine bleeding within 5 days after progesterone administration is a normal response to the progestin and Premarin®/Provera® challenge tests.
2. In the progestin challenge test, failure to experience uterine bleeding is consistent with primary uterine failure as in Asherman's syndrome or the absence of endometrial development from lack of or blocked estrogenic stimulation as a result of hypothalamic–pituitary failure or dysfunction (e.g., overexercise in athletes), postcontraceptive syndrome, premature ovarian failure, hyperprolactinemia, androgen excess, and, less commonly, primary hypothyroidism.
3. Failure to respond to the Premarin®/Provera® test indicates primary uterine failure.

Notes

1. The progestin and Premarin®/Provera® challenge tests assess the integrity of the gonadotropin–ovarian–uterine axis.
2. Women who experience prolonged amenorrhea associated with overexercise or other causes of hypoestrogenism are at risk for almost irreparable loss of bone mineral and increased rate of stress fractures.
3. In women, Cushing's syndrome or psychological stress with an increase in cortisol secretion and chronically suppressed hypothalamic gonadotropin-releasing hormone (GnRH) will result in anovulatory cycles.
4. Episodic decrements in GnRH pulse generation are likely to result in luteal inadequacy or oligoovulation. The more complete the suppression of GnRH pulses (both in frequency and amplitude), the more likely is the reduction in endometrial thickness, anovulation, and reproductive compromise.

ICD-9 Codes

Conditions that may justify this test *include but are not limited to*:

Uterus
626.0   absence of menstruation
661.0   primary uterine disorder

Suggested Reading

Berga SL. Stress and amenorrhea. *Endocrinologist* 1995; 5:416–21.

Morcos RN, Leonard MD, Smith M, Bourguet C, Makii M, Khawli O. Vaginosonographic measurement of endometrial thickness in the evaluation of amenorrhea. *Fertil Steril* 1991; 55:543–6.

Patterson DF. Menstrual dysfunction in athletes: assessment and treatment. *Pediatr Nurs* 1995; 21:227–9, 310.

Speroff L, Glass RH, Kase NG. Amenorrhea. In: *Clinical Gynecologic Endocrinology and Infertility*, 3rd ed. Speroff, L (Ed). 1983. Williams & Wilkins, pp 145–9.

Vutyavanich T, Rugpao S, Uttavichai C, Tovanabutra S, Ittipankul W. Secondary amenorrhea at Maharaj Nakhon Chiang Mai Hospital. *J Med Assoc Thai* 1989; 72:160–6.

9.1.2 Ovulation Determination by Colorimetric (Dipstick) Test for Luteinizing Hormone (LH) in Urine

Indications for Test

Colorimetric dipstick tests to detect urinary [LH] are indicated:

- As a self test by the patient in self assessment of fertility or most fertile time of cycle
- To increase the chances of pregnancy by indicating the optimal time for intercourse or artificial insemination

Procedure

1. Three days before the anticipated LH surge at midcycle, determine the semiquantitative urine [LH] by immersing an LH dipstick into a first morning void urine specimen using either:
   - A one-step LH dipstick test (e.g., OvuQuick® One-Step, ClearPlan Easy®, or SureStep™), or
   - A multistep LH dipstick test (e.g., ClearPlan Fertility Monitor® or OvuQuick®)
2. Repeat the test every 12 hours on a series of random urine samples (typically, 6 to 10) and record the estimate of [LH] for all tests by comparison of the dipstick color change for each of the patient's urine samples against the color chart provided with the test strips that indicates the scale of intensity of color observed and the corresponding estimate of [LH].

Interpretation

1. A marked increase in estimated [LH] compared to a previous sample marks the point of the LH surge and an ovulatory event.
2. The sensitivity and specificity of this test for timing of ovulation depend on the specific dipstick test method (single or multistep) used; however, both single- and multistep urine dipstick LH methods are equally effective when used correctly (>90% agreement with quantitative serum LH measurements) in identifying the timing of the LH surge.
3. The most fertile period occurs within 1 to 2 days after the LH surge as marked by the peak in LH concentration estimated by the highest color intensity of the test strips observed during the monitoring of LH concentration in serial urine specimens.

Notes

1. Correlation between urinary dipstick LH concentration and diagnosis of ovulation based on transvaginal ultrasound examination of the ovary (see Test 9.5.2) approaches 100%.
2. Determinations of basal body temperature (BBT), salivary ferning, or changes in cervical/vaginal mucous (spinnbarkeit) as measures of ovulation are ≤33% as accurate as serial measures of urinary [LH] by dipstick done at midcycle.
3. Cervical mucous viscoelasticity with its inverse relationship to mucus penetrability over the course of the menstrual cycle is widely variable and tends to peak only in the ovulatory phase.
4. Detection of ovulation by vaginal mucous characteristics perceived by the patient or by salivary ferning were reported to have sensitivities for identifying the LH surge of only 48.3% and 36.8%, respectively (Guida et al., 1999).

ICD-9 Codes

Conditions that may justify this test *include but are not limited to*:

V22.2   Pregnancy, incidental

Suggested Reading

Behre HM, Kuhlage J, Gassner C, Sonntag B, Schem C, Schneider HP, Nieschlag E. Prediction of ovulation by urinary hormone measurements with the home use ClearPlan Fertility Monitor: comparison with transvaginal ultrasound scans and serum hormone measurements. *Hum Reprod* 2000; 15:2478–82.

Eichner SF, Timpe EM. Urinary-based ovulation and pregnancy: point-of-care testing—a review. *Ann Pharmacother* 2004; 38:325–31.

Guida M, Tommaselli GA, Palomba S, Pellicano M, Moccia G, Di Carlo C, Nappi C. Efficacy of methods for determining ovulation in a natural family planning program. *Fertil Steril* 1999; 72:900–4.

Nielsen MS, Barton SD, Hatasaka HH, Stanford JB. Comparison of several one-step home urinary luteinizing hormone detection test kits to OvuQuick. *Fertil Steril* 2001; 76:384–7.

Wolf DP, Blasco L, Khan MA, Litt M. Human cervical mucus. IV. Viscoelasticity and sperm penetrability during the ovulatory menstrual cycle. *Fertil Steril* 1978; 30:163–9.

9.1.3 Premenstrual Dysphoric Disorder (PMDD): Use of a Diagnostic Questionnaire Based on Criteria for the Diagnosis of PMDD from the Diagnostic Statistical Manual of Mental Disorders (DSM-IV)

Indications for Test

Use of a diagnostic questionnaire in the evaluation of women for PMDD is indicated:

- When disabling premenstrual symptoms occur in cycling women
- In women undergoing effective antidepressant therapy, such as selective serotonin reuptake inhibitor (SSRI) or GnRH with add-back estrogen therapy, who may be candidates for ovariectomy as a method to control PMDD

Procedure

1. Administer the screening "Menstrual Experience Questionnaire" (Q9.1 in Appendix 2).
2. Review all positive responses with patients to verify their responses.
3. Record other patient symptoms not covered by the questions in the questionnaire.
4. In cases with qualitatively severe signs and symptoms of PMDD as well as 10 or more positive responses based on Q9.1, refer the patient for psychologist/psychiatrist evaluation.

Interpretation

1. Diagnosis of PMDD is established the patient provides a minimum of 5 verified positive responses out of the 11 questions on the Menstrual Experience Questionnaire.

Notes

1. PMDD is not clearly associated with a well-identified set of hormonal disturbances different from those seen in normal ovarian function and in premenstrual syndrome (PMS).
2. Interactions between normal changes in menstrual cycle hormones and serotonin reuptake is thought to underlie both PMDD and PMS consistent with the hypothesis that brain serotonin level plays a major role in modulating sex-steroid-driven behavior.
3. The choice of treatment of apparent PMDD with SSRIs vs. a more aggressive approach using ovariectomy, usually after a trial of GnRH therapy with add-back estrogen but not progesterone therapy, depends on adequately differentiating severe PMDD from PMS.
4. Typically, PMDD symptoms appear in the luteal phase and diminish when the follicular phase begins then disappear during the week after menses.
5. From epidemiologic surveys, as many as 75% of women with regular menstrual cycles experience some symptoms of PMS. From 3 to 10% of these cases may have PMDD.

ICD-9 Codes

Conditions that may justify this test *include but are not limited to*:

> 625.4   premenstrual tension syndromes/premenstrual dysphoric disorder (PMDD)

Suggested Reading

Eriksson E, Andersch B, Ho HP, Landen M, Sundblad C. Diagnosis and treatment of premenstrual dysphoria. *J Clin Psychiatry* 2002; 63(Suppl 7):16–23.
Steiner M, Pearlstein T. Premenstrual dysphoria and the serotonin system: pathophysiology and treatment. *J Clin Psychiatry* 2000; 61(Suppl 12):17–21.

9.2   Female Infertility, the Menopause Transition, and the Polycystic Ovary Syndrome

9.2.1   Progesterone, Luteinizing Hormone (LH), and Follicle-Stimulating Hormone (FSH) Testing in Female Infertility and Menopause

Indications for Test

Measurement of LH and FSH is indicated:

- As part of the basic infertility evaluation
- For assessment of the integrity of the female gonadal–pituitary axis when there is evidence for or suspicion of a loss of estrogenization
- For possible centrally mediated amenorrhea or reduced fertility after cancer chemotherapy
- To diagnose early stages of ovarian involution or gonad failure before 40 years of age

Measurement of progesterone may be indicated:

- With a history of unexplained infertility or repeated miscarriages
- If luteal phase defect or inadequate luteal phase is suspected based on the main symptom of short or irregular menstrual cycles

Procedure

1. Obtain the patient's menstrual history. If no menses have occurred for at least 4 months, obtain a random blood specimen for LH and FSH testing.
2. Use a urinary LH test kit (see Test 9.1.2) to estimate the precise time of the midcycle ovulatory LH surge.
3. Seven days after midcycle, obtain a set of blood samples for progesterone testing or arrange to have an endometrial biopsy done. Do not perform either test if time of midcycle is uncertain.
4. Because progesterone is released in a pulsatile fashion, a single blood sampling may give a nondiagnostic result. Obtain three blood samples about 30 to 60 minutes apart in the midluteal phase to overcome the pulsatile nature of progesterone secretion, recognizing that this procedure is usually less expensive, painful, and inconvenient than an endometrial biopsy.
5. Perform an endometrial biopsy on the 12th day of a 14-day luteal phase as an alternative to progesterone testing.

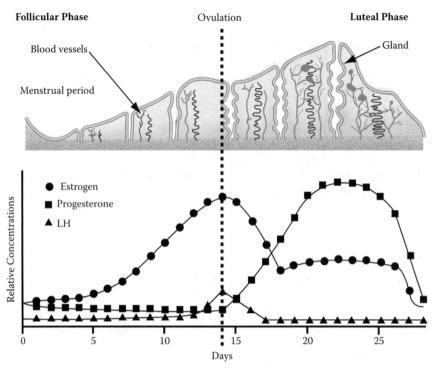

Figure 9.1

Changes in estrogen, progesterone, and luteinizing hormone (LH) during normal phases of the menstrual cycle and their effect on the endometrium. Rising levels of estrogen during the follicular phase thicken the endometrial lining of the uterus. The follicular phase varies in length, and the secretory (or luteal) phase lasts a predictable 14 days (i.e., the life span of the corpus luteum). Following ovulation, the mature ovarian follicle forms the corpus luteum, which produces progesterone. If pregnancy occurs, the production of progesterone from the corpus luteum continues for 7 weeks, directly related to the tonic release of LH from the pituitary gland, after which rising levels of human chorionic gonadotropin (hCG) from the placenta take on the LH function. If pregnancy does not occur, menses begins with the fall in progesterone levels as the corpus luteum involutes.

Interpretation

1. Serum concentrations for progesterone, LH, and FSH are dependent on the age of the patient, phase of menstrual cycle (Figure 9.1), and assay used (Table 9.1).
2. In a nonovulatory female, high [LH] and [FSH] suggest gonadal failure.
3. Women treated with conjugated equine estrogens have lower [LH] and [FSH] due to the negative feedback of estrogens on hypothalamic secretion of GnRH, but their levels remain above the premenopausal reference interval.
4. A high-dose hook effect can occur in some immunometric assays at high [LH] (e.g., ≥250 IU/L) and [FSH] (e.g., ≥350 IU/L).
5. Serum [progesterone] < 10 ng/mL in pooled sera or in 2 out of 3 specimens a week prior to the start of menstruation or 7 days after the LH surge is generally accepted as a diagnosis of luteal-phase defect.
6. Luteal-phase defects can be found in up to 30% of menstrual cycles of otherwise healthy women, with 6 to 10% of women who are fertile demonstrating an inadequate luteal phase.
7. Hypocholesterolemia, hyperprolactinemia, and hypothyroidism can all induce a luteal-phase defect.

Notes

1. The usefulness of gonadotropin and prolactin levels in the evaluation of infertility in a patient with normal menstrual cycles has not been demonstrated. The concept of luteal-phase defect is controversial, and its role in infertility is uncertain.
2. Infertile women with anorexia or bulimia have low [LH] and [FSH], indicating that improper nutrition affects ovulatory function via interference with hypothalamic–pituitary hormone secretion.

TABLE 9.1
Representative Serum Concentrations of Estradiol (E2), Luteinizing Hormone (LH), and Follicle-Stimulating Hormone (FSH)[a]

Group		E2 (pg/mL)	LH (IU/L)	FSH (IU/L)
Prepubertal (males and females)		<1.5	<1.0	<3
Adult males		0.8–3.5	1.0–9.0	1–13
Females				
Premenopausal		<400	—	—
Postmenopausal		<35	12–55	30–120
Menstruating females (phase and day)[b]				
Follicular	–12	11–69	1–18	2–12
	–4	63–165	—	—
Midcycle	–1	146–526	20–80	4–36
Luteal	+2	33–150	0.5–18	1–9
	+6	68–196	—	—
	+12	36–133	—	—

[a] Values provided are intended to serve as a guideline only; analyte values and units will vary by assays used in various laboratories.

[b] By day in cycle relative to ovulatory LH peak.

3. Anti-Müllerian hormone (AMH) is more sensitive and specific than FSH in identification of women in menopause transition (see Test 9.2.5) and is a useful adjunct to antral follicle count when estimating ovarian reserve.

4. An FSH/LH ratio at the start of stimulation with gonadotropins after pituitary downregulation is inversely correlated with a mature oocyte yield. Thus, for women undergoing assisted reproductive procedures, a baseline FSH/LH ratio ≥ 3 was shown to predict a poor response cycle (i.e., mature oocyte yield of ≤4 after administration of a GnRH agonist) (Ho et al., 2005).

5. FSH LH, estradiol, inhibin B, and AMH, as biochemical markers of ovarian reserve, cannot adequately predict poor outcomes in assisted reproduction, but they are of more help in predicting outcomes than assessments based on chronological age alone (Lutchman-Singh et al., 2005).

6. Inadequate secretion of LH results in a decrease in the substrate hormone [androstenedione] from the ovarian theca followed by a decrease in [estradiol] and subsequently lower [progesterone]. A suboptimal LH surge at ovulation causes deficient progesterone due to inadequately luteinized granulosa cells.

ICD-9 Codes

Conditions that may justify this test *include but are not limited to*:

253.2	pituitary hypogonadism
256.39	ovarian failure
626.0	absence of menstruation
628	infertility, female
752.7	indeterminate sex/pseudohermaphroditism

Suggested Reading

de Koning CH, Popp-Snijders C, Martens F, Lambalk CB. Falsely elevated follicle-stimulating hormone levels in women with regular menstrual cycles due to interference in immunoradiometric assay. *J Assist Reprod Genet* 2000; 17:457–9.

Ho JY, Guu HF, Yi YC, Chen MJ, Ho ES. The serum follicle-stimulating hormone-to-luteinizing hormone ratio at the start of stimulation with gonadotropins after pituitary down-regulation is inversely correlated with a mature oocyte yield and can predict "low responders." *Fertil Steril* 2005; 83:883–8.

Lutchman-Singh K, Davies M, Chatterjee R. Fertility in female cancer survivors: pathophysiology, preservation and the role of ovarian reserve testing. *Hum Reprod Update* 2005; 11:69–89.

Peters AJ, Wentz AC. Luteal phase inadequacy. *Semin Reprod Endo* 1995; 13:162–71.

Resch M, Szendei G, Haasz P. Eating disorders from a gynecologic and endocrinologic view: hormonal changes. *Fertil Steril* 2004; 81:1151–53.

Tremellen KP, Kolo M, Gilmore A, Lekamge DN. Anti-Müllerian hormone as a marker of ovarian reserve. *Aust NZ J Obstet Gynaecol* 2005; 45:20–4.

9.2.2   Female Estrogens—Estrone (E1), Estradiol (E2), and Estriol (E3)—in the Evaluation of Fertility and Ovarian Function

Indications for Test

Measurement of female estrogen levels are indicated in nonpregnant females:

- With amenorrhea or the absence of a normal menstrual cycle
- Who are undergoing ovulation induction and monitoring using E2 testing only
- With possible ovarian failure before age 40
- With precocious puberty (E2 testing only)

Procedure

1. Obtain a blood, urine, or saliva specimen for determination of [E2] and, as necessary, for determination of [E1] and/or [E3] (see Specimen Collection Protocol P2 in Appendix 2).
2. Use pooled specimens to detect [E1] and [E2] < 150 ng/L (Nelson et al., 2004).
3. Obtain a blood specimen for LH and FSH testing as a routine part of any fertility or menopause evaluation.

Interpretation

1. Normal values for E2 are dependent on the age of the patient, phase of menstrual cycle, and E2 assay used (Table 9.1)
2. In the noncastrate, a relatively low but detectable [E2] of 5 to 20 ng/L may be found for up to the first 10 years after menopause.
3. Females not on oral or transdermal estrogen replacement therapy (ERT) with well-established postmenopausal states typically have elevated [LH] (>10 IU/L) and [FSH] (>20 IU/L).
4. Although conjugated equine estrogens (CEEs) are not measured by the E2, E1, or total estrogen assays, treatment with any of these agents may be expected to reduce gonadotropin levels.
5. [LH] or [FSH] > 200 IU/L should raise suspicion of a gonadotropin-secreting tumor, usually within the pituitary.
6. An [E2] > 200 pg/mL for 2 days prior to the LH surge is required for ovulation.
7. High [E2] (>100 pg/mL) and normal [FSH] (<10 IU/L) on days 2 to 4 of the menstrual cycle are predictive of ovulatory dysfunction and poor response to treatments for infertility.
8. If the [LH] and [FSH] reported by the laboratory are unexpectedly low or are not consistent with the patient's clinical and other laboratory findings, consider the high-dose hook effect as the possible cause of this discrepancy. Request that laboratory personnel perform repeat testing for LH and/or FSH using suitable dilutions of the patient's serum sample.

Notes

1. Circulating concentrations of E2, LH, and FSH vary in a complex negative and positive feedback manner during the normal menstrual cycle in a predictable fashion.
2. E2 production is stimulated by LH, which in turn is stimulated by GnRH.

3. Chronically elevated [E2] has an inhibitory effect on LH production at the level of the pituitary.
4. Measurement of [E2] in blood or urine is of limited use in the diagnosis of menopause or evaluating the status of a high-risk pregnancy.
5. [E2] increases several hundredfold during a normal pregnancy, with its measurement having no known clinical usefulness in this condition.
6. The production of E2 from the ovaries in early postmenopause is substantial (15 to 20 mcg/day).
7. Estrogen replacement therapy should not be expected to suppress [LH] and [FSH] to normal or raise the [E2] to >200 pg/mL if bioidentical E2 is used as ERT. Little to no evidence (Boothby et al., 2004) supports individualized bioidentical hormone dosing based on salivary hormone concentrations.
8. Among women undergoing coronary angiography for suspected myocardial ischemia, the use of cholesterol-lowering agents or lower cholesterol levels themselves are not associated with significantly lower total [E2], bioavailable [E2], [E1], or progesterone.
9. [E2] (71 ± 52 pg/mL) were not significantly lower among premenopausal women with very low LDL cholesterol levels compared to [E2] (88 ± 67 pg/mL) in women with higher LDL cholesterol levels ($p = 0.32$).
10. In spite of the expected increase in conversion of androstenedione to E1 with obesity, there is no significant hyperestrogenemia in obese females, indicating that obesity-related estrogen is minor and statistically undetectable compared to E1 from normal ovarian sources.
11. Follicular cyst fluid contains high [E3] with levels independent of phase of ovulation and [E2].
12. In "triple test" studies, only maternal serum [hCG], not [AFP] or unconjugated [E3], was higher in pregnant *in vitro* fertilization patients as compared to those with spontaneous pregnancies.
13. Retrospective studies have associated oral E2 replacement therapy at a 1- to 2-mg/day dose with increased risk of breast cancer possibly related to concomitant urinary [E1] 5 to 10 times the upper limit of the reference interval for [E1] in serum from premenopausal women.

ICD-9 Codes

Conditions that may justify this test *include but are not limited to*:

256	ovarian dysfunction
256.39	ovarian failure
259.1	precocious sexual development and puberty (E2 testing only)
626.0	absence of menstruation/amenorrhea

Suggested Reading

Bairey Merz CN, Olson MB, Johnson BD, Bittner V, Hodgson TK, Berga SL, Braunstein GD, Pepine CJ, Reis SE, Sopko G, Kelsey SF. Cholesterol-lowering medication, cholesterol level, and reproductive hormones in women: the Women's Ischemia Syndrome Evaluation (WISE). *Am J Med* 2002; 113:723–7.

Bar-Hava I, Yitzhak M, Krissi H, Shohat M, Shalev J, Czitron B, Ben-Rafael Z, Orvieto R. Triple-test screening in *in vitro* fertilization pregnancies. *J Assist Reprod Genet* 2001; 18:226–9.

Boothby LA, Doering PL, Kipersztok S. Bioidentical hormone therapy: a review. *Menopause* 2004; 11:356–67.

Nelson RE, Grebe SK, OKane DJ, Singh RJ. Liquid chromatography–tandem mass spectrometry assay for simultaneous measurement of estradiol and estrone in human plasma. *Clin Chem* 2004; 50:373–84.

Su P, Zhang XX, Chang WB. Development and application of a multi-target immunoaffinity column for the selective extraction of natural estrogens from pregnant women's urine samples by capillary electrophoresis. *J Chromatogr B Analyt Technol Biomed Life Sci* 2005; 816:7–14.

Vermes I, Bonte HA, v d Sluijs Veer G, Schoemaker J. Interpretations of five monoclonal immunoassays of lutropin and follitropin: effects of normalization with WHO standard. *Clin Chem* 1991; 37:415–21.

Zumoff B, Strain GW, Kream J, O'Connor J, Levin J, Fukushima DK. Obese young men have elevated plasma estrogen levels but obese premenopausal women do not. *Metabolism* 1981; 30:1011–4.

9.2.3   The LH, FSH, LH/FSH Ratio, and Androgens in the Evaluation of Female Infertility and Diagnosis of Polycystic Ovary Syndrome (PCOS)

Indications for Test

Calculation of the LH/FSH ratio and total testosterone (TT) and delta 4-androstenedione (Δ^4AD) testing are indicated in females when:

- Both infertility and androgenic changes are associated with ovarian abnormalities (i.e., ovulatory dysfunction, hirsutism, hyperandrogenemia, and polycystic ovaries).

Procedure

1. Obtain the patient's menstrual history. Assess for degree of hirsutism, recognizing that visual scales are subjective with significant interobserver variation (refer to Test 9.4.1).
2. Examine the ovaries bimanually and image them with transvaginal ultrasound (see Test 9.5.2) to assess for masses and cysts.
3. Obtain a blood sample for [FSH] and [LH].
4. In patients suspected of PCOS, obtain a blood sample for TT and Δ^4AD testing.

Interpretation

1. LH/FSH ratios in serum from normal premenopausal women vary widely.
2. In general, patients with PCOS have a chronically elevated LH/FSH ratio.
3. An LH/FSH ratio > 3, obtained using assays for LH and FSH available before 1990 for diagnosing patients with PCOS, is not valid when current monoclonal assays are used to quantify [LH] and [FSH].
4. In 1991, [LH] by radioimmunoassay (RIA) gave consistently higher results than [LH] obtained using an immunoradiometric assay (IRMA) and, consequently, higher LH/FSH ratios compared to those obtained when [LH] was quantified by IRMA.
5. The high-dose hook effect may occur in some immunometric assays at [LH] ≥ 250 IU/L but not in samples containing [FSH] up to 350 IU/L.
6. The reference interval for [Δ^4AD] in premenopausal women using commonly available assays is 65 to 270 ng/dL; in postmenopausal women, [Δ^4AD] is usually <150 ng/dL (Marshall et al., 1977).
7. Up to 15% of patients with PCOS have elevated [Δ^4AD] (>270 ng/dL) even if the [TT] is not elevated.

Notes

1. Normally, Δ^4AD is produced in approximately equal amounts by both the adrenal glands and ovaries.
2. Prolactin levels may be mildly elevated (15 to 50 ng/mL) in 25 to 30% of patients with PCOS.
3. [FSH] is a poor predictor of the onset of menopause.
4. Acne or androgenic alopecia cannot be used reliably as a clinical sign of hyperandrogenism.

ICD-9 Codes

Conditions that may justify this test *include but are not limited to*:

256.39   ovarian failure
628   infertility, female

Suggested Reading

Azziz R, Carmina E, Dewailly D, Diamanti-Kandarakis E, Escobar-Morreale HF, Futterweit W, Janssen OE, Legro RS, Norman RJ, Taylor AE, Witchel SF. Position statement: criteria for defining polycystic ovary syndrome as a predominantly hyperandrogenic syndrome: an Androgen Excess Society Guideline. *J Clin Endocrinol Metab* 2006;91; 4237–45.

Fauser BC, Pache TD, Lamberts SW, Hop WC, de Jong FH, Dahl KD Serum bioactive and immunoreactive luteinizing hormone and follicle-stimulating hormone levels in women with cycle abnormalities, with or without polycystic ovarian disease. *J Clin Endocrinol Metab* 1991; 73:811–7.

Marshall DH, Crilly RG, Nordin BEC. Plasma androstenedione and oestrone levels in normal and osteoporotic postmenopausal women. *Br Med J* 1977; Nov 5:1177–79.

Vermes I, Bonte HA, v d Sluijs Veer G, Schoemaker J. Interpretations of five monoclonal immunoassays of lutropin and follitropin: effects of normalization with WHO standard. *Clin Chem* 1991; 37:415–21.

9.2.4 GnRH Agonist (Buserelin) Stimulation Test (BST) for Diagnosis of PCOS

Indications for Test

The BST is indicated for:

- Outpatient evaluation of obese and hirsute women with possible PCOS

Procedure

1. On day 4 or 5 of spontaneous or progestin-induced menses, obtain a morning baseline blood specimen sample for measurement of [LH], [FSH], LH/FSH ratio, 17-hydroxyprogesterone (17OH-Prog), estradiol (E2), and delta 4-androstenedione [Δ^4AD].
2. After blood collection, inject a GnRH agonist (1 mg buserelin s.q.)
3. Twenty-four hours later, obtain a second blood sample for determination of the same analytes tested in the baseline blood specimen.
4. Calculate changes in hormone levels before and after administration of GnRH agonist.

Interpretation

1. In women with PCOS vs. healthy control subjects, changes in hormone levels between baseline and 24-hour blood specimens after buserelin injection were as follows (Hagag et al., 2000):
   - LH/FSH ratio rose from 1.5 to 8.3 (PCOS) vs. 0.8 to 2.8 (control subjects).
   - [17OH-Prog] rose on average from 102 to 357 ng/dL (3.1–10.8 nmol/L) (PCOS) vs. 50 to 102 ng/dL (1.5–3.1 nmol/L) (control subjects).
   - [E2] rose on average from 73 to 403 pg/mL (267–403 pmol/L) (PCOS) vs. 60 to 138 pg/mL (222–508 pmol/L) (control subjects).
   - [Δ^4AD] rose from 200 to 304 ng/dL (7–10.6 nmol/L) (PCOS) vs. 106 to 163 ng/dL (3.7–5.7 nmol/L) (control subjects).

Notes

1. The specificity and positive predictive value of the BST were 100%, but its sensitivity and negative predictive values were lower, at 50 to 60%.
2. Objective criteria for the definitive diagnosis of PCOS have not been standardized; therefore, the interpretive hormone data presented above are intended to be representative of the changes in hormone levels that can occur in women with PCOS vs. healthy control subjects. Note the similar [17OH-Prog] response to injection of the GnRH agonist triptorelene as with buserelin (Figure 9.2).
3. The presence of multiple ovarian cysts, high basal levels of LH, or LH hyper-responsiveness to GnRH are not constant or specific for PCOS.
4. The hormone response to administration of 1 mg buserelin may be affected by obesity but not insulin resistance.

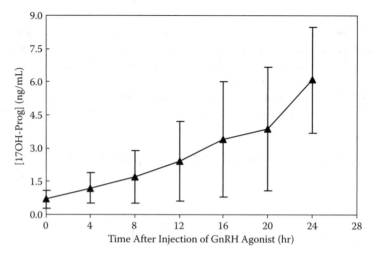

Figure 9.2
17-Hydroxyprogesterone (17OH-Prog) response to the gonadotropin-releasing hormone (GnRH) agonist triptoreline in patients with polycystic ovarian disorder (PCOD).

5. After administration of GnRH agonist leuprolide acetate, ovulatory women without clinical/biochemical hyperandrogenism, but with polycystic-appearing ovaries (PCOs), had response patterns in LH and 17OH-Prog that were intermediate between ovulatory women with normal appearing ovaries and women with PCOS.

6. About a third of ovulatory women with PCOs have a high-density lipoprotein (HDL) concentration < 35 mg/dL and a reduced glucose lowering response to insulin suggestive of mild insulin resistance.

ICD-9 Codes

Conditions that may justify this test *include but are not limited to*:

255.2	adrenogenital disorders/virilization
256.4	polycystic ovaries
278.00	obesity
704.1	hirsutism

Suggested Reading

Chang PL, Lindheim SR, Lowre C, Ferin M, Gonzalez F, Berglund L, Carmina E, Sauer MV, Lobo RA. Normal ovulatory women with polycystic ovaries have hyperandrogenic pituitary–ovarian responses to gonadotropin-releasing hormone-agonist testing. *J Clin Endocrinol Metab* 2000; 85:995–1000.

Hagag P, Ben-Shlomo A, Herzianu I, Weiss M. Diagnosis of polycystic ovary disease in obese women with a 24-hour hormone profile after buserelin stimulation. *J Reprod Med* 2000; 45:171–8.

9.2.5   Anti-Müllerian Hormone (AMH) in the Determination of Ovarian Reserve and the Status of the Menopause Transition

Indications for Test

Measurement of [AMH] is indicated for:

- Identification of the transition from normal ovulation to ovarian failure, as in menopause or in women with symptoms of menopause or perimenopause
- Fertility evaluation and quantitative determination of ovarian follicular reserve preliminary to *in vitro* fertilization procedures in non-PCOS patients

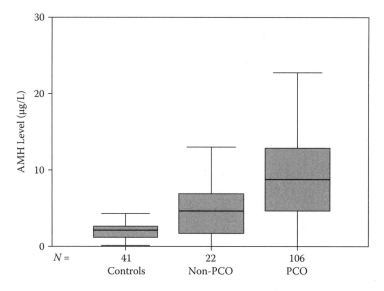

Figure 9.3

Changes in serum concentrations of anti-Müllerian hormone (AMH) in normo-ovulatory (control group) and anovulatory women—polycystic ovarian disorder (PCOD) and nonpolycystic ovarian groups—of reproductive age. AMH concentrations are increased in normogonadotropic anovulatory infertile women with or without PCO disorder compared to normally fertile control women. (From Laven, J.S. et al., *J. Clin. Endocrinol. Metab.*, 89, 318–323, 2004. With permission.)

Procedure

1. Note the patient's age, symptoms of hot flashes, length and predictability of menstrual cycle, and any stigmata of polycystic ovary syndrome (PCOS), including obesity, hirsutism, or infertility.
2. Obtain a blood sample for [AMH] regardless of the patient's phase of the menstrual cycle.

Interpretation

1. [AMH] (median; range) was significantly elevated ($p < 0.001$) in normogonadotropic anovulatory infertile women (7.6 mcg/L; 0.1–40.0) compared with controls (2.1 mcg/L; 0.1–7.4) and even more elevated ($p < 0.0001$) in the subset of patients with PCOS (9.3 mcg/L; 1.8–40.0) (Figure 9.3).
2. Baseline [AMH] does not predict response to *in vitro* fertilization procedures in PCOS patients.
3. [AMH] correlates negatively with age in control subjects ($r = -0.465$; $p = 0.005$).
4. [AMH] remains relatively static at a median of 2.8 to 3.5 mcg/L (20–25 pmol/L) from 18 to 29 years of age and tends to fall rapidly from age 30 to 37, reaching 50% (i.e., ~1.4 mcg/L or 10 pmol/L) of the level seen prior to age 30 (Figure 9.4).
5. Poor response to *in vitro* fertilization procedures, suggestive of a diminished ovarian reserve, is associated with reduced baseline [AMH].
6. AMH helps predict the occurrence of menopause transition within a 4-year interval. Adding the results of inhibin B measurement improves this prediction (refer to notes 4 and 5 below).

Notes

1. AMH is a member of the transforming growth factor beta (TGFβ) family of growth and differentiation factors and has an inhibitory effect on primordial follicle recruitment as well as on the responsiveness of growing follicles to FSH.
2. AMH is a homodimeric disulfide-linked glycoprotein with a molecular weight of 140 kDa. As mapped by Cohen-Haguenauer and reported in 1987, its gene is located on the short arm of chromosome 19 in humans, band 19p13.3. In males, AMH is strongly expressed in Sertoli cells from testicular differentiation up to puberty.

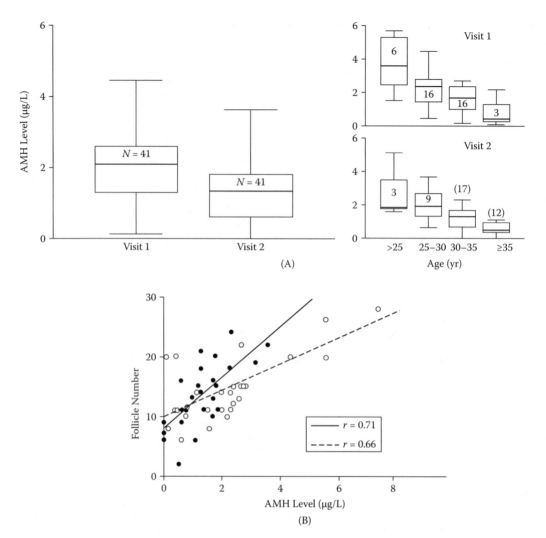

Figure 9.4
(A) Serum concentrations of anti-Müllerian hormone (AMH) in a group of healthy women between 2 visits, 2.6 ± 1.7 years apart (left panel), and as a function of the age of these women at each visit (right panel). (B) Correlation of AMH levels with number of antral follicles (O, visit 1; ●, visit 2). (From de Vet, A. et al., *Fertil. Steril.*, 77. 357–362, 2002. With permission.)

3. Because of its unique production by follicles during their various stages of development, AMH is a marker of ovarian follicular status.
4. In normo-ovulatory, infertile women, repeated measurements of [AMH] at cycle day 3 were more reproducible (intraclass correlation coefficient, 0.89; 95% CI, 0.83–0.94) than inhibin B (0.76; 0.66–0.86; $p < 0.03$), estradiol (0.22; 0.03–0.41; $p < 0.0001$), FSH levels (0.55; 0.39–0.71; $p < 0.01$), and early antral follicle counts (0.73; 0.62–0.84; $p < 0.001$) (Fanchin et al., 2003, 2005).
5. FSH, LH, estradiol, inhibin B, and AMH, as biochemical markers of ovarian reserve, cannot adequately predict poor outcomes in assisted reproduction, but they are of more help in predicting outcomes than assessments based on chronological age alone.
6. [AMH] in 119 patients was positively correlated with the number of antral follicles ($r = 0.77$; $p < 0.01$) and the number of oocytes retrieved ($r = 0.57$; $p < 0.01$) after GnRH agonist stimulation (van Rooij et al., 2002).
7. Using an AMH cutoff value of 1.13 mcg/L (8.1 pmol/L), AMH assessment predicted poor ovarian reserve on a subsequent *in vitro* fertilization cycle, with a sensitivity of 80% and a specificity of 85% (Tremellen et al., 2005).

8. [AMH] is superior to [FSH] in identifying women with reduced ovarian reserve. Note that only minimal changes in [FSH] have been observed in the 30 to 37 age group.

9. For women undergoing assisted reproductive procedures, a baseline FSH/LH ratio of ≥3 was predictive of a poor response cycle (i.e., mature oocyte yield of ≤4).

10. In female cancer treatment survivors, total antral follicle count by transvaginal ultrasound (see Test 9.5.2) appeared to be the most discriminatory test of ovarian reserve, followed by ovarian volume (Lutchman-Singh et al., 2005).

11. The menopause transition may be defined as a mean cycle length of <21 days or >35 days or as a mean cycle length of 21 to 35 days but with the next cycle not predictable within 7 days during the prior 6 months.

12. [AMH] in PCOS patients was significantly higher than in healthy women (5.49 ± 2.26 SD and 1.93 ± 0.51 mcg/L, respectively; $p = 0.001$). No significant changes in [AMH] occurred after six cycles of oral contraceptive therapy in either group (Somunkiran et al., 2007).

13. [AMH] was 65% lower in late-reproductive-age obese women compared to nonobese women (Freeman et al., 2007).

ICD-9 Codes

Conditions that may justify this test *include but are not limited to*

Ovary
256.31   premature menopause
256.39   other ovarian failure
256.4   polycystic ovary syndrome (PCOS)

Other
V49.81   menopause
627.2   postmenopause

Suggested Reading

de Koning CH, Popp-Snijders C, Martens F, Lambalk CB. Falsely elevated follicle-stimulating hormone levels in women with regular menstrual cycles due to interference in immunoradiometric assay. *J Assist Reprod Genet* 2000; 17:457–9.

de Vet A, Laven JS, de Jong FH, Themmen APN, Fauser BC. Anti-Müllerian hormone serum levels: a putative marker for ovarian aging. *Fertil Steril* 2002; 77:357–62.

Fanchin R, Schonauer LM, Righini C, Guibourdenche J, Frydman R, Taieb J. Serum anti-Müllerian hormone is more strongly related to ovarian follicular status than serum inhibin B, estradiol, FSH, and LH on day 3. *Hum Reprod* 2003; 18:323–7.

Fanchin R, Taieb J, Lozano DH, Ducot B, Frydman R, Bouyer J. High reproducibility of serum anti-Müllerian hormone measurements suggests a multi-staged follicular secretion and strengthens its role in the assessment of ovarian follicular status. *Hum Reprod* 2005; 20:923–7.

Freeman EW, Gracia CR, Sammel MD, Lin H, Lim LC, Strauss JF 3rd. Association of anti-Müllerian hormone levels with obesity in late reproductive-age women. *Fertil Steril* 2007; 87:101–6.

Ho JY, Guu HF, Yi YC, Chen MJ, Ho ES. The serum follicle-stimulating hormone-to-luteinizing hormone ratio at the start of stimulation with gonadotropins after pituitary down-regulation is inversely correlated with a mature oocyte yield and can predict "low responders." *Fertil Steril* 2005; 83:883–8.

Laven JS, Mulders AG, Visser JA, Themmen AP, De Jong FH, Fauser BC. Anti-Müllerian hormone serum concentrations in normoovulatory and anovulatory women of reproductive age. *J Clin Endocrinol Metab* 2004; 89:318–23.

Lutchman-Singh K, Davies M, Chatterjee R. Fertility in female cancer survivors: pathophysiology, preservation and the role of ovarian reserve testing. *Hum Reprod Update* 2005; 11:69–89.

Somunkiran A, Yavuz T, Yucel O, Ozdemir I. Anti-Müllerian hormone levels during hormonal contraception in women with polycystic ovary syndrome. *Eur J Obstet Gynecol Reprod Biol* 2007; 134:196–201.

Tremellen KP, Kolo M, Gilmore A, Lekamge DN. Anti-Müllerian hormone as a marker of ovarian reserve *Aust NZ J Obstet Gynaecol* 2005; 45:20–4.

van Rooij IA, Broekmans FJ, te Velde ER, Fauser BC, Bancsi LF, de Jong FH, Themmen AP. Serum anti-Müllerian hormone levels: a novel measure of ovarian reserve. *Hum Reprod* 2002; 17:3065–71.

van Rooij IA, Tonkelaar I, Broekmans FJ, Looman CW, Scheffer GJ, de Jong FH, Themmen AP, te Velde ER. Anti-Müllerian hormone is a promising predictor for the occurrence of the menopausal transition. *Menopause* 2004; 11(6 Pt 1):601–6.

Visser JA, de Jong FH, Laven JSE, Themmen APN. Anti-Müllerian hormone: a new marker for ovarian function. *Reproduction* 2006; 131:1–9.

9.3   Pregnancy

9.3.1   Human Chorionic Gonadotropin (hCG) and Alpha-Fetoprotein (AFP) in Pregnancy and Trophoblastic Diseases

Indications for Test

Testing for hCG is indicated for:

- Early detection of pregnancy prior to surgery, radiologic procedure, or other therapeutic or diagnostic intervention
- Monitoring the progress of a threatened miscarriage

Testing for AFP is indicated:

- In conjunction with testing for hCG + unconjugated estriol (uE3) (i.e., triple screen of AFP + hCG + uE3) and dimeric inhibin A (DIA) (i.e., quad screen of AFP + hCG + uE3 + DIA) in the routine assessment of maternal risk for fetal open neural tube defects (open spina bifida or anencephaly) and Down syndrome
- When multiple gestations, hydrocephalus, cystic hygromas in Turner syndrome, or a variety of other congenital anomalies are seen on prenatal anatomic ultrasound testing

Procedure

1. hCG testing:
    - Obtain a blood specimen or a concentrated, nonacidified first morning urine specimen for hCG testing.
    - Order qualitative or quantitative hCG testing of urine or serum as appropriate. Use the qualitative hCG assay for faster turnaround time of test results.
    - Obtain serial blood specimens for hCG testing if the initial test is found to be negative on the first day of a missed menstrual period.
    - Obtain a blood sample for hyperglycosylated hCG testing in the presence of a suspected trophoblastic neoplasm during pregnancy.
    - Obtain a blood sample for hCG free beta subunit in nontrophoblastic tumor patients
2. AFP testing:
    - In pregnant women, refer the patient to an obstetrician who will obtain a maternal blood specimen for AFP (i.e., as part of a triple or quad screen) testing between 15 and 20 weeks' gestational age for routine maternal screening for a fetus affected with an open neural tube defect and/or Down syndrome.
3. Do not rely exclusively on the results of *serum* hCG testing for the diagnosis of trophoblastic or nontrophoblastic lesions.

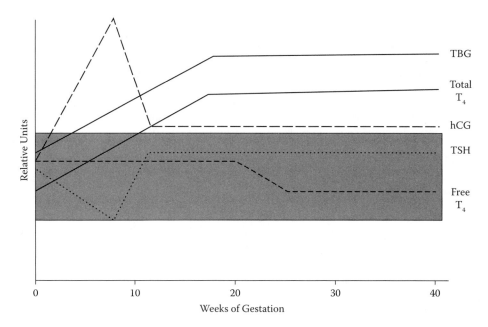

Figure 9.5
Human chorionic gonadotropin (hCG) and thyroid hormone and binding protein (TBG) changes during pregnancy (shaded area indicates reference interval for each hormone and TBG). *Note:* hCG originates from the placenta and within the first trimester relates inversely to TSH levels.

4. Confirm serially elevated *serum* hCG results with a *urine* hCG test. When the results of serum and urine hCG tests are discordant, consider the possibility of human anti-mouse antibodies (HAMAs) in the patient's serum as a potential cause of the discordance.
5. When laboratory test results are inconsistent with the patient's clinical or other laboratory and nonlaboratory test findings, consult with doctoral-level laboratory personnel on procedures to resolve possible HAMA interference before making a final diagnosis or recommending treatment.

Interpretation

1. hCG testing:
   - The expected [hCG] in serum or urine from a nonpregnant female is negative (qualitative assay) or less than the lower limit of detection (LLD; typically <3 IU/L) for the assay used to quantify [hCG].
   - Figure 9.5 shows the expected serial changes in hCG, thyroxine-binding globulin (TBG), and other hormones during a normal pregnancy.
   - If the serum [hCG] is $>10^5$ IU/L and on serial testing rises to $>3 \times 10^5$ IU/L, the differential diagnosis includes trophoblastic tumors (e.g., choriocarcinoma), lung cancer, hepatoma, or dysgerminoma in patients who have associated gynecomastia, precocious puberty, or markedly elevated [TSH] (i.e., >1000 IU/L).
   - False-positive or false-negative hCG test results may occur in serum samples containing human anti-mouse antibodies.
   - hCG of pituitary origin may be detected in normal menopausal females in low levels of 2 to 32 IU/L. To confirm the benign pituitary origin of these levels of hCG, estrogen hormone replacement therapy, given for 2 weeks, will result in an [hCG] of <2 IU/L (Cole et al., 2007).
2. AFP testing, as part of a triple or quad screen:
   - Consult with a referring obstetrician to obtain the results of pregnancy assessment, anatomic change or abnormality on fetal ultrasound, and maternal serum screening so as to correlate test results with the patient's clinical status.

Notes

1. Using *quantitative* hCG assays, hCG levels are detectable in serum or urine within 3 to 4 days of implantation of a conceptus.
2. *Qualitative* serum or urine assay requires 10 days to 2 weeks or up to 2 to 3 days before the first missed menstrual period to become positive (i.e., [hCG] > 20–25 IU/L).
3. Using an extremely sensitive assay for hCG, 10% of clinical pregnancies were undetectable on the first day of missed menses. In practice, an even larger percentage of clinical pregnancies may be undetected by use of current test kits for hCG (Wilcox et al., 2001).
4. In assisted reproductive technologies pregnancies, comparison of a single serum [hCG] with appropriate reference ranges enables approximately 40% of nonviable pregnancies to be identified with a high positive predictive value. Repeated measurements did not contribute further to the predictive value (Alahakoon et al., 2004).

ICD-9 Codes

Conditions that may justify this test *include but are not limited to*:

634	spontaneous abortion
V22.2	pregnancy, incidental

Suggested Reading

Alahakoon TI, Crittenden J, Illingworth P. Value of single and paired serum human chorionic gonadotropin measurements in predicting outcome of *in vitro* fertilisation pregnancy. *Aust NZ J Obstet Gynaecol* 2004; 44:57–61.

Bogart MH, Jones OW, Felder RA, Best RG, Bradley L, Butts W, Crandall B, MacMahon W, Wians FH Jr, Loeh PV. Prospective evaluation of maternal serum chorionic gonadotropin levels in 3248 pregnancies. *Am J Obstet Gynecol* 1991; 165:663–7.

Cole LA, Sasake S, Muller CY. Normal production of human chorionic gonadotropin in menopause. *N Engl J Med* 2007; 356:1184–86.

Rajesh R, Thomas SV. Prenatal screening for neural tube defects. *Natl Med J India* 2001; 14:343–6.

Smikle CB, Sorem KA, Wians FH Jr, Hankins GDV. Measuring quantitative serum human choriogonadotropin: variations in levels between kits [review]. *J Repro Med* 1995; 40:439–42.

Strickland DM, Butzin CA, Wians FH Jr. Maternal serum alpha-fetoprotein (MSAFP) screening: further consideration of low-volume testing. *Am J Obstet Gynecol* 1991; 164:711–4.

Wians FH Jr, Hankins GDV, Yeomans ER, Hammond TL, Bradley WP, Dev J. Maternal serum alpha-fetoprotein (MSAFP) screening: the Wilford Hall experience. *J Mil Med Lab Sci* 1990a; 19:168–77.

Wians FH Jr, Hankins GDV, Yeomans ER, Hammond TL, Strickland DM. Guidelines for establishing a maternal serum alpha-fetoprotein (MSAFP) screening program. *Milit Med* 1990b; 155:468–72.

Wians FH Jr, Hankins GDV, Yeomans ER, Bradley WP, Dev J. Calculation of Down syndrome risk using maternal age and alpha-fetoprotein concentration. *J Mil Med Lab Sci* 1991; 20:40–5.

Wilcox AJ, Baird DD, Dunson D, McChesney R, Weinberg CR. Natural limits of pregnancy testing in relation to the expected menstrual period. *JAMA* 2001; 286:1759–61.

9.4   Female Androgens

9.4.1   Androgen Excess in Females: Tests for Total (TT), Free (FT), and Bioavailable (BioT) Testosterone; Androstenedione; Sex Hormone Binding Globulin (SHBG); Dehydroepiandrosterone (DHEA); and DHEA Sulfate (DHEAS)

Indications for Test

Measurement of androgen levels in females is indicated in women:

- With hirsutism, with or without an ovarian mass, and other signs of hyperandrogenism
- With recent, unexplained changes in libido
- On androgen or estrogen therapy who demonstrate symptoms of excessive sex steroid effects

Procedure

1. On physical exam, seek evidence for acne, an androgen-dependent pattern of hair distribution, temporal balding, increased muscle mass, or clitoromegaly before initiating testing.
2. Obtain a semiquantitive assessment of the degree of androgen-related hirsutism by use of the Ferriman–Gallwey scoring technique if a research-level inquiry is to be made (AACE Hyperandrogenism Guidelines, 2001; refer to Table 1, Figure 1, in this article).
3. Obtain history of estrogen therapy and recent pregnancy, which might increase sex hormone binding globulin (SHBG), or androgen therapy, which, in turn, might decrease SHBG.
4. Identify changes in libido, sexual satisfaction, quality of life, and bone mineralization.
5. Obtain a random blood specimen for total testosterone (TT), androstenedione and DHEA, or DHEAS testing.
6. If factors that might alter SHBG levels, such as sex steroid therapy, are present, obtain blood samples for free testosterone (FT) or bioavailable testosterone (BioT) testing.
7. Initiate evaluation for an androgen-secreting tumor at a [TT] > 200 ng/dL.

Interpretation

1. [TT] in women varies not only with the menstrual cycle but also with age, race, and body mass index (Santoro et al., 2005). The normal [TT] in women is <40 ng/dL (1.4 nmol/L) as shown in detailed studies in which patients with menstrual abnormalities, hirsutism, acne, or other effects of androgens are excluded.
2. Typical TT levels in Tanner stage 2 and 5 females are <30 ng/dL and 10 to 40 ng/dL, respectively.
3. During pregnancy, estradiol occupies a substantial number of SHBG-binding sites, so [SHBG], as determined by immunoassay, is falsely elevated, which in pregnancy sera results in a calculated [FT] that is lower than [FT] measured by equilibrium dialysis.
4. Reference intervals for [FT] vary with the assay and laboratory. Examples of [FT] reference intervals for healthy women reported by various referral laboratories are 0.2 to 5 pg/mL, 1.3 to 6.8 pg/mL, and 3.3 to 13.0 pg/mL.
5. The true upper limit of the normal range for [FT] is probably closer to 8 pg/mL when only samples from patients without androgen excess (i.e., menstrual abnormalities, hirsutism, acne, or other effects of androgen excess) are included in the study population.
6. [TT] > 200 ng/dL should be considered suspicious for an androgen-secreting tumor, usually of ovarian origin.
7. [TT] > 300 ng/dL is practically diagnostic of adrenal or ovarian tumors or exogenous androgen administration.
8. The reference interval for [DHEAS] is 80 to 330 mcg/dL. [DHEAS] > 700 mcg/dL should be considered suspicious for an adrenal androgen-secreting tumor.

Notes

1. Increased sex drive may occur at higher levels of androgen.
2. Hypertrichosis and alopecia, which are not androgen dependent, should be differentiated from hirsutism and temporal balding, which may be androgen dependent. The more elaborate assessment of the degree of androgen-related hirsutism available using the Ferriman–Gallwey scoring technique is usually not necessary for this determination.
3. The Ferriman–Gallwey scoring technique is most useful for serial determinations of change in hirsutism following interventional therapy. The Ferriman–Gallwey score goes up to 76 and represents data from 19 different sites.
4. Virilization is an advanced hyperandrogenic state marked by clitoromegaly.
5. Rapid onset of hirsutism and other signs of hyperandrogenism should prompt screening for androgen-producing tumors using selected imaging techniques.

6. TT concentrations during female puberty correlate poorly with Tanner stage and are of little use as an adjunct to Tanner stage determination. With onset of puberty at about age 10 years, [TT] increases 20-fold in boys and 2-fold in girls, whereas androstenedione increases 10-fold in boys and 5-fold in girls (Elmlinger et al., 2005).

7. When unscreened clinical samples have been used to derive reference intervals appearing on lab reports, expect the interval to be shifted to higher levels than those from a well-selected population of healthy, nonhirsute women.

8. The ovaries contribute approximately half the circulating plasma [TT] in premenopausal women.

9. In both sexes, [androstenedione] decreases continuously during aging.

10. A large percentage of [TT] in women is derived from the peripheral conversion of adrenal [DHEA] and [DHEAS] (Arlt et al., 1998) which are hormones not subject to pituitary feedback control.

11. The mean (±SD) serum [FT] increased from a baseline of 1.2 ± 0.8 pg/mL (4.2 ± 2.8 pmol/L) during placebo treatment to 3.9 ± 2.4 pg/mL (13.5 ± 8.3 pmol/L) and 5.9 ± 4.8 pg/mL (20.5 ± 16.6 pmol/L) during treatment with 150 and 300 mcg of transdermal testosterone per day, respectively (Shifren et al., 2000).

12. Testosterone excess in women lowers [SHBG], which raises the [FT] and contributes to the strong correlation of 1/[SHBG] with [FT].

13. Severe insulin resistance results in compensatory hyperinsulinemia that stimulates ovarian androgen production. Acanthosis nigricans, a dermatologic manifestation of severe insulin resistance, is an epiphenomenon that marks this condition.

ICD-9 Codes

Conditions that may justify this test *include but are not limited to:*

Ovary
256.39   ovarian failure
256.4   polycystic ovaries/hyperandrogenism

Other
704.1   hirsutism
799.81   decreased libido

Suggested Reading

AACE Hyperandrogenism Guidelines. American Association of Clinical Endocrinologists medical guidelines for clinical practice for the diagnosis and treatment of hyperandrogenic disorders. *Endocr Practice* 2001; 7:121–34.

Arlt W, Justl HG, Callies F, Reincke M, Hubler D, Oettel M, Ernst M, Schulte HM, Allolio B. Oral dehydroepiandrosterone for adrenal androgen replacement: pharmacokinetics and peripheral conversion to androgens and estrogens in young healthy females after dexamethasone suppression. *J Clin Endocrinol Metab* 1998; 83:1928–34.

Bachmann G, Bancroft J, Braunstein G, Burger H, Davis S, Dennerstein L, Goldstein I, Guay A, Leiblum S, Lobo R, Notelovitz M, Rosen R, Sarrel P, Sherwin B, Simon J, Simpson E, Shifren J, Spark R, Traish A. Female androgen insufficiency: the Princeton consensus statement on definition, classification, and assessment. *Fertil Steril* 2002; 77:660–5. Also, Lobo R. *Contemp OB/GYN* 2003; Suppl.

Barbieri RL, Hornstein MD. Hyperinsulinemia and ovarian hyperandrogenism: cause and effect. *Endocrinol Metab Clin North Am* 1988; 17:685–703.

Braunstein GD. Androgen insufficiency in women: summary of critical issues. *Fertil Steril* 2002; 77(Suppl 4):S94–9.

Eggert-Kruse W, Kruse W, Rohr G, Muller S, Kreissler-Haag D, Klinga K, Runnebaum B. Hormone profile of elderly women and potential modifiers. *Geburtshilfe Frauenheilkd* 1994; 54:321–31.

Elmlinger MW, Kühnel W, Wormstall H, Döller PC. Reference intervals for testosterone, androstenedione and SHBG levels in healthy females and males from birth until old age. *Clin Lab* 2005; 51:625–32.

Ferriman DM, Gallwey JD. Clinical assessment of body hair growth in women. *J Clin Endocrinol* 1961; 21:1440–47

Lobo RA. Hirsutism, alopecia, and acne. In: *Principles and Practice of Endocrinology and Metabolism*, Becker KL et al. (Eds). 1990. Lippincott Williams & Wilkins, pp. 834–848.

Loy R, Seibel MM. Evaluation and therapy of polycystic ovarian syndrome. *Endocrinol Metab Clin North Am* 1988; 17:785–813.

Santoro N, Torrens J, Crawford S, Allsworth JE, Finkelstein JS, Gold EB, Korenman S, Lasley WL, Luborsky JL, McConnell D, Sowers MF, Weiss G. Correlates of circulating androgens in mid-life women: the study of women's health across the nation (SWAN). *J Clin Endocrinol Metab* 2005; 90:4836–45.

Shifren JL, Braunstein GD, Simon JA, Casson PR, Buster JE, Redmond GP, Burki RE, Ginsburg ES, Rosen RC, Leiblum SR, Caramelli KE, Mazer NA. Transdermal testosterone treatment in women with impaired sexual function after oophorectomy. *N Engl J Med* 2000; 343:682–8.

Young RL, Goldzieher JW. Clinical manifestations of polycystic ovarian disease. *Endocrinol Metab Clin North Am* 1988; 17:621–35.

9.4.2 Androgen Deficiency in Females: Tests for Total (TT), Free (FT), and Bioavailable (BioT) Testosterone; Androstenedione; Sex Hormone Binding Globulin (SHBG); Dehydroepiandrosterone (DHEA); and DHEA Sulfate (DHEAS)

Indications for Test

Measurement of androgen levels in females is indicated in women:

- With recent, unexplained changes in libido
- Receiving androgen or estrogen therapy who demonstrate symptoms of diminished sex steroid effects

Procedure

1. Obtain history of estrogen therapy and recent pregnancy, which might increase sex hormone binding globulin (SHBG), or androgen therapy, which, in turn, might decrease SHBG.
2. Identify changes in libido, sexual satisfaction, quality of life, and bone mineralization. Identify onset of menopause.
3. Obtain a random blood specimen for total testosterone (TT), androstenedione (A), and DHEA or DHEAS testing.
4. If factors that might alter SHBG levels, such as sex steroid therapy are present, obtain blood samples for bioavailable testosterone (BioT) in preference to free testosterone (FT) testing.
5. If androgen insufficiency is highly suspected, obtain blood sample for SHBG testing and calculate the free androgen index (FAI) = [TT]/[SHBG], or simply proceed to measurement of the BioT.

Interpretation

1. [TT] in women varies not only with the menstrual cycle but also with age, race, and body mass index (Santoro et al., 2005). The normal [TT] in women is <40 ng/dL (1.4 nmol/L), as shown in detailed studies in which patients with menstrual abnormalities, hirsutism, acne, or other effects of androgens are excluded.
2. Neither the clinically available platform nor conventional chemiluminescence assays for measurement of [TT] are reliable in the low range, as both lack sensitivity. Immunoassay of [TT], after extraction and LC-MS/MS, is a better method to test for androgen deficiency (Bachmann et al., 2002).
3. Typical TT levels in Tanner stage 2 and 5 females are <30 ng/dL and 10 to 40 ng/dL, respectively.
4. Reference intervals for [FT] vary with the assay and laboratory. Examples of [FT] reference intervals for healthy women reported by various laboratories are 0.2 to 5 pg/mL, 1.3 to 6.8 pg/mL (4.5–23.6 pmol/L), and 3.3 to 13.0 pg/mL.

5. The true lower limit of the normal range for [FT] is probably closer to 2 pg/mL when only samples from patients without complaints of disturbances in libido (see note 1 below) are included in the study population.
6. By definition, oophorectomy and normal adrenal status mark the androgen-deficient woman.
7. Androgen insufficiency may be diagnosed if [TT] falls into the lowest quartile of the reference interval for normally cycling, reproductive-age women.
8. [A], assayed by LC/MS, varies with menstrual cycle (reference intervals: luteal phase, 30–235 ng/dL; midcycle, 60–285 ng/dL). During the follicular phase, the ovaries and adrenals contribute equally to the circulating [A]. Postmenopausal [A] intervals are 20 to 75 ng/dL.

Notes

1. Sexual complaints, particularly lack of libido, unexplained fatigue, diminished sense of well-being, and dysphoric mood with decreased sexual receptivity and pleasure suggest androgen insufficiency (Basson et al., 2001).
2. Ovariectomized patients experience an accelerated decline in [DHEA] which is 50% greater than that in noncastrates after natural menopause. A lower [DHEA] may be expected after menopause-related estrogen deficiency, and a relative increase in androgens from adrenal sources is associated with increased abdominal obesity and adverse alterations in metabolic cardiovascular risk profile.
3. Androstenedione is normally converted to testosterone and dihydrotestosterone and originates from both the adrenal cortex and gonad.
4. In 261 premenopausal women who reported a decrease in satisfying sexual activity relative to their younger years and had a morning [FT] < 3.8 pmol/L (1.1 pg/mL), a daily transdermal dose of testosterone improved self-reported sexual satisfaction by a mean of 0.8 self-reported satisfactory sexual events (SSEs) per month. The rate of SSEs with higher and lower testosterone doses did not differ from that with placebo (Davis et al., 2008).
5. Many unanswered or incompletely answered questions in the diagnosis of female androgen deficiency persist (Braunstein et al., 2002). These include:
   • Which androgens (T, A, or DHEA) best reflect the androgen status of women?
   • What form of testosterone (BioT, TT, or FT) should be measured and how?
   • Do androgen levels fall after menopause and, if so, at what point in years does this occur?
   • What effect does oophorectomy have on androgen levels?
   • What is the relationship between androgens and sexual dysfunction or changes in libido?
   • What constitutes androgen insufficiency syndrome?
   • What is the range of conditions associated with androgen insufficiency?
   • How should a patient with suspected androgen insufficiency be evaluated?
   • How should androgen replacement therapy be monitored?

ICD-9 Codes

Conditions that may justify this test *include but are not limited to:*

Ovary
256.39   ovarian failure
256.4   polycystic ovaries/hyperandrogenism

Other
799.81   decreased libido

Suggested Reading

Bachmann G, Bancroft J, Braunstein G, Burger H, Davis S, Dennerstein L, Goldstein I, Guay A, Leiblum S, Lobo R, Notelovitz M, Rosen R, Sarrel P, Sherwin B, Simon J, Simpson E, Shifren J, Spark R, Traish A. Female androgen insufficiency: the Princeton consensus statement on definition, classification, and assessment. *Fertil Steril* 2002; 77:660–5. Also, Lobo R. *Contemp OB/GYN* 2003; Suppl.

Basson R. Female sexual response: the role of drugs in the management of sexual dysfunction. *Obstet Gynecol* 2001; 98:350–3.

Braunstein GD. Androgen insufficiency in women: summary of critical issues. *Fertil Steril* 2002; 77(Suppl 4):S94–9.

Davis S, Papalia M-A, Norman RJ, O'Neill S, Redelman R, Williamson M, Stuckey BGA, Wlodarczyk J, Gardner K, Humberstone A. Safety and efficacy of a testosterone metered-dose transdermal spray for treating decreased sexual satisfaction in premenopausal women. *Ann Intern Med* 2008; 148:569–77.

Santoro N, Torrens J, Crawford S, Allsworth JE, Finkelstein JS, Gold EB, Korenman S, Lasley WL, Luborsky JL, McConnell D, Sowers MF, Weiss G. Correlates of circulating androgens in mid-life women: the study of women's health across the nation (SWAN). *J Clin Endocrinol Metab* 2005; 90:4836–45.

9.5 Neoplasms of the Female Reproductive Tract

9.5.1 CA-125 Testing in Ovarian Cancer

Indications for Test

CA-125 testing is indicated:

- In postmenopausal women with a family history or other risk factors for ovarian cancer and signs and symptoms including bloating, urinary urgency, and increasing abdominal size
- Upon findings of an adnexal mass in a nonmenstruating female

Procedure

1. Perform a bimanual, rectovaginal pelvic exam to check for obvious adnexal masses.
2. Check for ascites, pleural effusion, lymphadenopathy, and an omental "cake," or mass, in the mid-upper abdomen.
3. Obtain an ultrasonographic study of the liver and transvaginal images of the ovaries if an adnexal mass or fullness is found clinically.
4. If ultrasonography of the liver shows cirrhosis, do not order CA-125 testing, as results may be falsely positive.
5. In general, avoid CA-125 testing in premenopausal women and never do CA-125 testing when the patient is menstruating.
6. If a CA-125 test is indicated, obtain the appropriate blood specimen for testing.
7. If the CA-125 level is elevated, double check the stage of the patient's menstrual cycle and obtain another blood specimen for repeat CA-125 testing to confirm the initial test results.
8. If the initially elevated CA-125 level is confirmed by repeat testing, obtain pelvic ultrasound of the adnexa within the first 7 to 8 days (follicular phase) of her menstrual cycle.

Interpretation

1. In healthy pre- and postmenopausal women, the [CA-125] is typically <35 U/mL; however, in premenopausal women, higher [CA-125] may occur in the absence of ovarian cancer (Table 9.2) because [CA-125] is elevated during menses and in patients with hepatic cirrhosis.
2. If a pelvic mass is present and an elevated baseline [CA-125] increases rapidly on repeat testing, epithelial ovarian cancer is likely.
3. [CA-125] is elevated in 60% of patients with pancreatic cancers, 20 to 25% of patients with other solid tumors, and 80% of all patients with epithelial ovarian cancers.
4. At CA-125 cutoff values of 35 U/mL and 65 U/mL, the diagnostic sensitivity, specificity, and accuracy of CA-125 in the diagnosis of ovarian cancer were 86%, 76%, 79% and 83%, 95%, 92%, respectively

TABLE 9.2
CA-125 Levels and Menopausal Status

Menopausal Status	CA-125 Concentration (U/mL)		
	Normal	Indeterminate	Elevated
Postmenopausal (age, >50 years)	<35	—	>35
Premenopausal (age, <50 years)	<35	35–200	>200

(Trimble et al., 1994). Therefore, a CA-125 cutoff value of 35 U/mL is not sufficiently sensitive to warrant use of CA-125 alone as a screening test for the detection of ovarian cancer. While lower CA-125 cutoff values might identify women at higher risk of developing ovarian cancer, the lower specificity for diagnosis using these values restricts their clinical utility.

5. Serial CA-125 testing may increase the sensitivity and specificity of this test.
6. Serial CA-125 levels which plateau or decrease usually indicate benign conditions.
7. In patients with a history of ovarian cancer, three progressively rising serum CA-125 concentrations, even within the normal range (<35 U/mL), at 1- to 3-month intervals are associated with a high likelihood of tumor recurrence.
8. CA-125 is not an accurate marker of ovarian cancer in premenopausal women, and 50% of patients with stage 1 ovarian cancer will have an isolated [CA-125] within the reference interval.

Notes

1. CA-125 is primarily a tumor marker for epithelial ovarian cancer.
2. In patients with ascites, cirrhotic liver disease should be sought out prior to CA-125 testing.
3. There is a greater risk of developing ovarian cancer in women with a positive family history, difficulty becoming pregnant, and a normal menopause with hot flashes. The risk for the disease is diminished with multiparity, a history of dysmenorrhea, tubal ligation, and hysterectomy.

ICD-9 Codes

Conditions that may justify this test *include but are not limited to*:

787.3   bloating
788.63   urinary urgency, and increasing abdominal size
V10.8   personal history of malignant neoplasm of the ovaries

Suggested Reading

Gadducci A, Capriello P, Bartolini T, Barale E, Cappelli N, Facchini V, Fioretti P. The association of ultrasonography and CA-125 test in the preoperative evaluation of ovarian carcinoma. *Eur J Gynaecol Oncol* 1988; 9:373–6.

Helzlsouer KJ, Bush TL, Alberg AJ, Bass KM, Zacur H, Comstock GW. Prospective study of serum CA-125 levels as markers of ovarian cancer. *JAMA* 1993; 269:1123–6.

Paley PJ. Ovarian cancer screening: are we making any progress? *Curr Opin Oncol* 2001; 13:399–402.

Trimble EL. The NIH Consensus Conference on Ovarian Cancer: screening, treatment, and follow-up. *Gynecol Oncol* 1994; 55(3 Pt 2):S1–3. Also, NIH Consensus Statement. Ovarian cancer: screening, treatment and follow-up. *NIH Consensus Statement*. 1994; 12:1–30.

Wilder JL, Pavlik E, Straughn JM, Kirby T, Higgins RV, De Priest PD, Ueland FR, Kryscio RJ, Whitley RJ, Nagell J. Clinical implications of a rising serum CA-125 within the normal range in patients with epithelial ovarian cancer: a preliminary investigation. *Gynecol Oncol* 2003; 89:233–5.

9.5.2 Imaging of the Ovaries with Transvaginal Ultrasound (TVUS)

Indications for Test

TVUS of the ovaries is indicated for:

- Anatomic identification of polycystic ovaries (PCOs) and ovarian tumors
- Identification of an androgen-secreting ovarian tumor in cases of hirsutism or virilization
- Assessment of cases of congenital gonadal dysgenesis (e.g., Turner syndrome)

Procedure

1. Use an intravaginal probe, capable of 3-mm resolution or better, for imaging of adenexal spaces.
2. Access the services of a skilled sonographer to measure all three dimensions of the ovary (length, width, and thickness).
3. Calculate ovarian volume using the prolate ellipsoid formula ($L \times H \times W \times 0.523$).
4. Perform TVUS as part of annual screening for cancer in higher risk individuals with a family history of ovarian cancer.

Interpretation

1. Normal, postpubertal, premenopausal ovarian dimensions ($L \times H \times W$) are 2.5–5.0 cm × 1.5–3.0 cm × 0.6–1.5 cm.
2. The ovaries are usually, but not necessarily, enlarged in women with PCOS.
3. A statistically significant decrease in ovarian volume occurs with each decade of life from age 30 to age 70. The upper limit of normal for ovarian volume is 20 cm^3 in premenopausal women and 10 cm^3 in postmenopausal women.
4. The criteria for identification of PCO by TVUS are at least 10 follicles, 2 to 8 mm in diameter each, arranged in a peripheral pattern, and associated with an increased amount of stroma relative to follicles.
5. In addition to ovarian cysts, polycystic ovary syndrome (PCOS) is characterized by a state of chronic anovulation manifested by hirsutism, obesity, abnormal menses, and increased serum LH/FSH ratios and androgens (usually testosterone).
6. As many as 40% of patients with PCOS may have normal size and appearance of the ovaries on TVUS.
7. In a 10-year study, 14,469 high-risk women with a first- or second-degree relative who had ovarian cancer were monitored with repeat TVUS. Of these, 180 had persistent ovarian abnormalities, 17 of whom had cancer. Of these 17 women, only four were detectable by other means. Four patients developed cancer after negative screening (van Nagell et al., 2000).
8. The sensitivity, specificity, and negative predictive value of TVUS for ovarian cancer were 81%, 99%, and 99.9%, respectively (van Nagell et al., 2000).
9. TVUS screening, when performed annually, is associated with a decrease in ovarian cancer stage at detection and a decrease in case-specific ovarian cancer mortality but does not appear to be effective in detecting ovarian cancer in which ovarian volume is normal.
10. The overall sensitivity and accuracy of TVUS in the diagnosis of adnexal masses are 76% and 83%, respectively; for magnetic resonance imaging, sensitivity and accuracy are 49% and 70%, respectively (Jain et al., 1993).

Notes

1. TVUS has a greater sensitivity than the transabdominal approach for the diagnosis of ovarian cysts but is insufficient to establish a diagnosis of PCOS.
2. TVUS is a better modality than magnetic resonance imaging for the assessment of suspected pelvic masses.
3. TVUS for general ovarian cancer screening has not been validated as being cost effective.

ICD-9 Codes

Conditions that may justify this test *include but are not limited to*:

255.2	adrenogenital disorders/virilization
256.4	polycystic ovaries (PCOs)
704.1	hirsutism
758.6	gonadal dysgenesis (e.g., Turner syndrome)

Suggested Reading

Jain KA, Friedman DL, Pettinger TW, Alagappan R, Jeffrey RB, Sommer FG. Adnexal masses: comparison of specificity of endovaginal US and pelvic MR imaging. *Radiology* 1993; 186:697–704.

Pavlik EJ, De Priest PD, Gallion HH, Ueland FR, Reedy MB, Kryscio RJ, van Nagell JR. Ovarian volume related to age. *Gynecol Oncol* 2000; 77:410–2.

van Nagell JR, De Priest PD, Reedy MB, Gallion HH, Ueland FR, Pavlik EJ, Kryscio RJ. The efficacy of transvaginal sonographic screening in asymptomatic women at risk for ovarian cancer. *Gynecol Oncol* 2000; 77:350–6.

9.5.3 Transvaginal Ultrasound (TVUS) Imaging of the Uterus for Endometrial Stripe Measurement and Assessment of Endocrine-Related Anatomic Abnormalities

Indications for Test

TVUS of the uterus for measurement endometrial stripe and assessment of uterine anatomy is indicated in women:

- When there is suspicion of endometrial cancer
- With persistent intermenstrual or postmenopausal spotting regardless of their hormone replacement therapy (HRT) status

Procedure

1. Use an intravaginal probe for imaging of the uterine cavity.
2. Measure the thickness of the endometrial stripe.
3. Document flow velocity in the uterine and ovarian arteries.

Interpretation

1. An endometrial stripe by ultrasound of 4 to 6 mm indicates no serious pathology.
2. An endometrial stripe < 4 mm is evidence against endometrial cancer in the postmenopausal woman treated with estrogens.
3. A *pulsatility index* (PI) of <2 may indicate endometrial cancer (Bourne et al., 1991). The PI is derived from flow velocity waveforms recorded from both uterine arteries and from within a tumor.
4. The sensitivity of TVUS for endometrial disease was 98.2% and 82.0% if cutoff points for endometrial thickness of 2 mm and 4 mm, respectively, were used (Van den Bosch et al., 1995).
5. No occurrence of endometrial hyperplasia or adenocarcinoma was found in 345 normal postmenopausal women on HRT when endometrial thickness was <9 mm. No correlation was found between various forms of HRT and the occurrence of endometrial hyperplasia or adenocarcinoma (Bonilla-Musoles et al., 1995).

Notes

1. TVUS is preferred over abdominal ultrasound for estimation of endometrial thickness.
2. TVUS is not a replacement for the endometrial biopsy but may direct the need for one.

ICD-9 Codes

Conditions that may justify this test *include but are not limited to*:

182   malignant neoplasm of body of uterus/endometrial cancer
626   hypomenorrhea

Suggested Reading

Bonilla-Musoles F, Ballester MJ, Marti MC, Raga F, Osborne NG. Transvaginal color Doppler assessment of endometrial status in normal postmenopausal women: the effect of hormone replacement therapy. *J Ultrasound Med* 1995; 14:503–7.

Bourne TH, Campbell S, Steer CV, Royston P, Whitehead MI, Collins WP. Detection of endometrial cancer by transvaginal ultrasonography with color flow imaging and blood flow analysis: a preliminary report. *Gynecol Oncol* 1991; 40:253–9.

Van den Bosch T, Vandendael A, Van Schoubroeck D, Wranz PA, Lombard CJ. Combining vaginal ultrasonography and office endometrial sampling in the diagnosis of endometrial disease in postmenopausal women. *Obstet Gynecol* 1995; 85:349–52.

Chapter 10

Gastroenteropancreatic (Noninsulinoma) Endocrine Tumors and Endocrine Tissue Tumor Markers

10.1 Gastrinomas

10.1.1 Gastrin: Basal Level as Screening Test for Gastrinoma or Zöllinger–Ellison Syndrome (ZES)

Indications for Test

Measurement of basal gastrin levels is indicated in patients with:

- Signs and symptoms as presented in "Signs and Symptoms of Gastroenteropancreatic (Noninsulinoma) Endocrine Tumors" (Table 10.1)
- A family history of multiple endocrine neoplasia (MEN 1) or ZES

Procedure

1. Instruct the patient to discontinue H2-receptor blockers (e.g., cimetidine, ranitidine) for 4 days or longer and proton pump blockers (e.g., omeprazole) for 10 days or longer prior to testing, as well as all topical antacids (e.g., sucralfate, liquid calcium-containing antacids) and anticholinergic agents for at least 24 hours prior to testing.
2. Obtain a fasting blood specimen for gastrin testing. If this initial gastrin result ≥ 100 pg/mL, obtain another blood specimen for confirmatory, repeat gastrin testing.
3. Upon finding an unequivocally elevated [gastrin], to locate its source and identify the presence of any abnormal or suspicious mass proceed with imaging studies:
   - Somatostatin receptor scintigraphy (SRS) of the abdomen, with a focus on pancreas imaging (see Test 10.4.1)
   - Multislice computed tomography (CT) or high-resolution magnetic resonance (MR) imaging of the abdomen with repeat imaging of any abnormal area suggested by somatostatin receptor scintigraphy
   - Endoscopic ultrasound to localize gastrinomas embedded within the duodenal wall if CT and MR imaging fail to localize tumor mass
4. Use abdominal angiography and the secretin stimulation of gastrin test (see Test 10.1.2) when all other localization techniques have failed to identify a gastrin-secreting tumor.

TABLE 10.1
Signs and Symptoms of Gastroenteropancreatic (Noninsulinoma) Endocrine Tumors

Clinical Syndrome	Paraneoplastic Signs and Symptoms Associated with Various Organs (% of Patients Affected)				
	Intestine	Stomach	Heart	Skin	Other
ZES	Jejunal ulceration, diarrhea	Large rugal folds, PUD	—	—	Abdominal pain, esophageal acid reflux
Carcinoid	Diarrhea (85)	—	RSHVD (40)	—	Vasomotor flushing (85), bronchoconstriction, wheezing
Glucagonoma	—	—	—	NME, cheilitis	Weight loss (98), DVT (30), DM (75–95), depression
WDHA	Watery diarrhea	Achlorhydria	—	—	Hypokalemia

Note: DM, diabetes mellitus; DVT, deep venous thrombosis; NME, necrolytic migratory erythema; PUD, peptic ulcer disease; RSHVD, right-sided heart valvular disease; VIPoma, vasoactive intestinal polypeptide secreting tumor; WDHA, watery diarrhea with hypokalemia and achlorhydria (VIPoma); ZES, Zöllinger–Ellison syndrome (gastrinoma).

Interpretation

1. A normal adult [gastrin] is <100 pg/mL, with >150 pg/mL occurring in patients with ZES and other conditions noted in Interpretation point 3 below. A serum [gastrin] of 100 to 150 pg/mL is indeterminate and may be associated with the conditions noted in Interpretation point 2 below.
2. Causes of elevated [gastrin], usually in the range of 100 to 500 pg/mL but occasionally >500 pg/mL, include retained antrum, achlorhydria, gastric outlet obstruction, gastric ulcer, hyperparathyroidism, renal insufficiency, insulin treatment, hepatic cirrhosis, and rheumatoid arthritis, as well as occasionally with antacid/gastric acid blocking therapy.
3. Serum [gastrin] may be as high as 10,000 pg/mL in patients with pernicious anemia, atrophic gastritis, or gastric cancer or those receiving treatment with H2 receptor or proton pump blockers.
4. In patients with ZES, only those with malignant disease had a serum [gastrin] > 1500 pg/mL. Half of all ZES patients with malignant disease have [gastrin] ≤ 1500 pg/mL.
5. SRS imaging (see Test 10.4.1) can help to identify a functional gastrinoma, but not all tumors will take up isotopically labeled somatostatin. Up to 50% of gastrinomas are not evident on preoperative imaging studies (Meko and Norton, 1995).
6. Use of atropine and other belladonna-related alkaloids may lower the basal gastrin level.

Notes

1. The duodenum is a major site for gastrinomas as it contains many G cells. Gastrin-secreting tumors of the duodenum may be less aggressive than those found in the pancreas, which does not contain G cells and must produce gastrin ectopically (Modlin and Lawton, 1994).
2. The best screening test for a gastrinoma is measurement of the fasting serum [gastrin].
3. Excess stomach acid resulting from a gastrinoma lowers intraluminal pH, damaging the walls of the intestine and leading to pancreatic enzyme inactivation, malabsorption, and steatorrhea.
4. Typically, gastrinomas are diagnosed in individuals between the ages of 50 and 70 years with a male/female preponderance of 1.5–2.1 and a delay in diagnosis of >5 years. MEN 1 patients tend to present at a younger age than those with sporadic onset ZES ($p < 0.0001$) (Roy et al., 2000).
5. In patients with large primary gastrin-secreting tumors of the pancreas, the development of liver metastases, especially when associated with bone metastases or Cushing's syndrome, and the extent of liver metastases are all important prognostic factors in patient survival (Chen et al., 1998; Yu et al., 1999).

6. Duodenal primary gastrinomas were found to have a significantly greater incidence of metastases (55%) and a significantly shorter disease-free interval (12 months) than pancreatic gastrinomas (22% and 84 months, respectively) (Norton et al., 1992).

7. With increased awareness of duodenal tumors, diagnosis and effective treatment of gastrinomas can be accomplished at surgery in 80 to 90% of patients (Meko and Norton, 1995).

8. In one series, all gastrinomas were found to secrete chromogranin A (Nobels et al., 1997), suggesting that this test is a useful tumor marker for residual disease, response to therapy, and disease recurrence (Nobels et al., 1998; Schurmann et al., 1992).

9. No particular MEN 1 gene mutation correlates with the clinical characteristics of patients with gastrinomas and its malignant potential (Goebel et al., 2000).

10. Patients with MEN 1 presented less frequently with gastrointestinal pain and bleeding and more frequently with nephrolithiasis.

11. Since 1980, successful antacid secretory therapy has dramatically decreased the rate of surgery for controlling stomach acid secretion and has led to patients presenting with less severe symptoms and fewer ZES-related complications (Roy et al., 2000).

ICD-9 Codes

Conditions that may justify this test include *but are not limited to*:

Ulcers

531	gastric
532	duodenal

Other

227	benign neoplasm of other endocrine glands and related structures/ZES
251.5	abnormality of secretion of gastrin
787.1	gastroesophageal reflux disease
789.0	abdominal pain

Suggested Reading

Chen H, Hardacre JM, Uzar A, Cameron JL, Choti MA. Isolated liver metastases from neuroendocrine tumors: does resection prolong survival? *J Am Coll Surg* 1998; 187:88–93.

Deveney CW, Deveney KE. Zöllinger–Ellison syndrome (gastrinoma): current diagnosis and treatment. *Surg Clin N Am* 1987; 67:411–22.

Ectors N. Pancreatic endocrine tumors: diagnostic pitfalls. *Hepato-Gastroenterol* 1999; 46:679–90.

Goebel SU, Heppner C, Burns AL, Marx SJ, Spiegel AM, Zhuang Z, Lubensky IA, Gibril F, Jensen RT, Serrano J. Genotype/phenotype correlation of multiple endocrine neoplasia type 1 gene mutations in sporadic gastrinomas. *J Clin Endocrinol Metab* 2000; 85:116–23.

Howard TJ, Stabile BE, Zinner MJ, Chang S, Bhagavan BS, Passaro E. Anatomic distribution of pancreatic endocrine tumors. *Am J Surg* 1990; 159:258–64.

Lamers CB, Buis JT, van Tongeren J. Secretin stimulated serum gastrin levels in hyperparathyroid patients from families with multiple endocrine adenomatosis type I. *Ann Int Med* 1977; 86:719–24.

Mansour JC, Chen H. Pancreatic endocrine tumors. *J Surg Res* 2004; 120:139–61.

Meko JB, Norton JA. Management of patients with Zöllinger–Ellison syndrome. *Ann Rev Med* 1995; 46:395–411.

Modlin IM, Lawton GP. Duodenal gastrinoma: the solution to the pancreatic paradox. *J Clin Gastroenterol* 1994; 19:184–8.

Nobels FR, Kwekkeboom DJ, Coopmans W, Schoenmakers CH, Lindemans J, De Herder WW, Krenning EP, Bouillon R, Lamberts SW. Chromogranin A as serum marker for neuroendocrine neoplasia: comparison with neuron-specific enolase and the alpha-subunit of glycoprotein hormones. *J Clin Endocrinol Metab* 1997; 82:2622–28.

Norton JA, Doppman JL, Jensen RT. Curative resection in Zöllinger–Ellison syndrome: results of a 10-year prospective study. *Ann Surg* 1992; 215:8–18.

Roy PK, Venzon DJ, Shojamanesh H, Abou-Saif A, Peghini P, Doppman JL, Gibril F, Jensen RT. Zöllinger–
    Ellison syndrome. Clinical presentation in 261 patients. *Medicine* 2000; 79:379–411.
Schurmann G, Raeth U, Wiedenmann B, Buhr H, Herfarth C. Serum chromogranin A in the diagnosis and
    follow-up of neuroendocrine tumors of the gastroenteropancreatic tract. *World J Surg* 1992:16:697–701.
Seregni E, Ferrari L, Stivanello M, Dogliotti L. Laboratory tests for neuroendocrine tumours. *Q J Nucl Med*
    2000; 44:22–41.
Yu F, Venzon DJ, Serrano J, Goebel SU, Doppman JL, Gibril F, Jensen RT. Prospective study of the clinical
    course, prognostic factors, causes of death, and survival in patients with long-standing Zöllinger–Ellison
    syndrome. *J Clin Oncol* 1999; 17:615–30.
Zöllinger R, Ellison E. Primary peptic ulceration of the jejunum associated with islet cell tumors of the
    pancreas. *Ann Surg* 1955; 142:709–23.

10.1.2 Secretin Stimulation of Gastrin Test in the Event of Elevated Basal Gastrin Levels

Indications for Test

The secretin stimulation test is indicated:

- When the serum [gastrin] in a single, properly collected blood sample is elevated in the absence of other factors known to cause a rise in [gastrin]
- To help further establish a diagnosis of ZES in patients with suggestive signs and symptoms (Table 10.1) and an equivocal basal [gastrin] (i.e., <100 pg/mL)

Procedure

1. Instruct the patient to discontinue H2-receptor blockers (e.g., cimetidine, ranitidine) for 4 days or longer and proton pump blockers (e.g., omeprazole) for 10 days or longer prior to testing, as well as all topical antacids (e.g., sucralfate, liquid calcium-containing antacids) and anticholinergic agents for at least 24 hours prior to testing.
2. After an overnight fast by the patient, obtain a blood specimen for determination of the basal serum [gastrin].
3. Administer secretin (2 mcg/kg) as an i.v. bolus.
4. Obtain blood specimens for gastrin testing at 2, 5, 10, 15, and 30 minutes after the injection of secretin.
5. In patients with a positive secretin stimulation test and negative or equivocal imaging studies for localizing the site of the gastrinoma, consult with an interventional radiologist on performing selective sampling for gastrin following catheterization of veins from the pancreas and the duodenum, intra-abdominal arterial injection of secretin, and collection of blood samples from these veins for gastrin testing.

Interpretation

1. In healthy subjects and patients with non-tumor-related peptic ulcer disease or antral G-cell hyperplasia, serum [gastrin] will not rise to >50% of the baseline [gastrin] after secretin injection.
2. A rise in [gastrin] to >200 pg/mL over baseline at any time point after secretin injection is considered diagnostic for ZES. In ZES, this peak level usually occurs at 10 minutes.
3. Seventy five percent of gastrinoma patients had a positive response to secretin by 5 minutes, 95% by 10 minutes, and 100% by 15 minutes. Only 6% of patients had a positive gastrin response at 2 minutes.
4. Secretin tends to inhibit gastrin release by normal gastrin-secreting G cells.
5. False-positive secretin stimulation tests have been reported in 5% and false negatives in 10 to 15% of patients with ZES.
6. False-negative secretin stimulation tests may occur in hypocalcemic patients.

Notes

1. The secretin stimulation test, with a diagnostic sensitivity and specificity of 90% for detecting gastrinomas, has superseded the 2-hour gastric acid output test (Delvalle and Yamada, 1995) in which basal gastric acid hypersecretion of >15 mEq/hr is consistent with the diagnosis of a gastrinoma.
2. In patients with a negative secretin stimulation test, the calcium infusion test, used for stimulation of gastrin as well as insulin, may be positive. The calcium infusion test (Frucht et al., 1989) should be reserved for patients with a negative secretin test and a strong clinical suspicion of ZES. A rise in serum [gastrin] of more than 50% over baseline after administration of calcium gluconate has a sensitivity of 74% for gastrinomas.
3. There is controversy as to whether pancreatic or duodenal gastrinomas are more common. By a two to one margin, pancreatic endocrine tumors (PETs) appear to predominate and duodenal tumors seem to be less aggressive (Mansour and Chen, 2004).
4. Although major morbidity occurs as a result of excess secretion of gastrin and other hormones from PET, all of these tumors have some degree of malignant potential.

ICD-9 Codes

Conditions that may justify this test include *but are not limited to*:

251.5   abnormality of secretion of gastrin

Suggested Reading

Delvalle J, Yamada T. Zöllinger–Ellison syndrome. In: *Textbook of Gastroenterology*, Yamada T et al. (Eds). 1995. Lippincott, p. 1430.

Deveney CW, Deveney KE. Zöllinger–Ellison syndrome (gastrinoma): current diagnosis and treatment. *Surg Clin North Am* 1987; 67:411–22.

Frucht H, Howard JM, Slaff JI, Wank SA, McCarthy DM, Maton PN, Vinayek R, Gardner JD, Jensen RT. Secretin and calcium provocative tests in the Zöllinger–Ellison syndrome: a prospective study. *Ann Intern Med* 1989; 111:713–22.

Giurgea I, Laborde K, Touati G, Bellanne-Chantelot C, Nassogne MC, Sempoux C, Jaubert F, Khoa N, Chigot V, Rahier J, Brunelle F, Nihoul-Fekete C, Dunne MJ, Stanley C, Saudubray JM, Robert JJ, de Lonlay P. Acute insulin responses to calcium and tolbutamide do not differentiate focal from diffuse congenital hyperinsulinism. *J Clin Endocrinol Metab* 2004; 89:925–9.

Mansour JC, Chen H. Pancreatic endocrine tumors. *J Surg Res* 2004; 120:139–61.

Maton PN, Gardner JD, Jensen RT. Diagnosis and management of Zöllinger–Ellison syndrome. *Endocrinol Metab Clin North Am* 1989; 18:519–43.

Mullan, MH, Gauger PG, Thompson NW. Endocrine tumours of the pancreas: review and recent advances. *Aust NZ J Surg* 2001; 71:475–82.

10.2   Carcinoid Tumors

10.2.1   5-Hydroxyindoleacetic Acid (5-HIAA): Random Urine, Timed Urine, and Blood Test for Carcinoid Syndrome

Indications for Test

Urinary or blood 5-HIAA testing is indicated in patients:

- Having signs and symptoms consistent with carcinoid syndrome as presented in Table10.1

TABLE 10.2
Foods and Medications That Can Affect
5-Hydroxyindoleacetic Acid (5-HIAA) Test Results

Foods Causing False Positives	Medications with Various Effects
Bananas	Monoamine oxidase inhibitors (MAOIs), serotonin reuptake inhibitors (SRIs)
Pineapples	Nicotine
Kiwi	L-Dihydroxyphenylalanine (L-dopa)
Eggplant	Salicylates, acetaminophen
Pecans	Glycerylguaiacolate
Walnuts	5-Fluorouracil (5-FU), streptozotocin, melphalan
Avocadoes	Phenothiazines, phenobarbital

Procedure

1. For at least a 1-week period, ensure that patients have not eaten any foods containing 5-hydroxy-tryptamine or taken any medications that interfere with the urinary 5-HIAA assay (Table 10.2).
2. Instruct the patient to go to the laboratory to receive either a random or 24-hour urine collection container specific for the 5-HIAA test that you have ordered (i.e., random or 24-hour). The 24-hour urine collection container must contain 25 mL of $6N$ HCl or 15 g of boric acid.
3. Have patient collect a timed, 12-hour overnight-collected urine sample as a practical alternative to a 24-hour urine collection.
4. Obtain a fasting blood sample for 5-HIAA testing in addition to 24-hour urine for the 5-HIAA test.

Interpretation

1. Age-dependent reference values for urinary [5-HIAA] in a random or timed urine sample from healthy individuals are:
   - 2 to 10 years, ≤12 mg/g creatinine or ≤8 mg/24 hr
   - >10 years, ≤10 mg/g creatinine or ≤6 mg/24 hr
2. Urinary [5-HIAA] > 30 mg/24 hr is diagnostic of carcinoid syndrome.
3. Depending on the assay for quantifying urinary [5-HIAA], falsely high or low values may occur in the presence of interfering drugs, including serotonin reuptake and monoamine oxidase inhibitors, phenothiazines, phenobarbital, isoniazide, L-dopa, nicotine, 5-fluorouracil, streptozotocin, melphelan, salicylates, acetaminophen, and glycerylguaiacolate.
4. There is no clear correlation between symptoms of carcinoid syndrome and [5-HIAA] in serially collected urine samples.
5. At a *plasma* [5-HIAA] cutoff value of 118 nmol/L, this test demonstrated diagnostic sensitivity, specificity, and efficiency for the diagnosis of carcinoid syndrome of 89%, 97%, and 93%, respectively (Carling et al., 2002).

Notes

1. Carcinoid syndrome is an uncommon disorder resulting from tumors of the enterochromaffin cells.
2. 5-HIAA is the principal metabolite of serotonin (5-hydroxytryptamine), which, in turn, is synthesized from tryptophan.
3. 5-HIAA is overproduced by midgut (i.e., ileal) carcinoid tumors metastatic to the liver.
4. Foregut carcinoid tumors infrequently cause the clinical syndrome characterized by vasomotor flushing, diarrhea, and right-sided heart valve disease.
5. Large fluctuations in urinary 5-HIAA excretion may be associated with more severe diarrheal episodes.
6. The [5-HIAA] in a 12-hour overnight-collected urine sample correlated well ($r = 0.81$) with values found in a 24-hour collection from the same individuals.

7. High variability in urinary 5-HIAA excretion throughout the day occurs in about 50% of carcinoid patients, with increased concentrations found in morning collections ($p = 0.0074$) and lower concentrations in the evening ($p = 0.0034$). In others, 5-HIAA excretion is flat throughout the day (Zuetenhorst et al., 2004).

ICD-9 Codes

Conditions that may justify this test include *but are not limited to*:

397.0   diseases of tricuspid valve
782.62   flushing
786.07   wheezing
787.91   diarrhea

Suggested Reading

Carling RS, Degg TJ, Allen KR, Bax ND, Barth JH. Evaluation of whole blood serotonin and plasma and urine 5-hydroxyindole acetic acid in diagnosis of carcinoid disease. *Ann Clin Biochem* 2002; 39(Pt 6):577–82.

Zuetenhorst JM, Korse CM, Bonfrer JM, Peter E, Lamers CB, Taal BG. Daily cyclic changes in the urinary excretion of 5-hydroxyindoleacetic acid in patients with carcinoid tumors. *Clin Chem* 2004; 50:1634–39.

10.2.2 Chromogranin A (CgA) Testing in Carcinoid Tumor Patients

Indications for Test

CgA testing is indicated:

- When a patient present with signs and symptoms of carcinoid syndrome (Table 10.1)
- Upon making the diagnosis of carcinoid tumor and monitoring status of carcinoid tumor activity after treatment
- For gathering supportive evidence for diagnosis of other neuroendocrine tumors including pheochromocytoma, medullary thyroid cancer (MTC), islet-cell tumors, pituitary tumors, gastrointestinal amine precursor uptake and decarboxylation (APUD) tumors, gastrinomas, small-cell lung cancer, and advanced prostate cancer

Procedure

1. Obtain history of patient use of acid blockers (i.e., proton pump and H2 receptor blockers). Defer testing for CgA if the patient used these drugs within the last 2 weeks (4 to 5 half-lives of the drug).
2. Evaluate the patient's renal and hepatic status by obtaining blood and urine samples for estimating glomerular filtration rate (GFR) and measuring creatinine clearance, bilirubin, alanine aminotransferase (ALT), and aspartate aminotransferase (AST). Do not test for CgA in patients with significant and fluctuating degrees of hepatic or renal impairment.
3. In patients with no laboratory evidence of renal or hepatic impairment or with stable renal and hepatic conditions, obtain a blood sample for CgA testing from a patient who has been supine for at least 20 minutes.
4. If the blood specimen cannot be assayed immediately for [CgA], freeze at least two separate samples of serum for storage and assay later on.
5. If serial CgA testing is planned, ensure that all testing is performed using the same CgA assay and that renal and hepatic statuses are monitored at the same time as [CgA].
6. Dilute and remeasure serum samples containing [CgA] in excess of about 600 ng/mL, as a hook effect may be present that is interfering with the accuracy of the test.

Interpretation

1. The reference interval for plasma [CgA] has been reported to be 10 to 50 ng/mL but may go up to ≤225 ng/mL using other assays in patients with normal renal function.
2. [CgA] rises in renal insufficiency with essentially all patients in stage 4 to 5 renal failure having an elevated CgA level.
3. Reference intervals and individual patient results differ significantly between different CgA assays, making direct comparisons nearly impossible.
4. A linear relation exists between [CgA] and carcinoid tumor burden beyond that for blood levels of serotonin, which plateau at a platelet storage capacity of about 5000 ng/mL.
5. Mildly elevated [CgA] was found in 7% of normal controls, whereas markedly elevated levels (>300 ng/mL) occur in 2% compared to 40% of patients with neuroendocrine tumors (Nobels et al., 1997).
6. Carcinoid tumors always secrete CgA, but elevated, circulating levels may be found in only 80 to 90% of patients with foregut and midgut carcinoids. In those with hindgut carcinoids, elevated [CgA] is found in nearly 100% of cases.
7. Varying percentages of patients with other neuroendocrine tumors have elevated CgA, including those with gastrinomas (100%), pheochromocytomas (89%), nonfunctioning tumors of the endocrine pancreas (69%), and MTC (50%).

Notes

1. CgA is a 439-amino-acid protein and member of the granin family of polypeptides normally found in secretory granules along with a variety of tissue-specific secretory hormones.
2. Carcinoid tumors characteristically secrete CgA along with serotonin and a variety of other modified amines and peptides.
3. CgA is coreleased with catecholamines from the adrenal medulla and from a wide variety of other peripheral and central nervous system tissues.
4. CgA release is unaffected by clonidine, metoprolol, phentolamine, or tyramine therapy.
5. Carcinoids arise from both the foregut (stomach, duodenum, pancreas, respiratory tract) and the hindgut (colon, rectum) in about 15% of cases. The majority of tumors (about 70%) arise from the midgut (jejunum, ileum, and appendix).

ICD-9 Codes

Conditions that may justify these tests *include but are not limited to*:

Neoplasms
162.9   malignant, of the bronchus and lung
193   malignant, of thyroid gland
211.7   benign, of the islets of Langerhans

Other
255.6   medulloadrenal hyperfunction
259.2   carcinoid syndrome
277.3   pituitary adenoma
397.0   diseases of tricuspid valve
782.62   flushing
786.07   wheezing, bronchoconstriction
787.91   diarrhea

Suggested Reading

Bajetta E, Ferrari L, Martinetti A, Celio L, Procopio G, Artale S, Zilembo N, Di Bartolomeo M, Seregni E, Bombardieri E. Chromogranin A, neuron specific enolase, carcinoembryonic antigen, and hydroxyindole acetic acid evaluation in patients with neuroendocrine tumors. *Cancer* 1999; 86:858–65.

Barakat MT, Meeran K, Bloom SR. Neuroendocrine tumours. *Endocr Relat Cancer* 2004; 11:1–18.

Deftos LJ. Chromogranin A: its role in endocrine function and as an endocrine and neuroendocrine tumor marker. *Endocr Rev* 1991; 12:181–7.

Ganim RB, Norton JA. Recent advances in carcinoid pathogenesis, diagnosis and management. *Surg Oncol* 2000; 9:173–9.

Haller DG. Endocrine tumors of the gastrointestinal tract. *Curr Opin Oncol* 1994; 6:72–6.

Lamberts SW, Hofland LJ, Nobels FR. Neuroendocrine tumor markers. *Front Neuroendocrinol* 2001; 22:309–39.

Mansour JC, Chen H. Pancreatic endocrine tumors. *J Surg Res* 2004; 120:139–61.

Nobels FR, Kwekkeboom DJ, Bouillon R, Lamberts SW. Chromogranin A: its clinical value as marker of neuroendocrine tumours. *Eur J Clin Invest* 1998; 28:431–40.

Schurmann G, Raeth U, Wiedenmann B, Buhr H, Herfarth C. Serum chromogranin A in the diagnosis and follow-up of neuroendocrine tumors of the gastroenteropancreatic tract. *World J Surg* 1992; 16:697–701.

Taupenot L, Harper KL, O'Connor DT. The chromogranin–secretogranin family. *N Engl J Med* 2003; 348:1134–49.

Tomassetti P, Migliori M, Simoni P, Casadei R, De Iasio R, Corinaldesi R., Gullo L. Diagnostic value of plasma chromogranin A in neuroendocrine tumours. *Eur J Gastroenterol Hepatol* 2001; 13:55–8.

10.3   Other Hormone-Producing Gastrointestinal Tumors

10.3.1   Glucagon: A Test for the Presence of a Glucagonoma

Indications for Test

Glucagon measurement is indicated in patients with:

- The 4D syndrome (Table 10.1)
- Hypoaminoacidemia and normocytic anemia (90% of cases), as well as hyperglycemia, rash, cheilitis, venous thrombosis, weight loss, and neuropsychiatric phenomena

Procedure

1. Obtain a fasting blood sample for glucagon determination.
2. Abdominal imaging studies are indicated if [glucagon] is >500 pg/mL. Because glucagonomas tend to be bulky, screen with abdominal ultrasonography before using more advanced imaging techniques (i.e., multislice CT or high-resolution MR).
3. If the [glucagon] is <200 pg/mL in a patient with necrolytic migratory erythema (NME) dermatitis (Kahan et al., 1977), obtain a blood specimen for amino acid testing.

Interpretation

1. The reference interval for [glucagon] is 50 to 200 pg/mL.
2. All patients with a glucagonoma have an elevated fasting [glucagon], usually well in excess of 1000 pg/mL, with a mean level of 2110 pg/mL (Stacpoole, 1981).
3. Very few patients with a glucagonoma have been described with [glucagon] in the range of 200 to 500 pg/mL.
4. Elevated fasting serum [glucagon] is usually associated with low plasma amino acid levels.
5. False elevations of [glucagon] can occur in other conditions such as diabetes mellitus, Cushing's syndrome, hepatic cirrhosis, pancreatitis, chronic kidney disease, acromegaly, obesity, and starvation, but levels rarely exceed 500 pg/mL in any of these conditions.
6. Glucagonomas are typically >4 cm at the time of diagnosis. They are usually found within the body and tail of the pancreas and can be readily imaged with current methodologies.

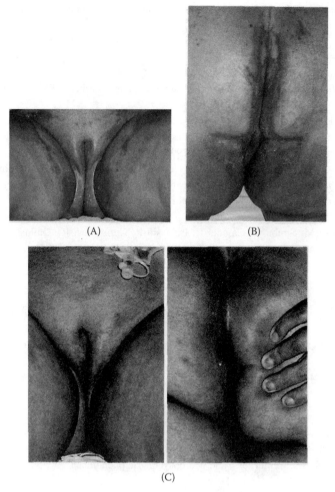

Figure 10.1

(A,B) Necrolytic migratory erythema with saprophytic superinfection in 31-year-old with a 5.2 × 3 × 4-cm glucagonoma arising from the pancreas. Glucagon levels fell from >2600 pg/mL to <100 pg/mL after surgical removal of mass. (C) Rash resolved in less than 1 week after surgery. Imaging studies showed no evidence of metastases after surgery.

7. An abdominal multislice CT (M-CT) scan with intravenous contrast will localize most (86%) of the typically large glucagonoma tumors as well as hepatic metastases.

8. Tumors too small to be seen by M-CT scan may be best visualized by endoscopic ultrasound, which has a sensitivity of 82% and a specificity of 95% for detection of glucagonoma (Rosch et al., 1992).

Notes

1. Determination of [glucagon] is strongly recommended in diabetes patients in whom an necrolytic rash (i.e., NME) fails to respond to antifungal or steroid therapy (Figure 10.1).

2. Detailed descriptions of approximately 300 cases of glucagonoma syndrome have appeared as of 2003 (Mansour and Chen, 2004), with an annual incidence estimated to be 1/20,000,000 (Stacpoole, 1981) and a prevalence of ≥13.5/20,000,000 (Echenique-Elizondo et al., 2004).

3. Pseudoglucagonoma syndrome characterized by NME dermatitis and one or more other manifestations of the syndrome, but no pancreatic tumor, is extremely rare.

4. Weight loss is not an invariant finding in the glucagonoma patient (Dons and Cashell, 1988).

5. The dermatitic rash, including cheilitis, found in patients with a glucagonoma may appear similar to the skin manifestations of kwashiorkor, pellagra, toxic epidermal necrolysis, or zinc deficiency and is probably related to decreased levels of circulating amino acids.

ICD-9 Codes

Conditions that may justify this test include *but are not limited to*:

Psychological illness
298.9   psychosis
311   depressive disorder, unspecified

Other
250.X   diabetes mellitus
270.7   other disturbances of straight-chain amino-acid metabolism
280.0   blood loss (chronic)
453.40   venous embolism and thrombosis of lower extremity
691.0   necrolytic migratory erythematous rash
783.21   weight loss, abnormal

Suggested Reading

Becker SW, Kahn D. Cutaneous manifestations of internal malignant tumors. *Arch Dermatol Syphiol* 1942; 45:1069–80.

Dons RF, Cashell AW. A recurrent intertriginous rash responsive to topical as well as surgical therapy: necrolytic migratory erythema with saprophytic superinfection. *Arch Dermatol* 1988; 124:431, 434.

Echenique-Elizondo M, Valls AT, Orúe JLE, Martinez de Lizarduy I, Aguirre JI. Glucagonoma and pseudo-glucagonoma syndrome. *J Pancreas* 2004; 5:179–85.

Gantcheva ML, Broshtilova VK, Lalova AI. Necrolytic migratory erythema: the outermost marker for glucagonoma syndrome. *Arch Dermatol* 2007; 143:1221–22.

Higgins GA, Fischman AB. The glucagonoma syndrome: surgically curable diabetes. *Am J Surg* 1979; 137:142–8.

Kahan RS, Perez-Figaredo RA, Neimanis A. Necrolytic migratory erythema: distinctive dermatosis of the glucagonoma syndrome. *Arch Dermatol* 1977; 113:792–7.

Mallinson CN, Bloom SR, Warin AP, Salmon PR, Cox B. A glucagonoma syndrome. *Lancet* 1974; 2:1–5.

Mansour JC, Chen H. Pancreatic endocrine tumors. *J Surg Res* 2004; 120:139–61.

Norton JA, Kahn CR, Schiebinger R, Gorschboth C, Brennan MF. Amino acid deficiency and the skin rash associated with glucagonoma. *Ann Int Med* 1979; 91:213–5.

Rosch T, Lightdale CJ, Botet JF et al. Localization of pancreatic endocrine tumors by endoscopic ultrasonography [comment]. *N Engl J Med* 1992; 326:1721–26.

Smith AP, Doolas A, Staren ED. Rapid resolution of necrolytic migratory erythema after glucagonoma resection. *J Surg Oncol* 1996; 61:306–9.

Stacpoole PW. The glucagonoma syndrome: clinical features, diagnosis, and treatment. *Endocr Rev* 1981; 2:347–61.

Vinik AI, Moattari AR. Treatment of endocrine tumors of the pancreas. *Endocrinol Metab Clin North Am* 1989; 18:483–518.

Wermers RA, Fatourechi V, Kvols LK. Clinical spectrum of hyperglucagonemia associated with malignant neuroendocrine tumors. *Mayo Clin Proc* 1996a; 71:1030–38.

Wermers RA, Fatourechi V, Wynne AG, Kvols LK, Lloyd RV. The glucagonoma syndrome: clinical and pathologic features in 21 patients. *Medicine* 1996b; 75:53–63.

10.3.2 Somatostatin: A Test for the Presence of a Somatostatinoma

Indications for Test

Somatostatin measurement is indicated when a patient presents with both a pancreatic or intestinal tumor and an inhibitory syndrome, including:

- Diabetes mellitus
- Steatorrhea with gastrointestinal hypochlorhydria

- Cholelithiasis
- Presence of a pancreatic tumor

Procedure

1. Obtain a fasting blood specimen for [somatostatin].

Interpretation

1. The reference interval for [somatostatin] is <100 pg/mL (61.1 pmol/L).
2. Patients with an inhibitory syndrome due to a pancreatic somatostatinoma have a markedly elevated [somatostatin], usually >1000 pg/mL (611 pmol/L).
3. Patients with somatostatinomas of the small intestine generally have far lower [somatostatin] levels than those with pancreatic somatostatinomas and are frequently asymptomatic.
4. Somatostatin secretion from tumors other than somatostatinomas may occur. These include small-cell lung cancer, medullary thyroid cancer, or pheochromocytoma. Measurement of [somatostatin] may be used as a tumor marker and index of response to antineoplastic therapy.

Notes

1. Somatostatin, a tetradecapeptide, inhibits the secretion of insulin, glucagon, gastrin, and growth hormone as well as cholecystokinin (CCK)-mediated secretion of pancreatic enzymes, intestinal absorption, and gastric acid secretion (Patel, 1992).
2. The median age at diagnosis of a somatostatinoma is 50 (range, 26–84 years), with an equal gender distribution (Harris et al., 1987).
3. Most somatostatinomas are solitary with about 75% of somatostatinomas located in the pancreas and 66% located in the head of the pancreas.
4. Measuring [somatostatin] has not been as sensitive for detection of intestinal as it has been for pancreatic somatostatinomas (Mao et al., 1995).
5. Somatostatinomas are often discovered incidentally during evaluation or operation for an unrelated complaint or problem (Norton, 1997).
6. Seventy-five percent of somatostatinomas are metastatic at diagnosis, with a size typically >5 cm (Snow and Liddle, 1995). The relatively large size of these tumors at the time of diagnosis makes multislice CT scanning an ideal first choice for imaging (Konomi et al., 1990), with somatostatin analog scintigraphy (see Test 10.4.1) as a second choice (Angeletti et al., 1998).

ICD-9 Codes

Conditions that may justify this test include *but are not limited to*:

250.XX   diabetes mellitus
536.8   dyspepsia and other specified disorders of function of stomach
574   cholelithiasis
579.4   malignant neoplasm pancreas
751.7   anomalies of pancreas
787.91   diarrhea, steatorrhea

Suggested Reading

Angeletti S, Corleto VD, Schillaci O, Marignani M, Annibale B, Moretti A, Silecchia G, Scopinaro F, Basso N, Bordi C, Delle FG. Use of the somatostatin analogue octreotide to localise and manage somatostatin-producing tumours. *Gut* 1998; 42:792–4.

Etzrodt H, Rosenthal J, Schroder KE, Pfeiffer EF. Radioimmunoassay of somatostatin in human plasma. *Clin Chim Acta* 1983; 133:241–51.

Ganda OP, Weir GC, Soeldner JS, Legg MA, Chick WL, Patel YC, Ebeid AM, Gabbay KH, Reichlin S. Somatostatinoma: a somatostatin-containing tumor of the endocrine pancreas. *N Engl J Med* 1977; 296:963–7.

Harris GJ, Tio F, Cruz AB. Somatostatinoma: a case report and review of the literature. *J Surg Oncol* 1987; 36:8–16.

Jensen RT, Norton A. Endocrine neoplasms of the pancreas. In: *Textbook of Gastroenterology*, Yamada, T et al. (Eds). 1995. Lippincott, p. 2131.

Konomi K, Chijiiwa K, Katsuta T, Yamaguchi K. Pancreatic somatostatinoma: a case report and review of the literature. *J Surg Oncol* 1990; 43:259–65.

Mao C, Shah A, Hanson DJ, Howard JM. Von Recklinghausen's disease associated with duodenal somatostatinoma: contrast of duodenal versus pancreatic somatostatinomas. *J Surg Oncol* 1995; 59:67–73.

Norton JA. Somatostatinoma and rare pancreatic endocrine tumors. In: *Textbook of Endocrine Surgery*. Clark OH, Duh QY (Eds). 1997. Saunders, p. 626.

O'Brien TD, Chejfec G, Prinz RA. Clinical features of duodenal somatostatinomas. *Surgery* 1993; 114:1144–7.

Patel Y. General aspects of the biology and function of somatostatin. In: *Basic and Clinical Aspects of Neuroscience*, Well C, Muller E, Thorner M (Eds). 1992. Springer, p. 1.

Rosch T. Functional endocrine tumors of the pancreas: clinical presentation, diagnosis and treatment. *Curr Prob Surg* 1990; 27:309–77.

Snow N, Liddle R. Neuroendocrine tumors. In: *Gastrointestinal Cancers: Biology, Diagnosis, and Therapy*, Rusygi A (Ed). 1995. Lippincott-Raven, p. 585.

10.3.3 Vasoactive Intestinal Peptide (VIP): A Test for the Presence of a VIP-Secreting Non-Beta, Islet-Cell Tumor (VIPoma)

Indications for Test

Vasoactive intestinal peptide measurement is indicated when:

- Making the diagnosis of a VIPoma based on clinical signs and symptoms (Table 10.1) associated with low serum potassium [K+]

Procedure

1. Obtain a fasting blood specimen for measurement of [VIP] and [K+].
2. Proceed with imaging study of the pancreas if the [VIP] > 150 pg/mL in the presence of fasting diarrhea >1 L per day.

Interpretation

1. The reference interval for plasma [VIP] is <190 pg/mL.
2. The mean plasma [VIP] in 29 patients with a VIPoma was 956 pg/mL (O'Dorisio et al., 1989).
3. A VIPoma is likely to be present if the [VIP] > 150 pg/mL in the presence of fasting diarrhea > 1 L per day. Ninety percent of VIPomas are found in the pancreas.
4. Patients with a VIPoma may have normal [VIP] when they are not symptomatic.
5. Octreotide therapy may suppress the circulating [VIP], but chronic kidney disease, hepatic cirrhosis, exercise, and diuretic or laxative abuse may raise it.

Notes

1. VIP, a 28-amino-acid protein, stimulates the secretion and inhibits the absorption of sodium, chloride, potassium, and water within the small intestine and increases bowel motility. These actions lead to secretory diarrhea, hypokalemia, and dehydration. Stool volumes of >3 L per day occur in 70% of these patients (Mekhjian and O'Dorisio, 1987).

2. VIPomas are rare tumors occurring with a frequency of 1 per 10 million per year. They arise most often in the pancreas but may appear in the colon, bronchus, adrenals, liver, and sympathetic ganglia (Friesen et al., 1987).

3. Adults usually present between ages 30 and 50 years, whereas children can present with a VIPoma between 2 and 4 years of age.

4. VIPomas are typically solitary and >3 cm in diameter, with 75% located in the tail of the pancreas. Metastases are found in 60 to 80% of patients at the time of diagnosis (Smith et al., 1998).

5. Up to 40% of VIPoma patients have hypercalcemia. Achlorhydria and hypochlorhydria (30% of cases) result from the inhibitory action of VIP on gastric acid secretion.

6. Neurotensinomas remain rare and truly difficult to separate from the symptom complex produced by VIP excess. The additional symptoms of edema, hypotension, cyanosis, and flushing should alert one to the possibility of a neurotensin-secreting tumor (Vinik et al., 1987).

7. Multidetector CT scan imaging will identify a VIP-secreting tumor in almost all proven cases. Somatostatin receptor scintigraphy (see Test 10.4.1) may also be a useful localizing test.

ICD-9 Codes

Conditions that may justify this test include *but are not limited to*:

275.42	hypercalcemia
276.51	dehydration
276.8	hypokalemia
536.0	achlorhydria
787.91	diarrhea

Suggested Reading

Bloom SR, Yiangou Y, Polak JM. Vasoactive intestinal peptide secreting tumors: pathophysiological and clinical correlations. *Ann NY Acad Sci* 1988; 527:518–27.

Friesen SR. Update on the diagnosis and treatment of rare neuroendocrine tumors. *Surg Clin North Am* 1987; 67:379–93.

Mekhjian HS, O'Dorisio TM. VIPoma syndrome. *Sem Oncol* 1987; 14:282–91.

Norton JA. Somatostatinoma and rare pancreatic endocrine tumors. In: *Textbook of Endocrine Surgery*. Clark OH, Duh QY (Eds). 1997. Saunders, p. 626.

O'Dorisio TM, Mekhjian HS, Gaginella TS. Medical therapy of VIPomas. *Endocrinol Metab Clin North Am* 1989; 18:545–56.

Smith SL, Branton SA, Avino AJ, Martin JK, Klingler PJ, Thompson GB, Grant, CS, van Heerden JA. Vasoactive intestinal polypeptide secreting islet cell tumors: a 15-year experience and review of the literature. *Surgery* 1998; 124:1050–5.

Verner JV, Morrison AB. Islet cell tumor and a syndrome of refractory watery diarrhea and hypokalemia. *Am J Med* 1958; 25:374–80.

Vinik AI, Strodel WE, Eckhauser FE, Moattari AR, Lloyd R. Somatostatinomas, PPomas, neurotensinomas. *Sem Oncol* 1987; 14:263–81.

10.3.4 Pancreatic Polypeptide (PP): A Test for the Presence of a Gastrointestinal or PP-Secreting Tumor (PPoma)

Indications for Test

Determination of the circulating [PP] is indicated:

- On finding a tumor in the pancreas or gastrointestinal tract, usually of a large or bulky size
- If the presence of a carcinoid tumor is suspected
- In cases of hepatomegaly, abdominal pain, and jaundice secondary to biliary tract obstruction by a large intrabdominal mass

Procedure

1. Obtain a fasting blood sample for measurement of [PP] and [CgA].

Interpretation

1. A plasma [PP] > 1250 pg/mL (300 pmol/L) is a positive screening test for a PPoma in patients less than 60 years of age. Expect about 2% false positives and about 55% false negatives at this cutoff level.
2. Elevations in [PP] can be seen in chronic kidney disease, old age, alcohol abuse, certain infections (e.g., mumps), chronic inflammatory disorders, acute diarrhea, diabetes mellitus, hypoglycemia, chronic relapsing pancreatitis, and the postprandial state.
3. Normally, [PP] rises in response to just the chewing of food consistent with vagal nerve stimulation while vagotomy or atropine blocks its secretion.
4. Elevated [PP] can be seen in 29 to 50% of patients with carcinoid syndrome, even if the tumor is not within the pancreas.
5. If the fasting [PP] is elevated on two or more determinations, an *atropine suppression test* (AST) may be used to differentiate between normal physiologic secretion of PP vs. secretion from a PPoma; however, the AST is potentially hazardous and has not received widespread acceptance (Adrian et al., 1986).
6. Low levels of PP occur when there is extensive pancreatic destruction.
7. An increased sensitivity in the detection of gastroenteropancreatic (GEP) tumors (>94%) may be achieved when measurement of [CgA] is combined with [PP] (Panzuto et al., 2004).

Notes

1. PP is secreted by approximately half of all pancreatic endocrine tumors (PETs) and may be of value as a biochemical marker for a wide variety of PETs. Mucinous pancreatic adenocarcinomas tend not to contain PP positive cells (Tamiolakis et al., 2003).
2. Very few tumors have been described that secrete PP alone. None of these has been associated with any clinical syndrome specifically related to PP secretion.
3. Although diarrhea may be a feature of their presentation, PPomas are usually silent and, because of their large size and malignant nature, tend to present with hepatomegaly, abdominal pain, and jaundice related to biliary obstruction (Vinik et al., 1987).

ICD-9 Codes

Conditions that may justify this test include *but are not limited to*:

.157.4	malignant neoplasm pancreas
576.9	unspecified disorder of biliary tract
782.4	jaundice, unspecified, not of newborn
789.0	abdominal pain
789.1	hepatomegaly
789.3	abdominal or pelvic swelling, mass, or lump

Suggested Reading

Adrian TE, Uttenthal LO, Williams SJ, Bloom SR. Secretion of pancreatic polypeptide in patients with pancreatic endocrine tumors. *N Engl J Med* 1986; 315:287–91.

Panzuto F, Severi C, Cannizaro R, Falconi M, Angeletti S, Pasquali A, Corleto VD, Annibale B, Buonadonna A, Pederzoli P, Delle Fave G. Utility of combined use of plasma levels of chromogranin A and pancreatic polypepetide in the diagnosis of gastrointestinal and pancreatic endocrine tumors. *J Endocrinol Invest* 2004; 27:6–11.

Schwartz TW. Pancreatic polypeptide: a hormone under vagal control. *Gastroenterology* 1983; 85:1411–25.

Tamiolakis D, Simopoulos C, Kotini A, Venizelos I, Jivannakis T, Papadopoulos N. Pancreatic polypeptide in the human pancreas: expression and quantitative variation during development and in ductal adenocarcinoma. *Acta Medica (Hradec Kralove)* 2003; 46:9–14.

Vinik AI, Strodel WE, Eckhauser FE, Moattari AR, Lloyd R. Somatostatinomas, PPomas, neurotensinomas. *Sem Oncol* 1987; 14:263–81.

10.4 Gastroenteropancreatic Tumor Imaging

10.4.1 Somatostatin Receptor Scintigraphy (SRS) Imaging with Somatostatin Analogs

Indications for Test

Somatostatin receptor scintigraphy is indicated:

- As the initial imaging modality of choice when attempting to localize suspected neuroendocrine gastroenteropancreatic (GEP) tumors containing somatostatin receptors (particularly the type 2 and type 5 receptor subtypes)

Procedure

1. Withhold octreotide, if used as therapy, for at least 2 days prior to study. Longer withdrawal may be necessary.
2. Infuse 3 mCi (111 MBq) of pentreotide (OctreoScan®) intravenously for SRS planar imaging.
3. Obtain SRS image of whole body with additional focus on selected areas (i.e., mid-abdomen) at 4 hours and at 24 hours.
4. If an upper abdominal (e.g., GEP tumor) is suspected, infuse 6 mCi (222 MBq) of OctreoScan® intravenously for single-photon emission computed tomography (SPECT).
5. To exclude a false-positive study from physiological uptake as well as clearly identify source of abnormal isotope uptake, obtain conventional ultrasound, multislice CT, or high-resolution MR imaging studies of the areas in question in conjunction with SPECT.

Interpretation

1. Images of concentrated uptake on scan suggest possibility of somatostatin receptor-bearing tumor.
2. Overall, SRS has sensitivity of at least 80% for detection of all PETs other than insulinomas.
3. The sensitivity of SRS for detection of GEP tumors is high but its specificity is low.

Notes

1. Indium-111-pentetreotide (^{111}In-DTPA-D-Phe1-octreotide) is an analog of somatostatin in which the radiolabel ^{111}In is chelated via diethylenetriaminepentaacetic acid (DTPA) to the eight-amino-acid residue octreotide (OctreoScan®; Mallinckrodt, Inc., 888-744-1414, www.imaging.mallinckrodt.com).
2. This test may be used to image tumors of the pancreas (glucagonoma, 100%; insulinoma, 42%), lung (small cell, 100%), stomach (gastrinoma, 100%), liver (carcinoid, 89%), gut (VIPoma, 100%), adrenal (pheochromocytoma, 86%; paraganglioma, 100%), thyroid (medullary carcinoma, 67%), and breast. The sensitivity of detection of these tumors by OctreoScan® is shown here as percentages based on data on file at Mallinckrodt.
3. Occult gastrinomas, difficult to visualize anatomically at surgery or by CT or MR imaging, may be found within an area defined by the junction of the cystic and common bile ducts superiorly, the junction of the second and third portions of the duodenum inferiorly, and the junction of the neck and body of the pancreas medially (Stabile et al., 1990).

4. Intraoperative gamma detection of radiolabeled peptides may allow the localization of occult tumors that contain specific peptide receptors (Woltering et al., 1994).
5. The sensitivity of planar SRS is less than with SPECT. When these two techniques were compared in a study of 52 patients with known tumors, planar imaging detected primary lesions in 13 patients and hepatic metastases in 21 patients, whereas SPECT found 31 primary lesions in 27 patients and hepatic metastases in 28 patients (Schillaci et al., 1999).
6. SPECT showed a significantly higher per lesion sensitivity (89.6%) than CT, MR, ultrasound, and angiography (72.6%) which was, in turn, significantly higher than planar SRS (53.3%) (Schillaci et al., 2003).
7. The sensitivity of planar SRS for detection of gastrinomas was 58%, equaling the sensitivity for ultrasound, angiography, and MR combined (Gibril et al., 1996). In another study, 40% (16 of 40) of metastatic PET lesions not found on transabdominal ultrasound, CT scanning, or MR were identified by SRS (Scherubl et al., 1993).
8. More lesions with positive or inconclusive uptakes will be detected with SRS than will be found with other morphological imaging techniques (e.g., 51 vs. 27 lesions in series from Saga, 2005), but many physiological uptakes will be detected as well.
9. SRS sensitivity was found to be 100% for limited GEP tumors vs. 72% for advanced tumor disease, whereas the chromogranin A sensitivity was 43% for limited vs. 57% for advanced disease (Cimitan et al., 2003).

ICD-9 Codes

Conditions that may justify this test include *but are not limited to*:

195   malignant neoplasm of other and ill-defined sites
751.7   anomalies of pancreas

Suggested Reading

Cadiot G, Bonnaud G, Lebtahi R, Sarda L, Ruszniewski P, le Guludec D, Mignon M. Usefulness of somatostatin receptor scintigraphy in the management of patients with Zöllinger–Ellison syndrome. *Gut* 1997; 41:107–14.

Cimitan M, Buonadonna A, Cannizzaro R, Canzonieri V, Borsatti E, Ruffo R, De Apollonia L. Somatostatin receptor scintigraphy versus chromogranin A assay in the management of patients with neuroendocrine tumors of different types: clinical role. *Ann Oncol* 2003; 14:1135–41.

Gibril F, Jensen RT. Diagnostic uses of radiolabelled somatostatin receptor analogues in gastroenteropancreatic endocrine tumours. *Dig Liver Dis* 2004; 36(Suppl 1):S106–20.

Gibril F, Reynolds JC, Doppman JL, Chen CC, Venzon DJ, Termanini B, Weber HC, Stewart CA, Jensen RT. Somatostatin receptor scintigraphy: its sensitivity compared with that of other imaging methods in detecting primary and metastatic gastrinomas—a prospective study [comment]. *Ann Intern Med* 1996; 125:26–34.

Gibril F, Doppman JL, Reynolds JC, Chen CC, Sutliff VE, Yu F, Serrano J, Venzon DJ, Jensen RT. Bone metastases in patients with gastrinomas: a prospective study of bone scanning, somatostatin receptor scanning, and MRI in their detection, their frequency, location, and effect of their detection on management. *J Clin Oncol* 1998; 16:1040–53.

Krenning EP, Kwekkeboom DJ, Bakker WH, Breeman WAP, Kooij PPM, Oei HY et al. Somatostatin receptor scintigraphy with [^{111}In-DTPA-D-Phe1]- and [^{123}I-Tyr3]-octreotide: the Rotterdam experience with more than 1000 patients. *Eur J Nucl Med* 1993; 20:716–31.

Reisine T, Bell GI. Molecular biology of somatostatin receptors. *Endocr Rev* 1995; 16:427–42.

Rufini M, Calcagni M, Baum R. Imaging of neuroendocrine tumors. *Sem Nucl Med* 2006; 36:228–47.

Saga T, Shimatsu A, Koizumi K, Ichikawa T, Yamamoto K, Noguchi S, Doi R, Ishibashi M, Machinami R, Nakamura K, Sakahara H, Endo K. Morphological imaging in the localization of neuroendocrine gastroenteropancreatic tumors found by somatostatin receptor scintigraphy. *Acta Radiol* 2005; 46:227–32.

Scherubl H, Bader M, Fett U, Hamm B, Schmidt-Gayk H, Koppenhagen K, Dop FJ, Riecken EO, Wiedenmann B. Somatostatin-receptor imaging of neuroendocrine gastroenteropancreatic tumors [comment]. *Gastroenterology* 1993; 105:1705–10.

Schillaci O, Corleto VD, Annibale B, Scopinaro F, Delle FG. Single photon emission computed tomography procedure improves accuracy of somatostatin receptor scintigraphy in gastro–entero–pancreatic tumours. *Ital J Gastroenterol Hepatol* 1999; 31(Suppl 2):S186–9.

Schillaci O, Spanu A, Scopinaro F, Falchi A, Corleto V, Danieli R, Marongiu P, Pisu N, Madeddu G, Delle Fave G, Madeddu G. Somatostatin receptor scintigraphy with ^{111}In–pentetreotide in non-functioning gastroenteropancreatic neuroendocrine tumors. *Int J Oncol* 2003; 23:1687–95.

Stabile BE, Morrow DJ, Passaro E. The gastrinoma triangle: operative implications. *Am J Surg* 1984; 147:25–31.

Woltering EA, Barrie R, O'Dorisio TM, O'Dorisio MS, Nance R, Cook DM. Detection of occult gastrinomas with iodine 125-labeled lanreotide and intraoperative gamma detection. *Surgery* 1994; 116:1139–47.

Woltering EA, O'Dorisio MS, Murphy WA, Chen F, Drouant GJ, Espenan GD, Fisher DR, Sharma C, Diaco DS, Maloney TM, Fuselier JA, Nelson JA, O'Dorisio TM, Coy DH. Synthesis and characterization of multiply-tyrosinated, multiply-iodinated somatostatin analogs. *J Pept Res* 1999; 53:201–13.

10.4.2 Ultrasonography (USG), Multislice Computed Tomography (M-CT), Magnetic Resonance (MR), and Venous Sampling for Imaging and Localization of Gastroenteropancreatic Endocrine Tumors Including Insulinomas

Indications for Test

Use of one or more of the above methods to localize pancreatic endocrine tumors is indicated to:

- Reduce the operative time required to search for mass lesions
- Limit damage or injury to normal tissue upon surgical removal of a pancreatic tumor (Huai et al., 1998; Kuzin et al., 1998)

Procedures

Ultrasonography

1. Access the services of a skilled and experienced ultrasonographer.
2. Proceed with transabdominal USG, the most widely available, low cost, and noninvasive imaging procedure.
3. Using conscious sedation and the help of an experienced and skilled endoscopist, proceed with endoscopic ultrasound (EUS) to visualize smaller PETs located in the head of the pancreas or duodenum. Note that the distal pancreas cannot be seen with EUS.

Multislice CT

1. If no tumor is seen using USG, proceed with M-CT scanning to obtain both arterial and portal venous phase images to identify intrapancreatic and hepatic lesions.
2. Use water as an oral contrast agent to help define duodenal tumors such as gastrinomas when using M-CT; otherwise, use M-CT with iodinated contrast agents to localize PETs in patients with normal renal function (Fidler and Johnson, 2001).

High-Resolution MR Imaging

1. As an alternative to M-CT with contrast, proceed with MR imaging, recognizing its inherent lower sensitivity for detection of PETs but its greater safety in chronic kidney disease patients.
2. As an alternative to M-CT and MR, consider proceeding directly to surgery with the use of extremely sensitive intraoperative USG imaging, recognizing that the pancreas must be fully mobilized for the surgeon to employ this modality.

Venous Sampling

1. In specialized situations, with the assistance of an experienced invasive radiologist, have percutaneous transhepatic portal venous sampling (PTPVS) performed upon placement of a sampling catheter percutaneously through the liver into the portal vein and then into the splenic and/or superior mesenteric veins (Bottger and Junginger, 1993).
2. In the specialized situation of gastrinoma localization, employ the services of an experienced interventional radiologist to selectively infuse secretin into the gastroduodenal, splenic, and common hepatic arteries via the femoral artery and obtain multiple venous samples for gastrin.

Interpretation

1. On USG, most PETs appear as hypoechoic, well-circumscribed masses within the pancreas. Conversely, liver metastases appear as hyperechoic lesions.
2. When compared to M-CT scanning and MR, EUS has a greater sensitivity for detecting tumors smaller than 3 cm (Muller et al., 1994; Rosch et al., 1992). This sensitivity is as high as 93% for detecting PETs of all sizes (Proye et al., 1998).
3. The sensitivity of M-CT is directly related to the size of the tumor. For tumors < 1 cm, sensitivity is <10%. For tumors 3 to 4 cm, the sensitivity increases to 40% (Orbuch et al., 1995). For tumors such as VIPomas, glucagonomas, somatostatinomas, and nonfunctioning tumors usually diagnosed at a larger size (>3 cm), the sensitivity is much greater and approaches 100% (King et al., 1994).
4. Using gadolinium as contrast, MR imaging identified no lesions < 1 cm, 50% of lesions between 1 and 2 cm in diameter, and all neoplasms > 3 cm (Boukhman et al., 1999).
5. Intraoperative USG can identify 75 to 90% of insulinomas without palpation. In addition, this technique defines the relationship between the tumor and surrounding structures for preoperative planning (Bottger and Junginger, 1993; Zeiger et al., 1993).
6. The sensitivity of PTPVS for localization of insulinomas and gastrinomas is over 80%.

Notes

1. The sensitivity of transabdominal ultrasound is, at best, 66 to 80%, with a slightly higher sensitivity for hepatic metastases. In actual practice, most centers report sensitivities of 40 to 60% for detection of PETs using this technique (Angeli et al., 1997; Galiber et al., 1998).
2. The hypervascular nature of most PETs makes contrast-enhanced M-CT imaging more sensitive than unenhanced imaging for detection of these tumors.
3. Using MR, PETs demonstrate low signal intensity on T1-weighted images and high signal intensity on T2-weighted images. Using MR with short time inversion recovery (STIR) sequences has increased the sensitivity for detection of hepatic metastases up to 83%.
4. Arterial infusion of secretogogs with venous sampling of hormones is an invasive, time-consuming, and operator-dependent procedure. An experienced interventional radiologist must be in attendance for successful use of this procedure.

ICD-9 Codes

Conditions that may justify this test include *but are not limited to*:

157.4   malignant neoplasm pancreas
211.7   neoplasms of the islets of Langerhans

Suggested Reading

Angeli E, Vanzulli A, Castrucci M, Venturini, M, Sironi, S, Zerbi A, Di, C, Pozza G, Del Maschio A. Value of abdominal sonography and MR imaging at 0.5 T in preoperative detection of pancreatic insulinoma: a comparison with dynamic CT and angiography. *Abdom Imaging* 1997; 22:295–303.

Bottger TC, Junginger T. Is preoperative radiographic localization of islet cell tumors in patients with insulinoma necessary? *World J Surg* 1993; 17:427–2.

Boukhman MP, Karam, JM, Shaver J, Siperstein AE, DeLorimier AA, Clark OH. Localization of insulinomas. *Arch Surg* 1999; 134:818–22.

Fidler JL, Johnson CD. Imaging of neuroendocrine tumors of the pancreas. *Int J Gastrointestinal Cancer* 2001; 30:73–85.

Galiber AK, Reading CC, Charboneau JW, Sheedy PF, James EM, Gorman B, Grant CS, van Heerden JA, Telander RL. Localization of pancreatic insulinoma: comparison of pre- and intraoperative US with CT and angiography. *Radiology* 1988; 166:405–8.

Huai, JC, Zhang W, Niu HO, Su ZX, McNamara JJ, Machi, J. Localization and surgical treatment of pancreatic insulinomas guided by intraoperative ultrasound. *Am J Surg* 1998; 175:18–21.

Kuzin NM, Egorov AV, Kondrashin SA, Lotov AN, Kuznetzov NS, Majorova, JB. Preoperative and intraoperative topographic diagnosis of insulinomas. *World J Surg* 1998; 22:593–7.

Muller MF, Meyenberger C, Bertschinger P, Schaer R, Marincek B. Pancreatic tumors: evaluation with endoscopic US, CT, and MR imaging. *Radiology* 1994; 190:745–51.

Proye C., Malvaux P, Pattou F, Filoche B, Godchaux, JM, Maunoury, V, Palazzo L, Huglo D, Lefebvre J, Paris JC. Noninvasive imaging of insulinomas and gastrinomas with endoscopic ultrasonography and somatostatin receptor scintigraphy. *Surgery* 1998; 124:1134–43.

Rosch T, Lightdale CJ, Botet JF, Boyce GA, Sivak MV, Yasuda K, Heyder N, Palazzo L, Dancygier H, Schusdziarra V. Localization of pancreatic endocrine tumors by endoscopic ultrasonography [comment]. *N Engl J Med* 1992; 326:1721–6.

Semelka RC, Custodio CM, Cem BN, Woosley JT. Neuroendocrine tumors of the pancreas: spectrum of appearances on MRI. *J Magn Reson Imaging* 2000; 11:141–8.

Zeiger MA, Shawker TH, Norton JA. Use of intraoperative ultrasonography to localize islet cell tumors. *World J Surg* 1993; 17:448–54.

10.5   Tumor Markers Secreted from Endocrine Tissue

10.5.1   Prostate-Specific Antigen (PSA) as a Tumor Marker for the Presence of Prostate Cancer (PCa) and Its Recurrence

Refer to Test 8.3.4.

10.5.2   Chromogranin A (CgA) Testing in Patients with Neuroendocrine Tumors Other than Carcinoid

Indications for Test

CgA testing is indicated when:

- Gathering supportive evidence for the diagnosis of a neuroendocrine tumor—pheochromocytoma, medullary thyroid cancer (MTC), islet-cell tumors, pituitary tumors, gastrointestinal amine precursor uptake and decarboxylation (APUD) tumors, gastrinomas, small-cell lung cancer, and advanced prostate cancer—if other tests for those tumors are equivocal
- Seeking a marker to monitor the status of neuroendocrine tumor after treatment
- Attempting to identify patients with a large neuroendocrine tumor burden or metastatic disease (Bajetta et al., 1999)

Procedure

Refer to Test 10.2.3.

Interpretation

1. CgA levels are usually independent of tumor functionality or hormone secretory capacity and may be elevated when hormone markers of the tumor are not present.
2. Elevated plasma CgA levels predict pheochromocytoma with 83% sensitivity and 96% specificity if the creatinine clearance is within normal limits.
3. CgA levels tend to correlate well with overall catecholamine production and may predict pheochromocytoma tumor bulk.
4. Cg A has a relatively low specificity (67%) for the diagnosis of gastrinoma.
5. Varying percentages of patients with other neuroendocrine tumors have elevated CgA, including those with gastrinomas (100%), pheochromocytomas (89%), nonfunctioning tumors of the endocrine pancreas (69%), and MTC (50%).
6. Out of 33 neuroendocrine tumor patients, CgA was elevated in 30 (91%) with levels (mean ± SEM) dependent on tumor location, as shown below (Schurmann et al., 1992):
   - Pancreas, 7068 ± 3008 ng/mL ($n = 8$)
   - Ileum, 5381 ± 1740 ng/mL ($n = 18$)
   - Stomach, 529 ± 179 ng/mL ($n = 7$)

Refer also to Test 10.2.3.

Notes

1. Other neuroendocrine tumor markers include plasma free metanephrine in pheochromocytoma, calcitonin in MTC, and gastrin in pancreatic gastrinoma.
2. Proton pump inhibitors will stimulate secretion of CgA from normal neuroendocrine cells unless such drugs are discontinued for four to five drug half-lives.
3. CgA testing is unreliable in patients with significant or fluctuating degrees of renal or hepatic impairment.

Additional notes regarding CgA may be found in Test 10.2.3.

ICD-9 Codes

Conditions that may justify these tests *include but are not limited to*:

162	malignant neoplasm of trachea, bronchus, and lung
193	thyroid cancer (MTC)
211.7	neoplasms of the islets of Langerhans
227.3	pituitary adenomas
600.2	benign localized hyperplasia of prostate

Suggested Reading

Bajetta E, Ferrari L, Martinetti A, Celio L, Procopio G, Artale S, Zilembo N, Di Bartolomeo M, Seregni E, Bombardieri E. Chromogranin A, neuron specific enolase, carcinoembryonic antigen, and hydroxyindole acetic acid evaluation in patients with neuroendocrine tumors. *Cancer* 1999; 86:858–65.

Barakat MT, Meeran K, Bloom SR. Neuroendocrine tumours. *Endocr Relat Cancer* 2004; 11:1–18.

Deftos LJ. Chromogranin A: Its role in endocrine function and as an endocrine and neuroendocrine tumor marker. *Endocr Rev* 1991; 12:181–7.

Hagn C, Schmid KW, Fischer–Colbrie R, Winkler H. Chromogranin A, B, and C in human adrenal medulla and endocrine tissues. *Lab Investig* 1986; 55:405–11.

Haller DG. Endocrine tumors of the gastrointestinal tract. *Curr Opin Oncol* 1994; 6:72–6.

Hsiao RJ, Parmer RJ, Takiyyuddin MA, O'Connor DT. Chromogranin A storage and secretion: sensitivity and specificity for the diagnosis of pheochromocytoma. *Medicine (Baltimore)* 1991; 70:33–45.

Lamberts SW, Hofland LJ, Nobels FR. Neuroendocrine tumor markers. *Front Neuroendocrinol* 2001; 22:309–10.

Mansour JC, Chen H. Pancreatic endocrine tumors. *J Surg Res* 2004; 120:139–61.

Nobels FR, Kwekkeboom DJ, Bouillon R, Lamberts SW. Chromogranin A: its clinical value as marker of neuroendocrine tumours. *Eur J Clin Invest* 1998; 28:431–40.

Schurmann G, Raeth U, Wiedenmann B, Buhr H, Herfarth C. Serum chromogranin A in the diagnosis and follow-up of neuroendocrine tumors of the gastroenteropancreatic tract. *World J Surg* 1992:16:697–701.

Taupenot L, Harper KL, O'Connor DT. The chromogranin-secretogranin family. *N Engl J Med* 2003; 348:1134–49.

Tomassetti P, Migliori M, Simoni P, Casadei R, De Iasio R, Corinaldesi R., Gullo L. Diagnostic value of plasma chromogranin A in neuroendocrine tumours. *Eur J Gastroenterol Hepatol* 2001; 13:55–8.

Tricoli JV, Schoenfeldt M, Conley BA. Detection of prostate cancer and predicting progression: current and future diagnostic markers. *Clin Cancer Res* 2004; 10:3943–53.

Chapter

11

Tests for Nutritional Status

11.1 Body Composition Analysis and Nutritional Assessment

11.1.1 Measurement of Oxygen Consumption (VO_2) and Resting Metabolic Rate (RMR) by Indirect Calorimetry and Calculation of Total Energy Expenditure (TEE)

Indications for Test

Measurement of VO_2 and RMR by indirect calorimetry (IC) with calculation of TEE is indicated:

- When making a determination of a patient's energy requirements
- When recommending caloric intake during acute or chronic disease states and weight-management programs

Procedure

1. Review the patient's weight history for any recent extreme changes in weight (>15 lb or 7 kg per month). Defer study of individuals who have undergone recent, marked change in weight as major shifts in fluid balance are likely to have occurred.
2. Ideal candidates for testing are those with a stable calorie intake for the last 7 to 14 days.
3. Instruct the patient on the following pretest requirements:
   - No eating or calorie intake for at least 4 hours before the test and no prolonged fasting (i.e., >18 hours)
   - No aerobic or strength training exercise for at least 4 and ideally 8 hours before the test
   - At least 10 to 15 minutes of quiet rest time before the test
   - No caffeine or nutritional supplements for at least 4 hours before the test
   - No ephedra, Ma Huang, pseudoephedrine, or other cardiac stimulants for at least 4 hours before the test
   - No smoking or use of nicotine for at least 1 hour before the test
   - No acute therapy (e.g., antibiotics) or transient, acute illness <24 hours before testing
4. Obtain the accurate height (cm), weight (kg), age (date of birth), and gender of the patient.
5. Determine the patient's usual activity level (athletic, moderately active, or sedentary) (Table 11.1).

TABLE 11.1
Activity Factors for Use
in Calculating the Thermic
Effect of Exercise

Activity	Activity Factor
Confined to bed	1.2
Ambulatory, low activity	1.3
Average activity	1.5–1.75
Highly active	2.0

Source: Adapted from Zeman, F.J., Ed., *Clinical Nutrition and Dietetics*, Macmillan New York, 1991; Long, C.L., in *Nutritional Assessment*, Wright, R.A. and Heymsfield, S.B., Eds., Blackwell Scientific, Boston, 1984.

6. Determine the time of the patient's last food intake and physical activity. Expect under-reporting of caloric intake and over-reporting of physical activity in weight-management clients for a variety of reasons, including:
   - Inaccurate portion assessment or incomplete recall
   - Psychosocial motivation or unconscious process of denial
7. During the test period, the patient must be comfortably recumbent or in a reclining chair in a quiet, preferably darkened room. Minimize all activity during the test. Even blepharospasm (i.e., repeated eye blinking) can lead to a falsely elevated VO_2 measurement.
8. Measure VO_2 and RMR by having the patient breathe into an indirect calorimeter or CO_2/O_2 analyzer (e.g., MedGem®, Microlife, Dunedin, FL; New Leaf®, Angeion Corp., St. Paul, MN) while the patient is semirecumbent for 5 to 10 minutes with a tight seal of the mask around the nose and mouth.
9. Refer to the IC device for readout of the values for VO_2 and RMR, a measure of calorie expenditure at rest (kcal/day).
10. Calculate the TEE by multiplying the RMR value by an activity factor obtained from Table 11.1.
11. When possible, calculate the respiratory quotient (RQ), defined as the ratio of carbon dioxide production (VCO_2) to oxygen consumption (VO_2):

$$RQ = VCO_2/VO_2$$

Note that the MedGem® instrument uses a fixed RQ value of 0.85 in its calculation of RMR.

12. Use a commercially available computer program to calculate the patient's caloric requirements for weight maintenance, gain, or loss.
13. Use the following approximate equivalents to convert nutrients to calories:
    - 1 gram carbohydrate = 4 calories.
    - 1 gram fat = 9 calories.
    - 1 gram protein = 4 calories.
14. Per current U.S. Public Health Service (USPHS) recommendations, provide the patient with advice on appropriate intake of calories distributed as carbohydrate, fat, and protein.

Interpretation

1. Various disease processes and physiologic conditions associated with increased energy expenditure and increased energy requirements result in:
   - Direct stimulation of the sympathetic nervous system (i.e., flight or fight reactions)
   - Increased oxygen delivery or increased motor activity (e.g., exercise)
   - Elevated body temperature
   - Indirect promotion of uncoupled or inefficient metabolism
   - Cytokine mediator release

TABLE 11.2
Relative Contribution of Different Tissues to Resting Metabolic Rate as a Function of Body Composition

Tissue	Body Weight (%)	Resting Metabolic Rate (%)
Heart, liver, kidneys, and brain	5–6	60–70
Skeletal muscle	30–40	16–22

Source: Elia, M., in *Energy Metabolism: Tissue Determinants and Cellular Corollaries*, Kinney, J. and Tucker, H., Eds., Raven Press, New York, 1992, pp. 61–77. With permission.

2. Underfeeding, which promotes use of endogenous fat stores, tends to cause decreases in the RQ, whereas overfeeding results in lipogenesis, which should cause increases in the RQ.

3. Marked increases in VCO_2 in response to overfeeding may cause respiratory compromise in patients with limited pulmonary reserve. Note that in this situation there will be marked increases in the respiratory quotient.

4. A serial, progressive pattern of inappropriately low RMR estimates for the degree of illness may imply impending septic shock and help to identify hospitalized patients at risk for organ failure and increased mortality.

5. Untreated hypothyroidism may lower the RMR by 20 to 30%, while subclinical hypothyroidism or inadequately treated hypothyroidism may lower the RMR by 10 to 20%.

6. In massively obese adolescents, the RMR for fat-free mass (FFM) might be expected to fall from the beginning of the weight-loss period and remain suppressed as long as energy restriction and weight reduction continues (Tounian et al., 1999).

7. Expect constitutionally lean children to have a low RMR, probably adaptive in nature. In obese children, expect RMR values to increase in proportion to the FFM (Tounian et al., 2003).

8. The RMR will be falsely low in individuals who have undergone a marked loss in weight over the previous 1 or 2 months.

9. Based on the patient's TEE value, an adjustment in current or prescribed caloric intake can be made to maintain the patient's body tissues or to induce a weight change of 0.5 to 2.0 lb/week until the patient achieves his or her ideal body weight.

10. Use the TEE (in kcal/day) to design a balanced diet around proper proportions of nutrients, usually distributed as 55 to 65% carbohydrates, 15 to 30% fat, and 10 to 20% protein.

11. Ordinarily, a weight-loss program should not reduce calorie intake to <1200 kcal/day or induce a calorie deficit of >500 kcal/day below the TEE.

Notes

1. It takes a calorie deficit of about 3500 calories to lose 1 pound of body fat. Ordinarily, recommend a deficit of between 150 and 250 kcal/day. Initially, it is best to accomplish this deficit by increased energy expenditure with little or no reduction in total calorie intake.

2. In obesity, muscle mass makes a significantly greater contribution to the RMR (13 kcal/kg/day) than does fat mass (4.5 kcal/kg/day), but, in aggregate, the heart, kidneys, liver, and brain contribute the most (200 to 400 kcal/kg/day) (Table 11.2).

3. Weight cycling (the losing and regaining of large amounts of weight, or "yo-yo dieting") does not significantly affect measurement of RMR, body composition, or body fat distribution (Wadden et al., 1996).

4. Unfortunately, with or without correcting for the metabolism of protein, the RQ derived from IC has low sensitivity (38.5–55.8%) and specificity (85.1–72.2%) as an indicator of over- or underfeeding. RQ measurements have only a weak, albeit significant, correlation to the actual percentage of calories provided/required.

5. The lowest energy expenditure is associated with sleeping. In mechanically ventilated critically ill patients, even chest physical therapy may be associated with metabolic increases of up to 35% over values during sleep.

6. Over a course of 3 to 5 days, clinically stable ventilator patients have more stable RMR measurements (range, 4–18%) than those who are more acutely ill (range, 37–56%), thus prompting more frequent measurements in the seriously ill population.

7. In ventilator patients, if the variation in VO_2 is low (CV $\leq$ 9.0%), the best estimate of RMR can be obtained by minute-to-minute repeat testing of VO_2 over intervals of $\geq$30 minutes. Greater variations in VO_2 require that repeat testing be done over intervals of $\geq$60 minutes for accurate estimation of RMR in these patients.

8. Using IC (MedGem®), the RMR measured in normal subjects (1607 $\pm$ 37 kcal/day) was significantly higher (5.1%) than the RMR from Sensormedics SM-2900 metabolic cart measurements (1529 $\pm$ 39 kcal/day), although the two were highly correlated ($r = 0.92$; $p < 0.0001$). A portion of the elevated RMR measured by the MedGem® instrument may have resulted from the slightly different body position used during testing with this instrument.

9. Gender specific Harris–Benedict equations yield crude estimates of RMR as follows:
   - Men (kcal/day) = 66 + (6.3 $\times$ weight in pounds) + (12.9 $\times$ height in inches) – (6.8 $\times$ age in years)
   - Women (kcal/day) = 655 + (4.3 $\times$ weight in pounds) + (4.7 $\times$ height in inches) – (4.7 $\times$ age in years)

10. Schofield height/weight equations for estimation of RMR in a mixed population of obese and nonobese children and adolescents, particularly boys, are preferred. Use of the FAO/WHO/UN University equation in girls and the Schofield weight equation in nonobese children yields better estimates of RMR than the Schofield height/weight equations.

11. In young Caucasian women who are normal weight, overweight, or obese, the Owen's, Bernstein, and Robertson–Reid equations, respectively, yield estimates of RMR similar to those derived from IC.

12. When estimating RMR, Owen's equation provides the best compensation for metabolic adaptation, occurring during therapeutic or spontaneous energy restriction.

13. In the critically ill, there is only a moderate correlation between measured RMR using IC and that predicted using the Harris–Benedict ($r = 0.57$) or Aub–DuBois ($r = 0.59$) formulas. The range of RMR, measured by IC, varies widely (70–140%) from that predicted by any of the aforementioned equations.

14. Estimates of VO_2 may be obtained using the metabolic cart (Sensormedics SM-2900) or Douglas bag techniques, both of which are more cumbersome and expensive than IC but useful for hospitalized patients on respirators.

15. Algorithms used to calculate body surface area and estimate metabolic rate remain untested. Little evidence supports their use in this role (Gibson and Numa, 2003).

16. The measurement of caloric requirements based on RMR, rather than RQ, may help to optimize nutritional support in critically ill patients.

17. Measured values for RMR will improve the adequacy of estimates of total parenteral nutrition (TPN) caloric requirements as opposed to use of empiric equations for calculating RMR. Use of these equations tends to result in inaccurate estimates of caloric requirements and results in overfeeding.

18. Critically ill patients require a relatively large volume of enteral (i.e., tube) feeding. Although overfeeding tends to occur during TPN, enteral nutritional support is frequently inadequate because of problems with gastrointestinal intolerance and interruptions in feeding.

19. Based on measured RMR, 41% of hospitalized patients were underfed and 27% were overfed. In contrast, when long-term, total parenteral or enteral feeding patients were studied by measurement of their RQ, 41.5% were overfed (received >110% of required calories), whereas 34.2% were underfed (received <90% of required calories).

20. Use of published guidelines to calculate, rather than measure, the RMR resulted in overfeeding of 1076 $\pm$ 660 kcal/day. A substantial cost savings accrues from use of IC to measure energy needs as it gives a lower, more appropriate number of calories to TPN patients.

ICD-9 Codes

Conditions that may justify this test include *but are not limited to*:

Thyroid disorders
242.9   thyrotoxicosis
244.9   hypothyroidism

Nutrition disorders
263.9   unspecified protein-calorie malnutrition
278.00   obesity
783.1   abnormal weight gain
783.2   abnormal loss of weight and underweight

Suggested Reading

Choban PS, Flancbaum L. Nourishing the obese patient. *Clin Nutr* 2000; 19:305–10.

Elia M. Organ and tissue contribution to metabolic rate. In: *Energy Metabolism: Tissue Determinants and Cellular Corollaries*, Kinney J, Tucker H (Eds). 1992. Raven Press, pp. 61–77.

Foster GD, McGuckin BG. Estimating resting energy expenditure in obesity. *Obes Res* 2001; 9(Suppl 5):367S–374S.

Foster GD, Knox LS, Dempsey DT, Mullen JL. Caloric requirements in total parenteral nutrition. *J Am Coll Nutr* 1987; 6:231–53.

Foster GD, Wadden TA, Mullen JL, Stunkard AJ, Wang J, Feurer ID, Pierson RN, Yang MU, Presta E, Van Itallie TB et al. Resting energy expenditure, body composition, and excess weight in the obese. *Metabolism* 1988; 37:467–72.

Gibson S, Numa A. The importance of metabolic rate and the folly of body surface area calculations. *Anaesthesia* 2003; 58:50–5.

Haugen HA, Melanson EL, Tran ZV. Variability of measured resting metabolic rate. *Am J Clin Nutr* 2003; 78:1141–45.

McClave SA, Snider HL. Understanding the metabolic response to critical illness: factors that cause patients to deviate from the expected pattern of hypermetabolism. *New Horiz* 1994; 2:139–46.

McClave SA, McClain CJ, Snider HL. Should indirect calorimetry be used as part of nutritional assessment? *J Clin Gastroenterol* 2001; 33:14–9.

McClave SA, Spain DA, Skolnick JL, Lowen CC, Kieber MJ, Wickerham PS, Vogt JR, Looney SW. Achievement of steady state optimizes results when performing indirect calorimetry. *J Parenter Enteral Nutr* 2003a; 27:16–20.

McClave SA, Lowen CC, Kleber MJ, McConnell JW, Jung LY, Goldsmith LJ. Clinical use of the respiratory quotient obtained from indirect calorimetry. *J Parenter Enteral Nutr* 2003b; 27:21–6.

Melanson EL, Coelho LB, Tran ZV, Haugen HA, Kearney JT, Hill JO. Validation of the BodyGem Hand-Held Indirect Calorimeter. *Int J Obes Relat Metab Disord* 2004; 28:1479–1484.

Perseghin G. Pathogenesis of obesity and diabetes mellitus: insights provided by indirect calorimetry in humans. *Acta Diabetol* 2001; 38:7–21.

Rodriguez G, Moreno LA, Sarria A, Fleta J, Bueno M. Resting energy expenditure in children and adolescents: agreement between calorimetry and prediction equations. *Clin Nutr* 2002; 21:255–60.

Siervo M, Boschi V, Falconi C. Which RMR prediction equation should we use in normal-weight, overweight and obese women? *Clin Nutr* 2003; 22:193–204.

Tounian P, Frelut ML, Parlier G, Abounaufal C, Aymard N, Veinberg F, Fontaine JL, Girardet JP. Weight loss and changes in energy metabolism in massively obese adolescents. *Int J Obes Relat Metab Disord* 1999; 23:830–7.

Tounian P, Dumas C, Veinberg F, Girardet JP. Resting energy expenditure and substrate utilisation rate in children with constitutional leanness or obesity. *Clin Nutr* 2003; 22:353–7.

Wadden TA, Foster GD, Stunkard AJ, Conill AM. Effects of weight cycling on the resting energy expenditure and body composition of obese women. *Int J Eat Disord* 1996; 19:5–12.

Weinsier RL, Nagy TR, Hunter GR. Do adaptive changes in metabolic rate favor weight regain in weight-reduced individuals? An examination of the set-point theory. *Am J Clin Nutr* 2000; 72:1088–94.

Weissman C, Kemper M, Damask MC, Askanazi J, Hyman AI, Kinney JM. Effect of routine intensive care interactions on metabolic rate. *Chest* 1984; 86:815–8.

Weissman C, Kemper M, Askanazi J, Hyman AI, Kinney JM. Resting metabolic rate of the critically ill patient: measured versus predicted. *Anesthesiology* 1986; 64:673–9.

Weissman C, Kemper M, Hyman AI. Variation in the resting metabolic rate of mechanically ventilated critically ill patients. *Anesth Analg* 1989; 68:457–61.

11.1.2 Body Composition Analysis (BCA): Lean Body Mass and Percent Body Fat (%BF) by Bioelectrical Impedance Absorptiometry (BIA) and Dual-Energy X-Ray Absorptiometry (DXA); Determination of Body Surface Area (BSA)

Indications for Test

Body composition analysis is indicated:

- When monitoring changes in body composition
- In patients undergoing major changes in diet or exercise activity
- During interventions that are known to alter body composition such as growth hormone treatment or the use of weight-gain or weight-loss drugs such as appetite stimulants or suppressors

Measurement of BSA may be indicated:

- In estimating the appropriate dose of a medication and the creatinine clearance (see Test 5.2.1)

Procedure

1. Note the interval since ingestion of the last meal. Assess if patient has been fasting or not.
2. Instruct patients to remove heavy objects from their clothes and take off their shoes and socks.
3. Use a commercial BIA device (e.g., Tanita®) capable of determining the patient's body weight, lean body mass (LBM), body mass index (BMI), fat-free mass (FFM), total body water (TBW), and %BF.
4. When the electrodes have been wiped clean with alcohol (not soap), instruct the patient to stand on them with bare feet.
5. Identify whether or not the patient is athletic and initiate the BIA measurement procedure by entering the patient's age, sex, and measured height and subtracting an estimate of the weight of their clothing (usually 2 to 4 lb). Make a notation of the estimated clothing weight used.
6. If a DXA device is to be used for measurement of %BF, LBM, and TBW, the appropriate computer software, different from that required for measurement of bone mineral density (see Test 7.1.1), must be obtained.
7. BSA may be calculated with the formula:

$$BSA\ (m^2) = W^{0.5378} \times H^{0.3964} \times 0.024265$$

where W = weight in pounds, H = height in inches, and m = meters
8. Alternatively, simply refer to a height and weight nomogram for estimation of BSA.

Interpretation

1. Although good correlations are reported between BIA and DXA for measuring LBM and BF, the following tendencies have been reported:
    - BIA tends to underestimate LBM and overestimate %BF compared to DXA, particularly in children and males, probably due to different assumptions about the constants used in the equations for calculating estimates of these parameters.
    - In diabetics, BIA overestimates (by 5%) the %BF in females but underestimates (by 10%) the %BF in males compared to DXA.
    - The recent ingestion of a meal leads to an additive decrease in the calculated value for BF by both methods.
2. In BIA, meal ingestion leads to an additive decrease in impedance and thus up to a 9 to 10% decrease in the calculated %BF from the beginning to the end of the day, with the most pronounced drop in %BF occurring after the first meal but lasting only 2 to 4 hours after each meal.

3. If the patient's BMI is <18, an anorectic eating disorder should be suspected. When making the diagnosis of such a disorder, use biochemical and psychological testing (e.g., see Tests 11.2.4 and 11.3.1).
4. In contrast to measurement of body composition by BIA, DXA is more precise and reproducible; however, measurement of BCA by DXA testing remains a nonreimbursed research technique.
5. The threshold for an increased risk of morbidity and mortality has been identified at a %BF (by BIA) of >28% for men and >32% for women.
6. Methods used to approximate BSA have little evidence to support their use as an estimate of metabolic rate but may help in calculating doses of medications.

Notes

1. Assessment of %BF helps determine which patients have weight changes related to changes in muscle mass rather than body fat.
2. Assessment of %BF, TBW, and FFM may be useful in designing and monitoring the effectiveness of therapeutic diets.
3. There are good correlations between %BF measured by DXA and BIA independent of age or BMI.
4. The assessment of BSA and LBM may be useful in deciding dosages of medications or test agents.
5. Active acromegaly is characterized by a decreased %BF and lower levels of leptin, both of which increase after the marked decrease in growth hormone following pituitary surgery (Bolanowski et al., 2002).
6. BIA depends on the conduction of electricity through the body and a measurement of the resistance to the flow of current usually from one leg/electrode to the other.
7. Methods other than BIA are widely available to assess %BF and LBM, including skin-fold thickness measurement, underwater weighing, and anthropomorphometry.
8. Underwater weighing requires a wave-free pool, an estimate of pulmonary dead-space volume, and a weighing scale, as well as a patient able to tolerate submersion. The Bod Pod® technique is similar to underwater weighing but is based on air displacement rather than water displacement.
9. Body fat estimation by skin-fold thickness measurement requires a calibrated caliper, a table of skin-fold thicknesses related to age and %BF, and an experienced operator.
10. Anthropomorphometry involves measurement of the neck, waist, and hip circumferences of women and the neck and abdomen circumferences of men to the nearest 0.25 inch. The %BF is then derived from a table of circumferences vs. height, in which circumferences are calculated as (abdomen – neck circumference) for men and [(waist + hip) – neck circumference] for women.
11. Anthropomorphometry is invalid in patients with goiters and in some bodybuilders. The U.S. armed forces abandoned its use after 2005.
12. Estimates of LBM, as derived from measurements of total body potassium and nitrogen by isotopic methods, are cumbersome, imprecise, poorly reproducible, and not widely available.
13. Near-infrared interactance progressively underestimates %BF as the degree of adiposity increases, particularly in extremely obese women (BMI > 50 kg/m^2).

ICD-9 Codes

Conditions that may justify this test include *but are not limited to*:

Thyroid disorders
242.9   thyrotoxicosis
244.9   hypothyroidism

Nutrition disorders
263.9   unspecified protein-calorie malnutrition
278.00   obesity
307.1   anorexia nervosa
783.1   abnormal weight gain
783.2   abnormal loss of weight and underweight
783.6   bulimia

Other
253.3   pituitary dwarfism
E947.0   diuretic toxicity

Suggested Reading

Bolanowski M, Nilsson BE Assessment of human body composition using dual-energy x-ray absorptiometry and bioelectrical impedance analysis. *Med Sci Monit* 2001; 7:1029–33.

Bolanowski M, Milewicz A, Bidzinska B, Jedrzejuk D, Daroszewski J, Mikulski E. Serum leptin levels in acromegaly: a significant role for adipose tissue and fasting insulin/glucose ratio. *Med Sci Monit* 2002; 8:CR685–9.

Elia M, Parkinson SA Diaz E. Evaluation of near-infrared interactance as a method for predicting body composition. *Eur J Clin Nutr* 1990; 44:113–21.

Gibson S, Numa A. The importance of metabolic rate and the folly of body surface area calculations. *Anaesthesia* 2003; 58:50–5.

Klimis-Tavantzis D, Oulare M, Lehnhard H, Cook RA. Near-infrared interactance: validity and use in estimating body composition in adolescents. *Nutr Res* 1992; 12:427–39.

Slinde F, Rossander-Hulthen L. Bioelectrical impedance: effect of 3 identical meals on diurnal impedance variation and calculation of body composition. *Am J Clin Nutr* 2001; 74:474–8.

Tsui EY, Gao XJ, Zinman B. Bioelectrical impedance analysis (BIA) using bipolar foot electrodes in the assessment of body composition in type 2 diabetes mellitus. *Diabet Med* 1998; 15:125–8.

11.1.3 Body Mass Index (BMI) Calculations

Indications for Test

Calculation of BMI is indicated in patients with:

- Central obesity and increased waist circumference who are not bodybuilding athletes
- Clinical or physical signs of conditions associated with obesity such as Cushing's syndrome, Prader–Willi syndrome, or diabetes mellitus (DM)
- Cardiovascular disease (CVD) risk factors

Procedure

1. Measure the patient's height (H) and weight (W), and calculate BMI according to formulas endorsed by the American Dietetics Association or USPHS:

$$BMI = \frac{704.5 \times W \text{ (lb)}}{H^2 \text{ (in.)}}$$

or

$$BMI = \frac{W \text{ (kg)}}{H^2 \text{ (m)}}$$

2. In addition to BMI, use nonstretchable tape to measure waist circumference when assessing a patient's risk for CVD.

Interpretation

1. Obesity may be defined as a BMI > 27.3 for women and > 27.8 for men, depending on ethnicity.
2. All Asian populations have a higher %BF at a lower BMI compared to Caucasians. Asians have a desirable BMI 3 to 4 points lower than that for Caucasians of similar %BF (Table 11.3).

TABLE 11.3
Body Mass Index (BMI) Target Criteria: World Health Organization (WHO) Ethnicity-Blind vs. Proposed Asian Ethnicity-Specific BMI Classifications

Classification of Weight	Body Mass Index (kg/m²)	
	Proposed Asian Criteria[a]	WHO Criteria[b]
Underweight	<18.5	<18.5
Normal range	18.5 to <23	18.5 to <25
Overweight	23 to <25	25 to <30
Obese	>25	>30

[a] In 2004.

[b] Prior to 2004.

Source: WHO Expert Consultation, *Lancet,* 363, 157–163, 2004.

3. Asians tend to have obesity-related complications at lower BMI than individuals of European ancestry and therefore should be considered obese at a BMI $\geq$ 26 and overweight at a BMI $\geq$ 24.
4. The grading of obesity by BMI follows the guidelines:
   - Grade I obesity, BMI = 30 to 34, inclusively
   - Grade II obesity, BMI = 35 to 39, inclusively
   - Grade III obesity, BMI $\geq$ 40
5. Some authorities advocate an upper limit of normal for the BMI of 25 with values between 26 and 30 defined as overweight.
6. Visceral adiposity—waist circumference of >102 cm or >40 in. for men and >88 cm or >34.5 in. for women—contributes to the clustering of other CVD risk factors, such as hypertension, insulin resistance/type 2 diabetes mellitus (T2DM), and dyslipidemia, within individual patients.

Notes

1. Circular slide charts are popular methods of estimating BMI. The accuracy of these charts at the extremes of BMI varies considerably.
2. A tabular presentation of values for BMI that permits rapid assessment of a patient's underweight (BMI < 18.5) or overweight (BMI > 25) status is available in Bailey and Ferro-Luzzi (1995).
3. At BMI levels > 25, visceral adiposity compounds the CVD risk of hypertension and dyslipidemia in both men and women and T2DM in women in particular.
4. On average, adults with a coronary artery event tend to have been underweight at birth, are thin at 2 years of age, and put on weight rapidly thereafter. This pattern of growth was associated with insulin resistance in later life. The risk of a coronary artery event strongly relates to a rapid tempo of childhood (ages 2 to 11 years) increase in BMI (Barker et al., 2005).

ICD-9 Codes

Conditions that may justify this test include *but are not limited to*:

250.X	diabetes mellitus
255.0	Cushing's syndrome
429.2	cardiovascular disease, unspecified
759.81	Prader–Willi syndrome
783.1	abnormal weight gain

Suggested Reading

Bailey KV, Ferro-Luzzi A. Use of body mass index of adults in assessing individual and community nutritional status. *Bull World Health Organ* 1995; 73:673–80.

Barker DJP, Osmond C, Forsén TJ, Kajantie E, Eriksson, JG. Trajectories of growth among children who have coronary events as adults. *N Engl J Med* 2005; 353:1802–9.

Deurenberg P, Deurenberg-Yap M, Guricci S. Asians are different from Caucasians and from each other in their body mass index/body fat percent relationship. *Obes Rev* 2002; 3:141–6.

Frankel HM. Determination of body mass index. *JAMA* 1986; 255:1292.

Janssen I, Katzmarzyk PT, Ross R. Body mass index, waist circumference, and health risk: evidence in support of current National Institutes of Health Guidelines. *Arch Int Med* 2002; 162:2074–79.

Ko GT, Wu MM, Tang J, Wai HP, Chan CH, Chen R. Body mass index profile in Hong Kong Chinese adults. *Ann Acad Med Singapore* 2001; 30:393–6.

Shirai K. Obesity as the core of the metabolic syndrome and the management of coronary heart disease. *Curr Med Res Opin* 2004; 20:295–304.

11.1.4 Waist-to-Hip Circumference Ratio (WHCR) and Waist Circumference (WC) or Girth as Measurements of Obesity and Cardiovascular Disease (CVD) Risk

Indications for Test

Calculation of WHCR and WC is indicated in:

- Cases of central obesity with a body mass index (BMI) above the ethnic norm (see Test 11.1.3)
- Identification of patients at higher risk for CVD
- Justification for the use of ICD Code 277.7 (dysmetabolism)
- Diabetes (DM) patients who may be above their ideal body weight
- Patients with CVD risk factors

Procedure

1. Locate the waist as the smallest circumference of the torso (not necessarily at the umbilicus). In obese patients without a waist, measure the girth as the smallest circumference in the horizontal plane between the 12th rib and the iliac crest (i.e., small of the back to the front, usually one inch above the umbilicus).
2. Measure the girth with the waist area of the patient exposed, standing relaxed, arms at the sides, and feet together. Use a Gulick tape measure (nonstretchable tape) and do not compress the skin. Have patient inhale, then exhale deeply. Measure girth after the exhalation.
3. Locate the hip as the maximum posterior extension of the buttocks. In the obese with an anterior abdominal wall that sags into this area, include the sagging abdominal wall, but not the panniculus, in the measurement.
4. Measure the hip circumference with patients in their underwear, standing tall, relaxed, and arms at their sides. Do not use stretchable tape or compress the skin.
5. Divide the waist circumference by the hip circumference to obtain the WHCR:

$$\text{WHCR} = \frac{\text{Waist circumference}}{\text{Hip circumference}}$$

Interpretation

1. A WHCR > 0.8 in women and > 0.95 in men has been associated with increased risk of CVD, DM, and hypertension (Table 11.4).

TABLE 11.4
Waist–Hip Circumference Ratio (WHCR) as an Isolated Indicator of Cardiovascular Risk in Men and Women

Men	Women	Risk
≤0.95	≤0.80	Low
0.96–1.00	0.81 to 0.85	Moderate
>1.00	>0.85	High

2. WC in excess of 86.5 cm (34 in.) for females and 96.5 cm (38 in.) for males are considered criteria to screen for DM.
3. The International Diabetes Federation (IDF) proposed in 2005 that the key feature of dysmetabolism is central obesity, defined as a WC ≥ 80 cm (32.6 in.) for female Europids and ≥ 94 cm (37 in.) for male Europids, with an even lower, ethnic-specific WC cutoff to be applied to Chinese, Japanese, and South Asian individuals. Refer to Table 3.15.
4. In a joint research project, scientists from the Medical College of Wisconsin, Columbia University, Nagoya University Graduate School of Medicine in Japan, and Verona University Medical School in Italy identified the WC for reduced CVD in men as ≤ 89 cm (35 in.) and in women as ≤84 cm (33 in.).
5. WC performs better as an index of trunk fat mass than does WHCR or the conicity index (Taylor et al., 2000) in children ages 3 to 19 years of age.

Notes

1. Increased WHCR ratios are associated with growth hormone deficiency.
2. In women, the relationship between the WHCR and the CVD endpoints of myocardial infarction, angina pectoris, stroke, and sudden death appears stronger than that for BMI.
3. Abdominal liposuction, with resultant lower WHCR, did not improve obesity-associated insulin insensitivity and cannot substitute for the metabolic benefits of weight loss (Klein et al., 2004).
4. The WHCR was the best anthropometric predictor of total mortality but was associated less consistently than BMI or WC with cancer incidence.
5. In men, significant associations were found between the WHCR and the occurrence of stroke ($p = 0.002$) and ischemic heart disease ($p = 0.04$).
6. In middle-aged men, the distribution of fat deposits may be a better predictor of CVD and death than the degree of adiposity.
7. In women, by both univariate and multivariate analyses, but not in men, the WHCR has been shown to be independent of age, BMI, smoking, hyperlipidemia, triglycerides, and systolic blood pressure as a predictor of CVD.
8. Measurement of WHCR is redundant at BMI ≥ 32, as the result will always be abnormal and the patient can be classified as at higher risk for CVD based on BMI alone.
9. There was no significant relationship between WCHR and intraabdominal fat, subcutaneous fat, or their ratio as determined by single-slice MR scan of the abdomen at the umbilicus. In studies of obesity, MR imaging of the abdomen remains a research tool.
10. The 2005 IDF guidelines defining dysmetabolism include an abnormal WC and any two of the following conditions:
    - Triglyceride levels ≥ 150 mg/dL (1.7 mmol/L)
    - High-density lipoprotein cholesterol (HDL-C) levels < 40 mg/dL (1.04 mmol/L) in males and < 50 mg/dL (1.29 mmol/L) in females
    - Systolic blood pressure ≥ 130 mmHg and diastolic blood pressure ≥ 85 mmHg
    - Fasting hyperglycemia (i.e., plasma glucose levels ≥ 100 mg/dL or 5.6 mmol/L) or a previous diagnosis of DM or impaired glucose tolerance

ICD-9 Codes

Conditions that may justify this test include *but are not limited to*:

Nutrition
429.2   cardiovascular disease, unspecified
783.1   abnormal weight gain
783.2   abnormal loss of weight and underweight

Other
250.XX diabetes mellitus
277.7   dysmetabolic syndrome
278.0   overweight and obesity

Suggested Reading

AACE/ACE Position statement on the prevention, diagnosis, and treatment of obesity. *Endocr Pract* 1998; 4:297–330.

Folsom AR, Kushi LH, Anderson KE, Mink PJ, Olson JE, Hong C-P, Sellers TA, Lazovich D, Prineas RJ. Associations of general and abdominal obesity with multiple health outcomes in older women. The Iowa Women's Health Study. *Arch Intern Med* 2000; 160:2117–28.

Gray DS, Fujioka K, Colletti PM, Kim H, Devine W, Cuyegkeng T, Pappas T. Magnetic-resonance imaging used for determining fat distribution in obesity and diabetes. *Am J Clin Nutr* 1991; 54:623–7.

Klein S, Fontana L, Young VL, Coggan AR, Kilo C, Patterson BW, Mohammed BS. Absence of an effect of liposuction on insulin action and risk factors for coronary heart disease. *N Engl J Med* 2004; 350:2549–57 (see comment and author reply in *N Engl J Med*. 2004; 351:1354–77).

Lapidus L, Bengtsson C, Larsson B, Pennert K, Rybo E, Sjostrom L. Distribution of adipose tissue and risk of cardiovascular disease and death: 12 year follow up of participants in the population study of women in Gothenburg, Sweden. *Br Med J (Clin Res Ed)* 1984; 289(6454):1257–61.

Larsson B, Svardsudd K, Welin L, Wilhelmsen L, Bjorntorp P, Tibblin G. Abdominal adipose tissue distribution, obesity, and risk of cardiovascular disease and death: 13 year follow up of participants in the study of men born in 1913. *Br Med J (Clin Res Ed)* 1984; 288:1401–4.

Taylor RW, Jones IE, Williams SM, Goulding A. Evaluation of waist circumference, waist-to-hip ratio, and the conicity index as screening tools for high trunk fat mass, as measured by dual-energy x-ray absorptiometry, in children aged 3 to 19 years. *Am J Clin Nutr* 2000; 72:490–5.

11.1.5 Ideal Body Weight (IBW) and Lean Body Mass (LBM) Determinations for Nutritional Assessment

Indications for Test

Calculation of gender-specific IBW and LBM based on height are indicated when:

- Making a rapid estimate of risk for malnutrition
- Estimating a desirable body weight for height
- Estimating caloric energy requirements, regardless of body frame size
- Assessing eligibility for weight-loss surgery

Procedure

1. To estimate IBW, use one of the following equations:
    - Devine (1974) equations:

$$\text{IBW}_{\text{men}} \text{ (kg)} = 50 \text{ kg} + (2.3 \text{ kg} \times \text{no. of inches above 5.0 ft in height})$$

$$\text{IBW}_{\text{women}} \text{ (kg)} = 45.5 \text{ kg} + (2.3 \text{ kg} \times \text{no. of inches above 5.0 ft in height})$$

or

$$IBW_{men} \text{ (lb)} = (106 \text{ lb for the first 5 ft of height})$$
$$+ (6 \text{ lb for each additional inch of height})$$
$$IBW_{women} \text{ (lb)} = (100 \text{ lb for the first 5 ft of height})$$
$$+ (5 \text{ lb for each additional inch of height})$$

- Robinson (1983) equations:

$$IBW_{men} \text{ (kg)} = 45.50 \text{ kg} + (2.3 \text{ kg} \times \text{no. of inches above 5.0 ft in height})$$
$$IBW_{women} \text{ (kg)} = 48.67 \text{ kg} + (1.65 \text{ kg} \times \text{no. of inches above 5.0 ft in height})$$

2. When using the Devine or Robinson equations to calculate a value for IBW on which to base the dosage (IBW_{dosage}) of hydrophilic drugs, use the formula:

$$IBW_{dosage} = IBW + [0.3 \times (ABW - IBW)]$$

where ABW is the actual body weight and 0.3 is a correction factor to account for the fact that ~30% of adipose tissue weight is water.

3. Calculate the %IBW using the formula:

$$\%IBW = (ABW/IBW) \times 100$$

4. Calculate LBM (kg) for men and women using the formulas:

$$LBM_{men} \text{ (kg)} = \left[1.10 \times \text{weight (kg)}\right] - 128\left[\text{weight}^2 \text{ (kg)} / \left(100 \times \text{height (m)}\right)^2\right]$$

$$LBM_{women} \text{ (kg)} = \left[1.07 \times \text{weight (kg)}\right] - 148\left[\text{weight}^2 \text{ (kg)} / \left(100 \times \text{height (m)}\right)^2\right]$$

Example—A man with weight = 173 lb and height = 6.0 ft:

$$\frac{173 \text{ lb}}{2.2 \text{ lb/kg}} = 78.6 \text{ kg}$$

$$6.0 \text{ ft} = 72.0 \text{ in.}$$

$$72 \text{ in.} \times \frac{2.54 \text{ cm}}{\text{in.}} \times \frac{100 \text{ cm}}{\text{m}} = 1.83 \text{ m}$$

$$LBM_{men} \text{ (kg)} = (1.10 \times 78.6) - 128\left[(78.6)^2 / (100 \times 1.83)^2\right] = 62.8 \text{ kg}$$

5. Alternatively, obtain the LBM graphically from widely available charts of height in inches vs. ABW in pounds.

Interpretation

1. Using %IBW and BMI (see Test 11.1.3), patients may be classified as follows:

%IBW	BMI (kg/m²)	Interpretation
<90	<18.5	Underweight
≥120	≥30.0	Obese

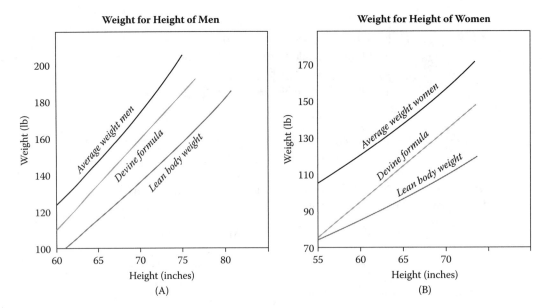

Figure 11.1
Nomograms of weight for height in (A) men and (B) women.

2. Estimates of healthy average BMI based on IBW using the Devine formulas correlate well with an estimated average BMI in men (23.0 kg/m^2) but underestimate the average BMI in women (20.8 kg/m^2) (Figure 11.1).
3. Compared with BMI percentiles, %IBW underestimates the severity of malnutrition in children with short stature and overestimates the severity of malnutrition in children with tall stature.
4. Substantial discrepancies have been observed when trying to classify individuals as underweight, normal weight, or obese by %IBW or BMI. These discrepancies may cause confusion when two or more indices are used simultaneously to classify the degree of obesity.

Notes

1. Do not use %IBW divided by height (%IBW$_{height}$):

$$\left[\frac{ABW/IBW}{Height\ (in.)} \right] \times 100$$

in lieu of %IBW alone, as results for %IBW$_{height}$ are notoriously inaccurate and variable both within and between observers, depending on the IBW equation used. Clearly, %IBW$_{height}$ is an unreliable measure of nutritional status.
2. The equations for estimating IBW shown above do not take into account body composition, frame size, age, or current weight and cannot be used to determine IBW for amputees.
3. A more accurate estimate of IBW can be obtained using equations derived from the plots of weight vs. height when different frame sizes (i.e., ectomorphs, mesomorphs, endomorphs) are taken into account.
4. Efforts to improve the accuracy of Devine's equations for calculating IBW using regression analyses of height and weight data resulted in equations similar to those published by Devine.
5. For the most part, formulas for IBW in men and women have been derived from the Metropolitan Life Insurance Company height and weight tables generated in the 1960 and 1970s and reflect the flaws inherent in that database.
6. Women are more frequently dissatisfied with their weight and see themselves as considerably heavier than they actually are, with a resultant higher risk for the development of an eating disorder and attempts to achieve a body weight that is ideal only to themselves.

ICD-9 Codes

Conditions that may justify this test include *but are not limited to*:

Nutrition
783.1   abnormal weight gain
783.2   abnormal loss of weight and underweight

Obesity
278.0   overweight and obesity
278.01 morbid obesity

Suggested Reading

Devine BJ. Gentamicin pharmacokinetics. *Drug Intell Clin Pharm* 1974; 8:650–655.
Pai MP, Paloucek FP. The origin of the "ideal" body weight equations. *Ann Pharmacother* 2000; 34:1066–69.
Poustie VJ, Watling RM, Ashby D, Smyth RL. Reliability of percentage ideal weight for height. *Arch Dis Child* 2000; 83:183–4.
Robinson JD, Lupkiewicz SM, Palenik L, Lopez LM, Ariet M. Determination of ideal body weight for drug dosage calculations. *Am J Hosp Pharm* 1983; 40:1016–19.
Tzamaloukas AH, Murata GH, Hoffman RM, Schmidt DW, Hill JE, Leger A, Macdonald L, Caswell C, Janis L, White RE. Classification of the degree of obesity by body mass index or by deviation from ideal weight. *J Parenter Enteral Nutr* 2003; 27:340–8.
Zhang Z, Lai HJ. Comparison of the use of body mass index percentiles and percentage of ideal body weight to screen for malnutrition in children with cystic fibrosis. *Am J Clin Nutr* 2004; 80:982–91.

11.2   Nutritional Status as Marked by Levels of Vitamins, Minerals, and Other Nutrients

11.2.1   Albumin (ALB), Total Protein (TP), and Prealbumin (PAB)

Indications for Tests

Albumin and TP tests, along with total cholesterol and creatinine, are indicated:

- In the general nutritional assessment of suspected cases of protein-calorie malnutrition (PCM) or nephrosis preliminary to PAB testing
- As a sensitive and cost-effective measure of malnutrition in patients who are hospitalized or have critical illness or chronic disease (e.g., fatty liver)

Prealbumin testing is indicated in patients:

- During prolonged hospitalization
- If nutritional compromise is in evidence or highly suspected

Procedure

1. Obtain the appropriate blood specimens for ALB, TP, total cholesterol, and creatinine testing followed by PAB testing.
2. Repeat measures of [PAB] up to twice a week in hospitalized or other nutritionally compromised patients, and note trends as interventions are applied.
3. If the serum [ALB] is <3 g/dL, request TP testing on a 24-hour urine collection to determine the TP excretion rate (see Specimen Collection Protocol P1 in Appendix 2).

TABLE 11.5
Interpretation of Prealbumin (PAB) Levels

PAB Concentration (mg/dL)	Implications
<5.0	Poor prognosis with severe malnutrition; hyperalimentation indicated
5.0–11.9	High risk for poor prognosis; nutritional intervention indicated
11.0–14.9	Prognosis guarded; monitor prealbumin frequently (at least twice a week)
15.0–35.0	Normal

Source: Prealbumin in Nutritional Care Consensus Group, *Nutrition*, 11, 170, 1995.

Interpretation

1. [PAB] correlates with hospital outcomes and helps to predict recovery from serious acute illnesses.
2. A serum [ALB] < 3 g/dL is evidence for severe PCM or urinary protein loss with possible nephrosis if the TP excretion rate is >1 g/24 hr.
3. Typical values for PAB and ALB in patients with mild, moderate, or severe PCM are shown in Table 11.5.
4. Lower [PAB] may occur with inflammatory disorders, cirrhosis, and protein-losing enteropathy or with protein malnutrition in malignancy in addition to nephrosis.
5. Using the Cox proportional hazards model, a [PAB] of 15 mg/dL was an independent predictor of mortality risk vs. higher values (relative risk [RR] = 4.48; $p < 0.01$), whereas PAB correlated ($p < 0.001$) more strongly with other nutritional markers (i.e., total cholesterol, apolipoprotein B, creatinine, and urea) than did serum [ALB].
6. Declining [PAB] during hospitalization indicates declining nutritional status and the need for prompt intervention to correct malnutrition to improve patient outcome and shorten hospitalization.
7. Falsely elevated [PAB] occurs as a result of acute alcoholic hepatitis and usually normalizes about one week after an alcoholic binge. Falsely elevated [PAB] also occurs with prednisone and progestational drug therapy.

Notes

1. ALB exists in a large body pool and has a half-life of 20 days vs. only 2 days for PAB; thus, [ALB] is relatively insensitive to short-term changes in nutrition, but [PAB] and other markers such as transthyretin and retinol-binding protein respond more quickly to such changes (Ihara et al., 2003).
2. Unlike ALB, PAB is not affected by hydration status and has a high ratio of essential to nonessential amino acids, making it a distinct marker for protein synthesis.
3. Many hospitalized elderly patients have nutrient intakes far less than their estimated maintenance energy requirements, contributing to an increased risk of mortality.
4. Zinc deficiency may lower [PAB]. Other vitamin and mineral deficiencies are not known to do so.

ICD-9 Codes

Conditions that may justify this test include *but are not limited to*:

263.9	unspecified protein-calorie malnutrition
583	nephritis and nephropathy

Suggested Reading

Beck FK, Rosenthal TC. Prealbumin: a marker for nutritional evaluation. *Am Fam Physician* 2002; 65:1575–78.

Bernstein L, Pleban W. Prealbumin in nutrition evaluation. *Nutrition* 1996; 12:255–9.

Brugler L, Stankovic A, Bernstein L, Scott F, O'Sullivan-Maillet J. The role of visceral protein markers in protein calorie malnutrition. *Clin Chem Lab Med* 2002; 40:1360–69.

Goldwasser P, Michel MA, Collier J, Mittman N, Fein PA, Gusik SA, Avram MM. Prealbumin and lipopro-tein(a) in hemodialysis: relationships with patient and vascular access survival. *Am J Kidney Dis* 1993; 22:215–25.

Ihara H, Matsumoto T, Shino Y, Hashizume N, Takase M, Nagao J, Sumiyama Y. Selective use of transthyretin and retinol-binding protein as markers in the postoperative assessment of protein nutritional status. *J Clin Lab Anal* 2003; 17:1–5.

Prealbumin in Nutritional Care Consensus Group. Measurement of visceral protein status in assessing protein and energy malnutrition: standard of care. *Nutrition* 1995; 11:169–71.

Sullivan DH, Sun S, Walls RC. Protein-energy undernutrition among elderly hospitalized patients: a prospective study. *JAMA* 1999; 281:2013–19.

11.2.2 Lymphocyte Count

Indications for Test

A lymphocyte count is indicated as a marker of nutritional status:

- In patients suspected of protein-calorie malnutrition (PCM)

Procedure

1. Obtain an anticoagulated blood specimen from the patient for differential assessment of the total white blood cell (WBC) count and the percent of each formed cellular element. A request for a complete blood count (CBC) is adequate for this determination.
2. Calculate the total lymphocyte count (TLC) = [(WBC count per mm^3) × (% lymphocytes)]/100.
3. For a more detailed assessment of immunocompetence, request lymphocyte subset analysis (i.e., number of T and B cells).

Interpretation

1. The TLC may be evaluated as follows:
   - Normal, 1500 to 3000/mm^3
   - Mild depletion, 1200 to 1500/mm^3
   - Moderate depletion, 800 to 1199/mm^3
   - Severe depletion, <800/mm^3
2. TLC < 1500 cells/mm^3 plus a serum albumin concentration < 3.5 g/dL and a serum transferrin concentration < 200 mg/dL are considered criteria for clinical malnutrition.
3. Higher death rates and other adverse outcomes occur in malnourished groups with low TLC.
4. In peritoneal dialysis patients, total body fluid volume status is usually the more important factor affecting the TLC than nutritional status (Ates et al., 2004).

Notes

1. Factors affecting the TLC, in addition to nutritional status, include inflammation, infection, stress, sepsis, medications, and various kinds of cancer and leukemias.
2. Hip-fracture patients with a low (<3.5 g/dL) [ALB] and a TLC < 1500 cells/mL when compared to patients with these analytes within the reference interval had much worse outcomes. Specifically, they were 2.9 times more likely to have a length of hospital stay greater than 2 weeks ($p = 0.03$), 3.9 times more likely to die within 1 year after surgery ($p = 0.02$), and 4.6 times less likely to recover their prefracture level of independence ($p < 0.01$) (Koval et al., 1999).
3. Iron overload and zinc deficiency reduce immunocompetence as reflected by lymphocyte subset analysis and decreased T and B cell counts.

4. Thirty-four percent of hospitalized patients had levels of lymphopenia likely to be associated with reduced cellular immunity (Bistrian et al., 1976).
5. In an institutionalized elderly population (Lukito et al., 2004):
   - Women had higher absolute counts of various lymphocyte subsets than men.
   - Serum zinc concentration in women was positively correlated with the absolute counts of CD3 (total T-cells), CD4 (T-helper cells), and CD19 (total B-cells).
   - Serum ferritin concentration in men was positively correlated with the number of CD8 (T-suppressor) and CD56 (natural killer) cells and with serum iron concentration.
   - After 22 months of observation, the mean number of CD4 cells of nonsurvivors (524×10^6 cells/L) was significantly lower than that of survivors (759×10^6 cells/L).
6. A negative value on the Rainey–McDonald Nutritional Index is consistent with malnutrition.
7. The Subjective Global Assessment and Nutritional Risk Index assessment is predictive of malnutrition and postoperative complications in patients undergoing major abdominal surgery (Sungurtekin et al., 2004).

ICD-9 Codes

Conditions that may justify this test include *but are not limited to*:

263.9   unspecified protein-calorie malnutrition

Suggested Reading

Ates K, Ates A, Kutlay S, Nergizoglu G, Karatan O. Total lymphocyte count in peripheral blood of peritoneal dialysis patients: relationship to clinical parameters and outcome. *J Nephrol* 2004; 17:246–52.

Bistrian BR, Blackburn GL, Vitale J, Cochran D, Naylor J. Prevalence of malnutrition in general medical patients. *JAMA* 1976; 235:1567–70.

Koval KJ, Maurer SG, Su ET, Aharonoff GB, Zuckerman JD. The effects of nutritional status on outcome after hip fracture. *J Orthop Trauma* 1999; 13:164–9.

Lukito W, Wattanapenpaiboon N, Savige GS, Hutchinson P, Wahlqvist ML. Nutritional indicators, peripheral blood lymphocyte subsets and survival in an institutionalised elderly population. *Asia Pac J Clin Nutr* 2004; 13:107–12.

Rai J, Gill SS, Kumar BR. The influence of preoperative nutritional status in wound healing after replacement arthroplasty. *Orthopedics* 2002; 25:417–21.

Sungurtekin H, Sungurtekin U, Balci C, Zencir M, Erdem E. The influence of nutritional status on complications after major intraabdominal surgery. *J Am Coll Nutr* 2004; 23:227–32.

11.2.3 Serum Creatinine (Creat) and Blood or Urine Urea Nitrogen (BUN)

Indications for Test

Serum creatinine and BUN levels may be indicated as markers of nutritional status:

- In cases of protein-calorie malnutrition or starvation
- Dehydration with hypovolemia

Procedure

1. Obtain a thorough patient history, particularly of age, gender, dietary intake of all forms of protein, thirst, diarrhea, fatigue, weakness, muscle cramps, and dizziness on standing.
2. Observe the patient for evidence of confusion, abdominal or chest pain, cyanosis, uremic odor, moisture of mucous membranes, and skin turgor.
3. Obtain a random blood specimen for serum [BUN] and [Creat], usually as part of an electrolyte panel including random [glucose]. Obtain a 24-hour urine collection for measurement of urinary urea nitrogen to make an estimate of nitrogen loss.

4. In the absence of diarrhea, divide the patient's 24-hour urinary urea nitrogen excretion by 0.85 to estimate total body nitrogen losses assuming that a minimum nitrogen loss of 2 g/day occurs through feces and perspiration.

Interpretation

1. The [BUN] correlates positively with the [Creat] as well as the degree of hydration of the patient.
2. The [BUN]/[Creat] ratio is normally ~10/1 but may rise to 20/1 or more in cases of hypovolemia with prerenal azotemia.
3. The level of serum [Creat] correlates positively with decline in renal status, increased protein intake, and muscle mass. It correlates negatively with age and decreases with vegetarian diets, female gender, and amputation of extremities.
4. The [Creat] does not correlate with obesity (fat mass), although there may be a weak correlation with BMI in nonobese individuals.
5. Tenting of the skin and dry mucous membranes tend to be poor markers of dehydration.
6. In chronic kidney disease (CKD) patients, very low protein diets are associated with stabilization of serum [Creat], which changes very little over time.
7. In patients with uncontrolled diabetes (random plasma [glucose] > 250 mg/dL or hemoglobin A_{1c} > 8%), hyperfiltration in the absence of dehydration may result in significant lowering of the serum [Creat], giving a false impression of normal kidney function (see Test 5.2.1).

Notes

1. A transient negative nitrogen balance may occur during the initial few weeks of a 30-g protein and 11.3-mmol (350 mg) phosphorus diet.
2. Nitrogen balances become positive or neutral, indicating no impairment in nutritional status with a very low protein diet over a period of 1 year in T2DM with CKD Stage 3 or 4 and proteinuria (Shichiri et al., 1991).
3. Medications that increase the serum [Creat] by reducing its tubular secretion include cimetidine, trimethoprim, and fibric acids other than gemfibrozil.
4. The reproducibility of values for the estimated glomerular filtration rate (eGFR), using various equations or creatinine clearance methods, is critically dependent on serum creatinine assay calibration. As a result of calibration differences, serum creatinine assays on the same samples were 0.23 mg/dL higher in the National Health and Nutrition Examination Survey III (NHANES III) than in the Modification of Diet in Renal Disease (MDRD) study (Coresh et al., 2002).
5. An elevated serum [Creat] is strongly associated with inadequate treatment of hypertension and the onset of CKD (Coresh et al., 2001).
6. The protein/creatinine ratio in spot urine specimens as well as values for eGFR obtained by equations using serum [Creat] help to identify children and adolescents with CKD (Hogg et al., 2003).
7. The prevalence of CKD in the U.S. adult population is around 11% (19.2 million) (Coresh et al., 2003). When classified by stage of CKD, the distribution of these patients in each stage of CKD were:

Stage 1 (persistent albuminuria with a normal GFR)	3.3%
Stage 2 (persistent albuminuria with a GFR of 60–89 mL/min/1.73 m²)	3.0%
Stage 3 (GFR, 30–59 mL/min/1.73 m²)	4.3%
Stage 4 (GFR, 15–29 mL/min/1.73 m²)	0.2%
Stage 5 (renal failure/dialysis)	0.2%
Total	11.0%

ICD-9 Codes

Conditions that may justify this test include *but are not limited to*:

Hydration
276.51   dehydration
276.52   hypovolemia

Other
263.9   unspecified protein-calorie malnutrition
994.2   effects of hunger/starvation

Suggested Reading

Coresh J, Wei GL, McQuillan G, Brancati FL, Levey AS, Jones C, Klag MJ. Prevalence of high blood pressure and elevated serum creatinine level in the United States: findings from the third National Health and Nutrition Examination Survey (1988–1994). *Arch Intern Med* 2001; 161:1207–16.

Coresh J, Astor BC, McQuillan G, Kusek J, Greene T, Van Lente F, Levey AS. Calibration and random variation of the serum creatinine assay as critical elements of using equations to estimate glomerular filtration rate. *Am J Kidney Dis* 2002; 39:920–9.

Coresh J, Astor BC, Greene T, Eknoyan G, Levey AS. Prevalence of chronic kidney disease and decreased kidney function in the adult U.S. population: Third National Health and Nutrition Examination Survey. *Am J Kidney Dis* 2003; 41:1–12.

Hogg RJ, Furth S, Lemley KV, Portman R, Schwartz GJ, Coresh J, Balk E, Lau J, Levin A, Kausz AT, Eknoyan G, Levey AS. National Kidney Foundation's Kidney Disease Outcomes Quality Initiative clinical practice guidelines for chronic kidney disease in children and adolescents: evaluation, classification, and stratification. *Pediatrics* 2003; 111(Pt 1):1416–21.

Shichiri M, Nishio Y, Ogura M, Marumo F. Effect of low-protein, very-low-phosphorus diet on diabetic renal insufficiency with proteinuria. *Am J Kidney Dis* 1991; 18:26–32.

11.2.4 Carotene

Indications for Test

A carotene level is indicated:

- As a marker of nutritional status in cases of suspected malnutrition based on low body weight or recent weight loss
- In individuals with abstinent anorexia or bulimic–purging psychological eating disorders

Procedure

1. Obtain a random blood sample for carotene testing.

Interpretation

1. Plasma levels of carotenoids, alpha-carotene, and beta-carotene are, at best, crude indicators of vegetable and fruit intake (Jansen et al., 2004).
2. Low levels of carotene are consistently found only in more extreme cases of inanition with malnutrition.
3. Transition from inadequate intake of carotenoids to the recommended intake found in five servings of fruits and vegetables per day can be documented by measurement of carotene levels before and after intervention.

Notes

1. A higher preconceptional intake of vegetable protein, fiber, beta-carotene, ascorbic acid, alpha-tocopherol, iron, and magnesium reduces the risk of offspring affected by orofacial clefts (Krapels et al., 2004).

2. A weak, positive association between beta-carotene intake and sperm concentration ($p = 0.06$) and progressive sperm motility ($p = 0.06$) has been observed (Eskenazi et al., 2005).

3. The estimated carotenoid intake from 24-hour diet recall and food frequency intake questionnaires was strongly associated with plasma concentrations of alpha-carotene, beta-carotene, and lutein (Natarajan et al., 2004). This corroborates the report of the effective use of 72-hour diet recall for estimation of carotenoid intake (Schroder et al., 2001).

4. Use blood levels of alpha-carotene, beta-carotene, and lutein to assess whether or not an increased intake to the recommended amount of five portions of fruits and vegetables per day has taken place (Brevik et al., 2004).

5. In patients seen preoperatively for minor surgery in Argentina, assessment of plasma levels of vitamins showed that 27% were deficient in vitamin C and 16% in vitamin A, but only 5% were deficient in carotenes (Zago et al., 1999).

6. Heavy consumers of soup in France had higher dietary intakes of folates, beta-carotene, and vitamin C. In these patients, soups contributed 12.5% of total intake of beta-carotene in men and 13% in women (Galan et al., 2003).

7. Plasma beta-cryptoxanthin appears to be the best indicator for fruit intake, as well as for the sum of vegetable, fruits, and fruit juice intake. Lutein levels best reflected vegetable intake (Jansen et al., 2004).

ICD-9 Codes

Conditions that may justify this test include *but are not limited to*:

307.1	anorexia nervosa
783.6	bulimia
V77.2	malnutrition

Suggested Reading

Brevik A, Andersen LF, Karlsen A, Trygg KU, Blomhoff R, Drevon CA. Six carotenoids in plasma used to assess recommended intake of fruits and vegetables in a controlled feeding study. *Eur J Clin Nutr* 2004; 58:1166–73.

Eskenazi B, Kidd SA, Marks AR, Sloter E, Block G, Wyrobek AJ. Antioxidant intake is associated with semen quality in healthy men. *Hum Reprod* 2005; 20:1006–12.

Galan P, Renault N, Aissa M, Adad HA, Rahim B, Potier de Courcy G, Hercberg S. Relationship between soup consumption, folate, beta-carotene, and vitamin C status in a French adult population. *Int J Vitam Nutr Res* 2003; 73:315–21.

Jansen MC, Van Kappel AL, Ocke MC, Van't Veer P, Boshuizen HC, Riboli E, Bueno-de-Mesquita HB. Plasma carotenoid levels in Dutch men and women, and the relation with vegetable and fruit consumption. *Eur J Clin Nutr* 2004; 58:1386–95.

Krapels IP, van Rooij IA, Ocke MC, West CE, van der Horst CM, Steegers-Theunissen RP. Maternal nutritional status and the risk for orofacial cleft offspring in humans. *J Nutr* 2004; 134:3106–13.

Natarajan L, Rock CL, Major JM, Thomson CA, Caan BJ, Flatt SW, Chilton JA, Hollenbach KA, Newman VA, Faerber S, Ritenbaugh CK, Gold E, Stefanick ML, Jones LA, Marshall JR, Pierce JP. On the importance of using multiple methods of dietary assessment. *Epidemiology* 2004; 15:738–45.

Schroder H, Covas MI, Marrugat J, Vila J, Pena A, Alcantara M, Masia R. Use of a three-day estimated food record, a 72-hour recall and a food-frequency questionnaire for dietary assessment in a Mediterranean Spanish population. *Clin Nutr* 2001; 20:429–37.

Zago L, Weisstaub A, Dupraz H, Godoy MF, Gasali F, Dirube C, Slobodianik NH, de Portela ML, Torino F, Rio ME. Nutritional status of surgical patients without apparent nutritional compromise. *Arch Latinoam Nutr* 1999; 49:1–7.

11.2.5 25-Hydroxyvitamin D [25(OH)D$_3$ Plus 25(OH)D$_2$]

Indications for Test

Measurement of 25-hydroxyvitamin D concentration [25(OH)D] is indicated:

- In the overall nutritional assessment of the individual
- When a patient is found to be hypercalcemic or at risk for osteomalacia

Procedure

1. Obtain the appropriate blood specimen for measurement of [25(OH)D] or vitamin D$_2$ plus D$_3$ by high-performance liquid chromatography (HPLC), gas chromatography–mass spectrometry (GC-MS), or tandem mass spectrometry (MS/MS).

Interpretation

1. Serum [25(OH)D] < 30 ng/mL (75 nmol/L) is associated with reduced calcium absorption, osteoporosis, and increased fracture risk.
2. Patients with overt vitamin D deficiency may be identified as having [25(OH)D] < 25 ng/mL, whereas those with vitamin D insufficiency have levels < 25 to 32 ng/mL.
3. A [25(OH)D] < 100 ng/mL is nontoxic.

Notes

1. An absence of seasonal variation and deficiency of 25(OH)D$_3$ in women can be attributed to patterns of lifestyle and to the traditional clothing of women, particularly among Muslims.
2. In a year-long study of 60-year-old postmenopausal Dutch women, a 29% decline in maximum 25(OH)D levels during the winter was not prevented by supplementation with vitamin D$_3$ in women with low bone mineral density (BMD) (Schaafsma et al., 2000).
3. Vitamin D-deficient Crohn's disease patients have higher serum alkaline phosphatase and parathyroid hormone levels without detectable differences in BMD compared to those with a 25(OH)D$_3$ level > 40 ng/mL (Siffledeen et al., 2003).
4. Lack of sunlight exposure, decreased iron and folate levels, and smoking status predict vitamin D deficiency.
5. Low intake of calcium and vitamin D with decreased 25(OH)D$_3$, especially in adolescents and elderly, pregnant, and lactating women, identifies these groups as at a higher risk for the development of disorders of bone metabolism (e.g., osteomalacia).
6. In 2007, the Food and Drug Administration (FDA) recommended that vitamin D$_3$ intake be raised to >800 units per day. Note that vitamin D$_3$ is the most active form of vitamin D.

ICD-9 Codes

Conditions that may justify this test include *but are not limited to*:

268.2   osteomalacia
275.42   hypercalcemia
V77.2   malnutrition

Suggested Reading

Heaney RP. Functional indices of vitamin D status and ramifications of vitamin D deficiency. *Am J Clin Nutr* 2004; 80(6 Suppl):1706–09S.

Koenig J, Elmadfa I. Status of calcium and vitamin D of different population groups in Austria. *Int J Vitam Nutr Res* 2000; 70:214–20.

Mirsaeid Ghazi AA, Rais Zadeh F, Pezeshk P, Azizi F. Seasonal variation of serum 25 hydroxy D_3 in residents of Tehran. *J Endocrinol Invest* 2004; 27:676–9.

Olkowski AA, Aranda-Osorio G, McKinnon J. Rapid HPLC method for measurement of vitamin D_3 and $25(OH)D_3$ in blood plasma. *Int J Vitam Nutr Res* 2003; 73:15–18.

Schaafsma A, Muskiet FA, Storm H, Hofstede GJ, Pakan I, Van der Veer E. Vitamin D(3) and vitamin K(1) supplementation of Dutch postmenopausal women with normal and low bone mineral densities: effects on serum 25-hydroxyvitamin D and carboxylated osteocalcin. *Eur J Clin Nutr* 2000; 54:626–31.

Siffledeen JS, Siminoski K, Steinhart H, Greenberg G, Fedorak RN. The frequency of vitamin D deficiency in adults with Crohn's disease. *Can J Gastroenterol* 2003; 17:473–8.

11.2.6 Cyanocobalamin (Vitamin B_{12}), Methylmalonic Acid (MMA), and Transcobalamin II (TC-II)

Indications for Test

Measures of $[B_{12}]$ and [MMA] are indicated when:

- Clinical signs of B_{12} deficiency, including macrocytic megaloblastic anemia, smooth tongue, and/or neuropathy, are found
- Screening vegetarian populations at high-risk for water soluble vitamin deficiencies

Measures of transcobalamin II (TC-II) unbound to vitamin B_{12} are indicated in:

- Childhood-onset megaloblastic anemia, usually in the offspring of consanguineous relationships
- Cases of megaloblastic anemia in which the serum $[B_{12}]$ is within the reference interval for healthy individuals

Procedure

1. Obtain a complete blood count for examination of the peripheral blood smear, leukocyte count (WBC), and determination of polymorphonuclear (PMN) cell morphology for the purpose of making a rapid provisional diagnosis of a vitamin B_{12}-deficient state.
2. Obtain a blood specimen for total $[B_{12}]$ testing.
3. If the $[B_{12}]$ is 100 to 400 pg/mL or in the lower quartile of the reference range, obtain a gastric parietal cell (GPC) antibody titer, [MMA], and homocysteine levels (see Test 4.1.3).
4. Obtain a blood specimen for serum or erythrocyte folate levels at the same time as the specimen for $[B_{12}]$, particularly if the red cell indices are normal, independent of the presence or absence of neuropathy.
5. Reserve direct measurement of TC-II for suspected cases of autosomal recessive TC-II deficiency or cases of unexplained macrocytic or megaloblastic anemia in which the serum $[B_{12}]$ is within the reference interval for healthy individuals.

Interpretation

1. Reference intervals for total $[B_{12}]$, serum folate, and erythrocyte folate levels are:
   - Vitamin B_{12}, >160 pg/mL
   - Serum folate, >4 ng/mL
   - Erythrocyte folate, 216 to 891 ng/mL
2. In general, total $[B_{12}]$ of >400 pg/mL is very good evidence for adequate B_{12} nutrition, whereas a $[B_{12}]$ of <100 pg/mL is unequivocal evidence for more severely deficient state.
3. A $[B_{12}]$ of 100 to 400 pg/mL is equivocal and prompts testing for [MMA] and homocysteine. An elevated level (i.e., three standard deviations above the mean) of either of these analytes is consistent

with B_{12} deficiency. Note that [MMA] may be elevated as a result of renal disease, thus limiting its use in making a definitive diagnosis of B_{12} deficiency.

4. In B_{12} deficiency, the PMNs tend to be multilobulated or hypersegmented.

5. The assay for total $[B_{12}]$ may overestimate bioactive $[B_{12}]$ if the binding protein haptocorrin is elevated or in patients (e.g., dental personnel) who have been chronically exposed to nitrous oxide, which oxidizes and inactivates B_{12}, which nonetheless remains detectable.

6. An elevated total WBC in patients with myeloid metaplasia or polycythemia vera may falsely elevate $[B_{12}]$.

7. Patients with hepatomas may have falsely elevated $[B_{12}]$ related to overproduction of vitamin B_{12} binding proteins.

8. Patients treated with metformin may have lower $[B_{12}]$ but rarely develop deficiency states.

9. Suspect partial TC-II deficiency in families with megaloblastic anemia and neurologic or mental disturbances even when the total $[B_{12}]$ is normal. In such patients, a low free TC-II level confirms this diagnosis.

10. Reference intervals for $[B_{12}]$ in nonpregnant women may not be applicable to the normal range for pregnant women (refer to note 3 below).

11. Gastric parietal cell antibodies reduce intrinsic factor secretion from the GPC of the stomach and may lead to pernicious anemia. The presence of GPC antibodies does not necessarily equate to cyanocobalamin deficiency or the diagnosis of pernicious anemia.

12. Substantial numbers of patients have subclinical B_{12} deficiency without clinical signs or symptoms. Isolated biochemical and hematologic abnormalities can be found in these patients.

Notes

1. Much of the measured vitamin B_{12} is bound to the carrier proteins haptocorrin and TC-II. Only the B_{12} bound to transcobalamin is functional.

2. An elevated unsaturated B_{12} binding capacity (UBBC), which reflects an elevated free TC-II, may occur in B_{12}-deficient patients with normal or near normal measured total $[B_{12}]$ if the haptocorrin level is elevated.

3. Serum $[B_{12}]$ decreases and UBBC increases steadily throughout pregnancy. In 35% of patients in the third trimester, the $[B_{12}]$ was <150 pmol/L (203 pg/mL), and in 68.6% the UBBC values were >15%. No pregnant patient was found to have an intracellular RBC vitamin B_{12} < 148 pmol/L (201 pg/mL) or a reduced hemoglobin or red blood cell count (Koebnick et al., 2002).

4. The absence of findings and symptoms of vitamin B_{12} and folate deficiency after Roux-en-Y gastric bypass (RYGB) suggests that these deficiencies are not clinically important postoperatively in most patients. In contrast, iron deficiency anemia is a potentially serious problem after RYGB (Brolin et al., 1998).

5. Megaloblastic anemia results from B_{12} and/or folate deficiency. Combined deficiencies commonly occur in nutritional disorders, especially alcoholism.

6. Out of 184 long-term intensive care unit (ICU) patients, 16 (9%) were iron deficient, 4 (2%) were vitamin B_{12} deficient, and 4 (2%) were folate deficient; thus, potentially correctable nutritional abnormalities were found in more than 13% of ICU patients (Rodriguez et al., 2001).

7. In older Chinese vegetarian women, the prevalence of vitamin B_{12} deficiency, with $[B_{12}]$ < 150 pmol/L (203 pg/mL) and [MMA] $\geq$ 0.4 µmol/L, was 42%. Another 32.8% had possible vitamin B_{12} deficiency with one or the other abnormality. However, macrocytic anemia did not increase significantly until the [MMA] became >1.0 µmol/L (Kwok et al., 2002).

8. Autosomal recessive TC-II deficiency is characterized by intracellular vitamin B_{12} deficiency.

9. Abnormalities of B_{12} binding proteins are rarely a cause of B_{12} deficiency in the aged; therefore, plasma TC-II bound by vitamin B_{12} should not be used as a marker of B_{12} deficiency in older people.

10. Hypochlorhydria in atrophic gastritis may result in enteral bacterial overgrowth and diversion of dietary vitamin B_{12} away from the host. Nonetheless, the ability to absorb crystalline vitamin B_{12} remains intact in older people with atrophic gastritis.

11. Atrophic gastritis results in low acid–pepsin secretion by the gastric mucosa, which in turn results in a reduced release of free vitamin B_{12} from food proteins.

ICD-9 Codes

Conditions that may justify this test include *but are not limited to*:

Anemia
281.1   vitamin B_{12} deficiency/veganism
281.9   macrocytic, megaloblastic

Other
356.9   unspecified neuropathy
529.4   atrophy of tongue papillae

Suggested Reading

Baik HW, Russell RM. Vitamin B_{12} deficiency in the elderly. *Annu Rev Nutr* 1999; 19:357–77.

Bibi H, Gelman-Kohan Z, Baumgartner ER, Rosenblatt DS. Transcobalamin II deficiency with methylmalonic aciduria in thRMR sisters. *J Inherit Metab Dis* 1999; 22:765–72.

Brolin RE, Gorman JH, Gorman RC, Petschenik AJ, Bradley LJ, Kenler HA, Cody RP. Are vitamin B_{12} and folate deficiency clinically important after Roux-en-Y gastric bypass? *J Gastrointest Surg* 1998; 2:436–42.

Koebnick C, Heins UA, Dagnelie PC, Wickramasinghe SN, Ratnayaka ID, Hothorn T, Pfahlberg AB, Hoffmann I, Lindemans J, Leitzmann C. Longitudinal concentrations of vitamin B_{12} and vitamin B_{12}-binding proteins during uncomplicated pregnancy. *Clin Chem* 2002; 48:928–33.

Kwok T, Cheng G, Woo J, Lai WK, Pang CP. Independent effect of vitamin B_{12} deficiency on hematological status in older Chinese vegetarian women. *Am J Hematol* 2002; 70:186–90.

Rodriguez RM, Corwin HL, Gettinger A, Corwin MJ, Gubler D, Pearl RG. Nutritional deficiencies and blunted erythropoietin response as causes of the anemia of critical illness. *J Crit Care* 2001; 16:36–41.

Savage DG, Lindenbaum J, Stabler SP, Allen RH. Sensitivity of serum methylmalonic acid and total homocysteine determinations for diagnosing cobalamin and folate deficiencies. *Am J Med* 1994; 96:239–46.

Teplitsky V, Huminer D, Zoldan J, Pitlik S, Shohat M, Mittelman M. Hereditary partial transcobalamin II deficiency with neurologic, mental and hematologic abnormalities in children and adults. *Isr Med Assoc J* 2003; 5:868–72.

van Asselt DZ, Thomas CM, Segers MF, Blom HJ, Wevers RA, Hoefnagels WH. Cobalamin-binding proteins in normal and cobalamin-deficient older subjects. *Ann Clin Biochem* 2003; 40:65–9.

11.2.7 Magnesium: Testing for Total [Mg] in Serum and Intracellular Red Blood Cell Ionized [Mg^{2+}]

Indications for Test

Measurement of [Mg] and/or [Mg^{2+}] is indicated in individuals with:

- Critical illness, malabsorption, or alcoholism
- Severe hyperglycemia or diabetic ketoacidosis (DKA)
- Hypocalcemia, preeclampsia, eclampsia, seizures, or cardiac dysrhythmias
- Congestive heart failure (CHF) or acute myocardial infarction (AMI)

Procedure

1. Recognize that 99% of the whole-body magnesium content is intracellular, and only about 60% of the [Mg] in serum is in the bioactive ionized form. Thus, accurate assessment of an individual's magnesium status, using serum measurements alone, is limited.

2. Measure total serum [Mg] as a screening procedure or in the clinical context of those at high risk for deficiency, as noted above, after obtaining a dietary history focused on intake of magnesium-containing foods (i.e., nuts, chocolate, shell fish, legumes, and whole grains).

3. For the most accurate assessment of magnesium deficiency, particularly in cases of insulin resistance, constipation, diarrhea, or fatigue, measure erythrocyte (red blood cell) intracellular free or ionized [Mg^{2+}] by phosphorus-31 nuclear magnetic resonance (^{31}P-NMR) spectroscopy.

4. Instead of using the NMR spectroscopy method for measurement of [Mg^{2+}], use the more available and cost-effective, albeit less accurate, ion-selective electrode method for determination of [Mg^{2+}] in red blood cells.

5. In high-risk patients for whom a critical decision is required as to replacement doses of magnesium, use the magnesium retention test (see Test 11.2.8) to provide treatment of deficiency and to make a diagnosis of the degree of deficiency.

Interpretation

1. [Mg^{2+}] in red blood cells, quantified by inductively coupled plasma–mass spectrometry (ICP-MS):
   - Healthy individuals, ≥1.5 to 3.1 mmol/L
   - Mg-deficient states, <1.0 mmol/L
2. Serum [Mg] reference intervals:
   - Healthy individuals, 1.6 to 2.4 mg/dL (females); 1.7 to 2.3 mg/dL (males)
   - Overt Mg deficiency, <1.7 mg/dL (0.70 mmol/L)
3. Overt hypomagnesemia ([Mg] < 1.2 mg/dL) is uncommon and identifies those with the most seriously deranged electrolyte disturbance.
4. Three quarters of patients with elevated plasma [Mg] > 2.4 mg/dL (1.0 mmol/L) may have chronic kidney disease (CKD). A total serum [Mg] of > 4.8 mg/dL (2.0 mmol/L) is rarely seen in hospitalized patients without CKD.

Notes

1. Circulating bioactive magnesium represents <1% of whole body levels.
2. Routinely determined [Mg] in serum is an insensitive index of magnesium deficiency.
3. The physiological functions of magnesium occur mainly intracellularly; therefore, measurement of intracellular [Mg^{2+}] is the best method for evaluating magnesium status.
4. Magnesium deficiency is linked to insulin resistance as well as to the use of diuretics, digoxin, and excessive alcohol intake.
5. Magnesium deficiency resulting from increased urinary excretion and decreased intestinal absorption is common in diabetes patients as well as in those given the misguided nutritional advice to avoid high-magnesium-containing foods such as chocolate and nuts because of their fat content.
6. In a diabetic, a dietary history of low intake of shellfish, legumes, and whole grains is sufficient to recommend magnesium supplements.
7. Patients with acute pancreatitis and hypocalcemia commonly have a whole-body magnesium deficiency despite a serum [Mg] within the reference range (Ryzen and Rude, 1990).
8. Magnesium deficiency is a common finding in ICU patients, resulting primarily from gastrointestinal or urinary magnesium losses. Underlying malnutrition and decreased dietary magnesium intake can be expected to accelerate magnesium depletion in ICU patients.
9. Parenteral magnesium therapy has a low risk of adverse effects, as suggested by the large and rapid dosing regimens used in many clinical studies (Crook, 1999).
10. No significant correlation was found among [Mg] in mononuclear cells, plasma, or erythrocytes (Elin and Hosseini, 1985).

ICD-9 Codes

Conditions that may justify this test include *but are not limited to*:

Cardiac illness
410	acute myocardial infarction
427	dysrhythmias in CHF
428.0	congestive heart failure (CHF)

Diabetic states
250.3   diabetic ketoacidosis (DKA)
790.29   severe hyperglycemia

Electrolytes
275.3   hypophosphatemia
275.41   hypocalcemia
276.8   hypopotassemia

Gastrointestinal
579.9   malabsorption
579.X   steatorrhea
787.91   chronic diarrhea

Other
642.X   mild, severe or unspecified preeclampsia/eclampsia
729.82   muscle cramps
780.39   convulsions
782.0   paresthesias
E904.1   lack of food/starvation
V11.3   alcoholism

Suggested Reading

Alfrey AC, Miller NL, Butkus D. Evaluation of body magnesium stores. *J Lab Clin Med* 1974; 84:153–62.

Crook M. A study of hypermagnesaemia in a hospital population. *Clin Chem Lab Med* 1999; 37:449–51.

Elin RJ, Hosseini JM. Magnesium content of mononuclear blood cells. *Clin Chem* 1985; 31:377–80.

Malon A, Brockmann C, Fijalkowska-Morawska J, Rob P, Maj-Zurawska M. Ionized magnesium in erythrocytes: the best magnesium parameter to observe hypo- or hypermagnesemia. *Clin Chim Acta* 2004; 349:67–73.

McLean RM. Magnesium and its therapeutic uses: a review. *Am J Med* 1994; 96:63–76.

Roberts LB. The normal ranges, with statistical analysis for seventeen blood constituents. *Clin Chim Acta* 1967; 16:69–78.

Ryzen E. Magnesium homeostasis in critically ill patients. *Magnesium* 1989; 8:201–12.

Ryzen E, Rude RK. Low intracellular magnesium in patients with acute pancreatitis and hypocalcemia. *West J Med* 1990; 152:145–8.

Ryzen E, Servis KL, De Russo P, Kershaw A, Stephen T, Rude RK. Determination of intracellular free magnesium by nuclear magnetic resonance in human magnesium deficiency. *J Am Coll Nutr* 1989; 8:580–7.

11.2.8 Magnesium [Mg] Retention Test

Indications for Test

The Mg retention test is indicated when:

- The serum magnesium concentration [Mg] > 1.7 mg/dL but more than one of the following conditions is found, suggestive of clinically significant ionized magnesium deficiency:
  - Paresthesias, muscle cramps, and/or weakness
  - Irritability, decreased attention span, impaired concentration
  - Chronic diarrhea, malabsorption syndromes, or steatorrhea
  - Diuretic, aminoglycoside, amphotericin B, or cisplatin therapy
  - Alcoholism, starvation, poorly controlled diabetes mellitus or diabetic ketoacidosis
  - Unexplained or refractory hypokalemia, hypocalcemia, or hypophosphatemia, particularly if associated with seizures
  - Variant angina or congestive heart failure
  - Ventricular dysrhythmias, possibly secondary to depleted K⁺ storage, which may be aggravated by Mg deficiency and refractory to correction with K⁺ supplementation

- Prolonged P–R and Q–T intervals on electrocardiogram (EKG)
- Supraventricular arrhythmias—premature atrial contractions (PACs), paroxysmal atrial tachycardia (PAT), or atrial fibrillation

Procedure

1. Obtain a baseline 24-hour urine sample for magnesium (preU$_{Mg}$) and creatinine (preU$_{Cr}$) testing.
2. Obtain a baseline blood sample for magnesium, calcium, phosphorus, and creatinine testing. Determine lean body mass (LBM) from Test 11.1.2.
3. Infuse Mg (0.2 mEq/kg LBM) in 50 mL of 5% dextrose in water i.v. over 4 hours (2.4 mg elemental Mg per kg LBM).
4. Monitor the patient's blood pressure, heart rate, and deep tendon reflexes every hour during the infusion. Stop the infusion if marked changes in the above parameters are observed.
5. Obtain a blood sample for magnesium, calcium, phosphorus, and creatinine at the end of the 4-hour infusion.
6. Obtain a post-infusion 24-hour urine sample for [Mg] (postU$_{Mg}$) and creatinine (postU$_{Cr}$) testing.
7. Calculate the percent magnesium retention as follows:

$$\text{Mg retention (\%)} = \left\{ 1 - \left[\frac{\text{postU}_{Mg} - \left(\frac{\text{preU}_{Mg}}{\text{preU}_{Cr}} \times \text{postU}_{Cr} \right)}{\text{Total Mg infused}} \right] \right\} \times 100$$

where the terms "pre" and "post" refer, respectively, to before and after infusion of MgSO$_4$.

Example:

Patient's LBM = 69 kg.
Concentration of MgSO$_4$ infused = 0.1 mmol/L.
Molecular weight of Mg = 24.3 mg/mmol.
Total elemental Mg infused = 69 × 0.1 × 24.3 = 167.67 mg.
Patient's 24-hour urine volume = 1799 mL = 17.99 dL.
preU$_{Cr}$ = 89.4 mg/dL.

Therefore,

$$\text{preU}_{Cr} \text{ (mg/kg)} = \frac{89.4 \text{ mg/dL} \times 17.99 \text{ dL}}{69 \text{ kg}} = 23.3 \text{ mg/kg}$$

$$\text{postU}_{Cr} = 23.0 \text{ mg/kg}$$

$$\text{preU}_{Mg} = 74.0 \text{ mg/day}$$

$$\text{postU}_{Mg} = 181.4 \text{ mg/day}$$

$$\text{preU}_{Mg/Cr} = \frac{74.0}{23.3} = 3.18$$

$$\text{preU}_{Mg/Cr} \times \text{postU}_{Cr} = 3.18 \times 23.0 = 73.05$$

$$\text{postU}_{Mg} - \left(\text{preU}_{Mg/Cr} \times \text{postU}_{Cr} \right) = 181.4 - 73.05 = 108.35$$

$$\frac{\text{postU}_{Mg} - \left(\text{preU}_{Mg/Cr} \times \text{postU}_{Cr} \right)}{\text{Total elemental Mg infused}} = \frac{108.35}{167.67} = 0.646$$

$$\text{Mg retention (\%)} = \left\{ 1 - \frac{\text{postU}_{Mg} - \left(\text{preU}_{Mg/Cr} \times \text{postU}_{Cr} \right)}{\text{Total elemental Mg infused}} \right\} \times 100$$

$$\text{Mg retention (\%)} = \left\{ 1 - \frac{181.4 - (3.18 \times 23.0)}{167.67} \right\} \times 100$$

$$= (1 - 0.646) \times 100 = 35.4$$

Interpretation

1. Mg retention reference interval is <25%.
2. Mg retention that is ≥25 to 50% suggests a possible magnesium deficiency.
3. Mg retention of >50% is virtually diagnostic of magnesium deficiency; the greater the retention, the greater the deficiency.

Notes

1. Mg retention correlates with skeletal muscle magnesium content.

ICD-9 Codes

Refer to Test 11.2.7 codes.

Suggested Reading

Reungjui S, Prasongwatana V, Premgamone A, Tosukhowong P, Jirakulsomchok S, Sriboonlue P. Magnesium status of patients with renal stones and its effect on urinary citrate excretion. *BJU Int* 2002; 90:635–639.

Ryzen E, Elbaum N, Singer FR, Rude RK. Parenteral magnesium tolerance testing in the evaluation of magnesium deficiency. *Magnesium* 1985; 4:137–47 (*erratum* in *Magnesium* 1987; 6:168).

11.2.9 Folate: Testing in Serum and Red Blood Cells (RBCs)

Indications for Test

Measurement of folate levels in serum, [S_{folate}], may be indicated in the assessment of:

- Nutritional deficiency
- Cardiovascular disease (CVD) risk
- Risk of fetal malformation with poor pregnancy outcome

Measurement of folate in erythrocytes [RBC_{folate}] is indicated as a marker of nutritional status in patients with:

- Megaloblastic anemia, neurologic problems secondary to malnutrition, or suspected nutritional vitamin deficiency
- Hyperhomocysteinemia, alcoholism, diabetes, or malabsorption syndromes

Procedure

1. When deciding which folate test to order (i.e., folate in RBC or serum), note that [RBC_{folate}] has higher sensitivity and specificity for detection of folate deficiency and malnutrition than does S_{folate}.
2. Obtain a blood specimen for S_{folate} testing. Laboratory personnel will ensure that the serum obtained from this specimen is protected from light and evaluate the serum for hemolysis. If the specimen is hemolyzed, lab personnel should reject the specimen and request that a fresh blood specimen be collected from the patient.
3. Obtain a blood specimen for RBC_{folate} testing if a deficiency of folate is suspected and the [S_{folate}] results are normal or in the low-end (i.e., quartile) of the reference interval.

Interpretation

1. Typical folate reference interval data:
   - [S_{folate}], >5.4 ng/mL
   - [RBC_{folate}], >280 ng/mL

2. Males tend to have lower serum folate levels (7.9 ± 3.1 ng/mL, or 18 ± 7 nmol/L) than females (9.7 ± 4.4 ng/mL, or 22 ± 10 nmol/L).

3. During treatment with high-dose methotrexate, [S_{folate}] drops to 0.2 to 3.1 ng/mL, prompting folate rescue therapy.

4. [S_{folate}] is positively correlated with dietary fiber consumption and is inversely correlated with vitamin B_{12} intake.

5. [RBC_{folate}] is a better measure of tissue folate stores than [S_{folate}], which may change rapidly with changes in diet.

6. The risk of myocardial infarction (MI) was threefold higher in individuals with [S_{folate}] in the lowest quartile (<2.4 ng/mL, or 5.4 nmol/L) compared with the highest quartile (>4.6 ng/mL, or >11.4 nmol/L) (Ma et al., 2001).

7. CVD patients have significantly lower [S_{folate}] and higher homocysteine (Hcy), vitamin B_{12}, and creatinine levels compared to non-CVD controls.

8. In a study of cognitive function, serum [Hcy] was negatively correlated with [S_{folate}]. Low folate levels largely accounted for the trend toward greater cognitive decline, observed over 7 years, in those with elevated [Hcy] (Kado et al., 2005).

9. [RBC_{folate}] is significantly lower in stroke patients with polymorphisms of the wild-type gene encoding 5,10-methylenetetrahydrofolate reductase (MTHFR), particularly the TT compared with the CT and CC MTHFR genotypes.

10. Compared to women with the wild-type CC genotype and [S_{folate}] > 3.3 ng/mL, women who have the TT genotype have a twofold greater risk of MI (odds ratio, 2.0; 95% CI, 1.0–3.7).

11. MTHFR polymorphisms, especially the 677C→T mutation, may contribute to vascular and neural tube birth defect risks, while reducing the risk of certain malignancies, such as colon cancer.

12. When combined together, folate deficiency, homozygous TT MTHFR gene mutations, and hyperhomocysteinemia are significant risk factors for intrauterine fetal death, habitual spontaneous abortion, thromboembolic disease in pregnancy, neural tube defects and fetal cardiac malformation, preeclampsia, placental abruption, and intrauterine growth retardation.

Notes

1. The principal dietary sources of folate are fruits, vegetables, and legumes.

2. Smokers have lower [S_{folate}], regardless of their dietary folate intake.

3. [S_{folate}] determined by competitive protein binding radioassay is 9% lower than results obtained with the reference method (i.e., isotope-dilution liquid chromatography–tandem mass spectrometry).

4. Microbiological assays that measure biologically active folates yield substantially lower estimates of folate concentration than immunometric assays.

5. MTHFR catalyzes the reduction of 5,10-MTHF to 5-MTHF or L-methylfolate, the main circulating form of folate, which serves as a methyl donor for remethylation of Hcy to methionine.

6. Some individuals are incapable of completely converting folic acid to its active form, L-methylfolate, because of gene polymorphisms, including the heterozygous 677C→T and homozygous T-to-T transition (TT) in a CC pair within the coding region of the wild-type gene encoding MTHFR. Therefore, MTHFR genotyping may be useful in individuals whose folate levels and clinical status are discordant.

7. High (>1 mg/day) folate and high (>500 mg/day) vitamin C intake appear to help those with both normal and heterozygous MTHFR genotypes maintain higher [S_{folate}].

8. Among the 293 Physicians' Health Study participants who developed MI during up to 8 years of follow-up compared to 290 control subjects, MTHFR polymorphisms were associated with higher [Hcy] but not with an increased risk of MI (Ma et al., 1996).

9. The periconceptional use of folic-acid-containing supplements reduces the first occurrence, as well as the recurrence, of neural tube defects and other poor pregnancy outcomes.

10. Women in whom adverse pregnancy outcomes are prevalent often consume diets that are low in vitamins and minerals, including folate.

11. Folate therapy may increase the risk of in-stent restenosis after coronary stenting in all patients except women, diabetics, and those with [Hcy] >15 mcmol/L. In these subgroups, folate therapy reduced the risk of restenosis (Lange et al., 2004).

12. In a population study of 601 individuals (ages 18 to 75 years) from the Canary Islands, mean [S_{folate}] and [RBC_{folate}] were 8.2 ng/mL and 214.3 ng/mL, respectively (Henriquez et al., 2004).

ICD-9 Codes

Conditions that may justify this test include *but are not limited to*:

429.2 cardiovascular disease
V22.2 pregnancy

Suggested Reading

Carmel R, Green R, Rosenblatt DS, Watkins D. Update on cobalamin, folate, and homocysteine. *Hematology (Am Soc Hematol Educ Program)* 2003; 62–81.

Delougheri TG, Evans A, Sadeghi A, McWilliams J, Henner WD, Taylor LM, Press RD. Common mutation in methylenetetrahydrofolate reductase: correlation with homocysteine metabolism and late-onset vascular disease. *Circulation* 1996; 94:3074–78.

Henriquez P, Doreste J, Diaz-Cremades J, Lopez-Blanco F, Alvarez-Leon E, Serra-Majem L. Folate status of adults living in the Canary Islands (Spain). *Int J Vitam Nutr Res* 2004; 74:187–92.

Icke GC, Dennis M, Sjollema S, Nicol DJ, Eikelboom JW. Red cell N^5-methyltetrahydrofolate concentrations and C677T methylenetetrahydrofolate reductase genotype in patients with stroke. *J Clin Pathol* 2004; 57:54–7.

Kado DM, Karlamangla AS, Huang MH, Troen A, Rowe JW, Selhub J, Seeman TE. Homocysteine versus the vitamins folate, B_6, and B_{12} as predictors of cognitive function and decline in older high-functioning adults: MacArthur Studies of Successful Aging. *Am J Med* 2005; 118:161–7.

Lange H, Suryapranata H, De Luca G, Borner C, Dille J, Kallmayer K, Pasalary, Scherer E, Dambrink JE. Folate therapy and in-stent restenosis after coronary stenting. *N Engl J Med* 2004; 350:2673–81.

Ma J, Stampfer MJ, Hennekens CH, Frosst P, Selhub J, Horsford J, Malinow MR, Willett WC, Rozen R. Methylenetetrahydrofolate reductase polymorphism, plasma folate, homocysteine, and risk of myocardial infarction in U.S. physicians. *Circulation* 1996; 94:2410–16.

Maruyama C, Araki R, Takeuchi M, Kuniyoshi E, Iwasawa A, Maruyama T, Nakano S, Motohashi Y, Nakanishi M, Kyotani S, Tsushima M. Relationships of nutrient intake and lifestyle-related factors to serum folate and plasma homocysteine concentrations in 30–69-year-old Japanese. *J Nutr Sci Vitaminol (Tokyo)* 2004; 50:1–8.

Pfeiffer CM, Fazili Z, McCoy L, Zhang M, Gunter EW. Determination of folate vitamers in human serum by stable-isotope-dilution tandem mass spectrometry and comparison with radioassay and microbiologic assay. *Clin Chem* 2004; 50:423–32.

Scholl TO, Johnson WG. Folic acid: influence on the outcome of pregnancy. *Am J Clin Nutr* 2000; 71(5 Suppl):1295–303S.

Tanis BC, Blom HJ, Bloemenkamp DG, van den Bosch MA, Algra A, van der Graaf Y, Rosendaal FR. Folate, homocysteine levels, methylenetetrahydrofolate reductase (MTHFR) 677C→T variant, and the risk of myocardial infarction in young women: effect of female hormones on homocysteine levels. *J Thromb Haemost* 2004; 2:35–41.

Vrentzos GE, Papadakis JA, Malliaraki N, Zacharis EA, Mazokopakis E, Margioris A, Ganotakis ES, Kafatos A. Diet, serum homocysteine levels and ischaemic heart disease in a Mediterranean population. *Br J Nutr* 2004; 91:1013–9.

11.2.10 Thiamine (Vitamin B_1), Riboflavin (Vitamin B_2), and Pyridoxine (Vitamin B_6)

Indications for Test

Thiamine, riboflavin, and pyridoxine are indicated as markers of nutritional status in:

- Renal failure patients on dialysis, alcoholics, or women who are pregnant or taking oral contraceptives and ingest high-protein, high-simple-carbohydrate diets with inadequate intake of fruits and vegetables or thiamine-fortified cereals

- Cases of refractory metabolic acidosis, hyperthyroidism, status after weight-loss surgery or anorexia (i.e., thiamine testing indicated)
- Cases of hyperthyroidism (riboflavin and pyridoxine testing) or hypothyroidism (riboflavin testing)

Procedure

1. Obtain the appropriate blood specimen for measurement of vitamin B_1, B_2, and/or B_6 levels.

Interpretation

1. Reference interval (HPLC) is 150 to 290 nmol/L; up to 38% of alcoholics were found to be erythrocyte thiamine deficient (90.8 nmol/L vs. 176 nmol/L in healthy controls; $p < 0.001$) (Herve et al., 1995).
2. Reference interval for plasma riboflavin is 2.3 to 14.7 mcg/dL (6.2–39.0 nmol/L).
3. Reference interval for plasma pyridoxine is 4 to 18 mcg/L (14.6–72.8 nmol/L).

Notes

1. Typically, erythrocyte transketolase (ETK), erythrocyte glutathione reductase (EGR), and erythrocyte aspartate aminotransferase (EAA) coenzyme activation assays are used to quantify vitamin B_1, vitamin B_2, and vitamin B_6 levels, respectively, in a red blood cell (RBC) hemolysate.
2. Direct measurement of thiamine and its phosphate esters was found to be a more sensitive and specific index of thiamine status than determination of ETK activity (Herve et al., 1995).
3. Direct measurement of thiamine (and its phosphate esters), B_2, and B_6 can be performed using high-performance liquid chromatography. In HPLC, the dioctylsulfosuccinate counter-ion facilitates unique retention of the pyridine-based vitamins (niacinamide and pyridoxine) and allows for concurrent measurement of both pyridoxal and riboflavin 5'-phosphate.
4. A score between 17 and 23.5 on the Mini Nutritional Assessment (MNA®) questionnaire (Guigoz et al., 1996) identified those at-risk frail, older persons with deficiencies of calcium, vitamin D, iron, vitamin B_6, and vitamin C for whom nutrition intervention might be effective (Vellas et al., 2000).
5. The prevalence of deficiency states for thiamine, riboflavin, and pyridoxine in hospital inpatients was found to be 21%, 2.7%, and 32%, respectively, with 49.2% of patients being deficient in one or more vitamin. The mean alcohol intake in this group was 9.7 units/week.
6. Deficiency states for thiamine, riboflavin, and pyridoxine were found in up to two thirds of parturient mothers and in up to one third of their infants (Sanchez et al., 1999).
7. The observed prevalence of deficiencies in an adult Mediterranean population was 6.4% for vitamin B_1 and 5.3% for vitamin B_2, whereas the expected prevalence, estimated from 48-hour food intake recall, was 6.2% (men) and 15.5% (women) (Mataix et al., 2003).
8. Long-term use of thiamine (100-mg), riboflavin (20-mg), pyridoxine (50-mg), folic acid (6-mg), and ascorbic acid (500-mg) supplements provides adequate vitamin levels in almost all dialysis patients (Descombes et al., 2000).

Illustration

See color photograph in Friedli, A. and Saurat, J.H., Oculo-orogenital syndrome: a deficiency of vitamins B_2 and B_6, *N. Engl. J. Med.*, 350(11), 1130, 2004.

ICD-9 Codes

Conditions that may justify this test include *but are not limited to*:

242.9	thyrotoxicosis
276.2	refractory metabolic acidosis
783.0	anorexia
E904.1	lack of food/starvation
V11.3	alcoholism
V22.2	normal pregnancy
V56.0	extracorporeal dialysis
V69.1	inappropriate diet and eating habits/excessively high protein or simple carbohydrate intake

Suggested Reading

Descombes E, Boulat O, Perriard F, Fellay G. Water-soluble vitamin levels in patients undergoing high-flux hemodialysis and receiving long-term oral postdialysis vitamin supplementation. *Artif Organs* 2000; 24:773–8.

Folkers K, Ellis J. Successful therapy with vitamin B_6 and vitamin B_2 of the carpal tunnel syndrome and need for determination of the RDAs for vitamins B_6 and B_2 for disease states. *Ann NY Acad Sci*, 1990; 585:295–301.

Guigoz Y, Vellas B, Garry PJ. Assessing the nutritional status of the elderly: the Mini Nutritional Assessment as part of the geriatric evaluation. *Nutr Rev* 1996; 54:S59–S65.

Herve C, Beyne P, Letteron P, Delacoux E. Comparison of erythrocyte transketolase activity with thiamine and thiamine phosphate ester levels in chronic alcoholic patients. *Clin Chim Acta* 1995; 234:91–100.

Jamieson CP, Obeid OA, Powell-Tuck J. The thiamin, riboflavin and pyridoxine status of patients on emergency admission to hospital. *Clin Nutr* 1999; 18:87–91.

Lakshmi AV, Ramalakshmi BA. Effect of pyridoxine or riboflavin supplementation on plasma homocysteine levels in women with oral lesions. *Natl Med J India* 1998; 11:171–2.

Mataix J, Aranda P, Sanchez C, Montellano MA, Planells E, Llopis J. Assessment of thiamin (vitamin B_1) and riboflavin (vitamin B_2) status in an adult Mediterranean population. *Br J Nutr* 2003; 90:661–6.

Planells E, Lerma A, Sanchez-Morito N, Aranda P, Lopis J. Effect of magnesium deficiency on vitamin B_2 and B_6 status in the rat. *J Am Coll Nutr* 1997; 16:352–6.

Sanchez DJ, Murphy MM, Bosch-Sabater J, Fernandez-Ballart J. Enzymic evaluation of thiamin, riboflavin and pyridoxine status of parturient mothers and their newborn infants in a Mediterranean area of Spain. *Eur J Clin Nutr* 1999; 53:27–38.

Suboticanec K, Stavljenic A, Schalch W, Buzina R. Effects of pyridoxine and riboflavin supplementation on physical fitness in young adolescents. *Int J Vitam Nutr Res* 1990; 60:81–8.

Tovar AR, Torres N, Halhali A, Bourges H. Riboflavin and pyridoxine status in a group of pregnant Mexican women. *Arch Med Res (Mexico)*, Summer 1996; 27:195–200.

Vellas B, Guigoz Y, Baumgartner M, Garry PJ, Lauque S, Albarede JL. Relationships between nutritional markers and the Mini Nutritional Assessment in 155 older persons. *J Am Geriatr Soc* 2000; 48:1300–9.

11.2.11   Potassium (K⁺)

Indications for Test

Measurement of serum K⁺ concentration [K⁺] is indicated as a marker of nutritional status in individuals with:

- Weakness, muscle cramps, and/or weight loss
- Arrhythmias and/or an abnormal electrocardiogram
- Renal disease and/or acidosis
- Recurrent vomiting, diarrhea, and/or an eating disorder

Procedure

1. Obtain a random blood specimen for K⁺ testing.

Interpretation

1. The usual reference interval for serum [K⁺] is 3.5 to 5.5 mmol/L; however, [K⁺] > 4.0 mmol/L is associated with a lower incidence of cardiac arrhythmia and a greater sense of well-being.
2. Chronically low [K⁺] may result in an insulin resistance syndrome and increased insulin requirements.

3. Chronically elevated [K$^+$] in the range of 5.5 to 5.9 mmol/L is usually well tolerated, but can result in serious cardiac bradyarrhythmia if acutely induced.
4. Patients with chronic kidney disease tend to have higher [K$^+$], whereas use of thiazide or loop diuretics will lower [K$^+$] and magnesium levels.

Notes

1. Ordinarily, the request for K$^+$ testing is processed as part of a panel of electrolytes, including sodium, chloride, and CO$_2$, as well as testing for creatinine and blood urea nitrogen.
2. In niacin-deficient alcoholics with anemia and hypoalbuminemia, the finding of lower [K$^+$] was associated with protein malnutrition, possibly related to pelagrin-associated diarrhea (Cunha et al., 2000).
3. In adolescent vegans, the diet-history method underestimated K$^+$ intake by 7% when compared with the 24-hour urine K$^+$ excretion method (Larsson et al., 2002).
4. Although central pontine myelinolysis is usually associated with rapid correction of hyponatremia and abnormal osmolality, preexistent and persistent hypokalemia can play a significant role in its pathogenesis (Sugimoto et al., 2003).
5. Patients with anorexia nervosa have a greater difference between the longest and the shortest QT interval on EKG than constitutionally thin and normal-weight women do. Oral K$^+$ supplementation reduces this difference (Franzoni et al., 2002).
6. Adoption of modern (Westernized) diets has led to a substantial decline in K$^+$ intake compared to the native diets of some populations having a traditionally higher intake of fruits and vegetables.
7. A 2.5- to 3.5-g daily dietary K$^+$ supply from fruits and vegetables (per the "5 to 10 servings per day" recommendation) is usually required for optimal nutritional health and maintenance of adequate whole-body K$^+$ stores.
8. Measurement of total body K$^+$ by whole-body counting revealed that 29.9% of males and 22.0% of females with cystic fibrosis were malnourished based on body cell mass, compared to identifying only 7.5% using standard anthropometric measurements (McNaughton et al., 2000).

ICD-9 Codes

Conditions that may justify this test include *but are not limited to*:

276.2	acidosis
427.9	arrhythmias
536.2	persistent vomiting
593.9	renal disease
729.82	muscle cramps
783.21	loss of weight
787.91	diarrhea

Suggested Reading

Cunha DF, Monteiro JP, Ortega LS, Alves LG, Cunha SF. Serum electrolytes in hospitalized pellagra alcoholics. *Eur J Clin Nutr* 2000; 54:440–2.

Demigne C, Sabboh H, Remesy C, Meneton P. Protective effects of high dietary potassium: nutritional and metabolic aspects. *J Nutr* 2004; 134:2903–6.

Franzoni F, Mataloni E, Femia R, Galetta F. Effect of oral potassium supplementation on QT dispersion in anorexia nervosa. *Acta Paediatr* 2002; 91:653–6.

Grunfeld C, Chappell DA. Hypokalemia and diabetes mellitus. *Am J Med* 1983; 75:553–4.

Larsson CL, Johansson GK. Dietary intake and nutritional status of young vegans and omnivores in Sweden. *Am J Clin Nutr* 2002; 76:100–6.

McNaughton SA, Shepherd RW, Greer RG, Cleghorn GJ, Thomas BJ. Nutritional status of children with cystic fibrosis measured by total body potassium as a marker of body cell mass: lack of sensitivity of anthropometric measures. *J Pediatr* 2000; 136:188–94.

Sugimoto T, Murata T, Omori M, Wada Y. Central pontine myelinolysis associated with hypokalaemia in anorexia nervosa. *J Neurol Neurosurg Psychiatry* 2003; 74:353–5.

11.2.12 Urinary Iodide (or Iodine) Excretion

Indications for Test

Urinary iodide excretion testing is indicated as:

- A marker of nutritional status in populations living in iodine-deficient areas with a high prevalence of endemic goiters (i.e., land-locked areas of Europe, Africa, and Asia)
- A means to assess the risk for thyroid goiter in residents of impoverished or remote, usually mountainous, inland regions
- An index of therapeutic or dietary-related iodine excess or depletion preliminary to radioiodine-related imaging or therapy in thyroid goiter or cancer patients

Procedure

1. For general nutritional assessment, measure fasting overnight urine iodide [I⁻] in population studies or 24-hour urine [I⁻] in individuals.
2. For patients requiring a low-iodine diet to obtain optimum thyroid imaging studies and response to radioiodine therapy, instruct the patient on a low-iodine diet for at least 2 weeks before the imaging study.
3. Collect a 24-hour urine specimen (see Specimen Collection Protocol P1 in Appendix 2) and measure urine iodide excretion after 10 days of the low-iodine diet.

Interpretation

1. A 24-hour urine iodide excretion of <80 mcg/day is consistent with severe iodine deficiency.
2. Very low or extremely deficient [I⁻] in urine is <10 mcg/dL, as suggested by the World Health Organization (WHO).
3. Ideally, iodine repletion is marked by the excretion of >150 mcg/L [I⁻] in urine with acceptable response to a low-iodine diet being marked by excretion of <100 mcg/day.

Notes

1. Urinary iodine excretion did not appear to be related to the presence of goiter or thyroid nodules (Brauer et al., 2005), but a correlation with decreasing iodide excretion (>90, 60–90, <60 mcg iodide per day) and increasing size of thyroid gland was observed (Thomson et al., 2001).
2. Categories of subjects who are at risk for iodine deficiency disorders were classified by the WHO International Committee for the Control of Iodine Deficiency Disorders in 1994.
3. As shown in Switzerland, dietary iodine insufficiency, possibly related to lower intake of fish and seafood, occurs frequently, such that 54.3% of Swiss Alpine children had urinary [I⁻] < 10 mcg/dL and 11.5% had levels under 5 mcg/dL (Zimmermann et al., 1998).
4. In Africa, iodine deficiency, which may affect up to 60 million persons, has been linked to multiple disorders including mental retardation, cretinism in newborns, goiter, deafness, miscarriage, lowered resistance to infectious diseases, and lowered intelligence (10 to 15 IQ points).
5. In countries with advanced food-processing industries, iodide is a widely used food preservative and deficiency states are uncommon; however, the availability of noniodized salt and avoidance of the use of iodine in the dairy and breadmaking industries have allowed the appearance of iodine deficiency in subpopulations of otherwise well-nourished societies.
6. In Australia in 1992 the urinary excretion of iodine was about 200 mcg/L. In 2002, it had dropped to 100 mcg/L as a result of the two factors listed in note 5 above.
7. Recommended iodine intakes range from about 120 mcg to <500 mcg per day. Intakes of 200 to 300 mcg are recommended during pregnancy and lactation.

ICD-9 Codes

Conditions that may justify this test include *but are not limited to*:

Health examination
V70.5   of defined subpopulations
V70.6   in population surveys

Other
193   thyroid cancer
240.9   goiter
241.0   thyroid nodule
V69.1   inappropriate diet and eating habits/deficient iodine intake

Suggested Reading

Barbaro D, Boni G, Meucci G, Simi U, Lapi P, Orsini P, Pasquini C, Piazza F, Caciagli M, Mariani G. Radioiodine treatment with 30 mCi after recombinant human thyrotropin stimulation in thyroid cancer: effectiveness for postsurgical remnants ablation and possible role of iodine content in L-thyroxine in the outcome of ablation. *J Clin Endocrinol Metab* 2003; 88:4110–5.

Brauer VF, Brauer WH, Fuhrer D, Paschke R. Iodine nutrition, nodular thyroid disease, and urinary iodine excretion in a German university study population. *Thyroid* 2005; 15:364–70.

Thomson CD, Woodruffe S, Colls AJ, Joseph J, Doyle TC. Urinary iodine and thyroid status of New Zealand residents. *Eur J Clin Nutr* 2001; 55:387–92.

Zimmermann MB, Hess S, Zeder C, Hurrell RF. Urinary iodine concentrations in Swiss schoolchildren from the Zurich area and the Engadine valley. *Schweiz Med Wochenschr* 1998; 128:770–4.

11.2.13 Iron: Testing for Serum Iron Concentration (SIC), Total Iron Binding Capacity (TIBC), and Ferritin

Indications for Test

Serum iron concentration, total iron binding capacity, and ferritin are indicated as markers of nutritional status in:

- Patients with fatigue, particularly those with anemia and microcytosis
- Cases of pernicious anemia, gastric surgery, or removal of terminal ileum
- Patients after weight-loss surgery
- Pregnancy, lactation, blood loss, menstruation, or iron sequestration (e.g., hemochromatosis, pulmonary hemosiderosis), chronic kidney disease (CKD)
- Adults with restless legs syndrome and children with symptoms of attention-deficit hyperactivity disorder (ADHD)

Procedure

1. Obtain a blood specimen for measurement of iron, TIBC, and ferritin in serum as a panel of tests in cases of anemia not clearly related to blood loss.
2. In patients with unexplained anemia and equivocal or discordant results upon testing for iron status (i.e., total SIC, TIBC, and ferritin), assess bone marrow iron stores by bone marrow aspirate and Prussian blue staining.
3. Obtain percent iron saturation (%TS), measurement of soluble transferrin receptor (sTfR), and the sTfR/ferritin index as supplemental tests in cases of unexplained anemia or as part of epidemiologic research studies.

Interpretation

1. Commonly accepted reference intervals for SIC, TIBC, and ferritin are:
   - SIC, >56 mcg/dL (10 μmol/L)
   - TIBC, <380 mcg/dL (68 μmol/L)
   - Ferritin, >20 mcg/L (levels > 100 mcg/L are toxic)
2. A serum [ferritin] cutoff value of ≤30 mcg/L or ≤10 mcg/L correctly identifies ~90% and 60%, respectively, of cases with iron deficiency anemia (IDA).
3. Ferritin values increase ~80 mcg/L for each 1% increase in transferrin saturation.
4. In general, a [ferritin] that is <20 mcg/L or >1.5 times the upper limit of the reference interval effectively identifies groups of individuals likely to have iron deficiency anemia or anemia of chronic disease (ACD), respectively
5. Deficient dietary iron intake is rare in the United States. Iron deficiency usually indicates blood loss.
6. SIC is lower and TIBC higher in iron-deficient athletes compared to nondeficient athletes, even in the absence of anemia.
7. A low SIC is commonly associated with anemia of celiac disease and may persist even after excluding gluten from the diet and giving routine doses of iron supplements.
8. Pica, a condition more common in some ethnic populations than others, may result in severe iron deficiency when substances such as clay are ingested, particularly during pregnancy.
9. Up to 84% of children with ADHD may have iron deficiency, with the severity of symptoms appearing to correlate with the severity of the deficiency. Taking an iron sulfate supplement 5 mg/kg/day or 80 mg/day has been shown to improve some measures of ADHD symptoms.
10. The differential diagnosis of microcytic anemia includes chronic infection, granulomatous disease, collagen vascular disease, malignancy, liver disease, CKD, and thalassemia. These conditions tend to be associated with a high SIC in the absence of blood loss.
11. Microcytic anemia, low SIC, low to normal TIBC, high serum ferritin, and increased bone marrow iron content with an absence of ringed sideroblasts results from primary defective iron reutilization. This disorder is refractory to iron therapy but responsive to danazol.
12. Even though sTfR values are significantly higher in patients with IDA than in those with ACD, the routine measurement of sTfR offers no advantage over TIBC for discriminating between people with biochemically defined iron deficiency anemia of chronic disease (Figure 11.2).

Notes

1. The gold standard method for evaluating body iron stores is histopathologic examination of a Prussian blue-stained bone marrow specimen.
2. Disorders of iron metabolism and impaired synthesis of protoporphyrin, heme, or globin lead to defective hemoglobin production and microcytosis.
3. Iron plays a role in dopaminergic-related brain functions contributory to ADHD. Preliminary evidence suggests possible improvement in ADHD as well as other mental and psychomotor functions with iron supplementation in iron-deficient individuals.
4. Folate and vitamin B_{12} deficiency is a major cause of nutritional anemia in kidney transplant recipients, of whom only ~10% have microcytic anemia from iron deficiency.
5. Using the sTfR/ferritin index, a significantly ($p < 0.01$) higher incidence of and more profound iron deficiency were observed in female (26%) than in male (11%) athletes (Malczewska et al., 2001).
6. Deficient iron in the substantia nigra with disruption of the dopaminergic system of the brain has been correlated with development of restless legs syndrome. Correction of an iron deficiency in the substantia nigra requires intervention sufficient to raise the serum ferritin level to >50 mcg/L (Sun et al., 1998).
7. Serum ferritin concentration, which may be elevated as a part of an acute-phase reaction, is an unreliable indicator of bone marrow iron stores.
8. In iron-deficient women of childbearing age, a high-iron diet produces a smaller increase in serum ferritin concentration than does iron supplementation (Patterson et al., 2001).
9. In hereditary hemochromatosis, male homozygotes for the C282Y mutation with ferritin levels ≥ 1000 mcg/L develop iron-overload-related disease in much larger proportions (28.4%) than do women (1.2%) (Allen et al., 2008).

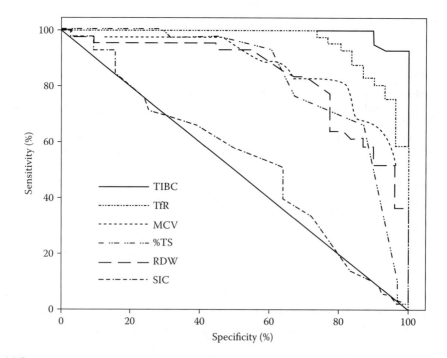

Figure 11.2

Receiver–operator characteristic (ROC) curves illustrating the diagnostic accuracy of various laboratory tests in distinguishing patients with iron deficiency anemia (IDA) from those with anemia of chronic disease (ACD). Curves (from upper right corner to diagonal) shown are for total iron binding capacity (TIBC), transferrin receptor (TfR), mean corpuscular volume (MCV), percent transferrin saturation (%TS), RBC distribution width (RDW), and serum iron concentration (SIC). The diagonal line represents the ROC curve for a test with no clinical value (i.e., area under the curve = 0.500). TIBC and TfR measurements provided the highest and similar discriminatory power for distinguishing between patients with IDA and those with ACD. (From Wians, F.H., Jr. et al., *Am. J. Clin. Pathol.*, 115, 112–118, 2001. With permission.)

ICD-9 Codes

Conditions that may justify this test include *but are not limited to*:

Anemias

280	iron deficiency
281.0	pernicious
285.1	acute posthemorrhagic

Other

275.0	disorders of iron metabolism
314.0	attention-deficit disorder
516.1	idiopathic pulmonary hemosiderosis
579.3	postsurgical nonabsorption
585	chronic kidney disease (CKD)
780.7	malaise and fatigue
V22.2	normal pregnancy
V24.1	lactating mother

Suggested Reading

Allen KJ et al. Iron-overload-related disease in HFE hereditary hemochromatosis. *N Engl J Med* 2008; 358:221–30.

Besa EC, Kim PW, Haurani FI. Treatment of primary defective iron-reutilization syndrome: revisited. *Ann Hematol* 2000; 79:465–8.

Kirschbaum B. Serial ferritin concentrations in hemodialysis patients receiving intravenous iron. *Clin Nephrol* 2002; 57:452–6.

Kotru M, Rusia U, Sikka M, Chaturvedi S, Jain AK. Evaluation of serum ferritin in screening for iron deficiency in tuberculosis. *Ann Hematol* 2004; 83:95–100.

Malczewska J, Szczepanska B, Stupnicki R, Sendecki W. The assessment of frequency of iron deficiency in athletes from the transferrin receptor-ferritin index. *Int J Sport Nutr Exerc Metab* 2001; 11:42–52.

Patterson AJ, Brown WJ, Roberts DC, Seldon MR Dietary treatment of iron deficiency in women of child-bearing age. *Am J Clin Nutr* 2001; 74:650–6.

Sun ER, Chen CA, Ho G, Earley CJ, Allen RP. Iron and the restless legs syndrome. *Sleep* 1998; 21:371–7.

Weiss G, Goodnough LT. Anemia of chronic disease. *N Engl J Med* 2005; 352:1011–23.

Wians FH Jr, Urban JE, Kroft SH, Keffer JH. Soluble transferrin receptor (sTfR) concentration quantified using two sTfR kits: analytical and clinical performance characteristics. *Clin Chim Acta* 2001a; 303:75–81.

Wians FH Jr, Urban JE, Keffer JH, Kroft SH. Discriminating between iron deficiency anemia and anemia of chronic disease using traditional indices of iron status versus transferrin receptor concentration. *Am J Clin Path* 2001b; 115:112–8.

11.2.14 Red Cell Indices

Indications for Test

Measurement of red cell indices—red blood cell (RBC) count, hemoglobin (Hb), hematocrit (Hct), mean corpuscular volume (MCV), mean corpuscular hemoglobin (MCH), mean corpuscular hemoglobin concentration (MCHC), RBC distribution width (RDW) or RBC mean index, reticulocyte count—and evaluation of RBC morphology are indicated in the assessment of nutritional status in:

- Cases of anemia
- Suspected nutritional vitamin or iron deficiency
- Evaluation of anemia in the absence of iron or vitamin deficiency (e.g., suspected thalassemia)

Procedure

1. Obtain an EDTA-anticoagulated whole blood specimen for routine complete blood count (CBC) with differential and review of a stained peripheral blood smear, including:
   - RBC count
   - Hb and Hct
   - Calculated indices: MCV, MCH, MCHC, RDW
   - Reticulocyte count
   - Evaluation of RBC morphology
2. Recognize limitations in the accuracy of percent reticulocytes values in acute anemias given the longer half-life of prematurely released reticulocytes.

Interpretation

1. Reference intervals:
   - Hb, 13 to 18 g/dL (8.1–10.2 mmol/L) (males); 12 to 16 g/dL (7.4–9.9 mmol/L) (females)
   - Hct, 42 to 52% (males); 37 to 48% (females)
   - RBC count, 4.42 to 5.56×10^{12}/L (males); 4.04–4.99×10^{12}/L (females)
   - MCV, 86 to 98 fL
   - MCH, 28 to 33 pg
   - MCHC, 263 to 375 g/L (males); 278–372 g/L (females)
   - RDW, 0.13 to 0.15
   - Reticulocyte count, <2%

2. Hb and Hct values are usually lower for females than males related to the effects of menstruation and androgen levels. Typically, a Hb < 10.0 g/dL and MCV ≤ 80 fL is seen in overt iron deficiency anemia.

3. The lowest acceptable Hb value in iron-treated pregnant women is proposed to be 11 g/dL (6.8 mmol/L) in the first trimester and 10.5 g/dL (6.5 mmol/L) in the second and third trimesters (Milman et al., 2000).

4. High RBC counts ($\geq 5.0 \times 10^{12}$/L) occur commonly in childhood anemia and do not help to differentiate iron deficiency from thalassemias in patients up to 48 months of age.

5. Microcytosis (MCV < 86 fL) is suggestive but not diagnostic of iron deficiency. Other possible explanations for a decreased MCV include thalassemic disorders; anemia of chronic disease, especially in elderly women; poisoning with lead, alcohol, or zinc; sideroblastic anemia; or copper deficiency.

6. Macrocytosis (MCV > 98 fL) may occur in deficiency of folic acid or vitamin B_{12}, severe hypothyroidism independent of pernicious anemia, alcoholic liver disease possibly via acetaldehyde adducts, or reticulocytosis following acute blood loss or in response to chemotherapy.

7. Distinctively greater MCV values occur with iron overload at diagnosis of hemachromatosis.

8. An elevated RDW may be found in polycythemia vera associated with elevated $[B_{12}]$ and microcytic erythrocytosis.

9. When reticulocytosis (reticulocyte count > 2%) occurs, only parameters for the mature RBC reflect nutritional iron status.

10. Normocytic anemias may occur in acute blood loss, hyper- or hypothyroidism, chronic kidney disease, spherocytosis, collagen vascular disease, and many other conditions, including chronic disease, bone marrow suppression, and autoimmune hemolytic anemia.

11. Compared to wild-type controls, increased mean Hb levels, Hct, MCV, MCH, and MCHC, but not RBC counts, occur in C282Y/C282Y hemochromatosis probands primarily via increased iron uptake and Hb synthesis by immature erythroid cells (Barton et al., 2000).

12. Hemoglobin, RBC counts, MCV, and MCH, but not MCHC, are significantly lower in well-nourished patients with sickle cell trait (Hb AS), but their reticulocyte counts and Hct values were found to be lower when compared to Hb AA controls (Pathak et al., 2003).

Notes

1. Mature erythrocyte indices are valid markers of iron status that remain independent of erythropoietic activity and are similar to whole RBC parameters only when the percentage of reticulocytes is low.

2. Microcytosis with anemia may be found in iron deficiency as well as in iron-replete conditions such as chronic disease, thalassemia, congenital sideroblastosis, and homozygous HbE disease.

3. Hematologic reference intervals for pregnant women should be derived from iron-replete subjects.

4. The indices of iron status may become disrupted in elderly patients with both acute and chronic inflammatory disorders. These indices cannot be used to reliably diagnose iron deficiency anemia in individuals with inflammatory conditions.

5. RBC and white blood cell (WBC) counts are lower in females with anorexia nervosa, with over 20% of cases being both anemic and leukopenic (Misra et al., 2004).

6. Possible age- and race-related trends for red cell indices, notably in RDW as well as in platelet count, and granulocyte and lymphocyte percentages have not been confirmed.

7. Values for TIBC and RDW are lower but values for ferritin and MCV are higher in patients with anemia of chronic disease than in those with iron deficiency anemia (Wians et al., 2001).

ICD-9 Codes

Conditions that may justify this test include *but are not limited to*:

280   iron deficiency anemias
282.49   thalassemia

Suggested Reading

Barton JC, Bertoli LF, Rothenberg BE. Peripheral blood erythrocyte parameters in hemochromatosis: evidence for increased erythrocyte hemoglobin content. *J Lab Clin Med* 2000; 135:96–104.

Bovy C, Gothot A, Krzesinski JM, Beguin Y. Mature erythrocyte indices: new markers of iron availability. *Haematologica* 2005; 90:549–51.

Cunietti E, Chiari MM, Monti M, Engaddi I, Berlusconi A, Neri MC, De Luca P. Distortion of iron status indices by acute inflammation in older hospitalized patients. *Arch Gerontol Geriatr* 2004; 39:35–42.

Milman N, Byg KE, Agger AO. Hemoglobin and erythrocyte indices during normal pregnancy and postpartum in 206 women with and without iron supplementation. *Acta Obstet Gynecol Scand* 2000; 79:89–98.

Misra M, Aggarwal A, Miller KK, Almazan C, Worley M, Soyka LA, Herzog DB, Klibanski A. Effects of anorexia nervosa on clinical, hematologic, biochemical, and bone density parameters in community-dwelling adolescent girls. *Pediatrics* 2004; 114:1574–83.

Pathak K, Kishore S, Anshu, Shivkumar VB, Gangane N, Sharma S. Study of haemoglobin S percentage and haematological parameters in sickle cell trait. *Indian J Pathol Microbiol* 2003; 46:420–4.

Tefferi A. Anemia in adults: a contemporary approach to diagnosis. *Mayo Clin Proc* 2003; 78:1274–78.

Weiss G, Goodnough LT. Anemia of chronic disease. *N Engl J Med* 2005; 352:1011–23.

11.3   Diagnosis of Eating Disorders and Nutrition-Related Diseases

11.3.1   SCOFF Questionnaire for Anorexia Nervosa (AN) and Bulimic or Binge Eating Disorders (BEDs)

Indications for Test

The SCOFF questionnaire is indicated when:

- There is evidence or history of a BED.
- A screening test for AN is required.
- BMI is <18 or patient is obese (BMI > 30).

Procedure

1. Administer the SCOFF questionnaire with its five questions (one each in five categories):
   Sick—Do you make yourself sick (vomit) because you feel uncomfortably full?
   Control—Do you worry that you have lost control over how much you eat?
   Oral—Have you recently lost more than 14 lb (6.3 kg) in a 3-month period as a result of decreased oral intake unrelated to specific illness?
   Fat—Do you believe yourself to be fat when others say you are too thin?
   Food—Would you say that food dominates your life?

Interpretation

1. For each "yes" response to the five questions, award 1 point.
2. A score of 2 indicates a positive result (i.e., a likely diagnosis of AN or other eating disorder).
3. Obese type 2 diabetes mellitus patients can exhibit binge eating behavior but not induce vomiting. This phenomenon has been called "diabulimia" and is characterized by the rapid eating of large quantities of food without any form of purging, including laxatives or fasting.

Notes

1. Depending on the severity of weight loss, the decision to hospitalize an individual for forced nutritional repletion or referral for psychiatric evaluation may be made if the SCOFF questionnaire is positive.
2. Binge eating disorder is characterized by low self-esteem; anxiety and depression; loss of sexual desire; feelings of guilt, shame, and disgust with self; suicidal thoughts; belief that weight loss will improve the quality of the patient's life; and feeling out of control when it come to eating.
3. Behaviors associated with BED include hiding food, going on many different diets, secretive eating patterns, and avoidance of social situations where food will be present.
4. Other features of BED include large fluctuations in weight, hypertension, hyperlipidemia, shortness of breath, lack of exercise tolerance, menstrual irregularities, and decreased mobility.
5. On examination, surreptitious vomiting may be suspected upon finding dental erosion from repeated exposure to acid gastric secretions, with ulcers and calluses seen on the dorsum of the hand caused by sticking a finger in the back of the throat to induce vomiting.

ICD-9 Codes

Conditions that may justify this test include *but are not limited to*:

307.50   eating disorder, unspecified
783.6   bulimia

Suggested Reading

Vale S. Better clinical management of anorexia nervosa in teens. *West J Med* 2000; 172:189–93, 365.

11.3.2 Dietary CAGE Questionnaire for Assessment of Saturated Fat and Cholesterol Intake

Indications for Test

The CAGE questionnaire may be indicated when:

- Screening individual patients whose dietary intake of fat and cholesterol exceeds recommendations for a healthy lifestyle per Adult Treatment Panel (ATP) III Guidelines for Therapeutic Lifestyle Changes (TLC)
- Deciding which patients might benefit most from intensive medical nutrition therapy

Procedure

1. Administer the CAGE questionnaire with its four questions (one each in four categories):
   Cheese—Do you prefer dairy foods with higher fat: milk with ≥2% fat, cream, whole-fat yogurt, ice cream?
   Animal fat—Do you prefer meats with higher fat content: fatty lunch meats, ground meat, sausage, frankfurters, bologna, salami, fried food, hamburgers?
   Got it away—Do you often eat meals purchased away from home: fast food, high-fat meals from restaurants?
   Extra eats—Do you often eat high-fat commercial products: cookies, candies (chocolate), pies, cakes, doughnuts, pastries?
2. Use a food log or obtain a 24- to 48-hour dietary intake recall history to assess the frequency of the above behaviors.
3. Provide the food log and dietary recall data to a nutritionist or dietary professional when making referrals for counseling.

Interpretation

1. Each positive response identifies an increased risk for high levels of atherogenic lipids (e.g., LDL), which may be modified by diet.
2. Positive responses to three or more questions should prompt referral of the patient to a nutrition/dietary professional.

Note

1. The use of this questionnaire has not been evaluated in terms of its impact on the results of the lipid tests (total cholesterol, triglycerides, HDL, and calculated LDL) in the standard lipid profile.

ICD-9 Codes

Conditions that may justify this test include *but are not limited to*:

V69.1   inappropriate diet and eating habits

Suggested Reading

National Institutes of Health Expert Panel. *Third Report of the National Cholesterol Education Program (NCEP) Expert Panel on Detection, Evaluation, and Treatment of High Blood Cholesterol in Adults (Adult Treatment Panel III)*, NIH Publ No 02–5215. 2002. National Institutes of Health, pp. 5, 7.

11.3.3 Three-Factor Eating Questionnaire (TFEQ) for Identification of Patients with Eating Behaviors Characterized by Dietary Restraint, Dietary Disinhibition, or Susceptibility to Hunger

Indications for Test

Administration of the TFEQ is indicated in (Stunkard and Messick, 1985):

- The psychological assessment of patients with maladaptive eating disorders
- Patients with obesity or excessive thinness resulting in either insulin resistance or risk for hypoglycemia
- The identification of chronic dieters
- The identification of patients most likely to respond to long-term use of appetite-suppressant medication (e.g., incretin mimetics) or behavioral modification therapy

Procedure

1. Administer the 51-item TFEQ (Q11.1 in Appendix 1) to the patient.
2. Calculate the score for each of the three factors (dietary restraint, disinhibition, and susceptibility to hunger) related to the evaluation of the patient's eating behavior by dividing the number of positive responses in each of the three factor categories by the maximum number of responses for each category shown below:
   - Factor 1, cognitive restraint of eating (control), 21 responses
   - Factor 2, disinhibited eating (lability), 16 responses
   - Factor 3, susceptibility to hunger before and after eating, 14 responses
   Example—16 positive responses to questions related to factor 1 yields a score of 76.2% [(16/21) × 100].

Interpretation

1. Scores less than 50% in each factor category indicate no particular eating tendency in any of the three factor categories.
2. A high score (>75%; positive responses on a minimum of 16 questions) in the factor I category indicates responsiveness to general information about caloric balance, nutrition, and traditional behavioral strategies for food stimulus control and prompt appropriate counseling and information transfer.
3. A high score (>75%; positive responses on a minimum of 12 questions) in the factor 2 category indicates responsiveness to interpersonal support or group approaches and prompt appropriate counseling in dealing with anxiety, depression, or loneliness.
4. A high score (>75%; positive responses on a minimum of 11 questions) in the factor 3 category indicates responsiveness to attributional techniques for coping with hunger or long-term use of appetite-suppressant medication such as Byetta® or Symlin®.
5. A low score (<25%; positive responses to fewer than 4 questions) in the factor 3 category should prompt appropriate counseling of the patient and avoidance of the use of weight-loss medications.
6. Even though vegetarians tend to have lower factor 1 category scores, they do not appear to be at increased risk for overeating disorders.
7. Women with high factor 2 and 3 category scores indicated a desire for and liking of simple carbohydrates (sweets, pastries served with coffee, fruit-based desserts) and high-fat condiments (butter and margarine) (Lahteenmaki and Tuorila, 1995).
8. Women with high factor 1 and low factor 2 scores tend to be unresponsive to manipulated palatability of food, whereas those with low factor 1 and high factor 2 scores are over-responsive and, thus, tend to be at greater risk of developing obesity (Yeomans et al., 2004).
9. Higher factor 1 scores tend to be positively associated in adults with preferences for healthy food groups such as green vegetables and negatively associated with a preference for french fries and sugar. Energy-dense foods containing higher percentages of fat were positively associated with uncontrolled eating (de Lauzon et al., 2004).

Notes

1. Other ways to assess nutritional status include the Rainey McDonald Nutritional Index (RMNI) in which a negative RMNI value is an indicator of malnutrition.
2. Most of the TFEQ questions related to factors 2 and 3 may be grouped into a global factor labeled "uncontrolled eating."
3. The original TFEQ had 18 items divided into three sections: 1, cognitive restraint; 2, uncontrolled eating; 3, emotional eating (Karlsson et al., 2000).
4. The revised TFEQ appears to explain disordered eating behavior better than the original TFEQ. The newer version involves three factors: 1, dietary restraint—strategic dieting behavior, attitude to self-regulation, and avoidance of fattening foods; 2, dietary disinhibition—habitual susceptibility, emotional susceptibility, situational susceptibility; 3, susceptibility to hunger—internal locus for hunger, external locus for hunger (Bond et al., 2001).
5. Factor 1 scores and serum leptin concentrations, higher levels of which are associated with diminished energy expenditure, showed a significant negative correlation ($r = -0.43$) in overweight preadolescent girls (Laessle et al., 2000). Girls who scored higher on factor 1 were found to have a lower energy intake than girls with uncontrolled eating and emotional eating behaviors (9164 kJ vs. 13,163 kJ, $p < 0.001$) (de Lauzon et al., 2004).

ICD-9 Codes

Conditions that may justify this test include *but are not limited to*:

277.7	dysmetabolic syndrome
278.00	obesity
783.22	underweight
V69.1	inappropriate diet and eating habits

Suggested Reading

Adami GF, Campostano A, Gandolfo P, Ravera G, Petti AR, Scopinaro N. Three-Factor Eating Questionnaire and Eating Disorder Inventory in the evaluation of psychological traits and emotional reactivity in obese patients. *J Am Diet Assoc* 1996; 96:67–8.

Bond MJ, McDowell AJ, Wilkinson JY. The measurement of dietary restraint, disinhibition and hunger: an examination of the factor structure of the Three-Factor Eating Questionnaire (TFEQ). *Int J Obes Relat Metab Disord* 2001; 25:900–6.

de Lauzon B, Romon M, Deschamps V et al. The Three-Factor Eating Questionnaire-R18 is able to distinguish among different eating patterns in a general population. *J Nutr* 2004; 134:2372–80.

Janelle KC, Barr SI. Nutrient intakes and eating behavior scores of vegetarian and nonvegetarian women. *J Am Diet Assoc* 1995; 95:180–8.

Karlsson J, Persson LO, Sjostrom L, Sullivan M Psychometric properties and factor structure of the Three-Factor Eating Questionnaire (TFEQ) in obese men and women: results from the Swedish Obese Subjects (SOS) study. *Int J Obes Relat Metab Disord* 2000; 24:1715–25.

Laessle RG, Wurmser H, Pirke KM. Restrained eating and leptin levels in overweight preadolescent girls. *Physiol Behav* 2000; 70:45–7.

Lahteenmaki L, Tuorila H. Three-factor eating questionnaire and the use and liking of sweet and fat among dieters. *Physiol Behav* 1995; 57:81–8.

Stunkard AJ, Messick S. The three-factor eating questionnaire to measure dietary restraint, disinhibition and hunger. *J Psychosom Res* 1985; 29:71–83.

Yeomans MR, Tovey HM, Tinley EM, Haynes CJ. Effects of manipulated palatability on appetite depend on restraint and disinhibition scores from the Three-Factor Eating Questionnaire. *Int J Obes Relat Metab Disord* 2004; 28:144–51.

11.3.4 Antigliadin Endomysial (EMA IgG and IgA) and Tissue Antitransglutaminase (TTG) IgA Antibody Testing in Diagnosis of Celiac Disease

Indications for Test

Antigliadin endomysial immunoglobulin G (IgG) and IgA and TTG IgA screening tests for gluten sensitivity are indicated in:

- Patients with vitamin deficiencies (e.g., vitamins D and B_{12}), particularly associated with gastrointestinal malabsorption
- Infants < 24 months of age with abdominal distension, pain, constipation with recurrent diarrhea, irritability, vomiting, anorexia, and/or weight loss
- Children with low bone density, short stature, delayed puberty, iron deficiency, puritic dermatitis herpetiformis, dental enamel hypoplasia, or epilepsy with occipital calcifications
- Patients with diarrhea or chronically loose stools and malabsorption, particularly if related to intestinal villous atrophy and crypt hyperplasia
- Hypothyroid patients who require very high doses of L-thyroxine therapy
- Most patients with osteoporosis (spine or hip T-scores < –2.5)
- Asymptomatic relatives of celiac disease patients
- Patients with type 1 diabetes mellitus (T1DM) or Down, Turner, or Williams syndromes

Procedure

1. Obtain measure of TTG IgA as a screening test for celiac disease, recognizing that celiac patients with IgA deficiency may be falsely negative for the typical TTG IgA elevation of their disease.
2. Obtain a blood specimen for quantitative IgA testing. If the results of this test indicate IgA deficiency, request endomysial antibody (EMA) IgG testing.

TABLE 11.6
Sensitivity and Specificity of Serologic
Screening Tests for Celiac Disease

Test	Sensitivity (%)	Specificity (%)
Tissue transglutaminase IgA (TTG-IgA)	90–98	94–97
Endomysial IgA antibodies (EMA-IgA)	85–98	97–100
Antigliadin IgA (AGA-IgA)	75–90	82–95
Antigliadin IgG (AGA-IgG)	69–85	73–90

Source: Adapted from Farrell, R.J. and Kelly, C.P., *Am. J. Gastroenterol.*, 96, 3237–3246, 2001.

3. Obtain a blood specimen for EMA and/or TTG IgA testing if the quantitative IgA test is normal.
4. If the results of EMA IgA or TTG IgA screening tests, when coupled with the patient's clinical and physical exam findings, are suggestive of celiac disease, obtain a jejunal biopsy for a definitive diagnosis.
5. Monitor bone density and TTG levels in patients with documented and treated celiac disease.

Interpretation

1. TTG results by enzyme-linked immunosorbent assay (ELISA) are sensitive and specific for identifying individuals with celiac disease; the test is relatively inexpensive and is not operator dependent, but it may be falsely negative in IgA deficiency and slightly less specific than EMA.
2. The EMA IgA antibody test is highly sensitive and specific for identifying individuals with celiac disease but is expensive, time consuming, and operator dependent, and it may be falsely negative in IgA deficiency.
3. IgA-deficient individuals are at higher risk for celiac disease. In IgA-deficient patients who are EMA IgG positive, proceed directly to jejunal biopsy.
4. In treated celiac patients, expect the bone density to rise and the TTG level to fall.

Notes

1. Transglutaminase 2 acts both as a deamidating enzyme that can enhance the immunostimulatory effect of gluten and as a target autoantigen in the immune response in celiac disease.
2. Celiac disease is closely associated with genes that code for human antigens DQ2 and/or DQ8 HLA haplotypes. The TTG and EMA tests are sensitive serum antibody tests for celiac disease with negative HLA-DQ types and are helpful in excluding its diagnosis (Hadithi et al., 2007).
3. The unequivocal diagnosis of celiac disease requires histopathologic evidence consistent with celiac disease and complete resolution of the patient's symptoms on a gluten-free diet.
4. Both IgA and IgG EMAs to the gluten protein gliadin have poor sensitivity and specificity for the identification of individuals with celiac disease (Table 11.6). EMA IgG testing should only be performed in individuals who are IgA deficient and positive for EMA IgA antibodies.
5. Celiac disease is associated with increased rates of anemia, osteoporosis, cancer, neurologic deficits, and additional autoimmune disorders, including T1DM and hypothyroidism.
6. Hypothyroid patients with celiac disease may require unusually high doses of thyroid hormone replacement, probably because of a small bowel absorption defect (Khandwala et al., 2005).
7. A strict, lifelong gluten-free diet (no wheat-, barley-, or rye-containing foods) is the only safe and effective treatment of celiac disease. An effective diet typically results in a lower TTG IgA titer.
8. IgG EMA was found in 12% of patients with multiple sclerosis and in 13% of blood donors. Antigliadin antibody (especially IgG isotype) can be a nonspecific finding, although blood donor screening suggests that a large number of individuals with celiac disease go undiagnosed.
9. Up to 8% of adults with unexplained iron deficiency anemia have celiac disease.

ICD-9 Codes

Conditions that may justify this test include *but are not limited to*:

Somatic disorders
783.0   anorexia,
783.21   loss of weight
783.43   short stature

Intestinal disorders
564.0   constipation
579.9   malabsorption
787.03   vomiting alone
787.7   abnormal feces
787.91   diarrhea
789.0   abdominal pain/infant colic

Other
250.X   diabetes mellitus (type1)
259.0   delay in sexual development and puberty
266.1   vitamin B_6 deficiency
275.42   hypercalcemia
280.9   iron deficiency anemia
520.4   disturbances of tooth formation
694.0   dermatitis herpetiformis
758.0   Down syndrome
758.6   gonadal dysgenesis

Suggested Reading

Alaedini A, Green PH. Narrative review: celiac disease—understanding a complex autoimmune disorder. *Ann Intern Med* 2005; 142:289–98.

Farrell RJ, Kelly CP. Diagnosis of celiac sprue. *Am J Gastroenterol* 2001; 96:3237–46.

Hadithi M, von Blomberg BME, Crusius JBA, Bloemena E, Kostense PJ, Meijer JWR, Mulder CJJ, Stehouwer CDA, Peña AS. Accuracy of serologic tests and HLA-DQ typing for diagnosing celiac disease. *Ann Intern Med* 2007; 147:294–302.

Horvath K, Hill ID. Anti-tissue transglutaminase antibody as the first line screening for celiac disease: goodbye antigliadin tests? *Am J Gastroenterol* 2002; 97:2702–4.

Khandwala KM, Chibbar R, Worobetz LJ. A case of celiac disease presenting with malabsorption of L-thyroxine. *Endocrinologist* 2005; 15:14–17.

Levine A, Bujanover Y, Reif S, Gass S, Vardinon N, Reifen R, Lehmann D. Comparison of assays for anti-endomysial and anti-transglutaminase antibodies for diagnosis of pediatric celiac disease. *Isr Med Assoc J* 2000; 2:122–5.

Pengiran Tengah CD, Lock RJ, Unsworth DJ, Wills AJ. Multiple sclerosis and occult gluten sensitivity. *Neurology* 2004; 62:2326–7.

Emerging or Historical Endocrine and Metabolic Tests with Potential Clinical Utility

12

Introduction

The clinical utility of the tests included in this chapter has not been well established; however, these tests have been and continue to be used in clinical practice and research by clinicians and investigators. Moreover, many of these tests have strong advocacy because they may offer greater discriminatory power than other available testing. Typically, the tests in Chapter 12 are more labor intensive or expensive or have less well-established reference ranges than alternative tests as indicated. Note that the clinical value of using one diagnostic test over another depends on whether that test improves patient outcomes beyond the outcomes achieved using the other test. If studies show that one test is safer or more accurate and specific than, but of similar sensitivity to, another test it may replace that test; however, if the safer test is more sensitive than the other test, it may lead to the detection of extra cases of disease. Results from treatment trials that enrolled only patients detected by a more sensitive test may not apply to these extra cases. Clinicians need to wait for results from randomized trials assessing treatment efficacy in cases detected by that test, unless they can be satisfied that the test detects the same spectrum and subtype of disease as the other test or that treatment response is similar across the spectrum of disease (Lord et al., 2006).

Suggested Reading

Lord SJ, Irwig L, Simes RJ. When is measuring sensitivity and specificity sufficient to evaluate a diagnostic test, and when do we need randomized trials? *Ann Intern Med* 2006; 144:850–5.

12.1 Tests for Thyroid Disorders

12.1.1 Thyrotropin-Releasing Hormone (TRH) Stimulation Test in Hypothyroidism

Possible Indications for Test

The TRH stimulation test may help to:

- Distinguish among primary, secondary or central, and tertiary or hypothalamic hypothyroidism
- Assess the effectiveness of thyroxin suppression therapy
- Identify patients in recovery from hypothyroidism
- Distinguish resistance to thyroid hormone from the elevated thyrotropin (TSH) levels associated with a TSH-secreting pituitary tumor

Contraindications to Testing

- Critical illness or therapy with dopamine or glucocorticoids at high doses that suppress TSH production

Alternatives to Testing

- TSH surge test (see Test 2.7.2) in children with short stature and possible central hypothyroidism
- Baseline sensitive TSH and free thyroxine (T_4) assay (see Tests 1.1.1 and 1.1.3) alone in diagnosis of hypothyroidism

Procedure

1. Obtain two baseline fasting blood samples, collected 20 minutes apart, for measurement of [TSH] and [free T_4].
2. Pool equal aliquots of serum from each sample.
3. Administer TRH (Protirelin®; 500 mcg as an i.v. bolus), after baseline sampling.
4. Obtain blood samples for TSH testing at 15, 30, 45, and 60 minutes after the administration of TRH.
5. Assay of the 30-minute and 60-minute samples for [TSH] is usually adequate for diagnosis.

Interpretation

1. Refer to "Distribution of Peak TSH Response to TRH (250 mcg) as a Function of Baseline TSH Concentration in 1285 Subjects" (Table 12.1) and "TRH-Stimulated TSH (sTSH) Test Results in Relation to Basal TSH and Free Thyroxine (T_4) Levels" (Table 12.2).
2. A TRH-stimulated [TSH] > 30 mcU/mL in all samples after baseline confirms primary hypothyroidism.
3. If the TRH-stimulated [TSH] exceeds 30 mcU/mL in one or more, but not all, timed samples after administration of TRH, a failing thyroid (i.e., a gland that requires excess TSH stimulation for adequate function) is suspect, even when the baseline [TSH] is within reference intervals.
4. Depressed or critically ill patients may show a significantly blunted TSH response (i.e., all serum samples have [TSH] < 30 mcU/mL following the administration of TRH), as well as a less blunted response when the patient is no longer depressed or critically ill. Such patients may or may not have hypothyroidism.

Notes

1. It is not possible to differentiate between pituitary and hypothalamic disease based solely on the patient's TSH response in the TRH test.
2. The TRH-stimulated TSH response, in terms of the fold-increase over the baseline [TSH], is more diagnostically useful than the absolute rise in [TSH] (Table 12.2).

TABLE 12.1
Distribution of Peak TSH Response to TRH (250 mcg) as a Function of Baseline TSH Concentration in 1285 Subjects

Baseline TSH Concentration (IU/L)	No. of Subjects	% of Subjects with a Peak TSH Concentration (IU/L) of			
		<20	20–24.9	25–29.9	>30
1.0–1.4	415	85.0	9.4	4.1	1.7
1.5–1.9	329	64.4	18.8	8.8	7.9
2.0–2.4	224	48.2	24.6	12.2	16.1
2.5–2.9	131	28.2	30.5	17.6	23.7
3.0–3.4	86	14.0	23.3	22.1	40.7
3.5–3.9	47	14.9	12.8	12.8	59.6
4.0–4.9	53	1.9	7.5	18.9	71.7

Note: TSH, thyroid-stimulating hormone (thyrotropin); TRH, thyrotropin-releasing hormone.

Source: Unpublished data courtesy of Texas Institute for Reproductive Medicine & Endocrinology, Houston, TX.

TABLE 12.2
TRH-Stimulated TSH (sTSH) Test Results in Relation to Basal TSH and Free Thyroxine (T_4) Levels

Basal sTSH (IU/L)	Free T_4 Concentration	Rise in sTSH (Fold Increase Over Baseline)	Peak sTSH (IU/L)	Thyroid Diagnosis
N	N	>3.0[b]	Typically <30	Euthyroid
H (5.0–15.0)	N or L	>3.0	>30	Hypothyroidism (early-primary)
VH (>15.0)	L	>3.0	>30	Hypothyroidism (established-primary)
L–LN	N or L[a]	<3.0	Typically <30	Hypothyroidism (pituitary-secondary)
L–LN	N or L	<3.0	<30[c]	Hypothyroidism (hypothalamic-tertiary)
<0.3	H	Negligible	n.s.	Hyperthyroidism
N or H	H	<3.0	Typically >30	TSH-secreting adenoma
N or H	H	>3.0	VH (>50)	Thyroid hormone resistance
Variable	Variable	>3.0	Typically <30	Nonthyroidal illness

[a] Euthyroid patients with acromegaly, depression, renal failure, or other chronic illness; those who have been exposed to excess glucocorticoids; or patients of advanced age may have this pattern.

[b] Mean rise, eightfold.

[c] Delayed to 45 to 60 minutes.

Note: H, high; L, low; LN, low–normal; N, within normal reference limits; VH, very high; n.s., not significantly different from baseline.

3. In patients with an intact hypothalamic–pituitary axis who are likely to have either hypo- or hyperthyroidism, there is no clear diagnostic advantage to TRH testing over measurement of a basal [TSH].

4. In 54 children with central hypothyroidism, a "normal" increase in [TSH] over baseline of 4.5 to 17.8 mcIU/mL after administration of TRH occurred in 23.3% of patients. A greater than normal (>17.8 mcIU/mL), absent/blunted (<4.5 mcIU/mL), or delayed TSH response to TRH was documented in 16.7%, 30%, and 30% of these patients, respectively (Mehta et al., 2003).

5. Protirelin® (Pyr–His–Pro–NH$_2$) order information: H-4915-GMP, 4003036 CAS No. 24305-27-9; Bachem Bioscience, Inc., 3700 Horizon Drive, King of Prussia, PA 19406; 1-800-634-3183 or 1-610-239-0300 (phone), 1-610-239-0800 (fax).

Suggested Reading

Brietzke S, Wians FH Jr, Jacobson JM, Ortiz G. Clinical significance of a slightly increased thyrotropin concentration. *J Clin Immunoassay* 1989; 12:224–8.

Hartoft-Nielsen ML, Lange M, Rasmussen AK, Scherer S, Zimmermann-Belsing T, Feldt-Rasmussen U. Thyrotropin-releasing hormone stimulation test in patients with pituitary pathology. *Horm Res* 2004; 61:53–7.

Mehta A, Hindmarsh PC, Stanhope RG, Brain CE, Preece MA, Dattani MT. Is the thyrotropin-releasing hormone test necessary in the diagnosis of central hypothyroidism in children? *J Clin Endocrinol Metab* 2003; 88:5696–703.

Schule C, Baghai TC, Tsikolata V, Zwanzger P, Eser D, Schaaf L, Rupprecht R. The combined T$_3$/TRH test in depressed patients and healthy controls. *Psychoneuroendocrinology* 2005; 30:341–56.

Snyder PJ, Utiger RD. Response to thyrotropin releasing hormone (TRH) in normal man. *J Clin Endocrinol Metab* 1972; 34:380–5.

Snyder PJ, Jacobs LS, Rabello MM, Sterling FH, Shore RN, Utiger RD, Daughaday WH. Diagnostic value of thyrotrophin-releasing hormone in pituitary and hypothalamic diseases: assessment of thyrotrophin and prolactin secretion in 100 patients. *Ann Intern Med* 1974; 81:751–7.

Spencer CA, Schwarzbein D, Guttler RB, Lo Presti JS, Nicoloff JT. Thyrotropin (TSH)-releasing hormone stimulation test responses employing third and fourth generation TSH assays. *J Clin Endocrinol Metab* 1993; 76:494–8.

Takasu N, Komiya I, Asawa T, Nagasawa Y, Yamada T. Test for recovery from hypothyroidism during thyroxine therapy in Hashimoto's thyroiditis. *Lancet* 1990; 336:1084–6.

Walsh JP, Ward LC, Burke V, Bhagat CI, Shiels L, Henley D, Gillett MJ, Gilbert R, Tanner M, Stuckey BGA. Small changes in thyroxine dosage do not produce measurable changes in hypothyroid symptoms, well-being, or quality of life: results of a double-blind, randomized clinical trial. *J Clin Endocrinol Metab* 2006; 91:2624–30.

12.1.2 Gentle Hair-Pull Test for Alopecia of Autoimmune, Endocrine, and Non-Endocrine Causes, Including Telogen Effluvium Following Stressful Events

Possible Indications for Test

The gentle hair-pull test may be indicated when:

- Hair loss is reported by the patient or observed by the examiner.
- The patient is documented to have experienced a recent, very stressful event, such as onset of Graves' disease or childbirth.

Contraindications to Testing

- Unacceptable level of pain and discomfort to the patient upon attempt to pull out hairs

Alternatives to Testing

- View photographs of the patient that objectively document alopecia.
- Collect the hair lost after a hair wash or removed upon combing the scalp for a standard ten strokes before treatment with a drug known to cause hair loss or a stressful event (e.g., surgery), and repeat collection at periodic intervals of weeks or months during therapy or after the event

Procedure

1. Obtain the patient's history of usual hair-care practices, use of any new products or recent changes in hair care, and any recently initiated new medications. Go to the current *Physician Desk Reference* or other objective source to identify which medications taken by the patient might have been associated with alopecia occurring with a frequency of >1% (e.g., pentosan polysulfate for interstitial cystitis).
2. Obtain a detailed history of symptoms of autoimmune disease, including thyroiditis, systemic sclerosis, lupus erythematosus, adrenal failure associated with adrenal antibodies, and rheumatoid arthritis.
3. Obtain a history of any recent stressful events, such as severe hyperthyroidism or hypothyroidism, surgery, accidents, critical illness, radiotherapy, or childbirth.
4. With thumb and forefinger, gently pull 8 to 10 closely grouped hairs from at least two separate areas of the scalp. Abort test if patient complains of moderate to severe pain with this maneuver.
5. Ideally, send any pulled hairs for examination by a dermatologist using light microscopy, particularly if the patient is getting frequent hair treatments (e.g., permanents) or is applying new hair-care products.

Interpretation

1. Normally, only two or fewer hairs can be removed easily per hair pull. Expect a maximum of one hair obtained for every six to seven hair pulls from adults, with fewer hair pulls required to obtain a maximum of one hair in adults with loose anagen syndrome or in children.
2. If four to six hairs come away, active shedding may be diagnosed and the test is positive.
3. If light microscopy is performed, broken hairs may be observed, especially if chemical or heat damage has occurred, and alopecia may be attributed to this damage.
4. Skin sensitivity to hair-care products may be responsible for a condition known as alopecia areata, which is characterized by patchy areas of hair loss.
5. If a hair pull test from apparently healthy scalp is abnormal, there is the possibility that hair loss may be progressing and may eventually encompass the entire scalp.
6. Autoimmune disorders may be associated with more generalized scalp alopecia, prompting an evaluation for the autoimmune diseases noted in Procedure point 2, above.
7. Major hair loss may occur during use of retroviral or cancer chemotherapeutic agents.
8. Telogen effluvium (i.e., major loss of hairs in the resting or telogen phase of the hair growth cycle) occurs within time periods of 3 to 6 months after major stressful events. In such cases, patients can be reassured that their hair loss is temporary and treatment for hair loss is not necessary.

Notes

1. The hair pull test has low diagnostic sensitivity and specificity for making any diagnosis, but it may help to validate the complaints of, or reassure, patients who complain of hair loss not obvious to the medical observer.
2. Evaluation of the degree of telogen effluvium (% = average number of hairs per hair pull × 100) affecting a selected area of the scalp, obtained by a trichogram, does not correlate with the severity of hair loss.
3. Obtaining a quantitative daily count of hairs lost after combing or showering is a cumbersome and nonstandardized method of assessing degree of hair loss but is popular with some patients.
4. Patients with longer hair may complain of hair loss more often than those with shorter hair because loss of longer hairs is more noticeable. Reassurance is more appropriate in these patients than special testing.

Suggested Reading

Guarrera M, Semino MT, Rebora A. Quantitating hair loss in women: a critical approach. *Dermatology* 1997; 194:12–6.

Olsen EA, Bettencourt MS, Cote NL. The presence of loose anagen hairs obtained by hair pull in the normal population. *J Invest Dermatol Symp Proc* 1999; 4:258–60.

12.1.3   Thyrotropin-Releasing Hormone (TRH) Stimulation Test in Thyroid Malignancies

Possible Indications for Test

The TRH stimulation test may be useful in:

- Identifying the degree of TSH suppression and making adjustments to thyroid hormone therapy given to thyroid cancer patients with:
  - High risk for recurrence
  - Metastatic disease
  - Detectable thyroglobulin levels at undetectable [TSH] measured using moderately sensitive assays (i.e., with a lower limit of detection [LLD] of 0.05–0.1 mcU/mL)
  - Interfering thyroglobulin (Tg) antibodies

Contraindications to Testing

- Critical illness or therapy with dopamine or glucocorticoids at high doses that suppress TSH production

Alternatives to Testing

- Use of an adequately analytically sensitive TSH assay (i.e., with a LLD of <0.05 mcU/mL) (see Test 1.1.1)
- Measurement of thyroglobulin concentration, [Tg]

Procedure

1. Obtain a baseline blood sample for TSH measurement by a moderately sensitive assay having a LLD of 0.01 to 0.05 mcU/mL. If [TSH] is detectable, then it is not necessary to perform a TRH stimulation test (see Test 12.1.1), and thyroid hormone dose may be adjusted as appropriate.
2. Measure [Tg] after TSH stimulation using thyrotropin infusion protocols before using this TRH stimulation test. If the [Tg] is above target (see Test 1.3.2) and the patient may be able to tolerate an increase in thyroid hormone suppression therapy dose, proceed with testing.
3. If [TSH] measured by a moderately sensitive assay is undetectable (<LLD), proceed with the TRH stimulation test:
   - Administer TRH (Protirelin®), 500 mcg as an i.v. bolus.
   - Obtain blood samples for [TSH] at 15, 30, 45, and 60 minutes after the administration of TRH. As a cost-effective alternative, collect blood samples for [TSH] at 30 and 60 minutes after administration of TRH, as the [TSH] at these two times points are usually adequate for diagnosis.
   - Quantify [TSH] in all samples using a TSH assay with better analytical sensitivity (i.e., lower value for LLD) than the initial TSH assay.

Interpretation

1. In treated thyroid cancer patients (Christ-Crain et al., 2002), expect close correlation ($p < 0.001$) between:
   - Basal and TRH-stimulated [TSH] when both are measured using moderately sensitive assays (LLD, 0.01–0.05 mcIU/mL; $r = 0.95$)
   - Basal and TRH-stimulated [TSH] when both are measured using the most sensitive assays (LLD, 0.005–0.01 mcU/mL; $r = 0.86$)
   - Basal [TSH] measured using the most sensitive assay and TRH-stimulated [TSH] measured using moderately sensitive assays ($r = 0.73$)
2. In metastatic thyroid cancer patients in whom a high degree of TSH suppression is the therapeutic goal, basal [TSH] determined by the most highly sensitive assays enables appropriate adjustment of thyroid hormone dose without resorting to a TRH stimulation test.

Notes

1. If only a lower sensitivity assay for TSH (LLD, 0.05–1.0) is available, it may be appropriate to monitor the degree of TSH suppression in high-risk thyroid cancer patients with a TRH stimulation test.
2. Adequately sensitive TSH assays document the effectiveness of thyroid hormone therapy in suppressing [TSH] to nonthyrotoxic levels. Thus, the TRH stimulation test is not necessary for this purpose.
3. Undetectable serum [Tg] during TSH suppression therapy with thyroxine in the treatment of a thyroid cancer patient does not exclude persistent thyroid cancer; however, TSH-stimulated serum Tg levels are undetectable in tumor-free differentiated thyroid cancer patients. Therefore, suppression of [TSH] to <0.1 mcU/mL does not appear to be necessary in patients with no evidence of active disease based on clinical evaluation and Tg data.
4. In patients with an intact hypothalamic–pituitary axis, there is no clear advantage to be gained by TRH stimulation testing over accurately measuring a basal TSH using a moderately sensitive assay in the diagnosis of either hypo- or hyperthyroidism.

Suggested Reading

Christ-Crain M, Meier C, Roth CB, Huber P, Staub JJ, Müller B. Basal TSH levels compared with TRH stimulated TSH levels for the diagnosis of different degrees of TSH suppression: diagnostic and therapeutic impact of assay performance. *Eur J Clin Invest* 2002; 32:931–7.

Gorges R, Saller B, Eising EG, Quadbeck B, Mann K, Bockisch A. Surveillance of TSH-suppressive levothyroxine treatment in thyroid cancer patients: TRH testing vs. basal TSH determination by a third-generation assay. *Exp Clin Endocrinol Diabetes* 2002; 110:355–60.

Kamel N, Gullu S, Dagci Ilgin S, Corapcioglu D, Tonyukuk Cesur V, Uysal AR, Baskal N, Erdogan G. Degree of thyrotropin suppression in differentiated thyroid cancer without recurrence or metastases. *Thyroid* 1999; 9:1245–8.

Pagano L, Klain M, Pulcrano M, Angellotti G, Pasano F, Salvatore M, Lombardi G, Biondi B. Follow-up of differentiated thyroid carcinoma. *Minerva Endocrinol* 2004; 29:161–74.

12.2   Tests for Pituitary Tumors

12.2.1   TRH Stimulation Test of Pituitary Gonadotropins in Pituitary Tumor Patients

Possible Indications for Test

The TRH stimulation test of pituitary gonadotropins may be indicated:

- To identify a potential serum tumor marker in patients with an apparently nonfunctional pituitary tumor such that the marker could be used to monitor therapy in these patients

Contraindications to Testing

- Elevated baseline serum tumor marker level (e.g., a pituitary hormone or hormone fragment as identified in Chapter 2) already identified in patients with a pituitary tumor
- Presence of very large pituitary macroadenomas, as pituitary apoplexy may result from injection of TRH

Alternatives to Testing

- Pituitary imaging studies with quantitation of tumor size before and after surgical resection of the tumor or radiotherapy to reduce the size of the tumor

Procedure

1. Obtain baseline blood samples for measurement of serum luteinizing hormone (LH), follicle-stimulating hormone (FSH), luteinizing hormone beta-subunit (LH-beta), and pituitary glycoprotein hormone alpha-subunit (PGH-alpha) levels at −30, −15 and 0 minutes. To eliminate the interference of
   hormone pulses and reduce the cost of testing, pool the serum from the three baseline blood samples.
2. Administer TRH as a 500 mcg i.v. bolus at time 0.
3. Obtain blood samples at 30, 60, and 90 minutes after injection of TRH, and measure the serum levels
   of the four hormones noted above. As an alternative to testing for all four hormones, it is cost effective
   to test an aliquot of serum from the baseline and post-TRH injection blood samples for [LH] and
   [FSH] only, while freezing the remaining serum for subsequent use if additional testing for the other
   hormones is deemed necessary.

Interpretation

1. A significant rise (defined below) over baseline at any time point after administration of TRH in the
   concentration of any of the four hormones is considered abnormal and evidence of a pituitary
   abnormality:
   * Women, >20% rise of [LH-beta] and/or >50% rise in [PGH-alpha], [LH], or [FSH]
   * Men, >25% rise of [FSH] and/or a >40% rise of [LH-beta], a >60% rise of [LH], or a >75%
     rise of [PGH-alpha]
2. This test has a low specificity for detection of glycoprotein hormone-secreting pituitary tumors.
3. Basal gonadotropins above 100 IU/L or PGH-alpha values > 7 ng/mL suggest the presence of a
   pituitary tumor.
4. False positives have been observed in patients with nonadenomatous pituitary masses.
5. The majority of nonsecreting pituitary macroadenomas arise from gonadotroph cells and can be
   recognized, even in postmenopausal women, by the serum LH-beta response to TRH.

Notes

1. Be alert for the onset of severe headaches during the TRH stimulation test as pituitary apoplexy has
   been associated with this test.
2. Pituitary tumors rarely secrete excess intact, bioactive LH or FSH unless stimulated by TRH.
3. Eleven of 16 women with adenomas not known to secrete intact pituitary hormones had significant
   increases in serum [LH-beta] in response to TRH. Of these, 3 had FSH responses, and 4 had LH
   responses. None of 16 age-matched, healthy women and none of 10 women with hormone-secreting
   macroadenomas had LH-beta, FSH, or LH responses to TRH (Daneshdoost et al., 1991).
4. This test appeared to be useful in preoperative identification of gonadotroph adenomas among other
   nonfunctioning pituitary adenomas, with 6 out of 7 tumors immunopositive for FSH and/or LH or
   their free subunits having a 50% rise in alpha-subunit over baseline after TRH (Gruszka et al., 2005).
5. TRH stimulation testing can identify gonadotroph adenomas in men with clinically nonfunctioning
   adenomas better than can basal concentrations of intact FSH and alpha-subunit FSH, alone or in
   combination, but not as well as in women (Daneshdoost et al., 1993).
6. FSH-beta testing is recommended only as a research procedure. In general, alpha-subunit testing is
   superior to beta-subunit testing. Note the apparently contradictory findings reported by different
   authors (Gil-del-Alamo et al., 1994; Somjen et al., 1997).

Suggested Reading

Daneshdoost L, Gennarelli TA, Bashey HM, Savino PJ, Sergott RC, Bosley TM, Snyder PJ. Recognition of
     gonadotroph adenomas in women. *N Engl J Med* 1991; 324:589–94.
Daneshdoost L, Gennarelli TA, Bashey HM, Savino PJ, Sergott RC, Bosley TM, Snyder PJ. Identification of
     gonadotroph adenomas in men with clinically nonfunctioning adenomas by the luteinizing hormone
     beta subunit response to thyrotropin-releasing hormone. *J Clin Endocrinol Metab* 1993; 77:1352–57.

Gil-del-Alamo P, Pettersson KS, Saccomanno K, Spada A, Faglia G, Beck-Peccoz P. Abnormal response of luteinizing hormone beta subunit to thyrotrophin-releasing hormone in patients with non-functioning pituitary adenoma. *Clin Endocrinol* 1994; 41:661–6.

Gruszka A, Kunert-Radek J, Pawlikowski M. Serum alpha-subunit elevation after TRH administration: a valuable test in presurgical diagnosis of gonadotropinoma? *Endokrynol Pol* 2005; 56:14–8.

Somjen D, Tordjman K, Kohen F, Baz M, Razon N, Ouaknine G, Stern N. Combined beta FSH and beta LH response to TRH in patients with clinically non-functioning pituitary adenomas. *Clin Endocrinol* 1997; 46:555–62.

12.2.2 Clomiphene Stimulation Test in Diagnosis of Gonadotropin Deficiency

Indications for Test

The clomiphene stimulation test may be indicated in:

- Adults only (not children) suspected of pituitary disorder affecting gonadotropins and/or gonadal failure
- Women in early menopause transition who wish definitive diagnosis of same or who are desirous of pregnancy in later life
- Preparation for *in vitro* fertilization to predict an individual's response to controlled ovarian hyperstimulation using the gonadotropin-releasing hormone (GnRH) agonist clomiphene citrate (CC) and human menopausal gonadotropins (hMGs) preliminary to retrieval of oocytes by follicular aspiration

Alternatives to Testing

- Baseline serum FSH and LH levels
- Daily collection of first morning voided urine for an entire menstrual cycle or up to 50 days (whichever comes first) with measurement of urinary LH and FSH, as well as estrone conjugates and pregnanediol glucuronide, to assess for evidence of luteal activity and day of luteal transition, if any
- Measurement of anti-Müllerian hormone (AMH) concentration (see Test 9.2.5)

Contraindications to Testing

- Non-adult status
- Liver disease
- Severe depression, as test has tendency to exacerbate it

Procedure

1. Perform testing at any time of the day or month in men or amenorrheic women; however, in menstruating women, start test on day 5 after the onset of menses.
2. Prior to the administration of clomophene citrate (CC), obtain a blood sample(s) for LH and FSH testing. Refer to Procedure point 6 for specimen collection protocol.
3. Administer CC, 100 mg p.o. daily for 5 days in women or for 7 days in men.
4. In women, obtain blood samples for LH and FSH testing on days 4, 7, and 10 and for progesterone testing on day 21.
5. In men, measure [LH] and [FSH] on days 7 and 10 after starting CC.
6. Pooling of serums from two or three blood specimens, obtained 15 minutes apart on each day of sample collection, improves the diagnostic sensitivity of the test.

Interpretation

1. By day 10 of the test, [LH] and [FSH] should rise by ≥30% and ≥20%, respectively, over baseline. In most patients with adequate GnRH secretory reserve, gonadotropin concentrations double by day 10.
2. Blunted LH responses (i.e., peak [LH] < 2 IU/L or <30% rise in [LH] over baseline) occur in prepubertal children and in patients with inadequate pituitary gonadotrope cell reserves or longer term loss of hypothalamic GnRH pulses, as may be seen in anorexia nervosa or hyperprolactinemia.
3. A serum progesterone level > 10 ng/mL in a woman on day 21 is consistent with induction of ovulation and normal luteal function.
4. The sum of [FSH] from the total of the four blood samples obtained before and after CC intake was best correlated with subsequent response to *in vitro* fertilization procedures. The upper limit of the reference interval for this sum, established in 26 patients who became pregnant, was 26 IU/L (Loumaye et al., 1990).
5. If the sum of [FSH] from all blood samples obtained before and after administration of CC was >26 IU/L, ovarian hypofunction was likely and successful response to *in vitro* fertilization procedures was unlikely (Loumaye et al., 1990).

Notes

1. This test evaluates the integrity of the entire hypothalamic–pituitary–gonadal axis.
2. Clomiphene, an antiestrogen, blocks the negative feedback of estradiol on production of GnRH in the hypothalamus and gonadotropins in the pituitary.
3. Clomiphene increases GnRH and gonadotropin secretion in both males and females.
4. The ovary may remain sensitive to elevated FSH in the early menopause transition and produce near-normal estrogen levels but less progesterone. The initiating event in the menopause transition appears to be the loss of inhibin negative feedback on FSH secondary to a diminished follicular reserve (Santoro et al., 2004).
5. Because FSH assay methodologies have changed considerably since 1990, the summation of [FSH] values and cutoffs, as reported by Loumaye et al. (1990), may not have validity using current assays.

Suggested Reading

Loumaye E, Billion JM, Mine JM, Psalti I, Pensis M, Thomas K. Prediction of individual response to controlled ovarian hyperstimulation by means of a clomiphene citrate challenge test. *Fertil Steril* 1990; 53:295–301.

Santen RJ, Leonard JM, Sherins RJ, Gandy HM, Paulsen CA. Short- and long-term effects of clomiphene citrate on the pituitary-testicular axis. *J Clin Endocrinol Metab* 1971; 33:970–9.

Santoro N, Lasley B, McConnell D, Allsworth J, Crawford S, Gold EB, Finkelstein JS, Greendale GA, Kelsey J, Korenman S, Luborsky JL, Matthews K, Midgley R, Powell L, Sabatine J, Schocken M, Sowers MF, Weiss G. Body size and ethnicity are associated with menstrual cycle alterations in women in the early menopausal transition: the Study of Women's Health across the Nation (SWAN) Daily Hormone Study. *J Clin Endocrinol Metab* 2004; 89:2622–31.

12.3   Tests for Hypoglycemia and Hyperglycemia

12.3.1   Oral Glucose Tolerance Test (oGTT) and the Hypoglycemic Index (HGI) for Diagnosis of Reactive Hypoglycemia (RH)

Possible Indications for Test

Determination of the HGI from an oGTT may be indicated in:

- Individuals with presumptive evidence for RH (i.e., symptoms of postprandial somnolence, shaky spells, or the dumping syndrome after carbohydrate loads)
- Diagnosing RH, regardless of cause

Alternatives to Testing

- Mixed meal tolerance test

Procedure

1. Insert a catheter into a large vein for frequent blood sampling.
2. After administration of a 75-g oral glucose load, wait 20 minutes before obtaining blood samples every 3 to 5 minutes for determination of [glucose] using a bedside capillary blood glucose monitoring device.
3. A preferable alternative to frequent blood sampling is to use subcutaneous continuous glucose monitoring (e.g., continuous glucose monitoring kits from MiniMed or DexCom) to precisely identify the lowest (nadir) [glucose] (Glu_{nadir}) obtained, usually within 4 hours of the oral glucose load. Recognize the lag time inherent in peripheral blood glucose testing.
4. Expect symptoms of hypoglycemia not only with a blood [glucose] < 60 mg/dL but also during a rapid fall in [glucose] into the normal range (e.g., 70–110 mg/dL).
5. Determine the change (decline, Δ) in blood [glucose] between the [glucose] in the blood sample obtained 90 minutes prior to the blood sample with the [glucose] representing the glucose nadir ($Glu_{nadir-90'}$): Δ (mg/dL) = ($Glu_{nadir-90'}$) − (Glu_{nadir}).
6. Calculate the HGI = Δ (mg/dL)/Glu_{nadir}.

Interpretation

1. In adults with a normal glucose tolerance test as well as non-overt or "early" type 2 diabetes mellitus (T2DM) without symptomatic hypoglycemia, HGI = 0.1 to 1.3 (95% confidence interval [CI]) (Cole et al., 1976).
2. In T2DM with symptomatic reactive hypoglycemia, HGI = 1.1 to 3.5 (95% CI) (Cole et al., 1976).
3. The rate of occurrence of RH may be as high as 50% in lean (body mass index [BMI] ≤ 25 kg/m^2) young women with polycystic ovarian syndrome (PCOS). Dehydroepiandrosterone sulfate (DHEAS) and prolactin levels tend to be lower in these RH patients.
4. RH in the fourth hour during an oGTT, together with a low DHEAS level, may be predictive of future DM in nonpregnant young women with PCOS even when they are not obese (Altuntas et al., 2005).

Notes

1. The use of the HGI, which reflects the rate of decline of the postprandial [glucose], may better identify symptomatic patients who are truly experiencing RH.
2. RH occurs within 2 to 5 hours following ingestion of nutrients and is predominantly associated with adrenergic rather than neuroglycopenic symptoms.
3. Causes of RH include alimentary-related hypoglycemia in patients with impaired glucose tolerance or functional idiopathic hypoglycemia or those who have had gastric surgery resulting in an increased rate of transit of liquid nutrients from the stomach to the small intestine (i.e., dumping syndrome).
4. Symptoms of hypoglycemia may be associated with a low absolute level of blood [glucose], as well as the rapidity of its rate of decline.
5. In patients who present with complaints of postprandial "hypoglycemia," the incidence of functional symptoms suggestive of, but not proven to be, hypoglycemia approaches 15%.
6. In individuals with more overt syndromes of insulin resistance and RH, Glu_{nadir} occurs later (i.e., fourth hour during an oGTT) compared to individuals who have only RH in which Glu_{nadir} is observed typically during the third hour, or at an earlier time point, of an oGTT.
7. Pregnant women who experience RH during a 100-g oGTT may have a significantly lower incidence of DM and low-birth-weight neonates (9.8%) than those who do not (28.6%) (Weissman et al., 2005).

Suggested Reading

Altuntas Y, Bilir M, Ucak S, Gundogdu S. Reactive hypoglycemia in lean young women with PCOS and correlations with insulin sensitivity and with beta cell function. *Eur J Obstet Gynecol Reprod Biol* 2005; 119:198–205.

Cole RA, Benedict GW, Margolis S, Kowarski A. Blood glucose monitoring in symptomatic hypoglycemia. *Diabetes* 1976; 25:984–8.

Weissman A, Solt I, Zloczower M, Jakobi P. Hypoglycemia during the 100-g oral glucose tolerance test: incidence and perinatal significance. *Obstet Gynecol* 2005; 105:1424–8.

12.3.2   75-Gram oGTT in the Research-Oriented Diagnosis of Nonpregnant Type 2 Diabetes Mellitus (T2DM) Patients

Possible Indications for Test

The 75-g oGTT may be indicated in the investigation of T2DM to:

- Make a research-level diagnosis of DM in a nonpregnant patient with a fasting plasma glucose (FPG) > 99 mg/dL (5.9 mmol/L) but < 126 mg/dL (7.0 mmol/L).
- Confirm the presence or absence of impaired glucose tolerance (IGT) in a high-risk individual (i.e., positive family history of DM, central obesity, hypertriglyceridemia).
- Ascertain whether or not glucose intolerance might be contributory to primary hypertriglyceridemia.

Contraindications to Testing

- In any patient with an established diagnosis of DM
- Hospital inpatient status, acute illness, immobilization, starvation, low-carbohydrate diet (<150 g/day), pregnancy, or drug therapy that impairs glucose tolerance, such as niacin, glucocorticoids, thiazide diuretics, beta blockers, or oral contraceptives

Alternatives to Testing

- Measurement of fasting and postprandial glucose, insulin, and hemoglobin A_{1c} concentrations

Procedure

1. Do not perform an oGTT if the fasting [glucose] on the day of testing is >125 mg/dL (6.95 mmol/L), as the patient may be overtly diabetic or not in a fasting state.
2. Clearly identify that the patient is ambulatory and consuming an unrestricted diet consisting of ≥150 g of carbohydrates per day for 3 days prior to the test.
3. Have the patient remain seated and do not allow the patient to smoke during the test.
4. Following an 8- to 14-hour overnight fast, obtain a fasting baseline venous blood sample for plasma glucose testing.
5. Administer 75 g of glucose orally over 5 minutes. For children of body weight <43 kg, administer 1.75 g glucose/kg up to a maximum of 75 g.
6. Obtain a venous blood sample for plasma [glucose] determination at 120 minutes after the administration of the oral glucose load.

Interpretation

1. Normal results for nonpregnant adults are a FPG concentration < 99 mg/dL (6.12 mmol/L) and a 2-hour post-glucose load plasma [glucose] < 140 mg/dL (7.8 mmol/L).
2. IGT is identified by a [FPG] > 99 mg/dL (5.9 mmol/L) and a 2-hour [glucose] > 140 mg/dL (7.8 mmol/L) but < 200 mg/dL (12.2 mmol) following the administration of a 75-g oral glucose load.

3. DM may be unequivocally diagnosed by the patient demonstrating one or more of the following three findings:
   - A random plasma [glucose] ≥ 200 mg/dL (12.2 mmol/L) in the presence of hyperglycemic symptoms (i.e., polyuria, polydipsia, polyphagia, weight loss, blurred vision, or fatigue)
   - [FPG] ≥ 125 mg/dL (6.95 mmol/L) on two separate occasions
   - A 2-hour glucose value ≥ 200 mg/dL (12.2 mmol) following a 75-g oral glucose load

Notes

1. This test is largely confined to research studies since the advent of the 2003 American Diabetes Association (ADA) standards for the diagnosis of T2DM based on [FPG].
2. The 100-g oGTT test is no longer used in nonpregnant patients as it has a much lower sensitivity and specificity for the diagnosis of DM than the 75-g oGTT.
3. A 2-hour postprandial glucose screen is usually adequate for detecting problems with glucose tolerance if an average meal with at least 70 to 80 g of carbohydrates is ingested.
4. If a 75-g oGTT in done during pregnancy, abnormal neonatal anthropometric features (macrosomia) may be associated with a [glucose] > 150 mg/dL at 16 to 20 weeks' gestation and >160 mg/dL at 26 to 30 weeks' gestation (Mello et al., 2003).

Suggested Reading

Mello G, Parretti E, Cioni R, Lucchetti R, Carignani L, Martini E, Mecacci F, Lagazio C, Pratesi M. The 75-gram glucose load in pregnancy: relation between glucose levels and anthropometric characteristics of infants born to women with normal glucose metabolism. *Diabetes Care* 2003; 26:1206–10.

12.3.3 Fasting Glucose/Insulin (Glu/Ins) Ratio vs. Homeostasis Model Assessment (HOMA) Formulas for Estimation of Insulin Sensitivity and Pancreatic Beta-Cell Function

Possible Indications for Test

Assessment of pancreatic beta-cell function and insulin resistance using formulas for the Glu/Ins ratio and HOMA may be indicated when:

- Evaluating research study groups of near-normal-weight individuals for insulin sensitivity
- Any combination of the following is present suggestive of insulin resistance or the so-called metabolic syndrome (prediabetes or early diabetes):
  - A fasting plasma glucose concentration > 99 mg/dL
  - Hypertension (BP, >120/>80 mmHg)
  - Gender-dependent low [HDL] and/or hypertriglyceridemia (>149 mg/dL)
  - Hemoglobin A_{1c} > 5.7%
  - Acanthosis nigricans (hyperpigmentation) around the neck and other intertrigonous areas
  - A large number of skin tags (>5) visible
  - Coronary artery disease

Contraindications to Testing

- In patients with overt or well-established diabetes mellitus and any patient treated with insulin
- In patients with established insulin deficiency, specifically type 1 diabetes mellitus (T1DM)
- When there is a research requirement for an accurate estimate of insulin resistance and β-cell deficit, neither of which can be obtained using the HOMA technique

Alternatives to Testing

- Intravenous glucose tolerance test (ivGTT)
- Euglycemic or hyperglycemic insulin clamp procedure
- Intravenous insulin tolerance test (ivITT) for determination of insulin sensitivity

Procedure

1. Following a 10- to 14-hour overnight fast, obtain a blood sample for both FPG and insulin determination.
2. Calculate the ratio of Glu/Ins = Glu (mg/dL)/Ins (μU/mL).
3. Calculate the insulin sensitivity index, S(i), using HOMA Formulas 1A or 1B:
   - Formula 1A—fasting insulin (μU/mL), fasting glucose (mg/dL):

$$(Ins \times Glu)/408.3$$

   - Formula 1B—fasting insulin (μU/mL), fasting glucose (mmol/L):

$$(Ins \times Glu)/22.5$$

4. Calculate the β-cell index using HOMA Formulas 2A or 2B:
   - Formula 2A—fasting insulin (μU/mL), fasting glucose (mg/dL):

$$(Ins \times 363)/(Glu - 63.5)$$

   - Formula 2B—fasting insulin (μU/mL), fasting glucose (mmol/L):

$$(Ins \times 20)/(Glu - 3.5)$$

Interpretation

1. A fasting Glu/Ins ratio < 6.0 is consistent with significant insulin resistance. Glu/Ins ratios between 6 and 10 suggest possible insulin resistance.
2. Using HOMA Formula 1, normal S(i) is defined as ≤1.0 and insulin resistance as S(i) > 1.0.
3. Normal %β-cell function by HOMA is a β-cell index ≥ 100%, with increasingly lower percentages correlated with insulin deficiency and diabetes.
4. The low precision of HOMA estimates of %β-cell function, imprecision of the insulin assay, pulsatility of insulin secretion, high proinsulin levels, stress, and the significant overlap of values for %β-cell function between normal subjects and those with insulin resistance limit the utility of this test in a clinical setting.

Notes

1. The low precision of %β-cell function estimates from the HOMA model stems from the high coefficient of variation (CV) of 31% for insulin resistance and 32% for β-cell deficit determinations.
2. Correlations (R_S) between estimates of insulin sensitivity by HOMA and those obtained by use of the euglycemic clamp procedure, fasting insulin concentration, and the hyperglycemic clamp procedure were 0.88 ($p < 0.0001$), 0.81 ($p < 0.0001$), and 0.69 ($p < 0.01$), respectively.
3. Estimates of deficient β-cell function obtained by the HOMA equation correlated with that derived using the hyperglycemic insulin clamp procedure ($r = 0.61$; $p < 0.01$) and with the estimate from the ivGTT ($r = 0.64$; $p < 0.05$) (Figure 12.1).
4. The euglycemic insulin clamp is another method of estimating insulin sensitivity (Dons, 1997).
5. A research alternative to HOMA is the continuous infusion of glucose with model assessment (CIGMA) procedure for evaluating insulin resistance and β-cell function which utilizes a continuous glucose infusion of 5 mg glucose per kg ideal body weight per minute for 60 minutes (Hosker et al., 1985), with measurement of plasma glucose and insulin levels.

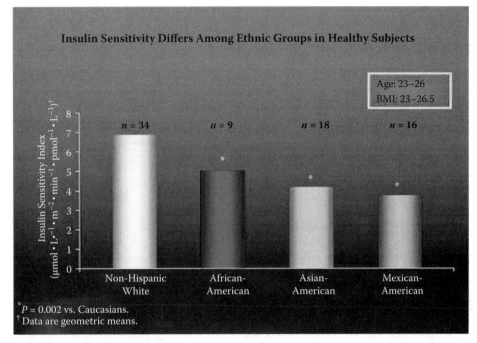

Figure 12.1

Insulin sensitivity (measured by hyperglycemic clamp technique) index (ISI) differs based on ethnicity. Glucose infusion (μmol/L per m[2] body surface area per minute) is shown as a function of insulin infusion (pmol/L) required to maintain a target glucose level. When more insulin is required to maintain a target glucose level, the ISI is lower in those with lower insulin sensitivity (or greater insulin resistance). (Adapted from Chlu, K.C. et al., *Diabetes Care*, 23, 1353–1358, 2000.)

6. Using the CIGMA approach to assessing insulin resistance and β-cell function, normal insulin resistance is defined as 1. The median resistance in a healthy reference population was 1.35 by CIGMA and 1.39 by glucose clamp. In DM patients, median resistance was 4.0 by CIGMA and 3.96 by glucose clamp. The CV for CIGMA determinations was 21% for insulin resistance and 19% for β-cell function.

7. Impaired β-cell function in patients with T2DM may be assessed by measurement of a proinsulin/insulin index in both the fasting basal and dynamic post-oral glucose load state (Tura et al., 2003).

8. Metabolically obese but normal-weight (MONW) women may be identified as having a higher percent body fat (%BF), lower fat-free mass, lower physical activity energy expenditure, and lower peak oxygen uptake (VO$_2$) associated with a %BF fat-adjusted insulin sensitivity index (ISI) of 2.1 vs. 0.93 for non-MONW women (Conus et al., 2004).

Suggested Reading

Chiu KC, Cohan P, Lee NP, Chuang LM. Insulin sensitivity differs among ethnic groups with a compensatory response in beta-cell function. *Diabetes Care*, 2000; 23:1353–58.

Conus F, Allison DB, Rabasa-Lhoret R, St-Onge M, St-Pierre DH, Tremblay-Lebeau A, Poehlman ET. Metabolic and behavioral characteristics of metabolically obese but normal-weight (MONW) women. *J Clin Endocrinol Metab* 2004; 89:5013–20.

Dons RF (Ed). *Endocrine and Metabolic Testing Manual*, 3rd ed. 1997. CRC Press, pp. 4-27–4-28.

Haffner SM, Miettinen H, Stern MP. The homeostasis model in the San Antonio Heart Study. *Diabetes Care* 1997; 20:1087–92.

Hosker JP, Matthews DR, Rudenski AS, Burnett MA, Darling P, Bown EG, Turner RC. Continuous infusion of glucose with model assessment: measurement of insulin resistance and beta-cell function in man. *Diabetologia* 1985; 28:401–12.

Levy JC, Matthews DR, Hermans MP. Correct homeostasis model assessment (HOMA) evaluation uses the computer program. *Diabetes Care* 1998; 21:2191–192.

Matthews DR, Hosker JP, Rudenski AS, Naylor BA, Treacher DF, Turner RC. Homeostasis model assessment: insulin resistance and beta-cell function from fasting plasma glucose and insulin concentrations in man. *Diabetologia* 1985; 28:412–9.

Tura A, Pacini G, Kautzky-Willer A, Ludvik B, Prager R, Thomaseth K. Basal and dynamic proinsulin–insulin relationship to assess beta-cell function during OGTT in metabolic disorders. *Am J Physiol Endocrinol Metab* 2003; 285:E155–62.

12.3.4   Determination of Glucagon-Like Peptide 1 (GLP-1) Deficiency

Possible Indications for Test

Measurement of [GLP-1] after an oGTT or mixed meal may be indicated when:

- Making a diagnosis of GLP-1 deficiency in type 2 diabetes mellitus (T2DM) patients
- Deciding to treat diabetics with GLP-1 analogs or dipeptidyl peptidase 4 (DPP-4) inhibitors

Contraindications to Testing

- Symptoms such as nausea and vomiting suggestive of delayed gastric emptying in the absence of mechanical obstruction (i.e., gastroparesis)
- GLP-1 resistance in which the baseline [GLP-1] is elevated
- Anorexia nervosa

Alternatives to Testing

- Identification of chronic hyperglycemia, particularly in the postprandial state of obese T2DM

Procedure

1. Obtain fasting blood sample for measurement of [GLP-1] at baseline. Elect to do this test if baseline [GLP-1] is not elevated (i.e., not >150 pg/mL).
2. Obtain blood samples for measurement of [GLP-1] at baseline (0 minutes) and at 30, 60, and 90 minutes after administration of a 75-g oral glucose load.
3. Alternatively, obtain blood samples for GLP-1 testing prior (0 minutes) to administering a 230-kcal mixed-content breakfast (60% of calories as carbohydrates) and at 30 and 60 minutes after the start of the meal.
4. Use computer-assisted analysis to measure the area under the curve (AUC) of [GLP-1] vs. time after the carbohydrate load from time 0 to 90 minutes.

Interpretation

1. Testing for postprandial GLP-1 remains investigational, as the reference interval for the timed GLP-1 response to an enteral glycemic load has not been established.
2. Mean integrated GLP-1 output (AUC) after a 75-g oGTT is about twice as high as the AUC after a mixed meal (230 kcal) in both normal and obese girls (Tomasik et al., 2002).
3. GLP-1 secretion may be lower in patients with metabolic syndrome and in sedentary individuals but is markedly reduced in patients with T2DM.
4. In patients with T2DM, the fasting plasma [GLP-1] (102 pg/mL) is similar to that of non-DM controls (97 pg/mL).
5. In patients with chronic hyperglycemia, a progressive desensitization of intestinal L-cells is induced, with consequent failure of GLP-1 to respond to stimulation by nutrient ingestion.

6. In patients with overt DM following chronic pancreatitis and loss of pancreatic glucagon secretory capability, plasma [GLP-1] was significantly elevated after an oGTT compared with the GLP-1 response to an oGTT in those with normal or impaired glucose tolerance.

7. Baseline fasting plasma [GLP-1] in adolescent girls with uncomplicated obesity or anorexia nervosa were similar (1.6–1.7 pmol/L) but significantly lower than in healthy, age-matched controls (2.6 pmol/L).

8. Secretion of GLP-1 is much lower in response to an oGTT and a mixed meal in girls with simple obesity or anorexia nervosa when compared to normal age-matched controls.

Notes

1. A major action of endogenous GLP-1 is to inhibit gastric emptying and glucagon secretion.

2. Regardless of the diagnosis of DM, states of insulin resistance, or insulin deficiency, chronic hyperglycemia appears to induce a progressive desensitization of intestinal L-cells with consequent decrease in the GLP-1 secretory response to a glycemic load.

3. About two thirds of the insulin response to an oral glucose or carbohydrate load is a result of the potentiating effect of the gut-derived incretin hormones, GLP-1, and glucose-dependent insulinotropic polypeptide (GIP). The effect of these hormones on insulin secretion is referred to as the *incretin effect*.

4. In patients with T2DM, there is a complete loss of the normal GIP-induced potentiation of second-phase insulin secretion. This occurs via an impaired GIP receptor even though secretion of GIP is not reduced.

5. Native GLP-1 is rapidly (<2 min) degraded in the circulation by cleavage at Ala(2), but linkage of acyl-chains to the alpha-amino group of His(1) or replacement of Ala(2) *in vitro* results in significantly increased biological effects of GLP-1 *in vivo* via decreased peptide degradation.

6. The remission, or honeymoon, phase of T1DM is associated with substantial recovery of β-cell function and marked improvement of endogenous insulin response to meals but little or no improvement in the insulin response to parenteral glucose, suggesting that incretins (GLP-1 and GIP) may play an important role in the regulation of glucose in this phase of DM.

7. The prolonged action of the GLP-1 analog exendin-4 has been shown to reduce the glycemic effects of a mixed meal in C-peptide-negative T1DM patients who were receiving intensive insulin therapy by delaying gastric emptying.

8. DPP-4 inhibitors delay the degradation of GLP-1, which results in an extension of the action of endogenous insulin and glucagon release suppression.

9. The AUC for [GLP-1] after an oGTT was significantly increased after physical activity compared to rest in lean subjects ($p = 0.05$) as well as in obese subjects after weight loss ($p < 0.05$), but not before weight loss (Adam and Westerterp-Plantenga, 2004).

10. GLP-1 completely suppresses glucagon at normal [glucose] (i.e., 77.5–90.1 mg/dL, or 4.3–5.0 mmol/L), but not at hypoglycemic glucose levels (i.e., ≤66.7 mg/dL, or ≤3.7 mmol/L). At plasma glucose concentrations < 77.5 mg/dL (< 4.3 mmol/L), the insulinotropic action of GLP-1 is negligible (Nauck et al., 2002). GLP-1 does not impair the overall counter-regulatory response to hypoglycemia except for a reduction in the growth hormone secretory response.

11. GLP-1 appears to play a protective role against cytokine-induced apoptosis and necrosis in β-cells through a protein kinase B-dependent signaling pathway (Li et al., 2005).

Suggested Reading

Adam TC, Westerterp-Plantenga MS. Activity-induced GLP-1 release in lean and obese subjects. *Physiol Behav* 2004; 83:459–66.

Dupre J. Glycaemic effects of incretins in type 1 diabetes mellitus: a concise review, with emphasis on studies in humans. *Regul Pept* 2005; 128:149–57.

Hiroyoshi M, Tateishi K, Yasunami Y, Maeshiro K, Ono J, Matsuoka Y, Ikeda S. Elevated plasma levels of glucagon-like peptide-1 after oral glucose ingestion in patients with pancreatic diabetes. *Am J Gastroenterol* 1999; 94:976–81.

Holst JJ, Gromada J. Role of incretin hormones in the regulation of insulin secretion in diabetic and nondiabetic humans. *Am J Physiol Endocrinol Metab* 2004; 287(2):E199–206.

Li L, El-Kholy W, Rhodes CJ, Brubaker PL. Glucagon-like peptide-1 protects beta cells from cytokine-induced apoptosis and necrosis: role of protein kinase B. *Diabetologia* 2005; 48(7):1339–49.

Lugari R, Dell'Anna C, Ugolotti D, Dei Cas A, Barilli AL, Zandomeneghi R, Marani B, Iotti M, Orlandini A, Gnudi A. Effect of nutrient ingestion on glucagon-like peptide 1 (7–36 amide) secretion in human type 1 and type 2 diabetes. *Horm Metab Res* 2000; 32:424–8.

Nauck MA, Heimesaat MM, Behle K, Holst JJ, Nauck MS, Ritzel R, Hufner M, Schmiegel WH. Effects of glucagon-like peptide 1 on counterregulatory hormone responses, cognitive functions, and insulin secretion during hyperinsulinemic, stepped hypoglycemic clamp experiments in healthy volunteers. *J Clin Endocrinol Metab* 2002; 87:1239–46.

Tomasik PJ, Sztefko K, Malek A. GLP-1 as a satiety factor in children with eating disorders. *Horm Metab Res* 2002; 34:77–80.

Tougas G, Eaker EY, Abell T, Abrahamsson H, Boivin M, Chen J, Hocking MP, Quigley EMM, Koch KL, Tokayer AZ, Stanghellini, V, Chen Y, Huizinga JD, Rydén J, Bourgeois I, McCallum RW. Assessment of gastric emptying using a low fat meal: establishment of international control values. *Am J Gastroenterol* 2000; 95:1456–62.

Xiao Q, Giguere J, Parisien M, Jeng W, St-Pierre SA, Brubaker PL, Wheeler MB. Biological activities of glucagon-like peptide-1 analogues *in vitro* and *in vivo*. *Biochemistry* 2001; 40:2860–9.

12.4   Tests for Markers of Cardiovascular Disease in Diabetes Mellitus Patients

12.4.1   Lipoprotein-Associated Phospholipase A_2 (Lp-PLA$_2$) as a Marker for Inflammation and Risk for Cardiovascular Disease (CVD)

Possible Indications for Test

Circulating Lp-PLA$_2$ as a marker of inflammation may be indicated in assessing risk for CVD and stroke in:

- Diabetes patients with hypertension
- Those patients who are more likely to have atherosclerotic disease based on family history and other risk factors but without elevations of low-density lipoprotein (LDL)

Alternatives to Testing

- Multiple other, better established lipid and nonlipid markers of CVD

Procedure

1. Obtain a blood sample for Lp-PLA$_2$ measured using enzyme-linked immunosorbent assay (ELISA) as developed by diaDexus, Inc. (San Francisco, CA).

Interpretation

1. Reference intervals:
   - Males, 131 to 376 ng/mL
   - Females, 120 to 342 ng/mL
2. Lp-PLA$_2$ values > 235 ng/mL may be useful in identifying individuals with a significant increased risk for CVD.

Atherosclerosis, LDL, and Lp–PLA$_2$

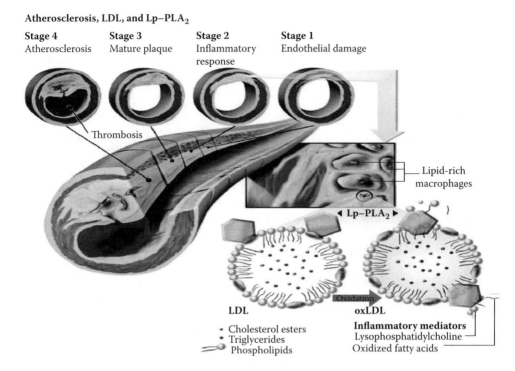

Figure 12.2
Lipoprotein-associated phospholipase A$_2$ (Lp-PLA$_2$) and the generation of inflammatory, atherogenic toxic mediators that promote the generation of lipid-rich macrophages. Lp-PLA$_2$, as a member of the phospholipase A$_2$ superfamily, hydrolyzes phospholipids and generates inflammatory, atherogenic mediators, including oxidized low-density lipoproteins (LDLs), lysophosphatidylcholine, and oxidized fatty acids.

3. Lp-PLA$_2$ correlates positively with LDL-C ($r = 0.36$) and negatively with HDL-C ($r = -0.33$) and does not correlate with CRP ($r = -0.05$).
4. Patients with a systolic blood pressure (SBP) > 130 mmHg and Lp-PLA$_2$ > 420 ng/mL have more than a 6-fold higher risk for cerebrovascular ischemia than do normotensives with a SBP < 113 mmHg and Lp-PLA$_2$ < 310 ng/mL.

Notes

1. Lp-PLA$_2$ is a member of the phospholipase A$_2$ superfamily, a family of enzymes that hydrolyze phospholipids (Figure 12.2).
2. Lp-PLA$_2$, also known as platelet-activating factor acetylhydrolase, is primarily associated with circulating LDL.
3. In the Atherosclerosis Risk in Communities (ARIC) study, an Lp-PLA$_2$ of 422 ng/mL was useful in identifying a higher risk of stroke and coronary heart disease (CHD) among patients with a baseline low-density lipoprotein (LDL) < 130 mg/dL.
4. Lp-PLA$_2$ may play an important proinflammatory role in the progression of atherosclerosis.
5. When traditional risk factors, including LDL-C, were taken into account, the association between Lp-PLA$_2$ and CHD was attenuated and did not achieve statistical significance.
6. For individuals with LDL-C below the median (130 mg/dL), Lp-PLA$_2$ and CRP were both significantly and independently associated with CHD.
7. Homozygosity for the Lp-PLA$_2$ activity-reducing A379V variant on the V379 allele was associated with lower risk of MI (odds ratio [OR], 0.56; 95%CI, 0.32–0.98). This lower risk was maintained after adjustment for lifestyle factors and levels of CRP, fibrinogen, and IL-6 (OR, 0.46; 95%CI, 0.22–0.93).

8. A variety of lipid-lowering agents of the statin class is known to reduce plasma levels of Lp-PLA$_2$ by about 20 to 25%.

9. Lp-PLA$_2$ levels are not affected by respiratory conditions, such as smoking, chronic obstructive pulmonary disease, asthma, or sinusitis, or by arthritic disease, including rheumatoid arthritis and osteoarthritis.

Suggested Reading

Abuzeid AM, Hawe E, Humphries SE, Talmud PJ. Association between the Ala379Val variant of the lipoprotein associated phospholipase A$_2$ and risk of myocardial infarction in the north and south of Europe. *Atherosclerosis* 2003; 168:283–8.

Ballantyne CM, Hoogeveen RC, Bang H, Coresh J, Folsom AR, Heiss G, Sharrett AR. Lipoprotein-associated phospholipase A$_2$, high-sensitivity C-reactive protein, and risk for incident coronary heart disease in middle-aged men and women in the Atherosclerosis Risk in Communities (ARIC) study. *Circulation* 2004; 109:837–42.

Dada N, Kim NW, Wolfert RL. Lp-PLA$_2$: an emerging biomarker of coronary heart disease. *Expert Rev Mol Diagn* 2002; 2:17–22.

Sudhir K. Lipoprotein-associated phospholipase A$_2$ clinical review: lipoprotein-associated phospholipase A$_2$, a novel inflammatory biomarker and independent risk predictor for cardiovascular disease. *J Clin Endocrinol Metab* 2005; 90:3100–5.

12.4.2 Tumor Necrosis Factor Alpha (TNFα) as a Marker of Intravascular Inflammation and Adiponectin Effects

Possible Indications for Test

Measurement of TNFα may be indicated in the:

- Assessment of risk for vascular disease in diabetes mellitus (DM) patients with hypertension and hyperlipidemia

Alternatives to Testing

- Markers of inflammation, including plasminogen activator inhibitor 1 (PAI-1), high-sensitivity C-reactive protein (hsCRP), and mean platelet volume (MPV)

Procedure

1. Obtain plasma or peripheral blood mononuclear cells for TNFα testing. Note that several different investigational assays for TNFα are available.

Interpretation

1. After adjustment for covariates, including percent body fat and waist-to-hip ratio, plasma [TNFα] was found to be positively related to [HbA$_{1c}$] and increased from baseline over time in elderly individuals with T2DM who had no change in BMI (Lechleitner et al., 2002).

2. Plasma [TNFα] increased significantly from 16.2 ± 9.6 pg/mL at baseline to 28.0 ± 13.8 pg/mL after 2 years ($p = 0.028$) concomitant with an increase in HbA$_{1c}$ from $6.4 \pm 1.2\%$ to $7.7 \pm 1.6\%$ ($p = 0.046$).

3. In a 6-week diet study, peripheral blood mononuclear cell TNFα activity rose by 86.3% ($p = 0.006$) or declined by 20% (p = not significant) on diets that either elevated advanced glycation endproducts (AGEPs) or resulted in 5-fold lower AGEPs.

Notes

1. There is a lack of a standardized assay for TNFα.
2. Visceral and subcutaneous adipose tissues, as well as lymphocytes, are major sources of cytokines or adipokines, including TNFα, interleukin-6 (IL-6), and PAI-1.
3. Reduction of adipose tissue mass through weight reduction in association with exercise reduces [TNFα], [IL-6], and [PAI-1] and increases [adiponectin].
4. Adiponectin inhibits the expression of TNFα-induced endothelial adhesion molecules, macrophage-to-foam cell transformation, smooth muscle cell proliferation, and TNFα expression in macrophages and adipose tissue.
5. Inflammation is a potential pathogenetic factor in atherosclerosis and insulin insensitivity associated with clinical markers of inflammation, including TNFα, hsCRP, small-particle LDL, and homocysteine.
6. Chronic inflammation appears to be a trigger for chronic insulin insensitivity and hence an increase in plasma [TNFα]. It has been suggested that overexpression of TNFα contributes to the development of insulin resistance and endothelial dysfunction.
7. In experimental systems using obese mice with homozygous null mutations at the TNFα or TNF receptor gene loci, genetic absence of TNF signaling in obesity significantly improved insulin receptor signaling capacity and consequently insulin sensitivity. These mutations were associated with lower PAI-1, transforming growth factor beta (TGFβ), lipid, and leptin production.
8. TNFα inhibits synthesis of adiponectin and enhances release of free fatty acids (FFAs) from adipose tissue.
9. Consistent with changes in TNFα, [CRP] rose by 35% in individuals on a diet that elevated AGEPs (H-AGEPD) and fell by 20% in individuals on an AGEP-lowering diet (L-AGEPD) ($p = 0.014$). Vascular cell adhesion molecule-1 (VCAM-1) declined by 20% on L-AGEPD ($p < 0.01$) and increased by 4% on H-AGEPD (Vlassara et al., 2002).

Suggested Reading

Aldhahi W, Hamdy O. Adipokines, inflammation, and the endothelium in diabetes. *Curr Diab Rep* 2003; 3:293–8.

Borst SE. The role of TNF-alpha in insulin resistance. *Endocrine* 2004; 23:177–82.

Dandona P, Aljada A. A rational approach to pathogenesis and treatment of type 2 diabetes mellitus, insulin resistance, inflammation, and atherosclerosis. *Am J Cardiol* 2002; 90(5A):27–33G.

Farmer JA, Torre-Amione G. Atherosclerosis and inflammation. *Curr Atheroscler Rep* 2002; 4:92–8.

Grimble RF. Inflammatory status and insulin resistance. *Curr Opin Clin Nutr Metab Care* 2002; 5:551–9.

Hotamisligil GS. Molecular mechanisms of insulin resistance and the role of the adipocyte. *Int J Obes Relat Metab Disord* 2000; 24(Suppl 4):S23–7.

Lechleitner M, Herold M, Dzien-Bischinger C et al. Tumour necrosis factor-alpha plasma levels in elderly patients with type 2 diabetes mellitus: observations over 2 years. *Diabet Med* 2002; 19:949–53.

Ouchi N, Kihara S, Funahashi T, Matsuzawa Y, Walsh K. Obesity, adiponectin and vascular inflammatory disease. *Curr Opin Lipidol* 2003; 14:561–6.

Ruan H, Lodish HF. Insulin resistance in adipose tissue: direct and indirect effects of tumor necrosis factor-alpha. *Cytokine Growth Factor Rev* 2003; 14:447–55.

Vlassara H, Cai W, Crandall J, Goldberg T, Oberstein R, Dardaine V, Peppa M, Rayfield EJ. Inflammatory mediators are induced by dietary glycotoxins, a major risk factor for diabetic angiopathy. *Proc Natl Acad Sci USA* 2002; 99:15596–601.

12.4.3   Plasminogen Activator Inhibitor-1 (PAI-1) as Serine Proteinase Inhibitor (SERPIN) in Patients with Thrombotic Vascular Disease

Possible Indications for Test

Measurement of PAI-1 enzyme may be indicated in the assessment of thrombosis risk in patients with:

- Postmyocardial infarction status
- Obesity, diabetes, insulin resistance, or cardiometabolic syndrome
- Renal disease
- Inflammatory lung disorders
- Neoplastic processes

Alternatives to Testing

- Multiple, better established lipid and nonlipid markers of risk for cardiovascular disease and thrombosis

Procedure

1. Obtain the appropriate blood sample for PAI-1 antigen testing relevant to the referral laboratory where this testing will be performed. Carefully follow detailed directions for sample collection, as specimen processing factors (time to cell separation after phlebotomy and storage temperature of sample) are critical.
2. Because PAI-1 antigen concentration or enzyme activity may be elevated if the specimen used for testing is contaminated with platelets, always obtain a repeat blood sample for PAI-1 testing, along with an EDTA specimen for a platelet count, if the first PAI-1 test result is elevated.

Interpretation

1. Reference intervals for PAI-1 are specimen and method specific (e.g., mass immunoassays vs. enzyme activity assays). The reference interval for the one-step enzyme immunoassay (EIA) for PAI-1 is 13.2 to 88 ng/mL (Nieuwenhuizen et al., 1995), whereas the reference interval for the Imubind® PAI-1 ELISA is 2 to 47 ng/mL (citrated plasma) ($n = 167$ healthy individuals).
2. Values for PAI-1 antigen concentration in human platelet-poor plasma range between 4 and 43 ng/mL (mean 18 ± 10 ng/mL) with a high correlation ($r = 0.80$) to PAI-1 enzyme activity (Thogersen et al., 1998).
3. Patients with recurrent deep vein thrombosis (DVT) typically have higher PAI-1 antigen levels (44 ± 20 ng/mL) than individuals without DVT.
4. PAI-1 antigen levels are elevated during the third trimester of pregnancy.

Notes

1. Plasminogen activator inhibitor-1 (PAI-1) is a SERPIN whose major function is to negate plasminogen activation and impair fibrinolysis.
2. Accelerated atherothrombotic processes are attributable not only to lipid and cardiovascular abnormalities but also to specific inflammatory states that lead to increased plasma levels of PAI-1.
3. An activated renin–angiotensin–aldosterone system (RAAS) significantly contributes to the upregulation of the PAI-1 gene via a receptor-mediated mechanism.
4. PAI-1 is overexpressed in adipose tissue and induced by angiotensin II.
5. Artificially high PAI-1 values may occur if samples are contaminated with platelets and undergo more than one freeze–thaw cycle.

Suggested Reading

Alessi MC, Juhan-Vague I. Contribution of PAI-1 in cardiovascular pathology. *Arch Mal Coeur Vaiss* 2004; 97:673–8.

Declerck PJ, Alessi MC, Verstreken M, Kruithof EK, Juhan-Vague I, Collen D. Measurement of plasminogen activator inhibitor 1 in biological fluids with a murine monoclonal antibody based enzyme-linked immunosorbent assay. *Blood* 1988; 71:220–5.

Huber K, Christ G, Wojta J, Gulba D. Plasminogen activator inhibitor type-1 in cardiovascular disease: status report 2001. *Thromb Res* 2001; 103(Suppl 1):S7–S19.

Juhan-Vague I, Alessi MC. PAI-1, obesity, insulin resistance and risk of cardiovascular events. *Thromb Haemost* 1997; 78:656–60.

Kohler HP, Grant PJ. Plasminogen-activator inhibitor type 1 and coronary artery disease. *N Engl J Med* 2000; 342:1792–801.

Nieuwenhuizen W, Laterveer R, Hoegee-de Nobel E, Bos R. A one-step enzyme immunoassay for total plasminogen activator inhibitor-1 antigen in human plasma. *Blood Coagul Fibrinolysis* 1995; 6:268–72.

Thogersen AM, Jansson JH, Boman K, Nilsson TK, Weinehall L, Huhtasaari F, Hallmans G. High plasminogen activator inhibitor and tissue plasminogen activator levels in plasma precede a first acute myocardial infarction in both men and women: evidence for the fibrinolytic system as an independent primary risk factor. *Circulation* 1998; 98:2241–47.

12.4.4   Mean Platelet Volume (MPV) and Measures of Platelet Activation as Indicators of Risk for Cardiovascular Disease (CVD) in Diabetes Mellitus (DM) Patients

Possible Indications for Test

Determination of MPV and other measures of platelet abnormalities may be indicated:

- To identify high-risk patients, particularly DM patients with unstable angina and/or electrocardiographic changes or non-Q-wave myocardial infarction
- In patients at high risk for vascular events to identify candidates for appropriate and effective antiplatelet treatment (e.g., aspirin, clopidogrel)
- When a patient experiences a thrombotic event while receiving aspirin or other anticoagulant therapy

Alternatives to Testing

- Multiple other, better established lipid and nonlipid markers of CVD

Procedure

1. Consult with clinical laboratory personnel to obtain the appropriate blood samples in the correct collection tubes for measurement of MPV and determination of platelet aggregation parameters.
2. Identify the resources required for determining other indices of platelet function, such as plasma soluble P-selectin concentration, urine 11-dehydrothromboxane B_2, glycoprotein IIb/IIIa receptors, platelet surface P-selectin, platelet–monocyte aggregates, and microparticles as supplemental to measurement of MPV and determination of platelet aggregation.

Interpretation

1. MPV is twofold higher (14.2 ± 2.2 fL) in T2DM patients than in non-DM patients (7.1 ± 1.2 fL). Among T2DM, MPV is higher in those who have microvascular complications, such as retinopathy (15.8 ± 1.3 fL) or microalbuminuria (15.6 ± 1.2 fL).
2. Higher MPV values may indicate a higher risk for CVD events in those known to be high risk based on other factors, but data are insufficient to support the ability of MPV to predict the risk of a vascular event in otherwise healthy adults.
3. The reference interval for platelet aggregation determined using the Platelet Function Analyzer-100™ (PFA-100™) instrument (Dade Behring; Leiderbach, Germany) has been reported (Grundmann et al., 2003) as 74 to 165 seconds. However, reference intervals established locally by the laboratory performing this testing and reported with the results of this testing are always preferred over product insert or literature data.

4. Abnormal results for platelet aggregation studies are not consistently predictive of vascular risk.
5. Platelet indices other than measures of their aggregation capability and MPV have no established role in CVD risk assessment.

Notes

1. Because the results of platelet function studies have not been shown to correlate well with clinical outcomes, proceed with aspirin treatment of all DM with the expectation of benefit in the majority of cases with or without detectably abnormal platelet studies.
2. A significant negative relationship exists between MPV and fasting triglyceride (TG) level.
3. The MPV response to low-dose collagen *in vitro* has been proposed as an indicator of platelet propensity to activation with confirmatory studies required (Park et al., 2002).
4. Because of their cost, time-consuming complexity, and requirement for specialized equipment, studies of platelet function (e.g., using the PFA-100™ or flow cytometry) are unlikely to be useful in estimating CVD risk in population studies involving large numbers of patients.
5. In addition to assessment of MPV, image analysis on flow cytometry can be used to estimate mean corpuscular volume (MCV) and detect platelet agglutinates and aggregates.

Suggested Reading

Andersen K, Hurlen M, Arnesen H, Seljeflot I. Aspirin non-responsiveness as measured by PFA-100 in patients with coronary artery disease. *Thromb Res* 2002; 108:37–42.

Eikelboom JW, Hirsh J, Weitz JI, Johnston M, Yi Q, Yusuf S. Aspirin-resistant thromboxane biosynthesis and the risk of myocardial infarction, stroke, or cardiovascular death in patients at high risk for cardiovascular events. *Circulation* 2002; 105:1650–5.

Endler G, Klimesch A., Sunder-Plassman H et al. Mean platelet volume is an independent risk factor for myocardial infarction but not for coronary artery disease. *Br J Haematol* 2002; 117:399–404.

Grundmann K, Jaschonek K, Kleine B, Dichgans J, Topka H. Aspirin non-responder status in patients with recurrent cerebral ischemic attacks. *J Neurol* 2003; 250:63–6.

Gum PA, Kottke-Marchant K, Welsh PA, White J, Topol EJ. A prospective, blinded determination of the natural history of aspirin resistance among stable patients with cardiovascular disease. *J Am Coll Cardiol* 2003; 41:961–5.

Jagroop IA, Mikhailidis DP. Mean platelet volume is an independent risk factor for myocardial infarction but not for coronary artery disease. *Br J Haematol* 2003; 120:169–70.

Kubota F. Analysis of red cell and platelet morphology using an imaging-combined flow cytometer. *Clin Lab Haematol* 2003; 25:71–6.

Mammen EF, Comp PC, Gosselin R et al. PFA-100 system: a new method for assessment of platelet dysfunction. *Sem Thromb Hemost* 1998; 24:195–202.

Papanas N, Symeonidis G, Maltezos E, Mavridis G, Karavageli E, Vosnakidis T, Lakasas G. Mean platelet volume in patients with type 2 diabetes mellitus. *Platelets* 2004; 15:475–8.

Park Y, Schoene N, Harris W. Mean platelet volume as an indicator of platelet activation: methodological issues. *Platelets* 2002; 13:301–6.

Schleinitz MD, Heidenreich PA. A cost-effectiveness analysis of combination antiplatelet therapy for high-risk acute coronary syndromes: clopidogrel plus aspirin versus aspirin alone. *Ann Intern Med* 2005; 142:251–9.

Tsiara S, Elisaf M, Jagroop IA, Mikhailidis DP. Platelets as predictors of vascular risk: is there a practical index of platelet activity? *Clin Appl Thromb Hemost* 2003; 9:177–90.

Yilmaz MB, Ozeke O, Akin Y, Guray U, Biyikoglu SF, Kisacik HL, Korkmaz S. Platelet aggregation in left ventricular thrombus formation after acute anterior myocardial infarction: mean platelet volume. *Int J Cardiol* 2003; 90:123–5.

12.4.5 Measurement of Carotid Artery Intima–Media Thickness (CIMT) as a Test for Atherosclerotic Disease and Risk of CVD

Possible Indications for Test

Measurement of CIMT may be indicated:

- As a surrogate marker of CVD risk particularly in diabetes mellitus (DM) patients with elevated blood pressure (BP) and history of smoking

Alternatives to Testing

- Multiple other surrogate markers of atherosclerotic vascular disease and CVD risk such as high-sensitivity C-reactive protein (hsCRP), LDL concentration, and coronary artery calcium scoring

Procedure

1. Measure the thickness from the outer wall to the inner luminal wall of the carotid artery immediately below the carotid bulb using B-mode carotid ultrasound images.

Interpretation

1. CIMT is thickest in the carotid bulb and increases linearly with age, most rapidly in the bulb.
2. With age, composite CIMT values increase most slowly in Caucasian females and most rapidly in Caucasian males. Nomograms of CIMT percentiles between the ages of 25 and 40 years are available in 5-year increments (Stein et al., 2004).
3. Sample size estimates project that 268 to 462 research subjects are needed to detect CIMT changes ≥ 0.010 mm/year in response to treatment interventions (Stein et al., 2004).
4. A median coefficient of variation of repeated baseline measures of CIMT is approximately 0.5% (range, 0.1–0.7%)

Notes

1. This test is technically complex, poorly reproducible, and known to be associated with variability of intimal thickness independent of changes in lipid levels.
2. Age-related differences in CIMT can be used to plan epidemiological and clinical trials investigating the prevalence of atherosclerosis and results of antiatherosclerotic interventions.
3. The severity of CIMT appears to be an independent predictor of cerebral vascular and coronary events. In asymptomatic individuals >45 years old, CIMT measurement can add incremental information to traditional CVD risk-factor assessment and identifies more atherosclerosis than predicted by the Framingham risk assessment (Smith et al., 2000).
4. CIMT appeared to be a useful surrogate marker of CVD risk in indigenous Australian subjects, correlating better than brachial artery flow-mediated vasodilation with established CVD risk factors (Chan et al., 2005). In this study, using nonparametric analysis:
   - Mean CIMT was significantly ($p = 0.049$) greater in DM vs. non-DM.
   - Significant correlations between CIMT and age ($r = 0.64$; $p < 0.001$), systolic BP ($r = 0.47$; $p < 0.001$), and nonsmoking ($r = -0.30$; $p = 0.018$) were found in non-DM.
   - Modest correlations between CIMT, age ($r = 0.36$; $p = 0.009$), and duration of DM ($r = 0.30$; $p = 0.035$) were found in DM.
   - Significant associations between CIMT and age ($t = 4.6$; $p < 0.001$), systolic BP ($t = 2.6$; $p = 0.010$), and HbA$_{1c}$ ($t = 2.6$; $p = 0.012$), smoking ($t = 2.1$; $p = 0.04$), and fasting LDL-cholesterol concentration ($t = 2.1$; $p = 0.04$) were found by linear regression analysis.

5. Hemoglobin A_{1c} was the only significant correlate of CIMT ($r = 0.35$; $p = 0.01$) in DM when age, sex, smoking, and history of CVD were taken into account.

6. The CIMT/lumen diameter ratio at lower CIMT levels appears to reflect arterial remodeling and does not have to be taken into account when determining the hazard ratio for risk of acute myocardial infarction (Bots et al., 2005).

7. Higher CIMT and levels of circulating inflammatory markers (i.e., CRP, IL-6, IL-18, and urinary catecholamines) occur in hypertensive patients associated with a morning BP peak vs. those without a morning peak in BP, possibly contributing to their increased CVD risk (Filippo et al., 2005).

8. In ARBITER 3, 57 patients treated with a statin plus extended-release niacin for 2 years had both an increase in HDL (23%) and a change in CIMT from 0.899 mm to 0.829 mm ($p = 0.055$) (Taylor et al., 2006).

Suggested Reading

Bots ML, Grobbee DE, Hofman A, Witteman JC. Common carotid intima–media thickness and risk of acute myocardial infarction: the role of lumen diameter. *Stroke* 2005; 36:762–7.

Chan L, Shaw AG, Busfield F, Haluska B, Barnett A, Kesting J, Short L, Marczak M, Shaw JT. Carotid artery intimal medial thickness, brachial artery flow-mediated vasodilation and cardiovascular risk factors in diabetic and non-diabetic indigenous Australians. *Atherosclerosis* 2005; 180:319–26.

Marfella R, Siniscalchi M, Nappo F, Gualdiero P, Esposito K, Sasso FC, Cacciapuoti F, Di Filippo C, Rossi F, D'Amico M, Giugliano D. Regression of carotid atherosclerosis by control of morning blood pressure peak in newly diagnosed hypertensive patients. *Am J Hypertens* 2005; 18:308–18.

Smith SC Jr, Greenland P, Grundy SM. Prevention Conference V: beyond secondary prevention—identifying the high-risk patient for primary prevention. *Circulation* 2000; 101:111–16.

Stein JH, Douglas PS, Srinivasan SR, Bond MG, Tang R, Li S, Chen W, Berenson GS. Distribution and cross-sectional age-related increases of carotid artery intima-media thickness in young adults: the Bogalusa Heart Study. *Stroke* 2004; 35:2782–7.

Taylor AJ, Lee HJ, Sullenberger LE. The effect of 24 months of combination statin and extended release niacin on carotid intima-media thickness: ARBITER 3. *Curr Med Res Opin* 2006; 22:2243–50.

12.4.6   Pulse-Wave Analysis as a Measure of Arterial Wall Stiffness in Patients with Atherosclerosis

Possible Indications for Test

Measurement of the pulse-wave velocity (PWV) may be indicated in patients with:

- Highly suspect or overt atherosclerotic disease
- Risk factors for cardiovascular disease (CVD), including intermittent claudication, diabetes mellitus, hypertension, hyperlipidemia, hyperhomocysteinemia, older age, smoking, or nonhealing foot ulcer

Alternatives to Testing

- Multiple other, better established nonlipid markers of CVD

Procedure

1. Using a commercially available PWV system (e.g., Colin VP-1000®), initiate pulse-rate monitoring by electrocardiogram and obtain phonocardiogram and pulse-volume recordings in a well-hydrated patient.

2. Calculate the brachial–ankle pulse-wave velocity (baPWV) by dividing the distance between the brachial and ankle arteries (measured by tape measure) by the pulse-wave transit time (PwTT).

3. As an alternative to the use of the VP-1000® device, apply a noninvasive, pencil-like pulse-wave probe (the SphygmoCor® or tonometer) to the radial artery of the right wrist to estimate arterial wall stiffness. The IntelliHeart™ pulse-wave analysis device is available in Australia (IM Medical, Ltd.; 12th Floor, 484 St. Kilda Road, Melbourne VIC 3004; 613-9860-0900 [phone]; 613-9860-0999 [fax]).

Interpretation

1. Arterial stiffness increases with duration of hypertension, DM, age, and atherosclerosis as a continuous variable. No specific PWV cutoff value for disease has been established. Reference intervals for a healthy population are available but not for an aging population as of this printing.
2. Patients with end-stage renal disease on dialysis (CKD Stage 5) have increased aortic stiffness as determined by measurement of an increased aortic PWV.
3. Increasing aortic PWV values are strongly associated with the presence and extent of atherosclerosis and constitute a predictor of CVD risk in hypertensive patients as well as a strong independent predictor of all-cause mortality, mainly from CVD, in dialysis patients.
4. Arterial wall stiffness is associated with reduced arterial flow volume in the lower extremities of DM and is reflected by a decrease in lower extremity PWV.
5. Large differences between the PWV (IntelliHeart™) of the lower extremities mark the limb more affected by atherosclerotic disease. The baPWV (VP-1000®) is increased in DM but decreased in the affected legs of those with obstructive PAD. Widening of the right-left difference in baPWV may help identify an extremity with more advanced obstructive peripheral artery disease (PAD).

Notes

1. Atherosclerotic hardening of the arteries is associated with a thicker arterial wall (see Test 12.4.5), increased stiffness of the artery, and a consequent increase in PWV.
2. PWV values are influenced by atherosclerosis of the artery as well as blood density.
3. A significant difference from the normal rate of age-related decline in vascular stiffness in elastic arteries of nondiabetic compared with diabetic arteries has been observed. Arteries in DM appear to age at an accelerated rate at an earlier age, eventually reaching a functional plateau.
4. The sensitivity and specificity of absolute values of PWV as predictors for severity of atherosclerotic disease are unknown.
5. The odds ratio of being in a high CVD mortality risk group (>5% at 10 years) for patients with PWV values in the upper quartile was 7.1 (95% CI, 4.5 to 12.3) (Blacher et al., 1999a,b).
6. Erectile dysfunction (ED) can be caused by reduced arterial and arteriolar inflow and may be related to increased PWV; however, other causes of ED include failed transmission of neural signals to and from the spinal cord secondary to peripheral and autonomic neuropathy, impaired sinusoidal endothelial cell nitric oxide (NO) release, and failure of the elastic fibers of the corpora to relax secondary to their stiffening from protein (collagen fiber) glycosylation.
7. The IntelliHeart™ pulse-wave analysis device is approved for use in Australia, with availability elsewhere pending.

Suggested Reading

Blacher J, Asmar R, Djane S, London GM, Safar ME. Aortic pulse wave velocity as a marker of cardiovascular risk in hypertensive patients. *Hypertension* 1999a; 33:1111–7.

Blacher J, Guerin AP, Pannier B, Marchais SJ, Safar ME, London GM. Impact of aortic stiffness on survival in end-stage renal disease. *Circulation* 1999b; 99:2434–39.

Blacher J, Guerin AP, Pannier B, Marchais SJ, London GM. Arterial calcifications, arterial stiffness, and cardiovascular risk in end-stage renal disease. *Hypertension* 2001; 38:938–42.

Cameron JD, Bulpitt CJ, Pinto ES, Rajkumar C. The aging of elastic and muscular arteries: a comparison of diabetic and nondiabetic subjects. *Diabetes Care* 2003; 26:2133–8.

Kilo S, Berghoff M, Hilz M, Freeman R. Neural and endothelial control of the microcirculation in diabetic peripheral neuropathy. *Neurology* 2000; 54:1246–52.

Lehmann ED, Gosling RG. Lowering of blood pressure and artery stiffness. *Lancet* 1997; 349(9056):955–6.

Sleight P. Lowering of blood pressure and artery stiffness. *Lancet* 1997; 349(9048):362.

Stansberry KB, Hill MA, Shapiro SA, McNitt PM, Bhatt BA, Vinik AI. Impairment of peripheral blood flow responses in diabetes resembles an enhanced aging effect. *Diabetes Care* 1997; 20:1711–16.

Stansberry KB, Peppard HR, Babyak LM, Popp G, McNitt PM, Vinik AI. Primary nociceptive afferents mediate the blood flow dysfunction in non-glabrous (hairy) skin of type 2 diabetes: a new model for the pathogenesis of microvascular dysfunction. *Diabetes Care* 1999; 22:1549–54.

Suzuki E, Kashiwagi A, Nishio Y, Egawa K, Shimizu S, Maegawa H, Haneda M, Yasuda H, Morikawa S, Inubushi T, Kikkawa R. Increased arterial wall stiffness limits flow volume in the lower extremities in type 2 diabetic patients. *Diabetes Care* 2001; 24:2107–14.

Yokoyama H, Shoji T, Kimoto E, Shinohara K, Tanaka S, Koyama H, Emoto M, Nishizawa Y. Pulse wave velocity in lower-limb arteries among diabetic patients with peripheral arterial disease. *J Atheroscler Thromb* 2003; 10:253–8.

12.4.7   Lipid Analyses by Proton Nuclear Magnetic Resonance (NMR) Spectroscopy (NMR LipoProfile®) or Vertical-Spin Density-Gradient Ultracentrifugation (VAP®) as Tests Predictive of Cardiovascular Disease (CVD) in Diabetes Mellitus (DM) Patients

Possible Indications for Test

Either VAP® (Vertical Auto Profile) or NMR LipoProfile® lipid analysis may be indicated following a conventional screening lipoprotein panel (fasting total cholesterol, triglycerides, HDL, and calculated LDL) to:

- Decide whether abnormal or desirable levels of fractionated LDL, HDL, and VLDL are present.
- Assess effectiveness of measures used to treat lipid abnormalities.
- Evaluate high-risk individuals with DM, a family history of premature vascular disease, occurrence of CVD events while being treated for a lipid disorder, or borderline lipid panel results that fail to completely normalize with appropriate treatment.

Alternatives to Testing

- Standard lipid profiles

General Procedure

1. Identify any secondary causes of lipid abnormalities such as uncontrolled DM, untreated hypothyroidism, chronic kidney disease (CKD), obstructive liver disease, nephrotic syndrome, and medications such as estrogen, diuretics, steroids, and nonselective beta blockers that might affect lipid levels.
2. After 10- to 12-hour fast by the patient, obtain a blood sample, separating the serum from the cells within 45 minutes of venipuncture.
3. Submit serum in a serum gel separator tube or a plastic transport tube, keeping the sample refrigerated or on a frozen cool pack.
4. Specimens may be sent frozen in a plastic transport tube when total storage and shipping time is more than 72 hours from the time of collection.
5. Analyze the serum sample for its lipoprotein subclass profile by vertical density-gradient ultracentrifugation (VAP®) or lipoprotein particle size and number by NMR spectroscopic analysis of proton spin resonance (NMR LipoProfile®) as specified below.

NMR LipoProfile® Procedure

1. Refrigerate or freeze serum sample prepared as described above.
2. Send sample for lipid fractionation using NMR spectroscopic methods to estimate particle sizes and numbers of particles across the LDL, HDL, and VLDL spectrums.

Interpretation of Results of Proton NMR Spectroscopy

1. Unlike the VAP®, the NMR LipoProfile® test directly measures the number of lipoprotein particles and their size.
2. Refer to "Reference Intervals for the Concentration and Size of Various Lipoprotein Particles Measured by Nuclear Magnetic Resonance Spectroscopy (NMR)" (Table 12.3) and "Analytical Imprecision in the Quantification of Various Lipoprotein Particles by NMR" (Table 12.4) for values expected for a large population and the relative precision of NMR fractionation.
3. Significantly stronger associations exist between lipid fractions measured by the NMR LipoProfile® and CVD endpoints such as myocardial infarction, angina, coronary heart disease (CHD) death, stroke, carotid atherosclerosis, coronary lumen diameter, and coronary calcification, as well as microvascular endpoints such as retinopathy and nephropathy than those associated with standard lipid profile or apolipoprotein fractions.

VAP® Procedure

1. Refrigerate or freeze serum sample prepared as described in the General Procedure, above.
2. Directly analyze lipoproteins by 1-hour vertical density-gradient ultracentrifugation followed by measurement of the cholesterol distribution using an online VAP-II analyzer.
3. Measure triglycerides enzymatically.
4. Identify the dominant LDL-R subclass pattern based on Rf (relative position) of the major LDL peak in the density gradient and qualitatively plot along a spectrum of particle densities, with "A" designated the most buoyant and "B" the most dense.
5. Lipid values reported include:
   - Total cholesterol (TC)
   - Triglycerides (TG)
   - High-density lipoproteins (HDLs), including HDL_2-C (large, buoyant), HDL_3-C (small, dense), and total HDL-C
   - Total non-HDL-C (a surrogate marker for total Apo-B) = LDL + VLDL
   - Total non-HDL-C/total HDL-C ratio
   - $VLDL_{1,2}$-C (large buoyant), $VLDL_3$-C (small remnant), and total VLDL-C
   - Lp(a) cholesterol [Lp(a)-C]
   - Intermediate-density lipoprotein cholesterol (IDL-C)
   - Real (measured, not calculated) LDL cholesterol (LDL-R-C)
   - Total LDL-C = LDL-R-C + Lp(a)-C + IDL-C
   - Homocysteine
   - High-sensitivity C-reactive protein (hsCRP)

Interpretation of Results of VAP® by Ultracentrifugation

1. Refer to "Therapeutic Target Values for Lipoproteins Measured by VAP®" (Table 12.5).
2. The VAP® provides a nonquantitative assessment of LDL particle size in contrast to the NMR LipoProfile® test.
3. The extent of VAP® lipid abnormalities may correlate with or suggest an increase in CVD risk.

Notes

1. The VAP® test analyzes the standard, direct, and indirectly measured lipid indices as well as non-HDL cholesterol and the nonlipids homocysteine and hsCRP.
2. Epidemiologic studies of LDL-C, upon which most therapeutic recommendations have been developed, use the calculated (or indirect) LDL-C value obtained using the Friedewald formula when the TC level is <400 mg/dL:

$$LDL\text{-}C = [TC] - [HDL\text{-}C] - [TG/5]$$

where [TG/5] provides an estimate of the [VLDL-C].

TABLE 12.3

Reference Intervals for the Concentration and Size of Various Lipoprotein Particles Measured by Nuclear Magnetic Resonance Spectroscopy (NMR)

Lipid Particles or Lipid	Men (n = 4054)			Women (n = 3317)			Total (n = 7371)		
	Mean ± SD	Median	Range	Mean ± SD	Median	Range	Mean ± SD	Median	Range
[VLDLp] (nmol/L)									
Total	84.8 ± 67.0	71.5	17.2–162.9	68.1 ± 62.5	54.6	8.3–141.0	77.3 ± 65.5	64.0	12.3–154.0
Large[a]	3.4 ± 8.9	0.8	0.1–8.5	2.5 ± 5.8	0.6	0.1–6.4	3.0 ± 7.7	0.7	0.1–7.6
Medium	51.9 ± 55.2	36.8	4.3–114.9	47.3 ± 4.9	26.3	2.2–87.4	46.2 ± 52.2	31.2	3.0–103.8
Small	29.5 ± 26.3	24.4	1.4–61.6	26.5 ± 25.7	20.4	0–58.9	28.1 ± 26.0	22.8	0.5–60.3
[LDLp] (nmol/L)									
Total	1535 ± 490	1468	972–2195	1489 ± 487	1419	949–2118	1514 ± 489	1445	961–2161
IDL	28 ± 43	7	0–86	26 ± 45	8	0–86	27 ± 44	5	0–86
Large	339 ± 241	297	70–657	524 ± 289	496	172–912	422 ± 279	381	99–792
Small	1169 ± 542	1122	516–1886	938 ± 564	870	242–1698	1065 ± 564	1021	370–1818
Medium to small	256 ± 542	246	119–402	212 ± 123	200	63–371	236 ± 121	228	88–390
Very small	913 ± 433	874	393–1483	727 ± 447	675	172–1329	829 ± 449	793	280–1425
[HDLp] (nmol/L)									
Total	28.1 ± 6.7	27.8	19.9–36.5	33.4 ± 7.7	33.0	24.2–43.5	30.5 ± 7.6	30.1	21.2–40.3
Large	5.3 ± 3.5	4.6	1.6–10.1	9.1 ± 4.9	8.3	3.5–16.1	7.0 ± 4.6	6.0	2.2–13.6
Medium	2.3 ± 3.4	0.9	0–6.8	3.1 ± 4.0	1.5	0–8.8	2.7 ± 3.7	1.1	0–7.8
Small	20.5 ± 5.3	21.6	14.0–26.9	21.2 ± 6.1	21.0	13.7–28.8	20.8 ± 5.7	20.8	13.9–27.8
Lipoprotein particle size (nm)									
VLDL	52.3 ± 13.2	49.1	41.1–65.9	55.2 ± 16.7	50.5	42.0–73.9	53.6 ± 15.0	49.7	41.5–69.4
LDL	20.4 ± 0.8	20.3	19.5–21.5	21.0 ± 0.9	21.0	19.9–22.3	20.7 ± 0.9	20.6	19.6–21.1
HDL	8.7 ± 0.4	8.7	8.3–9.3	9.0 ± 0.4	9.0	8.5–9.6	8.9 ± 0.4	8.8	8.4–9.5
Lipids (mg/dL)									
TG	157 ± 135	123	57–281	134 ± 104	106	54–237	146 ± 122	115	55–261
VLDL	119 ± 34	84	20–241	91 ± 103	62	13–190	106 ± 122	75	16–220
HDL-C	40 ± 14	38	25–57	54 ± 18	52	34–79	46 ± 17	43	27–70

Note: HDL-C, high-density lipoprotein cholesterol; HDLp, high-density lipoprotein particles; IDL, intermediate-density lipoprotein; LDLp, low-density lipoprotein particles; TG, triglycerides; VLDLp, very-low-density lipoprotein particles.

Source: LipoScience's NMR LipoProfile® data.

TABLE 12.4
Analytical Imprecision in the Quantification of Various Lipoprotein Particles by NMR

NMR Lipoprotein Parameter	Typical Range of Particle Diameters (nm)	NMR Particle Size Measurement Imprecision		
		Mean (nm)	SD (nm)	CV (%)
VLDL				
VLDL particles, total	27 to >60	96.4	4.1	4.3
Large VLDL or chylomicrons	>60	4.6	0.2	4.6
Medium VLDL	35–60	30.2	1.8	5.9
Small VLDL	27–35	61.6	4.6	7.5
LDL				
LDL particles, total	18 to 27	1404	51.5	3.7
IDL	23–27	39	10.7	27.4
Large LDL	21–23	490	33.5	6.8
Small LDL, total	18–21	876	79.2	9.1
Medium small LDL	19–21	199	17.2	8.6
Very small LDL	18–19	676	63.3	9.4
HDL				
HDL particles, total	7 to 13	29.3	0.4	1.5
Large HDL	9–13	9.1	0.4	4.6
Medium HDL	8–9	2.8	0.8	27.5
Small HDL	7–8	17.4	0.7	4.1
Particle sizes (nm)				
VLDL		53.7	1.06	2.0
LDL		21.1	0.13	0.6
HDL		9.0	0.04	0.4
Calculated lipids (mg/dL)				
Total triglycerides		147	1.6	1.1
VLDL triglycerides		107	1.6	1.4
HDL-cholesterol		48	0.5	1.1

Note: CV, coefficient of variation; HDL, high-density lipoproteins; IDL, intermediate-density lipoproteins; LDL, low-density lipoproteins; NMR, nuclear magnetic resonance; SD, standard deviation; VLDL, very low-density lipoproteins.

Source: Sniderman, A.D. et al., *Lancet*, 361, 777–780, 2003. With permission.

3. Quantification of small, dense vs. large, fluffy particle fractions of HDL-C, LDL-C, and VLDL-C particles may enhance the prognostic value of the lipid panel.

4. The standard lipid screening profile in conjunction with an assessment of other cardiovascular risk factors allows a calculation of a patient's long-term risk of CVD morbidity and mortality based on data from the Framingham study.

5. A series of studies, including the Cardiovascular Health Study, Women's Health Study, PLAC-1, VA-HIT, and Framingham Offspring Study, have shown that LDL-P by proton NMR is a better predictor of future CVD than LDL-C.

6. In children and young adults from Bogalsusa, Louisiana, VAP® lipid data did not help to predict abnormalities found on carotid intima–media thickness (CIMT) testing.

7. The Berkely lipid analysis program is based on VAP® test results.

TABLE 12.5
Therapeutic Target Values for Lipoproteins Measured by VAP®

Lipoprotein	Target Value, mg/dL (mmol/L)	Notes
HDL$_2$-C	>10 (0.26)	Large, buoyant particles
HDL$_3$-C	>30 (0.78)	Small, dense particles
Total HDL-C		
M	>40 (1.04)	
F	50 (1.30)	
Total non-HDL-C when CVD risk		
Low	<160 (4.14)	
High	<130 (3.37)	
Very high	<100 (2.59)	
Total non-HDL-C/HDL-C ratio	<4	
VLDL$_{1,2}$-C	<20 (0.52)	Large, buoyant particles
VLDL$_3$-C	<10 (0.26)	Small, remnant particles
Total VLDL-C	<30 (0.78)	
Lp(a)-C	<10 (0.26)	
IDL-C	<20 (0.52)	
LDL-R-C	<100 (2.59)	
Total LDL-C	<130 (3.37)	[Lp(a)-C] + [IDL-C] + [LDL-R-C]
Atherogenicity and buoyancy of LDL-C subclasses[a]		
LDL$_1$	Least buoyant; most atherogenic (B-position)	
LDL$_{2, 3, and 4}$	Intermediate (A/B position)	
LDL$_5$	Most buoyant; least atherogenic (A-position)	

[a] LDL-C subclasses are reported qualitatively as an Rf value (i.e., the relative position of the LDL-C subclass compared to the major LDL-C peak in the density gradient) within a spectrum of LDL-C subclass patterns ranging from A (least atherogenic) to B (most atherogenic)

Note: CVD, cardiovascular disease; HDL-C, high-density lipoprotein cholesterol; IDL-C, intermediate-density lipoprotein cholesterol; LDL-R-C, real (measured, not calculated) low-density lipoprotein cholesterol; Lp(a)-C, lipoprotein little a cholesterol; VAP, Vertical Auto Profile; VLDL-C, very-low-density lipoprotein cholesterol.

Suggested Reading

NMR Lipoprotein Profile

Beisiegel U, St Clair RW. An emerging understanding of the interactions of plasma lipoproteins and the arterial wall that leads to the development of atherosclerosis. *Curr Opin Lipidol* 1996; 7:265–8.

Blake GJ, Otvos JD, Rifai N, Ridker PM. Low-density lipoprotein particle concentration and size as determined by nuclear magnetic resonance (NMR) spectroscopy as predictors of cardiovascular disease in women. *Circulation* 2002; 106:1930–37.

Freedman DS, Otvos JD, Jeyarajah EJ, Barboriak JJ, Anderson AJ, Walker JA. Relation of lipoprotein subclasses as measured by proton nuclear magnetic resonance (NMR) spectroscopy to coronary artery disease. *Atheroscler Thromb Vasc Biol* 1998; 18:1046–53.

Kral BG, Becker LC, Yook RM, Blumenthal RS, Kwitterovich PO, Otvos JD, Becker DM. Racial differences in low density lipoprotein particle size in families at high risk for premature coronary heart disease. *Ethn Dis* 2001; 11:325–37.

Kuller LH, Arnold A, Tracy R, Otvos JD, Burke G, Pstay B, Siscovick D, Freedman DS, Kronmal R. Nuclear magnetic resonance (NMR) spectroscopy of lipoproteins and risk of coronary heart disease in the Cardiovascular Health Study. *Arterioscler Thromb Vasc Biol* 2002; 22:1175–80.

Lamarche B, Tchernof A, Moorjani S, Cantin B, Dagenais GR, Lupien PJ, Despres JP. Small, dense low-density lipoprotein particles as a predictor of the risk of ischemic heart disease in men. *Circulation* 1997; 95:69–75.

Mackey RH, Kuller LH, Sutton-Tyrrel K, Evans RW, Holubkov R, Matthews KA. Lipoprotein subclasses and coronary artery calcium in postmenopausal women from the Healthy Women Study. *Am J Cardiol* 2002; 90(Suppl):71–76i.

Otvos JD. Measurement of triglyceride-rich lipoproteins by nuclear magnetic resonance (NMR) spectroscopy. *Clin Cardiol* 1999; 22(6 Suppl):II21–7.

Otvos, JD, Jeyarajah EJ, Bennett DW. Krauss RM. Development of a proton NMR spectroscopic method for determining plasma lipoprotein concentrations and subspecies distribution from a single, rapid measurement. *Clin Chem* 1992; 38:1632–38.

Otvos JD, Jeyarajah EJ, Cromwell WC. Measurement issues related to lipoprotein heterogeneity. *Am J Cardiol* 2002; 90(Suppl):22–29i.

Packard CJ. Understanding coronary heart disease as a consequence of defective regulation of Apolipoprotein B metabolism. *Curr Opin Lipidol* 1999; 10:237–44.

Rosenson RS, Otvos JD, Freedman DS. Relations of lipoprotein subclass levels and low-density lipoprotein size to progression of coronary artery disease in the pravastatin limitation of atherosclerosis in the coronary arteries (PLAC-I Trial). *Am J Cardiol* 2002; 90:89–94.

Sniderman AD, Furberg CD, Keech A, van Lennep JER, Frohlich J, Jungner I, Walldius G. Apolipoproteins versus lipids as indices of coronary risk and as targets for statin treatment. *Lancet* 2003; 361:777–80.

Soedamah-Muthu SS, Chang Y-F, Otvos J, Evans RW, Orchard TJ. Lipoprotein subclass measurements by nuclear magnetic resonance (NMR) spectroscopy improve the prediction of coronary artery disease in type 1 diabetes: a prospective report from the Pittsburgh Epidemiology of Diabetes Complications Study. *Diabetologia* 2003; 46:674–82.

Weissberg PL, Rudd, JH. Atherosclerotic biology and epidemiology of disease. In: *Textbook of Cardiovascular Medicine*, 2nd ed., Topol, EJ (Ed). 2002. Lippincott Williams & Wilkins, p. 6.

VAP®

Garber DW, Kulkarni KR, Anantharamaiah GM. A sensitive and convenient method for lipoprotein profile analysis of individual mouse plasma samples. *J Lipid Res* 1994; 41:1020–26.

Kulkarni KR, Garber DW, Marcovina SM, Segrest JP. Quantification of cholesterol in all lipoprotein classes by the VAP-II method. *J Lipid Res* 1994; 35:159–68.

Kulkarni KR, Garber DW, Jones MK, Segrest JP. Identification and cholesterol quantification of low density lipoprotein subclasses in young adults by VAP-II methodology. *J Lipid Res* 1995; 36:2291–302.

Kulkarni KR, Marcovina SM, Krauss RM, Garber DW, Glasscock AM, Segrest JP. Quantification of HDL2 and HDL3 cholesterol by the Vertical Auto Profile-II (VAP-II) methodology. *J Lipid Res* 1997; 38:2353–64.

Tzou WS, Douglas PS, Srinivasan SR, Chen W, Berenson G, Stein JH. Advanced lipoprotein testing does not improve identification of subclinical atherosclerosis in young adults: the Bogalusa Heart Study. *Ann Intern Med* 2005; 142:742–50.

12.4.8   Apolipoprotein A (ApoA) and ApoA-I/ApoB Ratio Testing

Possible Indications for Test

Measurement of apolipoprotein A, along with apoliproteins B (ApoB), A-I (ApoA-I), A-IV (ApoA-IV), and triglycerides (TG), may be indicated:

- When the [HDL-C] is extraordinarily low (<25 mg/dL) in the presence or absence of atherosclerotic cardiovascular disease (ASCVD)
- In the diagnosis of Tangier disease
- For evaluation of the risk of ASCVD in nondiabetes (non-DM) and diabetes (T1DM and T2DM) patients

Alternatives to Testing

- Standard assay for HDL (see Test 4.2.6)

Procedure

1. Instruct patient not to consume any alcohol for 24 hours before testing.
2. Obtain a >12-hour fasting blood sample for ApoA, ApoB, ApoA-I, ApoA-IV, and TG testing.
3. Obtain a nonfasting or random blood sample if only the ApoA-I/ApoB ratio is to be calculated.
4. Calculate the ApoA-I/ApoB ratio.

Interpretation

1. Reference intervals for [ApoA-I] are:
   - Males, 88 to 180 mg/dL
   - Females, 98 to 210 mg/dL
2. Patients with an ApoA-I/ApoB ratio of >1.6 are considered to be at lower risk for ASCVD.
3. A low ApoA-I/ApoB ratio (<1.0) is an efficient predictor of ASCVD.
4. In T2DM patients, increased ApoA-IV levels are found, mainly related to hypertriglyceridemia and, to a lesser extent, lower [HDL-C].
5. Tangier disease patients have less than 1% of the ApoA-I found in reference plasma.

Notes

1. Studies on human ApoA-I (Milano) carriers have found that they are protected from ASCVD in spite of their markedly reduced [HDL-C].
2. In clinical studies, total [ApoA] is commonly measured, whereas in research studies, [ApoA-I] and [ApoA-IV] are usually measured.
3. In the Bogalusa Heart Study, low [ApoA-I] and a significantly higher ApoB/ApoA-I ratio (mean 0.62 ± 0.22) in children were positively related to the parental incidence of myocardial infarction (Srinivasan et al., 1995).
4. A raised ApoB/ApoA-I ratio (odds ratio 3.25 for top vs. lowest quintile; population attributable risks 49.2% for top four quintiles vs. lowest quintile) is significantly related to acute myocardial infarction ($p < 0.0001$) (Yusuf et al., 2004).
5. In a case-control cohort of apparently healthy persons not receiving lipid-lowering therapy and without overt DM, the ApoB/ApoA-I ratio was more closely associated with future ASCVD events than was the TC-HDL-C ratio. The two measures appeared to be equivalent in their ability to discriminate between persons with and those without CVD events (van der Steeg et al., 2007).
6. ApoA-IV plays a role in TG-rich lipoprotein metabolism by facilitating reverse cholesterol transport and cholesteryl ester transfer protein (CETP) activity.
7. ApoA-IV is genetically polymorphic in humans and composed of two major isoforms (ApoA-IV$_1$ and ApoA-IV$_2$).
8. The molecular variant of ApoA-I—ApoA-I (Milano) or ApoA-I (M)—has a Cys for Arg substitution. ApoA-I (M) forms dimers with a prolonged half-life, compared to monomers, and a more effective cholesterol removing function.

Suggested Reading

Chiesa G, Sirtori CR. Recombinant apolipoprotein A-I (Milano): a novel agent for the induction of regression of atherosclerotic plaques. *Ann Med* 2003; 35:267–73.

Contois J, McNamara JR, Lammi-Keefe C. Reference intervals for plasma ApoA-I determined with standardized commercial immunoturbidimetric assay: results from the Framingham Offspring Study. *Clin Chem* 1996; 42:507–14.

Estonius M, Kallner A. How do conventional markers of lipid disorders compare with apolipoproteins? *Scand J Clin Lab Invest* 2005; 65:33–44.

Srinivasan SR, Berenson GS. Serum apolipoproteins A-I and B as markers of coronary artery disease risk in early life: the Bogalusa Heart Study. *Clin Chem* 1995; 41:159–64.

van der Steeg WA, Boekholdt SM, Stein EA, El-Harchaoui K, Stroes ESG, Sandhu MS, Wareham NJ, Jukema JW, Luben R, Zwinderman AH, Kastelein JJP, Khaw K-T. Role of the apolipoprotein B–apolipoprotein A-I ratio in cardiovascular risk assessment: a case–control analysis in EPIC-Norfolk. *Ann Intern Med* 2007; 146:640–8.

Verges B. Apolipoprotein A-IV in diabetes mellitus. *Diabete Metab* 1995; 21:99–105.

Yusuf S, Hawken S, Ounpuu S, Dans T, Avezum A, Lanas F, McQueen M, Budaj A, Pais P, Varigos J, Lisheng L. Effect of potentially modifiable risk factors associated with myocardial infarction in 52 countries (the INTERHEART study): case-control study. *Lancet* 2004; 364(9438):937–52.

12.5 Tests for Hyperglycemic Complications of Diabetes Mellitus

12.5.1 Heart Rate Variability (HRV) Tests for Diabetic Autonomic Neuropathy (DAN)

Possible Indications for Testing

Use of HRV tests* may be indicated in:

- Any diabetes mellitus (DM) patient as a baseline and for follow-up assessment of DAN
- Patients with uncontrolled or "brittle" DM
- Patients with hypoglycemic unawareness and long-standing poor blood glucose control who may have catecholamine hyporesponsiveness to hypoglycemia or pituitary or adrenal insufficiency and are about to initiate an intensive insulin therapy program
- The presence of autonomic neuropathic symptoms including unexplained resting tachycardia, orthostatic symptoms or hypotension, constipation, gastroparesis, erectile dysfunction, gustatory sweating, night blindness, or urinary bladder dysfunction with detrusor reflex decompensation (see Q5.1 in Appendix 1)
- Assessment of degree of parasympathetic vs. sympathetic autonomic nervous system (ANS) dysfunction
- Assessment of response to therapy in patients with overt cardiovascular autonomic neuropathy (CAN), whether or not they are diabetic, as manifest by resting tachycardia (>100 bpm) or orthostasis (a fall in systolic blood pressure > 20 mmHg upon standing) without an appropriate heart rate response
- Planning an exercise program for individuals with DM who are about to embark on a moderate- to high-intensity exercise program, especially those at high risk for underlying cardiovascular disease (Zinman et al., 2004)

---

* HRV tests in the assessment of autonomic nervous system function include: (1) heart rate (R–R interval) variation with deep breathing, (2) heart rate response to Valsalva maneuver, and (3) immediate heart rate response to standing (30/15 ratio).

Contraindications to Testing

- Patients with a working pacemaker
- Patients who have eaten a large meal within the previous 30 to 60 minutes
- Patients afflicted with significant arrhythmias such as atrial fibrillation if a peak heart rate on expiration (E) to peak heart rate on inspiration (I) ratio (E/I ratio) is to be calculated
- Patients with proliferative retinopathy of DM or recent retinal hemorrhage if Valsalva is to be performed
- Patients with advanced pulmonary disease (e.g., emphysema)

Alternatives to Testing

- Refer to Section 5.3.1, Semmes–Weinstein Monofilament (S-WMF) Semiquantitative Test for Neurosensory Deficits in Diabetes Mellitus (DM) Patients; Section 5.3.2, Vibratory Sensation Testing in the Screening and Diagnosis of Peripheral Neuropathy in DM Patients; Section 5.3.3, Calculation of Gastric Emptying Rate by Conventional Radioscintigraphic Measurement of the Clearance of ^{99m}Tc-Albumin from the Stomach in Diabetes Mellitus Patients; and Section 5.3.4, Assessment of Erectile Function (EF) in Male DM Patients with Possible Neuropathy, Vasculopathy, and/or Hypogonadism.

Patient Preparation for HRV Testing

1. Initiate testing in a fasting patient or one who has had no more than a light meal at least an hour or more prior to testing.
2. Have patient empty bladder and bowels as appropriate in anticipation of performance of the Valsalva maneuver before testing.
3. Have the patient remove any metal objects from across the chest, particularly underwire brassieres.
4. Seat the patient in a comfortable chair with sufficiently firm back support. The chair should be stationary and not swivel, tilt, rock, or have wheels or casters.
5. Begin three-lead electrocardiogram monitoring, applying leads in a V-pattern to both sides of upper chest and to the abdomen with the patient in the sitting position.
6. Utilize a blood pressure cuff linked to a computer with a program (e.g., Ansar) for measuring both systolic (SBP) and diastolic (DBP) blood pressure at predefined intervals (typically, every 3 to 4 minutes).
7. Instruct the patient to remain as still as possible with feet flat on the floor. The patient should breathe freely at a relaxed, normal pace for 5 minutes and should not talk but should be instructed to indicate to the technician any discomfort or stress immediately. This lead-in period allows the patient to relax and let the respiratory system come to baseline.

General Notes on HRV Testing

1. Knowledge of early autonomic dysfunction can encourage patients and physicians to improve the metabolic control of DM and to use therapies, such as ACE inhibitors and beta blockers, which are proven to ameliorate CAN.
2. Based on metaanalyses, autonomic neuropathy patients have double the relative risk of silent myocardial ischemia, cardiac dysrhythmias, and sudden death regardless of duration of DM (Vinik et al., 2003).
3. There is increased 5-year mortality in DM subjects with autonomic neuropathy (27%) compared to those without (5%).
4. The presence of autonomic neuropathy may limit an individual's exercise capacity and increase the risk of an adverse cardiovascular event during exercise.
5. Men with the short form [Glu(9)/Glu(9)] of the three-amino-acid deletion (12Glu9) polymorphism in the alpha(2B)-adrenergic receptor gene have general depression of autonomic tone with vagal

ANS Test Results			
ANX 3.0 Copyright @2005 **ANSAR**	Patient: Normal, Patient Gender: Female Medications: Medical History: Number of ectopic beats: 0	Test Date: 4/2/2009 DOB: 1/28/1965 Age: 44	Physician: ANSAR Height: Weight (lbs): 145

Initial Baseline (Resting)	Value	Interpretation
Mean Heart Rate (HR) (expected, 60 to 100):	65	Normal
Range (max-min) Heart Rate (expected, 10 to 50):	13	Normal
Sympathetic Modulation LFA (expected, 0.5 to 10):	0.87	Normal
Parasympathetic Modulation (RFA) (expected, 0.5 to 10):	1.92	Normal
Sympathovagal Balance (LFA/RFA) (expected, 0.4 to 3.0):	0.45	Normal
Blood Pressure (expected, Systolic: 90 to 130; Diastolic: less than 85):	102/59	Normal

Deep Breathing	Value	Interpretation
Parasympathetic Response (RFA): (expected, 55.84 to 197.99; values are age and baseline adjusted)	39.62	Borderline low
Respiration Frequency (FRF) (expected, 0.09 to 0.15 Hz):	0.12	Normal
Range (max-min) Heart Rate (expected, 15 to 50 bpm):	28	Normal
E/I Ratio (expected, 1.2 to 1.6):	1.29	Normal
Blood Pressure (expected, decrease from initial baseline):	112/64	Normal

Valsalva	Value	Interpretation
Sympathetic Response (LFA): (expected, ×59.83 to ×212.12; values are age and baseline adjusted)	×45.76	Borderline low
Parasympathetic Response (expected, less than 1.67× [RFa of initial baseline]):	4.11	High*
Range (max-min) Heart Rate (expected, 15 to 50 bpm):	28	Normal
Valsalva Ratio (expected, 1.2 to 1.6):	1.31	Normal
Blood Pressure (expected, increase from initial baseline):	112/63	Normal

Stand	Value	Interpretation
Mean Heart Rate (expected, >10% but <30 beats increase from initial baseline):	72	Normal
Range (max-min) Heart Rate (expected, 15 to 50 bpm):	23	Normal
Sympathetic Response (LFA) (expected, [1.2 to 5] × [LFa of initial baseline]):	0.59	Low
Parasympathetic Response (RFA) (expected, decrease from initial baseline):	0.71	Normal
30:15 Ratio (expected, 1.15 to 1.5):	1.27	Normal
Blood Pressure (expected, increase from initial baseline):	105/71	Normal

Summary Diagnostic Implications:

There is an unexpected rise in BP from baseline to deep breathing. A high parasympathetic response to Valsalva challenge indicates possible paradoxic parasympathetic syndrome (PPS). Low sympathetic response during stand indicates sympathetic withdrawal (SW) which suggests orthostasis.

SW along with an increase in systolic BP (0 to 20 mmHg) from initial basline indicate a possible orthostatic intolerance.

* Paradoxic parasympathetic syndrome: an unexpected increase in parasympathetic activity during a sympathetic challenge.
NOTE: Results must be considered in the context the patient's medical history, current medications, diagnoses, symptoms, etc.

Physician's Signature: _____   Date: _____

Figure 12.3
Sample autonomic nervous system activity report (courtesy of Ansar Group, Inc.). Note reference intervals for autonomic nervous system (ANS) tests.

activity being especially impaired over time. In addition, this ANS disturbance has been associated with a reduced basal metabolic rate and is accentuated by central obesity (Sivenius et al., 2003).

6. HRV testing using the Ansar system evaluates noninvasively, simultaneously, and quantitatively both the parasympathetic and sympathetic limbs of the ANS (Figure 12.3).

7. The ability to interpret serial results of HRV testing requires accurate, precise, and reproducible procedures that use established physiologic maneuvers. Noninvasive autonomic function tests help to make a sensitive determination of early-onset CAN in 15 minutes.

8. Commercial devices such as the Ansar ANX 3.0, a portable system requiring a workspace of no more than 15 square feet, are available for HRV testing.

Heart Rate Variability with Deep Breathing and Measurement of Variation in R–R Interval and High-Frequency HRV Spectral Response (RFa)

Procedure

1. Instruct and guide the patient in normal, shallow breathing 5 seconds in and 5 seconds out slowly and smoothly without holding breaths for a period of 1 minute.
2. After the above, initiate six cycles of accentuated, deep breathing. Use a 5-second metronome to guide the patient in the performance of slow rhythmic breathing or use hand signals and calm voice directives to guide the patient's breathing.
3. Use the Ansar computerized BP and HR monitoring program to obtain and calculate values for the following:
   - Longest R–R interval for each E/I cycle and the mean longest R–R interval for the six cycles
   - Shortest R–R interval for each E/I cycle and the mean shortest R–R interval for the six cycles of deep breathing
   - R–R variation calculated based on both time interval and beats per minute (bpm):

   $$(\text{R–R variation})_{\text{interval}} = (\text{R–R interval})_{\text{mean longest}} - (\text{R–R interval})_{\text{mean shortest}}$$

   $$(\text{R–R variation})_{\text{bpm}} = \text{HR}_{\text{mean fastest}} - \text{HR}_{\text{mean slowest}}$$

   - E/I ratio from peak heart rate on expiration divided by peak heart rate on inspiration
   - Low and high respiratory frequency domains or frequency areas based on power spectral analyses of a series of successive R–R intervals and the corresponding ventilation pattern for each one, obtained by recording and analyzing short (i.e., 4-second or 15-second) R–R sequences
   - Shallow and deep breathing low (0.04–0.15 Hz) and high (>0.15 Hz) respiratory-related frequency areas (LFa and RFa, respectively)
   - Sympathovagal balance ratio (SBR) = LFa/RFa

Interpretation

1. E/I ratio reference intervals are shown in "Cardiovascular Tests for Diabetic Autonomic Neuropathy (DAN) in Adults < Age 65" (Table 12.6).
2. In assessment of the parasympathetic response to deep breathing, an E/I ratio of:
   - <1.2 is consistent with mild, acute ANS dysfunction (e.g., DAN).
   - 1.2 to 1.6 is consistent with unimpaired ANS function.
3. The Ansar test results report (Figure 12.3) provides power spectral analysis reference intervals for RFa, a marker of vagal or parasympathetic activity. The baseline reference interval for RFa is 0.5 to 10. The response of RFa to deep breathing, as well as Valsalva, is age and baseline adjusted.
4. If the high respiratory-related frequency area (RFa) response to deep breathing is:
   - Elevated (above the *upper* limit of the reference interval), pulmonary problems may be present and interfere with test interpretation and autonomic dysfunction may or may not be present.
   - Low (below the *lower* limit of the reference interval), mild, acute autonomic dysfunction is suggested.
   - <0.1, then the patient may be at high-risk for sudden death.
   - >10.0, along with LFa >10.0, then cardiac instability is suggested.
5. Using heart rate response to deep breathing, the SBR can be used to estimate parasympathetic and sympathetic tone and guide intervention with pharmacologic agents to reverse or prevent orthostasis secondary to DAN.
6. If SBR is:
   - <0.4, then parasympathetic excess is suggested.
   - 0.4–3.0, no imbalance in sympathetic or parasympathetic function is suggested.
   - >3.0, then sympathetic excess is suggested.
7. In case of ectopic heart beats, refer to "Interpretation of Heart Rate Variability (HRV) Test Results When Cardiac Ectopy Occurs" (Table 12.7) for interpretation of HRV deep breathing parameters.

TABLE 12.6
Cardiovascular Tests for Diabetic Autonomic
Neuropathy (DAN) in Adults <Age 65

Test	Interpretation		
	Expected	Borderline	Abnormal
Valsalva ratios			
Longest to shortest R–R interval	≥1.21	1.11–1.21	≤1.10
Slowest to fastest HR	≥1.50	—	<1.50
Deep breathing			
HR difference (bpm)	≥15	11–14	≤10
R–R interval variation (msec)	>30	20–30	<20
E/I ratio	1.2–1.6	1.1–1.2	<1.1 or >1.6
HR response (30/15 ratio)			
$R-R_{30}/R-R_{15}$	1.15–1.5	1.04–1.15	<1.04
SBP response after standing			
SBP difference	≤10	11–29	≥30
DBP response with handgrip			
DBP difference (mmHg)	≥16	11–15	≤10

Note: DBP, diastolic blood pressure; E/I, expiration/inspiration ratio; HR, heart rate; R–R, mean distance between R-waves by electrocardiography; $R-R_{30}/R-R_{15}$, R–R interval after 30 beats after standing/R–R interval after 15 beats after standing; SBP, systolic blood pressure;

Source: Ewing, D.J. et al., *Diabetes Care*, 8, 491–498, 1985. With permission.

TABLE 12.7
Interpretation of Heart Rate Variability (HRV)
Test Results When Cardiac Ectopy Occurs

Ectopy Occurs Only During Challenges	Interpretation
At rest (baseline)	Baseline for comparison is obscured; use an average of the other two baselines if possible.
With deep breathing (baseline)	A parasympathetic trigger to the arrhythmia may be present.
During Valsalva, stand baseline, or initial stand challenge	A sympathetic trigger to the arrhythmia may be present.
Only at the very start of testing	Possible "white coat" hypertensive syndrome may be occurring.
2 to 3 minutes into stand challenge	POTS may be present.
SBR (when ectopy occurs throughout testing only the resting SBR ratio is valid)	
<0.4 (low)	A parasympathetic component to the arrhythmia may be present.
0.4–3.0 (normal)	There is no neurogenic component to the ectopy.
>3.0 (high)	A sympathetic component to the arrhythmia may be present.

Note: SBR, sympathovagal balance ratio = LFa/RFa; LFa/RFa ratio, shallow breathing respiratory-related frequency area/deep breathing respiratory-related frequency area; POTS, postural orthostatic tachycardia syndrome.

Notes on Heart Rate Variability with Deep Breathing

1. If patient has significant arrhythmias such as atrial fibrillation, the E/I ratio will not be valid; however, other Ansar test results (e.g., SBR) will be valid in such patients and provide clinically relevant information. Note that cardiac ectopy and use of ANS-altering drugs will alter test results but are not contraindications to testing as the effects of these phenomena can be taken into account and permit valid interpretation of data.

2. The E/I ratio is a more indirect measure of parasympathetic activity than the spectral domain HRV measure of RFa.
3. Unusually high variation in the E/I ratio may be seen in tachypnea or primary pulmonary disorders and has no particular significance in terms of neuropathy or lack of it.
4. Based on dark-adapted pupil size, capillary latency time, and RR variation during beta-adrenergic blockade, the ANS may be impaired in 24 months after the diagnosis of poorly controlled DM.
5. The LFa is influenced by sympathetic nervous system activity, whereas the RFa is generally considered a marker of vagal or parasympathetic activity.
6. It is not unusual for the low-frequency HRV component (LFa), associated with BP regulation, and the high-frequency component (RFa), reliably determined only by characterizing range and median value of respiratory rate and tidal depth, to be totally superimposed on each other and, consequently, be nondiagnostic. It has been shown that there is a need for ventilatory control when assessing short-term resting HRV in children (Williams and Lopes, 2002).

Heart Rate Response to Valsalva Maneuver with Calculation of the Valsalva Ratio

Procedure

1. Use a computerized BP and HR monitoring program (e.g., Ansar) to obtain values for heart rate response during a Valsalva maneuver.
2. Enlist the assistance of an experienced instructor to instruct and guide the patient in the performance of a series of short Valsalva maneuvers, with short periods of normal breathing in between, as follows:
   - The first Valsalva should be held for 15 seconds (this allows for the computation of the Valsalva ratio).
   - After the first Valsalva, instruct the patient to relax for 15 seconds and assess the patient's general status.
   - If the patient is recovered (i.e., "feeling fine") after the first Valsalva maneuver, have the patient perform another Valsalva maneuver for 7 or 8 seconds.
   - Let the patient rest for 7 or 8 seconds after the second Valsalva maneuver.
   - Repeat the above two steps until three more Valsalva maneuvers (five total) are performed as tolerated by the patient.
3. If an individual patient has difficulty complying with the Valsalva maneuver or in achieving reproducibility in its performance, allow several practice sessions before committing to formal testing.
4. To obtain optimal results and reproducible data, instruct the patient in the performance of a Valsalva maneuver sufficient to generate 400 torr for 15 seconds using a calibrated aneroid manometer.
5. The computer program will provide values for:

$$(\text{Valsalva ratio})_{\text{Interval}} = (\text{R–R interval})_{\text{longest}}/(\text{R–R interval})_{\text{shortest}}$$

$$(\text{Valsalva ratio})_{\text{bpm}} = (\text{HR})_{\text{slowest}}/(\text{HR})_{\text{fastest}}$$

Interpretation

1. Valsalva ratio reference intervals are shown in "Cardiovascular Tests for Diabetic Autonomic Neuropathy (DAN) in Adults <Age 65" (Table 12.6).
2. In the assessment of the sympathetic response to Valsalva maneuvers, a Valsalva ratio of:
   - <1.2 is suggestive of moderate acute ANS dysfunction.
   - 1.2 to 1.6 is consistent with a functional ANS.
   - >1.6 suggests the need for evaluating the patient for hypertension or preclinical hypertension.
   - >2 may not correlate with any abnormality of blood pressure.
3. Age-dependent changes in Valsalva ratio reference intervals may exist and require further study.
4. In cases of ectopic heart beats, see Table 12.7 for interpretation of HRV indices with the Valsalva maneuver.

Notes on Heart Rate Response to Valsalva Maneuver

1. Use of the Ansar system of measurement (ANX 3.0) of HRV parameters requires only a simple Valsalva maneuver in which the patient is instructed to hold his breath and bear down. This may result in variability of HRV response, dependent on the effort made by the patient in performance of the maneuver. Note that this variability is not of critical importance because even submaximal effort by the patient in performing Valsalva maneuvers is sufficient to stimulate a sympathetic response detectable by the Ansar system.
2. The Valsalva stress test evaluates the sympathetic ANS.
3. Cardiac denervation syndrome is a severe manifestation of cardiovascular autonomic neuropathy and is manifested by a fixed heart rate (usually, ≥ 80 to 90 bpm) that normally does not change during a Valsalva maneuver, stress, exercise, sleep, or postural change.
4. Enhancement of cardiac vagal activity, as marked by HRV, might beneficially influence the prognosis in patients with heart disease by decreasing myocardial oxygen demand, reducing sympathetic activity, and decreasing susceptibility of the ventricular myocardium to lethal arrhythmia.
5. In studies of HRV and left ventricular ejection time, vagal effects became progressively stronger, with increasing sympathetic background activity, demonstrating the predominance of parasympathetic control of human heart rate with the implication that changes in cardiac activity resulting from changes in sympathetic control cannot be interpreted accurately unless concurrent vagal activity is assessed (Uijtdehaage and Thayer, 2000).

Notes on Paradoxic Parasympathetic Syndrome (PPS)

1. PPS is a pernicious, dynamic ANS imbalance in which the parasympathetics respond in tandem with the sympathetics following a sympathetic (e.g., Valsalva stress) challenge; hence, there is the paradox of a parasympathetic response to a sympathetic stimulus. The result is to cause the sympathetics to work harder to respond to the instantaneous effects of a stress event or metabolic disturbance. If PPS is present, detection of sympathetic withdrawal on ANS testing may be limited.
2. PPS is an abnormal parasympathetic response to a sympathetic stimulus (the paradox) and was found in about 70% of a 10,000-patient study database (Ansar data on file).
3. The ANS imbalance represented by PPS appropriately prompts interventional therapy.
4. To regulate the ANS and reestablish a proper dynamic balance of neurohumoral forces, it is appropriate to introduce a sympatholytic beta blocker without alpha blocking effects (e.g., carvediol) or an alpha-adrenergic sympathomimetic (e.g., midodrine) as safe and physiologic interventions to lower heart rate and BP.
5. Other features of PPS include:
   - Presentation as excess sympathetic activity (i.e., elevated BP and/or pulse)
   - Cardiovascular destabilization occurring as a result of attempted pharmacologic reduction in sympathetic excess as a consequence of less inhibition of the parasympathetics which forces the remaining sympathetics to over-respond to instantaneous changes in metabolic demand
   - Extremely uncontrolled DM in which there is a paradoxical increase in heart rate and BP in response to nonselective beta blockers or renin–angiotensin system-blocking agents, respectively
   - Increased sympathetic activity as well as enhanced parasympathetic activity associated with increased intracranial pressure in patients with acute-phase subarachnoid hemorrhage
6. The following questionnaire may be used to screen for PPS. Three or more positive responses are highly suggestive of PPS:
   - Do you have restless leg syndrome or night-time leg edema?
   - Do you have difficulty falling asleep over intervals of several hours or wake frequently during the night?
   - Are you a premenopausal (35 to 45 years old) female and get hot flashes?
   - Do you frequently get dizzy when you stand up?
   - Do you frequently have cognitive difficulties (memory, function) in the morning?
   - Do you have any stomach or intestinal upset (heartburn, frequent diarrhea or constipation)?
   - Do you frequently awake with a morning migraine or headache?

Immediate Heart Rate Response (HRR) to Standing (30/15 Ratio)

Procedure

1. To ensure that orthostatic hypotension (consistent with sympathetic withdrawal), if present, is fully revealed, have the patient stand from a seated position in an urgent (<5 second) manner without assistance, if possible.
2. If the patient needs help getting up or remaining upright, assistance to do so should be provided. Alternatively, a tilt table may be used to establish upright status of the patient.
3. On arising, let patient stand quietly to see if syncope results.
4. Permit patient to lean on something (e.g., cane, walker, or counter), if needed, to remain upright for up to 5 minutes if possible.
5. If a patient is unable to stand, instruct the patient to sit up from a supine position and hold a proper, seated posture for 5 minutes. Assist the patient in sitting up if necessary. To better evaluate Ansar results in these patients, administer the paradoxic parasympathetic syndrome (PPS) questionnaire (see Notes).
6. When the Ansar computer system is used, the electrocardiogram is marked at the time the patient becomes upright and the computer program provides values for the R–R intervals between the 15th and 30th heartbeats after the patient stands or becomes upright:

$$30/15 \text{ ratio} = (R–R)_{30}/(R–R)_{15}$$

and for low (0.04–0.15 Hz) and high (>0.15 Hz) respiratory-related frequency areas (LFa and RFa, respectively) from short (e.g., 4 second or 15 second) R–R sequences.

Interpretation

1. Table 12.6 shows reference intervals for HRR and SBP differences on standing.
2. In healthy individuals, there is a rapid increase in heart rate on standing from a supine or seated position which peaks at about the 15th beat. This is normally followed by vagal-mediated slowing, most pronounced at about the 30th beat.
3. Normally, the RFa in response to standing is less than baseline.
4. In the assessment of the HRR upon standing, a 30/15 ratio of:
   - <1.04 is overtly abnormal, establishing a diagnosis of advanced autonomic neuropathy with significant sympathetic withdrawal (i.e., orthostasis and chronic autonomic dysfunction).
   - 1.04 to 1.15 is borderline abnormal, suggesting that the syndrome of sympathetic withdrawal may be present.
   - 1.15 to 1.50 is consistent with normal ANS function.
5. To interpret 30/15 ratio results in case of ectopic heartbeats, refer to Table 12.7.

Notes on the HRR to Standing

1. The HRR response to standing evaluates the function of the efferent aspect (vagus nerve) of the ANS baroreceptor reflex.
2. An abnormal decrease in the sympathetic response to upright posture (i.e., sympathetic withdrawal) can be an early indicator of clinically significant orthostasis.
3. The severity of cardiac autonomic neuropathy correlates inversely with the maximal increase in heart rate after exercise ($r = {-}0.68$; $p <0.001$) (Kahn et al., 1986).

Suggested Reading

Boulton AJ, Vinik AI, Arezzo JC, Bril V, Feldman EL, Freeman R, Malik RA, Maser RE, Sosenko JM, Ziegler D, American Diabetes Association. Diabetic neuropathies: a statement by the American Diabetes Association. *Diabetes Care* 2005; 28:956–62.

Cammann H, Michel J. How to avoid misinterpretation of heart rate variability power spectra? *Comput Methods Programs Biomed* 2002; 68:15–23.

Cryer PE. Hypoglycemia-associated autonomic failure in diabetes. *Am J Physiol Endocrinol Metab* 2001; 281:E1115–21.

DCCT Research Group. The effects of intensive diabetes therapy on measures of autonomic nervous system function in the Diabetes Control and Complications Trial (DCCT). *Diabetologia* 1998; 41:416–23.

Ewing DJ, Martyn CN, Young RJ, Clarke BF. The value of cardiovascular autonomic function tests: 10 years experience in diabetes. *Diabetes Care* 1985; 8:491–8.

Ewing DJ, Boland O, Neilson JM, Cho CG, Clarke BF. Autonomic neuropathy, QT interval lengthening, and unexpected deaths in male diabetic patients. *Diabetologia* 1991; 34:182–5.

Freeman R, Saul P, Roberts M, Berger RD, Broadbridge C, Cohen R. Spectral analysis of heart rate in diabetic autonomic neuropathy. *Arch Neurol* 1991; 48:185–90.

Genovely H, Pfeifer MA. RR-variation: the autonomic test of choice in diabetes. *Diabetes Metab Rev* 1988; 4:255–71.

Grundy SM, Benjamin IJ, Burke GL, Chait A, Eckel RH, Howard BV, Mitch W, Smith SC, Sowers JR. Diabetes and cardiovascular disease: a statement for healthcare professionals from the American Heart Association. *Circulation* 1999; 100:1134–46.

Kahn JK, Zola B, Juni JE, Vinik AI. Decreased exercise heart rate and blood pressure response in diabetic subjects with cardiac autonomic neuropathy. *Diabetes Care* 1986; 9:389–94.

Kawahara E, Ikeda S, Miyahara Y, Kohno S. Role of autonomic nervous dysfunction in electrocardiographic abnormalities and cardiac injury in patients with acute subarachnoid hemorrhage. *Circ J* 2003; 67:753–6.

Maser RE, Mitchell BD, Vinik AI, Freeman R. The association between cardiovascular autonomic neuropathy and mortality in individuals with diabetes: a meta-analysis. *Diabetes Care* 2003; 26:1895–1901.

O'Brien IA, McFadden JP, Corrall RJ. The influence of autonomic neuropathy on mortality in insulin-dependent diabetes. *Q J Med* 1991; 79:495–502.

Pfeifer MA, Weinberg CR, Cook DL, Reenan A, Halter JB, Ensinck JW, Porte D. Autonomic neural dysfunction in recently diagnosed diabetic subjects. *Diabetes Care* 1984; 7:447–53.

Risk M, Bril V, Broadbridge C, Cohen A. Heart rate variability measurement in diabetic neuropathy: review of methods. *Diabetes Technol Ther* 2001; 3:63–76.

Routledge HC, Chowdhary S, Townend JN. Heart rate variability: a therapeutic target? *J Clin Pharm Ther* 2002; 27:85–92.

Sivenius K, Niskanen L, Laakso M, Uusitupa M. A deletion in the alpha2B-adrenergic receptor gene and autonomic nervous function in central obesity. *Obes Res* 2003; 11:962–70.

Task Force of the European Society of Cardiology and the North American Society of Pacing and Electrophysiology. Heart rate variability: standards of measurement, physiological interpretation and clinical use. *Circulation* 1996; 93:1043–65.

Uijtdehaage SH, Thayer JF. Accentuated antagonism in the control of human heart rate. *Clin Auton Res* 2000; 10:107–10.

Valensi P, Sachs RN, Harfouche B et al. Predictive value of cardiac autonomic neuropathy in diabetic patients with or without silent myocardial ischemia. *Diabetes Care* 2001; 24:339–43.

Vinik AI, Erbas T. Recognizing and treating diabetic autonomic neuropathy. *Cleve Clin J Med* 2001; 68:928–44.

Vinik AI, Erbas T, Pfeifer MA, Feldman EL, Stevens MJ, Russell JW. Diabetic autonomic neuropathy. In: *Ellenberg & Rifkin Diabetes Mellitus*, 6th ed., Porte D, Sherwin RS, Baron A (Eds). 2003a. McGraw-Hill, pp. 789–804.

Vinik AI, Freeman R, Erbas T. Diabetic autonomic neuropathy. *Semin Neurol* 2003b; 23:365–72.

Vinik AI, Maser RE, Mitchell BD, Freeman R. Diabetic autonomic neuropathy. *Diabetes Care* 2003c; 26:1553–79.

Williams CA, Lopes P. The influence of ventilatory control on heart rate variability in children. *J Sports Sci* 2002; 20:407–15.

Ziegler D. Cardiovascular autonomic neuropathy: clinical manifestations and measurement. *Diabetes Rev* 1999; 7:300–15.

Zinman B, Ruderman N, Campaigne BN, Devlin JT, Schneider SH. Physical activity/exercise in diabetes (position statement). *Diabetes Care* 2004; 27(Suppl 1):S58–62.

12.6   Tests for Determining Adrenal Status

12.6.1   Low-Dose Dexamethasone Suppression Test (DST) in Combination with the Ovine Corticotropin-Releasing Hormone (oCRH) Stimulation Test in the Differential Diagnosis of Cushing's Syndrome (CS)

Possible Indications for Test

The low-dose DST followed by the oCRH stimulation test (DST-CRH) may be indicated to:

- Differentiate CS caused by pituitary adenocorticotroph (ACTH-secreting) adenomas and some bronchial carcinoids from adrenal (cortisol-secreting) tumors and the majority of ectopic ACTH-producing cancers other than bronchial carcinoids in patients with established CS
- Distinguish patients with the appearance of CS (i.e., pseudo-Cushing's syndrome) from those with true CS originating from ACTH-secreting or cortisol-secreting lesion

Procedure

1. Obtain informed consent before performance of the invasive DST-CRH test.
2. Inform patient that an unpleasant episode of flushing may occur in about 15% of patients after injection of oCRH (Acthrel®; Ferring Pharmaceuticals, Inc., Tarrytown, NY) and that hypotension is a possible, but rare, occurrence at the usually recommended doses of oCRH.
3. Before performing this test, discontinue for 2 or more weeks the patient's use of any drugs that might inhibit or alter steroidogenesis.
4. Administer 0.5 mg dexamethasone p.o. every 6 hours for 2 days (8 doses) beginning at noon and ending at 6 a.m. Two hours after the last dose of dexamethasone, obtain two blood samples 15 minutes apart for baseline determination of both [cortisol] and [ACTH] in the pooled sample from both blood collections.
5. When testing in the outpatient setting, obtain a blood sample for measuring the circulating level of dexamethasone to confirm compliance with test agent administration.
6. Weigh the patient (in kg) and calculate the dose of oCRH (1 mcg oCRH per kg body weight).
7. Administer the dose of oCRH as an i.v. bolus over 1 minute at time 0 immediately after blood sampling as above.
8. Obtain blood samples for cortisol and ACTH testing at 15, 30, and 60 minutes after administration of oCRH.

Interpretation

1. An increase in [ACTH] of >50% or an increase in cortisol >20% over baseline at any time point after oCRH administration suggests CS caused by either an ACTH-producing, CRH-responsive, bronchial carcinoid or pituitary adenoma. Lesser increases in ACTH and cortisol occur in normal patients and in patients with pseudo-Cushing's syndrome as per "Interpretation of ACTH and Cortisol Changes During the CRH Stimulation Test" (Table 12.8).
2. Patients harboring adrenal cortisol-secreting tumors or who have iatrogenic CS will fail to experience a significant change in [ACTH] or [cortisol] after oCRH injection.
3. Patients with an ectopic ACTH syndrome caused by a noncarcinoid tumor who may or may not have a high basal [ACTH] and [cortisol] will have no ACTH or cortisol responses to oCRH.
4. After the DST-CRH, a +15-minute post-oCRH serum cortisol of >1.4 mcg/dL (38.6 nmol/L) reliably excludes a pseudo-Cushing's state related to alcoholism, obesity, or stress (100% specificity, sensitivity, and diagnostic accuracy) (Yanovski, 1995).
5. A peak [cortisol] of >20 mcg/dL after administration of oCRH indicates adequate (or excessive) corticotroph and adrenocortical function.

TABLE 12.8
Interpretation of ACTH and Cortisol Changes During the CRH Stimulation Test

Δ Cortisol (%)[a]	Peak Cortisol Concentration (mcg/dL)	Δ ACTH Concentration (%)	Possible Diagnoses
—	<20	—	2° Hypocortisolism
—	>20	—	Euadrenal Cushing's syndrome
>20	—	>50	Pituitary tumor Bronchial carcinoid
n.s.	—	n.s.	Adrenal tumor Ectopic ACTH production Exposure to exogenous glucocorticoids

[a] Δ, peak change in cortisol or ACTH concentration; % = {([H]$_{peak}$ − [H]$_{basal}$)/[H]$_{basal}$} × 100, where [H] = either cortisol or ACTH hormone concentration in basal and peak specimens.

Note: ACTH, adrenocorticotropic hormone; CRH, corticotropin-releasing hormone; —, no interpretive guidelines available; n.s., no significant change from basal concentration.

Notes

1. Ovine CRH is the pituitary corticotroph stimulating agent of choice because it has a longer half-life than human CRH (hCRH).
2. The diagnostic accuracy of the DST-CRH for CS is significantly greater than the accuracy of either test alone ($p < 0.01$).
3. Assess the integrity of the adrenocortical–pituitary corticotroph axis with DST-CRH only after first establishing the diagnosis of glucocorticoid excess based on UFC test results (see Test 6.2.2) or loss of diurnal cortisol secretion (see Test 6.2.1).
4. A 50% increase in ACTH *alone*, independent of the cortisol response, after the administration of oCRH identified 86% of patients with ACTH-dependent CS, whereas a 50% increase in cortisol *alone* identified only 61% of patients with ACTH-dependent CS (Yanovski, 1995).
5. A single cortisol value measured in blood after 2 days of low-dose DST and 15 minutes after oCRH stimulation was 100% specific and 100% sensitive for the diagnosis of CS (Yanovski et al., 1993).
6. In the DST-CRH test, a more than 30% suppression of serum cortisol after low-dose DST and/or more than a 20% increase in cortisol after CRH had significantly higher rates of sensitivity (97%) and specificity (94%) than either the high-dose DST or the CRH test alone in the differential diagnosis of ACTH-dependent Cushing's syndrome (Isidori et al., 2003).

Suggested Reading

Isidori AM, Kaltsas GA, Mohammed S. Discriminatory value of the low-dose dexamethasone suppression test in establishing the diagnosis and differential diagnosis of Cushing's syndrome. *J Clin Endocrinol Metab* 2003; 88:5299–306.

Nieman LK, Oldfield EH, Wesley R. A simplified morning ovine corticotropin-releasing hormone stimulation test for the differential diagnosis of adrenocorticotropin-dependent Cushing's syndrome. *J Clin Endocrinol Metab* 1993; 77:1308–12.

Yanovski JA. The dexamethasone-suppressed corticotropin-releasing hormone test in the differential diagnosis of hypercortisolism. *Endocrinologist* 1995; 5:169–75.

Yanovski JA, Cutler GB, Chrousos GP, Nieman LK. Corticotropin-releasing hormone stimulation following low-dose dexamethasone administration. A new test to distinguish Cushing's syndrome from pseudo-Cushing's states. *JAMA* 1993; 269:2232–38.

12.6.2   Metyrapone Test for Central Hypoadrenalism (CH)

Possible Indications for Test

The metyrapone test may be indicated:

- When both baseline plasma [ACTH] and [cortisol] testing and low-dose ACTH stimulation testing have failed to yield a definitive diagnosis of CH in a potentially adrenal-insufficient patient who has no evidence of cardiac or cerebrovascular disease on cardiovascular evaluation.
- As a simpler, less labor-intensive, and safer alternative to the insulin tolerance test (ITT) in diagnosis of CH and assessment of the hypothalamic–pituitary–adrenal axis (HPAA).

Contraindications to Testing

- Patients with cardiac disease (e.g., arrhythmias, angina, congestive heart failure) and other conditions for whom hypotension may be unacceptably hazardous
- Risk of life-threatening hypotensive shock that may occur upon blockade of cortisol synthesis by metyrapone, a risk that is minimized if the patient is recumbent overnight
- Overt hypoadrenalism with primary adrenal insufficiency associated with adrenal insult, markedly elevated [ACTH], and hyperpigmentation.

Alternatives to Testing

- Low-dose cosyntropin stimulation test (see Test 2.1.2) if baseline plasma [ACTH] and [cortisol] are equivocal
- Clinical diagnosis of cardiovascular disease in a patient who is overtly hypoadrenal based on a history of insult to pituitary, physical findings (e.g., hypotension), and other chronic hypoadrenal findings and symptoms exclusive of hyperpigmentation (see Table 6.2).

Procedure

1. Plan to perform this test in a well-supervised or hospital setting. Monitor blood pressure before the patient arises, and note symptoms of limb paresthesias, dizziness, fainting, disorientation, nausea, and vomiting.
2. Proceed with testing only after an older (>50 years) patient has received evaluation and clearance by a cardiologist.
3. Administer metyrapone tablets, 30 mg/kg orally, at midnight.
4. Obtain blood samples at between 0800 and 0930 hours for measurement of both 11-deoxycortisol and plasma cortisol.

Interpretation

1. A normal [11-deoxycortisol] (≥200 nmol/L) and normal plasma [cortisol] of ≥7.2 mcg/dL (200 nmol/L) indicate a normal HPAA.
2. A low [11-deoxycortisol] (<200 nmol/L) and low plasma [cortisol] of <7.2 mcg/dL (200 nmol/L) indicate CH.
3. A low [11-deoxycortisol] (<200 nmol/L) and normal plasma [cortisol] of ≥7.2 mcg/dL (200 nmol/L) indicate an invalid, nondiagnostic test. Hypoadrenalism is still a possibility in patients with this pattern of response.
4. A normal [11-deoxycortisol] (≥200 nmol/L) and low plasma [cortisol] of <7.2 mcg/dL (200 nmol/L) occur rarely. Patients with this pattern may have partial CH.

Notes

1. Metyrapone is not readily available in the United States but may be obtained on special request (Ciba/Novartis).
2. The overnight metyrapone test and high-dose ACTH stimulation test (see Test 6.1.1) are discordant more than 50% of the time in secondary adrenal insufficiency, prompting use of a low-dose ACTH test (see Test 2.1.2) in these patients.
3. A concordance of 100% between the overnight oral metyrapone, Test 6.1.1, and ITT is seen in patients with primary adrenal insufficiency, with Test 6.1.1 being the preferred one for this disorder.
4. In the most current study, reference cutoff levels for CH were established using the mean less 2 SD of the log-transformed data from 21 normal subjects, with a subnormal response to metyrapone defined as an [11-deoxycortisol] of <200 nmol/L. For a test to be accepted as yielding a valid subnormal response, the simultaneous plasma [cortisol] had to be <7.2 mcg/dL (200 nmol/L) (Fiad et al., 1994).

Suggested Reading

Fiad TM, Kirby JM, Cunningham SK, McKenna TJ. The overnight single-dose metyrapone test is a simple and reliable index of the hypothalamic–pituitary–adrenal axis. *Clin Endocrinol* 1994; 40:603–9 (comment in *Clin Endocrinol* 1994; 41:695).

12.7 Test for Bone Density in the Peripheral Skeleton

12.7.1 Radiographic Absorptiometry (RA) of Hand Bones

Possible Indications for Test

Estimation of peripheral bone density in the hand by RA may be indicated when:

- Large-scale screening for detection of more advanced bone loss is desired.
- Dual-energy x-ray absorptiometry (DXA) is not available or adequate to service large numbers of patients.
- It is necessary to estimate the severity and progression of rheumatoid arthritic disease.

Alternatives to Testing

- Bone mineral density (BMD) by central DXA (spine and hip) (see Test 7.1.1)
- Measurement of bone turnover and resorption markers in serum and plasma in monitoring responses to therapy for osteoporosis (see Test 7.2)

Procedure

1. Obtain a radiograph of the nondominant (usually left) hand using a 50- to 60-kVp, 300- to 400-mM, and 0.5- to 0.8-second exposure along with an aluminum wedge reference device.
2. Use nonscreened industrial x-ray film (i.e., Kodak® XTL-2) with a focus film distance of 105 cm.
3. Obtain two separate hand exposures to ensure that no film-specific defects, such as streaks or scratches, affect the results.
4. When the x-ray has been taken, scan the developed film onsite with a commercial desktop scanner, such as scanners manufactured by CompuMed (5777 W. Century Blvd., Suite 128, Los Angeles, CA 90045; 310-258-5000; osteo@compumed.net); contact CompuMed for information on the acquisition of scanner devices and the software analysis programs available.

5. Use a software analysis program (e.g., OsteoGram®) to calculate bone mineral content in arbitrary units (BMC-AU) and bone mineral density in arbitrary units (BMD-AU) from computer scan of the images from three phalanges.

Interpretation

1. RA estimates of bone mineral density are precise and accurate for the peripheral skeleton but cannot be relied upon to reflect changes in the spine (axial skeleton).
2. There is no method of translating BMD or bone mineral content (BMC) arbitrary units into the Z-values, T-values, and g/cm² units obtained from DXA.
3. Only in rheumatoid arthritis patients has it been suggested that RA-based BMD-AU values distinguish severity and progress of disease in the hand better than changes in BMD by DXA in the lumbar spine and total femur.

Notes

1. Patients with height loss, even if severe, may have a normal BMD-AU value on screening for osteoporosis and experience a delay in appropriate therapy.
2. RA may increase the cost of testing for bone mineral content, as its usefulness for serial assessment of bone density is limited and individuals with an abnormal RA often require DXA testing.
3. By the time osteoporosis can be visualized by a radiologist on a plain radiograph, at least 30% of the total bone mass may have been lost.
4. At best, RA is an attempt to use conventional radiography for detection of degrees of bone loss < 30%, but does not reliably detect deficits < 20%.

Suggested Reading

Bottcher J, Malich A, Pfeil A, Petrovitch A, Lehmann G, Heyne JP, Hein G, Kaiser WA. Potential clinical relevance of digital radiogrammetry for quantification of periarticular bone demineralization in patients suffering from rheumatoid arthritis depending on severity and compared with DXA. *Eur Radiol* 2004; 14:631–7.

Versluis RG, Petri H, Vismans FJ, van de Ven CM, Springer MP, Papapoulos SE. The relationship between phalangeal bone density and vertebral deformities. *Calcif Tissue Int* 2000; 66:1–4.

Yang SO, Hagiwara S, Engelke K, Dhillon MS, Guglielmi G, Bendavid EJ, Soejima O, Nelson DL, Genant HK. Radiographic absorptiometry for bone mineral measurement of the phalanges: precision and accuracy study. *Radiology* 1994; 192:857–9.

Yates AJ, Ross PD, Lydick E, Epstein RS. Radiographic absorptiometry in the diagnosis of osteoporosis. *Am J Med.* 1995; 98(2A):41–47S.

12.8 Male Reproductive System Test

12.8.1 Human Chorionic Gonadotropin (hCG): 15-Day Multiple-Injection hCG Stimulation Test in Boys with Intraabdominal Testes

Possible Indications for Test

Human chorionic gonadotropin stimulation of total testosterone (TT) and estradiol secretion with repeated injections over a 15-day period may be indicated:

- When attempting to determine if functioning testicular tissue is present intraabdominally in boys without scrotal testes and as a test of Leydig cell reserve in a cryptorchid patient
- As a follow-up test if the basal and stimulated [TT] after the single dose hCG test are inconclusive

TABLE 12.9
Testicular Response to Stimulation with Repeated Doses
of Human Chorionic Gonadotropin (hCG) in Males

Testicular Status	Testosterone Concentration, [T] (ng/dL)	
	Baseline	Stimulated
Normal Leydig cell reserve	Variable	>200 over baseline [T]
Decreased Leydig cell reserve[a]	Variable (typically <200)	<200 over baseline [T]
Limited to absent Leydig cell reserve[a,b]	Varies (<200)	≤Baseline [T]
Leydig cell failure[b]	<100	<100

[a] Hypogonadism, Kleinfelter's syndrome.

[b] Functionally or anatomically anorchid.

Alternatives to Testing

- Single-dose hCG test (see Test 8.1.2)
- Measurement of bioavailable testosterone (see Test 8.1.4), follicle-stimulating hormone (FSH), and luteinizing hormone (LH) (see Test 8.2.2)

Procedure

1. Obtain three separate baseline blood specimens 15 minutes apart. Order pooled TT and estradiol tests on these three baseline specimens (i.e., laboratory personnel will separate the serum from all three specimens and pool them together to perform TT and estradiol testing on the pooled serum specimen).
2. Administer 1000 units of hCG on day 1, on day 3 or 4, on day 7 or 8, and on day 10 from baseline.
3. On day 14 or 15, obtain three blood specimens collected 15 minutes apart and order pooled TT and estradiol testing on these specimens.

Interpretation

1. Refer to "Testicular Response to Stimulation with Repeated Doses of hCG in Males" (Table 12.9).
2. Only in normal pubertal boys, not prepubertal or Kallmann's patients, will the serum estradiol increase significantly after multi-dose hCG.

Notes

1. The preferred hCG stimulation test in prepubertal boys with possible hypothalamic hypogonadism is the single, not multiple, hCG injection test, with determination of [TT] 3 to 4 days later.
2. *In vitro*, interstitial testicular cells from patients with androgen insensitivity syndrome (AIS) respond to hCG added to the growth medium with about a twofold increase in testosterone secretion.

ICD-9 Codes

Conditions that may justify this test include *but are not limited to*:

752.5   undescended and retractile testicle/cryptorchism

Suggested Reading

Chemes H, Cigorraga S, Bergada C, Schteingart H, Rey R, Pellizzari E. Isolation of human Leydig cell mesenchymal precursors from patients with the androgen insensitivity syndrome: testosterone production and response to human chorionic gonadotropin stimulation in culture. *Biol Reprod* 1992; 46:793–801.

Dunkel L, Perheentupa J, Sorva R. Single versus repeated dose human chorionic gonadotropin stimulation in the differential diagnosis of hypogonadotropic hypogonadism. *J Clin Endocrinol Metab* 1985; 60:333–7.

Kolon TF, Miller OF. Comparison of single versus multiple dose regimens for the human chorionic gonadotropin stimulatory test. *J Urol* 2001; 166:1451–54.

12.9   Female Reproductive System Tests

12.9.1   Dexamethasone Suppression Test (DST) Combined with Gonadotropin-Releasing Hormone (GnRH) Agonist Stimulation for Determination of the Source of Androgen Production in Females

Possible Indications for Test

The DST, followed by GnRH agonist stimulation, for determination of the source of androgen production in the proven absence of an ovarian or adrenal tumor may be indicated in female patients if:

- Unexplained signs or symptoms of increased androgen production (i.e., moderate to severe acne, evidence of virilization such as clitoral hypertrophy, abnormal nonovulatory basal body temperature charts) are noted after imaging studies of the adrenal and ovaries fail to reveal the presence of tumors in either of these locations.

Procedure

1. Obtain a baseline blood specimen for total testosterone (TT) and dehydroepiandrosterone sulfate (DHEAS) testing (see Test 9.4.1). Calculate the patient's body mass index (BMI).
2. If the patient has a baseline [TT] that is ≥150 ng/dL (≥5.2 nmol/L), use high-resolution abdominal CT scanning and transvaginal ultrasound of the ovaries (see Test 9.5.2) to search for evidence of a tumor.
3. If an ovarian or adrenal mass is found, obtain repeat baseline [TT] and DHEAS for confirmation of abnormality in androgen secretion. Do not proceed with a DST in the event of an ovarian or adrenal mass.
4. If no mass is found, obtain repeat the blood sample for TT and DHEAS testing to confirm the presence or absence of hyperandrogenism.
5. If the repeat baseline [TT] is <150 ng/dL (<5.2 nmol/L) but >40 ng/dL (1.4 nmol/L), perform a DST during any phase of the menstrual cycle as follows:
   - Administer dexamethasone (0.5 mg p.o. q.i.d.) for 4 days (long test)
   - Obtain blood samples for serum cortisol, TT, free testosterone (FT), and DHEAS testing at 48 hours and/or 96 hours after the first dose of dexamethasone administered at 0800 hours on day 1
6. Follow the DST with a buserelin stimulation test (BST) (see Test 9.2.4) by injecting a GnRH agonist (e.g., triptoreline 100 mg or buserelin 1 mg s.c.) and obtain blood specimens for serum 17-hydroxyprogesterone (17OH-Prog) testing at 0800 (preinjection baseline), 2000, 2400, and 0800 hours.
7. Perform an ACTH stimulation test (refer to Tests 6.1.1 and 6.7.1) to determine 17OH-Prog production in patients identified as having an adrenal source of androgens, as in patients with 21-hydroxylase deficiency.

Interpretation

1. The results of the DST are shown in "Source of Androgen Production Based on Hormone Response During Dexamethasone Suppression Test (DST)" (Table 12.10) and of the BST in "Distinguishing Between Adrenal-Related, Idiopathic Nonadrenal, Non-Ovary, and Ovary-Related Hirsutism Using 17-Hydroxyprogesterone (17OH-Prog) Concentration after GnRH Agonist (i.e., Buserelin) Injection" (Table 12.11).

TABLE 12.10

Source of Androgen Production Based on Hormone Response During Dexamethasone Suppression Test (DST)

Time (hr)[a]	Hormone Response During DST	Most Likely Source of Androgen Production
48	$[TT]_{@48hr} < 50\%\ [TT]_{@0hr}$ and $[DHEAS]_{@48hr} < 50\%$ of $[DHEAS]_{@0hr}$	Adrenal glands[b]
96	$[FT]_{@96hr} > ULN$	Ovaries or an ectopic androgen-secreting tumor
96	$[DHEAS]_{@96hr} <$ (or only slightly >) ULN	Ovaries or an adrenal androgen-secreting tumor resistant to dexamethasone suppression

[a] After administration of dexamethasone. Note that a [cortisol] < 5 mcg/dL at 48 hours and/or 96 hours after the first dose of dexamethasone indicates patient compliance with the dexamethasone dosing regimen.

[b] Usually seen in patients with a body mass index (BMI) > 30.

Note: [DHEAS], dehydroepiandrosterone sulfate concentration; [FT], free testosterone concentration; [TT], total testosterone concentration; ULN, upper limit of normal.

TABLE 12.11

Distinguishing Between Adrenal-Related, Idiopathic Nonadrenal, Non-Ovary, and Ovary-Related Hirsutism Using 17-Hydroxy-Progesterone (17OH-Prog) Concentration after GnRH Agonist (i.e., Buserelin) Injection

Time from Injection	17OH-Prog Concentration (ng/mL ± SD)		
	ARH (n = 15)	NARH (n = 20)	ORH (n = 42)
0800 (0 hours)	1.8 ± 1.2	0.3 ± 0.1	0.7 ± 0.4
2000 (12 hours)	2.2 ± 0.9	1.5 ± 1.0[a]	2.4 ± 1.8[a]
2400 (16 hours)	2.6 ± 1.6	1.9 ± 0.8[a]	3.4 ± 2.6[a]
0400 (20 hours)	2.6 ± 1.8	2.2 ± 0.8[a]	3.9 ± 2.8[a]
0800 (24 hours)	2.4 ± 0.9	1.8 ± 0.7[a]	6.1 ± 2.4[a]

[a] $p < 0.01$.

Note: ARH, adrenal-related hirsutism; BST, buserelin stimulation test; GnRH, gonadotropin-releasing hormone; NARH, nonadrenal, non-ovary-related hirsutism; ORH, ovary-related hirsutism; SD, standard deviation.

Source: Bidzinska, B. et al., *Przegl. Lek.*, 57, 393–396, 2000. With permission.

2. Patients with adrenal-related hirsutism may be differentiated from those with idiopathic nonadrenal hirsutism by noting the early vs. late peak in [17OH-Prog] after GnRH agonist injection as well as the higher vs. lower baseline [17OH-Prog] at 0800 hours (0 hours) that occurs in these patients, respectively.

Notes

1. Typically, the adrenals are the major source of androgen secretion in women with idiopathic hirsutism.
2. Body weight correlates with adrenal androgen (DHEAS and testosterone) concentration in nonhirsute women and with androstenedione in hirsute women.

3. The ovaries are the major source of testosterone and androstenedione in women with polycystic ovary syndrome (PCOS), but the adrenal glands contribute substantially to these levels.

4. On serial baseline measurement of [TT] and [DHEAS], a virilizing adrenal adenoma is likely when levels of these hormones reach or exceed 200 ng/dL and 6600 ng/mL, respectively (Ho Yuen and Mincey, 1983).

5. Virilizing adrenal adenomas may produce both androgens and estrogens.

6. Ovarian vein sampling yielded false-positive results, suggesting the presence of tumor in 2 out of 8 cases of PCOS with [TT] ranging from 72 to 202 ng/dL (2.5–7.0 nmol/L). Thus, adrenal and ovarian vein sampling were of limited usefulness in the management of women with hirsutism or adrenal/ovarian neoplasia (Kaltsas et al., 2003b).

7. At 48 hours into the low-dose (2 mg administered) DST, none of 17 patients with androgen-secreting tumors obtained a >40% reduction or normalization of elevated or marginally elevated [TT] (Kaltsas et al., 2003a).

8. Hair growth rates correlated best with changes in serum [androstenedione]. Androstanediol glucuronide, derived primarily from adrenal precursors, was a better marker of adrenal androgen secretion than was DHEAS (Rittmaster and Thompson, 1990).

9. An ovarian origin of androgen excess was identified by GnRH agonist stimulation of 17OH-Prog in 8 out of 27 hirsute patients (30%) and by lack of dexamethasone suppression in 4 of these cases (22%). Only 4 of 8 patients showed concordant results by both tests (Re et al., 2002).

10. 17OH-Prog hyper-responds to GnRH agonist in cases of nontumorous ovarian hyperandrogenism as in the 17OH-Prog (ng/mL) response after GnRH agonist administration (see Test 9.2.4) in women with PCOS (Bidzinska et al., 2000).

Suggested Reading

Bidzinska B, Tworowska U, Demissie M, Milewicz A. Modified dexamethasone and gonadotropin-releasing hormone agonist (Dx-GnRHa) test in the evaluation of androgen source(s) in hirsute women. *Przegl Lek* 2000; 57:393–6.

Bouallouche A, Brerault JL, Fiet J, Julien R, Vermeulen C, Cathelineau G. Evidence for adrenal and/or ovarian dysfunction as a possible etiology of idiopathic hirsutism. *Am J Obstet Gynecol* 1983; 147:57–63.

Ehrmann DA, Rosenfield RL. Clinical review 10: an endocrinologic approach to the patient with hirsutism. *J Clin Endocrinol Metab* 1990; 71:1–4.

Ho Yuen B, Mincey EK. Role of androgens in menstrual disorders of nonhirsute and hirsute women, and the effect of glucocorticoid therapy on androgen levels in hirsute hyperandrogenic women. *Am J Obstet Gynecol* 1983; 145:152–7.

Ho Yuen B, Moon YS, Mincey EK, Li D. Adrenal and sex steroid hormone production by a virilizing adrenal adenoma and its diagnosis with computerized tomography. *Am J Obstet Gynecol* 1983; 145:164–9.

Kaltsas GA, Isidori AM, Kola BP, Skelly RH, Chew SL, Jenkins PJ, Monson JP, Grossman AB, Besser GM. The value of the low-dose dexamethasone suppression test in the differential diagnosis of hyperandrogenism in women. *J Clin Endocrinol Metab* 2003a; 88:2634–43.

Kaltsas GA, Mukherjee JJ, Kola B, Isidori AM, Hanson JA, Dacie JE, Reznek R, Monson JP, Grossman AB. Is ovarian and adrenal venous catheterization and sampling helpful in the investigation of hyperandrogenic women? *Clin Endocrinol* 2003b; 59:34–43.

Re T, Barbetta L, Dall'Asta C, Faglia G, Ambrosi B. Comparison between buserelin and dexamethasone testing in the assessment of hirsutism. *J Endocrinol Invest* 2002; 25:84–90.

Rittmaster RS, Thompson DL. Effect of leuprolide and dexamethasone on hair growth and hormone levels in hirsute women: the relative importance of the ovary and the adrenal in the pathogenesis of hirsutism. *J Clin Endocrinol Metab* 1990; 70:1096–102.

Rosenfield RL, Barnes RB, Ehrmann DA, Toledano AY. The value of the low-dose dexamethasone suppression test in the differential diagnosis of hyperandrogenism in women. *J Clin Endocrinol Metab* 2003; 88:6115–6.

12.9.2   Clomiphene Challenge Test (CCT) for Ovulatory Failure

Possible Indications for Test

The CCT may be indicated in the:

- Evaluation of the infertile cycling female
- Prediction of future fecundity in females under consideration for advanced fertility enhancement

Alternatives to Testing

- Anti-Müllerian hormone (AMH) (see Test 9.2.5)

Procedure

1. Obtain a blood sample for FSH testing on day 2 to 3 of the menstrual cycle during menses.
2. Administer clomiphene (100 mg p.o. on days 5 through 9 of the menstrual cycle).
3. Obtain a blood sample for FSH on day 9 or 11.

Interpretation

1. A baseline [FSH] > 13 mIU/mL is consistent with diminished ovarian reserve.
2. In women with adequate ovarian reserve, the [FSH] on day 9 or 11 of the CCT is nearly always <16 mIU/mL (mean ± SD, 11.5 ± 4.9 mIU/mL).
3. In women with diminished ovarian reserve, the [FSH] on day 9 or 11 of the CCT is nearly always >26 mIU/mL (mean ± SD, 8.9 ± 13.8 mIU/mL).
4. Approximately 10% of the patients in the general infertility population have an abnormal CCT, with the incidence of abnormal results increasing with age >30 years.
5. An abnormal CCT occurs with a higher frequency in patients who would otherwise be diagnosed with unexplained infertility and is prognostic of decreased long-term pregnancy rates.

Notes

1. The CCT evaluates ovarian follicular reserve. It is postulated that disparity between normal estradiol secretory capacity of the granulosa and diminished capacity to secrete inhibin might explain the inappropriately high FSH levels observed in infertility.
2. Clomiphene increases gonadotropin secretion and subsequent estradiol (E2) production from the ovaries, but the E2 response during the CCT did not predict ovarian responsiveness or pregnancy rates in patients undergoing assisted reproduction treatment.
3. If E2 production from the ovaries is adequate, it overcomes clomiphene inhibition and suppresses the pituitary production of gonadotropins. Thus, an elevated FSH before or after administration of clomiphene indicates a reduced number of ovarian follicles.
4. The GnRH stimulation test is limited by its lack of sensitivity in discrimination between delayed puberty and hypogonadotropic hypogonadism in children.

Suggested Reading

Kettel LM, Roseff SJ, Berga SL, Mortola JF, Yen SS. Hypothalamic–pituitary–ovarian response to clomiphene citrate in women with polycystic ovary syndrome. *Fertil Steril* 1993; 59:532–8.

Navot D, Rosenwaks Z, Margalioth EJ. Prognostic assessment of female fecundity. *Lancet* 1987; 2(8560):645–7.

Scott RT, Illions EH, Kost ER, Dellinger C, Hofmann GE, Navot D. Evaluation of the significance of the estradiol response during the clomiphene citrate challenge test. *Fert Steril* 1993a; 60:242–6.

Scott RT, Leonardi MR, Hofmann GE, Illions EH, Neal GS, Navot D. A prospective evaluation of clomiphene citrate challenge test screening of the general infertility population. *Obstet Gynecol* 1993b; 82:539–44.

12.9.3 Assessment of Female Sexuality in the Menopause Transition: Use of the Short Personal Experiences Questionnaire (SPEQ)

Possible Indications for Test

Assessment of female sexuality status may be indicated:

- When a female patient is in the menopause transition
- When complaints about libido are voiced or problems with sexual functioning are elicited from the perimenopausal patient who has no evidence of thyroid or other hormone imbalance
- As part of sex-steroid-related problems in studies of female populations

Procedure

1. Administer the SPEQ (Q12.1 in Appendix 1).
2. Review responses with the patient for verification.
3. Carefully maintain the privacy of data collected.
4. Obtain a blood sample for E2 determination in the perimenopause patient (refer to notes 5 and 6).

Interpretation

1. A score of 7 or less indicates sexual dysfunction.
2. SPEQ measures different domains of female sexual functioning as follows:
   - Libido (sexual interest)
   - Sexual responsivity (sexual arousal, enjoyment, orgasm)
   - Dyspareunia (vaginal dryness and dyspareunia)
   - Frequency of sexual activities
   - Partner-related factors (i.e., feelings for partner and partner problems with sexual performance)
3. Factors that explain over 50% of the variance in both frequency of sexual activities and sexual response are (Dennerstein et al., 2004):
   - Her prior level of functioning
   - Any change in partner status (gaining a new partner has a very positive effect but losing a partner has a negative effect on sexual functioning)
   - Feelings for her partner
4. A menopause-specific decline in sexual functioning correlates with a decline in [E2]. Specifically, a decline in [E2] affects sexual interest, arousal, enjoyment, orgasm, and dyspareunia. No significant effect of [E2] on frequency of sexual activities was observed (Dennerstein et al., 2002).

Notes

1. The most important factors affecting a middle-aged woman's sexual functioning are her prior level of functioning and relationship factors; these must be carefully assessed prior to considering the role of hormonal factors.
2. Using cognitive–behavioral interventions for the treatment of climacteric syndrome, significant improvements were observed in overall score of sexuality from the McCoy Female Sexuality Questionnaire ($p < 0.02$), administered before and after intervention. Improvements in anxiety ($p < 0.01$), depression ($p < 0.02$), partnership relations ($p < 0.02$), hot flashes ($p < 0.01$), and cardiac complaints ($p < 0.01$) were found using other measures. No changes were found in sexual satisfaction and stressfulness of menopause symptoms after intervention (Alder et al., 2006).
3. A major decline in a woman's sexual functioning occurs as she passes through the menopausal transition, based on the Melbourne Women's Midlife Health Project study. In this study, 42% of the premenopausal group (still menstruating within the last 3 months) had scores lower than 7 on the SPEQ. By the eighth follow-up year, the percentage of women (now postmenopausal) with such low scores was 88% (Dennerstein et al., 2002).

4. Significant declines in libido, sexual responsivity, and frequency of sexual activities and a significant increase in dyspareunia occur as women reach the late menopause transition (experienced after at least 3 months of amenorrhea) and a further decline with postmenopause. A woman's feelings toward her partner also change and become less passionate after entry into the late menopause transition (Dennerstein et al., 2001a).

5. If depression and treatment with antidepressants, known to have a major impact on sexual functioning, are not present and no partner relationship changes have occurred, hormonal factors should be studied in women whose decline in sexual functioning is clearly related to the menopausal transition.

6. The late menopause transition coincides with a steep decline in [E2], but no corresponding change occurs in total testosterone (Burger et al., 1999, 2000). In fact, there is a small but significant increase in the free testosterone index at this time secondary to a decline in SHBG (see Tests 9.4.1 and 9.4.2).

7. Alternatives or supplements to the SPEQ are the modified Sabbatsberg Sexual Self-Rating Scale and the Psychological General Well-Being Index (Davis et al., 2008).

8. The McCoy Female Sexuality Questionnaire has acceptable reliability, internal consistency, and apparent face and content validity, as well as considerable evidence of construct validity as demonstrated in more than a half dozen studies (McCoy et al., 2000).

Suggested Reading

Alder J, KE Besken, U Armbruster, R Decio, A Gairing, A Kang, J Bitzer. Cognitive–behavioural group intervention for climacteric syndrome. *Psychother Psychosom* 2006; 75:298–303.

Burger H, Dudley E, Hopper J, Groome N, Guthrie JR, Green A, Dennerstein L. Prospectively measured levels of serum FSH, estradiol and the dimeric inhibins during the menopausal transition in a population-based cohort of women. *J Clin Endocrinol Metab* 1999; 84:4025–30.

Burger HG, Dudley EC, Cui J, Dennerstein L, Hopper JL. A prospective longitudinal study of serum test-osterone, dehydroepiandrosterone sulfate, and sex hormone-binding globulin levels through the menopause transition. *J Clin Endocrinol Metab* 2000; 85:2832–38.

Davis S, Papalia M-A, Norman RJ et al. Safety and efficacy of a testosterone metered-dose transdermal spray for treating decreased sexual satisfaction in premenopausal women. *Ann Intern Med* 2008; 148:569–77.

Dennerstein L, Lehert P. Modelling mid-aged women's sexual functioning: a prospective, population-based study. *J Sex Marital Ther* 2004; 30:173–83.

Dennerstein L, Dudley E, Burger H. Are changes in sexual functioning during midlife due to aging or menopause? *Fertil Steril* 2001a; 76:456–60.

Dennerstein L, Lehert P, Dudley E. Short scale to measure female sexuality: adapted from McCoy Female Sexuality questionnaire. *J Sex Marital Ther* 2001b; 27:339–51.

Dennerstein L, Randolph J, Taffe J, Dudley E, Burger H. Hormones, mood, sexuality and the menopausal transition. *Fertil Steril* 2002; 77(Suppl 4):S42–48.

McCoy NL. The McCoy Female Sexuality Questionnaire. *Qual Life Res* 2000; 9: 739–45.

McCoy NL, Matyas J. Oral contraceptives and sexuality in university women. *Arch Sex Behav* 1996; 25:73–9.

12.10   Endocrine Tumor Test

12.10.1   Serotonin (5-Hydroxytryptamine) in Urine and Platelets: Tests in Carcinoid Tumor Patients

Possible Indications for Testing

Serotonin (5-hydroxytryptamine, or 5-HT) levels in urine and platelets may be indicated when:

- Urinary 5-hydroxyindoleacetic acid (5-HIAA) levels are normal in a patient with clinical signs and symptoms of carcinoid syndrome (i.e., vasomotor flushing and diarrhea) (see Table 10.1).
- The presence of a tumor suspicious for carcinoid has been found, particularly in foregut and midgut regions.

Alternatives to Testing

- Measurement of 5-HIAA and chromogranin A (see Tests 10.2.1 and 10.2.2)

Procedure

1. Be sure that patients have not recently taken any food containing 5-HT (e.g., plantains) or drugs that can affect the test results.
2. For measurement of urinary [5-HT], have patient collect an acidified 24-hour urine specimen (see Specimen Collection Protocol P1 in Appendix 2).
3. For measurement of platelet [5-HT], obtain the appropriate anticoagulated blood specimen for isolation of platelets and frozen storage of platelet aliquots (1×10^9 platelets per aliquot) until testing for platelet 5-HT level is performed.
4. Measure chromogranin A in blood samples from patients with carcinoid syndrome and a hindgut (colorectal) tumor.

Interpretation

1. Ordinarily, carcinoid tumors are diagnosed biochemically on finding of an increased urinary excretion of 5-HIAA by the patient; however, urinary and platelet [5-HT] provide complementary information.
2. The reference interval for urinary [5-HT] is 25 to 66 μmol/mol creatinine.
3. The reference interval for platelet [5-HT] is <500 ng/10^9 platelets.
4. Typically, hindgut carcinoid tumors do not produce 5-HT in excess.

Notes

1. [5-HT] in urine or platelets may help to identify the rare carcinoid tumor not detected by [5-HIAA].
2. Platelet [5-HT] may be useful in the follow-up of patients with carcinoid tumors that produce 5-HT in excess and may be more sensitive than urinary [5-HIAA] or [5-HT] in detecting tumors with a low level of 5-HT production.
3. Platelet [5-HT] is useful in the diagnosis of foregut carcinoid tumors or the rare patient with carcinoid syndrome but a normal urine or plasma [5-HIAA].

Suggested Reading

Anderson GM, Young JG, Cohen DJ, Schlicht KR, Patel N. Liquid-chromatographic determination of serotonin and tryptophan in whole blood and plasma. *Clin Chem* 1981; 27:775–6.

Kema IP, de Vries EG, Schellings AM, Postmus PE, Muskiet FA. Improved diagnosis of carcinoid tumors by measurement of platelet serotonin. *Clin Chem* 1992; 38:534–40.

Appendix 1.
Questionnaires

Q1.1
Thyroid Symptom Questionnaire (TSQ)

1. On a scale of from 1 to 10, what level of fatigue do you experience?   1 2 3 4 5 6 7 8 9 10
(Circle your level, with 10 being the highest level of fatigue)

	Yes	No	Not Applicable
2. Have you become *more* SHORT OF BREATH with less exertion than usual?	□	□	□
3. Have you developed problems with or an *increased* problem with depression?	□	□	□
4. Do you think you are losing *more* hair than usual?	□	□	□
5. Has your memory gotten *worse* lately—more than you might expect?	□	□	□
6. Have you or others noticed a swelling or swellings in your neck?	□	□	□
7. Have you become *more* unsteady on your feet?	□	□	□
8. Have you experienced an *increase* in muscle aches?	□	□	□
9. Have you experienced *recent* visual disturbances or change in disturbance?	□	□	□
10. Have you experienced any *recent* leakage of fluid from the breast(s)?	□	□	□
11. Have you developed an anemia (low red blood cell count) *recently*?	□	□	□
12. Do you experience *increased* discomfort or unpleasant awareness in your neck?	□	□	□
13. Have you experienced a *recent* decreased interest in sex/loss of libido?	□	□	□
14. Do you prefer warmer or cooler room temperatures?	□ Cooler	□ Warmer	□ Neither
15. Have you experienced a *recent* change in appetite?	□ Increase	□ Decrease	□ No change
16. Have your bowels/stool changed *recently*?	□ Looser	□ More constipated	□ No change

17. On a scale from 1 to 10 (highest), what level of lethargy do you experience?   1 2 3 4 5 6 7 8 9 10

18. On a scale of from 1 to 10 (highest), what level of weakness do you experience?   1 2 3 4 5 6 7 8 9 10

	Yes	No	Not Applicable
19. Have you experienced *recent or unexpected* weight GAIN?	□	□	□
20. Do you think your face has gotten *more* puffy or puffier than usual?	□	□	□
21. Have you experienced a *recent* decrease in perspiration?	□	□	□
22. Do you think your skin is drier or that you aneed more skin lotions *recently*?	□	□	□
23. Has you skin developed a more yellowish color *recently unrelated to the sun*?	□	□	□
24. Do you think your nails are growing out slower or becoming *more* brittle?	□	□	□
25. Have you experienced numbness in fingers of both hands *recently* or had a diagnosis of carpal tunnel syndrome?	□	□	□
26. Have you been diagnosed with sleep apnea?	□	□	□
27. Have your menstrual periods become *heavier* and/or *longer* than usual?	□	□	□
28. Have you been told that you:	□ Snore	□ Snore more loudly	□ I do not snore

Q1.1 (cont.)
Thyroid Symptom Questionnaire (TSQ)

	Yes	No	Not Applicable
29. Are you *more* irritable or *more* easily irritated than usual?	☐	☐	☐
30. Have you had *increasing* difficulty getting to sleep or staying asleep?	☐	☐	☐
31. Do you have *more* difficulty climbing stairs because of *recent* leg weakness?	☐	☐	☐
32. Have you experienced *recent or unexpected* weight LOSS?	☐	☐	☐
33. Have you experienced a *recent increase* in perspiration?	☐	☐	☐
34. Does your skin seem to itch *more* than usual?	☐	☐	☐
35. Do you think your nails are growing out *faster*?	☐	☐	☐
36. Have you experienced *increased* palpitations in the chest?	☐	☐	☐
37. Have your menstrual periods become *shorter* and/or *more frequent* than usual?	☐	☐	☐
38. Have your hands become *more* shaky or tremulous?	☐	☐	☐

Q5.1
Autonomic Neuropathy Symptom Questionnaire

Questions	Yes	No	Don't Know/ Not Sure
1. Do you have difficulty with *sexual function*?	☐	☐	☐
Men: How did you score on the IIEF-5?	☐ 1–4 ☐ 5–7	☐ 8–11	☐ 15–20 ☐ 21–25
2. Do you have a problem with excessive *perspiration*, not including with exercise?	☐	☐	☐
If "Yes," does excessive perspiration occur with ☐ meals ☐ anytime ☐ both?			
If "No," is your ability to perspire reduced when you are warm or exercising? Do you tend to become overheated?	☐	☐	
3. **Diabetics:** Are you sometimes unaware when your blood sugar is falling to a very low level (have a "hypo")?	☐	☐	☐
If "Yes," are you usually unaware when your blood sugar is getting low?	☐	☐	
If "No," is your blood sugar often more than 200 mg/dL?	☐	☐	
4. Do you have to *urinate* more than 5 or 6 times per day or rush to the bathroom to urinate on an urgent basis?	☐	☐	
If "Yes," do you lose control of the urine, and become incontinent?	☐	☐	
If "No," do you have difficulty starting or maintaining urine flow?	☐	☐	
5. Do you frequently have problems with *constipation, diarrhea*, or both occurring alternately?	☐	☐	
If "Yes," do you lose control of your bowels on occasion?	☐	☐	
If "No," do you use laxatives on a regular or frequent basis?	☐	☐	
6. Do you often get *faint, dizzy,* or *lightheaded* when you stand up?	☐	☐	☐ Yes to all 3
If "No," are you taking medication for blood pressure?	☐	☐	
7. Do you often have *nausea* with meals?[a]	☐	☐	
If "Yes," does this lead to *vomiting* on occasion?	☐	☐	
If "No," do you get cramping or abdominal pains after meals on more than an occasional basis?	☐	☐	
8. Do you get full or *bloated* easily when eating or after meals?[a]	☐	☐	
If "Yes," have you had gastroparesis or stomach atony?	☐	☐	☐

[a] If you are not a gastroparesis patient and are being treated with an incretin agent such as Symlin® (amylin analog) or Byetta® (GLP-1 mimetic), omit responses to Questions 7 and 8.

Note: GLP, glucagon-like polypeptide; IIEF-5, International Index of Erectile Function, Five Screening Questions.

Q5.2
Diabetic Peripheral and Sensory Neuropathy Screening Questionnaire

Sign, Symptom, Complaint	Yes	No
1. My feet tingle.	☐	☐
2. I feel pins and needles in my feet.	☐	☐
3. I have burning, stabbing, or shooting pains in my feet.	☐	☐
4. My feet are very sensitive to touch. (e.g., it hurts to have the bed covers touch my feet).	☐	☐
5. My feet hurt at night.	☐	☐
6. My feet and hands seem to get very hot or very cold.	☐	☐
7. My feet are numb and feel dead.	☐	☐
8. I don't feel pain in my feet, even when I have blisters or injuries.	☐	☐
9. I can't feel my feet when I'm walking.	☐	☐
10. The muscles in my feet and legs are weak.	☐	☐
11. I'm unsteady when I stand or walk and I have fallen down several times.	☐	☐
12. I have trouble feeling heat or cold in my feet or hands.	☐	☐
13. I have open sores (or ulcers) on my feet and legs. These sores heal very slowly or not at all.	☐	☐
14. It seems like the muscles and bones in my feet have changed shape.	☐	☐
15. Other symptoms that trouble me:	☐	☐

Source: Adapted from an educational program of the American Diabetes Association.

Q5.3
International Index of Erectile
Function-5 (IIEF-5) Questionnaire

Questions for IIEF-5 [Corresponding IIEF-15 Question]

Over the last 4 weeks …

1. How do you rate your confidence that you could get and keep an erection? [Q15]

2. When you had erections with sexual stimulation, how often were your erections hard enough for penetration? [Q2]

3. During sexual intercourse, *how often* were you able to maintain your erection after you had penetrated your partner? [Q4]

4. During sexual intercourse, *how difficult* was it to maintain your erection to completion of intercourse? [Q5]

5. When you attempted sexual intercourse, how often was it satisfactory for you? [Q7]

Responses to questions are scored as follows:

0 = No sexual activity, did not attempt intercourse.

1 = Very low, almost never or never, extremely difficult.

2 = Low, rarely, a few times (much less than half the time), very difficult.

3 = Moderate, occasionally, sometimes (about half the time), difficult.

4 = High, most times (much more than half the time), slightly difficult.

5 = Very high, almost always or always, not difficult, very high.

Source: Rosen, R.C. et al., *Int. J. Impot. Res.*, 11, 319–326, 1999. With permission.

Q5.4
International Index of Erectile Function-15 (IIEF-15) Questionnaire

Over the last 4 weeks...

1. *How often* were you able to get an erection during sexual activity?

2. When you had erections with sexual stimulation, *how often* were your erections hard enough for penetration?

3. When you attempted sexual intercourse, *how often* were you able to penetrate your partner?

4. During sexual intercourse, *how often* were you able to maintain your erection after you had penetrated your partner?

5. During sexual intercourse, *how difficult* was it to maintain your erection to completion of intercourse?

6. *How many times* have you attempted sexual intercourse?

7. When you attempted sexual intercourse, *how often* was it satisfactory for you?

8. *How much* have you enjoyed sexual intercourse?

9. When you had sexual stimulation or intercourse, *how often* did you ejaculate?

10. When you had sexual stimulation or intercourse, *how often* did you have the feeling of orgasm or climax?

11. *How often* have you felt sexual desire?

12. *How would you rate* your level of sexual desire?

13. *How satisfied* have you been with your overall sex life?

14. *How satisfied* have you been with your sexual relationship with your partner?

15. *How do you rate* your confidence that you could get and keep an erection?

Responses to questions are scored as follows:

0 = No sexual activity, did not attempt intercourse.

1 = Very low, almost never or never, extremely difficult.

2 = Low, rarely, a few times (much less than half the time), very difficult.

3 = Moderate, occasionally, sometimes (about half the time), difficult.

4 = High, most times (much more than half the time), slightly difficult.

5 = Very high, almost always or always, not difficult, very high.

Source: Rosen, R.C. et al., *Int. J. Impot. Res.*, 11, 319–326, 1999. With permission.

Q5.5
Sexual Encounter Profile (SEP) Questionnaire

Over the last 4 weeks …

1. Do you have the ability to achieve some erection?
2. Do you have successful penetration?
3. Do you have successful intercourse?
4. Are you satisfied with erection hardness?
5. Are you satisfied overall with your sexual experience?

Q8.1
Androgen-Deficient Aging Males (ADAM) Questionnaire

1. Do you have decrease in sex drive?
2. Do you have a lack of energy?
3. Has your strength or endurance decreased?
4. Have you lost height (more than 1 inch)?
5. Are you enjoying life less?
6. Are you sad or grumpy?
7. Are your erections less strong?
8. Have you noticed a recent deterioration in your ability to play sports or perform exercise workouts in the gym?
9. Do you fall asleep after dinner?
10. Has your work performance decreased lately?

Scoring: If the answer is "yes" to questions 1 or 7 or at least three questions other than 1 or 7, this screening test is positive and indicative of possible hypogonadism.

Q8.2
Daily Assessment of Mood Score Questionnaire

Negative Mood Statements

1. Much of the time I'm sad for no real reason.
2. I certainly feel useless at times.
3. I worry so much that at times I feel like I am going to faint.
4. I'm often so worried and nervous that I can barely stand it.
5. Often I can't understand why I have been so irritable and grouchy.
6. Some days I feel more irritated than usual.
7. I get nervous and irritated easily.
8. I get angry and want to take it out on myself.
9. I easily become impatient with people.
10. I have little control over my anger.
11. People are afraid of my temper.
12. Sometimes my temper explodes and I completely lose control.

Positive Mood Statements

1. Most days I am happy.
2. More days than not, I feel energetic.
3. I very seldom have spells of the blues.
4. I'm almost always a happy and positive person.
5. I rarely get in a bad mood.
6. At times, I am all full of energy.
7. I am happy most of the time.
8. Most days I feel very little or no annoyance.
9. I seldom feel anxious or tense.
10. I am usually calm and not easily upset.
11. I am not easily angered.
12. I wake up fresh and rested most mornings.

Determining a positive or negative mood score: (1) Award 1 point for each "yes" response to the negative mood statements (1–12). Add the total number of points and divide by the number of statements answered. Then assign a negative sign to this score. (2) Award 1 point for each "yes" response to the positive mood statements (1–12). Add the total number of points and divide by the number of statements answered. Then assign a positive sign to this score. (3) Sum the positive and negative scores to arrive at the overall strength of the individual's mood (mood scores range from –7 to +7).

Q 9.1
Menstrual Experience Questionnaire

Respond "Yes," "No," or "?" (don't know) to the following questions. If you are not sure about whether or not you experience the symptom noted or the symptom occurs very infrequently, circle the question mark. Note that all questions pertain to the *last half of your cycle* or the *two weeks before onset of menses*.

Yes	No	?	1.	Does your mood become markedly depressed with feelings of hopelessness and self-deprecation?
Yes	No	?	2.	Do you feel marked tension and anxiety? Do you feel "keyed up" or "on edge"?
Yes	No	?	3.	Do you experience marked mood swings with emotional or affective lability; do you become tearful, experience an increased sensitivity to rejection, or suddenly become very sad?
Yes	No	?	4.	Do you experience persistent or marked anger and irritability which leads to an increase in interpersonal conflicts?
Yes	No	?	5.	Do you experience a decreased interest in your usual activities (e.g., job, hobbies, friendships)?
Yes	No	?	6.	Do you experience difficulty concentrating?
Yes	No	?	7.	Do you experience fatigue, lack of energy, or lethargy?
Yes	No	?	8.	Do you experience marked changes in appetite with food cravings or overeating?
Yes	No	?	9.	Do you experience sleep disturbances with trouble getting to sleep and staying asleep, sleeping an excessive number of hours, or sleeping during the day when you would otherwise be awake?
Yes	No	?	10.	Do you experience a sense of being overwhelmed or out of control?
			11.	Do you experience a combination of physical symptoms including:
Yes	No	?		Breast tenderness and/or swelling?
Yes	No	?		Sense of bloating with weight gain?
Yes	No	?		Headache?
Yes	No	?		Muscle and/or joint pain?

List other symptoms not covered by the questions above in the space below:

Q11.1
Three-Factor Eating Questionnaire (TFEQ)

One point is given for each item in Part I and for each item in Part II. The underlined True/False answer is beside the number of the factor that it measures. The direction of the question in Part II is determined by splitting the responses at the middle. If an item is labeled "+" a response above the middle is given one point, and *vice versa* for a "−" label; for example, anyone scoring 3 or 4 on the first item in Part II (item 37) would receive 1 point. Anyone scoring 1 or 2 would receive a 0.

Part I				Factor Number
1.	When I smell or see a delicious, well-prepared main-course food item such as a meat dish, I find it very difficult to keep from eating even if I have just finished a meal.	<u>T</u>	F	2
2.	I usually eat too much at social occasions, such as parties and picnics.	<u>T</u>	F	2
3.	I am usually so hungry that I eat more than three times a day.	<u>T</u>	F	3
4.	When I have eaten my quota of calories, I am usually good about not eating any more.	<u>T</u>	F	1
5.	Dieting is so hard for me because I just get too hungry.	<u>T</u>	F	3
6.	I deliberately take small helpings as a means of controlling my weight.	<u>T</u>	F	1
7.	Sometimes things just taste so good that I keep on eating even when I am no longer hungry.	<u>T</u>	F	2
8.	Because I am often hungry, I sometimes wish that while I am eating an expert would tell me that I have had enough or that I can have something more to eat.	<u>T</u>	F	3
9.	When I feel anxious, I find myself eating.	<u>T</u>	F	2
10.	Life is too short to worry about dieting.	T	<u>F</u>	1
11.	Because my weight goes up and down, I have gone on reducing diets more than once.	<u>T</u>	F	2
12.	I often feel so hungry that I just have to eat something.	<u>T</u>	F	3
13.	When I am with someone who is overeating, I usually overeat, too.	<u>T</u>	F	2
14.	I have a pretty good idea of the number of calories in common food.	<u>T</u>	F	1
15.	Sometimes when I start eating, I just can't seem to stop.	<u>T</u>	F	2
16.	It is not difficult for me to leave something on my plate.	T	<u>F</u>	2
17.	At certain times of the day, I get hungry because I have gotten used to eating then.	<u>T</u>	F	3
18.	While on a diet, if I eat food that is not allowed I consciously eat less for a period of time to make up for it.	<u>T</u>	F	1
19.	Being with someone who is eating often makes me hungry enough to eat also.	<u>T</u>	F	3

Q11.1 (cont.)
Three-Factor Eating Questionnaire (TFEQ)

Part I				Factor Number
20.	When I feel blue, I often overeat.	T̲	F	2
21.	I enjoy eating too much to spoil it by counting calories or watching my weight.	T	F̲	1
22.	When I see a real delicacy, I often get so hungry that I have to eat right away.	T̲	F	3
23.	I often stop eating when I am not really full as a conscious means of limiting the amount that I eat.	T̲	F	1
24.	I get so hungry that my stomach often seems like a bottomless pit.	T̲	F	3
25.	My weight has hardly changed at all in the last 10 years.	T	F̲	2
26.	I am always hungry so it is hard for me to stop eating before I finish the food on my plate.	T̲	F	3
27.	When I feel lonely, I console myself by eating.	T̲	F	2
28.	I consciously hold back at meals in order not to gain weight.	T̲	F	1
29.	I sometimes get very hungry late in the evening or at night.	T̲	F	3
30.	I eat anything I want, any time I want.	T	F̲	1
31.	Without even thinking about it, I take a long time to eat.	T	F̲	2
32.	I count calories as a conscious means of controlling my weight.	T̲	F	1
33.	I do not eat some foods because they make me fat.	T̲	F	1
34.	I am always hungry enough to eat at any time.	T̲	F	3
35.	I pay a great deal of attention to changes in my figure.	T̲	F	1
36.	While on a diet, if I eat a food that is not allowed, I often then splurge and eat other high-calorie foods.	T̲	F	2

Part II

Directions: Please answer the following questions by circling the number above the response that is appropriate to you.

37.	How often are you dieting in a conscious effort to control your weight?	1 Rarely	2 Sometimes	3 Usually	4 Always	+1
38.	Would a weight fluctuation of 5 pounds affect the way you live your life?	1 Not at all	2 Slightly	3 Moderately	4 Very much	+1
39.	How often do you feel hungry?	1 Only at mealtimes	2 Sometimes between meals	3 Often between meals	4 Almost always	+3
40.	Do your feelings of guilt about overeating help you to control your food intake?	1 Never	2 Rarely	3 Often	4 Always	+1

(continued)

Q11.1 (cont.)
Three-Factor Eating Questionnaire (TFEQ)

Directions: Please answer the following questions by circling the number above the response that is appropriate to you.

41.	How difficult would it be for you to stop eating halfway through dinner and not eat for the next 4 hours?	1 Easy	2 Slightly difficult	3 Moderately difficult	4 Very difficult	+3
42.	How conscious are you of what you are eating?	1 Not at all	2 Slightly	3 Moderately	4 Extremely	+1
43.	How frequently do you avoid "stocking up" on tempting foods?	1 Almost never	2 Seldom	3 Usually	4 Almost always	+1
44.	How likely are you to shop for low-calorie foods?	1 Unlikely	2 Slightly unlikely	3 Moderately likely	4 Very likely	+1
45.	Do you eat sensibly in front of others and splurge alone?	1 Never	2 Rarely	3 Often	4 Always	+2
46.	How likely are you to consciously eat slowly in order to cut down on how much you eat?	1 Unlikely	2 Slightly likely	3 Moderately likely	4 Very likely	+1
47.	How frequently do you skip dessert because you are no longer hungry?	1 Almost never	2 Seldom	3 At least once a week	4 Almost every day	−3
48.	How likely are you to consciously eat less than you want?	1 Unlikely	2 Slightly likely	3 Moderately likely	4 Very likely	+1
49.	Do you go on eating binges though you are not hungry?	1 Never	2 Rarely	3 Sometimes	4 At least once a week	+2

50. On a scale of 0 to 5, where 0 means no restraint in eating (eating whatever you want, whenever you want it) and 5 means total restraint (constantly limiting food intake and never giving in), what number would you give yourself:

0	1	2	3	4	5	+1
Eat whatever you want, whenever you want it	Usually eat whatever you want, whenever you want it	Often eat whatever you want, whenever you want it	Often limit food intake, but often "give in"	Usually limit food intake, rarely "give in"	Constantly limiting food intake, never "giving in"	

51.	To what extent does this statement describe your eating behavior: "I start dieting in the morning, but, because of any number of things that happen during the day, by evening I have given up and eat what I want, promising myself to start dieting again tomorrow."	1 Not like me	2 Little like me	3 Pretty good description of me	4 Describes me perfectly	+2

Q12.1
Short Personal Experiences Questionnaire (SPEQ)

Please answer the following questions in terms of your current experience (in the last month) by circling the appropriate number. "Sexual activity" includes such behaviors as self-stimulation or masturbation, foreplay (arousal with partner), and actual intercourse.

		Not at all				A great deal	
A.	How enjoyable are sexual activities for you?[a]	1	2	3	4	5	6
B.	How often during sex activities do you feel aroused or excited?[a]	1	2	3	4	5	6
C.	Do you currently experience orgasm (climax) during sex activity?[a]	1	2	3	4	5	6
D.	How much passionate love do you feel for your partner?	1	2	3	4	5	6
E.	Are you satisfied with your partner(s) as a lover?	1	2	3	4	5	6
F.	Do you currently experience any pain during intercourse?	1	2	3	4	5	6
G.	Does your partner(s) experience difficulty in sexual performance?	1	2	3	4	5	6

H. About how many times have you had sexual thoughts or fantasies (e.g., daydreams) during the last month?[a]

   0   Never
   1   Less than once a week
   2   Once or twice a week
   3   Several times a week
   4   Once a day; sometimes twice
   5   Several times a day

I. About how may times during the past month have you had any sexual activities?[a]

   0   Never
   1   Less than once a week
   2   Once or twice a week
   3   Several times a week
   4   Once a day; sometimes twice
   5   Several times a day

J. Do you have a current sexual partner(s)?

   ☐ Yes
   ☐ No

K. What is your sexual preference?

   ☐ Heterosexual (male partner)
   ☐ Bisexual (both male and female partners)
   ☐ Lesbian/gay (female partner)

[a] Items included in composite scale score.

Source: Dennerstein, L. et al., *J. Sex Marital. Ther.*, 27, 339–351, 2001. With permission. (The McCoy Female Sexuality Questionnaire was originally published in McCoy, N. and Matyas, J., *Arch. Sex. Behav.*, 25, 73–79, 1996.)

Appendix 2.
Specimen Collection Protocols

Risks with Venipuncture

The risks of having blood drawn include:

- Excessive bleeding
- Fainting or feeling lightheaded
- Hematoma (blood accumulating under the skin)
- Infection (a slight risk any time the skin is broken)

Veins and arteries vary in size from one patient to another and from one side of the body to the other. Obtaining a blood sample from some people may be more difficult than from others

SCP-1. Urine Sample Collection: Patient Instructions

Patient Instruction Protocol for Collection of a 24-Hour Urine

- *Obtain the urine specimen container or containers* from the laboratory. Present the request form for a urine test to the laboratory.
- *Discard* the *first* morning urine specimen (between 6:00 a.m. and 7:00 a.m.) on the first morning of the 24-hour urine collection period.
- After discarding the first morning urine, *collect every drop* of urine passed into a suitable clean widemouth receptacle and pour into the container issued by the laboratory. Note that females should be issued a "hat" or special plastic collection device in which to pass the urine which is then to be poured into a narrow mouthed container.

- *Collect* the final urine sample the following morning at the same hour (between 6:00 a.m. and 7:00 a.m.).
- *Keep* the 24-hour urine collection container cool in either a refrigerator or an ice chest with ice throughout the collection period. (The ice chest or a written sign may be kept in the bathroom as a reminder that a urine collection is in progress.)
- Be sure to *label* the 24-hour urine collection containers with identification labels including name and identity number. The label must state the date and time the collection began and ended.
- *Submit* the 24-hour urine collection container accompanied by the request form to the laboratory promptly after the collection is complete.
- A blood sample may be required following completion of the 24-hour urine collection. Notice of this requirement should be given at the time the urine container is provided.
- *If a blood sample is required* by the laboratory, complete the urine collection on a day when the blood sample can be obtained.

Random Urine Sample Collection Protocol

- To collect a urine sample for protein or pH determination, males should wipe the head of the penis clean. Females need to wipe between the vaginal lips (labia) with soapy water and rinse well. Ideally, a special clean-catch kit that contains a cleansing solution and sterile wipes is used as directed.

Procedure

- Instruct the patient to first urinate a small amount into the toilet bowl to clear the urethra of any contaminants.
- Have the patient use a clean or sterile container to collect the urine sample.
- Have the patient collect at least 1 ounce of urine.
- Have the patient remove the container from the urine stream without stopping the flow and without touching any skin surfaces.
- For infants or individuals unable to assist with urine collection, clean and dry the genital area and attach a collection device to collect the urine. Be sure the collection device is attached securely to prevent leakage. After urinatation, place the urine (at least 20 cc) in a sterile container as soon as possible.

SCP-2. Saliva Sample Collection: Patient Instructions

Patient Instruction Protocol for Collection of Saliva Samples

- *Obtain* the appropriate containers for collection of saliva. Note that the Salivette® device has replaced widemouthed plastic cups or jars, which are no longer recommended as they allow evaporation and altered concentration of the sample.
- *Notify the physician if there is any bleeding inside your mouth and if this bleeding is a chronic condition.*
- If necessary, *stimulate salivation* with citric acid crystals, Crystal Light®, or sugar-free Kool-Aid® powder (i.e., about 1/8 tsp or the volume of a thumb tip) to obtain an adequate volume of saliva.
- Prior to collecting a saliva specimen, *do not:*
  - Eat or drink 30 minutes prior to collection.
  - Apply creams or lotions that contain steroids to the skin or mucous membranes immediately prior to collection.
  - Engage in activities that cause the gums to bleed, including brushing or flossing teeth, as blood in the saliva will invalidate the test.
- *Collect clear saliva* at both 0800 hours (8:00 a.m.) and 2300 hours (11:00 p.m.) as indicated.

- *Clearly label* each saliva container with the actual time of collection to the nearest minute. Use 24-hour clock times and large "AM" and "PM" labels to help lab personnel avoid confusion when entering the specimen collection time into a computerized laboratory information system. *Be sure that each container is correctly labeled* with your name, the physician's name, and the date and time of collection.
- *Store* specimens collected at home in the freezer. *Return* the carefully labeled specimens to a lab within one day of collection.
- *Tell the laboratory clerk to enter the time of collection* as the time you wrote on the specimen container, not the time you got to the lab.

SCP-3. Semen Sample Collection: Patient Instructions

Patient Instruction Protocol for Collection of Semen Samples

- *Consent* to collection of a semen specimen by masturbation.
- If this preferred method is unacceptable, as may be the case when religious beliefs, practice, or tradition forbid it, *request* a nonrubber, nonplastic perforated or imperforated pouch placed over the erect penis for use during intercourse to collect a specimen of seminal fluid.
- Note that under no circumstances should standard prophylactics or condoms be used to collect sperm as they may contain chemicals that are spermicidal. Expect a suboptimal analysis using any rubber or plastic product to collect seminal fluid.
- *Abstain* from ejaculating for 2 to 4 days (optimal interval, 3 days) prior to seminal fluid collection. Do not submit a specimen if the interval since the last ejaculation exceeds 5 days, as semen parameters start to change (degrade) at this point.
- *Obtain* a sterile glass jar and proceed on the morning of a prearranged day to a private area for specimen collection.
- *Keep the seminal fluid warm* (body temperature) and present the specimen for analysis within less than 30 minutes.
- *Be prepared* to submit a second specimen within 2 weeks, as there is marked day-to-day intraindividual variability in seminal fluid parameters, and repeat analysis is a standard part of any infertility evaluation.

SCP-4. Blood Sample Collection Protocol: Growth Hormone (GH) Testing

- Insertion of a blood sampling catheter must be done at least 15 minutes before a blood sample is obtained for measurement of GH.
- The protocol for the collection of blood samples during the arginine–GHRH (ARG–GeRef) infusion for the stimulation of GH in growth hormone deficiency test (Test 2.3.4) is shown below:

Specimen Collection Protocol During GH Stimulation Testing

| Time (min) | Obtain Blood Samples for | | Action |
	Glucose	GH	
0	✓	✓	Inject GeRef followed by L-arginine infusion
30	✓	✓	Stop L-arginine infusion
60	✓	✓	
90	✓	✓	

- Determination of insulin levels is optional at the times noted in the test protocol above, as is blood sampling at 120 minutes.

SCP-5. Blood Sample Collection Protocol:
Bilateral Inferior Petrosal Sinus Sampling (BIPSS)

- As with all invasive and complex procedures, be fully prepared and well organized in obtaining informed consent, supplies for obtaining multiple blood samples, and pathology assistance well in advance of the scheduled test date.
- Schedule that patient for BIPSS testing when an experienced invasive neuroradiologist is available to insert the bilateral inferior petrosal sinus (PS) catheters and adequate personnel are available to assist the radiologist in the radiology suite with the collection of multiple timed blood samples from each catheter prior to and after the administration of the stimulatory test agent.
- When performing BIPSS after the administration of ovine corticotropin-releasing hormone (oCRH) to exclude ectopic ACTH syndrome (Test 6.3.2), use the following procedure.
- Obtain the patient's weight (in kg) and calculate the dose of oCRH (i.e., Acthrel®, corticorelin ovine triflutate for injection; 1 mcg/kg body weight).
- Be sure that syringes (30 3-mL and 5 5-mL), blood collection tubes (27 EDTA and 3 red-top), a multi-event timer, and an adequate number of vials (i.e., 1 100-mcg vial per 100 kg body weight) of oCRH are available prior to beginning the procedure.
- Prepare a 50-mcg/mL solution of oCRH immediately before use by reconstituting each 100-mcg vial of oCRH with 2 mL of sterile saline.
- Insert two peripheral vein (PV) catheters, one for administration of oCRH and one for blood sampling.
- With the patient consciously sedated, have the neuroradiologist cannulate both the right and left petrosal sinuses.
- When both inferior PS (IPS) catheters are in place, obtain blood samples for [ACTH] from each PS catheter and a PV catheter at −15 and −5 minutes prior to injection of oCRH. Measure cortisol concentration, in addition to ACTH, in the −15-minute PV blood sample.
- At time 0, administer solution of 1 mcg/kg oCRH i.v. over 1 minute through a PV line.
- Simultaneously obtain blood samples (in EDTA tubes) for [ACTH] from the left PS (LPS) and right PS (RPS) catheters and the PV catheter at 0 (baseline), 1, 3, 5, 10, and 15 minutes after the start of oCRH infusion. In addition, obtain a blood sample (red-top tube) for cortisol testing at 5 minutes and 15 minutes after initiation of the oCRH infusion.
- A summary of timed specimen collection is shown below:

Specimen Collection Protocol During BIPSS Testing

Time Before or after Administration of oCRH (min)	Blood Specimen Collection Tube	Source of Blood Specimen			Analyte Tested
		PV	RPS	LPS	
−15	RTT	✓	—	—	Cortisol
	EDTA	✓	✓	✓	ACTH
−5	EDTA	✓	✓	✓	ACTH
0	EDTA	✓	✓	✓	ACTH
1	EDTA	✓	✓	✓	ACTH
3	EDTA	✓	✓	✓	ACTH
5	RTT	✓	—	—	Cortisol
	EDTA	✓	✓	✓	ACTH
10	EDTA	✓	✓	✓	ACTH
15	RTT	✓	—	—	Cortisol
	EDTA	✓	✓	✓	ACTH
30	EDTA	✓	✓	✓	ACTH
Total number of tubes ($n = 30$)		12[a]	9	9	

[a] Three red-top tubes (RTTs) and nine ethylenediaminetetraacedtic acid (EDTA) tubes.

Note: ACTH, adrenocorticotropic hormone; BIPSS, bilateral petrosal sinus sampling; LPS, left petrosal sinus; oCRH, ovine corticotropin-releasing hormone; PV, peripheral vein; RPS, right petrosal sinus.

SCP-6. Blood Sample Collection Protocol: Measurement of Potassium [K⁺] in Nonhemolyzed Blood

Use the following procedures to avoid hemolysis and falsely elevated $[K^+]$:

- If possible, discontinue angiotensin-converting enzyme inhibitors (ACEIs) and potassium-binding resins for at least 2 weeks.
- Discontinue all diuretics for at least 4 weeks and spironolactone for 6 weeks before measuring the serum $[K^+]$.
- Avoid fist clenching during collection, and wait at least 10 seconds after tourniquet release before drawing back on the syringe and collecting blood sample. Collect blood slowly using a syringe and needle. Do not use vacuum tubes.
- Carefully empty contents of syringe by letting the blood drip down the side of the centrifuge tube.
- Ensure separation of cells from fluid within 30 minutes of collection.

SCP-7. Blood Sample Collection Protocol and Report Form: Adrenal Vein Sampling

Procedure

Assemble the materials required:

Materials Required for Adrenal Vein Sampling

Provided by Radiology	*Provided by Pathology*
0.25 mg (250 mcg) ACTH as cosyntropin 1 vial, 5% dextrose in water (D5W) 250 mL i.v. solution	10 red-top or serum separation tubes (SST) blood collection tubes 1 test tube rack
10 5-mL disposable syringes	10 patient-specific (name and identification numbers) premade labels (two sets)

After catheterization of the adrenal veins and during blood sampling:

- Do not contaminate the sterile field. Wait to be handed all blood samples. Do not touch the radiologist's glove.
- Place blood samples in separate, appropriately labeled SST (red-top) tubes containing no anticoagulant.
- Wear a lead apron and do not turn your back during fluoroscopy without proper shielding.
- Agree with the radiologist ahead of time if the samples will be given to you with or without a needle on the syringe. If there is a needle, it is easy to inject the sample through the stopper of the tube, but a risk of needle sticks must be accepted. If there is no needle, you will need to loosen the SST stoppers ahead of time.

See sample form, next page.

Laboratory Evaluation of Adrenal Gland Lateralization of Aldosterone Production	**University Hospital - St Paul** **5909 Harry Hines Boulevard** **Dallas, TX 75390-9200** Phone: 214-879-3888 Fax: 214-879-6311

Patient's Name	Medical Rec No.	Physician/Mail Code	Date of Surgery	Interventional Radiologist

Specimen Source	During ACTH Infusion [Aldo], ng/dL	[Cortisol], ug/dL	A/C Ratio $(\times 10^{-3})$	$C_{RAV\ and\ LAV} \geq 3C_{IVC}$	Adrenal Vein A/C Ratio Dominant (D)	D/IVC	Nondominant (ND)	ND/IVC	$\frac{(A/C)_{Dominant}}{(A/C)_{Nondominant}}$
RAV	1000.0	1395.0	0.72	Yes	11.47	8.55	0.72	0.53	16.00
LAV	5300.0	462.0	11.47	Yes					
IVC	55.0	41.0	1.34	*Yes*	☐ Overall AVS successful?				
PV	52.0	34.4	1.51	*Yes*	☐ $C_{PV} \geq 20$ ug/dL?				

Baseline w/o ACTH			$(C_{RAV}/C_{IVC}) >10$	Yes	**IMPORTANT:** As always, clinicians must combine the information from AVS with other patient-specific information (e.g., imaging findings) when interpreting the results of AVS provided by this report. In addition, an experienced endocrinologist has reviewed, accepted, or modified appropriately the interpretive information provided in this report.
PV-B		25.0	$(A/C)_{RAV} > (A/C)_{IVC,PV}$	Yes	

Interpretive Notes:
AVS is considered successful if **BOTH** C_{RAV} and $C_{LAV} > 3C_{IVC}$

If $(A/C)_{Dominant}/(A/C)_{Nondominant} \geq 4$ and $(A/C)_{Nondominant} < (A/C)_{IVC}$ and dominant (A/C) value is from *RAV*, then laterality is *right* adrenal gland.

If $(A/C)_{Dominant}/(A/C)_{Nondominant} \geq 4$ and $(A/C)_{Nondominant} < (A/C)_{IVC}$ and dominant (A/C) value is from *LAV*, then laterality is *left* adrenal gland.

$(A/C)_{LAV} > (A/C)_{IVC,PV}$		Yes
$(A/C)_{LAV} / (A/C)_{RAV} <3$		No
$(A/C)_{LAV} / (A/C)_{IVC} < 3$		No

Diagnosis	Criteria	Criteria Met?
APA or PAH	$(A/C)_{Dominant}/(A/C)_{Nondominant} \geq 4$ **AND** $(A/C)_{Nondominant} < (A/C)_{IVC}$	Yes
Laterality [Right (RAGA) or Left (LAGA) Adrenal Gland Adenoma]		*LAGA*
BAH	$(A/C)_{Dominant}/(A/C)_{Nondominant} < 4$; BOTH $(A/C)_{LAV}$ and $(A/C)_{RAV} > (A/C)_{IVC}$	

Abbreviations: A, aldosterone; [Aldo], aldosterone concentration; C, cortisol; RAV, right adrenal vein; LAV, left adrenal vein; IVC, inferior vena cava; PV, peripheral vein; PV-B, peripheral vein-baseline; D, dominant; ND, nondominant; RAGA, right adrenal gland (aldosterone-producing) adenoma; LAGA, left adrenal gland (aldosterone-producing) adenoma; APA, aldosterone-producing adenoma; PAH, primary adrenal hyperplasia; BAH, bilateral adrenal hyperplasia; AVS, adrenal vein sampling; AG, adrenal gland; N/A, not applicable.

Anatomy of the Adrenal Glands

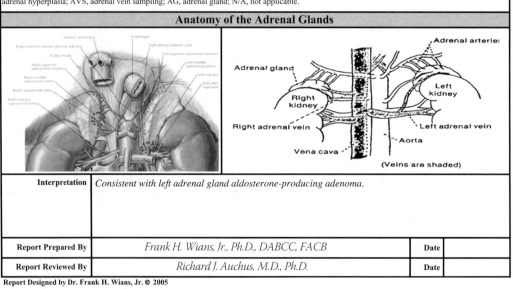

Interpretation	*Consistent with left adrenal gland aldosterone-producing adenoma.*

Report Prepared By	*Frank H. Wians, Jr., Ph.D., DABCC, FACB*	Date	
Report Reviewed By	*Richard J. Auchus, M.D., Ph.D.*	Date	

Report Designed by Dr. Frank H. Wians, Jr. © 2005

SCP-8. Blood Sample Collection Protocol: Selective Venous Sampling for Localization of Hyperfunctioning Parathyroid Glands

- Assemble materials required:
  - 30 EDTA (lavender-top) tubes
  - Test tube rack
  - Waterproof labels with sufficient space on the labels to enter:
    - Tube number
    - Patient's name
    - Collection time
    - Anatomical location from which the blood sample added to each tube was obtained
- Have a diagram of the vascular anatomy of the neck available to indicate the location, by tube number, from which each blood sample was obtained.
- Coordinate this procedure with a skilled interventional radiologist in a medical center and be prepared to personally assist in the fluoroscopy radiology suite.
- Have a bucket of ice and test tube rack available with which to store all tubes after blood collection prior to their delivery to the clinical laboratory at the end of the procedure.
- Have the interventional radiologist insert, under fluoroscopic guidance, a series of at least four catheters into the venous system draining the four quadrants of the thyroid. If you will be assisting the radiologist during specimen collection, minimize your x-ray exposure time by entering the radiology suite only after this time-consuming task is nearly accomplished. After catheterization of the parathyroid veins and during blood sampling:
  - Do not contaminate the sterile field. Wait to be handed all blood samples. Do not touch the radiologist's glove.
  - Place blood samples in separate, appropriately tubes.
  - Wear a lead apron and do not turn your back during fluoroscopy without proper shielding.
- Agree with the radiologist ahead of time if samples will be given to you with or without a needle on the syringe. If there is a needle, it is easy to inject the sample through the stopper of the tube, but a risk of needle sticks must be accepted. If there is no needle, you will need to loosen the stoppers ahead of time.
- Be sure that the exact anatomic area of blood collection by the catheter tip is marked with a wax pencil on a hard copy of the fluoroscopic image. In addition, you or a radiology technician must mark the anatomic location from which each numbered blood sample was obtained to create a general collection site map.
- As the radiologist obtains each blood sample, confirm the anatomic location from which the blood sample was obtained. Assay for intact parathyroid hormone (PTH) immediately using an operating-room-based automated immunoassay analyzer (Tosoh) or transfer the contents of the syringe into the appropriately labeled EDTA tube. Gently swirl and invert the tube to thoroughly mix the blood with the EDTA in the tube and prevent clots.
- Promptly deliver appropriately chilled (e.g., in ice) blood-containing tubes to the clinical laboratory for PTH testing, ensuring that all tubes are correctly and legibly labeled with the information indicated above.

SCP-9. Blood Sample Collection Protocol: Measurement of Testosterone in Blood

- Identify patient with a borderline low [TT] of <200 ng/dL on a single screening sample.
- Arrange specialized assistance (i.e., metabolic nurse) to obtain accurately timed specimens.
- Via an indwelling catheter, collect three blood specimens 20 minutes apart.
- Order pooled TT testing on these blood specimens.

Note: This protocol may be applied to measurement of any analyte with highly variable moment to moment changes in concentration.

Index